Metals and the Royal Society

Dedicated to our wives,
Phyllis and Ann,
who have have provided support and shown patience
in accepting the time we have spent in writing this book.

Metals and the Royal Society

D. R. F. West DSc, FIM and
J. E. Harris FIM, FREng, FRS

'The business and design of the Royal Society is – to improve the knowledge of natural things, and all useful Arts, Manufactures, Mechanick practices, Engynes and Inventions by Experiments – (not meddling with Divinity, Metaphysics, Moralls, Politicks, Grammar, Rhetorick or Logick)....All to advance the glory of God, the honour of the King....the benefit of his kingdom, and the general good of mankind'.

Robert Hooke, 1663.

'in proportion to the success with which the metallurgic art is practised in this country will the interests of the whole population, directly or indirectly, in no inconsiderable degree be promoted'.

John Percy,
from a lecture at the (Royal) School of Mines, London, 1851.

Book 568
Published in 1999 by
IOM Communications Ltd
1 Carlton House Terrace
London SW1Y 5DB

IOM Communications Ltd
is a wholly-owned subsidiary of
The Institute of Materials

ISBN 1 86125 028 2

Typeset by IOM Communications Ltd

Printed and bound in the UK by
The Universtiy Press, Cambridge

Contents

Foreword

Of the world's continuously-existing scientific academies, the Royal Society is the oldest. Its origins date back to 1645 when an assembly of scholars began to meet in London to discuss experimental philosophy (one source of their inspiration being the writings of the courtier and philosopher, Francis Bacon, who had died nineteen years earlier). Robert Boyle christened the group 'the invisible college'. During the Protectorate (1653–59) some of the group continued to meet at Gresham College, where they attended lectures by Lawrence Rooke and Christopher Wren, while others, including Robert Boyle and John Wilkins, went to Oxford and met in Wilkins's rooms at Wadham.

After the Restoration, the meetings were resumed in London where, on 28th November 1660, after one of Wren's lectures at Gresham College, the assembly withdrew to Rooke's apartment in the College and discussed 'a designe of founding a College for the Promoting of Physico-Mathematicall, Experimentall Learning'. Those present at that meeting, were: William, Viscount Brouncker, Robert Boyle, Alexander Bruce, Sir Robert Moray, Sir Paul Neile, John Wilkins, Jonathan Goddard, William Petty, William Ball (Balle), Lawrence Rooke, Christopher Wren and Abraham Hill. These, together with William Croone, can be taken as the original Founder Fellows.

Sir Robert Moray informed his friend, Charles II, of the proposed formation of the Society and the King 'did well approve of it, and would be ready to give encouragement to it'. Morley presented this good news at the second meeting of the Society on 5th December 1660. On 15th July 1662 the King granted 'The Royal Society' its first Charter, and a second Charter was granted on 22 April 1663 when he made the grant of the Arms. The motto chosen by the Society, *Nullius in verbia*, was an expression of determination to verify all statements by an appeal to facts, in the best Baconian tradition. In this second Charter the Society was referred to as 'The Royal Society of London for promoting Natural Knowledge'.

The King presented the Society with a handsome Mace and the Founders Book, which is still signed by every newly-elected Fellow (currently there are approximately 1200 FRSs together with 100 Foreign Members). At that time the King himself was in straightened circumstances so there was no grant of government funds. This absence of official aid gave the Society a degree of independence (and hence to be free, as it was quaintly put, 'to follow the whims of the afternoon'), but it is also true that acute shortages of cash have curtailed the Society's activities on a number of occasions throughout the succeeding years.

Of course the purpose of the regular meeting of the Society was to enable natural philosophers *to communicate*, a procedure which had previously been achieved mostly by the exchange of letters. An international epistolatory network had in fact been created during the first half of the 17th century, and the focal point and enablers for this 17th-century 'internet' were the 'intelligencers'. The most important of these was Martin Mersenne, a French monk whose main contact in England was Samuel Hartlib, though in due course Theodore Haak and Henry Oldenberg also became involved in these activities. Following the creation of the Royal Society, Oldenburg became its joint secretary and in 1665 started to publish *Philosophical Transactions*, a scientific periodical which reduced the need for the exchange of letters.

As well as creating one of the world's oldest scientific journals, Oldenberg put the recording of thc Royal Society's affairs on a firm basis, and the creation over the succeeding years of the Society's huge archives and collection of published works is due in no small part to his initiative. As well as records of proceedings a growing feature of the Society's archives and publications over the years has been comprehensive descriptions of its Fellows' lives, activities and achievements. These biographies, in the form of Notes, Obituaries and Memoirs, provide a unique source of information on the evolution of almost all scientific, medical and engineering disciplines during the past three and a half centuries. To avoid being swamped with detail it is necessary to chose a discipline, or sub-discipline, and trace its development through the comprehensive literature.

Both authors of this volume are practising metallurgists and the core of the approach to their writing has stemmed from their identification and study of a *metallurgical* thread running through the activities of the Society. This thread has related to the Society acting collectively and actively (such as when one of its Working Parties advised the government on, for example, corrosion of ships) and indirectly by electing to the Society scientists or engineers whose work has a strong metallurgical component. In the latter case information on this metallurgical activity eventually appears in the individual FRS's biographical memoir.

During the first two centuries of the life of the Society metallurgical strands can be traced through the work of physicists, chemists, mathematicians and engineers. It was not until the second half of the 19th century, stimulated largely by the application of science to the large-scale production of steel, that metallurgy began to appear as a separate discipline, having hitherto been in effect largely a sub-branch of chemistry with strong industrial associations. At about the same time the essential crystalline nature of metallic materials was widely recognised, and the extensive use of reflected light microscopy had excitingly opened up the world of microstructure. The full implications of the fact that metals consist of polycrystalline aggregates were recognised for the first time. The science of thermodynamics developed apace and the use of equilibrium phase diagrams grew in importance. As the distinguished metallurgist and historian of science, C.S. Smith, has written: 'the structure-property relationship has been the central theme of the last century of metallurgy and its statement constitutes the metallurgist's major contribution to science', see Appendix 8, Gen. Ref. 22.

X-ray diffraction, as discovered and widely applied from the early years of the 20th century, permitted quantitative evaluation of the crystalline state, not only in metals but also for a whole spectrum of other materials, inorganic, organic and biological. The advent of the electron microscope, together with advances in the theory of the solid state, added a new dimension to the study of metallurgy. The transition from the dominant role of empiricism in metallurgy to science-based development has been a vital feature of the past half century, and materials' artifacts with specific structures and properties can now be designed on an increasingly secure basis.

As far as the structure of the book itself is concerned, the first chapter sets the scene and discusses the origins of man's use of metals during several millennia preceding the creation of the Royal Society. It is then shown that the mining of metal ores has stimulated developments in technology, and the assaying and use of metals by alchemists played an important part in creating the conditions within which the scientific revolution could take root.

The majority of the remaining chapters are thematic, concentrating on particular subject areas, such as platinum metallurgy, minting of coins, metallurgy and electrical supply, the manufacture of iron and steel, and so on. These cover the scientific and technological contributions of more than five hundred Fellows out of the total of more than eight thousand elected to the Society since its formation. (Sadly no women, whose chief work has been metallurgical, are included in the above number; indeed women were not elected to the Fellowship until 1945). Reference is also made to the contributions of scientists who were not elected to the Society.

The title of this book *Metals and the Royal Society*, if interpreted literally, could imply a severe restriction on the subjects which could be covered in the text. In fact a liberal view has been taken and the boundaries assumed to be very permeable, particularly in chapters, 17, 18 and 19. Chapter 17 in fact specifically deals with developments in *non-metallic* materials, particularly during the last fifty years, illustrating the development of the broad field of materials science, technology and engineering, of which metallurgy now forms an integral part. The work of more than one hundred Fellows is described including, it is pleasing to report, several women scientists. The materials studied include semiconductors, superconductors, structural ceramics, polymers, composites and biological substances.

With the possible exception of the period when Sir Joseph Banks was President, physics has been regarded by the Fellows of the Royal Society as the senior discipline from which all others evolved. As the 19th century grew to a close, the great Lord Kelvin thought that physics had few more secrets to be revealed but along came Becquerel with his discovery of radioactivity and the resultant development of atomic science has so dominated the 20th century that it has become known as the 'Atomic Century'.

As well as the many nuclear discoveries made by its Fellows and Foreign Members, the Royal Society, as an organisation, has been intimately involved in the development of atomic science in Great Britain. The Society encouraged, by publishing their early papers, both Rutherford and Bohr even before they were involved in nuclear matters and it was in his 1920 Royal Society Bakerian Lecture that Rutherford first predicted the existence of the neutron. During the 1930s exodus of Jewish scientists from Europe, the Royal Society played an active part in finding them academic positions and helping them in other ways to settle in Great Britain. These scientists, including Peierls, Frisch, Simon, Kurti and Rotblat, made great contributions to the British nuclear programme. After the start of hostilities the Society facilitated the creation of, and acted as host for, the MAUD Committee, the body which co-ordinated Britain's early atom bomb programme.

Nuclear physics has also provided great insight into the structural, chemical and physical properties of metals. In addition, the technological manifestations of atomic science, nuclear weapons and atomic power stations, generated a multitude of metallurgical problems the solutions to which has involved the undivided attention of a large proportion of

Britain's metallurgical talent for almost a third of a century. The subsequent diaspora of these metallurgists, particularly into the universities, has greatly enhanced the strength of the metallurgy and material science disciplines within these academic institutes. It is a truism that physical metallurgy owes more to atomic energy than atomic energy owes to physical metallurgy. Taking all this into account, it is not surprising that Chapter 18, 'Metals and Atomic Energy', has turned out to be the longest in the book.

The development of atomic power follows what could almost be described as a classic British pattern. The technology was based on brilliant fundamental scientific discoveries, many of British origin, and the dedication and skill of the pioneering technologists brought early success – remarkably, during the 1970s, Britain generated more nuclear electricity than the combined total of the rest of the non-communist world. Then, a combination of unwise government and industry-based decisions, and international factors, first stalled the progress of the industry which then started to decline.

Britain's pattern of failure to build commercial success upon scientific excellence started about 1870 and Britain's relative industrial weakness was vividly illustrated at the outbreak of war in 1914 when it was discovered that the electrical and steel industries were only half the size of Germany's. Britain would have lost the war by 1916 were it not for massive imports of steel from America. Much the same happened during the 1939–45, so apparently Britain was unable to learn effectively from experience, and this is borne out by the continued decline in the international industrial league from 1945 to the present time. The cause of this malaise lies deep in Britain's imperial past, religious heritage, and in the education system and social structure. During all of this period the Royal Society has been *ipso facto* Britain's National Academy of Science, which forces the question 'can the Society escape some responsibility for the continuing failure of Britain's industrial policy?'.

This and other questions are addressed in chapter 19, the last of the 'thematic' chapters. It is suggested that a revolution is needed in the government's understanding of scientific issues and it is hinted that impending revisions of the constitution of the House of Lords, should ensure an effective balance of representation of the Royal Society and other scientific and technological bodies.

In a considerable Section following the thematic chapters, at the end of the book, brief biographies are given of past Fellows and Foreign Members mentioned in the text, together with some biographical notes of selected contemporary Fellows. Additional information, for example, on the recipients of Royal Society Awards, a bibliography and references are presented in a number of Appendices.

D. R. F. West
J. E. Harris

The Royal Society

The Royal Society is an independent academy promoting the natural and applied sciences. Founded in 1660, The Society has a dual role, as the UK academy of science, acting nationally or internationally, and as the provider of a broad range of services for the scientific community. It responds to individual demand with selection by merit, not by field. The Society's objectives are:

- To promote and recognise excellence in science and its application.
- To provide independent, authoritative advice, notably to the UK Government, on science and engineering-related matters and to inform public debate, through studies, submissions and reports.
- To support and encourage research and its application, through research fellowships and grants to individual scientists, and to disseminate the results of research through meetings, lectures, exhibitions and publications.

Fellows are elected for their contributions to science, both in fundamental research resulting in greater understanding, and also in leading and directing scientific and technological progress in industry and research establishments. A maximum of forty new Fellows may be elected annully.

Up to six Foreign Members are elected annually, and are selected from among persons of the greatest eminence, for their scientific discoveries and attainments, but who are not eligible for election as Fellows of the Society (owing to citizenship and residency restrictions).

Acknowledgements

The authors wish to express their thanks to the many individuals and also to those organisations, whose co-operation, assistance and advice have made possible the writing and production of this book:

At the Institute of Materials, the head of the Books Department, Mr Peter Danckwerts has provided extensive and invaluable work, not only in implementing the major task of producing the book, but also through his expert advice and encouragement. Also Mr Simon Crump has made a major and essential contribution in the production work.

Ms Hilda Kaune, the Institute's Archivist, has provided photographs and information which have contributed substantially; Mrs Maureen Johnson also assisted with information.

At the Royal Society the staff of the Library, in particular Mrs Mary Samson as Archivist, have been always ready to assist with access to reference sources and to help with advice and information. Mrs Sandra Cummings has given provided valuable help, especially concerning photographs, and also Mr Rupert Baker and Mr Paul Byrne have provided information; in the Publishing Department the co-operation of Mr Phillip Hurst and Mr John Taylor has been very helpful.

At Imperial College, the Archivist, Mrs Ann Barrett has given substantial assistance. The Science Museum Library at South Kensington and Bromley Central Library have also been important resources, including the assistance of the staff.

Professor Robert Cahn FRS, Dr James Charles FREng and the late Professor Bryan Coles FRS have provided expert comments and advice on major sections of the book.

In the provision of advice and information on particular sections of the book, and photographs we have greatly benefited from the assistance given by:

Dr John A. Bisshop; Professor W. Bonfield FREng, CBE; Dr James Charles FREng; Mr I. Cottingham; Professor B. Dyson; Dr. A.L. Greer; Professor D.B. Holt; Professor Sir Robert Honeycombe FRS, FREng; Professor D. Inman; Dr Frank A.J.L. James; Dr A. Jones; Professor T. Lindley; Mrs I. McCabe; Professor Sir Ronald Mason FRS; Mr S.W.K. Morgan; Mr A.C. Partridge; Professor D.W. Pashley FRS; Dr B. Smith; Professor G.D.W. Smith FRS; Professor B.C.H. Steele; Professor E. Tanner; Dr D. Temple; Mr P.R. West; Professor A. Willoughby and Professor A.H. Windle FRS.

Photographs and other illustrations have been provided by various organisations (see following section); thanks are expressed to individuals from these organisations who have assisted in these arrangements: Miss C. Arrowsmith; Mrs S. Ashton; Miss S. Barnard; Mr I. Cottingham; Monsieur D. Devriese; Mr E. Jenkins; Mr K. Moore; Ms Lynne Nakauchi; Dr P. Thwaites.

Notwithstanding the excellent advice received, the authors remain responsible for decisions concerning the scope of the book, including the choice of Fellows and Foreign Members to be included, and the selection of information, both scientific and biographical.

The preparation of brief biographical accounts of the lives of Fellows and Foreign Members of the Royal Society has been based largely on the main references listed in the Bibliography (namely *The Royal Society Obituary Notices and Biographical Memoirs*; the *Dictionary of Scientific Biography* and the *Dictionary of National Biography*), together with *Who's Who*. (See Appendix 8). The authors have depended extensively on these texts, both for the biographies and the thematic chapters, particularly in the case of Biographical sources of the Royal Society for 20th century Fellows. Without the excellent biographical writings of the individual authors of our sources, sometimes very comprehensive in their scope, this present book could not have been written, and the present authors express their grateful acknowledgement.

D. R. F. West
J. E. Harris

Acknowledgements for Illustrations

Careful efforts have been made to seek necessary permissions for the use of illustrations which are copyright material. If there remain omissions in this respect, apologies are offered, with the intention of making appropriate acknowledgements in any future edition, if information is supplied to the authors or photographers.

Photographs are reproduced by kind permission and acknowledgements as follows:

Many of the photographs of Fellows included in the Biographical Memoirs of the Royal Society were taken by Mr Godfrey Argent (Godfrey Argent Studio); for a number of other photographs taken by Walter Stoneman, Walter Bird and Baron Studios, the copyright is held by Godfrey Argent Studio. The authors wish to express their particular thanks to Mr. Godfrey Argent for his permission for the following photographs to be reproduced in this book;

Godfrey Argent
Andrade; Charnley; Constant; Coslett; Cook; Kurti; Oatley; Orowan; Polanyi; Randall; Wallis; Weck; Whittle.
Walter Stoneman
Akers; Anderson, Sir John; Baldwin; Bairstow; Chadwick; Darwin, C.G.; Haworth; Jones, H.; Lees; Lennard-Jones; Lindemann; Massey; Pfeil; Raynor; Richardson, O.W.; Stoner; Thompson, G.P.; Van Vleck; Whittle.
Walter Bird
Bates; Boys; Bragg, W.L.; Finch; Fleck; Kronberger; Rideal; Roberts, G.; Southwell.
Baron Studios
Lockspeiser.

Lotte Meitner-Graf (copyright Anne Meitner)
Bohr; Laue.

Royal Society
Bairstow; Barlow, W.; Beilby; Bélidor; Bengough; Bergman; Bernouilli; Berthollet; Berzelius; Bone; Boyle; Bradley; Bragg, W.H.; Bridgman; Cavallo; Cavendish, H.; Children; Dalby; Dalton; Daniell; Davy, H.; Dewar; Dolland, G.; Dolland, J.; Edwards; Euler; Ewald; Ewing; Fairbairn; Faraday; Folkes; Foster, G.C.; Fowler; Franklin, B.; Frohlich; Gibbs; Gilchrist; Glazebrook; Graham, T.; Grove; Haak; Hadfield; Halley; Hardy; Haüy; Haworth; Hele-Shaw; Herschel, J.F.W.; Inglis; Jenkin, C.F.; Joliot; Joly; Joule; Kupffer; Langmuir; Leeuwenhoek; Lipson; McLennan; Mellor; Mendeleev; Miers; Mond, R;

Moxon; Nernst; Newton; Oldenberg; Oram; Parsons, W. (Rosse); Pearson, G.; Pearson, K.; Petavel; Pope; Priestley; Ramsden; Rayleigh; Richards, T.W.; Richardson, O.W.; Prince Rupert; Rutherford; Simon; Smeaton; Soddy; Sommerfeld; Taylor; Telford; Thénard; Thomson, G.P.; Thomson, J.J.; Thomson, W. (Kelvin); Townsend; Tyndall; Unwin; Watt, J.; Wheatstone; Wilson, A.J.C.; Wollaston; Yarrow; Young.

Institution of Civil Engineers
Baker, B.; Barlow, W.H.; Clark, W.T.; Cubitt; Hawkshaw; Mallet; Rennie, John; Rennie, Sir John; Vignoles.

Institution of Electrical Engineers
Ampère; Clark, J.L.; Crompton; Crookes; Ferranti; Fleming, J.H.; Forrest, J.S.; Gay-Lussac; Herschel, F.W.; Muirhead; Oersted; Swan; Varley, C.F.; Volta.

Council of the Institution of Mechanical Engineers
Anderson, W.; Armstrong, W.G.; Bailey, R.W.; Bramwell; Brunel, I.K.; Gough; Hinton; Kennedy; Lanchester; Locke; Parsons, C.A.; Stephenson R.; Robertson; Whitworth.

Institute of Materials
Abel; Allen, N.P.; Arnold; Bell; Bessemer; Cavendish, W.; Chesters; Desch; Dorey; Finniston; Frank; Goodeve; Hatfield; Heycock; Hume-Rothery; McCance; Mott; Neilson; Neville; Parsons; Petch; Portevin; Richardson, F.D.; Rosenhain; Samuelson; Siemens; Snelus; Sorby; Stead; Swinburne; Sykes; White;
Turbinia, (Chapter 8). Periodic Table, (Chapter 11). Illustrations of fractures of pressure vessel and ship, (Chapter 14).

Johnson Matthey and Co. Ltd
(Sources: (1) Donald McDonald and Leslie B. Hunt, *A History of Platinum and its Allied Metals*, (2) *Platinum Metals Review*.)
Bacon; Buffon; Callendar; Chevenix; Crell; Griffiths, E.H.; Guyton de Morveau; Ingenhouz; Johnson P.N.; Klaproth; Lavoisier; le Chatelier; Magnus; Matthey; Mond, L.; Odling; Pepys, W.H.; Powell; Pushkin; Rideal; Roberts-Austen; Smee; Ulloa; Vauquelin; Wohler.
(Chapter 4. Medallions made by Graham; production of standard metres and kilograms; X-section of iridium–platinum alloy.)

Birkbeck College, University of London
Bernal.

California Institute of Technology
Pauling.

University of Illinois and Dr William A. Bardeen
Bardeen, John.

Imperial College of Science, Technology and Medicine
Archives:
Carpenter; Foster, C. le N.; Gowland; Guthrie; Hunt; Percy; Playfair; Stokes; Thorpe; Tilden; Willis.
Physics Department (Nick Jackson):
Coles.

Centre for Advanced Space Studies, Lunar and Planetary Institute, Houston
Urey.

Royal Military College, Sandhurst
Barlow, P.

St John's College, University of Cambridge
Love.

University of Sheffield
Eshelby; Sucksmith.

Crown Copyright
Published with the permission of the Defence Evaluation and Research Agency.
Griffith, A.A.

Akzo Nobel Ltd
Lord Kearton.

ICI Chemicals and Polymers Ltd
Birchall.

NNC Ltd
Franklin, N.L.

Rolls-Royce Plc
Hooker; Whittle and gas turbine, (Chapter 14).

UKAEA
Cockcroft; Fermi; Frisch; Glueckhauf; London, H.; Marsham; Marshall; Peierls; Penney; Spence.

Fayer, (Photographer)(Acknowledgements)
Skinner.

John Edward Leigh, Cambridge, (Photographer)
Bowden; Evans, U.R.

Maull and Fox, (Photographers)
Roscoe

Professor D. Shoenberg, FRS
Kapitza; Mendelssohn.

Cliché Bibliothèque Nationale de France, Paris
Réaumur.

McGraw Hill
Reproduced with the kind permission of McGraw Hill (illustrations from S.P. Timoshenko: *History of Strength of Materials*, McGraw-Hill Publishing Co, London, 1953.)
Kirchoff; Lamb; Maxwell; Neumann; Poisson; Poncelet; Prandtl; Rankine.
Brittania bridge section, (Chapter 9); fatigue machine, (Chapter 14).

A.T. Balkema and Balkema Rotterdam
Rossmanith, H.P., (Ed.), *Fracture research in Retrospect – An anniversary volume in honour of G.R. Irwin's birthday*, ISBN 90 5410 679 4, 1997, 584 pp.
Irwin.

Chapman and Hall, London
(Acknowledgements for illustrations from: from Sir Robert Hadfield, *Faraday and his Metallurgical Researches*, 1931 and *Metallurgy and its Influence on Modern Progress*, 1925.)
Banks; Barrett; Boulton; Bottomley; Brande; Christie; de la Rive; Faraday; Gore; Hopkinson, B.; Hopkinson, J.; Papin; Preece; Priestley; Rumford; Wedgewood.

Note: acknowledgements to the sources of other illustrations are made in the text as appropriate.

CHAPTER 1

Metals and the Origins of the Royal Society

Introduction

This chapter begins with a *tour d'horizon* of man's use of metals from the earliest times up to the 17th century before paying especial attention to how the metallurgical and mining industry has been a factor in preparing the ground for the creation of the Royal Society. Next consideration is given to other factors which led to the scientific revolution in Western Europe culminating in the formation of the Royal Society.

Man's use of metals

'Native' metals

It is not known with any degree of precision when man first used metals although it is quite certain that it all started within the period now referred to, paradoxically in this context, as the Stone Age and it involved the use of those metals which can be found in the 'native' form – gold, silver and copper. The two former metals are not particularly useful for any utilitarian purpose whereas, of course, copper is extremely important in mans' hesitant progress towards civilised life.

Native copper is far more common than is usually supposed. For example, at the beginning of the twentieth century no less than 15% of the huge consumption of copper worldwide came from native copper deposits, mostly in the Lake Superior area of North America. Moreover, most significant deposits of copper ore also contain some native copper. Its use by man has a long history; a completely oxidised native copper bead discovered in Ali Kosh in Iran has been dated as being manufactured in the 7th millennium BC, for example. By far the largest number of native copper artifacts, dating from 3000 BC to AD 1400, have been found in the above-mentioned Lake Superior district.

Meteoric iron also found application in the lithic period, and as it is almost invariably alloyed with 4–26wt% nickel it can readily be distinguished from smelted iron in archaelogical artifacts. However, because of its very high strength and brittleness and very high softening and melting temperature, it is much more difficult even than native copper to manufacture into useful products. The earliest artifact, a bead, dates from 3,500 BC, but knives of natural iron have continued to be made and used by Eskimos until quite modern times, until the 19th century in fact.

Smelted metals

Turning now to the historically-speaking far more important product – smelted copper – the earliest examples have been found in Iran and date from about 3800 BC and this can be taken as the earliest start of the Bronze Age (or rather as it should be called in its early stages, the Impure Copper Age – deliberate manufacture of bronze by adding tin to a copper melt came rather later). The appreciable transitional period when lithic people occasionally used native copper artifacts has become known as the Chalcolithic Interregnum.

Smelting of copper ore offered the prospect of much more widespread availability of the metal and came into general use in about 3800 BC. The smelting of iron ore started about a millennium later – an iron artifact was found in the tomb of Tutankamen and the military might of the Hittites was attributed to their skill in making iron weapons. The iron was a form of sponge iron and could be hammered into useful shapes.

As far as the manufacture of steel objects is concerned, it is not known when steel was discovered (produced by carbonising sponge iron) but the Romans had steel weapons. This carbonising process was used at least until the 18th century when the crucible method of steel production was developed in England, and marked, at least as far as the West was concerned, the start of the Industrial Revolution. During the whole of the first one and a half millennia of the Christian era, there is little information on developments in manufacturing steel artifacts, such as armour, though the writing of Pliny (AD 23–79) contains some valuable information.

Some early sources concerning the beginning of metallurgy

Aristotle (384–322 BC) thought metallurgy a lowly function and he argued that philosophers were of the first rank and should not be subjected to unwholesome or disgusting employments such as bending over furnaces inhaling noxious steams or torturing animals or handling dead bodies. He went on: 'for such studies as the heating and mixing of bodies offer to inquisitive curiosity, the naturalists of Greece trusted to slaves and mercenary mechanics, whose poverty or avarice tempted them to work in metals or minerals'.

In recent times the distinguished archaeologist, Gordon Childe, considered that when lithic people acquired metallurgy they fell from grace and abandoned arcadian idealistic communism for a society based on magic, metallurgy and hereditary power. Presumably he had in mind that with mining came a form of capitalism and power remained in the hands of a few as mining and metallurgy skills were handed down in secret from father to son.

Childe may have got his inspiration from the *Book of Enoch* (compiled in the last two centuries BC) in which there is an account of some of the consequences when the fallen angel, Azazel, taught mankind metallurgy:

> And Azazel taught men to make swords, and knives and shields, and breastplates, and made known to them the metals and the art of working them, and bracelets, and ornaments,

> and the use of antimony, and the beautifying of the eyelids, and all kinds of costly stones, and all colouring tinctures. And there arose much godlessness, and they committed fornication, and they were led astray, and became corrupt in all their ways.

Staying with *Enoch*, as the end of the world approaches it appears that metals would be among the first casualties:

> And it shall come to pass in those days that none shall be saved,
> Either by gold or by silver,
> And none shall escape.
> And there shall be no iron for war,
> Nor shall one clothe oneself with a breastplate.
> Bronze shall be of no service,
> And tin shall not be esteemed,
> And lead shall not be desired.
> And all these things shall be destroyed from the surface of the earth,
> When the Elect One shall appear before the face of the Lord of Spirits.

In the Biblical writings of the Old Testament, metals receive frequent attention including their processing (e.g. refining and casting) and various peaceful and military applications. The first mention (*Genesis* Ch. 2, v. 11) refers to the River Pison flowing through the land of Havilah 'where there is gold'. The historian Charles Webster in his book *The Great Instauration: Science, Medicine and Reform 1626–1660* extrapolated from this statement to suggest that 'by exegetical ingenuity, Adam could be transformed from a simple gardener into a proficient metallurgist'. However, this is not justified from the Biblical record which does not refer to metallurgical activity until *Genesis* Ch. 4, v. 22, where, among Adam's descendants was Tubal-Cain, son of Lamech, 'who forged all kinds of tools out of bronze and iron'. This ancient period has been thought of as being marked by the development of the whole range of refining and alloying. In the period of conflict between the people of Israel and the Philistines, the latter were the first to use iron, thus gaining a military advantage. Later in the construction of the temple by king Solomon, gold, silver, brass and iron were used in large quantities. Huram, from Tyre, 'highly skilled and experienced in all kinds of bronze work' was responsible for casting a variety of bronze artefacts, such as pillars and wheels (I *Kings* Ch. 7).

Philosophers such as Francis Bacon (1561–1626) and Paracelsus (1493/4–1541) criticised the academic scholars, while miners and craftsmen were held up as models of virtue. Ultimate virtue was awarded to Solomon whose empire and wealth were based on technological and scientific foundations. It was not for nothing that Bacon's House of Science and Technology (upon which some believe the Royal Society was modelled) was called Solomon's House.

Metallurgy in the 16th and 17th centuries

Although by the early 1500s the mining and processing of metal ores was an international industry of great age and importance, it had not found a chronicler of note. This changed in 1540 with the publication by Vannocio Biringuccio (1480–c1539) of *De la Pirotechnia* which was the first printed work covering a considerable part of the mining and metallurgical industry. Biringuccio was well qualified to tackle this momentous writing task as he had spent his working life in the metallurgical industry, first as a manager of iron mines, then as an armourer, gunfounder and military engineeer, finally becoming director of the Papal foundry and munitions in Rome.

In 1556, sixteen years after *De la Pirotechna,* was published *De re Metallica* by Georg Bauer (1494–1555), better known by his Latin name Agricola. His famous book was published postumously, but it had a very considerable impact and was the definitive work on mining for a century or more. Agricola was an important natural philosopher as well as author. He opposed the idea that metals contained the seeds of their own development rather like plants and he objected to the alchemical linking of the planets with the metals, which argued that the metals were formed through the agency of a heavenly force on stones. Agricola considered that to the conventionally-accepted six metals (gold, silver, iron, copper, tin and lead) should be added mercury and bismuth and later he suggested including antimony. He also thought, contrary to the received wisdom of Aristotle, that many more metals would one day be discovered.

The central question to be addressed here is whether or not the existence of the mining and metallurgical industry had an initiating influence on the scientific revolution. This is a question to which there is not a certain answer, but there are three possibilities. In the first place, a number of the savants and philosophers who played an important role in the creation of the scientific revolution were strongly influenced by having contact with the mining industry either during their childhood or during their subsequent careers. This is certainly true of Agricola, his contemporary Paracelsus and, most importantly, Francis Bacon.

Secondly, and linked to the above, the skills of the craftsmen in metal work of all description tended to push the natural philosophers towards a more experimental approach to science. This was certainly true of Bacon, and William Gilbert (1544–1603), England's greatest 16th century natural philosopher, too has written of his profound respect for metal workers, particularly those who worked in iron. Gilbert wrote the following praise of iron: 'Its use exceeds that of all other metals a hundredfold; it is smelted daily; and there are in every village iron forges. For iron is foremost among metals and supplies many human needs, and they the most pressing.' It is also the case that as the metal ore deposits became progressively more exhausted, deeper mines had to be sunk and the concomitant more complicated technology stimulated scientific activity.

The final contribution of metallurgy to the scientific revolution (and hence the creation of the Royal Society) appears the most obscure, but is the most fundamental. It is to do with assaying, which inevitably accompanies mining and metallurgical activities. Assaying is based on the principle that a pure metal is a final entity which cannot be broken down further into component parts and this contradicts Aristotle's theory that all matter ultimately consists of four elements, earth, air, fire, and water. This demonstrable belief played a significant part in undermining Greek science and hence assisted in initiating the scientific revolution.

The birth of chemistry

The emergence of chemistry as an independent discipline can be traced to two major influences, mining, metallurgy and assaying as described in Biringuccio's *De la Pirotechnia* and Agricola's *De Re Metallica* and the iatrochemistry (medical chemistry) ideas of Paracelsus (1493–1541). The latter's main influence arose from his view that the business of chemistry was not the transmutation of metals, though he held this to be possible, but with the preparation and purification of chemical substances for use as drugs. After him, his version of chemistry became an essential part of medical training and for nearly a century doctors were divided into paracelsists and herbalists, who kept to the old herbalist drugs.

The influence of the metallurgical industry on the beginnings of chemistry ranked with that of Paracelsus and it came about in the following manner. Agricola's *De Re Metallica* made a big impression on the metallurgist and assayist, Lazarus Ercker (1530–1594) during the preparation of his seminal work *Treatise on Ores and Assaying* (published in 1574).* Ercker was hostile to alchemical ideas and extended his terms of reference to include not only assaying and refining techniques but also the preparation and properties of various chemicals necessary for this work, i.e. acids, salts and saltpetre. In turn his fellow native of Saxony, Andreas Libavius (1560–1616), relied heavily on Ercker's writings in the preparation of his book *Alchemia* which is generally regarded as the first chemical text book. In this book, published in 1597, Libavius drew together the techniques and preparations of all the the chemical arts and crafts and presented chemistry as an independent discipline.

The scientific revolution – other factors

Metallurgy was of course only one of many influences leading to the scientific revolution taking place in Western Europe during the 16th and 17th centuries. Perhaps the first question to consider is how it came about that the revolution occurred in Europe rather than in India or China which had much longer traditions in mathematics and technology respectively. There is no simple or generally-agreed answer to this question but clearly the structures of the different societies must have played a role. One essential for science to flourish is communication and in the nature of things this can occur more readily in

*In 1683 a translation from the German of Ercker's book under the title *Fleta Minor* was published, by Sir John Pettus. The Royal Society records refer to Sir John Pettus (c.1640–1698; FRS 1663) but mention is also made of another Sir John Pettus (1613–1690). Reference to these and other sources leaves doubt as to which Pettus was the translator. A modern translation of the book by A. S. Sisco and C. S. Smith was published by the University of Chicago Press, Chicago, in 1951.

urban environments. In contrast to China and India where vast areas were controlled from very few urban centres, very large numbers of towns existed in Western Europe; a thousand of these came into being from the Middle Ages onwards.

There was also a difference in the structure and control of the cities between East and West. Whereas in oriental societies the merchants were rather despised, within western cities they flourished, the extreme examples being the courts of the Medici in Florence and the Doge in Venice. The merchants tended to support the teaching and practice of science and mathematics because employment of those trained in these subjects tended to increase their profits. In London the most important merchant-benefactor was Sir Thomas Gresham (1519–1579) from whose estate was established in 1598 Gresham College; its professors and friends were part of the 'invisible college' which in due course became the Royal Society.

The wealth and size of the town or city was also of importance; to form a scientific group of sufficient size to enjoy creative discussions and joint enterprises there needed to be sufficient numbers of scientists with leisure to spare on such activities. This required wealthy citizens which in turn was only likely to be found within a large overall population. London fitted these requirement; it was wealthy through international trade and a census of 1631 indicated a total population of 130,000, which was about an order of magnitude more than any provincial city at that time. (A city which could support a prosperous theatre during the late Elizabethan and early Stuart times was likely to be able to afford a scientific society).

Instruments and the practical arts

Bacon placed great emphasis on the experimental approach to scientific discovery (indeed he himself caught a chill and died when carrying out an experiment to freeze a chicken) but experimental science owed a great debt to the skills of the craftsmen, as Bacon himself acknowledged. Throughout the Middle Ages huge strides had been made in the mechanical arts; those linked to mining and metallurgy have already been mentioned, to which can be added water engineering, shipbuilding, printing, casting and manufacturing guns, glass-making, making locks and many more. As well as mechanical skills the availability of instruments assisted the experimentalists. Of those instruments which were invented around the turn of the 17th century mention must be made of the telescope (c. 1608), microscope (c. 1592), barometer (c. 1643) and the pendulum clock (c. 1592). The art and practice of woodcuts and engraving were of importance in illustrating scientific texts.

Role of classical and medieval science

There are distinguished authorities who argue that 17th century science can be regarded as the second phase of an intellectual movement which began when the philosophers of the 13th century read and digested, and later printed, in Latin translation the great scientific authors of classical Greek and Islam. Even though much of the Greek science was eventually rejected, some was retained and became of central importance in the new philosophy, and other Greek ideas, though fundamentally flawed, were the stimulus for the evolution of theories of profound importance.

An example of an important rediscovery of important classical documents was the finding of a complete text of *De Rerum Natura* by Lucretius (c.99–55 BC) in the garden of a monastery in 1417. The finder was the humanist scholar, Poggio Bracciolini and this discovery of the idea of systematic atomism changed the then current concept of nature.

Some of the giants of the scientific revolution were still deeply committed to classical ideas. Nicolaus Copernicus of Cracow in Poland (1473–1543) was by no means a committed measurer or experimentalist in what later became known as the Baconian tradition. In fact he made very few astronomical measurements himself and indeed showed no enthusiasm for this activity throughout his life. It was the seeking of a description of the heavens which was logical and aesthetically satisfying that moved him to advocate the heliocentric system. Moreover, he might have been very reluctant to put forward these views had he not been aware that some of the ancient Greeks, notably Aristarchus of Samos (c 250 BC), had first advocated such a system.

The great genius Johannes Kepler (1571–1630) in building on Copernican ideas was actually motivated by wanting to find harmonic relationships in the orbits of the planets, the elusive 'music of the spheres' – astonishingly, his laws came close to confirming such an arrangement. (It is of course important to remember the difficulties and dangers of the times through which Kepler lived; he suffered from much religious persecution, because of his Lutheran background; he was expelled from Graz (twice) and it was only with extreme difficulty that he managed to save his mother from being burned as a witch.) England's greatest sixteenth century natural philosopher, William Gilbert (1540–1603), who saw the earth as a huge magnet, could still talk about the it having a 'soul'. In similar vein, a few decades later, Aristotelian ideas often influenced Bacon's writings and he used such words as 'humours' and spoke of 'natural' and 'violent' motions.

William Harvey (1578–1657) stated in one of his books that 'The authority of Aristotle has always such weight with me that I never think of differing from him inconsiderably' and he remained an Aristotelian throughout his life. It was the basis of Aristotle's philosophy that circular motion is the noblest form of motion and that the circular motion of the heavenly bodies is reproduced in the microcosm of living organisms. In his account of the mechanical pumping of the blood around the body by the heart, Harvey speaks of its circular motion in Aristotelian terms and likens its circulation to the cycle of evaporation and condensation (as rain) of the moisture of the earth.

Clearly old ideas and theories took a long time to die. Even after the formation of the Royal Society a number of its members still believed in alchemy and carried out experiments in attempts to validate the alchemical practices. Somewhat surprisingly both Boyle (1627–1691; FRS 1663) and Isaac Newton (later Sir Isaac, 1642–1727; FRS 1672) could be numbered amongst the most enthusiastic alchemists.

English puritanism and the scientific revolution

It is not being chauvinistic to claim that no country in the world can match England's contribution to the scientific revolution in the 17th century. To substantiate such a claim it is only necessary to list some of the pioneering giants – William Gilbert, Francis Bacon, William Harvey, Robert Boyle, Robert Hooke (1635–1703; FRS 1663) and Isaac Newton. This section explores the extent to which puritanism sowed the seeds, or at least prepared a fertile ground, for this revolution.

A logical starting point for a survey of the 17th century is the year 1600 itself which saw the first performance in London of Shakespeare's play *Hamlet,* Kepler starting to work with Tycho Brahe (1546–1601) in his Prague observatory and the Italian philosopher, Giordano Bruno (1548–1600), burned at the stake in Rome for, amongst other 'transgressions', championing Copernicus's heliocentric theory. The publication of Gilbert's *De Magnete* (concerning magnetism) also occurred in this year.

The performance of *Hamlet,* which some would claim to be the greatest play ever written, was only possible because there was a reasonably large, literate and sophisticated theatre-going audience available to appreciate its subtleties. This need for an educated and appreciative constituency is no less true for a new scientific philosophy and both audiences were becoming available in London at that time.

Kepler's collaboration with Brahe was one of pivotal importance in the history of science; the brilliant mathematician was given access to the most comprehensive astronomical data on planetary movements ever assembled by a Western scientist. As a result of this short-lived collaboration, Kepler constructed his laws of planetary movement which influenced Newton's thinking some six decades later and thus had a profound influence on the English scientific revolution.

The burning of Bruno illustrates a marked difference between catholic Rome and protestant London, which was far more tolerant of 'the new science' – though not more tolerant of more basic departures in religious beliefs; well over 200 English Catholics had been executed by the Elizabethans as traitors by 1600. Finally, Gilbert's *De Magnete* which proposed that the earth is a great spherical magnet, and reported on various substances which produce static electricity, was the first treatise on physical science which was entirely based on experimentation. It was a fitting prelude to Bacon's new philosophy of experimental science as outlined in his *Advancement of Learning* in 1603 and *Novum Organum* of 1620.

Gilbert was court physician to Queen Elizabeth I, whose reign was drawing to an end (both she and Gilbert died in 1603). During her long reign (1558–1603) England enjoyed a great expansion in international exploration and increase in trade and this activity was of itself a great stimulus for technology and the improvements of instruments, particularly those used in navigation. The metallurgical industry expanded and that great indicator of economic activity, the output and consumption of coal, increased exponentially.

The Queen's power had been threatened by a Catholic revival but the execution of Mary Queen of Scots in 1587, and the defeat of the Spanish Armada a year later, had effectively removed this danger. However, as often happens when an external threat is removed, internal tensions within the indiginous Protestant church multiplied and sowed the seeds for the Civil War of 1642. Puritan anger was already rife because of the brutal persecution of Presbytarians by John Whitgift, archbishop of Canterbury, in 1583. The lot of the Puritans improved somewhat during the reign of James I (1603–25) but their effective persecution by Charles I led to the outbreak of the Civil War in 1642, Charles's execution in 1649 and the eleven years of the strict Puritan rule of Oliver Cromwell.

It is true that very few scientific discoveries of note were made actually during the Puritan reign, and that it took the restoration and royal patronage to launch the Royal Society. Nevertheless, the Puritans derived their theology from the Bible as the revelation of God's word and his values; they also believed that the created world shows God's design which could be investigated through their scientific endeavour. In their interpretation of Old Testament prophecy, the verse in the book of *Daniel* (Ch. 12, v. 4) that 'knowledge shall be increased' may have encouraged them to scholarly activity. There is support for the idea first put forward by Robert Merton in 1938 that Puritan values – hard work, attention to detail, single mindedness, and working things our for oneself rather than relying on received authority – created an atmosphere conducive to scientific discovery. When Puritan virtues are added to Baconian principles of induction and reliance on experiments then a powerful framework for scientific endeavour is indeed created.

Scientific letters – the role of the intelligencers

The act of creation in science may be a lonely experience but once a theory is formulated, or an important scientific result obtained, it is essential to tell one's colleagues about it, wherever they may be located. If scientists fail to communicate how do they know their ideas are original or their experiments unique and how can they subsequently claim precedence for their thoughts and results? These days a scientist will write up his theory and experimental data in the form of a paper or shorter communication (often referred to as letters) which will be published in a scientific journal. But when considering times before the scientific journal was invented, the chances are that a scientist would write to an 'intelligencer', an often self-appointed collector and transmitter of scientific letters. They formed a network of scientific correspondents, and it is no exaggeration to say that the scientific revolution could not have taken place without their co-operation.

The 'father' of the intelligencers was the French Minim monk, Marin Mersenne (1588–1648), who after studying theology at the Sorbonne joined the Franciscan order of Minims whose monastery, near the Place des Vosges in Paris, he entered as a young man; he lived there the rest of his life. The monastery became a centre of scientific life for Paris and the cleverest men of the time visited him there. His correspondence reached from London to Tunisia, Syria and Constantinople

Mersenne had strong links with London. An English correspondent wrote asking for 'any new observations, magnetical, optical, mechanical, musical, mathematical' that Mersenne might have from Italy or from Paris, and at the same time reported that he in turn would soon be sending a short treatise on the Roman system of measurement and another on the pyramids of Egypt. Mersenne sent out letters giving details of Parisian

Theodore Haak

Henry Oldenburg

experiments with telescopes and a new definition of the problem of the cycloid and reported on the chemistry of tin.

One of Mersenne's contacts in England was another intelligencer, Samuel Hartlib (1600–1662). Hartlib's circle of friends included Robert Boyle, John Milton (1608– 1674), John Beale (c. 1603–1683; FRS 1663), Seth Ward (1617–1689; FRS 1663) and Sir William Petty (1623–1687; FRS 1663), and he corresponded with scientists all over Europe and as far-afield as New England. He was an idealist who imagined that much practical benefit would arise from scientific endeavour.

As Hartlib's work-load grew he delegated some of his letter-writing tasks to Theodore Haak (1605–1690) an exile from the Palatinate, who established frequent contact with Mersenne from whom he received a vivid account of scientific activity in France. Accounts from Mersenne of the success of holding regular scientific meetings in Paris persuaded Haak that similar meetings should be held in England. In fact Haak may have been responsible for organising the early scientific meetings at Gresham College, which in due time led to the formation of the Royal Society. Haak's seminal contribution to the creation of the Society deserves more recognition.

Hartlib also delegated his letter-writing activities to the remarkable Henry Oldenberg (c.1618–1677; FRS 1663), an immigrant from Germany who was fluent in English, French and Italian as well as in his native German. This fluency in many languages, together with his charm and tact, enabled him gradually to displace Hartlib as the chief 'intelligencer' in England and led, almost inevitably, to his election to the Royal Society in 1661 and his appointment as its first (joint) secretary. To his foreign correspondence was added a great deal of letter-writing to natural philosophers working in England who had scientific findings they wished to communicate to the new society. Letter writing was not without its hazards. In 1667 Oldenburg was imprisoned for a few months in the Tower of London for a comment of his in a scientific letter which the Secretary of State considered a criticism of his conduct of the Anglo-Dutch War. Scientific letters grew in importance and would in most cases be of interest to others besides the recipient. To meet this difficulty scientists in Paris would put their ideas in a letter to a friend, have it printed and then send out hundreds of copies.

Such developments stimulated Oldenburg in 1665 to publish the world's first scientific journal, *Philosophical Transactions*, which in due course became an important publication for disseminating the new science; it is still being published today. It is interesting that the journal was owned by Oldenburg himself and only became an official publication of the Society a hundred years later. The launching of the scientific journal is one of the Royal Society's (and Oldenburg's!) greatest contribution to science and it arose directly from the exchange of scientific letters.

The Formation and Growth of the Royal Society

Gresham College

The College, which opened its doors to scholars in 1598, played a central role in the formation of the Royal Society. It was founded under the terms of the will of Sir Thomas Gresham and was located in his former home in Bishopsgate Street in the City of London. Its seven professors, probably all of whom were Puritans, taught: rhetoric, divinity, music, physics, geometry, astronomy and law. Science and mathematics thrived at the College and the practical applications of the new knowledge was a feature of the courses. Links were forged with the naval dockyard across the river and particular attention was given to navigation and ship design.

In 1645 at the College, and in nearby venues, a scientific club was formed which was to constitute the nucleus of the Royal Society some fifteen years later.* This group included John Wallis (1616–1703; FRS 1663) and Francis Glisson (1597–1677; FRS 1663). Incidentally, tradition has it that these meetings were first held at the suggestion of Theodore Haak. Discussions were restricted to scientific topics such as the weight of the air and the circulation of the blood. As the club had no permanent home and met on an *ad hoc* basis, it was referred to by Boyle as an 'invisible college'.

One of the reasons those interested in natural philosophy felt the need to form such a scientific club was that there was no other suitable forum available to them for such discourses. At that time the universities had curricula firmly based on traditional scholasticism; textual and ecclesiastical authority held sway. The dons were no more expected to do research than present-day schoolteachers are expected to publish research papers.

*Note: Elections to the Royal Society began in 1660, but in the Record Book of the Society the first dates of election are shown under the year 1663. In this connection it should be noted that Lawrence Rooke, one of the Founder Fellows, died in 1662, and he is not included in the list of Fellows in the *Record Book*.

After the victory of the Parliamentarians in 1646, Royalist professors were dismissed from their chairs at Oxford and Cambridge and the promotion prospects of avowed Puritans improved considerably. Thus it was that John Wilkins (1614–1672; FRS 1663) became Warden of Oxford's Wadham College, Wallis was appointed Savilian professor of geometry and Jonathan Goddard (1617–1675; FRS 1663) assumed the role of Warden of Merton College.

Wilkins and Wallis formed an offshoot of the Gresham College Club at Oxford and regular meetings were held at the lodgings of William Petty and at Wadham College. Robert Boyle became an active participant in 1654 and later on Christopher Wren (1632–1723; FRS 1663) and Robert Hooke joined the group. From about 1650, coffeehouses were opened at Oxford and became focal points for leisurely philosophical discussion. They were frequented by those who were to become Fellows of the Royal Society; in fact they have been called 'the cradle of the Royal Society'. Following the Restoration, the centre of gravity moved back to Gresham College where, on November 28th 1660, following a lecture by Christopher Wren, the historically-important decision was made to form what eventually became the Royal Society. According to *The Royal Society Journal Book* volume 1:

> ...it was proposed that some course might be thought of to improve this meeting to a more regular way of debating things, and according to the manner of other countries, where there were voluntary associations of men in academies, for the advancement of various parts of learning, so that they might do something answerable here for the promoting of experimental philosophy.

The academies of 'other countries' they must have had in mind were France's Montmor Academy (formed 1648) and Italy's Accademia del Cimento (1657). The latter has the honour of being the first 'modern' scientific academy, but it was dissolved in 1667 leaving the Royal Society as being the academy of science with the longest continuous existence. Closely rivalling the Royal Society in longevity though is the Académie Royale des Sciences which was founded in 1666 in Paris. It was to some extent stimulated into existence by the Cimento and the Royal Society, though it was quite dissimilar in character to either of these two older academies, particularly in respect to its more intimate relationship with the State.

Before leaving the subject of foreign academies, mention should be made of the Accademia dei Lincei (for the lynx-eyed ones) founded in Rome by Prince Frederico Cesi (1585–1630) in 1603, which Galileo joined on coming to Rome in 1611; the academy closed finally following Cesi's death in 1630. A more durable academy is the Berlin Academy which was established in 1700 and partly modelled on the Royal Society and Academie Royale (though it finally more closely followed the French model).

Composition and growth of the Royal Society

Of the twelve who attended the historic Gresham College meeting of 28th November 1660, ten were Royalists, one, Wilkins, was equivocal and another, Goddard, had served

Cromwell. Four were professors at Gresham College: Wren, Lawrence Rooke (1622–1662), William Petty (c 1623–1687; FRS 1663) and Goddard. Sir Robert Moray (c. 1608–1673; FRS 1663) a friend of Charles II (1630–1685; FRS 1665) informed the king of these proceedings and returned with encouragement from his Majesty.

Moray's friendship with the king ensured the granting of Royal Charters in 15 July 1662 and 22 April 1663.* Charles also presented the Society with its Charter Book (which is still in use and which all new Fellows must sign) and a handsome mace to use on important occasions. Charles's support for the Society was invaluable and his genuine interest in science was beyond question. He had a laboratory at Whitehall, where he 'was far more active and attentive than at the council board' and he founded a Mathematical School for instruction in the science of navigation. The latter initiative was on the advice of Samuel Pepys (1633–1703; FRS 1665); navigation studies generally were a great stimulus to scientific activity in both the 16th and 17th centuries. The King personally took part in scientific activities; in early May 1661 he stayed up all night to view through his telescope an eclipse of Saturn.

In its early years the Society evidently wanted to keep its membership for the upper echelons of the Court and those who espoused the Royalist cause – barons and above were accepted without question and the fact that a quorum of 21 was needed to elect new Fellows ensured that the prejudices of the majority prevailed, i.e. that Royalist candidates were preferred. Of the 119 Fellows in the Society in June 1663, Lewis Feuer identified 68 as having Royalist political attitudes or leanings; there were only 12 recognised Parliamentarians in the Society.

Although Catholics were few in number amongst the Fellows, they were influential. The colourful Sir Kenelm Digby (1603–1665; FRS 1663), a prominent Catholic whose father had been executed because of his involvement in the Gunpowder Plot, was a Fellow of the Society, as was another Catholic, William Howard, Viscount Stafford (1614–1680; FRS 1665), who was executed in 1680 on the false charge that he was trying to raise a Papal army in England. The Catholic who did most for the Society was Henry Howard (sixth Duke of Norfolk, 1628–1684; FRS 1666); he allowed the Society to hold meetings in Arundel House after the great fire of London and he presented the Society with the major part of his very valuable library. Another prominent Catholic Fellow was Lord Charles Herbert, later Marquis of Worcester (FRS 1673). Charles II himself had Catholic leanings; it was said he regarded that religion as being most compatible with his recreations!

In the early years of the Society the numbers of Fellows was small; in the period 1663–65 the average was 145 and a peak of 170 was reached in the period 1671–75. For the period 1696–1700 the total averaged 153, including 24 foreign members. The designation 'Foreign Member' to describe those resident abroad, (as distinct from 'Ordinary Member') does not appear in the Royal Society Record until the year 1813, when Jöns Jacob Berzelius (1779–1848; FMRS 1813) is thus listed.

*The 1663 Charter referred to the title of the Society as 'The Royal Society of London for promoting Natural Knowledge'.

The Society was not a professional group, and scientific men constituted only about a third of the Fellowship, while the remainder were Fellows by reason of their position in society or wealth , and acted a patrons of science. A. Rupert Hall records that noblemen, landed gentry or courtiers constituted 41% of the early Fellows; a substantial number had no continued interests in the meetings of the Society and did not pay their subscriptions. A further 40% were professional men including those in the medical profession; the remainder included merchants (6%) and foreigners (10%). Very few Fellows earned a living from science at any period of their lives.

Financially, the Society experienced difficulties in the early years, partly because of the failure of some Fellows to pay their subscriptions, and their consequent removal from the Register. However, when the 18th century opened there were signs of increasing membership, and some leading Fellows had built up correspondence with many leaders in science on the Continent. By mid-century the Society had achieved increased prestige and as the century progressed further the Society became recognised as the most important scientific body in the country. The Government began to seek advice from the Council on scientific questions. Also during the century the average number of elections to Fellowship increased, reaching in the last four years of the 18th century an average total membership of 501 Ordinary Fellows and 82 Foreign Members.

The substantial increase in the elections of Foreign Members during the century had been such that by 1761 their numbers constituted about a third of the total and had become a cause of concern to the Society. Accordingly the Council proceeded to enact Statutes which progressively reduced their number and in 1823 it was limited to 50. In the Royal Society today the maximum number of elections each year as Foreign Members is six, as compared with a maximum of 40 Ordinary Fellows.

The start of the 19th century saw the Society in a flourishing state with improved finances and steadily increasing membership. As the mid-century approached a peak total number of nearly 770 members was achieved. However, in the earlier part of the century the proportion of Fellows who were scientists remained at about one third , and many of the other Fellows did little to promote natural knowledge. The scientific Fellows were concerned with this situation which they saw as prejudicial to the reputation of the Society. As a result of this concern, by mid-century statute changes were made, including the strengthening of the procedures for election and restricting the numbers of Fellows elected, which led to the Society becoming a scientific institution of the highest rank with an enhanced degree of professionalism. As part of this change the total number of Ordinary Fellows had decreased to about 450 by the end of the 19th century. This number remained roughly the same during the first half of the 20th century, but substantial growth occurred later until now at the end of the century the total of Ordinary Fellows is around 1200, together with around 100 Foreign Members.

CHAPTER 2

Metals and the Early Period of the Royal Society

Introduction

From the earliest days of the Royal Society, many of the Fellows, pre-eminent figures in chemistry and physics, played key roles in the development of fundamental understanding in these subject areas during the next approximately one and a half centuries. Their work thus contributed indirectly to the body of knowledge from which metallurgy evolved as a distinct discipline in the second half of the 19th century. Also, some Fellows carried out important work directly connected with metals in relation to their processing, structures and properties. A further important area was the application of metals in many fields including the manufacture of precision scientific instruments, the development of the steam engine and in large scale civil engineering projects; here also Fellows who were distinguished engineers were prominent.

By way of background to the review of the work on metals carried out by 18th and 19th century Fellows, the significance of alchemy in relation to metallurgy is briefly considered.

Alchemy's influence on science

As science advances, old ideas, theories and practices have to be discarded. By the end of the 18th century alchemy was in retreat but was not yet finished. Its origins are obscure. It arose in both the East and the West and in part appears to have been based on Aristotelian ideas. Alchemists sought the Philosophers' Stone which could turn base metals into gold and also become man's perfect medicine, the *elixir vitae*, the elixir of life.

The essence of alchemy is in fact transmutation, and not just chemical change but passing from sickness into health, old age into youthfulness, and even passing from an earthly to a supernatural existence. In other words it sought rather cheerful objectives – wealth, health, longevity and even immortality, which is perhaps why it was such a popular pursuit over so many years. Its links with metallurgy are very strong, indeed its name derives from the Arabic *alkimia* where *al* is the definite article and *kimia* is believed to come from the Greek *chyma,* meaning to fuse or cast a metal. The first alchemists were probably metallurgists. The miners and the metallurgists, like the agriculturists, were seen to accelerate the normal maturation of the fruits of the earth, in a magico-religious relationship with nature. (In primitive societies the metallurgist is often a member of an occult religious society.)

The metals gold, silver, iron, copper, lead, tin and mercury were almost certainly all known before the rise of alchemy. Sulphur, the 'stone that burns' was very important and was known since prehistoric times in native deposits and was also produced as a byproduct when 'roasting' metal-sulphide ores. (Sulphur as an agent for change was replaced in the 18th century by the hypothetical phlogiston.) The alchemists reacted these metals, where possible, with a number of corrosive salts and claimed especial properties for the corrosion products.

In spite of the mysticism, and unrealistic objectives, and sheer nonsense, alchemy played an important part in the evolution of chemistry. The preface to the book *The Elements of the Theory and Practice of Chemistry* by Pierre Joseph Macquer (1718–1784) published in English in 1758, contains the important observation:

> Alchemists are rightly discredited for believing that chemical experiments will produce gold, but those experiments though quite useless with regard to the end for which they were originally made, proved the occasion of several curious discoveries.

If anything, Macquer was understating the importance of the alchemists who, over a long period of time, examined and tested virtually every substance known to man, thereby uncovering a good deal of the basic knowledge of the properties of various chemicals and compounds. Among its special achievements was the discovery of the mineral acids, hydrochloric, nitric and sulphuric. As well as phosphorus, alchemists are thought to have been the first to identify bismuth, arsenic, zinc and antimony. Also alchemy involved genuine attempts to investigate experimentally the structure of matter, which led to the development of important techniques.

Francis Bacon (1561–1626) at the start of the 17th century had already provided the most perceptive summary of the contribution of alchemy to science:

> Alchemy may be compared to the man who told his sons that he had left them gold buried somewhere in his vineyard, where they by digging found no gold, but by turning up the mould about the roots of the vines, procured a plentiful vintage. So the search and endeavours to make gold have brought many useful inventions and instructive experiments to light.

Chemistry and metallurgy

The publication by Robert Boyle (1627–1691; FRS 1663) of *The Sceptical Chymist or Chymico-Physical Doubts and Paradoxes* (1661) was a keystone in the foundation that eventually led to modern science, emphasising the importance of experiment to validate theory. Boyle did not accept the Greek concept of 'four elements': fire, air, earth, water. Among his important ideas was that of an 'element' which cannot be decomposed into simpler substances.

Robert Boyle

Isaac Newton

Antoine-Laurent Lavoisier

Joseph Priestley

> I now mean by Elements... certain Primitive and simple... bodies; which not being made of any other bodies, or of one another, are the Ingredients of which those call'd perfectly mixed Bodies are immediately compounded and into which they are ultimately resolved.

He was the first to distinguish definitely a mixture from a compound. Using the air pump constructed by Hooke (1635–1703; FRS 1663), Boyle demonstrated the effects of the elasticity, compressibility and weight of the air, which led to Boyle's Law, reported in 1662, showing the relationship between volume and pressure of a fixed mass of gas at a given temperature. Boyle's experimental work also included the melting of some metals, using 'a small crooked pipe of metal and glass such as tradesmen ... call a blowpipe'. He speculated on the mechanism of solidification of bismuth from his observations of crystalline facets on a fractured cast sample of bismuth.

Isaac Newton (later Sir Isaac, 1642–1727; FRS 1672) also carried out extensive work in chemistry, and was influenced by the writings of Robert Boyle. One of the topical fundamental questions that had concerned the alchemists, and in which Newton was interested, was the nature and behaviour of metals. Commencing around 1669 Newton compiled a collection of books on alchemy and chemistry; his interests included the literature of mining and metallurgy, such as Georg Agricola's *De Re Metallica* (1621) and Alvaro Alonso Barba's *Art of Metals,* 1670 and 1674. Early in the late summer of 1668 at Trinity College Cambridge, he ordered his first parcel of chemicals, including the metals antimony, silver and mercury and he also built a furnace. Chemistry remained an interest, and he worked in his laboratory after the writing of the *Principia (Philosophiae Naturalis Principia Mathematica),* and up to the time he left Cambridge in 1696.

Newton's chemical notebooks at Cambridge record information on his pyrotechnical experiments. Among substances of interest to many chemists were antimony and its sulphide ore, stibnite, from which the metal was produced by an ignition method. Cast antimony showed on its surface needle-like crystals in a star-shaped configuration ('Star of Antimony'). When antimony was melted with other metals a 'regulus' often crystalline in appearance, was formed at the bottom of the crucible. In work carried out in the early 1680s Newton examined the crystalline features of the crucible-cooled regulus produced from antimony melted with various other metals. His experiments included melting various proportions of 'regulus of iron' with copper, and he concluded that higher proportions of regulus in the mixture were preferable, presumably for obtaining striking crystalline appearances. He also used 'tinglass' (bismuth) and 'spelter' (zinc) in producing a regulus.

The discoveries of one of the great chemists of the 18th century Joseph Priestley (1733–1804; FRS 1766) greatly influenced the development of the subject. He experimented with the gas produced during fermentation and found that the layer of 'fixed air' (carbon dioxide) over a brewing vat was the same as that discovered by another highly influential scientist Joseph Black (1728–1799) in 1756. Before Priestley's studies only three gases were known, air, carbon dioxide and hydrogen; he discovered ten more, including nitric oxide, nitrogen dioxide, nitrous oxide, and hydrogen chloride. In other

Henry Cavendish

Louis Guyton de Morveau

Martin Klaproth

Antoni van Leeuwenhoek

words, Priestley was to do for the isolation and discovery of gases what later Humphry Davy (later Sir Humphry, 1778–1829; FRS 1803) was to do for metals.

Priestley's most famous discovery, possibly the most important scientific discovery of the 18th century, took place on August 1st 1774 when he obtained a colourless gas by heating red mercuric oxide (calx of mercury), using a burning glass to concentrate the sun's rays. This gas could support the combustion of a candle flame more vigorously than air and he called it 'dephlogisticated air', because it was then believed that ordinary air when it could no longer support combustion did so because it was saturated with phlogiston. He also showed that in breathing it was again the dephlogisticated air (i.e. oxygen) which was consumed. Photosynthesis too attracted his attention – he demonstrated that green plants actually produced dephlogisticated air (oxygen) from fixed air (carbon dioxide).

On a visit to Paris in 1774 Priestley informed Antoine-Laurent Lavoisier (1743–1794; FRS 1788) of his findings. In a sense Lavoisier had been preparing himself for news of Priestley's discoveries to bring to fruition his own thoughts on combustion. Lavoisier, who had a more analytical mind than Priestley, was much more willing to abandon old concepts, such as the existence of phlogiston, which had outlived their usefulness. The idea that combustibles contained a substance which they lost on burning had a long history – it was identified as sulphur by the Arabs and by the Paracelsians. The model was extended by Johann Joachim Becher (1635–1682) and by Georg Ernst Stahl (1660–1734) who named this mysterious invisible substance, phlogiston. It was considered that the reason a metal 'burned' and crumbled to an ash was that phlogiston had departed from it. Contrariwise, if a metal oxide (the calx or ash) was heated with a substance rich in phlogiston, such as charcoal, a transfer of phlogiston would take place from charcoal to oxide and as if by magic a shiny metal would be produced.

It was taken that phlogiston had no mass and from a theoretical point of view there was nothing wrong with such an assumption – after all, heat, electricity and magnetism also enjoyed this quality of weightlessness. The difficulty came when the idea had to be reconciled with experiment. When metals were 'burned', or simply rusted in air, they were considered to have lost phlogiston, yet the product, the ash, had actually increased in weight. For this to occur the phlogiston would have had to have had a negative weight, and such an idea was hard to accept. It was this need to postulate negative weight that eventually disproved the phlogiston theory.

Before he received news of Priestley's monumental discovery Lavoisier had embarked on what is now seen as a classical series of experiments. He was an extremely good experimentalist and had access to very accurate weighing scales (a good illustration of the importance of high-quality instruments in the creation of the scientific revolution).

He weighed small quantities of tin and lead to a high degree of accuracy and placed one or other of the metals in a sealed flask which he heated until no more ash, or calx, would form. He then reweighed the flask and observed no change in weight, but on letting air into the flask he found a weight change. Lavoisier was uncertain how to interpret these results until he heard from Priestley himself an account of his discovery of 'dephlogisticated air'. He realised immediately the significance of these findings and invented the name 'oxygen' (acid-maker) to replace Priestley's appellation 'dephlogisticated air'. He also coined the word 'azote' for the 'non-vital air' (nitrogen) as the residue when the oxygen in normal air had been used up by combustion.

In 1783 Lavoisier published his findings that the combustion of hydrogen in oxygen produces water, not realising that Henry Cavendish (1731–1810; FRS 1760) had preceded him. He went on to burn a number of organic compounds in oxygen and determined their composition by weighing the carbon dioxide and water produced; these were the first experiments in quantitative organic analysis. In 1789 Lavoisier reproduced his major findings in his monumental book *Traité élementaire de chimie.* He dismissed the phlogiston theory and to the established list of chemical elements – carbon, sulphur, phosphorus and all the known metals – he added oxygen and hydrogen and 'azote' or nitrogen; his list contained 23 candidates which would be accepted today as genuine chemical elements.

The subject of chemistry hardly existed before Lavoisier in particular laid down the fundamental guidelines based on quantitative experimentation and gave an accurate definition of the term chemical element as 'the last point which analysis is capable of reaching'. In noting these revolutionary achievements in chemistry the part played by electricity must also be acknowledged, particular the capability of the phenomenon to produce compounds from fundamental chemical elements and its complementary ability to divide compounds into their component parts.

At the same time credit must also be given to Robert Boyle, Robert Hooke and John Mayow (1640–1679; FRS 1678) for coming close to demonstrating, more than a century earlier, that air contained something that was essential to burning and turned arterial blood red. Boyle referred to it as 'a little vital quintessence that serves to the refreshment and restauration of our vital spirits' and Mayow called it *spiritus nitroaereus.*

It is not surprising that some consider Boyle and not Lavoisier, as the 'Father of modern chemistry'. However, his ideas did not constitute a mature theory, and this, together with an absence of good equipment and inadequate experimental techniques, and a lingering hangover from Greek animism and alchemical mysticism, led almost inevitably to the 'dark century' as far as chemistry is concerned.

Another outstanding Fellow who made major contributions was Henry Cavendish, a scientist of independent means who worked in his own laboratory on a range of topics in chemistry and physics. His investigations of the gas (hydrogen) produced by the action of acids on metals, experiments in the field of heat, and the determination of the freezing points of various materials were relevant to metallurgy.

Many experiments were carried out for the Royal Society by Jonathan Goddard (c. 1617–1675; FRS 1663), one of the founder Fellows and a member of the first Council. By profession a physician, his career included appointment as professor of physic at Gresham College. The antiquarian John Aubery (1626–1697; FRS 1663) noted concerning Goddard that the Royal Society 'made him drudge, for when any curious experiment was to be done, they would lay the task on him'. Among Goddard's papers in *Philo-*

sophical Transactions was one (1678) entitled 'Experiments on refining gold with antimony'; this paper reported on a method of refining gold by melting it with antimony, but it proved to be less profitable than cupellation.

Richard Watson (1737–1816; FRS 1769), whose varied career included appointments as Professor of Chemistry at Cambridge University and as Bishop of Llandaff, had an interest in metallurgical matters. His *Chemical Essays*, of which the first volume was published in 1781, included much information on non-ferrous metallurgy production in England, for example on the smelting of zinc.

In France the interests of Louis Bernard Guyton de Morveau (1737–1816; FRS 1788) included metallurgy and mineralogy, and he was consulted by mines and foundries in Burgundy. His industrial interests included glassmaking. As a chemist his work included measurements of cohesion between mercury and various other metals. In Germany, Martin Heinrich Klaproth (1743–1817; FRS 1795) was a leader in analytical chemistry, working with minerals from all over the world. He discovered, or co-discovered, zirconium, uranium, titanium, strontium, chromium and cerium in the period 1789–1803. Nicholas Joseph de Jacquin (1727–1817; FRS 1788) whose interests included chemistry spent a period of his varied career at the Royal Hungarian Mining Academy, Schemnitz, as Professor of Practical Mining and Chemical Knowledge, particularly introducing experimental classes. Nicolas Vauquelin (1763–1829; FRS 1823), a Frenchman, analysed many minerals and in 1798 discovered a metal that he named chromium and an oxide that was later named beryllia.

Microscopy

Robert Hooke, a great microscopist, constructed one of the most famous of early compound microscopes. In 1665 he produced his most important book *Micrographia*, a scientific milestone and a masterpiece of 17th century science, which incorporated microscopical observations reported at Royal Society meetings. This was the first great work devoted to such observations; it contained profuse illustrations. The importance of microscopy for biological sciences was revealed. There were also some metallurgical observations, including the point of a needle and a razor; of the latter he said the parts that had been finished on a grinding stone appeared rougher, 'looking almost like a plow'd field, with many parallels, ridges, and furrows, and a cloddy, as 'twere, or an uneven surface'. Hooke observed (following Henry Power (1623–1668; FRS 1663)) the melting of sparks struck from steel by flint and explained this as due to the concussion heating of a very small part of the steel.

The scientific career of Antoni van Leeuwenhoek (1632–1723; FRS 1680) began in about 1671. This was the time when he constructed his first simple microscopes or magnifying glasses, based on the idea of the glasses that drapers used to inspect cloth. The instrument utilised a small glass lens, ground by hand, and he proceeded to produce lenses of increasing quality and in large numbers; it is estimated that he obtained lenses of resolution 1 micron and magnifying power of 500. His chief area of research was the

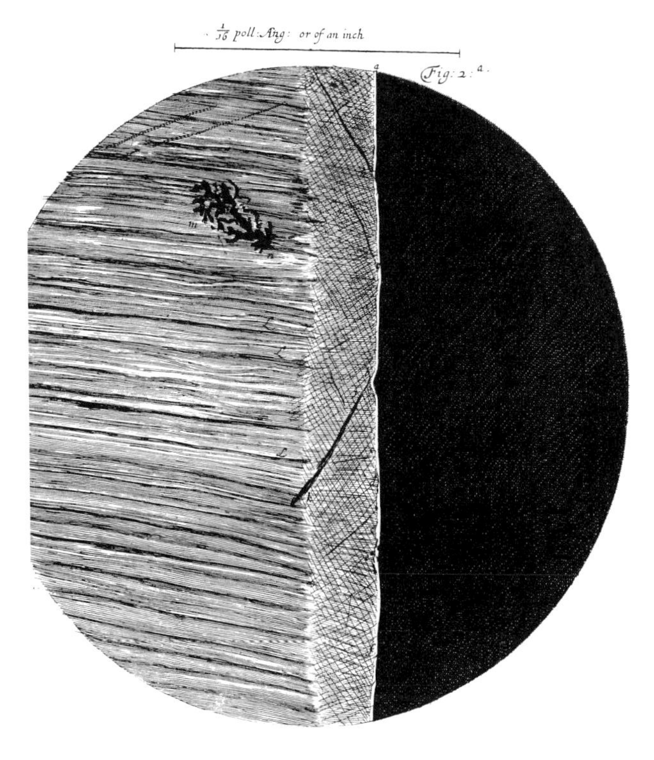

Edge of a razor as observed with a microscope, (Hooke, *Micrographia*, 1665).

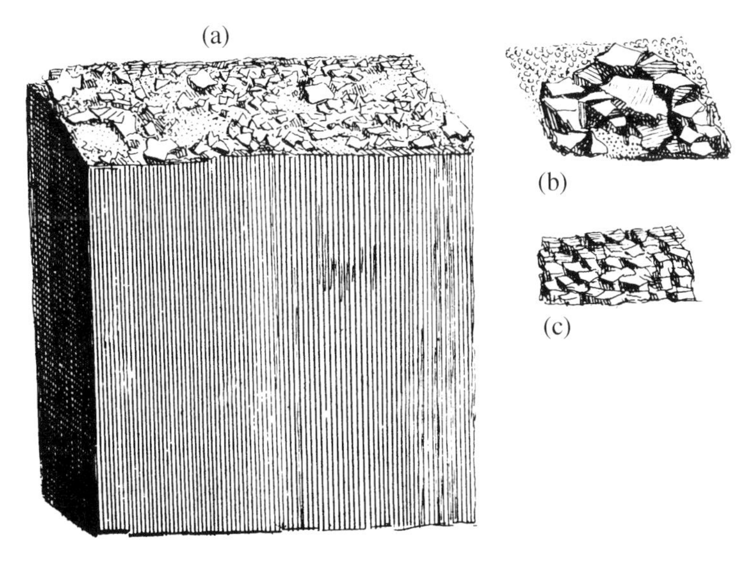

Fracture of wrought iron bar (a) natural size, (b) and (c) as appearing under a magnifying glass, (Réaumur, 1722).

microscopical investigation of organic and inorganic structures, in particular in the field of biology, in which he made important discoveries. In 1674, he recognised the nature of micro-organisms, on the basis of observations of bacteria and protozoa in water: subjects of other investigations included spermatozoa, blood vessels and blood, and plant anatomy.

In 1702 Leeuwenhoek reported in *Philosophical Transactions* surface observations of notches and holes at the edge of a razor, but these did not reveal the true structure of the metal. A year later he reported observations of a 'tree' (dendrite) of silver grown by placing a fragment of copper in silver nitrate solution, although he did not appear to recognise its crystalline nature. In 1709, in a *Philosophical Transaction* paper entitled 'Observations upon the edge of Razors etc.', he commented on features such as notches on razors. Concerning forged metals he referred to them as consisting

> of a great many small particles, the coarsest of which are always obvious when we come to break the mettals: and how often soever you melt any of these mettals and break them again after they are cold, you will always be able to discover the grainy particles thereof; but you will find them so strongly joyned and riveted in one another that they appear to be one body.

Henry Power published *Experimental Philosophy,* in three books on microscopy, atmospheric pressure, and magnetism (1664). The book on microscopy was the first book in English on this subject, and the first in any language to describe the nature of metals, with observations on mercury, gold, silver, iron, copper, tin and lead. Concerning the solids he wrote:

> Look at a polished piece of any of these Metals and you shall see them all full of fissures, cavities, and asperities, and irregularities; but least of all in Lead which is the closest and most compact Body probably in the world.

His observations on the fused globular nature of sparks struck by flint from steel led to Hooke's explanation of their origin in *Micrographia.*

Much was done to popularise the use of the microscope by Henry Baker (1698–1774; FRS 1741) who had wide-ranging interests in natural philosophy including the microscopical examination of water creatures and fossils. Among his writings on microscopy was *Employment for the Microscope* (1753), in which much space was devoted to salts, minerals and inorganic experiments generally, including observations on the growth of dendritic crystals or 'trees' of various metals.

Two other Fellows who examined metallurgical features, namely fracture surfaces, were René-Antoine Ferchault de Réaumur (1683–1757; FRS 1738), who used a magnifying glass and a microscope to examine ferrous alloys, and William Lewis (1708–1781; FRS 1745) who used a magnifying glass.

Temperature measurement

Newton published anonymously a work of metallurgical significance in *Philosophical Transactions,* a fundamental paper on thermometry. Employing his great experimental skills he made a glass thermometer, using linseed oil instead of alcohol, to extend the working range. He selected the freezing point of water and blood heat as 'fixed' points of 0 and 12 degrees. To deal with temperatures well above the boiling point of water he marked other points, including those for some low melting metals (lead and tin) using an arbitrary law of cooling, that allowed time measurements to be substituted for temperature measurements. From measurements of the cooling rate of a block of iron in the range of his thermometer, he extrapolated to the cooling rate at much higher temperatures; knowing the time for the block to cool from (say) dull red heat to the temperature of boiling water, he estimated the scale value to be assigned to red heat. Then, based on measurements of cooling times from red heat to the solidification temperatures of the metals, their melting points were assigned. 1° in Newton's scale is approximately equal to 3°C, but the relationship over the temperature range studied deviates from linearity.

Josiah Wedgwood (1730–1795; FRS 1783), whose work included studies of ceramic chemistry, had a great influence on pottery manufacture, contributing to the change from empiricism to greater dependance on scientific measurement and calculation; he also did work relevant to metallurgy. He was probably the first to develop a practical technique for measuring high temperatures; three papers on this were presented to the Royal Society, the first being published in 1773, which led to his election as a Fellow. In this first paper entitled 'An attempt to make a thermometer for measuring the higher degrees of heat, from a red heat up to the strongest that vessels made from clay can support', two types of pyrometer ('thermometer for strong fire') were reported, one based on the distinctive colour changes that occur when 'calces of iron with clay' were heated, and the other relying on the progressive contraction of clay at increasing temperatures. In the first method the colour of a disc heated with a particular batch of pottery was matched against a set of reference discs. For the second, and the better, of the two methods Wedgwood produced thermometer pieces relating to his scale, and procedures for the accurate measurement of contraction. His scale commenced at red-heat, and he reported 'that the greatest heat I have hitherto obtained in my experiments is 160°' (on his scale). In some collaborative experiments done at the Tower of London, using metals, the fusion temperatures were reported as: Swedish copper 27, silver 28, gold 32; brass 21; the welding heat of iron was given as 90–95, and cast iron melted at 130 (on his scale).

Physical and mechanical properties

Measurements of densities of alloys had long been used to obtain understanding of the nature of the interactions between metals, and in 1679 and 1680 the Royal Society investigated various combinations of copper, tin, lead and antimony in 50–50 proportions, with respect to densities and other features. Robert Boyle developed a hydrometer for measuring density in the testing of coins by comparison with standards. Francis Hauksbee

Joseph Wedgwood

Bernard de Belidor

George-Louis Buffon

Daniel Bernoulli

Leonhard Euler

(c.1666–1713; FRS 1705) a pioneer in the field of electricity, made accurate measurements of the specific gravity of a number of solids.

Quantitative measurements of the strength of metals had been made in earlier work before the founding of the Royal Society, for example in 1638 by Galileo Galilei (1564–1642). In the early years of the Royal Society, Hooke reported his important work on elastic behaviour which led to Hooke's law. Another Fellow, Sir William Petty (1623–1687; FRS 1663) presented a discourse to the Society in 1674 which included 'a new Hypothesis of springing or elastique Motions'. William Croone 1633–1684; FRS 1663), whose main research interests were in the medical field, was involved in tests on silver wires (1.5 and 4 mm diameter) carried out before the Royal Society; the outcome, however, did not appear to have been significant. Petrus van Musschenbroek (1692–1761; FRS 1734) and Georges-Louis Leclerc Comte de Buffon (1707–1788; FRS 1740) did important work on the mechanical testing of metals and other materials, in the context of building technology.

In the same field, Bernard Forest de Bélidor (1693?–1761; FRS 1726), whose scientific career involved the science of mechanics, wrote books on structural technology that were invaluable to architects, builders and engineers. Roger Joseph Boscovich (1711–1787; FRS 1761) born in Yugoslavia, a scientist of wide-ranging interests, gave mathematical form in 1758 to Newton's concept of forces between elementary particles, thus making a fundamental contribution for later developments in the mathematical theory of elasticity of crystals. The very important metallurgical research of Réaumur included hardness testing. Claude-Joseph Geoffroy (1685–1752; FRS 1715), in the area of fracture characteristics made an interesting investigation of copper–zinc alloys covering a range of compositions. John Theophilus Desaguliers (1683–1744; FRS 1714) in 1764, working on friction electric machines, pressure-welded lead spheres by twisting flattened and cleaned surfaces under load.

Perhaps best known in relation to the mechanical properties of materials was Hooke's work as mentioned above, reported in his pamphlet, *De potentia restitutiva* (1678) pronouncing the law of elasticity: the proportionality of strain to stress. In this work Hooke's law was demonstrated using several experimental situations: a long wire and also a helix of metal wire vertically suspended and loaded by weights on a scale pan with measurements made of extension; a watch spring coiled into a vertical spiral and loaded to produce angular rotation. A cantilever of dry wood loaded at the free end was found to follow the same law. Among Hooke's other contributions relevant to metallurgy was an outline of an experimental approach to crystallography; he realised that simple close-packed arrays of globular bodies would represent all of the shapes observed in alum crystals.

In the field of strength of materials the work of Buffon is of considerable interest. During his early period in Paris he was involved in investigations of the strength of timber for naval vessels, under the auspices of the naval authorities. In the course of his important work on timber, which included very large specimens, he also carried out tests on large metal specimens. In his testing equipment the load was transmitted through a

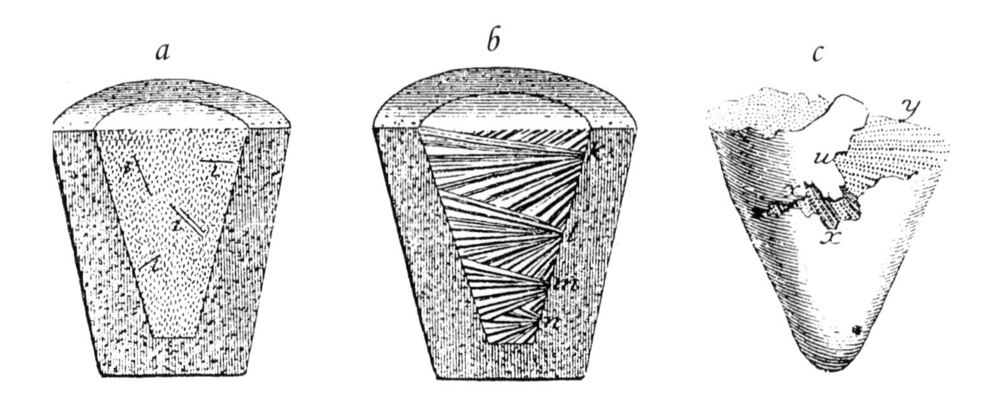

(a) and (b) Ingots of antimony solidified under various conditions and subsequently broken. (c) Lead ingot removed from crucible and, when just solid, broken by a hammer; showing intergranular fracture, (Réaumur, *Mem. Acad. Sc.*, 1724, 307–316).

forged iron shackle, which held the wooden log being tested. Possibly one of the shackles broke in service, and others were tested intentionally to provide a comparison; the shackle links had a sectional area of around 13 cm^2, and the average of four tests showed a breaking stress of about 6,000 lb/in^2 (41 MPa); this low value, indicating poor quality of the bar iron, was in contrast to wires of the same material these being twelve times as strong.

Petrus van Musschenbroek carried out a comprehensive series of accurate laboratory experiments to determine the behaviour of building materials, glass, timber and metals (including iron) under stress; the results were reported in 1729 in his *Physicae experimentales et geometricae.* In his tensile tests on glass he used dumb-bell shaped specimens, but instead of suspending them by cords as used by another previous worker he designed a new form of shackle for this and also for metal-wire specimens. His beams were square-section small rods, in most cases only ~0.4 cm^2 sectional area. The apparatus was designed for laboratory use, and there is no evidence that it was employed by those engaged in industrial practice. However, Musschenbroek obtained important results, including the fact that the resistance to buckling of a strut of a given section varies inversely as the length, anticipating the theoretical deduction by Leonhard Euler (1707–1783; FRS 1747) many years later.

Réaumur in his *Memoirs* gave what was probably the first account of commercial methods of mechanical testing of metals, including a description of a bend test using a wire or strip specimen. In his work on hardness testing, he extended the ancient test of whether a piece of metal can be filed, by using an ordered series of seven minerals as 'files', to carry out semi-quantitative scratch tests. He also used a test in which two short lengths of rod of triangular cross section were located across one another with a contact point along an edge; a blow was struck which produced an indentation at the contact point, the size of the indentation being used as an indication of hardness.

Of considerable metallurgical interest is the paper by Claude-Joseph Geoffroy in 1725 on alloys of copper and zinc, involving observations on the fracture characteristics. From microscopical observations he commented on the striae in a particular commercial alloy and on the variegated yellow and white striae seen in another alloy. He prepared alloys himself of various compositions, and found that the nature of the fracture varied with the composition. An alloy with equal proportions of brass and zinc produced a brilliant appearance and a fracture like glass. An alloy of higher zinc content (with two parts of brass and three of zinc) behaved similarly, but of a duller grain. At still higher zinc content, the metal was still brittle, but became dull and, bit by bit, developed grains with small facets, peculiar to zinc. In the direction of increasing copper, by the use of a higher brass content, the grain became larger, golden in colour, and then at even higher copper content, the alloy was ductile, with an ashy brown fracture with no fibres. From his description it can be concluded that he had covered the composition range of the gamma, beta, and alpha phases of the copper–zinc system, and related them to composition and serviceability.

Réaumur in 1724 reported observations on ingots of antimony solidified under various conditions and subsequently broken, and of a lead ingot broken near its melting point, showing intergranular fracture. In 1763 William Lewis reported experiments on gold–platinum alloys, having viewed the fracture with a magnifying glass and commenting

> the gold and platina appeared unequally mixed and several small particles of the latter were seen distinct; nor was the mixture entirely uniform after it had been again and again returned to the fire and suffered many hours of strong fusion.

In the field of fluids and phenomena associated with flow, important studies were made by Daniel Bernoulli (1700–1782; FRS 1750) which became of great importance in many fields of science and engineering, including metallurgy. His book *Hydrodynamica* was an outstanding publication presenting practical and theoretical aspects of equilibrium, pressure and velocity in fluids, it also gave the first attempt at a thorough mathematical interpretation of the behaviour of gases, based on the assumption that they consist of small particles.

Concerning thermal properties Benjamin Franklin (1706–1790; FRS 1756) suggested an experiment in which rods of different metals were heated at one end and wax rings placed at different positions along the rods, dropping off at times dependent on the thermal conductivity of the metals. Measurements of thermal expansion in relation to the production of precision instruments became an important practical activity as described below.

The work of Francis Hauksbee also included experiments on capillarity and surface forces; for example he studied the rise of fluids in tubes and between plates of glass and of brass.

Makers of precision instruments

In a significant field of practical applications, by the end of the 17th century London was the world's most important centre for making clocks and watches; knowledge of the mechanical and physical properties of materials such as brass and steel was inevitably of considerable practical importance in the construction of these mechanisms and of other precision scientific instruments. Hooke's interest in the field included his application of a spiral spring to regulate the balance of watches, and in 1675 Thomas Tompion (1639–1757) the leading man in the field, although not elected FRS, under the direction of Hooke made one of the first English watches with a balance spring. Among other famous names in this field were George Graham (c.1673–1751; FRS 1721), who in his earlier career had worked with Tompion, and also John Harrison (1693–1776) who was not elected a Fellow but was the recipient of the Copley medal in 1749.

Among the problems to be overcome in producing highly accurate timekeeping devices was the effect of change in pendulum length and rate of timekeeping resulting from temperature changes. Clock makers made measurements of the thermal expansion of metals, brass and steel for pendulums being important materials. In 1715 Graham began to work on this problem, including making experiments on the properties of metals. He designed the first precision dilatometer consisting of fixed end pieces between which the metal rod to be investigated was placed. A micrometer screw was located in a threaded hole in one of the end pieces. The specimens were heated to a series of constant temperatures and the positions of the micrometer screw in contact with the specimen were used to show the length change, indicated on a dial. Graham by 1722 had developed his mercury-compensated pendulum, which after careful testing he reported in 1726 in a Royal Society paper. The expansion of a steel pendulum was exactly compensated for by the expansion of mercury in a jar connected with it to maintain a constant vibrating length. To avoid the inconvenience of using a fluid he suggested the use of two different metals to achieve the required compensation, but he did not work out the problem. Graham's work extended beyond clocks and watches to scientific instruments, including barometers and astronomical equipment. He was also involved in the collaboration between the Royal Society and the French Academy of Sciences around 1741 on standard weights and measures. He arranged for the preparation of two brass rods graduated according to the standard yard kept in the Tower of London, for use in the programme of comparisons.

The work of John Harrison included the major achievement of the concept and production of a precise marine clock (in modern terms a chronometer) for determining longitude at sea. Before this, early in the 1720s, Harrison made a turret clock which needed no lubrication; he selected materials for this work, finding that the hardwood (lignum vitae) in combination with brass gave an excellent, oil-free bearing surface. In the mid-1720s he directed the construction by his brother James of a series of precision long-case clocks, smaller versions of the turret type. In this project he overcame the effect of change in temperature by inventing a 'gridiron' pendulum in which the bob was suspended by a series of parallel rods, alternately of steel and brass; these were arranged so that the

John Dollond

George Dollond

Tiberius Cavallo

Jesse Ramsden

John Smeaton

Frederick William Herschel

downward expansion of the steel resulting from a temperature increase was exactly compensated by the upwards expansion of the brass. Thus the effective pendulum length (the distance between the point of suspension and the centre of gravity) remained constant, leading to an accuracy of one second a month. Harrison used two of these regulators to test his other clocks. By 1730 he had a plan for his first marine timekeeper, and sought the advice of the the Astronomer Royal, then Edmund Halley(1656–1742; FRS 1678) who advised him to contact George Graham, from whom he received encouragement and support.

Over the years Harrison produced four designs of timekeeper (H1–H4). He invented a revolutionary new device, the bimetallic strip, for the temperature compensation in the spiral spring system in the H3 instrument version. Two flat strips, one of brass and the other steel, were riveted together; the higher expansion coefficient of the brass compared with steel caused the bimetallic strip to bend into a curved shape with the brass on the convex side. One end of the spring was fixed and the movement of the other end shortened the balance spring. In addition to the technical challenges, Harrison encountered substantial difficulties in satisfying the requirements of the Board set up to judge the acceptance of the instrument for the award of the prize that had been offered for success. However, eventually in 1733 he received the completion of the financial award, totally around £20,000. His son William (1728–1815; FRS 1765) was involved in the later stages of the project, and travelled on the voyage on which H4 was given a trial.

During the 18th century several other Fellows of the Royal Society also worked on measurements of thermal expansion. Musschenbroek in 1731 described a pyrometer (a term which he introduced derived from the Greek *pur* meaning fire). In the earlier literature the term was applied to instruments for studying expansion of materials on heating *viz* dilatometers. Musschenbroek's instrument consisted of a horizontal metal bar, with one end fixed and the other connected to a system of wheels to show the expansion when the bar was heated by a number of spirit lamps; the length change was transmitted by a mechanical system to a pointer on a scale enabling comparison of different metals to be made. John Smeaton (1724–1792; FRS 1753) in 1754 reported an improved version of a dilatometer (which he referred to as 'a new pyrometer') in which the elongation of the rod was amplified; the expansion of the specimen was compared with that of a brass bar which formed the base of the instrument. The bars could be immersed in a bath heated by lamps. Smeaton obtained data for a range of metals including iron, steel, lead, tin, zinc and copper. John Ellicott (c.1706–1772; FRS 1738) constructed equipment in which the expansion of a metal bar was compared with that of steel and also built a pendulum in which the different expansion coefficients of brass and iron were used to control the regularity of clocks. Jesse Ramsden (1735–1800; FRS 1786) who was prominent in precision mechanics using metals in the construction equipment such as precision balances and screw-cutting lathes, also made a high precision dilatometer (about 1784). The metal rod to be investigated was placed in a tank, with one of its ends resting on a fixed stud, with the other on a hinged spring; the length changes were determined by microscopical observation of the movement of the free end of the rod . Another Fellow, Tiberius Cavallo

(1749–1809; FRS 1779), gave a Bakerian Lecture in 1790 presenting a new type of pyrometer. This used a standard bar of glass for comparison with the metal being investigated; the mechanism used enabled the length change to be measured accurately, by means of a microscope and a micrometer.

Another among the group of skilled scientific instrument makers was John Whitehurst (1713–1788; FRS 1779) whose work included the making of clocks and of pyrometers, barometers and other instruments. Also in England John Senex (?–1740; FRS 1728) was chiefly known as a cartographer and globe maker. Edward Nairne (1726–1806; FRS 1776) had a shop in London as an 'optical, mathematical and philosophical instrument maker'; he constructed on the basis of plans prepared by Joseph Priestley, an electrical machine and also improved the astronomical apparatus at Greenwich. Also prominent as a maker of scientific instruments Edward Troughton (1753–1835; FRS 1810) included in his range of important products a precision balance, around 1797, which incorporated a large brass column.

Two other fellows, distinguished as instrument-makers, particularly in the field of optical equipment, were John Dollond (1706–1761; FRS 1761) and his grandson George Dollond (1774–1852; FRS 1819).

On the Continent another Fellow was an important figure in the field of horology. Ferdinand Berthoud (1727–1807; FRS 1764), a Swiss, working mainly in Paris developed an accurate and practical marine chronometer; this instrument incorporated improvements which have been largely retained in modern instruments. Berthoud also made a dilatometer with a device for amplifying small movements which became widely used in dilatometers subsequently.

Non-ferrous metals and alloys

The production of metallic mirrors of optimum shape and quality for reflecting telescopes was an important metallurgical activity in the early years of the Royal Society, in which Newton played a role. Copper–tin alloys (bronzes) had been in use for mirrors since antiquity, a typical composition being copper with approximately 25 wt% tin, sometimes with small additions of other elements such as arsenic. Newton experimented to find the best speculum metal; it was not possible to make an alloy that reflected as well as silvered glass, but he found that metal specula useful for astronomical purposes could be cast and polished. Even small mirrors required skill and patience to prepare. In 1672, a letter he sent to the Secretary of the Royal Society noted that speculum metal often contained small pores, visible only under the microscope, which wore away faster during the polishing of the mirror of the reflecting telescope and spoiled the image. Other workers were John Hadley (1682–1744; FRS 1717), who effected improvements in the telescope, and produced a 6 inch (15 cm) diameter mirror, and James Short (1710–1768; FRS 1737) who made improvements in metallic specula and was the first to produce a true parabolic shape.

The task of producing much larger mirrors was pursued in the latter part of the century by Frederick William Herschel (later Sir William, 1738–1822; FRS 1781). Herschel, who was born in Germany and came to England in 1757, had a distinguished career in astronomy, and recognised the need for instruments with considerable light-gathering power, for the investigation of very distant objects. Having found it necessary to carry out himself more of the work of grinding and polishing of large mirrors, he produced many telescopes for sale. His own favourite telescope had a mirror of 18 inches (~46 cm) diameter, but, in 1781, he set out to produce one of three foot (~0.9 m) diameter, and 30 foot (~9 m) focal length. The casting of the rough disk was beyond the capacity of the local foundries; accordingly he converted the basement of his house into a foundry, and experimented with various alloy compositions, but unfortunately leaking of the mould and cracking of the brittle metal prevented success. Several years later a 48-inch (~1.4 m) diameter mirror was cast in London.

Newton carried out other work on non-ferrous metals. In a *Philosophical Transactions* paper in 1672 he noted that bismuth mixed with bell metal made it white, but its fumes led to pores in the metal, while arsenic whitened the metal. A decade later in experiments on various alloys of antimony, copper, bismuth and zinc, he variously described their fractures as glassy, with glittering granulae, or steel-like. Newton also had an interest in discovering the alloy with the lowest melting point which could be prepared from the common metals. He investigated alloys produced from various proportions of lead, tin and 'tinglass' and found that ' lead two parts, tin 3 parts, tinglass 4 parts, melted together make a very fusible metal which will melt in the sun'.

Practical aspects of alloys and their processing also engaged the attention of Robert Hooke. He described the production of lead shot by pouring molten lead–arsenic alloy through a kind of colander into water, commenting that the skin that formed in the presence of arsenic contracted uniformly and gave a spherical shape to the product. This method of producing lead shot had been invented around 1650 by Prince Rupert of the Rhine (1619–1682; FRS 1665), a Royalist commander in the Civil War who had wide-ranging scientific and technological interests. The best 17th century account of bearings was written by Hooke; among his comments he referred to good performance of gudgeons of hardened steel, running in bell-metal sockets, providing that dust and dirt were excluded, and a constant supply of oil provided. In the area of crystallisation, Hooke observed hexagonal dendritic crystals of ice forming on urine, and noted that he had found similar features on the starred regulus of antimony. Similar, but much smaller figures appeared on the surface of lead containing arsenic and other impurities. He also commented on the green transparency of gold leaf either to the naked eye or to the microscope. He did not report microscopical observations of fractured metal, even though he had observed with the naked eye fractures in relation to solidification mechanisms.

Substantial metallurgical interests were held by Joseph Moxon (1627–1691; FRS 1678). His book on printing *Mechanick Exercises on the Whole Art of Printing* (1683) was the first full length manual on the subject, describing a technique of type founding which did not change materially until well into the 19th century. Type casting and the alloys used for this were among his special interests. He described a lead-antimony alloy resembling modern type metal made by fusing ~1.4 kg of antimony sulphide with an equal weight of

Prince Rupert

Joseph Moxon

Denis Papin

Claude Berthollet

iron and alloying the resulting material with ~11.4 kg of lead; this would have produced an alloy containing approximately 8 wt% of antimony. Also Moxon noted that the addition of a small amount of tin made the material more fluid for producing small letters.

Engraving was an ancient technique used by goldsmiths and silversmiths to embellish their products; lines were engraved using a sharp instrument called a burin. The technique could also be applied to engrave a metal plate, which after inking could be used for printing; this was the procedure used to make intaglioprint. Copper was commonly used for the plates, but iron, pewter, silver and zinc were sometimes used. The technique of mezzotinting, invented by Ludwig von Siegen (1609–?), was a system of intaglio work involving the formation of furrowed indentations on the plate crossing each other. In 1654 it was demonstrated to Prince Rupert in exile, who subsequently introduced and used it in England, producing the indentations with a knurled wheel.

A device for casting medallions was demonstrated to the Royal Society in 1684 by Denis Papin (1647– c.1712; FRS 1682) who was born in France and came to England in 1675; casting in vacuum followed by the application of air pressure improved the surface detail achieved in the casting.

Alloys from the copper–zinc system (brasses) had been used since Greek times, and brass manufacture became established in England during the reign of Elizabeth I. Brasses were produced by a 'cementation' type process in which pieces of copper were placed in crucibles, with charcoal and a zinc-containing ore, calamine. On heating to around 1000°C, for times up tp 20 hours, the zinc vapour formed from the ore and zinc diffused into the copper to produce brasses containing around 30 wt.% zinc, or more. An account of English brassmaking was given by Christopher Merret (1614–1695; FRS 1663), a physician and a founder member of the Royal Society. This metallurgical topic was included in a book on glassmaking, namely a translation by Merret in 1662 of the important book by Antonio Neri (1576–c.1614), *L'Arte vetraria*. This influential translation was made at the suggestion of Robert Boyle and at the instigation of the Royal Society. Merret added to the translation extensive observations of his own on aspects such as the design of glassmaking furnaces, types of glass being produced in England and materials used.

Brasses became important materials, for example for instrument makers. A range of compositions were of interest, as indicated by the investigations of Geoffroy, referred to above. In the 17th century Prince Rupert is said to have produced an alloy whose zinc content was higher than its copper content, referred to as 'Princes metal', with some decorative uses. In the latter part of the 18th century James Keir made an important contribution in the field of brasses. Keir (1735–1820; FRS 1785) was an industrial chemist who in 1775 began business as a glass manufacturer at Stourbridge. A year later he presented a paper to the Royal Society reporting observations on the crystallisation of glass during slow cooling. This paper included comments on the crystallisation rates and crystal forms; also he suggested that basaltic geological formations such as the Giant's Causeway may have been formed by crystallisation of molten rock during cooling. In 1778 he gave up his glass business to take charge of the Soho, Birmingham, Engineering Works of Matthew Boulton (1728–1809; FRS 1785) and James Watt (1736–1819; FRS

1785) in Birmingham during their absence on business. In the metallurgical field his name is associated with the invention in 1779 of an alloy, with the composition in the proportions by weight of copper, zinc and iron 50:37.5:5. Copper–zinc brasses containing 35 wt% zinc were more difficult to work than those of 10–20 wt% zinc content. Keir showed that by increasing the zinc content beyond 35 wt% an alloy was obtained that could be forged or worked either at red-heat or cold. He attempted to sell the alloy to the Admiralty for sheathing ships' ladders and for making bolts. In fact the material was not widely used, although some half a century later a somewhat similar alloy (Muntz metal, typically 40 wt% zinc) was patented by Muntz and used for ships' protection. Keir also wrote on experiments on the dissolution of metals in acids; his suggestions contributed to the discovery of the electroplating process.

In the field of the platinum group metals covering chemical, processing and physical metallurgy, Fellows of the Royal Society played prominent roles commencing in the 18th century. In addition, in connection with gold and silver as precious metals, and also base metals, in coinage, Fellows were substantially involved in the work of the Royal Mint (see Chapter 5).

Peter Woulfe (1727–1803; FRS 1785), a chemist and mineralogist, who had a laboratory in London, included in his interests tin in Cornwall; he was the first Bakerian Lecturer (1765), and spoke on a mineralogical theme. Among other Fellows interested in minerals and mining, Frank Nicholls (1699–1778; FRS 1728), a physician, examined the structure of metallic veins, and published observations on mines of Devon and Cornwall in *Philosophical Transactions* (1727–28) concerning the minerals of tin and other metals, including iron.

Also in *Philosophical Transactions* (1751) the Rev. William Henry (?–1768; FRS 1755) reported on a visit he had made to the Ballymurtogh mines in County Wicklow, Ireland. He described the principal works in a hill, with shafts giving access to a series of minerals at successive depths: an ironstone, followed by a lead ore, then silver and a rich copper ore. He discussed an accidental discovery that had been made at the mine, that when an iron shovel had been left in the stream of blue-coloured water that ran from the mine, it was found some days later to be encrusted with copper. One of the mine proprieters made some experiments from which he concluded that in the acidic solution iron dissolved and copper precipitated. Following this the miners dug long pits in which bars of iron were placed in the water. It was found that the bars soon contracted a 'copper rust', which progressively 'eat away' the iron. The copper was removed as a 'mud' which was dried and smelted to produce substantial quantities of high purity copper.

In the early years of the Royal Society an eminent physician Edward Brown (1642–1708; FRS 1667), who had wide ranging interests and travelled extensively, published in *Philosophical Transactions* (1670) a description of mines in Transylvania and Hungary. He reported on various metal sources including gold and silver. Concerning the gold mines at Chemnitz which had been worked nine hundred years, he described in detail the method of extracting gold by amalgamation.

A colourful character elected to the Royal Society was Rudolf Eric Raspe (1737–

1794; FRS 1769), author of *Baron Munchausen's Travels*. This book was written during a period from 1783–1786 when Raspe was assay master at some mines at Dulcouth in Cornwall, having been appointed by Matthew Boulton (1728–1809; FRS 1785).

Ferrous metallurgy – production, treatment and properties

Interests of Fellows of the Royal Society in ferrous metallurgy can be traced from the early period of the Society. They include efforts to understand the nature of irons and steels (including work by Robert Hooke). A number of Fellows were prominent in industrial developments in the production and treatment of ferrous materials, although the major advances in the 18th century, for example in iron smelting, were made by men who were not elected to the Royal Society. There were many small ironworks scattered through Britain located where there were local sources of iron ore, and wood for producing charcoal, water power for working bellows and tilt hammers. Blast furnaces produced pig iron (containing a total of several percent of other elements, notably carbon, silicon, manganese, sulphur and phosphorus). A great obstacle to development was the need to rely on charcoal as the reducing agent in the smelting process; the disadvantages included its expense and the limited weight of ore and limestone it could support in the blast furnace. In 1708 important work by Abraham Darby I (1677–1717) led to the successful smelting of iron ore at Coalbrookdale, using coke instead of charcoal, allowing greater weights of materials in the furnace and higher temperatures. There were difficulties in producing satisfactory wrought iron because of the brittleness of the coke-smelted pig iron. The puddling process patented in 1784 by Henry Cort (1740–1800) was a critical advance in producing wrought iron. The process involved melting pig iron and treating with iron ore to remove carbon and other elements by oxidation, followed by working the semi-solid product into shapes. Wrought iron products became crucially important in the development of civil engineering .

An interesting 17th century contribution to ferrous metallurgy was made by Prince Rupert, through his inventive ability and interest in military matters. In 1671 a warrant was issued

> for a licence to Prince Rupert at the yearly rent of 20s, of the sole exercise for fourteen years from 6 May last of his inventions of converting edged tools and other forged instruments and iron wire into steel; also of softening cast or melted iron so as to be wrought like forged iron, and also of tincturing copper upon iron; with commission to him to take oaths from all the workmen employed in the said arts not to divulge them.

In this and other ventures Prince Rupert was associated with Anthony Ashley Cooper (Lord Ashley, later Earl of Shaftesbury, 1621–1683; FRS 1663). In about 1678 Prince Rupert invented a procedure of 'annealing' iron ordnance in glassworks (in effect a malleabilising process) which may have improved the fracture resistance, but it was not adopted.

In the Sussex iron industry the Fullers of Brightling Park were among the ironmasters.

The furnace they built at Heathfield in 1693 was producing 200 tonnes of iron per year in 1717 and was central in the 18th century in their production of guns, some of which were exported; other products included rollers. John Fuller (FRS 1704) had two sons: John (died 1755; FRS 1727) and Rose (died 1777; FRS 1732).

Early work on smelting using multiple tuyeres in the furnace with charcoal was done by James Cockshutt (c. 1742–1819; FRS 1804). Cockshutt who was a civil engineer was associated with the Wortley Ironworks, Sheffield, and was often employed by John Smeaton. John Roebuck (1718-1794; FRS 1764) was, in effect, the founder of the iron industry in Scotland (see below).

Fellows in several countries in continental Europe also played important roles in industrially related work. In France Réaumur, a scientist of great diversity in his interests was a key figure in metallurgical developments, particularly in the ferrous field, while Buffon beginning in 1767 was also involved in iron production. Also, in France, Guyton de Morveau worked on blast furnace practice and the chemist Claude Louis Berthollet (1748–1822; FRS 1789) had an interest in processes for the production of iron and steel. In Sweden Torbern Olof Bergman (1735–1784; FRS 1765) carried out important work.

An interesting insight into the importance of the purity of iron ore is associated with a *Philosophical Transactions* paper (1751) by Peter Ascanius (FRS 1755), a Swedish scientist. As noted by Thomas Thomson (1773–1852; FRS 1811) in his book *History of the Royal Society* in the section on mineralogy, Swedish iron was superior to iron from other countries for the making of steel by the cementation process. Ascanius reported on a particular Swedish mine at Taborg which was in effect a substantial mountain of pure iron ore, which had been worked for nearly three hundred years.

Among other Fellows who took an interest in metal mining matters was Peter Simon Pallas (1741–1811; FRS 1764), a zoologist and traveller. From his travels in Siberia he reported in *Philosophical Transactions* (1776) his finding on a mountain ridge of a large mass of 'iron', previously discovered by Russian miners. Analysis by Martin Klaproth and others showed this to be an alloy of iron and nickel, and it was interpreted as being meteoric in nature.

Robert Hooke was interested in steels and his book *Micrographia* had a section on the hardening and tempering of steel, in relation to his observations on interference colours; he related the colours to those possessed by thin films of mica and other substances. He believed steel to be iron in which certain salts had been incorporated, in contrast to a common contemporary view that steel was the purest, most refined iron. Crucible cast steel did not become important in Europe until the development of the process in 1740 by Benjamin Huntsman (1704–1760) whose interests included the production of a material suitable for springs and pendulums of clocks. However, under the name of wootz, this type of alloy was produced in India, and Hooke was aware of it; he noted in his diary that

> steel was made by being calcined or baked with the Dust of charcole, and that bringing it up soe as to melt made the best steel after it had been wrought over againe, that it would be at first porous but upon working and hammering as fine as glasse.

Torbern Bergman

Antoine Réaumur

In 1694 Hooke showed that red hot iron became hotter and appeared to burn when a blast of air was directed on to it.

The various technological activities of Joseph Moxon included a substantial interest in steels. Among his publications were *Mechanick Exercises or the Doctrine of Handy-works applied to the art of smithing, joining, carpentry, turning, bricklaying etc.*This book, begun in around 1677, included interesting practical information on ferrous metallurgy, referring for example to topics such as case-hardening, fractures, and heat treatment. He compared types of steel from different countries in relation to their selection for various applications, such as watch springs and chisels. Advising blacksmiths on their selection of material he wrote

> Therefore when you chuse Iron, chuse such as it bows otf'nest before it break, which is an Argument of Toughness; and see it break sound within, be grey of Colour like broken Lead, and free of such glistering Specks you see in broken Antimony, no flaws or divisions in it; for these are arguments that it is sound and well wrought at the Mill ... The Rule to know good Steel by. Break a little piece of the end of the Rod and observe how it breaks; for good Steel breaks short off, all Gray, like frost work Silver. But in the breaking of the bad, you will find some grains of Iron shining and doubling in the Steel.

John Roebuck early in his career was a physician in Birmingham, but having retained his interest in chemistry he spent his spare time in chemical experiments, especially with a view to applying chemistry to some of the Birmingham industries. His inventions included an improved method of refining gold and silver in connection with the jewellery trade, and he set up a laboratory and a precious metal refinery working with a Birmingham merchant as a business associate. He contributed to Birmingham industrial life by becoming a consulting chemist for the local manufacturers. Of particular importance in process improvements was his introduction of the use of lead vessels instead of glass

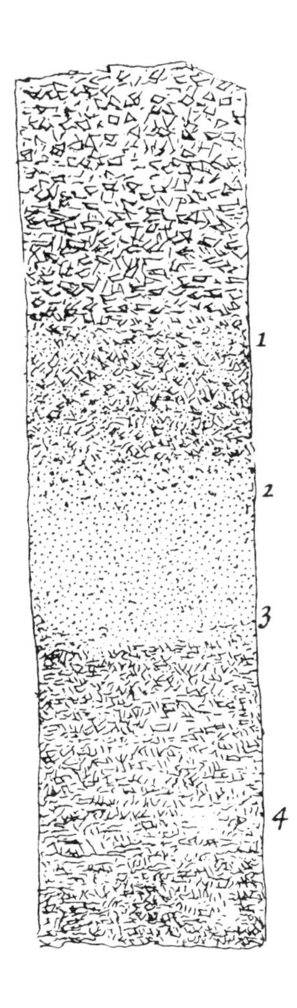

James Watt

Matthew Boulton

Steel bar quenched after heating only the top end to a high temperature; fracture viewed with a magnifying glass, gradient of grain size is seen below point 3 the structure has not been affected by the heating process. (Réaumur, 1722)

vessels in the manufacture of sulphuric acid. The first of the leaden chambers was erected by 1746, and encouraged by success he and his associate established a second sulphuric acid plant, near Edinburgh. He then became involved in the production of ceramics, and, also most importantly, he founded the Carron Ironworks near Falkirk, Stirlingshire, with several partners,including his three brothers. This latter venture was the real foundation of the Scottish iron industry, using a furnace in which coke replaced charcoal (1760), and a blowing engine installed by John Smeaton. Malleable iron was produced in 1762. George III granted the works a Royal Charter, and from 1779 ordnance known as 'carronades' were produced.

Antoine Réaumur was placed in charge by the Academy of Sciences of writing an industrial encylopedia. His investigation of the iron and steel industry in France was his most important and original contribution to industrial technology, and his findings were presented to the Academy in a series of memoirs in 1720, 1721 and 1722 . These were collected and published under the title *L'art de convertir le fer forgé en acier, et l'art d'adoucir le fer fondu, ou de faire des ouvrages de fer fondu aussi finis que de fer forgé.*

The French government took considerable interest in Réaumur's work, hoping that it would help the ferrous metals industry; the Regent played a part in enabling Réaumur to obtain information concerning the iron and steel industries of other countries, and also subsidised his researches by granting him a pension.

Among the qualities that Réaumur brought to his work were his mathematical ability, very keen powers of observation and lively imagination. An important feature of his metallurgical approach was that he examined fractures of steels and cast irons with the lens and the microscope, and used his observations in a practical and theoretical way, showing himself to be ahead of his time. He was interested in the changes accompanying the conversion of iron into steel in the cementation process which involved prolonged heating of iron in contact with charcoal leading to carbon diffusion and carbon enrichment. He was also concerned with the selection of iron for the carburising process on the basis of fracture studies. He adopted a quantitative approach to grain size in steel by reference to a set of standards. This enabled him to investigate the effect of heat treatment temperature, and to show an increase of grain size with increasing temperature. In his published results in 1722 this effect was clearly illustrated in a drawing, as seen through a magnifying glass, of the fracture surface of a steel bar, quenched after heating from one end only so as to produce a temperature gradient. Réaumur also considered the reason for the hardening of steel on quenching. Another area of his activity was in the tinplate industry and he reported on the industrial secrets of the process.

Réaumur also carried out very important work on cast iron technology, and was the originator in Europe of the annealing process of malleabilising white cast irons. The process was originally used in China in the last few centuries BC and its invention in England is usually credited to Prince Rupert (see earlier) in the context of improving the properties of ordnance. The principle of malleabilising is to form graphite from the cementite in a white cast iron. Réaumur used large kiln type furnaces with temperatures around 950–1000 °C . Heating the iron in bone-ash, a neutral environment, he produced 'black heart' iron, while, with an oxidising environment, iron oxide ('hammer scale') the product was 'white heart' iron. He used his fracture tests to distinguish between irons given various treatments and to select irons for various purposes, including castings for subsequent malleabilising, and for studying the malleabilising process. He observed with his lens the black dots in malleable iron, but did not know that these were graphite. Observations were also made on the white fracture of rapidly cooled cast iron and the relation of fracture to section thickness.

Bergman in his treatise *De praecipitatis metallicis* considered the phenomena observed when metals dissolve in acids. Following the work of Sven Rinman (1720–1792) on etching, involving observations on the differences in rates at which iron and steel dissolve, Bergman (*De analysi ferri,* 1781) made quantitative studies of carbon in various ferrous materials, measuring their rates of dissolution in acid and the gases evolved (mainly hydrogen). He carried out dry assaying (fusions in sealed crucibles), in which iron ore was reduced by the carbon in cast iron to form a metal and slag; the metallic iron produced could be mechanically worked, indicating that the carbon in the cast iron had reduced some of the oxide ore. In his wet assaying work on cast irons, he noted that the residues had the appearance of plumbago (graphite); higher proportions of residues from cast iron as compared with steel and wrought iron indicated higher carbon content of the former. The work of Bergman was significant in beginning to throw light on the differences between iron and steel in terms of their carbon content. In other work he found that iron was not the only magnetic metal, but that cobalt, nickel and manganese also showed magnetic behaviour.

Guyton de Morveau investigated the use of coal in place of charcoal in blast furnaces (1769), while in 1777 he described a flux for assaying iron ore. Other investigations concerned experiments on solidification and on etching. He noted that a sample of blister steel melted in a crucible and solidified under a slag developed a surface covered with regularly intersecting lines, like cross hatching, as observed with a magnifying glass; his illustrations indicate dendrite-like features. He produced illustrations of surfaces of various other solidified metals, including gold, copper, tin and platinum. In 1779 he examined a large plate of cast iron from the bottom of one of Buffon's furnaces. He polished metals using the lapidary's wheel. In 1786 he reported results of using etching to study differences between quenched and slowly cooled steel and cast iron, noting that a drop of nitric acid can be used to detect the differences between white and grey cast iron; he suggested that differences between iron, steel and cast iron were attributable to the presence of carbonaceous matter. Also in the same year Claude Berthollet (as a co-author) reported that cast iron, wrought iron and steel were different only in carbon content; an acid test was use on polished material e.g. on a steel the acid did not affect the carbonaceous matter, the latter being deposited and appearing as a black spot.

Metallurgy and the Industrial Revolution

Developments in metallurgy, particularly in the production of iron, played an essential role in the extraordinary advances in engineering beginning in the 18th century, for example in the development of steam power and in the large scale application of iron in civil engineering. Fellows of the Royal Society were prominent in the invention and development of steam engines. Denis Papin who assisted Robert Boyle with his experiments with the air pump, demonstrated to the Royal Society in 1679 a 'steam digester' (pressure cooker) for which he invented a safety valve. Thomas Savery (1650–1715; FRS 1706), a military engineer and inventor was the first to use fuel as a practical means of performing mechanical work. In 1698 he obtained a patent for an engine to raise water by the agency of fire and the following year demonstrated a working model at a Royal Society meeting. Also, of metallurgical interest, was his invention, patented in 1706, of a double hand bellows 'sufficient to melt any metals in an ordinary wood or coal fire, thus avoiding the need for assay furnaces'. Thomas Newcomen (1644–1729) built the first successful steam engine, which included a copper boiler and a brass cylinder. Martin Triewald (1691–1747; FRS 1731) who had come to England from Sweden in 1716 and assisted Newcomen with at least one of his engines, wrote an account of Newcomen's invention.

John Smeaton, often regarded as the founder of British civil engineering, whose career involved the construction of canals and and bridges, included in his work the improvement of the Newcomen steam engine; one aspect of this was the introduction of chains made from flat iron plates, joined by round pins. Smeaton was also consulting engineer at the Carron Company's Ironworks at Falkirk. The work of James Watt in the last third of the 18th century, by his radical design changes involving a separate condenser, transformed the potentialities of steam power, with far reaching effects in industrialisation. Close fitting between piston and cylinder was essential in Watt's engine and early cylinders supplied were not sufficiently accurate in bore to prevent steam leakage. However, the construction of a new type of boring machine by John Wilkinson (1728–1808) led to large cast-iron cylinders being made of sufficient accuracy. Iron was a critical material for many other parts of the engines, including pistons and connecting rods.The success of Watt's engine, patented in 1769, increased the demand for iron and extended the applications of heat power. His engine was developed, first in partnership with John Roebuck of the Carron Ironworks and then, commencing in 1773, with Matthew Boulton at the Soho works in Birmingham, where two full-scale engines were built, with cylinders of cast iron, a significant improvement. More patents were taken, and many engines were built and came into industrial use, e.g. to operate a tilt hammer at at the Bradley forge works of John Wilkinson .

In the sphere of civil engineering, important developments occurred both in terms of theory and in applications of materials and construction methods. The work of Leonhard Euler included the calculation of curves into which materials could be bent under the action of forces; in 1757 he derived an expression showing the load at which a long flexible column would become unstable. Charles-Augustin de Coulomb (1737–1806) also made important contribution to the theory of the strength of materials. In the early part of the century Benjamin Robins (1707–1751; FRS 1727) mathematician and engineer worked in both the civil and military spheres; he invented the ballistic pendulum for measuring the velocity of projectiles. Also in the military sphere Marc Isambard Brunel (1769–1849; FRS 1814) during his period as chief engineer of New York, erected an arsenal and cannon foundry where he introduced new machinery for casting and boring ordnance.

Fellows of the Royal Society were involved in the engineering applications of cast iron. John Smeaton used it in the construction of windmills, water wheels and pumps, while John Rennie (1761–1821; FRS 1798), who became a distinguished civil engineer, introduced cast iron pinions in place of wooden trundles into mill machinery early in his career. Cast iron was used in bridge building as early as 1779 when Abraham Darby III (1750–1789) began the construction of the bridge over the River Severn at Coalbrookdale. Thomas Telford (1757–1834; FRS 1827) designed many iron bridges and viaducts in his canal and road building projects and achieved an unequalled mastery in the structural use of cast iron. During the latter part of the century masonry was increasingly replaced by cast iron in civil engineering construction. An interesting example was the work of William Strutt (1756–1830; FRS 1817), a prominent mill owner, who developed an interest in iron through his objective to construct 'fire-proof' mill buildings; in some cases he used iron in conjunction with clay and plaster features of construction. Just after the turn of the century John Rennie, between 1801 and 1805 introduced cast irons and columns and roofs for dock warehouses. Cast iron played an important role in the development of the theory of construction, since as its use extended and failures occurred, attention was drawn to the need for improved understanding of its mechanical behaviour and casting technology.

An interesting part of the metallurgical story of the Industrial Revolution is found in the influence and activities of a group of eminent scientists, engineers and industrialists in the Birmingham area, who came to be known as the Lunar Society. The term 'Lunar' was not applied to them until around 1776, although the group began in 1765 and continued until 1791. The arrangements were informal and evolved to the holding of monthly meetings near the time of full moon – safer conditions for travel; no minutes of the meetings were taken, nor were publications produced.

The objective was to advance science and the arts and to apply science to industry and crafts. The group, which over the years consisted of only 14 members, was independent of the Royal Society, but nearly all of the members were elected as Fellows. Members participated in Royal Society activities by submitting papers, attending meetings and sponsoring new Fellows. The Lunar Society came to represent a high concentration of Fellows associated with industrial activities. Six of the members have been referred to in this chapter, namely: Matthew Boulton; James Keir; Joseph Priestley; Josiah Wedgwood; James Watt and John Whitehurst. Mention has been made of some of their individual activities and of some of their associations involving metallurgy, such as that of Boulton with Watt and Keir in the Soho Foundry, which had been found by Boulton.

Boulton was very influential in the group and his career illustrates his prominent position in British science and engineering of his time. His involvement with Watt in the development and production of steam engines is well known, but it is interesting to note also his substantial and wide-ranging metallurgical interests and activities. In connection with his steam engine work beginning around 1769 he was concerned with finding metallic materials that could be readily cast and machined, and could also resist wear and corrosion by steam; another aspect was the effect of mercury which was under consideration for use as a steam-seal. In ferrous metallurgy, Boulton had an interest in the Huntsman and the cementation steelmaking processes, and on a visit to the Carron Works he experimented with iron smelting. Other projects were the production of magnets, and the introduction of 'Sheffield plate' into Birmingham; the latter product, for use by silversmiths, was made by overlaying copper plates with silver plates and then hot-rolling. In the field of mining and metallurgical chemistry, Boulton was associated with the Cornish mines (as mentioned concerning his appointment of Raspe to an assaying post); Boulton himself became quite skilled in assay and smelting techniques. Yet another aspect of his work was the production of coinage. Thus, Boulton's career and influence, including his association with others in the Lunar Society, represent a combination of engineering and metallurgical achievements, linked with industrial sucess, constituting a significant contribution to the industrialisation of Britain.

The comprehensive account given by Robert Schofield in his book *The Lunar Society of Birmingham*, which incorporates some fascinating metallurgical insights, includes the following statements epitomising the significance of the achievements of the Birmingham group:

> 'the Lunar Society was a brilliant microcosm of that scattered community of provincial manufacturers and professional men who found England a rural society with an agricultural economy and left it urban and industrial'... 'Together they comprised a clearing-house for the ideas which transformed their country materially, socially and culturally within a generation.'

CHAPTER 3

Chemistry, Metallurgy and Materials Processing

Introduction

Metallurgical advances in extraction and refining and in other aspects of metal processing inevitably and dominantly stemmed from the major experimental and theoretical developments in chemistry during the 18th and 19th centuries. In the latter part of the 19th century the research of chemists in classifying the elements leading to the Periodic Table was a vital advance. Many Fellows of the Royal Society in England and Europe were prominent in various chemical fields. The activities of some of these Fellows included not only metallurgical themes, but also developments in the science and technology of the high-temperature processing and properties of non-metallic materials. This chapter reviews their contributions, together with the work of other Fellows with backgrounds in physics and engineering, ranging over metals, glasses, cements and ceramics, mainly in the period from around 1800 to the mid-20th century.

Contributions of chemists to metallurgy

Subsequent to the work of Antoine-Laurent Lavoisier (1743–1794; FRS 1788) in France in the 1780s, which showed that compounds consist of two or more elements, another French scientist, Joseph Louis Gay-Lussac (1778–1850; FMRS 1815), found that when hydrogen and oxygen were combined to form water they always did so in fixed proportions. Claude Louis Berthollet (1748–1822; FRS 1789) proposed that matter consists of small discrete particles, invisible to the naked eye and located a definite distance from each other; this distance was considered to increase with increasing temperature.

In England, one of the key figures was John Dalton (1766–1844; FRS 1822) who, in the first decade of the 19th century, proposed a theory that matter consists of atoms (from the Greek for 'that which cannot be divided'). Atoms of the same element were considered to be alike in every respect, mass, volume and chemical properties, while different elements had their own type of atoms and properties. He proposed that when different elements combine, the atoms join in simple proportions to form compound atoms (later called molecules). Dalton suggested and applied a method of determining the relative atomic weights of elements, compared to that of hydrogen. In 1803 he proposed his law of partial pressures. Dalton also devised a somewhat cumbersome list of symbols to denote the elements, which he continued to use throughout his life, although the system proposed in around 1814 by Jöns Jacob Berzelius (1779–1848; FMRS 1813) gained universal currency among chemists.

New elements were identified, in addition to the around twenty known to Dalton. Lavoisier had postulated the existence of many metals, but, early in the 19th century, the available metals were mainly those known from ancient times, namely iron, copper, lead, tin and mercury, extracted by smelting ores using wood or charcoal; also available were silver, produced by parting from lead, gold in elementary form separated by comminution and gravity concentration, and the platinum group metals. The task of discovering and isolating new elements, begun earlier, continued progressively into the 19th century, and by mid-century most of the metals had been discovered, although only a small number had been produced in a pure state, and engineers used mainly iron and a few copper-based alloys for constructional applications. However, many of the metallic elements eventually became of extensive importance in modern industry in the form of alloys.

Among distinguished chemists abroad, Berzelius in Stockholm was a brilliant experimentalist, the scope of whose work ranged widely including electrochemical theory, mineralogy, analytical chemistry, the theory of definite proportions and chemical equivalents. In analytical chemistry he introduced a variety of new methods and an improved degree of precision, while in mineralogy he established a chemically based classification. He discovered cerium, selenium and thorium, and other elements were discovered by students working with him; also he investigated the properties of zirconium and silicon.

Students of Berzelius included Friedrich Wöhler (1800–1882; FMRS 1854) who in 1826 among various projects, assisted in the discovery of silicon, boron and zirconium. In his work in Germany his finding of a method for obtaining nickel and cobalt free from arsenic from their ores proved of significant industrial importance. Wöhler also played a pioneer role in the discovery of aluminium and beryllium, and carried out work on the platinum-group metals. In 1850 he showed that copper-coloured crystals from blast furnace slag were a compound of titanium, carbon and nitrogen. He also had an interest in geological specimens and published many papers on minerals, meteorites and their analysis.

Robert Wilhelm Eberhard Bunsen (1811–1899; FMRS 1858) in Germany carried out important and wide-ranging research, including collaboration with others. His investigations in electrometallurgy included the preparation of some rare-earths of the cerium group; he devised a calorimeter to determine their specific heats, leading to the determination of atomic weights. Work with Henry Roscoe (1833–1915; FRS 1863) involved photochemical research. Bunsen's well-known burner was developed in the 1850s and soon displaced the blowpipe in the dry tests of analytical chemistry. In around 1860 Bunsen with Gustav Robert Kirchhoff (1824–1887; FMRS 1875)introduced the spectroscope and during the 1860s they developed spectroscopy, leading to their announcement of two new elements, caesium and rubidium. Other Fellows became active in spectroscopy including the study of metals. For example, George Downing Liveing (1827–1924; FRS

John Dalton

Jöns Berzelius

Joseph Gay-Lussac

Friedrich Wöhler

1879) and James Dewar (later Sir James, 1842–1923; FRS 1877) jointly carried out spectroscopical research between 1878 and 1900, including the subject of metal vapours.

Justus von Liebig (1803–1873; FMRS 1840), whose research was mainly in organic chemistry, established a high reputation which drew young chemists from all parts of Europe to work with him at Giessen. The State eventually decided to build him a large chemical laboratory outside Giessen, an important event, marking the start of a new way of training scientists. Liebig's approach was to teach chemistry without any reference to special applications, and to provide students with knowledge which they were later to apply; students progressed systematically from elementary operations to independent research through guidance from an established scientist. Many eminent investigators, teachers, and practical men for various industries came from his school.

An outstanding figure in England was Humphry Davy (later Sir Humphry, 1778–1829; FRS 1803; PRS 1820–1827) working at the Royal Institution in London. A development of great significance in relation to the future of metallurgy occurred in 1807 when he successfully electrolytically reduced compounds of sodium, potassium, barium, strontium and calcium to produce the metals using the newly developed voltaic cell (see Chapter 8). In the following year he isolated magnesium by an electrolytic process in which the metal was absorbed into a mercury cathode to form an amalgam. Working at the Royal Institution also, initially with Davy, a dominant scientist in the 19th century was Michael Faraday (1791–1867; FRS 1824) whose outstanding research activities spanned many fields of physics and chemistry, including electrochemistry, and had a major influence in the metallurgical field.

Other scientists who became Fellows of the Royal Society worked in the field of metallurgical chemistry e.g in the work carried out in England in the field of platinum metallurgy (see Chapter 4). In England, the work of Charles Hatchett (1765–1847; FRS 1797) included the analysis of a mineral, now known as columbite or niobite, which led to his reporting in 1801 that it contained a hitherto unknown metal, which he called columbium (now called niobium). Among others whose work was of metallurgical significance were a number of French scientists. Louis Jacques Thénard (1777–1857; FMRS 1824) in collaboration with Gay-Lussac reported in 1808 that they had prepared potassium by fusing potash with iron filings in a gun barrel; sodium was prepared similarly. Jean Baptiste André Dumas (1800–1884; FMRS 1840) investigated problems in metallurgy such as the preparation of calcium and the treatment of iron ores. Henri Victor Régnault (1810–1878; FMRS 1852), travelled extensively in Europe to study mining and metallurgical processes and was Director of the famous porcelain works at Sèvres; in addition to his important work in organic chemistry he carried out systematic investigations of the specific heats of many solids and liquids. Pierre Eugène Marcellin Berthelot (1827–1907; FMRS 1877) began his research on thermochemistry in 1864 by investigating heat changes in chemical reactions. He invented the bomb calorimeter in which a gas was mixed with excess oxygen, compressed to 20–25 atmospheres and then sparked.

Louis Thénard

Heinrich Magnus

Gustav Kirchhoff

Henry Roscoe

Also he introduced the terms 'exothermic' (heat evolved in a reaction) and 'endothermic' (heat absorbed). Hans Peter Jörgen Julius Thomsen (1826–1909; FMRS 1902) in Denmark carried out a thirty-year programme of thermochemical studies of chemical reactions, and also worked on electrochemistry. The wide-ranging research of the German scientist, Heinrich Gustav Magnus (1802–1870; FMRS 1863) included work on the reduction of oxides of cobalt, nickel and iron by hydrogen (published in 1825) and on the spontaneous inflammability of these metals in finely divided form.

The interest of Henri Louis Le Chatelier (1850–1936; FMRS 1913) in equilibria in chemical reactions, including those in the blast furnace (see Chapter 7) led to the principle that bears his name (published as a note in 1884) and was subsequently widely applied. In 1888 it was expressed in a simpler form: 'Every change of one of the factors of an equilibrium occasions a rearrangement of the system in such a direction that the factor in question experiences a change in a sense opposite to the original change.' (It is of interest to note that Joseph John Thomson (later Sir Joseph,1856–1940; FRS 1884) who graduated in mathematics at Cambridge in 1880, presented, in the same year, a thesis for a fellowship at Trinity College, in which he enunciated the same principle).

As chemistry progressed during the 19th century, similarity of chemical behaviour between various types of elements was observed, and various workers including William Odling (1829–1921; FRS 1859) tooks steps towards the recognition of a periodic relationship. However, it was not until 1869 that the Russian chemist Dmitri Ivanovich Mendeleev (1834–1907; FMRS 1892) presented the classification of the Periodic Table. When the 63 elements then known were listed in order of atomic weights he found that that those with similar properties appeared at regular intervals (or periods) and could be arranged in vertical columns. From the existence of gaps in the Table it was predicted that new elements remained to be found and the discovery of germanium, gallium and scandium confirmed this.

John Alexander Reina Newlands (1837–1898) had preceded Mendeleev in formulating a concept of periodicity of properties in the early 1860s, although his ideas were not accepted at the time. He was not elected a Fellow of the Royal Society, but eventually (1887) he was awarded the Davy Medal. Another chemist, Julius Lothar Meyer (1830–1895) produced a periodic law independently of Mendeleev which described the properties of the elements, but which did not predict the properties of as yet unknown elements; he also was a recipient of the Davy Medal, although not elected a Fellow. The interest of Johannes Robert Rydberg (1854–1919; FMRS 1919) in correlating physical and mechanical properties of the elements with their position in the Periodic Table led on to his work on spectrum analysis; in a publication in 1890 he presented a mathematical expression for the frequencies of spectral lines for the elements, which includes a constant, which became known as the Rydberg constant. In the field of X-ray spectroscopy Karle Manne Georg Siegbahn (1886–1978; FMRS 1954) carried out important work; his name has become a spectroscopic unit.

Among Fellows who carried out research on metallic elements and their chemistry, Sir Henry Enfield Roscoe (1833–1915; FRS 1863) did pioneering work on vanadium,

Dmitri Mendeleev

William Odling

Thomas Thorpe

Theodore Richards

involving collaboration with Thomas Edward Thorpe (later Sir Edward, 1845–1925; FRS 1876); Thorpe, with his students, made accurate measurements of the atomic weights (relative atomic masses) of titanium, silicon, gold, tin and radium. Theodore William Richards (1868–1928; FMRS 1919) in the USA was awarded the Nobel Prize for Chemistry for his accurate determinations of atomic masses of many elements. In 1913 he detected differences of atomic mass between 'ordinary' lead and lead extracted from uranium minerals, in which radioactive decay had occurred; this demonstrated the uranium decay series and provided confirmation of the proposal by Frederick Soddy (1877–1956; FRS 1910) of the concept of isotopes.

The electronic structures of the elements in relation to their positions in the Periodic Table are discussed in Chapter 11.

Non-metallic high-temperature processing – glasses, cements, ceramics

In the early part of the 19th century the British Government provided support for tackling the important problem of improving the quality, uniformity of properties and low porosity of optical glass for achromatic lenses for telescopes in order to compete with other countries. The background involved the Board of Longitude of which Sir Humphry Davy was a member. In 1824 Davy proposed that the President and Council of the Royal Society should appoint a committee to confer with members of the Board with a view to conducting such experiments as were thought necessary at the expense of the Board. The committee initially included Davy (chairman), William Thomas Brande (1788–1866; FRS 1809), George Dollond (1774–1852; FRS 1819), Charles Hatchett, Henry Kater (1777–1835; FRS 1814), Henry Warburton (1784–1858; FRS 1809) and Thomas Young (1773–1829; FRS 1794).

The committee asked the glass making firm of Pellatt and Green at the Falcon works in Southwark to construct a furnace and Michael Faraday (1791–1867; FRS 1824) to analyse the glass produced. In 1825 the committee appointed Faraday as a member and also set up a sub-committee comprising John Frederick William Herschel (later Sir John, 1792–1871; FRS 1813), Dollond and Faraday. Herschel in 1822 had suggested the use of lead borate for the production of a highly refractive optical glass; Herschel with James Smith had produced a glass of refractive index 1.866, although the material was too soft for optical use. Subsequently experiments were made using a furnace built at the Royal Institution and Faraday took over the glassmaking in 1827, commencing a period of arduous work. In 1828, in view of the defects in the glass arising from the use of clay crucibles, he began to use platinum for crucibles and items such as ladles and stirrers which came into contact with the molten glass. Faraday also used platinum powder, aiming to nucleate gas bubbles, which would separate from the glass thus reducing porosity. Samples of improved glass were prepared for examination by Dollond and Herschel, but did not prove to be of important practical use. Faraday's first Bakerian Lecture 'On the manufacture of glass for optical purposes' was presented in three sessions to the Royal Society in 1829. However, Faraday regarded the time spent on the project as pre-

John Joly

Charles Boys

Joseph Mellor

Derek Birchall

venting him from doing his own research. Nevertheless his discovery of the magneto-optical effect would not have been possible without the use of a piece of lead silico-borate glass.

John Hopkinson (1849–1898; FRS 1878) entered the service of Messrs Chance Brothers and Company, glass makers of Smethwick, near Birmingham in 1872 as engineering manager. An important part of his work was concerned with lighthouse illumination. Several decades later, in 1900,Walter Rosenhain (1875–1934; FRS 1913) who subsequently gained an international reputation for his metallurgical work, also entered the service of Chance Brothers when he left Cambridge University; he spent six years there and wrote a textbook on glass manufacture.

Sir Charles Algernon Parsons (1854–1931; FRS 1898) played an important role in the manufacture of optical glass in the 1920s. He purchased the Derby Crown Glass Company in 1922 and under the name of the Parsons Optical Glass Company he produced around 100 different types of glass for optical purposes and made many improvements in the manufacturing process. Later, in another company and plant set up near his turbine works he embarked on the making of large telescopes.

A number of distinguished chemists who became Fellows of the Royal Society carried out research in the field of glasses during the during the first half of the 20th century. Early in the 1914–18 war Great Britain faced problems in the production of glasses for special applications. By 1915 Herbert Jackson (later Sir Herbert, 1863–1936; FRS 1917) had determined formulae for a large number of glasses, including optical and heat resisting types. He made his knowledge available to glass manufacturers to assist them in dealing with production problems. He became expert in the effects of metallic oxides as colouring agents in ceramics enabling him to identify glazes and in this context he had an interest in oriental archaeological studies. Jackson also had an interest in a range of ceramic materials – refractories, pottery, porcelain and clays. Morris William Travers (1872–1961; FRS 1904) joined the firm of Baird and Tatlock when the 1914–18 war began; his work on building glass furnaces and in the production of scientific and other glassware was important for the war effort. In the field of glass technology William Ernest Stephen Turner (1881–1963; FRS 1938) was a pioneer. In 1904 he joined the academic staff of the Chemistry Department at Sheffield University where initially he taught a course on physical chemistry to students of metallurgy. Turner was associated with the setting up of a Technical Advisory Committee in 1914 to assist local industries and in addition to metallurgical problems glass soon became an important issue. In 1915 this led to the establishment of a department of glass technology in the university which became the leading centre for glass research.

Two of the great names in steel making in the 19th century (see Chapter 7): Charles William Siemens (later Sir William, 1823–1883; FRS 1862) and Henry Bessemer (later Sir Henry, 1813–1898; FRS 1879) also had an involvement with the glass industry. Siemens, in 1847, working with his younger brother Frederick (1826–1914) began investigations which led to the development of the regenerative gas furnace. The first practical application, before it was used in the steel industry was in 1861 to one of the glass making furnaces at the works of Messrs Chance Bros of Birmingham; fuel consumption in the glass making process was greatly reduced and the method was successfully introduced into the industry. In the field of sheet glass production in 1846, Bessemer, who had already been concerned with the production of samples of optical glass of large diameter, had the idea of passing molten glass between a pair of rolls. He wanted to carry out the melting in an open hearth furnace instead of a crucible and to run the glass from a slit at the bottom of the furnace. Although Bessemer's experiments did not lead to a commercial process he produced a long length of glass sheet and his idea foreshadowed later developments. More than a century later Lionel Alexander Bethune Pilkington (later Sir Alistair, 1920–1995; FRS 1969) is particularly associated with the float glass process at Pilkington Plc, Lancashire. Following his conception of the idea in 1952, seven years of development work led to the process in which a layer of molten glass is floated on a bath of molten tin in an inert atmosphere to prevent oxidation of the tin, giving a high quality surface finish to the glass.

Henri Le Chatelier, when he embarked on an academic career at the École des Mines in 1877 selected the field of cements for his research. This choice was influenced by a family connection with Vicat, the first investigator of hydraulic cements. Study of the literature showed Le Chatelier the unsatisfactory state of knowledge and he began his work by investigating the setting of plaster of paris, repeating an experiment carried out by Lavoisier; this led to a paper published in 1882. He proceeded then to study cements composed of calcium silicates and calcium aluminates, including experiments on setting in air,water and sea water. His view on the mechanism of setting involved the formation of a supersaturated solution from which products of reaction separated as interlocking crystals, giving rise to a hard mass. These investigations were reported in Le Chatelier's doctoral thesis in 1887 and he also carried out various studies of ceramics and glass.

An eminent military engineer, Charles William Pasley (later Sir Charles, 1780–1861; FRS 1816) who became director of the Royal Engineers' establishment at Chatham, Kent, published in 1836 the first edition of a work on *Limes, calcareous cements, mortar, stuccos and concretes, and on puzzolannas, natural and artificial water cements equal in efficiency to the best natural cements of England, improperly termed Roman cements, and an abstract of former authors on the same subject*. This work had begun two years previously and he reported discoveries based on his experiments, which quickly led to the manufacture in large quantities of artificial cements, including the Portland type.

In the 20th century cements and concretes were included in the research interests of Reginald Edward Stradling (later Sir Reginald, 1891–1952; FRS 1943) and William Henry Glanville (later Sir William, 1900–1976; FRS 1958), both of whom as part of their careers worked at the Building Research Station of the Department of Scientific and Industrial Research. Steel-reinforced concrete was a particular theme of Glanville's research, involving fundamental aspects such as adhesion of concrete to steel, creep and flow under load and stresses arising from shrinkage of the concrete.

John Joly (1857–1933; FRS 1892) worked on various aspects of inorganic materials, from the view point of geology and mineralogy, but including some aspects of metallurgical interest. He carried out investigations of the properties of minerals, for example specific gravity. In his experiments on the fusion of minerals he produced for the first time artificially crystals of calcium oxide and magnesium oxide in the oxy-hydrogen flame, and, using his 'meldometer', crystals of platinum and vanadium. In the 'meldometer' melting points were determined by supporting a small sample of the materials under investigation on an electrically heated platinum ribbon. Melting was observed with a microscope; the thermal expansion of the ribbon and hence its temperature were indicated by microscopical observation. In making measurements of volume changes associated with the fusion of rocks and minerals Joly introduced a method of removing the need for a containing vessel by utilising the surface tension of the materials undergoing melting.

A notable invention in the field of inorganic fibres was made by Charles Vernon Boys (later Sir Charles, 1855–1944; FRS 1888) in the 1890s. He was a highly skilled experimenter and in his work on the design of scientific instruments he invented a process for producing very fine quartz fibres. This invention was associated with his construction of a sensitive instrument (a 'microradiometer') for detecting infrared radiation; for this he required a galvanometer suspension, with characteristics superior to the silk, metals and glasses previously used. Boys had the idea of melting various siliceous minerals in the oxygen blowpipe and drawing fine fibres. Quartz fibres were found to be ideal: when twisted and released they showed no permanent set; also they had high strength, low expansion behaviour and insulating properties. Their production required a rapid drawing technique to avoid premature solidification of the molten material. Boys developed a process in which an oxy-hydrogen flame was used to melt a quartz rod, which was attached to a crossbow; the firing of the bow produced long uniform fibres of less than 1 micron in thickness. The strength was found to increase with a decrease in fibre diameter. Fused quartz fibres proved to be of major importance in experimental physics.

In the latter part of the 19th century and extending well into the 20th century, the chemist Joseph William Mellor (1869–1938; FRS 1927) was a distinguished worker in the field of ceramics. His wide-ranging investigations included: the plasticity of clays; the specific heats of firebricks at high temperatures; the fine grinding of ceramic materials; the constitution of the clay molecule and the crazing, peeling and durability of glazes.

John (Jack) Hugh Chesters (1906–1994; FRS 1969) was a leader in the field of ceramic refractory materials, associated particularly with the iron and steel industry (see Chapter 7). The innovative research of other Fellows on inorganic materials in the latter part of the 20th century, for example that of James Derek Birchall (1930–1995; FRS 1982) is discussed in Chapter 17.

CHAPTER 4

Precious Metals – the Platinum Group

Introduction

Platinum and its allied metals, palladium, osmium, iridium and rhodium, which are valued as 'precious' metals, have also played a vital part in many important technological advances, particularly during the past one and a half centuries. The history of the metallurgy of these elements, in which many Fellows of the Royal Society have played a role provides an interesting 'case study' illustrating many facets of fundamental and applied metallurgical development over a period of around 250 years.

Although it is very doubtful whether platinum was recognised as a separate metal in antiquity, traces of it have been found in artifacts from ancient Egypt. The best known example is a small strip of native platinum set on the surface of a box among many hieroglyphic inscriptions, made of gold on one side and of silver on the other. This artifact is dated to the 7th century BC. The French chemist, Pierre Eugène Marcellin Berthelot (1827–1907; FMRS 1877) was a pioneer in chemical archaeology, analysing metallic objects from ancient Egypt and Mesopotamia. In 1900 he was asked to investigate this artifact and found that one of the characters on the silver side differed considerably from the others. His tests showed that it was almost insoluble in aqua regia, and he concluded that it was a complex alloy containing several metals of the platinum group; there was no evidence that the Egyptian craftsman had noted a difference between this metal and the silver, and it had probably been mistaken for silver. Subsequent observations have been made of platinum metal inclusions in gold artifacts from ancient times, and it has been shown that they occur quite commonly in alluvial gold, often as alloys of more than one of the platinum group metals.

In South America platinum had been known in the 18th century as a white metal, in association with gold, under the name of 'platina' (or white gold). A surprising starting point for the beginning of scientific interest in Europe is found in the career of a young Spanish naval officer, Antonia de Ulloa (1716–1795; FRS 1746) an astronomer and mathematician who accompanied a French expedition to Quito in the 1730s. Through the seizure of their ship by a British naval vessel he was brought to London, where he was befriended by Martin Folkes (1690–1754; FRS 1714) then president of the Royal Society, and by William Watson (1715–1787; FRS 1741). Ulloa, following his election as a Fellow in 1746 returned to Spain, where in 1748 he published an account of the expedition, making reference to platina. William Brownrigg (1711–1800; FRS 1742) received samples of platinum from Charles Wood, who had obtained them in Jamaica and had done some experimentation. Brownrigg, after carrying out some experiments, passed the samples to Watson who, in 1750, communicated to the Royal Society 'several papers concerning a new semi-metal called platina'.

Torbern Olaf Bergman (1735–1784; FRS 1765) in Sweden had an interest in the classification and nomenclature of the elements, and he proposed the use of the name 'platinum' instead of the previously used 'platina', in line with various other metals for which he had adopted the Latin ending 'um'. In 1777 he presented evidence which conclusively refuted the view of Georges-Louis Comte de Buffon (1707–1788; FRS 1739) that platinum was not an individual metal.

Production and processing – early developments

The beginning of the 19th century saw the start of an important scientific collaboration between William Hyde Wollaston (1766–1828; FRS 1793) and Smithson Tennant (1761–1815; FRS 1785). Tennant, following studies at Edinburgh, moved to Cambridge in 1782 and then in 1785, while still an undergraduate he was elected FRS. He became friends with Wollaston, who was at Cambridge from 1782–1789. During 1784 Tennant had visited mines and chemical plants in Denmark and Sweden, where he had met leading chemists including Lorenz Florenz Friedrich von Crell (1745–1816; FRS 1788) and had become interested in the problem of producing platinum in malleable form. In London in 1800 Wollaston and Tennant formed an informal partnership, and one of their objectives was to develop malleable platinum; to this end they purchased a large quantity of alluvial platinum for their experiments. It was known that when platinum was dissolved in aqua regia a black residue remained, and they agreed that Tennant should examine the residue while Wollaston should focus on the soluble portion. As the work progressed it led to the identification of two new elements by Tennant, reported to the Royal Society in 1804, and named by him iridium and osmium.

Wollaston's work led to the discovery of two other new elements, the first of which he referred to as palladium in one of his notebooks in 1802. He did not immediately publish the results, but instead without revealing the authorship he advertised, with a handbill, that a new noble metal, palladium (or new silver) was available for sale through a London shop. This unorthodox procedure led to some scientific controversy in which a distinguished young Irish chemist, Richard Chenevix (c.1774–1830; FRS 1801) took part. Chenevix purchased the available stock of palladium and after carrying out experiments, he read a paper to the Royal Society in which he agreed that the material possessed the properties claimed for it, but that it was merely an alloy of platinum and mercury. Chenevix wrote to Nicolas Louis Vauquelin (1763–1829; FMRS 1823) in Paris and sent him a sample. Vauquelin was a first-class experimental chemist and analyst who carried out extensive research on platinum, and became involved in commercial enterprises of refining and fabrication; he also worked on other platinum group metals. Having checked the properties claimed in the hand-bill and found them to be correct Vauquelin did not find either platinum or mercury, and tentatively suggested that the material might be a new element. The controversy was entered into by several scientists in Europe, but in 1805 Wollaston read a paper to the Royal Society which put an end to the mystery.

Antonia de Ulloa

Martin Folkes

William Watson

Lorenz von Crell

In the meantime in 1804, three days after Tennant had reported his discovery, Wollaston reported to the Royal Society the discovery of another new element, designated rhodium. Subsequently Wollaston played the main role in the successful development of malleable platinum, but it was not until he was ill, and near his death in 1828, that he dictated a paper to his friend Henry Warburton (1784–1858; FRS 1809) revealing the process details, which Warburton read as a Bakerian Lecture. The process involved a powder route in which the conditions were optimised by purification by careful precipitation to produce a cake of the metal, hard enough to be handled, heated and forged. Another Fellow William Allen (1770–1843; FRS 1807), among his many interests, took up the refining and fabrication of platinum in 1805.

The melting of platinum (melting point of the pure metal ~1770 °C) was an important technological challenge and a number of scientists in England and abroad attempted to melt small quantities. William Lewis (1708–1781; FRS 1745) carried out a major series of experiments and reported them to the Royal Society in 1754 and 1757. This work included attempts to melt platinum with nearly every other metal then known. An alloy produced from a mixture of equal parts of platinum and gold was brittle, but could be worked after annealing. Lewis used a magnifying glass to view its fracture and commented on the structure of the partly melted alloy. He also discovered the precipitation of platinum from solution by sal ammoniac. Antoine-Laurent Lavoisier (1743–1794; FRS 1788) in 1772 was involved in experiments using one of the 'great burning glasses' in which a lens was used to focus the sun's rays; attempts to melt native platinum were unsuccessful. Later, in other experiments using a 1.8 m diameter lens, iron was melted, but the native platinum was not. Jan Ingenhousz (1730–1799; FRS 1769) melted a platinum wire by an electric discharge from a Leyden jar in an oxygen atmosphere in 1775. Lavoisier had achieved very small-scale melting in 1782 by directing a stream of oxygen into a hollowed out piece of charcoal in which a small quantity of platinum had been placed. Von Crell in Germany carried out some unsuccessful attempts at melting. Discharge from large galvanic cells was used by John George Children (1777–1852; FRS 1807) to melt platinum and iridium, and William Hasledine Pepys (1775–1856; FRS 1808) also took an interest in achieving melting.

Among researchers in various countries Louis Bernard Guyton de Morveau (1737–1816; FRS 1788) in France carried out extensive work which was important in establishing fabrication methods and properties.Following experiments by other workers on the production of platinum by solution in aqua regia and precipitation with sal ammoniac, he did further research in the mid 1770s. He heated the precipitate powder without fluxes, and with various fluxes including one containing arsenic; the effect of arsenic (as demonstrated by another worker) in lowering the melting point of platinum led to complete fusion but a brittle product. Heating the precipitate without fluxes did not give complete fusion, but the product was malleable and consolidation was aided by forging (1775). In 1784 Frank Karl Achard (1753–1821) had reported a method for making platinum crucibles involving the use of arsenic to achieve fusion, and subsequently removing the arsenic; de Morveau in 1785 took up this process and significantly improved it by modifying the

William Wollaston

Richard Chenevix

Nicolas Vauquelin

Jan Ingenhousz

flux composition. Continuing his interest in platinum metallurgy he investigated the possibility of platinum being used to adulterate gold, platinum being cheaper than gold at that date. He published a paper in 1803 on gold alloys to determine the effect of platinum on the colour and specific gravity; it was also mentioned that a drop of aqua regia could be used to detect a thin gold coating on a platinum coin by exposing the underlying grey metal.

In Russia, a member of the Russian Court, Count Apollos Mussin-Pushkin (1760–1805; FRS 1799) began work on platinum in 1797 and continued until his death in 1805, the main contribution being a method of refining the metal and rendering it malleable. Friedrich Wöhler (1800–1882; FMRS 1854) working in Germany on the platinum metals, reported in 1828 a method for the refining of iridium and osmium by converting the two metals to soluble chlorides that could be separated in solution; this process was used by refiners for many years.

Percival Johnson and George Matthey

Among the names of other Fellows of the Royal Society,working in England, those of Percival Johnson (1792 –1866; FRS 1846) and George Matthey (1825–1913; FRS 1879) occupy particularly prominent positions for their scientific and industrial work on platinum and its allied metals.

Percival Johnson had been apprenticed to his father, John Johnson, in London in his business of assaying and the buying and selling of precious metals; John, in 1786 had inherited the business, established in 1777 by his father, also named John Johnson. At the age of 25 in 1817 Percival set himself up independently as 'Assayer and Practical Mineralogist' in the City and subsequently in Hatton Garden in 1822. He rapidly rose to the highest eminence as an assayer and metallurgist, and his expertise was much sought after. Rival assayers used to send him compounds or minerals of a difficult and complicated nature for him to analyse for them. He was noted for the extreme accuracy of his assays and he, for the first time, reported the exact amount of gold and silver in the specimens submitted, whereas before the quantities had only been approximately stated. This led to difficulties with the bullion dealers who had been used to wider margins and therefore easier profits. He stated that he was willing, if required, to purchase all bars on the basis of his own assays, and this led him to take up the refining business; a gold refinery was built in Hatton Garden. His ability in this was soon publicly recognised and when gold bars from certain Brazilian mines came over in very large quantities and were refused at the Mint on account of brittleness, he was consulted and successfully undertook to refine and toughen them. He had earlier found that this gold contained palladium and other platinum group metals, and having succeeded in devising a separation process he introduced it commercially. Over a period of 20 years from 1832 he refined over a quarter of a million ounces of Brazilian gold, recovering large quantities of precious metal. It was not until 1846, when he was a candidate for election to the Royal Society that full details of the process were disclosed in a letter to the president of the Society. He

Apollos Mussin-Pushkin

William Pepys

Percival Johnson

George Matthey

also suggested in this letter uses for palladium as a protective coating for silver, as an alloy with silver for dental purposes and in a ternary alloy with copper and silver for instrument construction.

Among steps taken to promote the application of palladium, a ceremonial chain made from the metal was presented to King George IV. In addition to activity in palladium, Johnson, together with Thomas Cock (1787–1842) (who became his brother-in-law) had successfully extracted and refined the other metals of the platinum group shortly after the business was established, and the platinum group metals (except ruthenium, which was not discovered until 1844) were available to Michael Faraday (1791–1867; FRS 1824) and James Stodart (1760–1823; FRS 1821) (see Chapter 11) for their work on the alloying of steel. Johnson was also active in the mining field as a consultant and owner of mines, but his greatest success was in the platinum business as the first to successfully refine and manufacture the metal on a commercial scale, and playing a major part in extending its applications,

One of the sons of Thomas Cock, William John Cock (1813–1892), became a partner with Percival Johnson in 1837 and their collaboration resulted in an increase in the size of platinum ingots and sheets produced. The year 1838 saw a further significant development, when Percival concluded an agreement with a wealthy friend, John Matthey, involving capital for expanding the business and the apprenticeship of two of John's sons, George and Edward (1836–1918). George Matthey joined the firm at the age of 13 in 1838 while Edward joined 12 years later at the age of 14.

George began his work with the firm then known as Johnson and Cock, in the assay laboratory, and besides gaining an excellent training from Percival Johnson, he had the benefit of William Cock as a mentor. William Cock retired in 1838, owing to ill health, and George Matthey, now 20 years old, having shown a particular interest in the small scale platinum refining, was put in charge of these operations. He set about his task with great determination, and there began a remarkable period of persistent scientific endeavour together with an acute business sense, in which Matthey transformed a laboratory activity into an industrial enterprise and made platinum available for use throughout the world. Working long hours, six days a week, Matthey put in hand the preparation of larger, sounder, and more malleable ingots, and the introduction of more effective separation and refining techniques. By the end of the year he was supplying Michael Faraday – for his studies on magnetism – with wire and foil in both platinum and palladium, with rhodium, iridium and osmium in metallic form and also with a number of platinum group metal compounds.

In 1849, Albert, the Prince Consort (1819–1861; FRS 1840) suggested the holding of a Great Exhibition (1851) to promote the application of science to industry. There was opposition from industrialists, including Johnson, but Matthey persuaded his employer to take part; their display, including platinum crucibles, was awarded a prize medal. However, when Matthey saw the much larger platinum still displayed by a French competitor, his determination was strengthened to become pre-eminent in the platinum business. In 1850 Matthey obtained a new source of platinum from Russia by an agreement with one of the Russian mine-owners to make him sole refiner and selling agent. This achievement prompted Percival Johnson to take Matthey into partnership in 1851, the firm becoming Johnson and Matthey, and with Matthey largely in charge rapid progress was made, with improvements in the design and construction of platinum boiling flasks, and much larger platinum ware was exhibited in Paris in 1855.

An important technological development occurred in France in 1857 when Henri Etienne Saint-Claire Deville (1818–1881) and Henri Jules Debray (1827–1888; not elected FRS) developed a furnace made of limestone blocks and fired by a mixture of oxygen and coal gas, which could refine and melt platinum on a large scale. Matthey acquired the British rights in the same year and after several years work of further work at Hatton Garden satisfactory production was achieved. In 1862 Matthey and Deville cast a large ingot, weighing 3,215 ounces (100 kg) which was displayed at an exhibition in London, together with a large sulphuric acid boiler, (the first boiler had been made by Wollaston in 1805) and numerous other forms of platinum; these included tubes joined by a process of fusion welding using an oxy-hydrogen blowpipe which Matthey had developed in 1861 and which was adopted for boiler construction.

George Matthey's brother Edward, who had undertaken a course of metallurgical instruction at the Government School of Mines in Jermyn Street, became a junior partner in the company (which became Johnson Matthey and Co) in 1860; he concentrated on the gold and silver refining. The search for applications for platinum continued and in an International Exhibition in Paris in 1867 an exhibit comprising 15,000 ounces (~470 kg) of platinum showed what the company had achieved. This exhibition led to further developments on International Standards (metres and kilograms), using a platinum –10% iridium alloy proposed by Deville and Debray. Matthey played an important part in supplying the high-purity metals, in making large castings and in machining the bars; in 1874, in Paris under the supervision of Matthey and Henri Tresca three large castings were cut and remelted to ensure homogeneity, to produce a casting weighing 236 kilograms. Also a new method of refining was reported by Matthey to the Royal Society in 1879. International collaboration, involving among others Matthey, Deville and Debray gave rise to major developments in the production of large castings and forgings of platinum–iridium alloy.

By 1880 the platinum business was a prosperous activity; Matthey continued for many years to supervise the refining and melting, but was able to maintain and extend his contacts with the scientific community. He provided platinum metals, without charge to those studying their properties, for example pure wire for the first resistance thermometer produced by Hugh Longbourne Callendar (1863–1930; FRS 1894) in 1885. Matthey finally retired in 1909, following a career that had spanned just over seventy years. He had been elected FRS in 1879, his supporters including Sorby and Roberts-Austen, and was described as 'distinguished as a metallurgist, having special knowledge of the metals of the platinum group. The development of the platinum industry was mainly due to his efforts.'

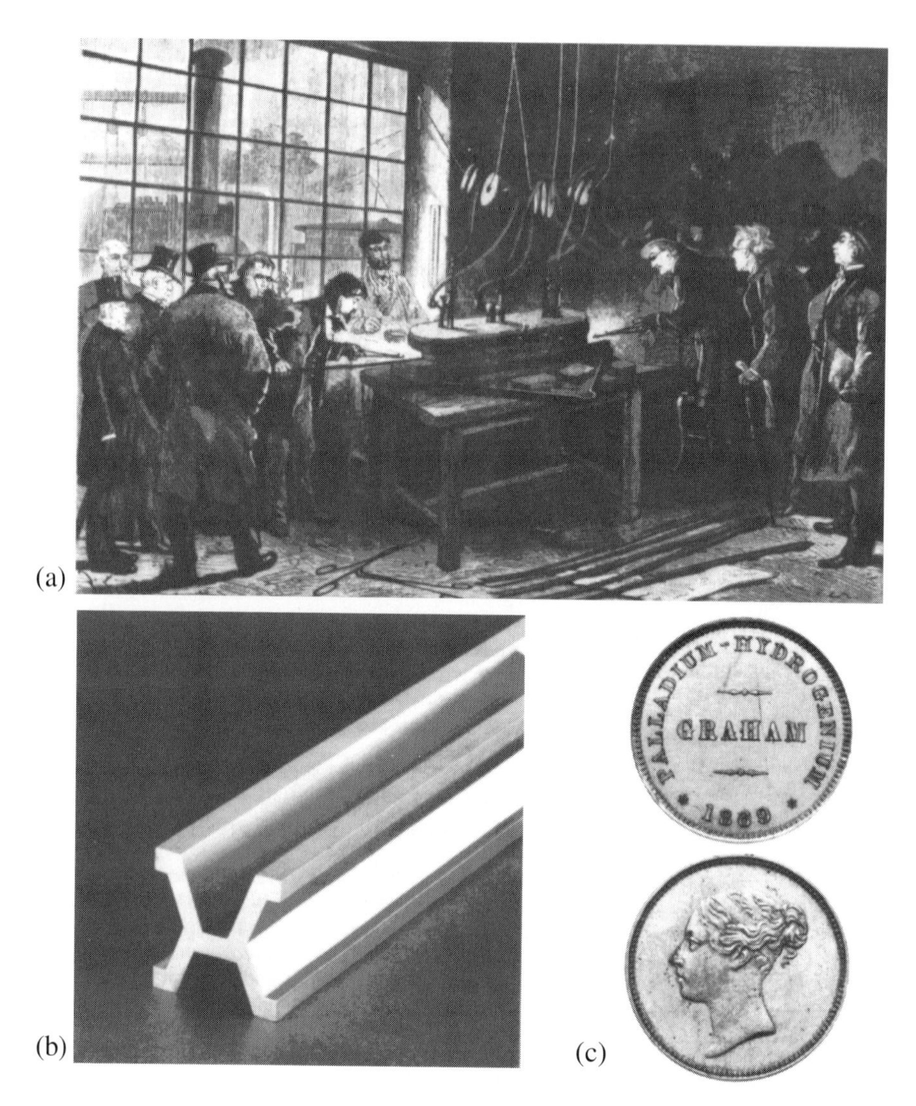

(a) Melting and casting process of high purity platinum–10% iridium ingot for the production of standard metres and kilograms. Three ingots were cut into pieces and remelted to achieve a 236 kg ingot. The process, which was carried out at the Conservatoire des Artes et Metiers in Paris, was conducted by George Matthey and Henri Tresca (Professor of Mechanics at the Conservatoire and Secretary of the French section of the International Metric Commission). Henri Deville and Henri Debray were present as technical advisors. (Engraving from the French magazine *L'Illustration*, 16 May, 1874.) (b) The ingot was subjected to forging and cold drawing to produce the X-section designed by Tresca aiming for maximum rigidity; problems were intially encountered in the drawing process in obtaining the requisite quality of the product, but eventually success was achieved. (Ref. Ch. 4.1.) (c) Medallions made by Graham at the Mint in 1869; he believed that the alloy used consisted of palladium and 'hydrogenium', the name he proposed for the form in which he considered hydrogen existed when absorbed. (Ref. Ch. 4.1.)

Applications and other developments

In the 19th century, Michael Faraday's name is significantly associated with the history of platinum; he said 'This beautiful, magnificent and valuable metal is very remarkable in its known special uses'. At the Royal Institution, in collaboration with James Stodart, a London cutler and maker of surgical instruments, Faraday made an investigation of the alloying of iron by additions of platinum, palladium, rhodium, iridium and osmium, with a practical aim of seeking improved materials for cutting or corrosion resistant applications; the results were published in 1820 and 1822 (see Chapter 11).

Also of interest at about the same time Robert Were Fox (1789–1877; FRS 1848) a scientific writer, who also carried out research including the field of metalliferous mining, published in 1819 a paper on 'Alloys of Platinum' in the periodical *Annals of Philosophy*. In this paper Fox reported observations from his experiments in which pieces of platinum were heated to a red heat e.g. with a blowpipe in contact with a piece of another metal. When equal amounts of platinum and tin were heated in this way they combined 'suddenly with great vehemence', with the emission of light and heat, producing globules of fused metal. It was noted that exclusion of air was desirable to achieve success. When antimony and platinum in contact were given prolonged heating, the antimony sublimed, and the remaining metal was malleable, consisting essentially of platinum. Heating zinc in platinum did not achieve alloying, but oxidation of the zinc.

The career of Martin Klaproth (1743–1817; FRS 1795) in Germany included an interest in ceramics. In the winter of 1788–89 he read a paper on 'On the use of platina in the decoration of porcelain' to the Berlin Academy, and exhibited samples. The metal, in powder form, mixed with a flux was applied with a brush, and then the object was heated in a furnace. For some years Klaproth was a consultant to the Berlin porcelain works. Later, probably around 1806, platinum was used for decorative purposes in the factory of Josiah Wedgwood (1730–1795; FRS 1783) in England.

The second decade of the 19th century saw the discovery of a phenomenon which was to have profound effects in the chemical industry of the future, and in which several Fellows of the Royal Society played a part. Sir Humphry Davy (1778–1829; FRS 1803) in 1817 reported the results of experiments on the combustibility of mixtures of coal gas and air, and noted that heated platinum wire, introduced into the gas mixture, became white hot. Without realising the potential significance of this effect, he had discovered the phenomenon of heterogeneous catalytic oxidation. His younger cousin Edmund Davy (1785–1857; FRS 1826) at the Royal Institution, followed up this work, including the use of finely divided platinum, and his results attracted the attention of Johann Wolfgang Döbereiner (1780–1849) in Germany, who by 1821 had repeated Edmund Davy's experiments and had appreciated the significant fact of the activity of the platinum rather than the action of the gas upon it. Döbereiner found that platinum in powder form would ignite a mixture of hydrogen and either air or oxygen at room temperature or below. News of the results was passed on to Louis Jacques in France, and he began to study, in collaboration with Pierre Louis Dulong (1785–1838; FRMS 1826) the effects of heated solid metals in inflammable gas mixtures. They reported in 1823 observations on several platinum group metals in bringing about the combination of hydrogen and oxygen. At about the same time Faraday repeated some of this work and reported his observations. Subsequent developments led eventually to the far-reaching importance of catalysis, a term which was introduced by Jöns Jacob Berzelius (1779–1848; FMRS 1813) in 1836, in industry and in which the platinum metals were to play a pre-eminent part.

In the 18th century Jan Ingenhousz played a role in encouraging scientists to take an interest in platinum and he had a set of waistcoat buttons made by Matthew Boulton (1728–1809; FRS 1785).

Platinum became an important material for crucibles, stirrers etc in the government sponsored work, in which Faraday and others were involved in the 1820s, aiming to improve the quality of glass for optical purposes. Platinum crucibles in the earlier years of the 19th century played a vital role in the development of analytical chemistry and several Fellows of the Royal Society were associated with this area. Other later applications for the platinum metals included temperature measurement (thermocouples and resistance thermometers) (see Chapter 10) and furnace heating elements.

In extraction metallurgy, Robert Wilhelm Eberhard Bunsen (1811–1899 ; FMRS 1858) in 1868 developed procedures for treating the residues of ores after platinum had been extracted, to separate palladium, ruthenium, iridium and rhodium.

In the field of batteries and their applications the research of Alfred Smee (1818–1877; FRS 1841) made significant contributions in electroplating of platinum and palladium. Also two other Fellows Thomas Hetherington Henry (c.1816–1859; FRS 1846) and Henry Beaumont Leeson (1800–1872; FRS 1849) were involved in this area.

An important application for platinum was its use for Grove cells – William Robert Grove (later Sir William, 1811–1896; FRS 1840). In 1839 he described, in a letter to *Philosophical Magazine*, an experiment on 'an important illustration of the combinations of gases by platinum', in which a galvanometer was permanently deflected when connected with two strips of platinum covered by tubes containing oxygen and hydrogen. In a further letter to the same journal in 1842, and in a private letter to Faraday, he described the first demonstration of a fuel cell. This utilised platinum foil coated with spongy platinum as the electrodes (acting as a catalyst for the combination of oxygen and hydrogen) and sulphuric acid as the electrolyte. There were alternate tubes of oxygen and hydrogen through each of which the platinum foil was passed to dip into dilute sulphuric acid, and the hydrogen and oxygen were 'burned' slowly to generate a voltage. A series of fifty pairs produced a brilliant spark between charcoal points, and decomposed hydrochloric acid and acidulated water. Stemming from this invention Ludwig Mond (1839–1909; FRS 1891) reported in 1889 work that he had done with Carl Langer (1859–1935) on designing a 'gas battery' involving the use of platinum foil coated with platinum black; however, the life of this cell proved to be too short.

In the 20th century platinum has continued to play an essential role in the development of fuel cell technology. Among current systems is the Ballard fuel cell, incorporating a polymer membrane with nano-sized platinum based electro-catalyst. In another important line of development Francis Thomas Bacon (1904–1992; FRS 1973) became

Thomas Graham

William Grove

Francis Bacon

Eric Rideal

interested in fuel cells in the early 1930s. Initially he repeated Groves' work, but then proceeded to replace the precious metal catalyst by nickel electrodes, with an alkaline solution (potassium hydroxide). Following a long and successful development programme stacks were produced with nickel electrodes of specially controlled porosity to work at elevated temperature; these were used in the Apollo space programme and continue to be used in the Space Shuttle.

Another Fellow who had an ongoing interest in fuel cells was Ulick Richardson Evans (1889–1980; FRS 1949); his first paper on this theme was co-authored with Eric Keightley Rideal (later Sir Eric, 1890–1974; FRS 1930) who subsequently worked with Bacon.

The interaction of hydrogen with platinum and palladium is a subject of considerable theoretical and practical interest. Thomas Graham (1805–1869; FRS 1836) in the mid-1860s found that when red-hot platinum was exposed to hydrogen, the gas was absorbed and retained for an indefinite time; in the case of palladium, five or six hundred times its own volume of hydrogen was absorbed , and certain palladium-silver alloys, made for Graham by George Matthey, behaved in a similar manner. Graham also showed for palladium that hydrogen could be occluded electrolytically by immersion in dilute sulphuric acid in contact with zinc. Graham's research laid the foundations for the development of modern equipment to generate high purity hydrogen and stimulated numerous subsequent investigations by other workers. Among such investigations Alfred René Jean Paul Ubbelhohde(1907–1988; FRS 1951) included in his research interests the interaction of hydrogen with palladium and with iron. X-ray and electrical resistivity measurements were made to investigate the effects of absorption of protium and deuterium by palladium; other aspects were the effects of hydrogen on the crystallographic and surface structures of palladium and iron, and on the elastic properties of palladium.

An event involving palladium and deuterium occurred in 1989 which caused extraordinary interest in the scientific world relating to nuclear fusion. Martin Fleischmann (1927– ; FRS 1986) and Stanley Pons (1943 –) at Utah reported that they had achieved a controlled nuclear fusion reaction by an electrolytic method under laboratory conditions at room temperature. Their method was based on the occlusion of deuterium by palladium ; they electrolysed water containing deuterium isotope using a palladium cathode and achieved heat in excess of that resulting from the passage of the electric current. It was suggested that 'cold fusion' had occurred associated with an effective high pressure of deuterium in the electrode. Considerable efforts to repeat the experiments were made in a number of laboratories including that at Harwell, but the 'fusion' effect has remained unresolved .

In the platinum metals industry in the 20th century, Alan Richard Powell (1894–1975; FRS 1953) played a major role, particularly through his researches including analytical chemistry, and his influence as Research Manager for Johnson Matthey and Co. Ltd for 36 years (see Chapter 6).

CHAPTER 5

Minting of Coinage

Introduction

The vital economic role of gold and silver role in the coinage of the realm provides a historical theme through the biographies of Fellows of the Royal Society who worked at the Royal Mint or at other mints. Since the founding of the Royal Society many Fellows have held important offices in the Royal Mint with crucial responsibilities. The history of the organisation of the Mint is complex, involving legal, economic and political factors, together with technical metallurgical work relating to the coin composition and properties. Over the years there have been various important offices, such as Warden, Master and Assayers, and a number of changes in the administrative structure have taken place since the early days of the Royal Society. For some of the Fellows who held posts at the Mint in the earlier years, for example Isaac Newton (later Sir Isaac, 1642–1727; FRS 1672), their scientific eminence was achieved in fields other than metallurgy. However, in the 19th century much of the metallurgical work of Thomas Graham (1805–1869; FRS 1836) and William Chandler Roberts-Austen (later Sir William, 1843–1902; FRS 1875) was done at the Mint. Other Fellows, particularly in the 18th century, were notable for their achievements in the administrative and political fields. This chapter does not present details of the history of the Mint, but focuses briefly on some aspects of metallurgical and engineering interest involving Fellows. Mention is made of copper coinage as well as precious metals; also reference is made to Fellows who held technical responsibilities in mints in other countries.

Metallurgy and mints

In the early days of the Royal Society, Henry Slingsby (1621–c.1688; FRS 1663) became Master of the Mint in 1663. James Hoare (?–1696; FRS 1664) whose association with the Mint was as Surveyor, Warden and Controller, retained his goldsmith's business at Cheapside, and was involved with his family in the founding of Hoare's Bank. Hoare had to appoint a deputy at each of the five temporary mints opened in county towns for the great recoinage of clipped and obsolete silver commenced in 1696; for the Deputy controller at Chester he chose Edmund Halley, the astronomer (1656–1742; FRS 1678). Hoare's son, James (?–1679; FRS 1669), was Joint Controller with his father at the Mint beginning in 1662.

Isaac Newton's 31 years as Master constituted an important period in the history of the Mint. His initial appointment as Warden in 1696, was procured by Charles Montagu (later Earl of Halifax, 1661–1715; FRS 1695; President 1695–1698), after 35 years at

Cambridge, and Newton continued to hold his fellowship and chair at Cambridge for another five years. It was a critical time for the Mint; there had been debasement of the silver coinage and there was a monetary crisis. Parliament had passed an Act at the beginning of the year deciding on recoinage by remelting worn-out silver coins and replacing them with new-minted coins. Newton's skill in numbers enabled him to understand without delay the accounting system, and also he demonstrated great administrative gifts. The recoinage was completed in the summer of 1698 and Newton's understanding of the operation of the Mint was such that he had virtually assumed the title of Master. As Warden he was responsible for certain disciplinary matters, including bringing clippers and counterfeiters to justice. On the death of Thomas Neale (1641–1699?; FRS 1664), the Master of the Mint, at the end of 1699, Newton was appointed to the post and Sir John Stanley (1663–1744; FRS 1698) became Warden.

Newton resigned both his posts in Cambridge in 1701, a year of marked minting activity, when he received nearly £3,500 as Master; his average annual income as Master over the years was probably about £1,650. For a period of some 7 years the administrative duties of the Mint were dominant features of his work. Among many activities he improved the standard of accuracy in the weight of coins, and tried to enforce high precision in provincially stamped hallmarks. Also during his period of office there occurred the substitution of gold for silver as the main national coinage and the Mint became involved in the purchase, storage and marketing of tin from the mines in the Duchy of Cornwall. He carried out a thorough study of the weights, fineness and value of foreign coins; he wrote on economic matters and became the nation's authority on all matters concerning the world's money. Some writers have suggested that Newton personally carried out certain assaying work for the determination of gold and silver, associated with a tradition (extant at the time of Roberts-Austen) that he used a particular furnace in the Mint; however, this suggestion appears to lack authoritative support.

Newton's successor as Master in 1727 was John Conduitt (1688–1737; FRS 1718), whose career involved service in the army, and as a member of parliament. He wrote an essay in 1730 on 'Observations on the present state of our gold and silver coins', showing great knowledge of the history of the currency and much care in experimental assaying.*

A heavy recall and recoinage of worn gold coins began in 1773. On the passage of an act for the better regulation of gold coinage in 1775 John Whitehurst (1713–1788; FRS 1779) was appointed as stamper of the money weights. In 1798 when the Privy Council appointed a committee 'to take into consideration the state of the coins of this realm' Henry Cavendish (1731–1810; FRS 1760) and Charles Hatchett (1765–1847; FRS 1797)

*In the period 1725–1727 Walter Cary, Clerk to the Privy Council (?–1757; FRS 1727) served as Warden of the Mint. In the years following the period of Conduitt's mastership, six Masters were elected Fellows, either before or during their period of office: the Hon Richard Arundel, MP, Master 1773 (?–1757; FRS 1740); Philip Stanhope KG, fifth Earl of Chesterfield, Master for four months during 1789–90 (1755–1815; FRS 1776); George Townshend, Earl of Leicester, Master 1790–94 (1755–1811; FRS 1781); Sir George Yonge, KB, Bart, MP, Master 1794–98 (1731–1812; FRS 1784); Robert Banks Jenkinson, Lord Hawkesbury, later Prime Minister, Master 1798–1801 (1770–1828; FRS 1794); Charles George Arden, Baron Arden, Master 1801–1892 (1756–1840; FRS 1786).

William Brande

Edmund Halley

John Frederick Herschel

John Rennie

conducted a detailed investigation of the 'comparative wear of gold. The main conclusion of their work was that the 'extraordinary loss which the gold coin of the kingdom is stated to have sustained within a certain limited time cannot be attributed to any important defect in the composition or quality of the standard gold'.

Also in 1798, another Fellow, Robert Bingley (?–1847; FRS 1809) was appointed as King's Assay Master, recommended by the Lords of the Treasury as being by 'study an able chemist'. Nearly 60 million pounds sterling of the new coinage introduced in 1817 (sovereigns and half sovereigns) were struck while he held this office up to 1835. He did valuable work in pointing out and correcting some errors which were liable to occur during gold assaying. He carried out experiments on platinum and its separation from gold in assaying.

John Rennie (1761–1821; FRS 1798) an eminent civil engineer, who was involved in many major projects on canals, docks, harbours and bridges was called upon to make an inspection of the Mint; he was associated with Matthew Boulton (1728–1809; FRS 1785) who had set up a mint at Soho, Birmingham. Boulton obtained many orders for copper coinage for foreign countries and British territories, and was entrusted with the production of the whole British and Irish copper coinage from 1797–1807; he also played a role in assisting the work of the Mint. Another Fellow, James Keir (1735–1820; FRS 1785) in 1797 undertook the production of a new copper coinage for Great Britain; he also supplied machinery to the new Mint at Tower Hill, for which John Rennie erected the buildings. George Rennie (1791–1866; FRS 1822) the elder son of John, whose engineering career included the manufacture of machinery, was appointed in 1818, on the recommendation of Sir Joseph Banks (1743–1820; FRS 1766, and President from 1778 to 1820) and James Watt (1736–1819; FRS 1785), as inspector of machinery and clerk of the irons (dies) at the mint; he held this post for nearly eight years.

In 1823 William Thomas Brande (1788–1866; FRS 1809) was consulted by the government on the manufacture of iron and steel, which included the objective of obtaining an improved metal for the dies used in the coinage. His report led to improvements and economies at the Mint and also to his appointment as Superintendent of the Die Department, and in 1854 to his becoming Chief Officer of the Coinage Department.

Sir John Frederick William Herschel (1792–1871; FRS 1813) was appointed Master of the Mint in 1850 and was involved in major re-organisation; having served for only five years he retired due to ill-health. Thomas Graham who was a non-resident assayer during Herschel's term of office, succeeded Herschel as Master, and presided until his death in 1869. He introduced various reforms and economies and supervised the major change from bronze to copper coinage; in his fundamental researches, he was helped by the resources available at the Mint.

In 1865 William Roberts-Austen was engaged by Graham as a personal assistant. Following Graham's death reorganisation at the Mint took place in accordance with the provisions of the Coinage Act of 1870 and the recommendations of a Royal Commission appointed to enquire into the administration as far back as 1848. A new post was created, that of 'Chemist of the Mint', to which Roberts-Austen was appointed in 1870; he, with the deputy master, then toured thirteen European mints to study their methods. Some of the assay work was done by non-resident assayers, but in 1882 this activity was concentrated in one office with Roberts-Austen as 'Chemist and Assayer', a post which he held until the time of his death; he also held the post of Acting Deputy Master for several months preceding his death.

Roberts-Austen carried out a wide range of investigations connected with the metallic currency during his 33 year association with the Mint; these were described in the memoranda he was required to supply each year for publication in the annual reports of the deputy-master and comptroller of the Royal Mint, regularly presented to Parliament. They included treatment of brittle gold, assaying by means of the spectroscope, liquation in alloys, density of gold–copper alloys, steel for the manufacture of dies, rate of wear of gold coins, casting, electrodeposition of iron, eutectic alloys, diffusion in metals, surface treatment of silver and bronze medals and bronzes for medals.

During the period when the Mint employed non- resident assayers several Fellows of the Royal Society served in this capacity: William Allen Miller (1817–1870; FRS 1845); August Wilhelm von Hofman (1818–1892; FRS 1851); and John Stenhouse (1809–1880; FRS 1848). Another, later link with the Mint was that of Guy Dunstan Bengough (1876–1945; FRS 1938) who, following his graduation in 1902, spent several months at the Mint gaining research experience with Roberts-Austen.

Fellows of the Royal Society also played a role in mints in other countries. Louis Bernard Guyton de Morveau (1737–1816; FRS 1788) was appointed by Napoleon in 1799 as administrator to the mints in France. Nicholas Louis Vauquelin (1763–1829; FMRS 1823), who held for a period the post of Professor of Assaying at the École des Mines, served as official assayer for precious metals for Paris. James Prinsep (1799–1840; FRS 1828) spent much of his career in mints in India and at Calcutta constructed a very accurate assay balance. Stanley Jevons (1835–1882; FRS 1872) received through Thomas Graham the appointment of Assayer to the Australian Royal Mint in Sydney. Before taking up the appointment he studied gold and silver assaying with Graham and at the Paris Mint to acquire the necessary skills and then went to Australia in 1854 where he worked very successfully for five years. When he returned to England his career turned to political economy and logic. In 1868 he read a paper to the London Statistical Society 'On the condition of the metallic currency of the United Kingdom' in which he recommended a re-coinage and the introduction of an international money. Previously, in order to estimate the age of the gold circulation in England, and the loss on light gold coins, he had used information supplied by bankers to make a census of over 165,000 gold sovereigns. William Gowland (1842–1922; FRS 1908) went to Japan in 1872 and worked for several years as Chemist and Metallurgist to the Imperial Mint at Osaka. In 1878 he became Assayer and Chief of the Foreign Staff and Adviser to the Imperial Arsenal.

CHAPTER 6

Non-Ferrous Extraction Metallurgy

Introduction

During the latter half of the 19th century tremendous advances occurred in chemical and engineering aspects of primary (extraction) metallurgy both in the methods and scale of processes for extracting the established metals and in invention and development relating to metals previously regarded as rarities. Chemistry provided basic understanding of reactions, new reduction procedures, accurate analytical techniques, electroforming and electroplating (stemming from Faraday's work), knowledge of refractory materials leading to higher temperature capability and resistance of furnace linings to attack, and methods of separating pure metallic oxides from their ores. Electrolysis came to play a major role in extraction metallurgy. From the engineering standpoint, in the 19th century, techniques were developed for handling large quantities of ores and for processing the minerals they contained.

The mechanisation of most industries depended on the use of steam and there were incremental improvements in design and in metal-working for boiler construction; then towards the end of the century the steam turbine was invented, one version being that produced by Charles Algernon Parsons (later Sir Charles, 1854–1931; FRS 1898). By the turn of the century the development of the turbine for electric power generation led to the electrometallurgical production of aluminium on a manufacturing scale. Electric furnaces came to play a wide-ranging role in both non-ferrous and ferrous metallurgy.

The 20th century has continued to see important developments in the extraction of many metals, and the production of metals in high purity states. These have proceeded alongside the exciting advances in alloy metallurgy, in terms of alloy composition and processing, which have made possible extraordinary engineering developments in many fields, such as aerospace (in which nickel-based and titanium-based alloys are vital) (see Chapter 14).

In extraction metallurgy a number of Fellows of the Royal Society played important roles, particularly in the iron and steel industry. In the non-ferrous field, Fellows have also been active, notably in relation to aluminium, nickel, copper and zinc, gold and silver and platinum and rare metals. John Percy (1817–1889; FRS 1847), although having a particular interest in ferrous metallurgy, was also concerned with non-ferrous metals. A century later Frederick Denys Richardson (1913 –1983; FRS 1968) was especially active in the metallurgical science of the iron and steel industry, but in the Nuffield Research Group in Extraction Metallurgy at Imperial College he was also involved with non-ferrous metals.

As a 19th century author John Percy was particularly significant and in 1861 he published the first volume of his great treatise on *Metallurgy,* dedicated to Michael Faraday (1791–1867; FRS 1824). This work, which Percy called the 'task of his life', developed into a series of volumes (1861–1880), which remained uncompleted. The first volume dealt with fuels, refractories, copper, zinc and brass. In his introductory chapter to this volume he dealt not only with metallurgical processes, but also with chemical and physical properties including aspects such as structural changes during solidification and working, and properties such as strength, toughness and fracture. The subject of the second volume, published in 1864 was iron and steel. He dealt not only with ironmaking and steelmaking, but also with physical and chemical properties; in his section on alloys he discussed a range of alloy additions, including results obtained by Faraday and James Stodart (1760–1823; FRS 1821). A volume on the metallurgy of lead was published in 1870, and the first part of the volume on silver and gold in 1880. A distinctive feature of his writing arose from the care with which he examined the relations of the metals to other elements and to each other. The treatise contained over 3,500 pages covering exact descriptions of metallurgical processes and detailed scientific discussion of the chemical problems involved. Percy made a practice of visiting many industrial works and was able to present detailed information based on the knowledge thus gained. He obtained, and fully acknowledged, the assistance he received, for example, from works' managers in the preparation of the treatise; his students were among those who assisted in this way.

Percy clearly recognised the importance of metallurgy in the national scene. In his inaugural address at the School of Mines in London, he referred to the fact that metallurgy was looked on as an empirical art, and he concluded this lecture by pointing out that 'in proportion to the success with which the metallurgic art is practised in this country will the interests of the whole population, directly or indirectly, in no inconsiderable degree be promoted.'

Aluminium

Humphry Davy (later Sir Humphry, 1778–1829; FRS 1803), who in 1807 produced sodium and potassium by electrolysis of the alkalis (see Chapter 8), attempted unsuccessfully to extract aluminium from alumina; it has been suggested, however, that he may have obtained the metal in very impure form. Jöns Jacob Berzelius (1779–1848; FMRS 1813) also tried unsuccessfully to produce aluminium from alumina. Hans Christian Oersted (1777–1851; FMRS 1821) in 1825 showed to the Academy of Sciences in Copenhagen a sample of a metal which he believed was aluminium. This had been prepared by using a potassium–mercury amalgam to reduce aluminium chloride; the mercury was removed by distillation, and a grey powder was left. Friedrich Wöhler (1800–1882; FMRS 1854), encouraged by Berzelius, carried out similar experiments, reported in 1827. Initially he used a potassium–mercury amalgam, before developing an improved method. This involved heating in a platinum crucible a small amount of potassium covered with aluminium chloride. Wöhler showed that the metal globules produced contained no potassium; he confirmed the low density of the metal and and studied a number of its reactions. In the mid-1850s Robert Wilhelm Eberhard Bunsen (1811–1899; FMRS 1858) prepared aluminium from its molten chloride.

In 1854 Henri St Claire Deville (1818–1881) reduced aluminium chloride with sodium and by the following year exhibited small bars of aluminium at the Paris exhibition. When information became known in London, experiments were made using the Deville process at the Royal School of Mines under John Percy. Among developments in England was the production of aluminium from around 1860 to 1874 at the Washington works of Bell Brothers in County Durham. Isaac Lowthian Bell (later Sir Lowthian, 1816–1904; FRS 1874), best known for his work in ferrous extraction metallurgy was one of the directors of the company and adapted and improved the Deville process. A mixture of ground bauxite and soda ash was heated to reduce the oxide as sodium aluminate from which pure alumina was precipitated; this was briquetted with salt and was chlorinated in a vertical retort and used to react with metallic sodium.

In the 1850s Henri Louis Le Chatelier (1850–1936; FMRS 1913) collaborated with Deville on the electrolysis of a molten salt mixture to produce aluminium. However it was much later that a successful development of an electrolytic process was achieved by Charles Martin Hall (1863–1914) in the USA and Paul Louis Toussaint Heroult (1863–1914) in France. Hall and Heroult worked independently and patented processes in 1886, which involved the electrolysis of alumina dissolved in molten cryolite (sodium aluminium fluoride). Industrial development led to large scale production of aluminium. In the UK the British Aluminium Company Ltd adopted the electrolytic process, based largely on a report by Lord Kelvin (formerly Sir William Thomson, 1824–1907; FRS 1851) who was a Director of the Company from 1898 to 1907.

Nickel

Among non-ferrous metals which became of great importance, both as a base for alloys and as an alloying element in steels, is nickel. In 1826 Friedrich Wöhler found a method for obtaining nickel and cobalt, free from arsenic, from their ores, thus making an important industrial advance. Another significant development was associated with Percival Norton Johnson (1792–1866; FRS 1846) a major figure in the platinum industry (see Chapter 4) who produced nickel silvers on a significant scale between 1829 and 1832. This was based on an arrangement with Geitner who had established in Germany a process for extracting reasonably pure nickel and a plant for producing nickel silvers. Johnson set up a small works by the Regent's Canal where a reverberatory furnace treated the nickel speiss from Germany containing nickel, iron and arsenic; this treatment removed the arsenic, and the sintered cake produced was sent to Hatton Garden. Further processing included melting with copper, charcoal dust and borax, and adding zinc; alloy bars with a composition of 18 parts of nickel, 55 of copper and 27 of zinc were cast into iron moulds. Some 60 years later the invention of nickel steels extended the market for nickel, and the firm of Henry Wiggin and Co. Ltd was producing nickel in Birmingham. Another Fellow of the Royal Society, David Forbes (1828–1876; FRS 1858) was involved during part of his career in nickel prospecting and smelting, and in a partnership in a nickel smelting works in Birmingham.

A major discovery, made in 1884 by Ludwig Mond (1839–1909; FRS 1891), led to a large scale new industrial process for extracting nickel from a gaseous compound formed with carbon monoxide. In one of Mond's projects concerning the recovery of chlorine wasted in the ammonia soda process, it was observed that nickel valves in the chlorine process were attacked by carbon monoxide. Mond, having moved to London in 1884, followed his great interest in research in the laboratory he set up in the stables of his London home. Here in 1889 a discovery was made that was to become of outstanding industrial significance. Mond's assistant, Carl Langer (1859–1935) heated finely divided nickel in carbon monoxide gas and noted that the colour of the carbon monoxide flame, burning as a jet from the furnace, changed from blue to a 'sickly green'. Mond was called to see this phenomenon and they found that a nickel 'mirror' was deposited on the surface of a white porcelain tile placed in the flame. The gas emerging from the furnace was found to contain a volatile compound of nickel, hitherto unknown, and named by them 'nickel carbonyl'. Lord Kelvin elegantly described the phenomenon 'as giving wings to nickel'. The carbonyl, when heated to about 180°C decomposes into pure nickel and carbon monoxide.

Mond recognised the significance of this reaction for selectively winning nickel from the Sudbury, Ontario ores and by 1892 he had set up a small model plant in his laboratory and had produced nickel from a number of ores. The essence of the process developed was to form nodules of pure nickel from small nickel pellets introduced into the gaseous atmosphere. By the end of 1892 Mond had established a pilot plant near Birmingham on land belonging to Henry Wiggin and by 1895 3,000 lb (1360 kg) of nickel per week was being produced from Canadian matte, in contrast to the small quantities available at the beginning of the century. The Mond Nickel Company was formed by the end of the century. The importance of nickel greatly increased with the development of nickel-containing alloy steels in the latter part of the century, while the 20th century has seen continued increase in importance with nickel playing a crucial role in high temperature technology, for example, forming a vital base of alloys for jet engine components. This carbonyl (Mond) process, named after its discoverer, led to important developments in the nickel industry. The career of Ludwig Mond and the careers of his two sons, both of whom became Fellows of the Royal Society, Sir Robert Ludwig Mond (1867–1938; FRS 1938) and Alfred Moritz Mond (first Baron Melchett, 1868–1930; FRS 1928) were closely linked with the development of the nickel industry in Britain.

Copper and Zinc

Copper and zinc and their alloys, such as the brasses, have occupied an important place in metallurgical history in England, including the period of the Royal Society prior to 1800 (see Chapter 2). Beginning in the 18th century South Wales became a prominent centre for copper production. Among those involved with this activity were John Vivian

Ludwig Mond

Robert Mond

William Crookes

Robert Hunt

who established his Hafod copper smelting works at Swansea in 1810 in partnership with his sons. The eldest son Richard Hussey Vivian (later Sir Richard, 1775–1842; FRS 1841) had a distinguished career, mainly in the army. The younger son, John Henry Vivian (1785–1855; FRS 1823), in 1823 published details of the copper smelting procedures at Hafod in a substantial paper in *Annals of Philosophy*. The eight procedures that he described included calcination, production of matte, melting and casting; the final stage of refining, or toughening, involved 'poling' to achieve the proper 'pitch' (see Chapter 15).

John Hawkshaw (later Sir John, 1811–1891; FRS 1855) who became a leading civil engineer, spent a period (1832–1834) in charge of some copper mines in Venezuela.

For the production of zinc, early in the 20th century from 1910–1915, Brunner–Mond, with the involvement of Robert Mond operated a chloride electrolysis process. Zinc oxide was heated in calcium chloride solutions discarded from the Solvay process. Carbon dioxide was passed into the slurry and the zinc chloride formed was purified and then electrolysed using rotating rotating iron discs as cathodes with carbon anodes; zinc powder was produced at the cathodes. Attempts to produced zinc by electrolysis of the molten chloride were unsuccessful.

Stephen Esslemont Woods (1912–1994; FRS 1974) made major contributions to extraction metallurgy, particularly to that of zinc. Woods joined the research department of the then National Smelting Company at Avonmouth, Bristol, direct from Oxford in 1936. At that time the company was producing zinc by the horizontal distillation process, but in 1933 it had begun to operate an additional vertical retort plant obtained under licence from the USA. To feed these plants the imported sulphide ore was roasted on sinter machines. The sulphur dioxide evolved was converted into sulphuric acid by a converter process with a vanadium catalyst. In 1929 a research department was formed to increase the efficiency of both zinc and acid plants to the maximum.

Woods, as a chemist, was allocated to work on acid plant problems and immediately made an impact. He carried out a detailed heat survey of the converters and showed that there was insufficient cooling between some of the stages, and maximum possible conversion was not being obtained. Adjustments were made to the heat exchangers and immediately the overall conversion efficiency was raised by over 1%. This improved operating profits and made a large reduction in atmospheric pollution. Woods, in addition to this work on the acid plant contributed to the solution of a number of metallurgical problems which beset the research department. From the early stages of its inception there had been a belief in the department that the formidable problems involved in the development of a blast furnace method, to obtain continuous production of zinc on a large scale and with improved fuel economy, could be overcome. In 1945 a colleague, J. Derham, suggested the use of a spray of molten lead as a shock chilling condensing medium for the gases, which went far in solving the main problem of the project, which was the formation of an oxidised product by the back reaction: $Zn + CO_2 = ZnO + CO$. As the molten lead containing zinc in solution was cooled, molten zinc floated to the top and was removed.

Clement le Neve Foster

Alan Powell

A series of furnaces of increasing size were built and showed the value of Derham's suggestion, but also disclosed a number of lesser problems and at this stage Woods made a number of contributions. By 1950 the project was heavily committed. Two large furnaces had been erected at Avonmouth and were operating intermittently. Whilst the blast furnaces were producing metal, operation was plagued by the build-up of zinc oxide in the furnace top and the flues leading to the condensers. The dreaded back reaction had not been completely overcome and the situation was serious. Several major modifications were rapidly tried without success. The situation was saved by a simple solution proposed by Woods. He reasoned that if further air was added to the gases leaving the furnace, equivalent by volume to 10–14% of that blown through the tuyeres, the rise in CO_2 content from 8 to 11% in the top gas, though having an adverse effect on equilibrium, would be more than offset by the rise in temperature of the gas (250ºC) that would result from the combustion of the CO. Some cooling in the critical zone between the exit from the furnace and the encounter with the shock-chilling lead spray, could then be tolerated. Tests proved almost immediately to be successful and at once became a standard part of furnace operation. It was a major factor in establishing the commercial viability of the process.

Gold and silver

In 1824 John Children (1777–1852; FRS 1807) discovered a method of extracting silver without the use of mercury; this method was purchased from him by several American companies. In 1829 another Fellow, Hugh Lee Pattinson (1796–1858; FRS 1852) discovered an easy way of separating the silver from a lead ore, but owing to lack of funds was not able to complete his researches. He was appointed manager to a lead smelting and refining works in 1831, where he perfected, and put into practice, his process for desilvering lead, and patented it in 1833. The profits enabled him subsequently, in a partnership, to establish chemical works at Felling and later at Washington, near Gateshead. Pattinson's process depended on the fact that when argentiferous lead is cooled from the liquid state the first crystals to form have a very low silver content, while the liquid is enriched in silver; in a series of repeated stages involving removal of the crystals in a perforated ladle, lead substantially enriched in silver was obtained and subjected to cupellation to obtain the silver. This process permitted the successful working of previously neglected lead ores. Pattinson also discovered and brought into use a method of separating magnesia from limestone rock, and a process for producing oxychloride of lead, a valuable pigment.

In 1835–36 Antoine-César Becquerel (1788–1878; FMRS 1837) constructed a plant for the electrolytic extraction of silver, lead and copper in France, although this did not operate on an economic basis.

John Percy presented a paper in 1848 on a method of extracting silver from its ores, depending on the solubility of the chloride in sodium thiosulphate; this was his most notable invention in the field of practical metallurgical processses, and led to the Von Patera process, used at Joachimsthal, and the Russell process developed and used in America, which was able to treat ores poor in silver.

Also in the metallurgy of precious metals, William Crookes (later Sir William, 1832–1919; FRS 1863) developed a process for extracting gold and silver, using an amalgam. This was reported in *A Treatise of Metallurgy* which he, in collaboration with Ernst Rohrig in 1868, wrote based on the German edition of Kerl's *Metallurgy*. This substantial three-volume work was intended to provide English metallurgists with contemporaneous experience of metallurgical operations acquired mainly from the European continent. The first volume dealt essentially with the extraction and refining of many non-ferrous metals. In the second volume copper and iron extraction were discussed, while the third volume dealt with steel and fuel, and included a historical review of pyrometry.

George Thomas Beilby (later Sir George, 1850–1924; FRS 1906) was active in extraction metallurgy, involving gold and also sodium. When the McArthur–Forrest process for the extraction of gold from its ores using potassium cyanide was introduced, Beilby invented a process of manufacturing cyanide from ammonia; this process was operated from 1890 to 1906 by the Cassel Gold Company.

In the sphere of mining Robert Stephenson (1803–1859; FRS 1849), in the early part of his engineering career, spent three years in the 1820s supervising the working of some gold and silver mines in Colombia, South America. Percival Johnson had a substantial activity in mining. He consulted and visited professionally many mines in England, Wales, Scotland and Ireland and elsewhere. He was involved with the then booming mines in Devon and Cornwall, concerned with lead, copper, tin and silver; he eventually became owner of several mines and introduced improvements in mineral treatments. The mining engineer Thomas Sopwith (1803–1879; FRS 1845) was associated with lead mining in Northumberland and Durham, and wrote a paper on the lead mines of England. Robert

Hunt (1807–1887; FRS 1854) held the appointment of Keeper of Mining Records for 37 years. His publications included *The History and Statistics of Gold* (1851)and a comprehensive study entitled *British Mining; a Treatise on the History, Discovery, Practical Develoment and Future Prospects of the Metalliferous Mines of the United Kingdom.* Clement le Neve Foster (later Sir Clement, 1841–1904; FRS 1892) was a graduate in mining and metallurgy from the Royal School of Mines, London, whose distinguished career lay in geological survey and mining. In 1868 his exploration work abroad included a survey of gold fields in Venezuela, while from 1869–72 he was an engineer to a gold mining company in Northern Italy.

Guy Dunstan Bengough (1876–1945; FRS 1938) in around 1914 spent a period as a metallurgist and assayer with a tin syndicate in Burma, making a extensive tour prospecting for minerals and learning the processes of extraction of gold and silver from their ores.

In more recent years, in the 1950s, another Fellow (Cecil Reginald Burch 1901–1983; FRS 1944) spent a period in Cornwall in which he developed an interest in tin mining and mineral technology and learned to pan for tin and gold. He invented a novel mineral classifier to deal with tin 'trailings' and pursued research in this field.

Less common metals

William Ramsay (later Sir William, 1852–1916; FRS 1888), whose name is mainly linked with his fundamental research on the inert gases, also carried out some applied, industrial type work. For example, he worked out a process to produce radium from residues containing pitchblende available at a mine in Cornwall. The mining company set up a small factory in London's east end with the title The British Radium Corporation, and Ramsay acted as advisor. Also Ramsay, working with a chemical manufacturer, used the process devised for pitchblende to treat a new mineral, termed thorianite, containing thorium oxide, to produce radium and barium.

At an early stage of his career Cecil Henry Desch (1874–1958; FRS 1923) was employed as a chemist at F. Kendall and Son in Stratford on Avon, where he did work aiming to prepare small quantities of the rare earths, particularly scandium and the development of a department for fine chemicals.

Alan Richard Powell (1894–1975; FRS 1953), research manager for Johnson Matthey and Co. Ltd for 36 years, made outstanding contributions to chemical and extraction metallurgy, particularly concerning platinum group metals and other 'less common' metals, e.g. rare earths. In 1919, Powell, with W.R. Schoeller published a text book on the *Analysis of Minerals and Ores of the Rarer Elements.* Powell also studied the chemistry and metallurgy of niobium and tantalum, expecting that their alloys would find special engineering applications. Another activity was the study of the extraction of zirconia from zircon.

For 14 years Powell was associated with Schoeller in researches on the estimation of the component metals of tantaloniobic minerals; this work was mainly carried out at the Sir John Cass Technical Institute in Aldgate, London. The objectives were to improve methods of attacking minerals containing tantalum and niobium in association with other metals, and to obtain good quantitative separation of the oxides of tantalum and niobium. An advance was achieved by the novel step of controlling the separation of tantalum and niobium solution of the oxides in ammonium oxalate with tannin. Powell was involved in the development of other analytical methods and one of his specialities became the separation of elements of very similar chemical characteristics.

Research on the production of platinum-group thermocouple materials of high stability led to work on the further refinement of all the rhodium residues in stock and on methods for refining, ruthenium, osmium, iridium and platinum. The laboratory also assumed the responsibility for precious metal analysis.The expert services of the analytical laboratory became vitally important and were made available to metallurgical scientists in the 1930s and 1940s engaged in experimental investigations relating to the theory of alloy formation involving elements for which which standard methods of separation and determination were not available. Considerable quantities of platinum-bearing rocks were discovered in South Africa in the latter part of the 1920s. In ores in which the precious metals were associated with sulphides of copper, nickel and iron, the extraction presented difficulties. The research laboratory in 1926 began to work on the urgent problem of finding a reliable and economic method for treating these ores. Within a year the problem had been solved by Powell and two of his colleagues with enormous economic consequences. A pilot plant was erected and then a full-scale plant, which by 1929 was dealing with large quantities of sulphide ore concentrates.

John Stuart Anderson (1908–1990; FRS 1953), whose main field of research was solid-state chemistry, while he was in Australia in an early part of his career, carried out collaborative investigations relevant to the Australian mineral industry. Using zircon and monazite concentrates, the lower halides of zirconium and hafnium were prepared with the objective of using these to effect separation of the metals.

Electric furnace technology

Electric furnace technology began to develop in the latter part of the 19th century, in relation to both non-ferrous and ferrous extraction metallurgy. Charles William Siemens (referred to as William, later Sir William, 1823–1883; FRS 1862) demonstrated the melting of iron using an electric arc furnace with horizontal carbon electrodes at the Paris Exhibition in 1879. Two years later Sebastian Ziani de Ferranti (1864–1930; FRS 1927) assisted Siemens in experiments with an electric furnace for making steel and in 1887 Ferranti developed an induction-heating furnace. Another Fellow, John Joly (1857–1933; FRS 1892) about 1887–88 developed an electric arc furnace, by which he reduced aluminium from topaz.

In France, using an electric arc furnace, Ferdinand Frederic Henri Moissan (1852–1907; FMRS 1905) made outstanding contributions in high temperature metallurgical chemistry, producing many non-ferrous metals and also important compounds of numerous elements. This work of Henri Moissan had its origin in the programme of research that he began in 1884 on compounds of fluorine. His experiments hoping to obtain carbon in the form of diamond from gaseous compounds of fluorine was unsuccessful, but he pursued the objective of obtaining artificial diamond. A meteorite from Canon Diablo, consisting mainly of metallic iron had been found to contain small crystals of diamond. Moissan theorised that the carbon had originally been dissolved in the iron in its original molten state and that during solidification great pressure had been generated in the interior of the meteorite. Having experimented without success with iron saturated with carbon at about 1000°C Moissan then decided to use higher temperatures to achieve greater solubility of carbon in iron, and it was for this purpose that he constructed an electric furnace the original version of which was demonstrated in 1892.

Moissan's furnace consisted of a rectangular block of lime in the centre of which a hole had been scooped. This block was covered with a rectangular lid; two grooves of circular section admitted the carbon poles which served as electrodes, and an arc was struck between the poles. Later, an electromagnet was use to deflect the arc downwards onto the material to be heated. Subsequently, to reduce cost, the furnace body was constructed of limestone. The crucible to be heated stood on magnesia, to avoid the rapid formation of calcium carbide and, for some purposes, crucibles were constructed of a grid of alternate slices of carbon and magnesia. By heating an inclined carbon tube in the arc, and feeding in at one end a mixture of an oxide such as chromium oxide and carbon the metal flowed out at the other end, and a continuous supply was obtained. The first experiment of crystallising carbon under pressure from iron was made using 200 g of Swedish iron, melted with sugar charcoal in a carbon crucible. The crucible was then plunged into a container of cold water. Although an explosion was feared no accident occurred then, nor during many subsequent experiments. The iron was dissolved in dilute hydrochloric acid; the residue, consisting mainly of carbon in various forms, was extracted, and graphite removed by various chemical treatments, including the use of concentrated acids. The final residue was floated in bromoform, in which some transparent dust of specific gravity 3 to 3.5 sank, while a black substance floated. The transparent particles scratched ruby, burned to carbon dioxide, and showed octahedral facets. However, the conclusion that diamond had formed was not confirmed by other experiments by distinguished workers some decades later; the product obtained by Moissan possibly consisted of transparent refractory minerals, such as spinel.

Moissan's electric furnace provided a great stimulus to the advance of high-temperature chemistry. Many products were prepared by him: metallic chromium, manganese, molybdenum, niobium, tantalum, tungsten, uranium, vanadium, zirconium and titanium and many rare-earth metals ; carbides of silicon (carborundum), lithium, calcium, barium, strontium, cerium, lanthanum, ytrrium, thorium, aluminium, manganese and uranium. He also prepared silicides of iron, chromium and borides of iron, carbon, silicon and the metals of the alkali earths. His last research dealt with the distillation of titanium in the arc furnace.

CHAPTER 7

Iron and Steel Production

Introduction

Advances in chemical and process engineering metallurgy concerning iron and steel production illustrate well the pioneering contributions in Britain of chemists, engineers, industrial inventors and entrepreneurs who became Fellows of the Royal Society. During the 18th century and the first half of the 19th century there were many small iron works scattered throughout Britain. The blast furnaces which produced pig iron provided the manufacturing base for the armaments in the Napoleonic wars. Puddling furnaces produced wrought iron, of low carbon content, and containing elongated slag particles. Steel, which was almost exclusively used for cutlery and tools, was made by the crucible process, or by welding 'cemented' bars to form 'shear steel'. In the crucible process, bar iron and added materials were melted in clay crucibles.

This chapter reviews the work of Fellows of the Royal Society, mostly in Britain during the 19th and 20th centuries, contributing to the revolutionary advances in iron and steel production which have been so influential in shaping the history of the modern world.

Ironmaking in the 19th century

In iron smelting technology in Britain, following earlier developments important work was done by James Neilson (1792–1865, FRS 1846) in the introduction of hot blast, and by Isaac Lowthian Bell (later Sir Lowthian, 1816–1904; FRS 1874) in advancing scientific understanding of the ironmaking process. Another Fellow who developed a major, innovative industrial activity was Josiah John Guest (later Sir John, 1785–1852; FRS 1830). At the Dowlais Ironworks in South Wales he tried the use of improved blowing engines and the use of hot blast, and was among the first ironmasters to undertake the rolling of heavy rails. The works expanded from producing around 5,000 tonnes of iron in 1806 to an annual production, including bars and rods, of around 100,000 tonnes by mid-century.

Robert Wilhelm Eberhard Bunsen (1811–1899; FMRS 1858) and Lyon Playfair (later Lord Playfair, 1818–1898; FRS 1848) collaborated in investigating the industrial production of iron in Germany and England. Bunsen's scientific interests in chemistry were wide-ranging; much of his work involved metallurgical topics, and he was also concerned with the application of experimental science to industrial problems. Between 1838 and 1846 he developed methods for the study of gases, while he was investigating the industrial production of cast iron in Germany, and also, in collaboration with Playfair in England. It was found that in the charcoal-burning German furnaces, over 50% of the heat of the fuel was lost in the escaping gases, while the coal-burning English furnaces were even more inefficient, losing more than 80% of the heat. In an 1845 paper 'On the gases evolved from iron furnaces with reference to the smelting of iron'. Bunsen and Playfair suggested techniques for recycling gases through the furnace, and discussed ways of recovering valuable escaping materials.

It was around 1835 that James Neilson commenced the work which resulted in the discovery of the value of the hot blast in the ironmaking process. Blast furnace practice at that time was directly opposite to this. From the fact that greater quantities and higher quality iron were produced by the blast furnace in winter than in summer, the ironmasters had concluded that this was due to the lower temperature of the blast in winter; various devices were used to artificially cool the blast. Neilson considered that the higher yield of the furnaces in winter could be explained, at least partly, by the higher moisture of the air in summer. The comparative inefficiency of the blast in a particular situation, in which the blowing-engine was located half a mile away from the furnace, led to his experiments and important invention. He concluded that the effects of distance between furnace and blowing-engine would be overcome by heating the blast by passing it through a red-hot vessel, to increase its volume and the work done by it.

Neilson carried out experiments and found that heated air in a tube surrounding a gas burner increased the illuminating power of the gas; he also discovered that by blowing heated air instead of air at ordinary temperature into an ordinary smith's fire a more intense heat resulted. He proposed the substitution of a hot blast for a cold blast to increase efficiency, but the ironmasters were reluctant to carry out trials with their furnaces; even those willing to permit an experiment objected to the alterations in the furnace arrangements which Neilson considered necessary. Eventually, however, the effects were properly tested at the Clyde Ironworks, and were so successful that a partnership was set up and patents were taken out in 1828. Neilson and others soon improved the equipment and the invention became widely used in Scotland and beyond. There was, however, resistance to the validity of the patent and refusals by ironmasters to take out licences. Litigation, with some successes for Neilson, ensued in Scotland and in England, and the patents expired in 1842.

Isaac Lowthian Bell at the age of 19, joined his father at the ironworks of the firm of Messr. Losh, Wilson and Bell at Walker, near Newcastle. In 1842 he supervised the erection of a blast furnace, and two years later a second furnace was installed with which experiments were made for smelting Cleveland ironstone. Also in 1844 Isaac with his brothers, Thomas and John, leased a blast furnace at Wylam-on-Tyne where the trials of Cleveland ore continued under his direction. In 1854 Bell's firm opened a works with three blast furnaces on the north bank of the Tees opposite Middlesbrough. There followed a period in which there were difficulties in the transport of ore and Bell was involved in promoting the construction of the Cleveland railway to bring the ironstone to the banks of the Tees. Parliamentary approval was only obtained after severe and expen-

James Neilson

Isaac Lowthian Bell

Lyon Playfair

Bernhard Samuelson

sive contests. During the trade depression that followed the Cleveland developments, Bell Bros acquired colliery properties and expanded their activities in this field.

Subsequently Bell's firm entered the field of steel production. Open-hearth furnaces were erected at Clarence, where steel was first made in 1889. After two years Bell and his partners negotiated with Messrs Dorman, Long and Co. who were among the first to manufacture rolled steel girders in Britain, leading to an amalgamation in 1899 and the building of important steel works at Clarence. A further industry was added later to the wide range of the firm's activities stemming from the discovery of rock salt 1200 feet (~ 365 m) below the surface on the south side of the river Tees. The firm of Bell Brothers in all its branches became in Lowthian Bell's lifetime a huge concern with some 6,000 employees in its mines, collieries, and ironworks.

Bell travelled abroad, and closely studied the conditions of iron manufacture, especially in America. He was outstanding as an authority on the blast furnace and the science involved in its operation, and was also a leader in general knowledge of chemical metallurgy. The results of Bell's experimental researches on blast furnace practice, in which he was assisted by Charles Romley Alder Wright (1844–1894; FRS 1881) were published in 1872 in his classical *Chemical Phenomena of Iron Smelting; an experimental and practical examination of the circumstances which determine the capacity of the blast furnace, the temperature of the air and the proper conditions of the materials to be operated upon.* In 1884 Bell's second great scientific treatise was published, *The Principles of the Manufacture of Iron and Steel.*

Bernhard Samuelson (later Sir Bernhard, 1820–1905; FRS 1881) played an important role as an 'ironmaster' from 1853 when he erected blast furnaces at South Bank, Middlesbrough to use the Cleveland iron. Ten years later he sold this plant and built new and extensive works near Newport in South Wales, where he developed his interest in the practical application of science and studied the construction of large blast furnaces. In 1887 the firm of Sir B. Sanderson and Co. was formed and by 1905 the pig iron production was around 300,000 tonnes per year. Sanderson also built the Brittania Ironworks in Middlesbrough in 1870. He experimented with the Siemens–Martin process to produce steel from the iron produced from the Cleveland ore, but this venture was not successful.

In France, Henri Louis Le Chatelier (1850–1936; FMRS 1913), from his interest in the conditions needed for equilibrium in chemical reactions, enunciated in 1884 the principle that bears his name. Important investigations were made by Le Chatelier and his pupils on the chemistry of the blast furnace. Work on gases at high temperatures included an examination of the reaction $2CO \leftrightarrow CO_2 + C$ which had been studied by Lowthian Bell, and the determination of the effect of temperature on the equilibrium. The influence of iron and its oxides in promoting the reaction was discovered; also an investigation was made of the equilibria between metallic iron, its oxides, and the oxides of carbon on the one hand or hydrogen and water vapour on the other. Other high-temperature work was continued in later years, and in 1900 Le Chatelier determined the conditions necessary for the synthesis of ammonia, including the use of iron as a catalyst; in 1912 reactions in steelmaking were studied.

Henry Bessemer

Charles William Siemens

Steelmaking in the 19th century

In steelmaking during the latter half of the 19th century the names of four Fellows of the Royal Society are particularly significant in fundamental metallurgical inventions: Henry Bessemer (later Sir Henry, 1813–1898; FRS 1879); Charles William Siemens (later Sir William, 1823–1883; FRS 1862); George James Snelus (1837–1906; FRS 1887), and Percy Carlyle Gilchrist (1851–1935; FRS 1891). John Percy (1817–1889; FRS 1847) was also an influential figure in the whole sphere of ferrous extraction metallurgy through his analysis of British ores, his role as a leader in higher education at the Royal School of Mines, London, and through his writings. Sir William Cavendish (Seventh Duke of Devonshire, 1808–1891; FRS 1829) also played an important role in the ferrous industry through his establishment of a large plant in the Barrow-in-Furness area, and as first president of the Iron and Steel Institute (1869–71).

The introduction of the Bessemer converter process in 1856 and of the Siemens open-hearth process ten years later, together with the improvements in blast furnace practice, provided the foundations for the modern steel industry of the world. The first Bessemer works to produce steel on a commercial scale was built at Sheffield, while Siemens carried out extensive experiments with an acid open-hearth furnace at Birmingham in the mid 1860s, and the first open-hearth steel plant was erected at Landore in South Wales in the latter part of the decade. The development of the basic process, patented in 1879, involving notably the work of Gilchrist and of his cousin Sidney Gilchrist Thomas (1850–1885), was a major advance, which made possible the production of steel from the extensive worldwide deposits of phosphoric ores.

The earliest basic Bessemer works to produce steel on a commercial scale were in

Newcastle and Staffordshire. Later the basic open-hearth process became a serious competitor. The British iron and steel industry was the largest in the world until the 1880s but by the time of the first World War the total size of the industry, and the average size of the plants was smaller than in the USA and Germany and the plants were less modern.

The beginning of the most important stage of Bessemer's career dates back to the Crimean war, when many inventors, including Bessemer, were turning their attention to the faults in the British artillery. One of his early proposals was to fire elongated shot from a smooth-bore gun and to obtain rotation by grooving the projectile. Although he received no encouragement from the British war office, the Emperor Napoleon invited him to Vincennes, where some experiments proved conclusively that the material then available for gun construction was too weak for this new projectile system. As a result, Bessemer began efforts to produce a stronger material by the melting of pig or cast iron with steel in a reverberatory or cupola furnace.This was the subject of the first of the long series of patents in steel manufacture that Bessemer took out from 1854 to 1869. The combination of cast iron and steel gave promising results but did not provide the required properties.

In his autobiography, Bessemer wrote 'My knowledge of iron metallurgy at that time was very limited.... but this was in one sense an advantage for me, for I had nothing to unlearn.'

While melting pig iron in a reverberatory furnace Bessemer found that some pieces exposed to the air blast on one side of the bath remained unmelted in spite of the intense heat; examination showed that these were shells of decarburised iron, the carbon having been burnt out by the blast. This observation led to experiments that formed the basis of three other patents in the period December 1855–February 1856; these concerned a process in which molten pig iron from the blast furnace or cupola was run into a large tipping vessel – the Bessemer converter – an air blast being introduced through tuyeres to pass up through the charge. In March 1856 another specification was filed, for the addition of some recarburising material to be added after the carbon and impurities had been burnt out, so as to obtain steel of the required carbon content. Between the middle of 1855 and the summer of 1856 Bessemer carried out many experiments at his laboratory at Baxter House, St Pancras, London, aiming to establish his process on an industrial scale.

The challenge was how to decarburise the charge completely and to keep it liquid by the combustion of the impurities in the molten iron by means of an air blast. The first converter consisted of a cylindrical chamber lined with fireclay, with a row of tuyeres near the bottom and an opening at the top for the discharge of the burning gases. The converter held 10 hundredweight (~ 500 kg) of molten metal, the air being admitted into the charge for about 10 minutes, producing a violent explosion of sparks and flames and melted slag, lasting some minutes. When this had subsided the charge was tapped, and the product was found to be a fully decarburised, malleable product. After many experiments the fixed converter was replaced by one mounted on trunnions.

Bessemer's experiments attracted much attention, leading to widespread enthusiasm when he read his famous paper entitled 'On the manufacture of Malleable Iron and Steel without Fuel' at the British Association meeting in Cheltenham in 1856. Within a month

Bessemer converter: (A) View of converter, (B) Converter in charging position, (C) Blowing position, (D) Pouring steel into casting ladle (E), (F) Ladle pouring into ingot mould, (G) Plan of tuyeres. (Ref. Ch. 7.1.)

Square bar of Bessemer steel: twisted cold (shown at Sheffield, 1861). (Ref. Ch. 7.1.)

Bessemer plant at Sheffield. (Ref. Ch. 7.1.)

of this meeting he had received £ 27,000 from ironmakers for licences to use the process. However, trials hastily made by the licensees, using irons with higher phosphorus content than that used by Bessemer led to failures. Consequently there was a strong adverse reaction, and Bessemer's reputation and the process were endangered. However, in 1858, having carried out further experiments at Baxter House, and having spent thousands of pounds in constructing new plant, Bessemer was able to show his numerous licensees why they had failed and how they could make first-class steel with certainty using irons produced from low-phosphorus ores. Violent opposition from the steel trade ensued, which Bessemer met by erecting his own works in Sheffield in 1859, and starting business as a steelmaker. Another critical factor in overcoming the difficulties in achieving a successful commercial process was a contribution by Robert Forester Mushet (1811–1891); he suggested and demonstrated the effectiveness of the addition of a proportion of manganese ore (spiegeleisen) to the melt. There was previous knowledge and a number of patents, showing that manganese was an important element in steelmaking and Mushet took out a patent soon after Bessemer commenced his process. Manganese acted as a deoxidising agent to deal with the problem of oxygen in solution in the molten steel.

Bessemer's invention revolutionised the commercial history of the world, and it was later said 'The invention takes its rank with the great events which have changed the face of society since the middle ages'.* Manufacture proceeded steadily on an increasing

*A. S. Hewitt, address to the Iron and Steel Institute, 1890.

William Cavendish
(Seventh Duke of Devonshire)

Percy Gilchrist

George Snelus

John Percy

scale, and within five years the Bessemer process had been adopted by all the steelmaking countries of the world. Although William Kelly (1811–1888) in the USA had pre-dated Bessemer in the discovery of a process using a blast of air to convert pig iron into steel, and had obtained a USA patent in 1857, it was the Bessemer process, with mechanical superiority of the converter, that led to the first great expansion of the American steel industry. Soon after 1859 Bessemer made a speciality of gun-making at Sheffield and manufactured some hundreds of weapons for foreign governments; however, opposition to the use of steel for ordnance in Britain led to a delay of 20 years before steel was used by the British services. Bessemer steel exhibited in London in 1862 illustrated the state of manufacture at the Sheffield works; it included locomotive boiler plate tubes, a 24-pounder gun, square steel bars and double-headed steel rails twisted cold into spirals; a 14-inch (~ 0.33 m) ingot and the crankshaft of a 250 horse-power engine. In 1879, the Bessemer process made a further vital advance, following the discovery of a means of removing phosphorus in the converter; steel manufacture was thereby greatly facilitated and made more economic in both England and America.

One of the largest Bessemer plants in Europe was established by the Seventh Duke of Devonshire to exploit the rich deposits of high grade haematite iron ores on his north Lancashire estate; Barrow-in-Furness developed into a port and a centre of ferrous production.

From 1865 onwards competition was experienced by the Bessemer process from the Siemens steelmaking process which became widely used in Britain. An early enthusiast in the then new science of thermodynamics, in 1847 Charles William Siemens had already begun work on the regenerative principle in the employment of heat to obtain increased economy in its practical use. In 1857 in conjunction with his younger brother and pupil Frederick Siemens (1826–1914) he commenced investigations with a view to using the principle of regeneration in the field of metallurgy. The results of their work, in which the elder brother had the greatest share, was the regenerative gas furnace. It was first practically applied in the works of Messrs Chance, of Birmingham, to one of their glass-smelting furnaces, and proved a complete success. The great economy of fuel, together with the great ease with which high temperatures could be maintained and regulated, soon led to its application in the production of iron and steel. Siemens did not remain content with the application of the furnace to the then known methods of ironmaking and steelmaking, but turned his attention to improving the methods of manufacture. In the steel works which he founded in Birmingham in 1865, he brought to a thorough practical success the open-hearth steelmaking process, which came to bear his name. In this process molten metal lay on the furnace hearth with the flames above it; the oxygen of the iron ore was used to remove carbon from the iron. A subsequent improvement of this became the Siemens–Martin process involving adaptations, introduced by the French engineer Pierre Émile Martin (1824–1915) in 1864; the use of fuel gas instead of solid fuel enabled temperatures high enough to melt scrap steel to be achieved. Within three or four years after this the Landore works, near Swansea, were organised on an extensive scale for production, and operated until about 1888. The new material –

mild steel – thus introduced, speedily proved its excellence for metallic structures in which strength and lightness were desired, and it began to replace iron in the construction of the highest classes of steamships and marine steam boilers.

Siemens also experimented with a 'direct' steelmaking process in which iron ore was reduced in a gaseous atmosphere in a rotating vessel at high temperature to produce impure metal for refining in the open-hearth furnace. Another innovative technique on which William Siemens worked was arc melting. In 1878 and 1879 he patented a small indirect arc furnace for melting. This equipment consisted of a crucible made from non-conducting material and electrodes which were laterally adjustable; one electrode was hollow to permit the introduction of a reducing or inert gas. It appears that Sebastian Ziani de Ferranti (1864–1930; FRS 1927), while working at the Siemens Brothers plant in Charlton, London, in 1881, assisted William Siemens with an electric furnace for steelmaking. Among Ferranti's research activities in electrical technology was the development of an induction heating furnace, patented in 1887.

The critical industrial problem of producing Bessemer steel from pig iron high in phosphorus was successfully solved by collaboration between Sidney Gilchrist Thomas (1850–1885) and his cousin Percy Carlyle Gilchrist resulting in a patent in 1879. However, George Snelus had conceived the possibility of eliminating phosphorus from molten pig iron by oxidation in a basically-lined vessel. Having taken out a British patent in 1872, he carried out experimental trials in a Bessemer converter and succeeded in almost entirely eliminating phosphorus from 3 to 4 tonne charges of molten phosphoric pig iron thus making the first specimens of 'basic' steel by the pneumatic process. There were some practical difficulties, however, which were not fully overcome, associated with the use of lime, and the final success was achieved by Thomas and Gilchrist. Another significant contribution made by Snelus in the field of metallurgical chemistry was in the analytical determination of phosphorus in steel; also, he took out in 1902 a patent for the manufacture of iron and steel in a basic lined rotary furnace.

Thomas originally intending a career in medicine was forced for financial reasons to give up this idea and took up a job as a court clerk. He devoted spare time to the study of natural science and in particular chemistry; in 1870 he attended a course of lectures at the Birkbeck Institute where he heard one of the lecturers, Mr George Chaloner, say 'The man who eliminates phosphorus by means of the Bessemer converter will make his fortune.' From that time Thomas became a constant reader on the subject and in 1872 and 1873 he studied at the Royal School of Mines, London, and obtained a first class award in advanced mineralogy and inorganic chemistry. Following this he began experiments and conceived the idea of rendering both the Bessemer and Siemens processes capable of removing phosphorus, based on research by French metallurgists who replaced the usual silica lining in furnaces, with one manufactured from limestone.

In 1878 the then manager of the Blaenavon Works in South Wales had appointed Thomas's cousin, Percy Gilchrist as Works Chemist. An important ironmaking industry had been established in Blaenavon, blast furnaces having been built a century earlier; at the time of Gilchrist's appointment a new site across the valley in Blaenavon had been opened.Thomas persuaded Gilchrist to carry out experiments during his leisure hours at Blaenavon to test his idea. The first patent in 1877 was for linings made by mixing limestone with a binder of sodium silicate, and these linings were used for the first trials at the Blaenavon works, where the manager Edward Martin was also involved. These linings perished rapidly and in 1878 they obtained a patent for subjecting certain types of limestone to a very high temperature to produce a dense material for making into lumps or bricks; this patent was of the greatest importance. At the Spring meeting of the Iron and Steel Institute in 1878 Thomas reported that he and Gilchrist and Martin had almost completely removed phosphorus from Cleveland iron in a Bessemer converter. Events then led to E. Windsor Richards of Bolckow, Vaughan and Company obtaining support from the directors of the company for the process to be given a trial. John Stead (1851–1923; FRS 1903) and Richards, both consultants to the Company, were involved with Thomas and Gilchrist in the practical development of the process. Experiments began in 1878 in Middlesbrough using as a fixed converter an old cupola lined with crushed dolomite with a sodium silicate binder, and tuyeres near the bottom of the sides of the cupola. The process involved the addition of spiegeleisen, containing manganese, in the ladle, and casting the steel into ingots. The analyses made by Stead showed that considerable amounts of phosphorus remained in the steel; he concluded that the secret of dephosphorisation was the prolongation of the blowing after the carbon was removed and sufficient iron oxide had been produced to oxidise the phosphorus. This view was not initially accepted by the inventors, but, after further failures in the trials, they followed Stead's suggestion and advised Richards to 'overblow' a charge, thus achieving success; the afterblow process was patented in 1879.

The essential conditions for the successful working of the basic Bessemer process were found to be: the use of a basic lining of correct composition with properties enabling it to adhere to the walls of the converter, the formation of a rich basic slag early in the process, and the use of an 'afterblow'. For many years the process was the primary means of producing cheap steel on a large scale, and the slag found an application as a fertiliser. In 1881 Thomas and Gilchrist set up their own steelworks, North Eastern Steel Company, at Middlesborough. This was sold to Dorman, Long and Company in 1903, when Gilchrist retired from active metallurgical work.

Of interest in the chemistry of steel refining in the Bessemer process was research in spectrography by Walter Noel Hartley (later Sir William, 1847–1913; FRS 1884), professor of chemistry at the Royal College of Science at Dublin. He took a series of photographs of spectra at intervals of half a minute during the complete course of a Bessemer melt and showed that both iron and manganese are largely vaporised.

John Percy began his very important work on iron and steel early in his career. The Great Exhibition of 1851 contained a very extensive and highly interesting series of British ores, collected by one of his friends, and Percy, in the same year, undertook to supervise their analysis. He was involved in the question of the need to remove phosphorus in the Bessemer steelmaking process, and both Thomas and Gilchrist were students of his. Over a period of more than 30 years, Percy made a collection of metallurgical specimens (totalling around 4,000 items) to illustrate points relating to the manufacture and use of metals; this unique collection was preserved and exhibited to the public as the 'Percy Collection'. Percy made a particularly important contribution to metallurgy as an author through his great treatise *Metallurgy* (see Chapter 6), which included detailed information on ferrous metallurgy.

Ironmaking and steelmaking in the 20th century

During the 20th century, improvements in blast furnace practice continued. The furnace size increased dramatically to a point where one modern furnace can supply up to around 10,000 tonnes of liquid iron per day in the very large, integrated steel works that have been built. In recent years there has been interest in potential substitutes for the blast furnace process, for example, reduction of iron ores at temperatures below the metal's melting point. For some 100 years the open-hearth and Bessemer-based processes produced most of the steel made. However, by the mid-20th century oxygen, available on a tonnage scale, came to be used universally in bulk steelmaking in place of air to remove carbon; e.g blowing oxygen into the top of liquid iron in a converter vessel; (the use of oxygen in steelmaking was not a new idea, Bessemer having referred to it in patents). Other important developments include the extensive use of electric furnace technology and improvements in refractory materials for furnace linings. Also, in place of conventional ingot production, continuous casting of steel became the established method in the 1950s. (Bessemer, recognising the advantages of such a process had, in 1856, filed a patent for equipment for producing steel sheet by pouring the liquid metal between two water-cooled rollers; however, it was not developed into an industrial process. In 1891 Bessemer described how a tundish with a series of holes in the base could be located above the roll gap to achieve uniform feed and constant width of strip.)

In the 20th century, Fellows of the Royal society have been responsible for many advances in the practice and theory of ironmaking and steelmaking and processing: Robert Abbott Hadfield (later Sir Robert, 1858–1940; FRS 1909); John Oliver Arnold (1858–1930; FRS 1912); William Bone (1871–1938; FRS 1905); Cecil Henry Desch (1874–1958; FRS 1923); William Herbert Hatfield (1882–1943; FRS 1935); Charles Alfred Edwards (1882–1960; FRS 1930); Andrew McCance (later Sir Andrew, 1889–1983; FRS 1943); Charles Goodeve (later Sir Charles, 1904–1980; FRS 1940); Montague (Monty) Finniston (later Sir Monty, 1912–1991; FRS 1969); Frederick Denys Richardson (1913–1983; FRS 1968) and John (Jack) Hugh Chesters (1906–1994; FRS 1969). Also Stanley Baldwin (later Earl Baldwin of Bewdley, 1867–1947; FRS 1927) who became Prime Minister of England spent twenty years, beginning in 1888, in the family steel business before entering national politics. The careers of these Fellows involve industry companies, research establishments and universities. In this chapter the emphasis is on activity relating to the primary production of iron and steel, while ingot production, mechanical working and alloy development (e.g. the work of Hadfield, Hatfield and Stead) are considered elsewhere (e.g. see Chapters 11 and 15).

At Sheffield University John Arnold was closely involved in the application of scientific knowledge to the manufacture of iron and steel. Essentially a chemist, Arnold gave much attention to analytical processes and was joint author of the book *Steelworks Analysis,* first published in 1891. He improved existing methods of analysis of steels and developed new techniques; the book demonstrated the application of advanced inorganic chemistry to practical metallurgical purposes.

Cecil Desch, who became well known for his work in physical metallurgy, had strong associations with the steel industry, beginning in the 1920s during his tenure of the chair of metallurgy at Sheffield University. When he moved to the National Physical Laboratory in 1932, one of the projects in which he took a special interest was the determination of the oxygen content of steels using the vacuum fusion process. When he retired from the NPL in 1939 he was appointed as a scientific adviser to the Iron and Steel Research Council, and in 1943 he joined the board of Messrs Richard Thomas and Co. Ltd, with the main task of establishing a research laboratory in the works at Ebbw Vale. However, before this laboratory was completed the company amalgamated with Baldwins Ltd to form Richard Thomas and Baldwins Ltd. Desch resigned his appointment and became scientific adviser to the Whitehead Iron and Steel Co.

William Bone, distinguished for his achievements in fuel science and technology, particularly combustion phenomena, had a close connection with the iron and steel industry beginning in his youth; fuel economy in the production of iron and steel was one of his interests. At the University of Leeds in the early 1900s his work included studies of reactions in the fuel bed of gas producers. During the period 1926–38 at Imperial College he carried out a collaborative programme of fundamental research on chemical reactions in the blast furnace, involving the oxides of iron and of carbon; the laboratory investigations were supplemented by the study of gas compositions and the state of reduction of ore in the interior of an operating blast furnace.

The career of Charles Edwards was closely involved with ferrous metallurgy, including production processes. In 1910 he took a post as metallurgist with Bolckow, Vaughan and Company and Dorman, Long and Company at Middlesborough. Here his interests included low carbon steels and he wrote papers on the merits of oxygen enrichment in blast furnace production. In 1920 when he became professor of metallurgy at the University College of Swansea he began involvement with the local industry of tin plate and black plate manufacture and his researches included subjects such as blowholes in steel ingots and the alloys of iron and tin.

Andrew McCance after graduation from the Royal School of Mines in 1910 began his career at Beardmores in Glasgow. In 1919, foreseeing a great future in alloy steelmaking, he started the Clyde Alloy Steel Company, Motherwell, supported financially mainly by Colvilles Ltd. In spite of this busy life during the early 1920s, it was during this period that McCance made his most important scientific contribution. This was to the physical chemistry of steelmaking which involves the removal of impurities by a reactive slag. At that time, much was known from practical observation, but much less was properly understood. McCance recognised that an interdependent collection of balanced reactions

Stanley Baldwin
(Earl Baldwin)

William Bone

Charles Edwards

Andrew McCance

were involved and that the balances altered with progress in the refining, with changes in slag composition or temperature of the bath, with the temperature change on pouring and casting, and with the deoxidizer additions made at that time. He put this on a quantitative basis, so the detailed understanding of the often complicated pattern of the reactions was much improved and practical observations became understandable. McCance began work on this topic at Beardmores and was not in a position to carry out personal experimental measurements. His principal paper (1925) represented nine years of work, mainly in spare time, drawing on the available published thermodynamic data and then working out the best figures that gave internal consistency in the group of reactions involved. With his very complete treatment of all the reactions, McCance took steelmaking chemistry up to the stage where it became necessary to consider the activities in the slag and some initial consideration of this aspect was given in his extensive review of 1938.

John Craig (later Sir John, 1874–1957) the chairman of both the Clyde Alloy Steel Company and of Colvilles, was particularly impressed by McCance and in 1930 invited him to become General Manager of Colvilles. This was the start of a partnership that continued until 1956 and shaped the development of the Scottish steel industry, with Craig as the major negotiator and McCance as the technical reorganiser. The decade of the 1930s saw many changes – political, industrial and technical. These included mergers of companies, the closure of uneconomic plant and the building of new plant. The integrated Clyde Iron-Clydebridge development came into operation at the beginning of World War II. The Government took control of the iron and steel industry in September 1939, and during the war, government policy was against developments in steel works, except for special purposes. McCance was fully occupied with the technical problems of keeping the plant working at full capacity and Colvilles became one of the biggest producers of armour plate for ships and tanks. After the war various developments proceeded, including the setting up of the Ravenscraig complex.

At the end of the war, Charles Goodeve was invited at the suggestion of Sir Andrew McCance to become director of the British Iron and Steel Research Association (BISRA). This was to be financed by the steel industry and the Department of Scientific and Industrial Research through the British Iron and Steel Federation. He inherited an embryo organization, and was determined to build up a research association for an industry which was then producing about 12 million tonnes of steel per annum. He saw the work of a research association in terms of the science and technology and in relation to the activities of universities and industrial laboratories. He regarded BISRA as complementary to, and not in competition with, the research and development groups within industry. The programmes were controlled by panels consisting of senior managers or technical men from industry; detailed work was controlled by more specialised committees and was carried out either in the BISRA laboratories or in works. The laboratories were set up in various locations: Battersea (London), Swansea, Sheffield and Teesside. Goodeve collected around him a keen group of engineers and scientists and stimulated them to act on their own initiative, giving them all the support he could. A profound effect on the steel industry came about, through the BISRA research programmes, and the forum for tech-

Charles Goodeve

Frederick Denys Richardson

John Chesters

Monty Finniston

nical discussion and collaboration between specialists in the separate steel companies.

The many major projects which contributed to improved understanding and practice covered areas such as blast furnaces, open hearth furnaces, arc furnaces and automatic gauge control in strip production. Goodeve developed operational research to promote efficiency and BISRA demonstrated the need for larger ore carriers and deeper berths, which the industry opposed. He maintained a good balance between short- and long-term research, and stimulated academic interest in metallurgical work by ensuring that his departments sponsored postgraduate research in universities. He was very much concerned with the future of Britain as a whole and he saw the iron and steel industry in the context of Britain's continuing development. Goodeve retired from BISRA in 1969, and with the nationalisation of the steel industry the Research Association was absorbed into the British Steel Corporation over the next few years.

Also at the end of the war Denys Richardson was invited by Goodeve to build up the BISRA chemistry department. He began by visiting various works to find out about ironmaking and steelmaking and was struck by the large number of reactions that occur and equilibria that are approached during the processing, and by the importance of oxygen potentials. He met Dr H.J. Ellingham, who informed him about two free energy diagrams (now known as Ellingham diagrams) that he had published in 1944 with Professor Cecil Dannatt who was at the Royal School of Mines, Imperial College; these diagrams displayed free energy changes for the formation of a substantial number of oxides and sulphides as a function of temperature. The order of stability of compounds over a range of temperatures was displayed, and Richardson saw their potential for clarifying reactions and equilibria. With colleagues, including Dr J.H.E. Jeffes, a number of new diagrams were prepared; these eventually covered oxides, sulphides,silicates and carbides, and nomograms were devised for reading off data such as the partial pressures of O_2 and SO_2, carbon activities, equilibrium gas ratios (CO/CO_2, H_2/H_2O, H_2/H_2S, CH_4/H_2) and CO pressures in the presence of carbon in equilibrium with oxide systems at any temperature. A special series of such diagrams were prepared to demonstrate the equilibria in the iron blast furnace, and many aspects of the shaft and hearth reactions were clarified for the first time. At the BISRA research laboratories in Battersea, an experimental chemical group was set up. In 1948 the Faraday Society arranged a discussion, organised by Denys Richardson,on the physical chemistry of process metallurgy. This was attended by most of the significant workers in the field; the Faraday discussion was a landmark, and showed the enormous scope for basic experimental work relating to the problems of metal production at high temperatures.

In 1950 Richardson took up a fellowship at the Royal School of Mines in the metallurgy department to be financed for a five year period by the Nuffield Foundation with support also from the College. A number of projects were started, marking the beginning of what became the Nuffield Research Group in Extraction Metallurgy, and which proved to be the world's major academic research group in this subject, with a strong team. Continued financial support was obtained and the work involving many research students and research colleagues covered many topics, relevant to ferrous and non-ferrous metallurgy. These included free energy diagrams in the analysis of metallurgical processes such as the vacuum degassing of steel; metal solutions; thermodynamics and structures of slag solutions; equilibria and kinetics involving solids; kinetics of reactions using levitated metal drops; bubbles, drops and interfacial phenomena. Richardson's textbook *The Physical Chemistry of Melts in Metallurgy* published in 1974 became a standard work. Richardson recognised the importance of engineering aspects of metal processing: fluid flow and heat and mass transfer, and this field was also developed.

The science and technology of ceramic refractory materials has played an ongoing, vital role in the iron and steel industry, and the name of Jack Chesters was prominent in this. In 1934 he was invited to form a refractories section in the new Central Research Laboratory for the United Steel Companies in Sheffield. This section aimed to obtain improved understanding of furnace performance and a reduction in the cost of refractories. Many valuable discoveries were made, including the development of sea water magnesia, stabilised dolomite and carbon refractories. In the period 1939–45 he made a major contribution to the war effort by securing supplies and perfecting strategic materials for iron and steelmaking. From the late 1940s to the early 60s most of his energies were devoted to leading research on furnace design and fuel efficiency which further enhanced his international reputation.

In 1970, only eighteen months before his retirement was due he was invited to become Director of what had been the British Iron and Steel Research Association (BISRA) and which was responsible for the Corporate Laboratories of the British Steel Corporation. It was a difficult time in which reorganisation and rationalisation were being envisaged, but he said that he had never fired a research man and did not intend to start. He accepted the post on his terms, and maintained them, although after his retirement changes inevitably followed. Then in 1976 Chesters became chairman of the Watt Committee on Energy. Chesters's publications included several books; for example his *Steelplant Refractories,* first published in 1944, brought all his work to that date together, and was recognised worldwide as the definitive source.

In 1966, the steel industry was about to be renationalised and the metallurgist Monty Finniston was invited to join the government committee planning the future organisation of the new corporation. This led to him joining the corporation in 1967 as a Deputy Chairman, becoming Chief Executive in 1971 and, finally Chairman in 1973. There were immense problems in bringing together the 14 steel companies with 39 steelmaking plants and numerous steel-working plants.The committee had a sound plan which involved essential reorganisation and modernisation and envisaged an expansion to an annual output of around 35 million tonnes over a ten-year period involving an estimated cost of £3 billion. Finniston initiated and made good progress with the implementation of the plan, but trouble came when the steel market was hit by the effects of the oil crisis and world depression. Also when a Labour government returned to power he fought for the right of management in a state-owned industry to be free from political intervention. In spite of Monty's impressive contributions to the industry, his contract with BSC was not renewed by the Government in 1976 and the massive drop in steel demand in 1975 made some of the planned developments unnecessary. With the election of a Conservative government in 1979, BSC's decisions included closing of obsolete works and that modernisation programme gave British Steel one of the best facilities in Europe.

The industrial production of iron and steel is of huge importance in Britain, not only in terms of the metallurgy involved, but also in relation to the economy and politics of the nation. This chapter has focussed on the role of Fellows of the Royal Society in iron and steel production, with little reference to the wider aspects of national interest. Interestingly, during the last three decades of the 20th century, no other Fellows apart from those referred to above have worked directly in ferrous ironmaking and steelmaking, although Fellows of the Royal Academy of Engineering have played key roles. In Chapter 19, however, wider issues stemming from science and technology and relating to the national economy are discussed, in the context of the role of the Royal Society in the 'Second Industrial Revolution'; further reference is made to steelmaking, particularly in the light of its history from the mid-20th century to the present day.

CHAPTER 8

Metallurgy and the Electrical Age

Introduction

Among the major scientific and technological advances that took place during the 19th century those concerning electricity and magnetism were particularly far-reaching in their significance; indeed, as this Chapter suggests, the 19th century might be called 'The Electrical Age'. The Electrical Society of London, founded in 1837, expressed the view that 'as a universal agent in nature and space, electricity takes the first rank in the temple of knowledge.' In relation to metallurgy the linking themes included the use of electricity in production and processing techniques and the vital role of metallic materials in the generation, distribution and application of electricity. This Chapter begins with consideration of early developments in magnetism and electricity as a background to reviewing aspects such as batteries and their metallurgical applications, electric lighting and the development of the electricity industry in the United Kingdom.

Early discoveries in magnetism and electricity

If one places two different metals in one's mouth a tingling sensation is felt. What has happened is that the metals have acted as electrodes with saliva the electrolyte, and the tingling is due to the generation of a minute electrical current; one has re-discovered galvanism! One of the greatest mysteries in the history of science is that in spite of the simplicity of the metals-in-mouth experiment it took mankind until the end of the 18th century to identify and make use of galvanism (Luigi Galvani (1737–1798) and Allessandro Volta (1745–1827; FRS 1791). This discovery has turned out to be of immense importance; it launched the Electrical Age which, in turn and with James Clerk Maxwell (1831–1879; FRS 1861) as the bridge-builder, initiated the scientific revolution of the 20th century and the Atomic Age.

It is not then surprising that the history of magnetism and electricity constitutes a major part of the history of physics. Metals in their turn have a symbiotic relationship with magnetism and electricity. It was the metal iron which could be magnetised and hence used for compass needles, and of course the first electrodes were metals and metals are uniquely-good conductors of electricity. The complement of this is that a number of metals were first isolated and identified by electrolysis.

Stepping backward now to the earlier times, there is some evidence that the magnetic properties of the mineral magnetite (Fe_3O_4), known as lodestone, were used in compasses as early as the 26th century BC by the Chinese. Such claims have, however, been challenged, some authorities being convinced that compasses were introduced to China as late as the 13th century AD, from Italy or the Middle East.

Magnets were mentioned in Greek writings from as early as 800 BC and from early times magnetite was mined in the Greek province of Magnesia. According to Lucretius (c 99–55 BC) 'magnet stones' attracted iron by atoms from the stone emanating from its surface striking and removing all the atoms of air between the stone and the iron which then closes the gap by being drawn into the vacuum:

> Why iron for all its strength
> Can be seduced, enticed;
> First from this stone must flow
> A multitude of atoms forms;
> Perhaps a surge which with its blows
> Parts all the air that lies between the iron and the stone.
> And when this space is emptied out,
> Becomes a void,
> Straightaway the atom particles of iron rush in
> In thronging numbers and together joined.

He elaborates further: 'Then all the air behind/ Propels and pushes it along.'

When contact is made between the iron and the stone 'they are held together by invisible hooks'

> And some things seem to cling, held together as if by hooks and eyes.
> This seems to be the case with magnet and with iron.

Thales of Miletus (640–548 BC) is credited with being the first to observe that when amber was rubbed with a dry cloth it acquired the property of attracting light objects, in other words it gained what would now be called a static electricity charge. (The word 'electricity' is derived from 'electron', the Greek word for amber.) In spite of this knowledge, the Greeks did not theorise to any considerable extent on the nature of magnetism and electricity. This is fortunate, according to Arthur Koestler, for had they done so they would have polluted magnetism and electricity with Aristotlean animism and delayed even further into the future man's understanding of the processes.

There followed a long period of time when little or no progress was made with the theory or application of magnetism or electricity. In 1269 Petrus Peregrinus (born c 1220), an engineer in the French army under Louis IX, published his *Epistola de Magnete* which was an account of his experiments with magnets and described a simple compass consisting of a piece of magnetized iron floating in water.

Right at the end of the 16th century William Gilbert (1544–1603), court physician to Elizabeth I, building on the pioneering work of the humble Wapping compass-maker,

Robert Norman, carried out extensive experiments on magnetism. He was very successful, confirming Norman's discovery of magnetic dip and fashioning a lodestone into a sphere and he held a small compass to its surface plotting lines of equal potential. By such experiments he concluded that the earth was itself a giant magnet with a magnetic pole near each geographic pole. He also showed that the strength of a lodestone can be increased by combining it with soft iron and that iron and steel can be magnetised by stroking them with a lodestone.

Gilbert went on to demonstrate that iron magnets lose their magnetism when heated to red heat (i.e. above the Curie Point) and cannot be remagnetised until they are cooled back to room temperature. So numerous were his discoveries that it was not until William Sturgeon (1783–1850) made the first electromagnet in 1825 and Michael Faraday (1791–1867; FRS 1824) began his studies that substantial new knowledge was added to Gilbert's magnetic discoveries. Gilbert also studied static electricity, showing that other materials besides amber generated a charge on being rubbed with a cloth.

By carrying out such a thorough experimental programme Gilbert anticipated the advocacy of the experimental method, in 1620, by his contemporary at court, Francis Bacon (1561–1626). Gilbert still retained Aristotlean ideas though and spoke of the 'soul' of the magnet – he stood like a Colossus astride Greek animism and modern (experimental) science.

Gilbert reported his findings in *De Magnete*, his book published in 1600. Thus the 17th century, the century of the scientific revolution, got off to a splendid start as far as magnetism and electricity is concerned, and further great progress might have been expected. It was not to be. Very few of the great scientists of the age contributed to the subject and when they did they usually got things wrong – the great Sir Isaac Newton (1642–1727; FRS 1671) erroneously deduced an inverse cube law for electrostatic attraction, for example. The fledgling Royal Society took relatively little interest in the subject.

The one outstanding advance of the century, as far as equipment was concerned, was the development of the first continuous static electricity generator by Otto von Guericke (1620–1682), Mayor of Magdeburg in Germany. He made a globe of sulphur which was rotated on an axle with a cloth rubbing on its surface. It was a very efficient apparatus with the globe strongly attracting light-weight objects. Inspired by some observations by the French astronomer, Jean Picard (1620–1682), Francis Hauksbee (c 1666–1713; FRS 1705), Robert Hooke's assistant at the Royal Society, made a machine similar to Guericke's except that instead of a ball of sulphur he used a glass sphere. When rotated and rubbed the glass sphere became charged and not only attracted objects but also glowed internally. No explanation for the illumination was forthcoming.

Another of Newton's disciples, Stephen Gray (1666–1736; FRS 1733) made the important discovery that electricity could flow from one place to another without any appearance of movement of matter, that is, it was weightless. Electricity could be generated in, and held in bodies such as glass or silk but could not flow through them. These he called *electrics* – what would be known today as insulators or dielectrics. In contrast, wet string, or more importantly metals, do conduct electricity, but could not be used to generate electricity. These were his *non-electrics* or what are now called conductors. The fact that metals are excellent conductors of electricity has been profoundly important, not only in the history of the technology of metals, but also in metallurgical science, for example, the concept of free electrons leading to the development of the electron theory of metals. The little-known figure of Stephen Gray deserves a place in the pantheon of metallurgical science.

Static electricity generators did not disappear with the advent of galvanism and the invention of the battery; perhaps the best known is the Wimshurst machine (1883) in which electrostatic charges were multiplied by induction and accumulation. Of more interest in the present context is the development of a hydroelectric machine by William Armstrong (later Lord Armstrong, 1810–1900; FRS 1846) the famous Victorian metallurgical engineer and arms manufacturer. Armstrong had heard that an operator of a colliery steam engine had observed that escaping steam from a boiler could give rise to an electrical charge. He exploited this discovery by constructing an insulated iron boiler with 46 jets through which steam could escape and the individual charges were accumulated and succeeded in producing sparks up to 56 cm long.

So much for generating static electricity, how about storage? Perhaps the most ubiquitous apparatus, certainly the best known, was the Leyden jar, a type of condenser developed in about 1745 in the town of Leyden; its origin is usually attributed to Petrus van Musschenbroek (1692–1761; FRS 1734). Benjamin Franklin (1706–1790; FRS 1756) acquired a Leyden jar shortly after its invention and soon discovered that the dry atmosphere of Philadelphia, where he was living at that time, made it ideal for carrying out experiments in electrostatics. It was Franklin who conceived of positive and negative electricity and in 1752 deliberately flew a kite in a thunderstorm and channelled some electricity to earth down the cord thereby demonstrating the electrical nature of thunderstorms. This was a highly dangerous experiment. Franklin kept the Royal Society informed of his discoveries in a series of letters; in fact the Society played a pivotal role as an accumulator and distributor of information on matters electrical during much of the 18th century.

Following his experiments with the kite, Franklin invented the lightning conductor and when he visited England during the 1760s he acted as consultant on fitting conductors to a number of notable buildings and monuments, including St Paul's Cathedral. During this period an argument broke out between so-called experts on whether lightning conductors should have sharp or blunt ends (it was rather like the argument which led to war in Swift's *Lilliput* between those who cut their eggs at the round end and those who chose the sharp end). Franklin fell out of favour with George III because of his support for the American rebels' cause and as he supported sharp ended conductors, George III ordered that the conductors fitted on Kew Palace should have blunt ends! This dispute led in due to course to the resignation of the President of the Royal Society, Sir John Pringle (1707–1782; FRS 1745). A contemporary wit summed up the controversy with this epigram:

While you, great George, for safety hunt,
And sharp conductors change for blunt,
The nation's out of joint.
Franklin a wiser course pursues,
And all your thunder fearless views,
By keeping to the point.

Franklin seemed to have a profound effect on many people. Joseph Priestley (1733–1804; FRS 1766) was persuaded by him from about 1767 onwards to concentrate more on science (which hitherto had been something of a hobby), particularly on electricity and optics. Franklin also persuaded him to write his important book, *History and the Present State of Electricity*. Thus inspired, Priestley went on to deduce that electrostatic charge is concentrated on the outer surface of a charged body and he proposed an inverse square law for charges. Priestley exploited his new tool – the electrical discharge from a friction machine or Leyden jar – to reunite oxygen and hydrogen to form water, an experiment repeated much more quantitatively by Henry Cavendish (1731–1810; FRS 1760).

The Electrical Age

It is difficult to overestimate Benjamin Franklin's contribution to electrical science, not least because, by his example or his persuasion, others were stimulated to look more deeply into the phenomenon. By the mid-1780s Luigi Galvani, an anatomy professor at the University of Bologna, had been studying 'animal electricity' for fifteen years and it was during an experiment to build on Franklin's work on atmospheric electricity that he hung a dissected frog to an iron railing by means of a brass hook. When the wind blew and the frog's legs touched the railing it went into spasm. It was already known that electricity caused muscles to contract and Galvani thought that the source of the electricity was the frog's body itself. In fact of course what Galvani had accidentally set up was an electric cell with brass and iron as electrodes and the frog's body fluids as electrolyte.

Fortunately, Allessandro Volta a professor of natural philosophy at Pavia University, interpreted Galvani's result correctly and this led him to develop his 'voltaic pile', consisting of discs of copper or silver separated from discs of zinc by flannel soaked in brine – i.e. the first battery. In contrast to the electrostatic machines and the Leyden jar, a steady and constant supply of low voltage electricity was now available for the experimentalist.

On the 20th March 1800 Volta wrote a long letter in French to Sir Joseph Banks (1743–1820; FRS 1766, President 1778–1820), describing his discoveries and giving details of how to construct his 'pile'. In some ways this is one of the supremely important letters in science, it could be regarded as instigating the 'Age of Electricity' and was an auspicious launching of the new 'scientific' century. It turned out that the new source of electrical power was almost as important in the evolution of metallurgy as it was in the development of the science of electricity itself. Inspired by Volta's letter, Anthony Carlisle (1768–

1840), a surgeon, and a chemist, William Nicholson (1753–1815) constructed a voltaic cell and decomposed water into its elements. This created a demand for such cells and the most successful manufacturer of these was William Cruickshank (1745–1800) whose product consisted of copper and zinc electrodes soldered together in pairs and immersed in dilute sulphuric acid inside a wooden trough.

At the newly-formed Royal Institution Humphry Davy (later Sir Humphry, 1778–1829; FRS 1803) acquired a Cruickshank cell and could be said to have created the science of electrochemistry during the first decade of the 19th century. He made early experiments on galvanism in 1800, and in 1801 his first communication to the Royal Society was an *Account of some galvanic combinations.* In 1806 Davy became sufficiently free from his various duties at the Royal Institution, (which included work on leather and printing and the assembly of a mineral collection) to return to research of his own choice on galvanic experiments. He investigated the chemical action of the voltaic battery and in 1807 reported that he had decomposed 'the fixed alkalis' (potash and soda) to produce ' two new inflammable substitutes' (potassium and sodium).

In his MS notebook dated Oct. 19th 1807 Davy wrote:

> When Potash was introduced into a tube having a platina wire attached to it so (sketch shown....) and fused into the tube so as to be a conductor.i.e. so as to contain just water enough though solid – and inserted over mercury, when the Platina was made neg – No gas was formed and the mercury became oxydated – and a small quantity of the alkaligen was produced round the plat: wire as was evident from its giving inflammation by the action of water — When the mercury was made the neg. gas was developed in great quantities from the pos. wire, and none from the Neg mercury and this gas proved to be pure *oxygene* Cap[l] Exp[t] proving the decomposition of POTASH.

Davy also discovered calcium, barium, strontium and magnesium, and isolated boron, and introduced the idea that an electrical force binds the atoms together in a chemical compound. However, he was not successful in producing aluminium (see Chapter 6).

Batteries and their development

In addition to the important work of Davy in the isolation of the light elements, many other Fellows in England pursued research in the field of the development and applications of batteries.

Among these were John George Children (1777–1852; FRS 1807) and his father George (1742–1818). George retired from his business in Tonbridge, Kent and devoted his energies and much of his money in aiding John in the construction of new and large galvanic batteries, which consisted of a number of cells containing plates of copper and zinc. In 1808 John contributed a paper to the Royal Society on the best way of constructing a voltaic apparatus for chemical research. He established a good laboratory at Tonbridge in which Humphry Davy made many experiments. Davy, William Allen (1770–1843; FRS 1807) and William Hasledine Pepys (1775–1856; FRS 1808) came to Tonbridge to carry out an experiment in August 1808 aiming to melt platinum wire by a battery larger

Benjamin Franklin

Allessandro Volta

Joseph Banks

Humphry Davy

than any previously constructed. Davy was encouraged by this to build a still larger battery at the Royal Institution, but Children went on to make an even larger version in 1813. Early in July of that year a group of 38 distinguished scientists, including various Fellows of the Royal Society: Davy, William Hyde Wollaston (1766–1828; FRS 1793), Charles Hatchett (1765–1847; FRS 1797) William Allen, William Hasledine Pepys (who, later – 1823 – constructed a large battery), William Thomas Brande (1788–1866; FRS 1809) and Henry Warburton (1784 ?–1858; FRS 1809) witnessed the melting of a small sample of iridium. This was achieved by means of 'the greatest galvanic battery that has ever been constructed', consisting of 20 pairs of copper and zinc plates, each 6 feet (~ 1.8 m) long by 2 feet 8 inches (~0.8 m), suspended from the ceiling, and then lowered into a tank containing over 900 gallons (~ 4,100 litres) of dilute nitric and sulphuric acids. Subsequently the battery was modified to increase its power and in 1815 a sample of pure iridium, weighing ~ 7 g was 'fused into an imperfect globule' by a high current.

Much activity took place in improving the performance of batteries. In France Antoine-César Becquerel (1788–1878; FMRS 1837) made a battery that could supply current at a reasonably constant electromotive force. John Daniell (1790–1845; FRS 1814) a great friend of Michael Faraday (1791–1867; FRS 1824), working in London in 1836 invented the 'constant battery' (a copper–zinc cell) which established the possibility of maintaining powerful and uniform quantities of electricity for required periods. The battery consisted of a copper cylinder immersed in a saturated solution of copper sulphate; there was an inner porous cell within which was fixed a zinc rod immersed in dilute sulphuric acid. This invention was significant in various practical applications including electroplating, gilding, etc. William Robert Grove (later Sir William, 1811–1896; FRS 1840) in 1839, devised a primary cell, which came to be called after him; this consisted of a number of small cells contained in glass vessels, the electrolytes being separated by means of broken-off bowls of clay tobacco pipes. It consisted of a zinc pole, dipping into dilute sulphuric acid and a platinum pole, dipping into concentrated nitric acid. By assembling a number of such units he was able to provide a useful source of constant and continuous current. Joseph Wilson Swan (later Sir Joseph, 1828–1914; FRS 1894) invented the cellular lead plate for secondary batteries.

Alfred Smee (1818–1877; FRS 1841) also produced a battery (zinc and silver in sulphuric acid) which was used for trade purposes. During the 1840s and 1850s Robert Wilhelm Eberhard Bunsen (1811–1899; FMRS 1858) made various improvements in the galvanic battery. In 1841 he made a carbon–zinc battery which he used to produce a very bright electric arc light; carbon was used as a cheaper alternative to platinum or copper. Josiah Latimer Clark (1822–1898; FRS 1889) invented a battery (the Clark cell) reported to the Royal Society in 1874.

An important development occurred in 1842 when William Grove devised the first fuel cell in which oxygen and hydrogen were combined using spongy platinum as electrodes in sulphuric acid, (see Chapter 4).

John Children

Alfred Smee

John Daniell

Auguste de la Rive

Electrometallurgical processing

In the early 1830s Michael Faraday discovered what came to be known as Faraday's laws of electrolysis, which described quantitatively the relationship between the extent of chemical decomposition of a conducting material and the amount of electricity that passes through it. The unit now known as the Faraday is the amount of electricity necessary to liberate 1 gram-equivalent of any element from conducting solutions or conducting molten salts. With guidance from William Whewell (1794–1866; FRS 1820) with whom he corresponded, Faraday introduced the terminology which came to be used in electrochemistry: electrolyte, electrolysis, electrode, anode, cathode, ion, anion and cation. When news of Faraday's fundamental discoveries spread across Europe it stimulated a plethora of creative activity, which led to major practical consequences in the industrial development of techniques such as electroplating, electrogilding and electroforming.

Alfred Smee was active in electrometallurgy, including the art of electrotyping. In 1840 he published *Elements of Electro-Metallurgy*, discussing the laws relating to the reduction of metals in different states, and description of the processes for plating platinum and palladium so that reliefs and intaglios in gold could readily be obtained.

Henry Beaumont Leeson (1800–1872; FRS 1849) filed a patent in 1842 concerning the electrodeposition of a wide range of metals, and suggesting the procedure of agitation of either the component to be plated or of the electroplating solution, so as to achieve smooth electrodeposits at high current densities.

Among early industrial applications of electrodeposition was the work of Henry Bessemer (1813–1898; FRS 1879) in copper plating white-metal castings (see Chapter 15) using an acid solution of the metal; the castings were laid on the bottom of a shallow zinc tray, and a saturated solution of sulphate and of nitrate of copper were poured in, resulting in a bright copper layer over the whole surface of the casting. Auguste de la Rive (1801–1873; FMRS 1846) experimented on electroplating without much success as early as 1825; subsequently (1839–40) from further work he described a method of using a solution of gold chloride, but this did not give sufficiently good adhesion or the deep yellow colour required by gilders. In 1843 Charles William Siemens (referred to as William, later Sir William, 1823–1883; FRS 1862) came to England from Germany on behalf of his brother Werner (Ernst Werner von Siemens, 1816–1892) with an electroplating process invented by Werner; the process was sold to Elkington Brothers, Birmingham who were pioneering and influential in the electroplating industry from around 1840.

George Gore (1826– 1908; FRS 1865) was very interested in electrodeposition, and in 1851 moved to Birmingham, already the chief centre of electroplate manufacture. From 1854 to 1863 he published a series of papers on the electrodeposition of metals; his many discoveries were important in the early days of the subject. Golding Bird (1814–1854; FRS 1846) in 1837, described the electrodeposition of nickel, lead, iron, copper, tin, zinc, bismuth, antimony, silver, manganese and silicon in crystalline form by the action of long continued currents at low voltage through solutions of metal salts; for example,

Michael Faraday

Faraday's laboratory at the Royal Institution as in Faraday's time (from Brande's *Manual of Chemistry*, 1819).

he used solutions of nickel chloride or sulphate to obtain a layer of nickel on a platinum cathode. At the Royal Mint William Chandler Roberts-Austen (1843–1902; FRS 1875) in his work relating to die manufacture investigated the electrodeposition of iron. John Edward Stead (1851–1923; FRS 1903) investigated the crystallisation of electrolytic iron.

The availability of Daniell's 'constant battery' played an important role in the transition of electrodeposition from science to successful economic application. In this battery the conditions enabled copper to be deposited as a firm metallic layer. Daniell and others noted that if the layer of deposited copper was stripped off it showed in relief the marks of any scratches on the original electrode. By 1838 it had been realised that there was a possibility of making practical use of this effect and in this same year in St. Petersburgh Academician M.H. von Jacobi, a German who had emigrated to Russia in 1834, invented electroforming (which he termed 'Galvanoplasty'). The process consisted of using a conducting material to produce a mould of an object to be copied, and then making the mould a cathode in an electrolytic bath with a solution of a metal salt as electrolyte. The electrodeposited metal film which formed on the mould surface almost perfectly reproduced the fine detail of the original object, and separating it from the mould resulted in the production of a replica of the highest quality, known as an electrotype.

Electrotyping was a copying process *par excellence* and provides an interesting illustration of a technique stemming from fundamental research carried out by distinguished scientists including Fellows of the Royal Society, which became of widespread importance in society. It came to be extensively used for making duplicate plates for relief or letter press printing, since by carefully choosing the metal used for the duplicates they could be made much more durable than the original and hence more suitable for long press runs without damaging expensive type of valuable halftones or linocuts. In Britain, the March 1839 issue of William Sturgeon's *Annals of Electricity* reported the developments in St. Petersburgh and a few months later a book printer, C.J.Jordan, described his achievements in copying engraved plates, set type, coins and medals.(*Mechanics Magazine*). In 1840 Sturgeon's *Annals of Electricity* published a 62 page article by Thomas Spencer, a Liverpool picture – framer, entitled *Instruction for the multiplication of works of art in metal by voltaic electricity, with an introductory chapter on electrochemical decomposition by feeble currents.* The article contained an illustration of the apparatus used by Spencer reproduced by means of an electrotype plate, probably the first published example of such a figure. These developments were to lead to an enormous growth in the publication of illustrated books and magazines, starting with the *Illustrated London News* (1842) and *Scientific American* (1845). No other invention has led to a greater access by the public to reproductions of works of art and hence the spread of culture.

In the 1860s electrodeposition for the reproduction of statues and other works of art for museums and art galleries became a large business. Some of the reproductions were huge – Messrs Elkington made 22 statues around two metres in height weighing as much as three tonnes. Some of the supporting figures associated with the Albert Memorial, situated in Kensington, are copper electrotypes. In 1872 the South Kensington Museum

Josiah Clark

George Gore

James Joule

Alexander Fleck

offered a catalogue of about 500 electrotype reproductions for sale to the public. The Victoria and Albert Museum currently possesses electrotype silver and gilt copper reproductions of many of the world's most famous pieces of the goldsmith's art; of particular note is an electrotype of the exquisite bronze doors of the Baptistry in Florence by Gilberti.

Among other areas of electrometallurgy James Prescott Joule (1818–1889; FRS 1850) carried out investigations to produce a range of mercury amalgams by electrochemical procedures. Using pressure to remove excess mercury he obtained amalgams of definite compositions, such as Pt + 2Hg, Cu + Hg, 2Zn + Hg and 7Sn + Hg. An electrolytic amalgam method was used by Leslie Fleetwood Bates (1897–1978; FRS 1950) to prepare high-purity manganese.

Robert Bunsen in 1852 commenced using electrochemical techniques to isolate pure metals in quantities sufficient for measurements of properties to be made. Chromium was prepared from a solution of the chloride and magnesium from the fused chloride; commercial manufacture of magnesium was undertaken, and the metal came into use as an illuminating agent. In the mid 1850s he prepared sodium and aluminium from their molten chlorides. With the assistance of Augustus Matthiesen (1831–1870; FRS 1861) he isolated lithium, barium, copper and strontium from their fused chlorides. In collaboration with others some rare earth elements of the cerium group were prepared.

In the field of electrometallurgy for the extraction of metals the development of a large scale process for aluminium was an outstanding industrial advance (see Chapter 8).

Another example associated with electrometallurgical processing was the production of sodium. George Beilby (later Sir George, 1850–1924; FRS 1906) in 1900, working with Castner, invented a process of cyanide manufacture which involved the use of sodium. Beilby joined the board of the Castner–Kellner Alkali Company at Runcorn, manufacturers of pure caustic soda, which was the raw material for sodium production by electrolysis of the molten caustic. He erected plant at Newcastle upon Tyne and continued throughout his life association with the operation of the works at Runcorn, Newcastle and Glasgow. In 1916 Alexander Fleck (later Baron Fleck of Saltcoats, 1889–1968; FRS 1955), stemming from his contacts with Beilby, joined the Castner–Kellner Alkali Company which at that time produced sodium using an old plant at Wallsend. Fleck analysed and modified the production process, including experiments with larger cells and methods for automatic handling, which formed the basis for the design of later plant. In the early 1920s Fleck's responsibilities extended beyond Wallsend and he correlated work on the direct extraction of sodium from mercury cell amalgam.

Faraday and the evolution of electric light

Davy made two major contributions to the development of the electric light. Firstly, to an astonished audience at the Royal Institution in 1808, he demonstrated his latest invention – the carbon electric arc. Secondly, in 1813 he offered Michael Faraday a position as laboratory assistant at the Royal Institution.

Davy, in relation to the spark produced between the extremities of Volta's pile, commented that when pieces of well-calcined carbon were used instead of metals the spark was larger and clear white in colour. Curiously, although he realised the potential of the electric arc as a heat source he did not consider it a practical means of illumination. Nevertheless, at least in theory, there existed by 1810 the two essential ingredients for a lighting system; a power source, the voltaic cell, and a lamp, the carbon electric arc. There were though two drawbacks, the low capacity and high expense of battery-power and the fact that the arc lamp, although suitable for large space or outside use, was too brilliant, noisy and smelly for domestic use. It took another seventy years, and some major developments in electrical science, and materials technology, for these two problems to be solved.

The starting point for the development of an electrical dynamo (and motor) was the discovery in 1819 by Hans Christian Oersted (1777–1851; FMRS 1821) that there was a connection between magnetism and electricity. During the winter of 1819/20 Oersted presented a lecture to some of his postgraduate students at the University of Copenhagen during the course of which, as a demonstration, he brought a compass close to a wire carrying an electrical current; the compass moved and a circular magnetic field around the wire could be traced.

Oersted's discovery stimulated a burst of activity amongst his fellow scientists in the western world, notably amongst these was André-Marie Ampère (1775–1836; FMRS 1827) at the École Polytechnique in Paris. This 'Newton of Electricity' within two months of the announcement of Oersted's findings reasoned that if a magnetic field could be produced by a passage of electricity then maybe naturally occurring magnetic materials, lodestones, had within them rotating subatomic electrical currents which were the source of the magnetism.

Bearing in mind that the discovery by Joseph John Thomson (later Sir Joseph, 1856–1940; FRS 1884) of the electron lay three-quarters of a century into the future, Ampère's prescience was remarkable. In 1913, when Niels Henrik David Bohr (1885–1962; FMRS 1926) described his model for the atom it was thought that the orbital motion of the electrons around the nucleus provided the Ampèrian circuits. (It later turned out that there was a more intimate connection between magnetism and electricity; in 1921 Owen Willans Richardson (later Sir Owen, 1879–1959; FRS 1913) demonstrated that it was the spin of an electron around its own axis which governed magnetic properties. As the electron represents the most fundamental unit of electric charge, Richardson's work marks the ultimate bringing together of electricity and magnetism thus completing a journey started by Oersted a hundred years earlier.)

Returning to the 1820s, in 1821 Davy reported that he had repeated and developed Oersted's experiment. A further ten years had to pass though before Faraday's crowning achievement, the discovery of electromagnetic induction in 1831. This phenomenon was the basis of the electro-motor and of the dynamo, and within a year Hippolyte Pixii in Paris had developed a hand-cranked d.c. generator.

Hans Oersted

André Ampère

Charles Wheatstone

William Thomson
(Lord Kelvin)

Lighthouses

By the mid-19th century up to 800 ships a year were being wrecked around Britain's coast so one of the main incentives for an improved light source was for use in lighthouses. In 1849 Floris Nollet (1794–1853) a Belgian physicist had an ingenious, if tortuous, idea for a light. It was to decompose water by electrolysis and then burn the hydrogen and oxygen to heat up a piece of lime to incandescence (known as 'Drummond' or 'limelight'). For this purpose Nollet had constructed a huge magneto-electric generator and an associate of his, the British engineer, Frederick Hale Holmes, had the idea of using the output of the generator to power a carbon arc light suitable for use in a lighthouse. Following preliminary trials at Blackwell and South Foreland (which Faraday attended) the first permanent installation was made in 1862 in the famous Wyatt lighthouse at Dungeness.

Self excitation

Holmes's magneto-electric generator at Dungeness weighed 5 tonnes yet only produced about 1kw! Its days were numbered – the much more efficient self-excited generators were developed just four years later, in 1866, independently by the Siemens brothers (William and Werner), Samuel Alfred Varley and Charles Wheatstone (later Sir Charles, 1802–1875; FRS 1836). The idea here was that the current produced by the generator itself is used to energise its own field windings; Charles William Siemens and Wheatstone both presented papers to the Royal Society on the same day in February 1867 describing the process, and demonstrating working models.

In 1871, Zenobe Theophile Gramme (1816–1901), a Belgian and another early associate of Nollet, adopted the ring armature and thereby produced a very efficient dynamo. Siemens in Germany responded by producing generators with drum armatures. Between them, by 1880, Gramme and Siemens produced about a thousand dynamos. These found application mostly in arc lighting, first of all in Paris (Gare du Nord, 1875, Grand Magasins du Louvre 1877) and then in London (Gaiety Theatre, Embankment, Holborn Viaduct, Billingsgate 1878). The Severn Railway Bridge at Sharpness was completed with the aid of an arc light supplied by a Gramme dynamo which was subsequently used to illuminate a football match in the Forest of Dean.

As far as metallurgical applications of the dynamos are concerned, in the period 1877–82, an extensive development on the application of electricity to mining was carried out, following the purchase of a copper mine in the Caucasus by Werner and Carl Siemens. Krug von Nidda (1810–1885), head of the Prussian State Mines, was also interested in the electrolytic refining of copper, and he purchased dynamos for a copper refining plant at Oker.

Turning now to transformers, the American Joseph Henry (1797–1878) must be credited with the discovery of self-induction by d.c. currents (this is the phenomenon which causes a spark when circuits are broken and which is responsible for inducing a current in the secondary windings of a transformer). The first practical use of transformers was, however, not illustrated till fifty years later by Lucien Gaulard and John Dixon Gibbs of Paris at a demonstration in London in 1883. The US patent rights for the Gualard and Gibbs system were purchased by Westinghouse who were responsible for the widespread adoption of transformers and d.c. distribution in America.

Electrical materials

For the first seventy five years of the electrical age, i.e. till about 1875, the technology depended mainly on traditional materials, notably copper which was universally used as a conductor. It was soon realised that its conductivity depended critically upon its purity, and investigations of the properties of copper and also of other metallic and non-metallic materials became an important activity.

The introduction and development of telegraphy and telephony in the latter part of the 19th century involved major contributions from a number of Fellows of the Royal Society. The partnership of Charles Wheatstone and William Fothergill Cooke (later Sir William, 1806–1879) which began in 1837 led to important inventions on the road to commercial development. Wheatstone, in 1834, had been appointed as Professor of Experimental Physics at King's College, London. In the same year at the College he carried out an important investigation to determine the velocity of transmission of an electrical discharge through a copper wire. The apparatus involved a total length of wire of around half a mile (~ 0.8 km) with three gaps; a rotating mirror device was used to measure the time interval between the sparks that were produced acrosss the gaps.

Among those involved in the major activities of cable laying under water was one of the most distinguished scientists of the 19th century,William Thomson (later Lord Kelvin; 1824–1907; FRS 1851). When the Atlantic Telegraph Company was formed in 1856 he was appointed a director. He carried out theoretical work concerning the conduction of copper through long insulated wires. Also, finding that the electrical conductivity of copper was significantly reduced by impurities, he set up a system of testing conductivity in the factory. In his Bakerian lecture of 1856 he reported results from his studies of properties of metals.

Pioneering experiments on insulation materials for underwater cables was carried out by Wheatstone in 1844 in Swansea Bay while he was staying with his friend Lewis Weston Dillwyn (1778–1855; FRS 1804) a Member of Parliament. He tried underwater cables insulated with rope soaked in tar and also worsted and marine glue, but neither were a success and submarine telegraphy had to wait for the introduction of gutta percha, a substance of vegetable origin similar to rubber. While it was extensively used for telegraph cables it was found unsuitable for current-carrying electrical wires and with the discovery of 'vulcanised' rubber by Charles Goodyear (1800–1860) in 1839, this was the preferred material for insulation for a number of years. However, it was not the complete solution, the copper wire itself was usually tinned and a layer of pure rubber inserted between the conductor and the vulcanised rubber to prevent chemical reaction between the copper and sulphur in the vulcanised rubber. The whole was then often encased in an outer lead sheath.

Cromwell Varley

Alexander Muirhead

William Preece

Richard Glazebrook

The work of Cromwell Fleetwood Varley (1828–1883; FRS 1871) was associated particularly with the Atlantic cable. Following the failure of the cable laid in 1858 a committteee was set up comprising Varley, Robert Stephenson (1803–1859; FRS 1849) and Sir William Fairbairn (1789–1874; FRS 1850) to consider various aspects of another cable-laying project. Varley conceived and constructed an 'artificial line' as an analogue of the working conditions of the cable. Using electric resistances to represent the copper conductor, together with capacitors representing induction effects, he was able to predict the speed of signalling.

When the electric telegraph was introduced Werner Siemens recognised its possibilities and in 1847, he co-founded the firm of Siemens and Halske (Johann Georg Halske) in Berlin. His brother William was appointed the London agent. Werner was the discoverer of the technique of insulating telegraph wires with gutta-percha, and this led on to the invention of submarine cables. In 1858 with the growth of their business in this field the brothers set up a factory at Millbank in England, which in 1866 was moved to Charlton. The successful cable laying and repairing activities of the Siemens Brothers firm were world wide and included the completion of the laying of four Atlantic cables. Many of the machines and processes were greatly improved by William, while some of them owed their existence to him. The cable-ship designed by him for the laying of the direct Atlantic cable in 1874, was named the *Faraday*. In his pioneering work in numerous and important inventions e.g relating to the application of electrical energy, William worked in association with Werner. In another field, that of transport William was one of the first to apply electric power to railway locomotion.

Another Fellow of the Royal Society, Robert Stirling Newall (1812–1889; FRS 1875) played an important role in the field of submarine cables through his invention of a manufacturing method for steel wire, his studies of gutta percha and the production of submarine cables strengthened by wire.

Among leading figures in telegraphy, Josiah Clark became chief engineer to the Electric and International Telegraph Company; his work included standards of electrical quantities and resistance. Alexander Muirhead (1848–1920; FRS 1904) developed a major industrial activity and also made precision measurements of resistance and capacity. William Henry Preece (later Sir William, 1834–1913; FRS 1881) played an important role in the introduction of wireless telegraphy and the telephone in Great Britain and made a significant measurement of the electrical resistance of pure iron. In 1883 Richard Tetley Glazebrook (later Sir Richard, 1854–1935; FRS 1882) was secretary of a British Association committee on electrical standards, and carried out many accurate measurements. Working with a link with William Thomson, in the field of telegraphy, Henry Charles Fleeming Jenkin(1833–1885; FRS 1865) made experiments on the resistance and insulation of electric cables, involving measurements on gutta percha; he also participated in work on electrical standards.

James Swinburne (later Sir James, 1858–1958; FRS 1906) was a leading electrical engineer, who made important contributions in areas such as the development of the theory of dynamo design and of alternating current measuring instruments. He was also

James Swinburne

Rookes Crompton

First transformer constructed of silicon steel (1873), patented by Hadfield (Ref. Ch. 8.11).

particularly interested in the application of physics and chemistry to industrial problems and recognised the need to develop new materials. His interests ranged widely and his numerous patents (he was associated with around 130 in the period 1883–1922) included metallurgical activities such as the treatment of sulphide ores and ores of antimony and arsenic, extracting metals, electric cells or batteries, electrodeposition of zinc and electrolysis of fused salts, working and soldering aluminium. He was concerned with incandescent lamps, lamp filaments and arc lamps, and for a period worked in association with Joseph Wilson Swan (later Sir Joseph, 1828–1914; FRS 1894). He set up a laboratory in London in 1894 and did consulting work. Around 1899 he suggested that lamp filaments and artificial silk might be made from viscose. In his presidential address to the Institution of Electrical Engineers in 1907, surveying the field of heavy electrical engineering, he commented that there was a great need for improvement in the insulation of cables, since at that time this was limiting the pressures and hence the distance of transmission.

Around 1902 Swinburne became aware of the product of the reaction between phenol and formaldehyde, and he formed a syndicate to proceed to commercial development of the resin product. In 1907 when he came to patent his product he found that he had been anticipated by only one day by the Belgian chemist Leo Hendrik Baekeland (1863–1944) working in the USA, with his invention of Bakelite, the first fully synthetic plastic. However, Swinburne was able to patent a lacquer, which he manufactured and which was used for the protection of brass and similar polished surfaces.

As far as transformers were concerned, they suffered from large hysteresis and eddy current losses in their iron cores. These losses became larger with increasing age and this necessitated periodically withdrawing the transformers from service and re-annealing the cores to restore their original magnetic properties. The solution to this problem had to wait until early in the 20th century when silicon steels were developed by Robert Hadfield (later Sir Robert, 1858–1940; FRS 1909). The magnetic properties of the new alloy, known as 'Stalloy', actually improved with age. Alloys based on Hadfield's original discovery are still in use and efficiencies in excess of 99% are regularly achieved in modern transformers (this is just as well bearing in mind the amount of heat that would otherwise be generated in these cores).

Although these developments and improvements were important enough the development of a practicable electric lamp filament was the main barrier to the adoption of electric lighting for most of the 19th century. This development and the background to it will now be described.

The development of the incandescent lamp

The carbon arc lights which were invented right at the start of the 19th century, and which were subsequently refined and developed, continued to have severe disadvantages. In the first place they were very powerful and only a few could be run off each generator. Moreover they gave off noxious fumes and were altogether quite unsuitable for illuminating shops, offices or homes, so the market for the new lighting was very limited. What was desperately needed was to 'subdivide' the electric light, that is to say to make smaller units which were clean and safe to operate. The high-resistance filament incandescent lamp seemed to be the ideal solution. The fact that the passage of an electric current through a conductor generated heat had been known since the 18th century – it had been observed for example that lightning conductors glowed red upon being struck by lightning and the current conducted to earth. Pioneering workers in the 1840s, such as W. E. Staite (1809–1854) had realised that the conductors (ie the filaments) would have to be protected from oxidation if they were to be raised to incandescent temperatures and to survive for long enough to be useful. Unfortunately vacuum technology at the time did not permit the production of a good enough vacuum.

It was the development of the mercury vacuum pump by Hermann Johann Philipp Sprengel (1834–1906) in 1865 which made the incandescent lamp a possibility. It was still necessary to develop a suitable filament. As early as around 1840, William Grove, experimented by passing an electric current through a helix of platinum in a glass vessel, representing the first filament lamp, but the life of the filament was short. It was eventually confirmed that platinum, the favourite material of the pioneers, had its fusion temperature too close to its incandescent temperature for it to be truly useful. Attention was focused on the carbon filament produced by 'carbonising', i.e. partially combusting fine threads of organic material.

Among many workers the two leading contenders in the race to produce a practical incandescent lamp were Joseph Swan in England and Thomas Alva Edison (1847–1931) in America. It is interesting to compare their methods of working. Swan was very much the lonely scientist; he had patiently worked on the problem for more than twenty years, mostly with just one assistant. In complete contrast, Edison had not started to work on the problem until 1878 when he began to use the very considerable resources at his laboratory at Menlo Park (one of the world's first industrial research laboratories).

Edison adopted a blitzkrieg approach, he and his associates tried to make filaments from numerous refractory elements, including iridium, boron, chromium, molybdenum and osmium. During these experiments he made the important discovery that if the filaments were heated while the vacuum pumps were still running, trapped oxygen on their surfaces was driven off with an associated improvement in subsequent performance. During his attempts to produce a carbon filament he carbonised a bewildering variety of organic materials including turnips, pumpkins, squash, egg plant, apples, Southern moss, palmetto, monkey grass, Mexican hemp, jute, manila fibre, coconut palm, coconut hair, human hair, spider's web, fish line, cotton, tar, celluloid, wood shavings, cardboard, cork and even the skin off a rice pudding! In all more than 6,000 experimental filaments were prepared.

In the period 1879–80, both Swan and Edison succeeded in producing practical carbon-filament incandescent electric lamps. Swan utilised carbonised parchmentised thread, that is to say ordinary cotton thread pre-treated in sulphuric acid. Edison found success with carbonized bamboo slivers and the bamboo filament lamp remained the standard incandescent lamp in America from 1881 until 1894.

During the commercial production stage Swan found difficulties with his carbon filaments because his starting material, the cotton thread, did not itself have uniform properties nor uniform diameter. He overcame this problem by producing 'squirted' filaments, made by injecting through a fine capillary a solution of nitrocellulose into a coagulant such as alcohol thereby producing reconstituted cellulose thread. This thread was far more uniform than any other 'natural' fibre previously available (he was in fact producing the first man-made fibre). The squirted cellulose thread could subsequently be carbonized to give a uniform carbon filament of extremely high quality.

Swan installed his lighting system in his own home at Gateshead in November 1880. Lord Salisbury had Swan lamps installed and powered by a converted sawmill on his estate. In December 1880 Sir William Armstrong had Swan lamps installed at his home, Cragside, at Rothbury.

Both Swan and Edison became lamp manufacturers and electrical entrepreneurs and eventually the Edison and Swan United Electric Light Company was formed. In 1881 Swan engaged James Swinburne as a young engineer, to establish a lamp factory in Paris, and by the end of the year France was being supplied with French produced Swan lamps; later, still associated with Swan, he started a lamp factory in Boston, USA.

Swan had the distinction of providing incandescent lamps for the world's first central electrical power station which supplied electrical power for both public and private (domestic) use. The date was October 1881 and the place was Godalming, a small town in Surrey. Initially seven arc lamps were used for street lighting and about 40 Swan incandescent lamps for lighting shops and domestic interiors. The power source was a waterwheel already installed in a mill previously used for leather dressing, and the generator was a Siemens d.c. machine excited by a separate d.c. machine. The enterprise was not economic and was discontinued in May 1884; nevertheless it was an historic landmark.

The first central steam-powered generator was Edison's station at Holborn Viaduct which operated from mid-January 1882. This was a much larger affair capable of supplying a thousand incandescent lamps with a total electrical load of 80 kilowatts. This was a prelude to Edison's Pearl Street station which opened in New York in September 1882. Many power stations and distribution centres soon followed but it is highly appropriate to refer here to the enterprise of one other entrepreneur and pioneer of d.c. supply (Edison favoured d.c.), the brilliant Sebastian (Pietro Innocenzo Adhemar) Ziani de Ferranti (1864–1930; FRS 1927), who became known as 'The English Edison'.

To set the scene for Ferranti's triumphs it is helpful to start with a performance of Gilbert and Sullivan's opera 'Patience' at the newly-opened Savoy Theatre on 5th October 1881. To Edison's chagrin, Swan had won a contract to illuminate the Savoy with his incandescent lamps, and for this had received much favourable publicity (though Edison's industrial 'spy', his UK manager Edward Johnson, had discovered that from a position behind the wings the electric lamps were augmented by gas lights!). In writing the libretto for 'Patience' Gilbert wished to pull fun at the then current Aesthete movement and its leader Oscar Wilde, who was caricaturised (in part) as Bunthorne, 'a fleshy poet'. A favourite haunt of the aesthetes was the Grosvenor Art Gallery in Bond Street owned by the aristocrat, Sir Coutts Lindsay. In the second act of the opera, Bunthorne describes another aesthete as 'a greenery-yallery, Grosvenor Gallery, Foot-in-the-grave young man'.

Joseph Swan

Sebastian Ferranti

Irving Langmuir

Walther Nernst

Perhaps assisted by this notoriety, Sir Coutts's business expanded so he decided to install a portable electrical generator at his gallery to provide improved lighting for his pictures. The new form of illumination impressed his neighbours who asked if they too might be supplied with electricity from his generators. This unexpected sideline expanded until Sir Coutts needed to take on extra staff, one of whom was a supplier of equipment, Sebastian Ferranti.

Ferranti came from a distinguished Venetian family; indeed an ancestor of his, Sebastino Ziani, was Doge in 1173 followed by his son, Pietro, in 1205; de Ferranti was added to the name in the 18th century. Ferranti's father came to England in 1859 and established a photographic studio in Liverpool. Having attended University College London, for a few months, Ferranti was employed in the experimental department of the Siemens works at Charlton. During this period Ferranti developed a new type of generator and an improved arc lamp. When it was found that William Thomson had invented a similar generator, Ferranti, still only 18 years old, formed a company with him (together with a solicitor called Francis Ince, his future father-in-law) to manufacture and market what had become known as the Ferranti–Thomson dynamo. This machine had a novel zig-zag armature and it yielded five times as much output as any other machine of equal size (output was measured in terms of the number of incandescent lamps that could be powered by the dynamo). A feature of Ferranti's research and development was his innovative use of materials for electrical conductors and insulators – a particularly noteworthy development of his was the use of oil-impregnated paper as insulator for copper conductors.

Under Ferranti's inspired technological management, Sir Coutt's Grosvenor Gallery station continued to grow until it supplied power to cover 300 customers over an area extending from Regent's Park to the river Thames, and from Knightsbridge to the Law Courts. Ferranti then had an idea which was simply breathtaking in its brilliance and audacity. This was to transfer the Bond Street activities to a huge purpose-built station eight miles away at Deptford. The planned capacity was 90 MW and was capable of supplying two million lamps, more than the projected demand for the whole of London. This is how it came about that starting with a desire to illuminate a few paintings a plan was conceived to light up all London with power from the largest central electrical station the world had ever known.

Ferranti incorporated into his Deptford station many developments which were years ahead of his time. Apart from using reciprocating engines instead of using turbines (see below) Deptford incorporated all the features of a station designed at least 50 years later. To minimise power losses he transmitted power at the then unprecedented pressure of 10,000 volts d.c. and he designed a main cable for this consisting of two concentric copper tubes of equal cross sectional area with the inner one carrying the current and the outer one earthed. The two tubes were insulated from each other by paper impregnated with ozokerite wax. This cable was satisfactory and remained in service till 1933.

A great disappointment to Ferranti and his company was the decision by the Board of Trade to restrict their area of operation to a much smaller level than they had been led to expect. In these circumstances Ferranti's grand design and large output could no longer be sustained; in August 1891 he resigned from the enterprise. The company subsequently ran into severe technological problems, the most severe being the corrosion and failure of many boiler tubes.

One of the principal differences between Ferranti's Deptford enterprise and a modern power station is the former's use of a reciprocating engine rather than a steam turbine. In the field of steam engine technology Britain had reigned supreme throughout the 100 years preceding the 1880s and for much of this time the Willans reciprocating steam engine (Peter Willans, 1851–1892) more that held its own against foreign competition. It is not then surprising that Ferranti chose Willans engines for Deptford, even though Charles Algernon Parsons (later Sir Charles, 1854–1931; FRS 1898) had patented his invention, the steam turbine, as early as 1884.

Charles Parsons and the steam turbine

The idea of allowing steam to impinge directly on to a turbine consisting of metal blades around the periphery of a wheel to produce a rotary movement had been in existence for a long time but it was Charles Parsons who brought the concept to fruition.

Parsons had an unusual background for an engineer; he was the younger son of William Parsons the third Earl of Rosse (1800–1867; FRS 1831), who had been a President of the Royal Society. As a young man of 19 Parsons had entered Cambridge in 1873 and, as there was no engineering faculty at that time, attended lectures on mechanisms and applied mechanics. In 1877 he became an apprentice at Sir William Armstrong and Co, and then became a junior partner in the Gateshead company, Clarke Chapman. In 1884 he took out his first turbine patent, based on the principle of developing power from the velocity of steam rather than from its static pressure. In his design steam was fed at high pressure into a series of turbines, each consisting of blades mounted on a rotating shaft.

In the 1884 design the turbine blades were of cast brass with slots cut to produce an integral component, but there were problems of porosity and failures often occurred. The integral design was subsequently abandoned and in 1891 a method was used of stamping strips from alpha brass to fit into grooves in the wheel; in 1896 strip blading rolled to an aerodynamic profile was introduced. However, as inlet temperature was increased above about 240°C, embrittlement of the brass occurred. With progress in turbine design all-steel blading was adopted in around 1918, and in 1925 a ductile stainless iron became standard to provide corrosion resistance. Increase in blade velocities in the exhaust stages led to water droplets in the wet steam causing erosion. Parsons designed an experimental rig to study this problem and concluded that hardness was the chief requirement to resist erosion; in 1926 he adopted a technique of brazing a hard shield on the appropriate section of the blades.

The first rotors consisted of a high carbon steel shaft ground to a good finish, and then, as forging procedures improved, larger size single-piece components became available and were used for the high pressure rotors. Turbine cases too evolved, from brass to cast

Parsons first turbo-dynamo (1884).
Output 7 1/2 kW, 18,000 rev/min.

Charles Parsons

The first turbine driven vessel *Turbinia*, designed by Parsons, attained a speed of 34 1/2 knots in 1897. Later in the same year Parsons demonstrated the vessel's performance at the Naval Review, which was held to celebrate Queen Victoria's Diamond Jubilee.

iron to cast steel. In 1889 Parsons founded C. A. Parsons and Co. and in 1890 the Forth Banks Power station at Newcastle became the first generating station in the world to use turbo-generator equipment. By 1912 over a third of the generators in operation and almost all the new ones being installed in public supply systems were turbines.

The employment of turbines, together with related improvements in boilerhouse plant and condensing equipment, greatly reduced the amount of coal needed for each kWh generated. Parson's invention has remained the basis of turbo-generator design. Later Parsons entered the field of marine propulsion and this led to the progressive adoption of turbines throughout the world for naval, and eventually also for merchant shipping.

In the closing decades of the century, while Ferranti and Parsons were laying down the foundations for the modern power station, important developments were also taking place in the design of the incandescent lamps. The main incentive here was the inadequacies of the carbon filament. These developments are now described.

Alternatives to the carbon-filament lamp

In the early 1880s, electric lighting using carbon filament lamps was expensive and its limited success was largely due to the inadequacies of gas lighting. At that time gas lamps were little more than holes in metal pipes where the incomplete combustion of the coal gas caused unburnt carbon particles to be raised to incandescent temperatures in the flame. (It is interesting that at that time both gas and electric lighting relied on the emission from incandescent carbon.)

It was the development of the incandescent gas mantle which re-established the gas industry as a powerful contender for the illumination market. The underlying physical phenomenon for the mantle was the fact that certain metal oxides selectively emit radiation in the visible range when heated, a principle which had been exploited since the mid nineteenth century. Here intense illumination required for some theatrical performances had been achieved by heating blocks of calcium oxide (lime) to a high temperature by means of a gas flame; it was known as 'limelight'. The developer of the most successful gas mantle, Carl Auer von Welsbach (1858–1929) found that the best mixture was a mixture of 99% thorium oxide and 1% cerium oxide applied to a silk or cotton fabric (which burned off on first use leaving a network of the two oxides).

Incandescent oxides were also exploited in the innovative 'Nernst' electric lights. Hermann Walther Nernst (1864–1941; FMRS 1932) discovered that some oxides which were insulators at low temperatures became quite good conductors when heated. His lamp, which became commercially available from 1900, consisted of rods of oxide which could be heated initially by a platinum resistance coil and thereafter raised to incandescent temperatures by the passage of an electrical current through the oxide itself. The lamp was 50% more efficient than carbon filament bulb and had the advantage that the incandescent oxide rod did not have to be protected by a vacuum. It is surprising that it did not find wider commercial success.

Apart from being expensive the carbon filament lamp had little development potential because its filament cannot be used much above 1600°C – it rapidly evaporates at higher temperatures. The obvious response was to develop metal filaments and the prime candidates were (approximate melting temperatures): osmium (3,000°C), tantalum (2,995°C) and, most promising of all, tungsten (3,410°C). The versatile von Welsbach developed an osmium filament lamp which was available for sale by 1899 and Siemens and Halske developed a tantalum filament. Thc ultimatc goal was a tungsten filament and this was achieved by William David Coolidge (1873–1975) in 1909 in the USA who succeeded in drawing tungsten into a wire by first producing tungsten powder and compressing, heating and hammering it.

Coolidge worked at the newly-formed General Electric Research Laboratory and later his colleague, Irving Langmuir (1881–1957; FMRS 1935), demonstrated that a longer and more efficient life for the tungsten filament could be achieved if it was wound into a spiral. Later further improvements were obtained by winding the coiled wire itself into a larger coil (the coiled-coil filament), and this arrangement is still employed in the modern electric light bulb. Because of the very high temperatures achieved in tungsten filaments grain growth occurred until single grains traversed the filament, and sliding along these boundaries restricted the filament life. An ingenious solution to this problem was the incorporation of volatile elements such as potassium, which formed vapour bubbles which pinned the boundaries and restricted grain growth. Langmuir's work included investigations of surface phenomena concerning the reaction of gases with heated tungsten filaments (see Chapter 11).

In England, research at the laboratories of the General Electric Company Ltd, Wembley made important contributions in fields such as the manufacture of lamps and valves. John Walter Ryde (1898–1961; FRS 1948), with his team, included in their work the development of the high pressure mercury vapour lamp in the early 1930s. Also in the 1930s John Turton Randall (later Sir John, 1905–1984; FRS 1946) developed luminescent powders for use in discharge tubes.

Another leader in the science of thermionic emission phenomena was Owen Willans Richardson. He introduced the concept of an evaporation potential barrier for electrons and received the Nobel Prize for Physics in 1928. The wide ranging theoretical work at Cambridge by Ralph Howard Fowler (later Sir Ralph, 1889–1944; FRS 1925) included thermionic and photo-electric emission of electrons from metals.

Rookes Evelyn Bell Crompton (1845–1940; FRS 1933), an engineer, whose career extended well into the 20th century was involved in the manufacture of electric lighting plant. A new pipe-casting plant at the Stanton Ironworks, owned by his brothers was the first industrial building to be so lit. Beginning in about 1885 James Swinburne worked for a period with Crompton in his dynamo works.

Returning to the closing years of the 19th century, it is noteworthy that most of the above developments of improved gas mantles and the use of metallic filaments in incandescent lamps did not take place in Britain, but abroad, mainly in Germany and the USA. What had happened to Britain's lead in electrical technology in particular and in engineering in general, and what part did the Royal Society take in arresting this decline?

John Ryde

John Randall

Owen Richardson

Ralph Fowler

Britain's international position

Towards the end of the 19th century Britain should have been in a uniquely strong position to take a world lead in the emerging field of electrical technology. Pioneering work by British physicists, from Davy and Faraday to Maxwell had laid down the foundation stones for the subject. Moreover, the inventive genius of Swan complemented the engineering vision of Ferranti and Parsons. To add to all this there was the wealth of the Empire, enormous coal reserves and a labour force skilled in technology as a consequence of Britain being the birthplace of the Industrial Revolution. In addition, much of the population was already concentrated into cities so it was ideally distributed to take advantage of the supply of electricity from central stations. As the new century approached there seemed no reason to suppose that Britain would not occupy as dominant a position in the era of electricity as it had previously enjoyed in the age of steam.

However, it did not happen. Scientific and engineering genius is no match for the ignorance and incompetence of the politicians responsible for devising the legal framework within which the industry must operate. Neither could earlier discoveries of monumental importance to science guard against lack of vision by some industrialists who fail to support research and development activities which are essential if scientific discoveries can be converted into profitable technology. One manifestation of this lack of vision was Swan's inability, at the start of the 1880s, to persuade British industry to support financially his plans to develop a metallic filament for his lamps. In fact British industry did not establish an industrial-based research laboratory in the field of lighting until 1919; this was the General Electric Company Laboratory initially at Hammersmith and later at Wembley.

Thus for the forty years that straddled the turn of the century, virtually no British-sponsored research into lighting was carried out and inevitably we were excluded from the the highly lucrative business of manufacturing and selling metal-filament electric lamps. Equally inevitably, three of the four major electrical manufacturing companies which became established in this country, Siemens, BTH and British Westinghouse, were subsidiaries of German or USA corporations. The fourth one, GEC, had been created by a German immigrant, Hugo Hirst (later Baron Hirst, 1863–1943) and owed its prosperity to an agreement with a German company, Auergesellshaft, to manufacture for the British market Welsbach's 'Osram' metal filament lamp.

To add insult to injury, we took no part in the development of polyphase a.c., for example, and we relied on American and German technology for our electrical transport system – the London Underground was almost entirely financed with American money and used American equipment. General Electric (US) was our major supplier of electric tram motors and provided virtually all the generators for the traction power houses.

Electricity supply was in an even worse state than the manufacturing industry. The power stations were far too numerous and inefficient. At the outbreak of the 1914–18 war there were about 500 stations in the UK but their total output was 25% lower than that supplied by a single company to the city of Chicago and, perhaps not surprisingly, the average cost per unit was twice that paid by the Americans. More importantly, the

UK's electricity industry was only half the size of Germany's and this placed this country in a grave disadvantage throughout the hostilities. The effect of shortages of electricity on the manufacture of munitions became critical by 1916 and from that time to the end of the war Lloyd George himself (1863–1945) became involved in increasing electric output by 50% during the following two years.

Strenuous efforts were made to improve the international standing of the UK in the world of electricity in the immediate post-war years, but these met with little success. In 1923 the total output of the UK, which supplied 45 million people was less than that of Metropolitan New York, which supplied 7 million. The annual consumptions of electricity in kwh/person was: Switzerland 700, Canada 612, Norway 493, USA 472, Sweden 364, France 147 and UK 139 (though it has to be said that nearly all those countries above Britain had the benefit of extensive hydro-electric stations).

Throughout this period there is little evidence of the Royal Society taking a creative interest in the British industrial scene; certainly it did not alert the government of the disastrous consequences of lack of investment in research and development in the electrical field. Where Fellows of the Society were invited to assess the industrial potential of the new technology and to guide the politicians in formulating legislation and controls, their advice was often less than helpful, as will now be explained.

The Royal Society and electrical legislation

Reference has already been made to the unfortunate example in the 18th century of the dispute involving George III and Benjamin Franklin concerning whether lightning conductors should have blunt or sharp ends, which eventually led to the resignation of its President of the Royal Society, Sir John Pringle.

Moving forward a century to the final quarter of the 19th century, the biggest disservice to the budding electrical industry by the scientific establishment was their enthusiasm for d.c. power, the adoption of which, because it could not readily be raised to high voltages in order to minimise distribution losses, severely restricted the area an individual generating station could supply (the insistence on at least partial d.c. supply severely crippled Ferranti's innovative plans at Deptford, for example). Supporters of d.c. included Sir William Thomson, Rookes Crompton, John Hopkinson (1849–1898; FRS 1878) and Alexander Blackie William Kennedy (later Sir Alexander, 1847–1928; FRS 1887)

Examples of leading scientists and engineers making unhelpful statements include Alexander Kennedy's prediction: 'In 15 years from now we shall not be having a three phase plant we shall be having a single phase' and William Henry Preece (electrical engineer to the Post Office) after predicting that currents in electrical supply cables would not increase so that they should not interfere with telephone lines, went on to express the view that in any case the telephone would not be widely adopted in Britain because, unlike America, 'we have a superabundance of messenger boys'. The leading scientists and engineers also tended to miss the tremendous potential of incandescent electric lighting.

A select committee was set up in 1879 'to consider whether it is desirable to authorise Municipal Corporations or other local authorities to adopt any scheme for lighting by electricity; and to consider how far and under what conditions, if at all, gas or other Public Companies should be authorised to supply light by electricity'. The chairman of the Committee was Lyon Playfair (later Lord Playfair, 1818–1898: FRS 1848). He received evidence from numerous scientists and engineers, including William Siemens, John Hopkinson (1849–1898; FRS 1878), and Sir William Thomson.

When Playfair started his deliberations, in 1879, the incandescent lamp was still in the development stage; nevertheless there had been sufficient claims for its potential, particularly from Thomas Edison, for the Select Committee to have taken it seriously. In fact Edison was dismissed as an American upstart and, astonishingly, Swan was not even invited to give evidence to the Committee. Thus the Committee assessed the immediate need for legislation on the basis of using electricity for no other purpose than to supply carbon arc lamps.

A clause in the Playfair report may be the origin of many of the subsequent difficulties the industry was to face:

> …a monopoly given to a private company should be restricted to the short period required to remunerate them for the undertaking, with a revisionary right in the municipal authority to purchase the plant and machinery on easy terms.

This recommendation was enshrined in the 1882 Electricity Lighting Act which gave public authorities the right to buy out at knock-down (i.e. scrap) prices private electricity generating companies after only a 21-year period of operation.

The severity of this limiting legislation took a little time to sink in. Immediately after the Act was passed, 69 provisional orders were granted during the remainder of 1882. However the number of orders reduced to four during 1883 and was zero in 1884. It was later revealed that none of these 73 provisional Acts actually led to the construction of an electrical station. In 1888 the period before a compulsory purchase order could be activated was increased from twenty one to forty two years but the Amending Act also introduced a new and damaging principle – the agreement of the Local Authority was to become necessary before a provisional order could be granted. The electricity supply industry was almost stifled at birth by inappropriate and unhelpful legislation, and for this existing Fellows of the Royal Society (or those who eventually were elected to the Society) must take their share of the blame.

Subsequent governments were either Conservative or Liberal. The latter continued to support municipal socialism as a matter of principle whereas the Conservatives were content to go along with it for they saw it (quite correctly) as a bulwark against fullscale nationalisation. Thus the doleful influence of the the 1882 Act continued well into the 20th century. Its principal effect was to encourage the formation of a multiplicity of small and uneconomic stations. Matters were made even worse when in an influential report a senior Board of Trade official, Major Marindin, (Sir Francis Arthur Marindin,

1838–1900) recommended that individual customers should be given a choice of a.c. or d.c. supply. In effect this advocated a duplication of supply in each area. In another part of his report Marindin severely restricted the area the Deptford station could supply in central London and this was one of the final nails in the coffin of Ferranti's splendid concept of a central electrical station for London.

Another intangible extrinsic factor was the British attitude to science and technology. British scientists tended to remain amateur and individualistic. In the 19th century when Oxford and Cambridge were indifferent to experimental science France established the École Polytechnique and Germany its Technology Colleges. The education system in Britain was geared to producing officers and administrators for the Empire. When Wilfred Moseley of South African Diamonds advocated a vocational slant to British education, Matthew Arnold the then Chief Inspector of British schools, was horrified and called Moseley's intervention. '*A miscalculation of what culture really is, and calculated to produce miners and engineers not citizens of light*'. It is true that in 1889 Parliament empowered local authorities to levy a two pence rate to establish technical colleges, but as this was to be voluntary, it had little effect.

It was little wonder that the emerging electrical industry, which should have been at the vanguard of technological change, was desperately short of trained staff and creative ideas. As a general rule engineers were excluded from the boardrooms of British companies. In 1890 the journal *Electrical Review* surveyed company directors in British electrical companies and found that only 20% of them were engineers (of the remainder 45% were 'gentlemen' (!), 18% were titled aristocrats, 9% were lawyers and 8% military men).

The true pioneers of the industry, Swan, Crompton, Ferranti and Parsons (all of whom became FRSs) did not shine in the world of big business. Swan failed in his attempts to encourage work on metal filaments, Crompton was gradually squeezed out of having a controlling interest in his companies, Ferranti preferred to pursue perfection than gain commercial viability and Parsons concentrated his meagre financial resources on the development of turbines for marine applications. The nearest Britain came to a successful engineer/entrepreneur was Charles Hesterman Merz (1874–1940), the chief architect of the hugely successful Newcastle Electric Supply Company.

There were of course other factors which held back Britain's electrical industry for so many years, the most important of which were a relatively efficient steam and gas industry. As far as steam is concerned, its success was two-edged; leaders emerged from prime movers for electrical generation, from Willans to Parsons, but Britain's efficient steam engines hindered the development of electrical traction and electrical drive for machinery.

In the case of the gas industry, Britain was the world's pioneer in gas technology and by the time electric lighting began to compete the industry was very efficient. For the whole of the 19th century in the UK electric lighting was much more expensive than gas lighting. The early ventures with arc lighting were short-lived because they were up to seven times more expensive than gas lighting, and carbon filament incandescent electric lighting was always at least twice as expensive as gas. It is true that electric lighting was cleaner and more intense, but even these advantages were greatly diminished with the development of the gas mantle. (In contrast, the American gas industry was rather inefficient, which is one of the reasons Edison could operate successfully his power station in Pearl Street, New York while his Holborn Viaduct station – the first central steam-driven station in the world – could not show a profit and was closed down after a couple of years' operation.)

It was not until the 1910s with the introduction of the metal filament electric lamps from abroad, that electric lighting could compete with gas on a cost basis (metal filament lamps for a given light output consumed about a third as much energy as carbon filament lamps). For street lighting, gas remained competitive until the 1950s – there are still examples of it to be found today, including near the Royal Society building in London.

In December 1923 Britain elected a minority Labour government which was very much aware of the poor state of the electrical industry, hindered as it was by an absence of national standards – at that time there were no less than seventeen different electrical frequencies in use. Prominent members of the Party, including Herbert Stanley Morrison (Baron Morrison of Lambeth, 1888–1965) saw wholesale nationalisation as the solution to the industry's ills, but such a course was not possible without Liberal support. The government had to satisfy itself by making a grant to the industry of ten million pounds to be used in an attempt to standardise the frequency. Standardisation would though have resulted in huge quantities of electrical equipment becoming redundant, and this would have been very expensive. One of the last acts of this short-lived government was to commission Merz and William McLellan to assess the true cost of standardisation.

Before Merz and McLellan could produce their report there was a general election and the Conservatives regained control with Stanley Baldwin (Earl Baldwin, 1867–1947; FRS 1927) returning to his former position as Prime Minister. For the only time in the history of electrical power this country had a Chief Minister who was well-acquainted with the problems of the industry. Baldwin had trained as a metallurgist and was a member of a family with widespread metallurgical connections but he also had a good grounding and appreciation of general technology.

Recognising the pressing need for radical reform of electrical generation and distribution Baldwin wanted to appoint William Douglas Weir (first Viscount Weir, 1877–1955), a distinguished industrialist and sometime Secretary of State for Air, as a Minister in his government. Weir turned down Baldwin's invitation but he did agree to chair a committee investigating how electricity was organised in Britain. The Committee reported in May 1925 and recommended the establishment of an independent body, to be known as the Central Electricity Board charged with a duty of establishing a 'gridiron' system of high-voltage transmission lines – this soon became known as the 'National Grid'. This was completed by 1933, and even before this, the influence of the grid was being felt – in the period 1929 to 1933, world output of electricity increased by 5% whereas in Britain it increased by 30%.

While the grid itself was a success, little progress was made in reducing the number of supply authorities. Even at the conclusion of the Second World War there were still 560 individual supply undertakings operating with 300 stations, half of which had outputs of less than 10 MW. Evidently all attempts to bring about an improvement in the structure of the industry by a process of evolution had failed and all that was left was full-scale nationalisation. In 1947 the Electricity Act was passed and the generation, transmission and distribution of electricity for public supply was transferred to public ownership.

An important subsequent landmark was the publication of the Herbert Report (Edwin Savory Herbert (Baron Tangley, 1899–1973) in 1956. This prepared the way for the formation in 1957 of the Central Electricity Generating Board (CEGB) which was to become the largest fully-integrated Electrical Utility in the world. The CEGB placed a great emphasis on science as evidenced by its investment in research and the origin of this policy can be traced to two sentences in the Herbert Report:

> Science is not now a thing apart but is intimately associated with all activities and aspects of industry. The electricity supply industry, by its very nature, should be in the forefront of the major industries organising and applying research and development on an extensive scale.

The CEGB certainly responded to Herbert's homily with enthusiasm, it built up a research organisation employing a thousand professionals (BScs and PhDs) located at three large central laboratories at Marchwood, Leatherhead and Berkeley. It might have been reasonable to ask what exactly these researchers could do to improve the technology, bearing in mind that conventional, coal-burning stations, which produced more than 80% of the electricity, had barely changed in basic design since the end of the last century.

In fact what kept the research officers busy was the consequences of a decision made by government before the creation of the CEGB, and that was to create in Britain the world's largest programme of construction of civil nuclear power stations. More or less by government decree, the CEGB was eventually required to construct sixteen Magnox reactors (see Chapter 18) at eight sites (with a further two in Scotland) to be followed by a programme of Advanced Gas-cooled Reactors and eventually a single Pressurised Water Reactor. At one time in the 70s, when all the Magnox stations had been commissioned, the UK was generating more nuclear electricity than the combined total of the rest of the non-communist world.

To launch this programme the distinguished engineer, Christopher Hinton (later Lord Hinton, 1901–1983; FRS 1954) who had been responsible for constructing the Windscale Piles and the Calder Hall Magnox reactors, was appointed the first chairman of the CEGB. Subsequently, he invited Leonard Rotherham (1913–; FRS 1963) to join his board as the member responsible for research and development; Rotherham had previously worked with Hinton on the Windscale and Calder Hall projects, and was an expert on the metallurgy of the Magnox fuel elements. During the manufacture of nuclear weapons and the development of civil reactors unique demands were made on materials and many Fellows of the Royal Society were involved in studying these problems both in the Atomic Energy Authority and the CEGB. These activities and achievements are described in Chapter 18 together with a brief review of the evolution of atomic science.

CHAPTER 9

Engineers and Their Metallic Materials

Introduction

The mechanical behaviour of metallic materials is of critical relevance to engineering practice in terms of aspects such as the shaping and fabrication, and also selection for service; these applied aspects are sometimes referred to as engineering metallurgy. Experience of the strength of metals dates back to the prehistoric use of metals, and their hardening by cold work and annealing to produce softening were used in ancient times. The field of strength of materials and mechanical testing was active in the earlier period of the Royal Society (see Chapter 2) and was of continued and increasing importance in the period of extraordinary achievements in large scale engineering construction in the 19th century, such as roads, railways, bridges, canals, tunnels and ships. Late in the 18th century cast iron had become an important material for large structures. It continued in widespread use in the first half of the 19th century, where its good compressive strength but low tensile strength was suitable for compression members such as columns. It then faded out as wrought iron and later, steel, were used with higher tensile strength, suitable for beams and members in tension. Around the mid-century widespread applications for wrought iron included rails, beams, boilers and ships; large components could be made from small beams welded together under a forge.

In their design and construction work, some of the eminent engineers of the 19th and 20th centuries, Fellows of the Royal Society, became closely involved in metallurgical matters concerning the metals and alloys on which their success depended. This chapter considers some aspects of their work in civil and mechanical engineering (mainly using ferrous materials), notably in the construction of railways and bridges, ship construction and armour; developments in mechanical testing are also briefly considered.

Metallurgy and 19th century engineering construction

The engineering work of Thomas Telford (1757–1834; FRS 1827), which began in the 18th century continued actively into the 19th century. In 1814 he was consulted on a scheme to bridge the River Mersey at Runcorn, and as a basis for his design of a large suspension bridge for this task he collaborated in a programme to determine the tensile strength of wrought iron. Peter Barlow (1776–1862; FRS 1823) and Bryan Donkin (1768–1855; FRS 1838) assisted in this programme of mechanical testing, using a special Bramah hydraulic press (Joseph Bramah, 1748–1814) at the Patent Chain Cable Works of William Brunton (1777–1851). Although this bridge project was abandoned Telford's subsequent achievements included the massive enterprise of the design and construction of the suspension bridge across the Menai Straits, separating the mainland of North Wales and the Isle of Anglesey; this bridge, completed in 1826, was part of the road from Shrewsbury to Holyhead to link with ships to Dublin. The suspension chains were to consist of iron links, each about 8 ft (~2.4m) long, wrought at Upton forge, near the Shrewsbury Canal. The links were then taken by canal to Shrewsbury for tensile testing on a machine devised by Telford. From the tests he estimated that each link should be capable of withstanding a load of ~88 tons (~880 kN) before fracture; however, he specified that tests should be limited to 35 tons (~350 kN) to avoid the risk of straining the links, having calculated the maximum loading in service to be only half of that value. For protection from corrosion by the salt environment, the links were heated in a stove and quenched in linseed oil, prior to painting.

The civil engineering activities of William Tierney Clark (1783–1852; FRS 1837) involved important bridge construction, including the use of cast iron, a material of which he had gained experience during the early part of his career when he worked at the Coalbrookdale Ironworks in Shropshire.

In the theory of strength of materials Thomas Young (1773–1829; FRS 1794) extended the work of Leonhard Euler (1707–1783; FRS 1746), and introduced into beam theory the term 'neutral axis'; his work included the loading of columns. Young also made an approximate determination of the crushing strength of cast iron in the form of a small prism (~ 3 × 3 × 6 mm) using a vice. George Rennie (1791–1866; FRS 1822) published a paper in *Philosophical Transactions* in 1818, describing experiments made on the strength of materials; part of the work concerned the crushing strength of a number of metals including cast iron and copper, while tests in tension included steel. Of considerable practical significance, Peter Barlow reported in 1817 on the strength of timber and other materials, providing essential data for engineering calculations on the basis of numerous experiments in the Woolwich dockyards. Later, experiments on the 'resistance' of iron which formed the basis of the design of the Menai suspension bridge were submitted by Telford for Barlow's examination. Barlow's treatise *The Strength of Materials* of which a new edition, revised by his sons Peter William Barlow (1809–1885; FRS 1845) and William Henry Barlow (1812–1902; FRS 1850), was published in 1867 included much information on the mechanical properties of iron.

William Fairbairn (later Sir William, 1789–1874; FRS 1850) and Eaton Hodgkinson (1789–1861; FRS 1841), a mathematician, carried out important research on mechanical properties. Hodgkinson began this work in 1822 and in 1824 published a paper on 'The transverse strain and strength of materials'. In a later paper (1830) on 'Theoretical and experimental researches to ascertain the strength and best forms for iron beams', data for cast irons were reported; the moduli of elasticity, elastic limits, and ultimate strengths showed different values for tension and compression. A major series of experiments was reported by Hodgkinson in a Royal Society paper in 1840 entitled 'Experimental researches on the strength of pillars of cast iron'.

During the 1830s a controversy arose as to whether iron produced by the 'hot-blast' process invented by James Neilson (1792–1865; FRS 1846) was inferior in properties to

Thomas Telford

Isambard Kingdom Brunel

Robert Stephenson

William Fairbairn

the 'cold-blast' product. Fairbairn and Hodgkinson were commissioned by the British Association for the Advancement of Science to carry out comparative tests. Hodgkinson made tension and compression tests; Fairbairn focused on bending tests but also experimented on the effect of long maintained loads. The specimens used by Fairbairn were cast bars 1.5 m in length and 2.5 cm square, mounted on supports and centrally loaded incrementally, with deflections reported to 0.025 mm; microscopical observations of fracture surfaces were made. In the tests on time dependence of properties (in effect 'creep': see Chapter 14) Fairbairn, from observations after 15 months of loading reported that fracture would occur sooner in the cold-blast iron than in the hot-blast material; in 1842 he reported data for a loading period of more than four years, and experiments extended up to seven years, although the results of the latter were not reported. Fairbairn, in describing in 1838 the deformation process, referred to 'the minute crystalline particles of the bars...' and 'whether those particles ... by slow, though imperceptible degrees, becoming hourly weaker, until the cohesive power is entirely destroyed and rupture takes place'. From his data Fairbairn concluded 'that a cast iron bar is capable of resisting for a series of years a force equal to 7/8 and sometimes 9/10 of the load that would break it'.

In later pioneering experiments, reported in 1856, Fairbairn investigated the effect of increasing test temperature on the tensile properties of iron; he used a lever testing machine with which a load of up to ~ 450 kN could be applied, the specimen being heated in a bath of water or oil. Soon after this work, Fairbairn, in the interesting context of the alloying behaviour of iron, reported on ' Experiments to determine the strength of some alloys of nickel and iron, similar in composition to meteoric iron.'

In the 19th century in the development of railway engineering in Britain several eminent engineers who played dominant roles became Fellows of the Royal Society; of particular note concerning the use of iron were the construction projects of Isambard Kingdom Brunel (1806–1859; FRS 1830), the son of Marc Isambard Brunel (1769–1849; FRS 1814), and Robert Stephenson (1803–1859; FRS 1849), the son of George Stephenson (1781–1848). Isambard Brunel had worked with his father on tunnelling and went on to many outstanding construction achievements in railways, bridges and ships. The careers of both George Rennie and his younger brother John (later Sir John, 1794–1874; FRS 1823) included periods in railway engineering. John Urpeth Rastrick (1780–1856; FRS 1837) as a civil engineer was much involved in railway engineering, and beginning at an early stage of his career, gained knowledge of cast iron, holding appointments in ironworks, in which he was involved in iron founding and in the manufacture of machinery.

Another eminent engineer involved in railway work, Charles Blacker Vignoles (1793–1875; FRS 1855) in 1837 introduced the flat-bottomed rail section which became standard. In engineering construction for the development of railways in Ireland Robert Mallet (1810–1881; FRS 1854) played a major role. Mallet in 1852 patented the buckled plate, consisting of plates of wrought iron of square or rectangular shape, formed to introduce curvature which provided an increased strength:weight ratio; this came into widespread use, for example in railway bridges.

For these and other engineers the availability of ferrous materials of suitable properties and form was of critical importance and in 1848 a Royal Commission was set up to report into the application of iron to railway structures. The six commissioners were all Fellows of the Royal Society: Lord John Wrottesley (1798–1867; FRS 1841), Robert Willis (1800–1875; FRS 1830), Capt. Henry James (1803–1877; FRS 1848), George Rennie, William Cubitt (1785–1861; FRS 1830) and Eaton Hodgkinson. Their report, published in 1849, included minutes of evidence taken, comprising the opinion of nearly all the leading engineers of that time, data on the strength of iron, and the effects of continuous and intermittent loads and of long-continued impacts, (see Chapter 14). Willis in collaboration with Henry James and Capt. Douglas Strutt Galton (1822–1899; FRS 1859) made experiments on the effect of moving loads to simulate trains on iron rails; George Gabriel Stokes (later Sir George, 1819–1903; FRS 1851) played a role in contributing to the mathematical analysis.

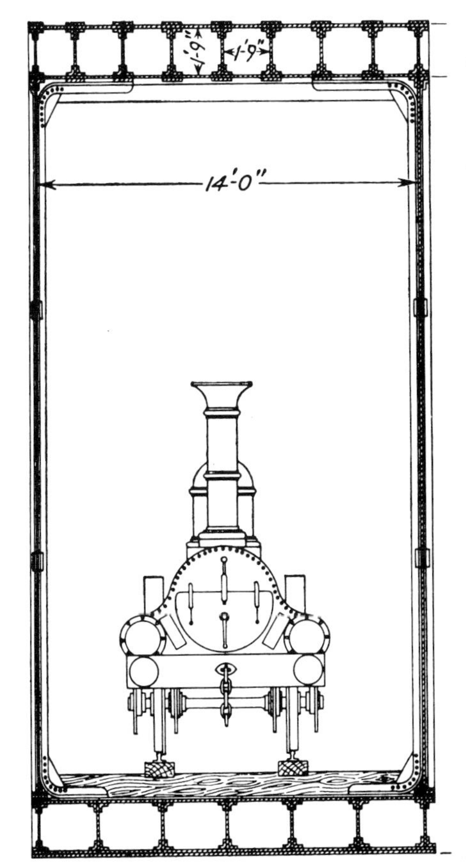

Britannia tubular-type bridge, North Wales; transverse section through the middle of the tube.

Late in the 1840s, Robert Stephenson faced the challenge of bridging the Menai Straits and the Conway River for the Chester–Holyhead railway line. A timely event a few years earlier was the publication by Henry Moseley (1801–1872; FRS 1839) of an important book, *The mechanical principles of engineering and architecture.* Moseley's work through his interest in mechanics had contributed to knowledge of the statics of masonry arches and to the development of the wrought iron girder bridge. Stephenson, with mathematical assistance from Eaton Hodgkinson and in collaboration with William Fairbairn at Fairbairn's shipyard at Millwall embarked on a classic series of experiments to determine the best form of tubular girders for the long span bridges to be built. Three forms of such tubes, to be constructed by riveting from wrought iron plates and angle sections were considered, of circular, oval and rectangular sections respectively. Experimental model tubes were made from wrought iron of the three shapes and of various dimensions; these were supported at the ends and loaded at midpoint to failure. The information thus obtained not only satisfied their specific objective, but also laid the foundations for modern structural engineering theory. A hollow rec-

Stephenson's Conway Bridge (from a contemporary print).

View from Conway Castle of Telford's Bridge and Stephenson's Bridge (1998).

William Clark

Peter Barlow

William Henry Barlow

Robert Willis

Sir John Rennie

Charles Vignoles

tangular section tube was selected large enough for the passage of a train; the top and bottom involved a cellular construction, while the sides consisted of stiffened single plate. Their work included the testing of a one-sixth scale model, ~26 m in length, which was progressively modified to improve its performance. The bridge was completed in 1850. Fairbairn also carried out a test on site of the first tube for the Conway bridge; this tube was 129 m long, maximum height 7.8 m and 4.5 m wide, and weighing 1300 tonnes; the tube was supported and a central load of 95 tons (~950 kN) was applied for about 24 hours, when a measurable deflection was noted. In the Britannia bridge across the Menai Straits each main tube weighed more than 1500 tonnes. The plates used for the construction of the tubes had to be straightened before they could be punched for riveting; cold rollers were sometimes used, or flattening with sledge hammers.

Another Fellow of the Royal Society, William Pole (1814–1900; FRS 1861), made a major contribution to the Brittania bridge project. This involved the scientific application of elastic beam theory to the practical task of achieving the best stress distribution when connecting the separate spans of the tubular girders. Also Charles Babbage (1791–1871; FRS 1816), one of pioneers of mechanical computation, had an interest in railway engineering and the construction of large tubular bridges, and corresponded with Fairbairn on the experiments with model tubes.

For Isambard Kingdom Brunel, as chief engineer to the Great Western Railway, the Tamar River was a formidable obstacle to the extension of the railway west of Plymouth into Cornwall from Devon. At Saltash, which was the most suitable location for a bridge, the river was 335 m wide. He met the challenge in the design in wrought iron of the last and greatest of his railway works – the Royal Albert bridge at Saltash which was opened in 1859. He used a cast iron caisson nearly 11 metres wide to construct a central pier built on rock 24 metres below the high water mark. For the two spans of the bridge he used two arched tubes of elliptical cross section. For this bridge project Brunel made use of the results of testing to destruction various forms of large wrought iron girders.

In his railway engineering work Brunel was expert in the use of wrought iron, masonry and timber. However, after experiencing a situation of the fracture of some cast iron girders on a bridge, as a result of a timber platform catching fire, he wrote' Cast iron girder bridges always give trouble... I never use cast iron if I can help it; but in some cases it is necessary, and to meet these I have had girders cast of a particular mixture of iron...'

Joseph Locke (1805–1860; FRS 1838), who with Brunel and Robert Stephenson constituted the triumvirate of leading railway engineers) also distrusted the use of cast iron in railway bridges and preferred to use masonry. However, in contrast, Robert Stephenson when using cast iron arches in the construction of the bridge over the River Tyne carried out thorough foundry experiments on various 'mixes' to achieve the best for the production of the iron, and tested the arches at the works.

William Henry Barlow was a civil engineer, who carried out many scientific researches, including work on the strength of beams. He was a proponent of the advantages of steel, noting that Britain was lagging behind some other countries, such as the USA, in its

William Cubitt

Joseph Locke

John Hawkshaw

Benjamin Baker

James Watt　Matt: Boulton　Thos. Telford

Robt Stephenson　Wm Fairbairn　B Baker

Signatures of some engineering Fellows from the Royal Society records.

application for structural engineering. Another eminent civil engineer John Hawkshaw (later Sir John, 1811–1891; FRS 1855) was also concerned with the application of steel for structural purposes, for example, for railway bridges. The Board of Trade exercised authority in this sphere and in the early 1870s set up a committee, involving Hawkshaw, Barlow and Colonel William Yolland (1810–1885; FRS 1859) to determine safe specifications. The recommendations in the committee's report included reference to the processing procedures for the steels and the loading conditions in service. Barlow and Yolland worked together again on the Court of Enquiry established following the Tay Bridge disaster that occurred in 1879.

In the era of steel Benjamin Baker (later Sir Benjamin, 1840–1907; FRS 1890), a civil engineer whose interests included the strength of materials, obtained early experience as an apprentice in the Neath Abbey steelworks. He was a partner in the construction of the Forth Bridge from 1883 to 1890, the first large bridge to be made completely of steel. By this time the steel industry was able to supply much larger plates and a variety of rolled sheet sections as compared with the wrought iron era. The steel for the bridge was produced by the Siemens open hearth process; plates as large as 30 × 5 feet (~ 9 × 1.5 m) were available for forming into large tubes. For this steel the tensile strength was around 33 tons/in^2 (~ 510 MPa) as compared with the value of 22 tons/in^2 (~ 340 MPa) for the wrought iron, only available in smaller sizes, as used earlier by Stephenson, thus leading to reduction in weight and cost.

During the 19th century the ship building industry saw the transition from sail to steam propulsion and from wooden hulls to metal construction. In the 1830s Fairbairn developed an interest in iron ships and carried out extensive experiments in 1838–39 on the strength of wrought-iron plates and their riveted joints. Among other famous engineers, Isambard Kingdom Brunel is remembered for his construction of large steamships. His *Great Britain,* launched in 1843 as a transatlanic liner was the largest vessel afloat at approximately 3,000 tonnes It was propeller driven, with a hull consisting of overlapped wrought iron plates 6 ft × 2.5 ft (~1.8 x 0.8 m). In 1852 there began the construction of the *Great Eastern,* conceived by Brunel. His associate, John Scott Russell (1808–1882; FRS 1849), a naval architect, was given the contract to construct the vessel at the Thames shipyard at Millwall, formerly owned by Fairbairn. In this project Brunel also consulted William Froude (1810–1879; FRS 1870) an engineer and naval architect, who had previously worked with him on railway construction. Froude carried out important studies of the motion of ships in waves, and its relation to ship design, which was later influential in the design of naval vessels. With a length of around 210 m and a beam of 25 m, the ship had an iron frame weighing ~ 6,300 tonnes. The 30,000 iron plates used in the construction were joined by 3 million rivets. There were two sets of engines, one driving paddles and the other a propeller; in addition masts and sails were provided. Construction was completed in 1857 and launching took place the next year with great difficulties. The *Great Eastern* made successful Atlantic crossings, but there were mishaps and accidents, and financially the ship was unsuccessful; however, it later achieved success in a role for cable laying, across the Atlantic and elsewhere in the world.

The *Great Britain* (length approx. 98 m).

The *Great Eastern* (length approx. 210 m).

Naval shipbuilding also saw important changes, particularly in the latter part of the century, with the impetus for development from the arms race with France. The *Warrior* in whose design John Scott Russell was involved, was built at the Thames Ironworks at Blackwall, and launched in 1860; it was, in effect the first ocean going, iron-hulled battleship incorporating an iron structure in frame and plating, a system of compartments giving increased watertight integrity and armour protection. The length was ~ 128 m, the beam ~ 18 m, with a displacement of ~ 9,360 tonnes. The guns included a novel type with a rifled barrel, involving steel tube with wrought iron coils as reinforcement fitted over it, with breech loading invented by Sir William George Armstrong (later Lord Armstrong, 1810–1900; FRS 1846). Much attention was devoted by Armstrong to problems in connection with the mounting and working of guns on ships.

The choice of the most effective material for armour plating for warships was an important technological matter, and as early as 1860, the Special Committee on Iron was set up by the government; the membership included John Percy (1817–1889; FRS 1847) as a metallurgist and William Fairbairn and William Pole as engineers. Four years of work led to four reports. A popular view at that time was that high hardness was the essential property required for armour plate; however, the brittleness of such material

William White

Henry Oram

Alfred Yarrow

Frederick Lanchester

e.g. a hard steel, led to fracture of the plating under the impact of the projectile. The Committee proposed the use of plates of wrought iron as a soft, tough material to absorb the energy of the impact. Some years later Charles William Siemens (later Sir William, 1823–1883; FRS 1862), serving on another committee supported the same approach to the selection of materials for armour plating.

Armstrong sought to improve armour plating and produced steel of high tensile strength and great toughness by heat treatment. The design and construction of cruisers was a special interest of his and the Elswick firm built several vessels of this class. In 1882 the shipbuilding firm of Messrs Mitchell and Swan joined forces with Armstrong's Company. The firm of Robert Abbott Hadfield (later Sir Robert, 1858–1940; FRS 1909) was involved in armour plating and developed special steels for armour-piercing projectiles; tests were made of resistance to armour-piercing projectiles and investigations of the phenomena of impact at high velocities.

Edward James Reed (later Sir Edward, 1830–1906; FRS 1876) was a leading naval architect involved in the introduction of armoured vessels. He became Chief Constructor of the Royal Navy in 1863, and introduced a new system of framing, and the *Bellerophon* was among the vessels he designed; later he worked with Sir Joseph Whitworth (1803–1887; FRS 1857) at his Manchester works. In the period 1871–79, when Reed was a consultant based in London, his chief assistant was Francis Elgar (1845–1909; FRS 1896); Elgar's distinguished career included a period as Director of Dockyards, and work in the shipbuilding industry involving the manufacture of armour plate in the development of vessels of iron and steel with heavy armaments, an important role was played by another naval architect William Henry White (later Sir William, 1845–1913; FRS 1888). The royal dockyards and private yards were engaged in this activity, and new methods of construction were being devised. White's career included a period as designer and manager at Armstrong's warship yard at Elswick, and subsequently as director of naval construction at the Admiralty. He made many improvements, embodying advances in machinery, gunnery and quality of materials. From 1908 to 1910 he was the first President of the Institute of Metals.

Henry John Oram (later Sir Henry, 1858–1939; FRS 1912) had a distinguished career in engineering in the Royal Navy; in 1907 he became Engineer-in Chief of the Fleet with the rank of Engineering Vice-Admiral. He played an important role in developments such as the introduction of water-tube boilers and of oil-fuel turbines. He served as president of the Institute of Metals from 1914 to 1916.

An eminent ship-builder and marine engineer, Alfred Fernandez Yarrow (later Sir Alfred, 1842–1932; FRS 1922) created the great engineering firm that carried his name. Starting with a small activity in London in 1868 building steam launches, and later torpedo boats, a new shipyard was opened on the Clyde in Scotland in 1907. Here, during the 1914–18 war, 29 destroyers were constructed. Yarrow was one of the first ship designers to make systematic experiments and speed trials, and to recognise the advantages of using high tensile steels and aluminium, leading to the reduction of the thickness of plates and of the weight of hulls and fittings.

The closing years of the 19th century saw crucial developments in road transport through the pioneering work of Frederick William Lanchester (1868–1946; FRS 1922). In 1894 following a period working for a company on gas engines in which he invented the pendulum governor and an engine starter, Lanchester decided to start out on his own account and formed a small private syndicate, setting up a workshop. His ambitious objective was to develop a 'real motor car' as distinct from a 'horseless carriage', incorporating features such as a noiseless and vibrationless engine and transmission, with springing and steering appropriate for higher speeds than those associated with horse-drawn vehicles. He recognised that this task would require innovative design and also new manufacturing techniques, and in the course of his work he produced a number of important inventions: these included the splined shaft which required a new machining technique; roller bearings with short rollers of extreme accuracy, and for this and other precision work he designed and patented what is now known as the centreless grinder. His first, experimental car was produced in 1906 and by the turn of the century cars were being produced for the market.

Lanchester was also a pioneer in the science of aeronautics, and in addition carried out some experimental work in aircraft construction. In 1911, in collaborative work, he designed an experimental aircraft with wings covered by an aluminium skin and a steel sheet fuselage; although undercarriage failure occurred in the first trial flight, many of his ideas were subsequently utilised in the construction of seaplanes used in anti-submarine activity during the 1914–18 war.

Looking back over the 19th century in the field of civil engineering, the great achievements of British engineers, including a number of Fellows of the Royal Society, were already clearly apparent during the first half of the period. It is interesting to highlight the mid-century opportunity for the display of British engineering skills provided by the construction of the Crystal Palace in Hyde Park, London, which was built to house the Great Exhibition, which opened on 1 May 1851. A Royal Commission had been set up in 1849 'for promoting arts, manufactures and industry by means of a great collection of works of art and industry of all nations'. The building was planned by Joseph Paxton (later Sir Joseph, 1803–1868) who was not an engineer. The detailed design and working drawings and the construction were done by Messrs Fox, Henderson and Co. of Smethwick. The building was 563 m long and 124 m wide and involved the use of ~ 700 tonnes of wrought iron, for girders etc., and 3,800 tonnes of cast iron for columns etc; the vast area of glass was provided by Messrs Chance Bros. The work of erection took only approximately six months, and the design was such that the building could be dismantled and moved to another location in Sydenham, south of London. The Commissioners, under the presidency of Prince Albert (1819–1861; FRS 1840) included William Cubitt and Robert Stephenson; John Scott Russell served as one of the secretaries to the Commission. Cubitt was chairman of the building committee whose members included Robert Stephenson and Isambard Kingdom Brunel. Various committees dealt with the arrangements for exhibits and the committee concerned with machines included in its membership Brunel and Sir John Rennie.

Mechanical testing machine used by Hodgkinson (*Phil. Trans.* 1840)

Among other Fellows associated with the vast project were Lyon Playfair (later Lord Playfair, 1818–1898; FRS 1848) who was appointed as a Special Commissioner, a vital role which involved many important aspects, including organisation and coordination. It was Playfair who proposed a classification scheme for the exhibits based on eight divisions, among which was metallurgy. Finally, five divisions were established of which one dealt with raw materials and included metallurgical operations; the division on manufacture encompassed metallic, vitreous and ceramic products. Overall, it is clear that metallurgical activities played a major role in the materials of construction and in the wide ranging exhibits displayed in the remarkable venture represented by the Great Exhibition of 1851.

Mechanical testing equipment

The common mechanical testing of metals was well established by the last quarter of the 19th century in Britain and abroad e.g. in the USA. Machines of substantial capacity were available earlier in the century. George Rennie in 1818 reported in a Royal Society paper a new machine able to apply a load of ~11 tons (~110 kN). Woolwich Dockyard had a 100 ton (~1000 kN) machine described by Peter Barlow in 1837. In 1840 Hodgkinson reported to the Royal Society on a machine for compression testing to determine the breaking strength of cast iron pillars up to several feet in length and presented data obtained. Subsequently he constructed an improved design of machine at University College London and Fairbairn made a machine at his works in Millwall.

In the latter part of the Victorian period tensile testing machines were built by many manufacturers, and accurate extensometers were produced including adaptability for high temperature testing. Among Fellows of the Royal Society who contributed in the field of mechanical testing Sir James Alfred Ewing (1855–1935; FRS 1887) and William Dalby (1862–1936; FRS 1913) developed extensometers. Dalby's device was an optical load-extension recorder, reported in 1912. The principle lay in applying a load to a test piece and a master-bar coupled together in series. The extension of the master-bar tilted a mirror and deflected a focused spot of light in one direction a distance proportional to the applied load, while the extension of the test piece tilted another mirror which deflected the light spot in a transverse direction, this latter displacement being proportional to the elongation of the test piece. This was a great advance on then available equipment; it provided precision, enabled rapid tests to be performed and opened up new fields of investigation on the mechanical properties of engineering materials. Dalby's publications included a book *Strength and structure of steel and other metals* (1928).

William Cawthorne Unwin (1838–1933; FRS 1886) also had a major interest in mechanical testing and properties of materials. In his earlier career he collaborated with Fairbairn. In 1883 he published a treatise entitled *The testing of materials for construction;* this included detailed descriptions of testing machines, extensometers and autographic recording apparatus, and experimental data for a range of materials. Unwin's investigations included the relationship of elongation to test bar size in the testing of steels and he was also concerned with the accuracy of extensometers.

For high-temperature tensile testing of iron Walter Rosenhain (1875–1934; FRS 1913) and J. C. W. Humfrey constructed apparatus at the National Physical Laboratory, reported in 1909. Direct electric current was used for heating the specimen and the apparatus was enclosed in a chamber which could be evacuated. The apparatus was subsequently modified for the monitoring of stress and strain, and temperatures up to 1100 °C could be achieved, measured accurately with a platinum type thermocouple. It was found

William Dalby

William Unwin

that the high temperature tensile strength of steel was markedly dependent on the rate of straining.

During the 1914–18 war, work in the laboratory of Joseph Ernst Petavel (later Sir Joseph, 1873–1936; FRS 1907), Professor of Engineering at Manchester University, included mechanical testing of materials.

In the first decade of the 20th century, tensile testing at cryogenic temperatures became an established technique. Sir Robert Hadfield (1858–1940; FRS 1909) and Sir James Dewar (1842– 1923; FRS 1877) were notable researchers in this area; for example, a paper presented by Hadfield in 1905 described apparatus available at the Royal Institution, London in which specimens of iron and various steels were immersed in liquid nitrogen and tested to fracture to provide data on tensile strength and ductility.

During the 19th and early part of the 20th century as the importance of service conditions involving repeated stress cycles or high temperatures became prominent, appropriate tests to determine fatigue and creep properties, and also fracture toughness were required; reference to some of these is made in Chapter 14.

CHAPTER 10

The Emergence of Physical Metallurgy

Introduction

In the history of metallurgical science one of the most influential advances was the application of optical microscopy to the investigation of the structure of metallic materials; this was associated particularly with Henry Clifton Sorby (1826–1908; FRS 1857). In the latter part of the 19th century, resulting from his work and that of a number of eminent metallurgists, improvements in the understanding of iron and steels occurred and over a period of several decades the view that metals consisted of aggregates of crystals was established. This chapter considers the emergence of the subjects of metallography and physical metallurgy, including the importance of techniques such as pyrometry and thermal analysis, which were essential in the investigation of phase transformations in materials in various conditions of thermal and mechanical treatment.

Metallography and physical metallurgy

By the last decade of the 19th century, the term 'metallography', which had been used as early as 1721 in a broad sense to signify the description of metals and their properties, had come into usage to describe a specific area of metallurgy. William Chandler Roberts-Austen (later Sir William, 1843–1902; FRS 1875) in his classic book *An Introduction to the Study of Metallurgy,* published in 1898, referred to the considerable attention given in (then) recent years to the study of the structure of metals and alloys, in which an essential feature was the use of optical microscopy. The book, which aimed to treat the subject as a whole, included a short chapter entitled 'Microstructure of metals and alloys' in which Roberts-Austen commented 'The subject of microscopic metallography has assumed much importance, as it will in the future become a powerful ally of chemical analysis...'; in this chapter he referred to aspects such as the preparation of specimens for microscopical observations and microscope illumination systems. Other major chapters in the book were concerned with the physical properties of metallic materials, alloys, and thermal treatments,including photographs of microstructures. The book also contained considerable information on recent research findings by a number of workers.

In 1910 Cecil Henry Desch (1874–1958; FRS 1923) also published a classic textbook, entitled *Metallography,* which saw a series of editions, including up-dating, the last being published in 1944. In his preface he commented that 'the study of metallic alloys by physical and microscopical methods has reached so great a development in recent years as to form a distinct branch of physical chemistry'. He defined the subject area in some detail, stating:

> Metallography may be defined as the study of the internal structure of metals and alloys, and of its relation to their composition, and to their physical, chemical and mechanical properties. It is a branch of physical chemistry since the internal structure depends on the physical and chemical conditions under which the solid metal or alloy is formed, and the study of the structure presents itself as a department of the study of equilibrium in heterogeneous systems. Whilst, however, physical chemistry concerns itself in general only with the nature and quantity of the phases in a system and with the transformations of energy which accompany chemical changes, metallography takes into account a further variable, namely the arrangement of the component particles. It is thus intimately connected with crystallography .... The study of structure has proved an indispensable auxilliary to chemical analysis in the scientific control of the metallurgical industries...

The book had a substantial content relating to phase diagrams, and also covered pyrometry, specimen preparation and microscopical examination, solidification, diffusion, physical properties, corrosion, plastic deformation, and the metallography of iron and steel, and other industrial alloys.

Early in the 20th century and to some extent synonymously with the term metallography, the term 'Physical Metallurgy' was coming into use , and was accepted and defined by Walter Rosenhain (1875–1934; FRS 1913) in his important book *An Introduction to Physical Metallurgy*, published in 1914. Rosenhain, noting that the term was not yet in widespread use, defined it as

> that great branch of knowledge of metals which has to a large extent grown up during the last fifty years – a branch which concerns itself with the nature, properties and behaviour of metals and of alloys as such, as distinct from the far older branch of metallurgy which deals with the reduction of metals from their ores.

He noted that the term 'Metallurgy' had been almost entirely confined to the latter meaning, and he suggested that the field of the reduction of metals and their refining might be termed 'Process' or 'Chemical Metallurgy', alongside the newer field of 'Physical Metallurgy.' He wrote

> thus defined and understood the scope of physical metallurgy is an exceedingly wide one, and one which brings it well over the border-land of several sister sciences – such as chemistry on the one side, physics on the other, and that branch of knowledge generally known as 'strength of materials' in yet another direction. Besides these, crystallography bears largely on our subject. But while our young science necessarily draws largely upon the resources of these her elder sisters, she is not without gifts in return.

Rosenhain presented his subject matter under two main headings, namely, 'The structure and constitution of metals and alloys' and 'The properties of metals as related to

Henry Sorby

William Roberts-Austen

Cecil Desch

Walter Rosenhain

Photomicrographs (Sorby 1864) (a) meteoric iron; (b) wrought iron armour plate – longitudinal section; (c) white cast iron, mag X7, Gen. Ref. 22, (Acknowledgments).

their structure and constitution'. The first of these covered topics such as microscopic examination, phase diagrams, physical properties of alloys, and typical alloy sytems, including a chapter on the iron–carbon system. In the second part of the book, mechanical testing received considerable attention, as did deformation, thermal treatments and casting; a chapter on defects and failures was also included.

The field of study involving the structure of metallic materials had grown out of the work of mineralogists and crystallographers early in the 19th century. However, before this optical microscopy had been used by several Fellows of the Royal Society; e.g. Henry Power (1623–1668; FRS 1663), Robert Hooke (1635–1703; FRS 1663) and René-Antoine Ferchault de Réaumur (1683–1757; FRS 1738). The microscopical observations made, including some on fractured surfaces, were however, of limited usefulness in obtaining fundamental understanding of the nature of metallic materials. Early in the 19th century (1817) John Frederic Daniell (1790–1845; FRS 1814) presented results in a paper entitled 'The mechanical structure of iron developed by solution' which showed the value of macroetching. His observations on ferrous samples included layers of spongy graphite on grey cast iron after immersion in dilute hydrochloric acid, a fibrous appearance on wrought iron and plate-like features on white cast iron; fractured surfaces of heat treated steel were also examined after immersion in acid. He reported 'The good qualities of steel seem to depend for different purposes upon a varying mechanical arrangement of its particles.'

A critical advance came with the work of Henry Sorby reported in the 1860s utilising the examination of polished and etched metallic samples by reflected light by microscopy at magnifications up to about 650 times. Having learnt the biological technique of preparing thin sections, Sorby, in 1849, applied it to rocks using polarised light in transmission. His investigations on a range of geological materials made important scientific contributions, although his work was rather slow to be widely appreciated; some geologists saw little importance 'in studying mountains with microscopes'. Sorby's interest had been stimulated by the observations on certain meteorites that had been reported as early as 1808 by Aloys Joseph Beck Edler von Widmanstätten (1754–1849), director of the Imperial Porcelain works in Vienna; he had observed the geometrical structure that appeared when the polished meteorite surfaces were etched, which subsequently came to be termed a 'Widmanstätten' structure, and which provided a clue to the internal crystalline nature of the material.

Thus, in 1863 Sorby turned his attention for a while to metals, beginning with meteorites (as metallic bodies having an easily visible microstructure). It proved necessary to adapt his technique to take account of the opaque nature of the metallic constituents of the meteorites and he prepared specimens to be viewed by reflected light and modified the microscope accordingly. From this he was led to study the structure of 'artificial' steels and iron, including an alloy which he prepared by melting iron and nickel which, when slowly cooled, showed a Widmanstätten structure. Specimens were prepared by a series of filing, grinding and polishing operations, followed by etching with dilute nitric acid. In 1863 he recorded in his diary observations on the structure of a ferrous sample, which he interpreted as a Widmanstätten structure. Early in 1864 he recorded structures by a process in which a relief-etched surface was inked and pressed to paper. He worked with a local photographer and prepared several photomicrographs of steel, which he displayed and discussed in 1864 when he presented a paper at the British Association meeting in Bath. This paper was only published in abstract, but represented the foundation of metallography. Among important conclusions, it became clear that metals were crystalline. Sorby referred to several constituents, which he considered when arranged in various ways would give rise to different types of iron and steel.

Sorby moved on to other fields, not returning to steel until 1882, and not publishing in detail until 1885; his main influence stemmed from his papers in 1885 and 1887. By that time interest in metal microstructures had been aroused by publications from workers abroad, who had not been aware of Sorby's earlier work. From his researches Sorby realised that a sequence of constituents appear as carbon is alloyed in increasing amounts with iron, and that the 'pearly constituent' (pearlite) consisted of lamellae of iron containing virtually no carbon, alternating with an iron–carbon compound; this constituent formed during slow cooling from a higher temperature at which a different constituent existed. Sorby also identified graphite and iron oxide in ferrous samples. Sorby's work greatly advanced the understanding of the structures of steels; in England it was not taken up rapidly by other workers, but by the end of the century it was being widely used both in England and abroad.

A number of other Fellows of the Royal Society, actively and effectively pursued microscopical studies, most of them beginning their work in the latter part of the 19th century and, particularly, but not exclusively, on steels: Thomas Andrews (1847–1907; FRS 1888); John Oliver Arnold (1858–1930; FRS 1912); William Roberts-Austen; William Gowland (1842–1922; FRS 1908); Henry Louis Le Chatelier (1850–1936; FMRS 1913); John Edward Stead (1851–1923; FRS 1903); Charles Thomas Heycock (1858–1931; FRS 1895); Francis Henry Neville (1847–1915 ; FRS 1897); Cecil Henry Desch;

Walter Rosenhain; Sir Henry Cort Harold Carpenter (1875–1940; FRS 1918); Sir Robert Abbott Hadfield (1858–1940; FRS 1909); and Albert Marcel Germain René Portevin (1880–1962; FMRS 1952).

Stead in 1898 reported important observations on the crystalline structure of iron and steel. He made detailed observations on the effects of heat treatment temperature and time on iron and various steels, including grain growth; he showed that 'burning' at very high temperatures involved incipient melting, sometimes with oxidation. Stead also introduced a method of heat tinting specimens by oxidation which became an accepted method for the microscopical examination of cast irons. He was strongly influential in his advocacy of the value of microscopy for research and control in the steel industry. Arnold sought to show industrialists the importance of microscopy in giving information for the control of manufacture and causes of failure, while Andrews in his lectures to engineering students at Cambridge in the early 1900s included the importance of microscopical examination. Gowland, when he became professor of metallurgy at the Royal School of Mines, London, in 1902, used the new methods of investigating steel, particularly microscopy.

Le Chatelier played an important role in contributing improvements in apparatus and techniques; he designed the inverted microscope (1900) an important development, and introduced the method of electrolytic etching. In 1901 he published the conclusions of a group of investigators, in a volume entitled *Contribution à l'étude des alliages*, which served for a long period as a book of reference for metallographers. Also he used dilatometry and measurements of electrical resistance as a means of following transformations in alloys with change of temperature. A number of workers, including Roberts-Austen, Stead and Desch, used photography to record microstructures, as an alternative to the meticulous sketching of the structural features. Rosenhain was also interested in the design of microscopes, and made important contributions to metallographic techniques. For example, the investigation of dental amalgams presented special difficulties in the microscopical examination because of the presence of liquid mercury at ordinary temperatures. Rosenhain devised special techniques for polishing,etching and photographing at temperatures of about –60 °C.

Concerning the basic nature of crystalline aggregates Daniell had speculated earlier in the century on the nature of intercrystalline boundaries. He was among the first to state clearly that a relationship between a regular internal structure and external shape is not necessary; also he speculated on the nature of a nucleus during solidification and commented on the similarity of structure between liquid and solid metals. Of special interest were his ideas on grain boundaries and the geometry of intercrystalline junctions. Joseph Antoine Ferdinand Plateau (1801–1883; FRS 1870), a Belgian physicist, carried out important work on surface tension effects using thin soapy films. This work was relevant to the later understanding of metallic grain structures. Lord Kelvin (formerly Sir William Thomson, 1824–1907; FRS 1851) also did classic work; using soap films he showed that the shape designated a tetrakaidecahedron (eight hexagonal sides and six square sides)

John Arnold

William Gowland

John Stead

Harold Carpenter

could be used to completely fill a space and to fulfil the interfacial energy balance requirements. These ideas were to be applied in the 20th century by Desch to metallic crystalline aggregates and later developed by Cyril Stanley Smith (1903–1991).

A significant event in the literature of physical metallurgy was the publication in 1939 of the comprehensive book *Metals* by Harold Carpenter and J. M. Robertson. The book was in two volumes, spanning nearly 1500 pages with the general objective of presenting the characteristics of metals and explaining their relationship to composition and heat treatment and to properties. An introductory chapter dealt with aspects such as the importance of metals, industrial metals and alloys and extraction and refining. In fifteen chapters the book presented a very wide coverage of themes in physical and mechanical metallurgy, including extensive reference to information from research papers by many workers. The first volume included consideration of microstructure, crystal structure, solidification, mechanical properties and mechanical testing, oxidation and corrosion, together with processing (casting, heat treatment and mechanical treatment). In the second volume there was extensive coverage of industrial ferrous alloys, dealing with the iron–carbon system, alloy steels and cast irons; non-ferrous metals and alloys, e.g. those based on copper, aluminium, nickel, and lead respectively were discussed.

Pyrometry and thermal analysis

Thermal analysis, the study of thermal changes that occur during the heating and cooling of alloys, was also proving to be an important complementary technique for the study of alloy constitution. In the essential area of pyrometry earlier work in the 18th century had been carried out by several Fellows: Petrus von Musschenbroek (1692–1761; FRS 1734), Tiberius Cavallo (1749–1809; FRS 1779) and Josiah Wedgwood (1730–1795; FRS 1783)(see Chapter 2). At the beginning of the 19th century, in 1803, Louis Bernard Guyton de Morveau (1737–1816; FRS 1788) in the context of his interest in the measurement of temperature in furnaces and kilns reported a pyrometer based on the expansion of a platinum rod. In 1821 Daniell also developed an instrument based on expansion of a platinum rod, encased in graphite, with a mechanical lever as a pointer. Then in the latter part of the 19th century other Fellows made important contributions in the field of electrical resistance pyrometry and thermocouples. James Prinsep (1799–1840; FRS 1828) introduced a pyrometric procedure using a graduated series of alloys, whose melting indicated temperature points.

Charles William Siemens (later Sir William, 1823–1883; FRS 1862) in 1860 suggested that temperatures could be measured from the change in electric resistance with temperature of a metal. He devised the first resistance thermometer, using insulated copper wire, to check the temperature of coils of submarine cables. Subsequently he substituted a coil of platinum for copper; the platinum wire was wound on an insulating fire clay core and was contained in an iron tube. The Bakerian Lecture which he gave in 1871 was entitled 'On the increase of electrical resistance in conductors with rise of temperature and its application to the measure of ordinary and furnace temperatures'. On the recommendation of the British Association, a committee, whose membership included

Henry Le Chatelier

George Carey Foster

Ernest Griffiths

Hugh Callendar

several Fellows of the Royal Society, was formed to test his pyrometer. One of these Fellows, George Carey Foster (1835–1919; FRS 1869) at University College, London, carried out experiments which indicated lack of reliability of the pyrometer's measurements; another committee member suggested that this might be due to contamination by silicon in a furnace atmosphere. The pyrometer lost favour for several years, but the skilled work of Ernest Howard Griffiths (1851–1932; FRS 1895) and Hugh Longbourne Callendar (1863–1930; FRS 1894) at Cambridge led to the development of successful, reliable instruments (see Chapter 16).

Callendar, in 1885, entered the Cavendish laboratory, where Joseph John Thomson (later Sir Joseph, 1856–1940; FRS 1884), quickly recognised his considerable gifts as a skilful experimenter, and assigned him a project on the accurate measurement of the electric resistance of platinum, and its variation with temperature, in the context of its use for temperature measurement. This work, which involved taking care to avoid strain or contamination of the wire, was carried out under difficult conditions; however, by sealing a platinum coil, wound on a piece of mica, inside the glass bulb of the air thermometer he was using as a standard, he developed a reliable formula relating change in resistance to temperature. Unknown to Callendar, research was about to begin by Griffiths at Sidney Sussex College, which led to him constructing a number of platinum resistance thermometers to assist the work of Charles Thomas Heycock (1858–1931; FRS 1895) and Francis Henry Neville (1847–1915; FRS 1897) on alloy constitution, which required accurate temperature measurement. Griffiths collaborated with Heycock and Neville in calibrating the instruments at several fixed points, including ice, steam, the boiling points of several organic compounds, and of sulphur; this led to the accurate determination of the melting points of various metals and alloys. In 1889 Griffiths, having become aware of Callendar's development of a platinum resistance thermometer, collaborated with him to determine temperatures up to 1100°C, with accurate and reliable devices. Through a manufacturing arrangement with the Cambridge Instrument Company the instruments were introduced into the iron and steel industry and other industries.

John Allen Harker(1870–1923; FRS 1910), who becamea member of staff at the National Physical Laboratory when it was established in 1899, was concerned with standards of temperature measurement using platinum resistance thermometers. In 1905 he reported on his construction of a furnace in which very high temperatures could be achieved by passing electric current through the furnace tube which consisted of a conducting ceramic; using this furnace he carried out a redetermination of the melting point of platinum.

In 1821 Thomas Seebeck (1770–1831) reported his important discovery of thermoelectricity, and this opened up a field of research in which a number of Fellows and Foreign Members of the Royal Society took part. Antoine-César Becquerel (1788–1878; FMRS 1837) in 1826 proposed the use of a thermocouple (platinum and palladium) for high temperature measurements. A difficulty was encountered by Henri Victor Regnault (1820–1878; FMRS 1852) in that variable results were found with an iron-platinum couple. However, an influential advance was made in 1888 by Henri Le Chatelier, namely of the platinum–platinum/rhodium thermocouple, as a highly accurate, dependable method of temperature measurement. In his thesis work on cements Le Chatelier found it necessary to devise a means of measuring high temperatures. He showed that, in addition to contamination by iron, an important source of error was the lack of uniformity of the wires, giving rise to local electromotive forces; by using a platinum wire with one of 90% platinum–10% rhodium alloy reliable results could be obtained, and this couple was generally adopted for work at high temperatures. Two days after the reading of the paper describing the new pyrometer, Ludwig Mond (1839–1909; FRS 1891) ordered two such thermocouples by telegram, and these, introduced into his works, were the first to be employed in industry. Robert Hadfield installed a Le Chatelier pyrometer in his works in 1891 for control of furnace temperatures. The calibration of the couples by using the melting-points of pure metals as fixed points, instead of by comparison with an air thermometer, was introduced at the same time. Le Chatelier was also a pioneer in the use of the optical pyrometer; he measured the emissivities of various solids, and obtained reasonable values for the temperature of the interior of a steel furnace.

Among those who adopted thermoelectric pyrometry for the study of alloys was Roberts-Austen, who also devised an automatic recording instrument, in which the movements of the reflected beam of a mirror galvanometer that detected the thermocouple response were recorded photographically. At the National Physical Laboratory, Rosenhain used a differential thermocouple technique and also adopted the 'inverse rate' method for determining heating and cooling curves for phase equilibria and transformation studies; he designed a gradient furnace to obtain very regular cooling, and a plotting chronograph, which simplified the taking of the curves; he also designed furnaces for very high temperatures.

Temperature measurement using thermocouples, both of the platinum and of the base metal types became an invaluable technique, in the whole field of metallurgy. The problem of instability of platinum/platinum–rhodium couples remained an issue, and in 1921 Alan Richard Powell (1894–1975; FRS 1953) as research manager for Johnson, Matthey and Co. Ltd, started investigations on the matter. It was shown that instability was due to impurities in the metals used for the manufacture of the thermocouples, and particularly to to the presence of iron in the rhodium. Suitable methods were developed for eliminating iron, and materials giving thermocouples of high stability were produced.

The development of physical metallurgy in the 20th century

Throughout the 20th century, physical metallurgy, with its focus on metallic structure and properties and their interrelationships, has played a major role not only in the metallurgical sphere, but in many fields of science and engineering.The following chapters of this book are largely concerned with the role of Fellows in various aspects of the development of physical metallurgy and of its achievements.

The basic techniques of optical microscopy and thermal analysis described above, supplemented by modifications such as the use of polarized light and differential scanning calorimetry, have continued to be essential up to the present day, in many applications. Alongside the use of these techniques, investigations of a wide range of physical and chemical properties, begun in the 19th century, have proceeded through the present century, giving deeper understanding of metallic and non-metallic materials. Electrical and magnetic behaviour have been increasingly important themes, including phenomena such as superconductivity and special magnetic materials. These, and other properties are reviewed in Chapter 11.

In 1912 the discovery of X-ray diffraction provided a new and invaluable tool to the physical metallurgist. Access to the structure of metals and alloys was no longer limited to those features revealed by optical microscopy, i.e. relating to the morphology of crystals in crystalline aggregates. X-ray diffraction enabled the details of crystal lattices to be obtained. Later advances came through electron diffraction and electron microscopy, resolving fine scale features, including defects such as dislocations; high resolutions down to atomic level have been achieved by electron microscopy and by techniques such as field ion microscopy. Understanding at the level of electron structure has been developed by techniques such as X-ray spectroscopy. Chapter 12 discusses advances in various techniques for structural investigations.

The field of mechanical behaviour, in which important pioneer work was done in the 19th century, has been a major, vital theme in the current century, particularly in the period following the 1939–45 war. Chapter 13 deals with this theme, including the concept of dislocations. The development of dislocation theory and experimental information obtained from advances in techniques led to a revolution in the understanding of mechanical behaviour. Thus physical metallurgists have been able to assist engineers in successfully meeting the many challenges in engineeering applications as discussed in Chapter 14. The importance of processing in controlling structure, which is currently a topic of major importance is considered in Chapter 15.

Chapter 16 considers various fundamental aspects of physical metallurgy concerned particularly with alloy phase equilibria and phase transformations, including, for example, the importance field of ferrous alloys. Advances in the 20th century are considered, including the development of modern alloy theory. Chapter 17 presents a brief review of contributions of Fellows in the various fields of non-metallic materials;this includes mention of the use of techniques of investigating structures and properties, which are vital in physical metallurgy and which have contributed to laying the foundations of materials science.

Chapter 18 presents an account of one of the most spectacular themes in 20th century science and engineering, namely the elucidation of atomic structure and the ensuing developments leading to the military and civil use of atomic energy. The crucial metallurgical challenges – relating to the chemical, physical and processing spheres – that have been confronted and overcome, provide an excellent illustration of the importance of metallic and non-metallic materials, and of the vital contributions made by Fellows and Foreign Members of the Royal Society.

CHAPTER 11

Physical and Chemical Properties of Metals and Alloys

Introduction

This chapter which is concerned with the properties of metals and alloys is an appropriate place to consider the fundamental question 'what is a metal?'. It turns out that it is rather difficult to arrive at a simple definition of a metal, but the question is considered here briefly in relation to the Periodic Table, atomic structure, types of bonding in materials, and properties. This forms a background for considering in some detail the way in which the 19th century, particularly the second half, saw the opening up of the major field of the detailed scientific studies of the properties and constitution of alloys, including notably of alloy steels. The chapter illustrates the role of Fellows of the Royal Society in alloy development beginning in the early part of the century. An associated stream of research during the century was conducted by distinguished physicists and chemists, many of whom were Fellows, involving investigations of physical properties of metal and alloys. Electrical and magnetic properties formed a major area, which with the development of cryogenics led to the discovery of superconductivity, and has been pursued actively throughout the 20 th century. Other aspects of properties; e.g. thermal, diffusion phenomena, behaviour at high pressures and surface behaviour are also considered. The phenomenon of radioactivity, which led to the modern industrial activity of atomic power, is reviewed in Chapter 18.

What is a metal?

Of the hundred odd elements in the Periodic Table, 80 % are defined as metals, the non-metals essentially occupying the right hand corner in a triangle bounded by a 'diagonal' running from boron to astatine (Table 1). The boundary is not clear cut and elements which are marginally non-metallic when crystalline, such as silicon, germanium, selenium and tellurium, become metallic when molten. Also high pressure and impurities can cause a change of state. Almost anything can become a metal if its atoms are brought close enough together. Small additions of impurities can cause non-metals to become metals, for example additions of phosphorus to silicon can do this (but above the trace levels necessary to cause silicon to become an extrensic semiconductor).

Thus it is difficult to state a simple definition of a metal. On the other hand most people would claim to be able to recognise a metal by visual and tactile inspection.

Table 1

PERIODIC TABLE OF THE ELEMENTS

IA																	
1 H 1	IIA		At.Wt Element At.No.									IIIA	IVA	VA	VIA	VIIA	4 He 2
7 Li 3	9 Be 4											11 B 5	12 C 6	14 N 7	16 O 8	19 F 0	20 Ne 10
23 Na 11	24 Mg 12	IIIB	IVB	VB	VIB	VIIB	VIII			IB	IIB	27 Al 13	28 Si 14	31 P 15	32 S 16	35.5 Cl 17	40 Ar 18
39 K 19	40 Ca 20	45 Sc 21	48 Ti 22	51 V 23	52 Cr 24	55 Mn 25	56 Fe 26	59 Co 27	59 Ni 28	64 Cu 29	65 Zn 30	70 Ga 31	73 Ge 32	75 As 33	79 Se 34	80 Br 35	84 Kr 36
85.5 Rb 37	88 Sr 38	89 Y 39	91 Zr 40	93 Nb 41	96 Mo 42	98 Tc 43	101 Ru 44	103 Rh 45	106 Pt 46	108 Ag 47	112 Cd 48	115 In 49	119 Sn 5o	122 Sb 51	128 Te 52	127 I 53	131 Xe 54
133 Cs 55	137 Ba 56	139 La 57	178.5 Hf 72	181 Ta 73	184 W 74	186 Re 75	190 Os 76	192 Ir 77	195 Pt 46	197 Au 79	201 Hg 80	204 Tl 81	207 Pb 82	209 Bi 83	210 Po 84	210 At 85	222 Rn 86
223 Fr 87	226 Ra 88	227 Ac 89															

Rare earths	139 La 57	140 Ce 58	141 Pr 59	144 Nd 60	147 Pm 61	150 Sm 62	152 Eu 63	157 Gd 64	159 Tb 65	182.5 Dy 66	165 Ho 67	167 Er 68	169 Tm 69	173 Yb 70	175 Lu 71
Actinides	227 Ac 89	232 Th 90	231 Pa 91	238 U 92	237 Np 93	242 Pu 94	243 Am 95	247 Cm 96	247 Bk 97					254 No 102	257 Lw 103

Metallic behaviour is associated with strength and ductility, opaqueness, having a characteristic lustre, and coolness to the touch, a consequence of high thermal conductivity. The latter is complemented by good electrical conductivity, and this can be very striking – conductivities up to 10^{10} times that of non-metals are quite common. It may be noted that semiconductors too can exhibit good electrical conductivity but this conductivity is thermally activated and hence, in contrast to metals, increases with increasing temperature. In fact, as absolute temperature is approached the conductivity of semiconductors becomes vanishingly small whereas that of metals remains high, and some become superconductors.

The physical properties of bulk metals depend on the 'collective' behaviour of the huge number of atoms present. Chemical properties, in contrast, are governed by the behaviour of the individual atoms reacting one at a time in concert or in sequence. As is well known, the chemical properties of metals are characterised by the ability to form strongly basic oxides and hydroxides and to react with acids to form salts.

In order to understand more fully the origins of the physical properties of bulk metals, and the chemical properties of individual atoms, it is necessary to consider the electronic structure of atoms, and a good starting point is the Rutherford/Bohr atomic model (Ernest

Table 2 Occupancies of the electron shells of the first twenty-six elements.

Atomic Number	Element	Symbol	Electronic Configuration K	L	M	N
1	Hydrogen	H	1			
2	Helium	He	2			
3	Lithium	Li	2	1		
4	Beryllium	Be	2	2		
5	Boron	B	2	3		
6	Carbon	C	2	4		
7	Nitrogen	N	2	5		
8	Oxygen	O	2	6		
9	Fluorine	F	2	7		
10	Neon	Ne	2	8		
11	Sodium	Na	2	8	1	
12	Magnesium	Mg	2	8	2	
13	Aluminium	Al	2	8	3	
13	Silicon	Si	2	8	4	
15	Phosphorus	P	2	8	5	
16	Sulphur	S	2	8	6	
17	Chlorine	Cl	2	8	7	
18	Argon	Ar	2	8	8	
19	Potassium	K	2	8	8	1
20	Calcium	Ca	2	8	8	2
21	Scandium	Sc	2	8	9	2
22	Titanium	Ti	2	8	10	2
23	Vanadium	V	2	8	11	2
24	Chromium	Cr	2	8	13	1
25	Manganese	Mn	2	8	13	2
26	Iron	Fe	2	8	14	2

Rutherford, later Lord Rutherford, 1871–1937; FRS 1903 and Niels Henrik David Bohr, 1885–1962; FMRS 1926).

Only three fundamental particles are involved: the proton with unit positive electrical charge, the neutron with zero charge, and the electron with negative electrical charge equal in magnitude to that of the proton. The proton's mass is almost identical to that of the neutron, each being about 1837 times that of the electron. The simplest atom is that of hydrogen, which consists of a nucleus of a single proton around which rotates a single electron. For all the other atoms the nucleus consists of both protons and neutrons with the electrons in the rotating cloud being equal in number, for a neutral atom, to the protons in the nucleus. The number of protons in the nucleus defines the atomic number of he element.

Bohr's brilliant modification to the relatively crude Rutherford atomic model was to assume that the motions of the electrons around the nucleus are governed by quantum theory rules. These require that the electrons could only move within fixed orbits or shells. The theoretical maximum capacities of the shells conform to the formula ($2\ n^2$) where n = 1 for the first (innermost) shell, and so on. This formula predicts maximum

Arnold Sommerfield

Linus Pauling

electron capacities of 2, 8, 18, 32, 50, 72 and 98, for shell numbers 1, 2, 3, 4, 5, 6 and 7 (these shells are sometimes designated K, L, M, N, O, P and Q respectively). In practice no shell contains more than 12 electrons. The occupancies of the electron shells of the first 26 elements are listed in order in Table 2.

The first feature to note is periodicity – sodium comes eight elements after its fellow alkali metal lithium, and eight elements before potassium. The three inert gases, helium, neon and argon come at eight element intervals, and chlorine eighth after fluorine, and so on. (This periodicity was of course the feature which stimulated Dmitri Ivanovich Mendeleev (1834–1907; FMRS 1892) to compile his table in 1869 – see Chapter 3 – it has been called the greatest achievement in chemistry of the 19th century. In like manner, the interpretation of the Table according to the electronic structure of the elements is one of the scientific highlights of the 20th century).

The principal interest here is how the electronic structure relates to the properties of metals. An examination of Table 2, or preferably of the complete Periodic Table, reveals that nearly all the 81 metals have either one or two electrons in their outermost shell; these, the most loosely bound electrons in most cases are known as valency electrons. When the number of valency electrons exceeds two, e.g. aluminium (3) and germanium (4), the metallic characteristics of the element become less marked.

For the majority of elements to have only one or two electrons in the outer shell seems incompatible with filling systematically and progressively the shells from the inside outwards with increasing atomic number. Indeed this does occur for elements 1 to 18 (i.e. as far as argon) and among these there are only five metals – representing only 28 % of the 79% known as metals.

This puzzle is resolved by considering in more detail the electronic structure of argon and those elements immediately following it. Argon has 8 electrons in its outer shell and is chemically inactive even though this M shell can accommodate up to 18 electrons. From this it is concluded that 8 electrons is a stable low energy configuration even when in an unsaturated shell – this is important in relation to bonding of the elements (see below).

As the electronic configuration of argon is 2.8.8, with an incompletely filled outermost shell, the next element above it in the table might have been expected to take the arrangement 2.8.9. In fact the new electron finds it energetically favourable to start the occupancy of the next shell in the sequence (the N shell) leaving the filling of the M shell till later. Hence element 19 becomes the monovalent alkali metal potassium with the configuration 2.8.8.1. There follows divalent calcium 2.8.8.2, but after this the M shell starts to fill with the formation of scandium 2.8.9.2, which presages the very important group known as the transition metals starting with titanium 2.8.10.2, and ending with copper 2.8.18.1.

In the unfilled shells of a number of these transitional metals, specifically iron, nickel and cobalt, some of the magnetic moments of the electrons are not mutually cancelled, and this gives rise to ferromagnetic properties – i.e. they can be readily magnetised and in some cases made into permanent magnets.

The lanthanides (rare earths) and actinides form additional transitional groups, from cerium to lutetium and from thorium to lawrencium respectively. By such 'back-filling' of inner shells while maintaining one or two electrons in the outer shells most of the elements in the Periodic Table have the electronic configuration of metals.

Many Fellows of the Royal Society were pioneers in the elucidation of the atomic and electronic structures of the elements and their work is not reviewed in detail here. From the experimental view point the results of X-ray spectroscopy played a vital role, for example, in the research of Manne Georg Siegbahn (1886–1978; FMRS 1954).

Bonding of elements

The inert gases owe their chemical inactivity to the fact that either their outermost shell is full, helium, 2, and neon, 2.8, for example, or is only partly filled but has the highly stable, low energy, 8-electron configuration viz: argon, 2.8.8; krypton, 2.8.18.8; xenon, 2.8.18.18.8 and radon, 2.8.18.32.18.8. Two of the important bonding processes, ionic and covalent bonding, involve the elements adopting electronic configurations identical to those of the inert gases. They do this by losing, gaining or sharing valency electrons.

The majority of elements are metals with only one or two valency electrons in their outermost shell. In many cases if these valency electrons are lost, say to another element, a positively charged ion is left with the electron configuration of an inert gas. For example, should sodium, 2.8.1 lose its valency electrons the resulting singly positively charged ion would have the highly stable configuration of neon, 2.8. Similarly, should calcium, 2.8.8.2 lose both its valency electrons, the result would be a doubly positively charged ion with the electronic configuration of argon, 2.8.8. For rather obvious reasons metals are known as electropositive elements. In contrast, non-metals, electronegative elements, achieve an inert gas configuration by acquiring one or more valency electrons. For example, fluorine, 2.7 adopts the electronic configuration of neon, 2.8 when it gains a valency electron.

Consider an ionic chemical reaction. If a sodium atom, 2.8.1 comes into contact with a chlorine atom, 2.8.7, and gives up its valency electron to the non-metal, the result is the release of energy and the formation of two stable ions, Na 2.8^{+} and Cl $2.8.8^{-}$. These then attract each other to form a stable ionic compound, sodium chloride, NaCl, common salt, which crystallizes into a regular cubic structure.

Metals are then electropositive elements usually with only one or two valency electrons which can achieve a low energy configuration more readily by losing, rather than gaining electrons. This can serve as a chemical definition of a metal.

In summary, if the outermost shell contains one or two valency electrons, the element is an electropositive metal, whereas with 6 or 7 valency electrons, the element is an electronegative non-metal. The question arises as to the situation with 3, 4 or 5 valency electrons in the outermost shell; for example lead 2.8.18.32.18.4 is a metal while silicon 2.8.4 is a non-metal.

Whereas an ionic bond is formed by transferring electrons to achieve a stable configuration, covalent bonds are formed by sharing the outermost electrons. For example, carbon which has a valency of four, crystallises as diamond by forming covalent bonds with four other carbon atoms each of which contributes one electron to form a stable 8 electron ring. Silicon, grey tin and germanium atoms show similar bonding.

Since the valency electrons are each involved in individual bonds and none is available for conducting electricity, covalent substances are insulators and non-electrolytes. However, consider as an example a phosphorus atom added to a a sample of silicon. Four of the 'impurity' five valency electrons will form covalent bonds with a corresponding number of silicon electrons, but the remaining electron, loosely attached and easily ionised off the phosphorus atom will be free to move through the silicon lattice. In contrast an atom of lower valency will generate an 'electron hole; which is also capable of breaking free and moving through the lattice. Both of these are 'impurity semiconductors', the basis of the transistor.

Covalent bonds also constitute the binding process for many types of molecules and compounds e.g. between carbon and hydrogen. Most organic compounds are completely covalent. Sometimes simple covalent molecules can be made to unite with molecules of the same type to form large chain-type structures (polymers) e.g. the gas ethyne C_2H_2 can be made to polymerise to form polythene.

Highly electropositive metals such as the alkali metals do not form covalent bonds, and this type of bonding is not very important generally as far as metals are concerned. However, covalent bonds form when weakly electropositive metals such as gold, silver and mercury, combine with weakly electronegative elements.

Two other weak types of bonding should also be mentioned for completeness, namely van der Waals and hydrogen bonding, but these need not be considered here in relation to metals.

Electronic structure and bonding have been major themes of investigation in the 20th century and among the many scientists who have contributed Gilbert Newton Lewis (1875–1946; FMRS 1940), a theoretical chemist in the USA began studies on the electronic theory of valency in 1916. Irving Langmuir (1881–1957; FMRS 1935) built on Lewis's work suggesting that chemical reactions depend on outer shells of electrons. Also in the USA the research of Linus Pauling (1901–1994; FMRS 1948), included bonding in metallic materials, such as the transition metals, in terms of electron orbitals; his book *The Nature of the Chemical Bond* (1939) was highly influential. In England the advances in theoretical chemistry of Charles Alfred Coulson (1910–1974; FRS 1950) relating to molecular systems included aspects relevant to metals.

The metallic bond and metallic properties

As discussed above the typical metal atom has far too few electrons in its outer shell for it to form a covalent bond with another metal and ionic bonding is only possible between metals and non-metals. Yet solid metals are very strong, indicating strong bonding between the atoms and again the valency electrons play a dominant role. The fundamental idea which was first put forward by Paul Karl Ludwig Drude (1863–1906) and Hendrik Antoon Lorentz (1853–1928; FMRS 1905) stemming from work begun in the last decade of the 19th century, is that the individual metal atoms give up their valency electrons to form a sort of gas or plasma of freely moving electrons surrounding a 'grid' of positively charged metal ions. They considered that the electrons obeyed the laws of classical mechanics and it was not until 1928 that Arnold Sommerfeld (1868–1951; FMRS 1926) applied quantum mechanics to the theory – this resolved a difficulty concerned with the specific heat of the free electron gas. Since that time there have been many developments in the theory which are not discussed here, but, although it is an oversimplification the central idea of the lattice of ions in a 'sea' of electrons is helpful.

The model readily accounts for many of the known physical properties of metals – their high electrical conductivity, for example, resulting from the relatively easy movement of electrons through the structure. The delicate balance of repulsion between neighbouring ions and attraction between ions and the sea of electrons is disturbed by an imposed stress, thereby stimulating a restoring force. From this metals are expected to exhibit a degree of elasticity, as indeed they do.

Metals typically consist of a huge number of crystals (crystals from the Greek 'clear ice'). During cooling from the liquid state metals normally crystallise, although with extremely rapid cooling rates amorphous (glassy) structures can be obtained in some metallic alloys. The size and shape of the crystals depends on the solidification conditions and also on subsequent mechanical and thermal treatments in the solid state. As

George Pearson

James Bottomley

William Barrett

James Ewing

John Hopkinson

Bertram Hopkinson

there are no directional chemical bonds, only an isotropic attraction of positively charged ions to a negatively charged field the metals generally crystallise in simple close-packed structures (see below). If a close-packed crystallographic plane in such a metal crystal is imagined, it will contain a lattice of identical ions so that sliding in that plane by one or more lattice spacings will return the symmetry, and hence should be readily possible. Such considerations lead to the predicition that metals will be ductile, which is in fact the case.

Similar considerations lead to the conclusion that any particular ion could be substituted by one of another element with minimum disturbance of the lattice. This is in accord with the observation that many metals readily form alloys with other metals over a wide range of compositions. In 'solid solutions' the solute atoms either take up positions on the solvent metal lattice, or if the solute atoms are sufficiently small they can occupy interstitial sites.

The nature of close packed structures is a key feature for the understanding of many metallic properties. The structures can be simulated by imagining the packing of the red balls in a triangle at the beginning of a game of snooker; the balls arrange themselves into a perfectly close-packed array in which (should the plane be infinitely extended) it would be obvious that each ball is surrounded by six others in a hexagonal pattern. Consider moving from this two-dimensional array by adding another layer of balls, placing them in the hollows between the balls in the first layer; the pattern of the first layer is reproduced but displaced to one side in relation to the lower layer. If the first layer is labelled A and the second layer B, there is a choice when a third layer of balls is added, since there are two sets of hollows. With one of these sets the balls in the third layer will be directly above those in the first layer, giving ABABAB... stacking. With the other set of hollows the sequence becomes ABCABC... in which the fourth layer is directly above the first.

Both of these structures exhibit maximum stacking density; whereas a random assembly of balls would have 36% free space, maximum packing density corresponds to 26% Each ball (atom) is in contact with 12 others i.e. has a coordination number of 12. Crystals with ABABAB stacking are known as hexagonal close-packed (hcp) whereas the ABCABC types are face-centred cubic (fcc). Examples of hcp metals are beryllium, cadmium, magnesium and zinc. Fcc metals include calcium, copper, gold, nickel, lead, platinum and silver.

There is a third simple crystal type which is also common among metals – it is body-centred cubic (bcc) in which the atoms exist at the corners of the basic cube with another in the centre of the body of the cube. It is not constructed from close packed planes so it is less closely packed than hcp or fcc; its free space is as high as 32% and its coordination number is only 8; however, it has six next neighbour atoms at distances slightly greater than that of the nearest neighbours. Examples of bcc metals are the alkali metals, lithium, sodium and potassium; and barium, chromium, molybdenum, vanadium, tungsten and tantalum.

Samuel Christie

Augustus Matthiesen

Robert Hadfield

William Hatfield

Some metals assume different crystal structures depending on the ambient temperature. This is particularly common among the transition metals; the best known example is iron, which <910°C and >1400°C is bcc, while it is fcc between these temperatures. Plutonium has no less than seven changes in crystal structure with increasing temperature. The phenomenon is referred to as allotropy or polymorphism.

The type of crystal lattice has an important influence on the plastic deformation behaviour of a metal. A simple picture is obtained by considering the process of slipping of close-packed planes of atoms over one another, but detailed understanding involves the consideration of the movement of defects termed dislocations. For example, an edge dislocation is associated with an uncompleted plane of atoms. The screw type of dislocation is important also in deformation and in crystal growth. The presence of crystal (grain) boundaries in polycrystalline metals is also of major importance in relation to mechanical properties. The subject of deformation is discussed more fully in Chapters 13 and 14.

Crystals also contain features termed point defects: vacancies and interstitials. The term vacancy is used when an atom is missing from a lattice site, while the term interstitial is used to describe an atom forced into an interstitial position in the lattice. Vacancies are a thermodynamically stable feature of a lattice, in contrast to dislocations. The equi librium concentration of vacancies increase exponentially with increase in temperature (for example, in copper at room temperature vacancies are 100,000 lattice spacings apart, whereas at 1007°C they are separated by only 1 atomic spacing. Vacancies play a key role in diffusion and also in the phenomenon of deformation by creep, which is discussed in Chapter 14.

In concluding this brief discussion of the properties of metallic materials it is appropriate to mention their optical behaviour. A beam of light falling on the surface of a metal would cause its free electrons to oscillate thereby absorbing energy from it at all wave lengths and so making the metal opaque. Subsequently the electrons would fall back to their lower energy states and emit nearly all the radiation they had received. This is a way of saying that light is almost totally reflected by nearly all metals giving rise to the familiar metallic lustre. Interestingly the only two metals which have readily recognisable, and very beautiful, colours were probably the first to be discovered by man – gold and copper. In the 19th century William Chandler Roberts-Austen (later Sir William,1843–1902; FRS 1875) included in his many interests the colours of metals and alloys in relation to his work at the Royal Mint and to art.

Alloy development – up to the mid-20th century

The latter half of the 19the century saw the opening up of the major field of the detailed scientific study of the structure,constitution and properties of alloys, including, importantly, alloy steels. However, in the earlier part of the century, some interesting systematic, albeit empirical work on alloy development had been done, notably by Michael Faraday (1791–1867; FRS 1824) and James Stodart (1760–1823; FRS 1821) on the alloying of steels. Stodart, a man of considerable scientific attainment and practical experience, was a maker of surgical instruments and a cutler, with his business at 401 Strand, London. The work done by the partnership of Faraday and Stodart, in which Faraday played the predominant role in the planning, conduct and presentation of the research programme, appears to have developed from the considerable interest that had arisen in 'wootz', a high-carbon steel from India, whose production involved a melting process. Sir Joseph Banks (1743–1820; FRS 1766) received a supply of this steel, late in the 18th century, and arranged for investigations of its nature and properties to be carried out by George Pearson (1751–1828; FRS 1791) and by Stodart. Pearson in a paper in 1795 referrred to Stodart forging a penknife from wootz, and finding it superior to any other steel then used in England; by 1820 Stodart was producing surgical instruments and other items from this material. These included the knife edges for the original invariable pendulum produced by Captain Henry Kater (1777–1835, FRS 1814), who presented a Bakerian lecture in 1820 'On the best kind of steel, and form, for a compass needle'.

At the time when the partnership work with Faraday began Stodart had previously published (in 1804 and 1805) papers on metallurgical topics,and it is noted that he tried the technique of heating steel in a salt bath. His work had included an interest in temper colours, a subject on which which Humphry Davy (later Sir Humphry, 1778–1829; FRS 1803) with Thomas Thomson (1773–1852; FRS 1811) had also reported. Stodart showed that the colour of the steel did not change with tempering when air was excluded, and therefore considered that the colours resulted from oxidation. Near the end of 1818 Stodart requested help from the Royal Institution with regard to the chemical aspects of the investigation of wootz, and Faraday, not yet director, and some thirty years younger than Stodart, had the task of analysing wootz samples and of 'duplicating' the material. He also had carried out chemical analysis for Josiah Guest (later Sir Josiah, 1785–1852; FRS 1830) of the famous Dowlais Ironworks in South Wales, and visited the works in 1819, thus obtaining some knowledge of ferrous metallurgical processing; he also made other visits including a copper works near Swansea. In the same year Faraday published two metallurgical papers, one on the separation of manganese from iron, and the other on an analysis of a wootz steel. In 1820 and 1822 respectively Faraday and Stodart published papers on 'Experiments on the alloys of steel made with a view to its improvement' and 'On the alloys of iron and steel' the latter being read before the Royal Society.

The work of Faraday and Stodart was concerned with synthetic wootz, and also with a substantial range of alloy steels, and included etching with acids and the use of heat tinting. They stated

> In proposing a series of experiments on the alloys of iron and steel with various other metals, the object in view was twofold: first, to ascertain whether any alloy could be artificially formed better, for purpose of making cutting instruments, than steel in its purest state; and second, whether any such alloys would, under similar circumnstances prove less susceptible of oxidation. New metallic combinations for reflecting mirrors were also a collateral object of research.

At the Royal Institution small ingots (or buttons) less than one pound in weight (< 450 g) were melted and cooled in the crucible in a coke fired furnace with bellows, described as a blast furnace. Wootz and other ferrous materials, including meteoritic iron, were used as the base material, while the wide range of additions included carbon, chromium, nickel, titanium, silver, gold and the platinum group metals (except ruthenium); manganese was not mentioned, and the later investigation of Faraday's specimens by Sir Robert Abbott Hadfield (1858–1940; FRS 1909) showed <0.1wt % of this element. Many metals of interest in modern alloy steels were difficult to obtain, but the noble metals were available, and William Hyde Wollaston (1766–1828; FRS 1793), provided the platinum group samples. Faraday appears to have been aware of the importance of accurate heat treatment control, referring, for example, to steel containing rhodium as requiring a higher tempering temperature than for the best wootz. Observations of the rusting behaviour of the steels showed some good results, for example, with iron–platinum, and tests were also made of resistance to acid attack. However, it seems that the work on additions of chromium was incomplete when the project was discontinued, so that its valuable role, which led many years later to the development of stainless steels by other workers, was not exploited. Faraday made some 'macroscopic' observations of samples using optical microscopy. In the later stages of the work by Stodart and Faraday, some of the alloys were made on a larger scale in the works of Sanderson in Sheffield, and various artefacts were produced by another Sheffield firm including razors containing small amounts of platinum group metal.

In 1863 John Percy (1817–1889; FRS 1847) wishing to make full analyses of Faraday's alloys, enquired as to where they were located, but found that Faraday did not know. However, Faraday had forgotten that he had packed and labelled a box of specimens in the Royal Institution. More than a century later than the preparation of these specimens, Sir Robert Hadfield obtained permission to make a detailed metallurgical investigation of the contents of the box; a few more specimens found in the Science Museum at South Kensington were also examined. The box contained 79 specimens, comprising alloys of iron with nickel, chromium, copper, gold, silver and the platinum metals. Since the whole series weighed just under 8 lb (~ 3.6 kg) tests had to be done on small quantities of material, but included chemical analyses, measurements of hardness, density, microstructure, corrosion resistance, and electrical and magnetic properties. The results were communicated to the Royal Society in 1931 and also in Hadfield's book *Faraday and his Metallurgical Researches* in the same year. Some specimens were forged into rods on which dilatometric and tensile tests could be made and some thermal analysis. From one platinum-containing specimen, strips were forged which were made into blades for twenty miniature pocket knives, which Hadfield presented to colleagues and distinguished persons. Hadfield regarded Faraday's work as of pioneering significance in the field of alloy steels.

Robert Hadfield's work was of particular significance in ferrous alloy metallurgy, opening up the major area of the detailed development of alloy steels. In 1882 he discovered the remarkable properties of steel containing certain certain proportions of manganese, which led to a patent in 1883. Arising from a problem in the works, Hadfield had initiated a sytematic study of alloys of iron with manganese, and with silicon respectively, aiming to produce a hard steel for applications such as tramway wheels and grinding discs. His detailed studies eventually covered manganese contents up to 36 wt%. In 1882 he found that increasing the manganese content beyond the usual limit, to 2.5–7.5 wt%, produced a brittle material, but a further increase up to 12–14 wt% manganese, with about 1% carbon, gave a steel which although apparently soft, was not abraded by a file; quenching from 900–1000°C softened the alloy, in contrast to the usual quench hardening effect associated with previously known steels. In a tensile test the steel drew out uniformly, in contrast to the local elongation or 'necking' that occurred in most metals. Also the material was non-magnetic. Progress in the application of manganese steel was slow at first, mainly owing to the difficulties associated with its working. The essential characteristic of the steel of becoming locally intensely hard when deformed, led to wide applications in railway and tramway crossings, in crushing and dredging machinery, etc. and its use in the 1914–1918 war for British military helmets.

Hadfield's work on manganese steels was typical of his metallurgical investigations, making an extensive series of alloys with increasing proportions of the added metal, and carrying out systematic tests to determine their mechanical,electrical and magnetic properties, thus providing a great mass of industrially important data.The next investigation of this kind which dealt with the alloys of iron and silicon did not aim to produce a very hard steel, but alloys were obtained (e.g. with ~3.5 wt% silicon), which later became important in transformers because of their valuable electrical and magnetic properties.

These two investigations were followed by others, concerned with alloys of iron with aluminium (1890), chromium (1892), nickel (1899), and tungsten (1903). None of these, however, yielded results as interesting as those for the alloys with manganese and silicon. One difficulty was that of obtaining the alloying metals in a form which did not involve the introduction of other elements in significant quantities. Chromium, for example, could only be obtained as ferro-chrome with a high proportion of carbon, leading to steels comparatively rich in that element; thus, although the high corrosion resistance of some of the chromium alloys was observed, the mechanical properties were not good.

From the time of Hadfield's discovery of the special features of the 12–14 wt% manganese steel, considerable interest in the properties of this and other alloy steels was shown by a number of eminent physicists. Hadfield himself was active in joint investigations or in providing materials for study. James Thomson Bottomley (1845–1926; FRS 1888) was the first to investigate the properties of the patented manganese steel. Confirmation of the non-magnetic properties was obtained, for example by John Hopkinson (1849–1898; FRS 1878), William Fletcher Barrett (later Sir William; 1844–1925; FRS 1899) and James Alfred Ewing (later Sir James, 1855–1935; FRS 1887), although Ewing showed that with the use of high fields the alloy could be magnetised. The work done by Barrett covered magnetic measurements and other physical properties of steels spanning manganese contents up to 21 wt%. In addition Barrett also made property measurements of some ternary and quaternary alloy steels produced by Hadfield containing elements such as silicon, aluminium, tungsten; those containing silicon or aluminium were found to have low magnetic hysteresis and high resistance. The observations on silicon-containing steels, together with the fundamental work of John Hopkinson on this type of material, was important in the development of these steels for use in transformers. Bertram Hopkinson (1874–1918; FRS 1910), the eldest son of John was among Hadfield's collaborators on investigations of the electrical and magnetic properties of various alloys.

At Sheffield in the university as Head of the Metallurgy Department, John Oliver Arnold (1858–1930; FRS 1912) played a role in the rapidly developing world of alloy steels, concerning the addition of vanadium as an element in tool steels and also of molybdenum as an alloy addition. Also he championed the cause of maintaining high quality in steel and was always ready to defend the local industry against attempts to obtain trade by supplying inferior cutlery products in place of the genuine material.

The wide-ranging researches of William Roberts-Austen included extensive investigations of structure-property relationships. A notable example was the fundamental work begun in 1890, at the request of the Institution of Mechanical Engineers, on the effects of small additions of certain elements on the mechanical and physical properties of the common metals and their alloys. The results of the work of the Alloys Research Committee were incorporated in a series of five reports (dated 1891, 1893, 1895, 1897 and 1899 respectively) from Roberts-Austen to the Institution, giving a mass of valuable information. Subjects on which experimental results were reported included: silver and gold and their impurities; the effects of arsenic; antimony and bismuth on copper; the effects of alloying aluminium with iron, copper and nickel; brasses; diffusion; the relationship between the melting point of alloys and the atomic volumes of the constituent metals; and treatments of certain steels. Following Roberts-Austen's death a sixth report was completed by William Gowland (1842–1922; FRS 1908). The work of this committee was transferred to the National Physical Laboratory in 1902. Henry Cort Harold Carpenter (later Sir Harold, 1875–1940; FRS 1918) continued the work leading to the production of the seventh report concerned with steel, and the eighth report which was a comprehensive study of copper–aluminium alloys. Subsequently Walter Rosenhain (1875–1934; FRS 1913) took over responsibility for the work and the ninth report dealt with the ternary alloys of copper, aluminium and manganese, while the tenth dealt with alloys of aluminium and zinc.

The phenomenon of age-hardening (precipitation hardening) of aluminium alloys was discovered by Alfred Wilm in Germany in 1906; he found that an aluminium–copper–magnesium alloy quenched from around 500°C and allowed to remain at room temperature for several days showed an increase in hardness. Wilm carried out further detailed experiments and published his results in 1911. Heat treatments of quenched aluminium alloys subsequently aged at room temperature or elevated temperatures became of great importance associated with an ongoing field of alloy development as the 20th century progressed.

In England the requirements of the 1914–18 war directed the attention of the investigators of the Alloys Research Committee particularly to the light alloys of aluminium,

James Dewar

John Fleming

Joseph Petavel

John Tyndall

and the eleventh report presented the results of a comprehensive study of a large number of these alloys. Several new aluminium based alloys were described, the most important of which was the 'Y' alloy, containing copper, nickel and magnesium and showing resistance to high temperatures.

In 1921 the Committee became the Alloys of Iron Research Committee, and there ensued a series of investigations of the constitution of the binary alloys of iron with other elements, using materials of the highest attainable purity, and involving a number of improvements in technique.

In the earlier part of the 20th century the name of William Herbert Hatfield (1882–1943; FRS 1935) became particularly associated with the development of the stainless steels. This followed the fundamental discovery made in 1912 by Harry Brearley (1871–1948), then employed by Messrs Firth, Sheffield, that a steel containing 12.8 wt% of chromium did not rust in air and was only attacked with difficulty by the acid reagents used to etch specimens for microscopy; this steel came into general use for cutlery applications. Brearley moved to the Sheffield firm of Brown Bayley's and a steel containing 12–14 wt% chromium and about 0.3 wt % carbon was developed. In the research and development on stainless cutlery steels carried out at that time one approach led to a ferritic steel, known as 'stainless iron' in which a low carbon content gave a combination of corrosion resistance with good ductility. Another approach led to Krupp of Essen patenting in 1912 a wide range of compositions of steels containing both nickel and chromium. Hatfield, as director at the Brown Firth research laboratories made a particular contribution by investigating thoroughly the effects of composition and treatment on the properties of austenitic steels containing nickel and chromium. He selected the steel with 18% chromium and 8% nickel as having the best properties, and this, under the trade name of 'Staybrite' adopted by his firm, and the more general designation '18:8', became the most important of the stainless steels. As it was intended to use the steel in a great variety of industries, it was necessary to examine, under various conditions of temperature, the chemical action of such acids and other corrosive agents as might be encountered. Hatfield published many papers in this area, and in 1928 noted that over 15,000 of these experiments had been made in his laboratory. He also carried out work in the field of corrosion problems arising in stainless steel weldments.

Hatfield was also responsible for producing two other steels with special properties. One was a non-magnetic alloy with a high coefficient of expansion, approaching that of aluminium alloys, so that it could be used with them in aero-engine valve sleeves.The second steel had great strength at high temperatures and proved useful for hot shears and similar applications for which resistance to thermal shock is needed. He also improved the process of nitriding.

Electrical and magnetic properties

Measurements of physical properties of metals, and also of alloys as a function of chemical composition played an important part in the development of metallurgical science and

John McLennan

Herbert Fröhlich

John Bardeen

Peter Kapitza

many Fellows of the Royal Society were involved. Of the wide range of physical properties those relating to electricity and magnetism formed a major theme, progressing from the early days of the 'electrical age' at the beginning of the 19th century. Here the contributions of Fellows on this theme are reviewed, including in the latter part of the century the phenomenon of superconductivity. It is also of interest to note a very significant discovery made early in the century, namely that reported in 1821 by Thomas Johann Seebeck (1770–1831) of thermoelectricity, or as he referred to it 'thermomagnetism.'

From experiments using various various combinations of metals, Seebeck had found that an electric current flows when two different metals are joined in a closed circuit with the two junctions held at different temperatures. This discovery stimulated great interest and experiments by others soon began, involving a number of Fellows in England and in Europe. These included Michael Faraday (1791–1867; FRS 1824), Hans Christian Oersted (1777–1851; FMRS 1821) and Heinrich Gustav Magnus (1802–1870; FMRS 1863). James Prescott Joule (1818–1889; FRS 1850) working in the late 1850s drew up a thermoelectric series of 51 common metals and alloys, including cast irons, wrought iron and steels. Seebeck's discovery led to the development of the crucial technique of measurements of high temperatures using thermocouples, (see Chapter 10). Later in the century Charles William Siemens (later Sir William, 1823–1883; FRS 1862) suggested another technique for temperature measurement based on changes of electrical resistance of metals with changing temperature.

An early example of investigations of electrical and magnetic properties is the work of Samuel Hunter Christie (1784–1865; FRS 1826) one of Faraday's friends. He was a mathematician and physicist who held an appointment at the Royal Military Academy, Woolwich; it appears that his research on the magnetic and electrical properties of metals (Bakerian lecture, 1833) involved some of Faraday's alloys. Faraday himself later did research on the magnetic properties of single crystals of several metals and his work on the magnetic anistropy of bismuth was reported in 1849. Dominique Francois Jean Arago (1786–1853; FMRS 1818) carried out research in the field of electricity and magnetism which included the discovery in 1822 at Greenwich of the dampening effects of metals on a compass needle. Several years later he showed that the rotation of a non-magnetic material, particularly copper, created a magnetic effect on a magnetised needle.

William Thomson (later Lord Kelvin, 1824–1907; FRS 1851), beginning in 1842, made fundamental studies of electrical and magnetic fields. Then from 1849 to 1859 he developed the discoveries and theories of Faraday on paramagnetism and diamagnetism into a fuller theory and used the terms magnetic permeability and susceptibility.

Among the varied research activities of John Tyndall (1820–1893; FRS 1852) were investigations of diagmagnetic behaviour in crystalline materials; his Bakerian Lecture in 1855 dealt with 'the nature of the forces by which bodies are repelled from the poles of a magnet...'. Tyndall also worked on the thermal properties of crystals.

A classical series of studies of certain physical properties, including the first systematic investigation of electrical conductivity (1859–1863) was carried out by Augustus

John Van Vleck

Edmund Stoner

Kurt Mendelssohn

Leslie Bates

Willie Sucksmith

Matthiesen (1831–1870; FRS 1861) working in a laboratory in his home at Torrington Place, London. The preparation of copper of the highest possible conductivity had become of great practical importance in relation to telegraphy. Matthiesen showed that discrepancies of previous observations and the low conductivity of certain samples assumed to be pure, resulted from the presence of trace amounts of impurities. He attempted to relate conductivity to the nature of the alloys, including the concept of 'solidified solutions' of one metal in another.i.e 'a most intimate mixture, such as would occur in the sudden conversion of a liquid into a solid… in fact, a perfectly homogeneous diffusion of one body into another… Even under the most powerful microscope it would be impossible to distinguish the components of a solid solution.'

Also in the area of electrical and thermal properties the work of Gustav Heinrich Wiedemann (1826–1899; FMRS 1884) with Rudolph Franz reported in 1853 that, at a constant not very low temperature, the electrical conductivity of metals is approximately proportional to their thermal conductivity (the Wiedemann–Franz ratio). Wiedemann also in the field of the relationship between magnetisation and mechanical phenomena carried out experiments on the torsion and magnetisation of steel and iron. Anders Jönas Ångström (1814–1874; FMRS 1870), whose name is associated with the unit of length in his research on spectroscopy, also demonstrated the proportionality of electrical and thermal conductivities in his early work.

George Carey Foster (1835 –1919; FRS 1869) used a Wheatstone bridge for the accurate measurement and comparison of standards of electrical conductivity and constructed a modified apparatus known as the Carey Foster bridge.

In the early 1920s Evan James Williams (1903–1945; FRS 1939), working at Swansea for his MSc, investigated the influence of small additions of metals on the electrical conductivity of mercury at various temperatures. Also he made a mathematical study of the movement of liquid metal conveying an electric current when located in an magnetic field.

Magnetic and electrical properties were a major interest of John Hopkinson (1849–1898; FRS 1878). His earliest important paper on the magnetic properties of metals, entitled 'Magnetization of iron' (1885) became a landmark in the development of the subject; it reported on the permeability, magnetic hysteresis and electrical resistance of a number of steels, including the manganese-containing steel invented by Hadfield and various cast irons. The magnetic measurements were made by a novel method, using samples in the form of short bars, enclosed in a massive yoke of soft iron. Subsequent research was concerned with the magnetic properties at various temperatures of nickel, iron and alloys of nickel and iron. For example, he found that in iron ~25 wt% nickel alloy, which was non-magnetic at room temperature, became strongly magnetic on cooling in a freezing mixture to around -50°C. Hopkinson introduced the term coercive force, and his name is attached to the peak of susceptibility close to the Curie point.

Magnetism was also the main field of research of James Alfred Ewing who was one of the first to observe the phenomenon of hysteresis. In 1881, when working on the effect of stress on the thermoelectric properties of metals, he discovered that the thermoelectric effect lagged behind the applied stress. He then studied the transient currents produced by twisting a magnetised wire and, finding a lag also, he introduced the term 'hysteresis' from the Greek 'to be late'. Turning to magnetism he found in 1882 that the area enclosed by the hysteresis loop was proportional to the work done during a complete cycle of magnetisation and demagnetisation. The work of Peter Barlow (1776–1862; FRS 1823) included a study of the dependence of the magnetic properties of iron on temperature. Also Thomas Andrews (1847–1907; FRS 1888) investigated the magnetic properties of iron and steel.

Success in liquifying gases opened up the field of cryogenics, and the properties of matter at low temperatures became an important field of research on gases. In England most of the early liquifaction work had been done at the Royal Institution, especially by Michael Faraday, who by 1845 had liquefied all the known gases, except oxygen, nitrogen, hydrogen, nitric oxide,carbon monoxide and methane. It was also the Royal Institution that became the centre of the experimental work in cryogenics carried out by James Dewar (later Sir James, 1842–1923; FRS 1877). Dewar was the first to succeed in liquefying hydrogen, and in 1898 he reached a temperature of 13 K. His principal interest was in the properties of matter, in the previously uninvestigated range of very low temperatures, and his pioneering work in this area was made possible by his invention in 1892 of the vacuum-jacketed flask. Between 1892 and 1895 Dewar collaborated with John Ambrose Fleming (later Sir John, 1849–1945; FRS 1892) in research on the electrical and magnetic properties of metals and alloys. The curves of resistance versus temperature indicated that for pure metals the resistance should vanish at absolute zero. They obtained accurate data in the temperature range from ~470 to 73 K on conduction, thermoelectricity, magnetic permeability and dielectric constants of metals and alloys. Joseph Ernst Petavel (later Sir Joseph, 1873–1936; FRS 1907) worked with Dewar and Fleming at the Royal Institution measuring physical properties at low temperatures, for example, the electrical resistance of bismuth in a magnetic field of pure mercury.

Robert Hadfield carried out important collaborative investigations of properties at cryogenic temperatures. In 1904 with Dewar he published a paper on the mechanical and other properties of iron and various alloys, including some of the non-ferrous type, in liquid air. In 1921 he was a co-author with Heike Kamerlingh Onnes (1853–1926; FMRS 1916) and H.R. Woltjer of the Leyden cryogenic laboratory of a paper on the magnetic properties of alloys of iron with nickel and manganese in liquid hydrogen (-253°C); then in 1933 with Wanda Johannes de Haas (1878–1960) he reported reported results on physical properties, particularly magnetic, of alloys of iron especially with nickel and manganese in liquid helium (-269 °C) Hadfield's involvement in low temperature investigations also included mechanical properties (see Chapter 13).

Investigations of the behaviour of gases and of the electrical properties of metals at very low temperatures were the major areas of research for Onnes who worked from 1882 at Leyden. He built apparatus for the liquifaction of air on a large scale, and liquified hydrogen in 1906 and helium two years later. Having prepared very pure mercury to reduce the effect of impurities, he found in 1911 a discontinuous decrease in resistance to zero, at a low temperature. This phenomenon of superconductivity (first called supraconductivity) was found for various metals, with different transition temperatures. He discovered various effects associated with the superconductive state including its disappearance in a magnetic field. He also studied the low temperature conductivity of lead, nickel and manganese – iron alloys.

As the 20th century progressed low temperature physics relating to electrical (notably superconducting phenomena), magnetic and other physical properties at very low temperatures down to <4 K became a major area of research. In the extensive experimental and theoretical investigations many scientists who became Fellows of the Royal Society were prominent. Many advances in the fundamental understanding of the solid state were made, not only relating to metals and alloys, but also in materials such as semiconductors (see Chapter 17). In England, major university centres for low temperature studies have been at Oxford (the Clarendon laboratory) and Cambridge (the Mond laboratory); important work on magnetism has also been done at Leeds. There have been major activities elsewhere, for example, on continental Europe and in the USA. Important advances in the development of new magnetic alloys have also been made.

Concerning superconductivity, Max von Laue (1879–1960; FMRS 1949) in Berlin wrote important papers relating to research by Walther Meissner (1882–1974). John Cunningham McLennan (later Sir John, 1867–1935; FRS 1915) at Toronto made experimental investigations of superconductivity effects, for example, in lead and lead–bismuth alloys. In the development of the theory of superconductivity. John Bardeen (1908–1991; FMRS 1973) played an important role. At the University of Illinois, Urbana he began a collaboration with Leon Neil Cooper (1930–) and John Robert Schrieffer (1931–) in 1957 which culminated in the joint award of a Nobel Prize for Physics in 1972. Herbert Frölhich (1905–1991; FRS 1951) had previously pointed out that electron–phonon interaction could produce a weak attraction between two electrons and that this was fundamental in the origin of superconductivity; the work of Bardeen, Cooper and Schrieffer provided evidence of electron pairing. The research of Kurt Alfred Georg Mendelssohn (1906–1980; FRS 1951) included the superconducting behaviour of alloys, and studies of thermal resistivity of a range of metals and semiconductors as a function of temperature and magnetic fields. Among other whose work involved superconductivity, Heinz London (1907–1970; FRS 1961) investigated thin metal films.

The research of John Martin Rowell (1935–; FRS 1989) in the USA in low-temperature physics has included superconducting tunnelling and its application; his pioneering experimental work has profoundly influenced the understanding of electrical transport in metals. Also Laszlo Solymar (1930–; FRS 1995) has worked on electrical properties of materials, including superconductor tunnelling. The work of Sir Martin Francis Wood (1927– ; FRS 1987) at Oxford Instruments plc is concerned with the technology of high magnetic fields and the development of industrial cryomagnetic equipment includes superconducting applications.

After World War II a contemporary Fellow, Alfred Brian Pippard (later Sir Brian, 1920–; FRS 1956) at the Cavendish laboratory, began to use microwaves to study su-

Frederick Lindemann
(Lord Cherwell)

Bryan Coles

perconductors, in particular conduction in a thin layer at the surface; the anomalous skin effect had been discovered by Heinz London (1907–1970; FRS 1961). Later, Pippard's research included magnetoresistance in metals, and he published a book on *Dynamics of conduction electrons* in 1964.

Also in the Cavendish laboratory Brian David Josephson (1940–; FRS 1970) in his research on superconductivity in 1962 set out to calculate the current due to quantum mechanical tunnelling across a thin strip of insulator between two superconductors. The prediction that an alternating current (a.c.) occurs in the barrier when a steady external voltage is applied became known as the Josephson a.c. effect. Also when a steady magnetic field is applied across an insulating barrier a steady current flows (the Josephson d.c. effect). The predictions were verified by other workers and important practical applications stemming from this work have included fast switching devices in computers. Josephson shared a Nobel Prize in Physics in 1973 with Leo Eskai (1925–) and Ivar Giaever (1929–).

In the field of magnetism Paul Langevin (1872–1946; FMRS 1928), a leading mathematical physicist carried out important work on paramagnetic and diamagnetic phenomena applying electron theory; he predicted the phenomenon of paramagnetic saturation, which was confirmed experimentally by Onnes. In early work Langevin discovered that X-rays liberate secondary electrons from metals.

In the USA John Hasbrouk van Vleck (1899–1980; FMRS 1967) from the 1920s working successively at the Universities of Minnesota, Wisconsin and Harvard, came to be considered as the father of modern magnetism. He applied quantum mechanics to the theory of electric and magnetic susceptibilities and also established the detailed theory of paramagnetism, including studies of rare earths, on a firm quantum mechanical basis. He developed all the fundamental formulae for the interactions between magnetic electrons and the surrounding ions in the crystal lattice. His research spanned widely across the field of magnetism, including ferromagnetism.

Van Vleck was a joint recipient of the Nobel Prize for Physics in 1977, his co-recipients being Philip Warren Anderson (1923–; FMRS 1980) who early in his career had worked with Van Vleck at Harvard and Sir Nevill Francis Mott (1905–1996; FRS 1936). The award was made 'for their fundamental theoretical investigations of the electronic structures of magnetic and disordered systems'. The research of Philip Anderson has ranged over a number of themes, including the development of theoretical treatments of antiferromagnets, ferroelectrics and superconductors, the Josephson effect, impurity atoms in metals, disordered structures, and low temperature properties of glasses, including spin glasses. Nevil Mott's many outstanding research contributions included work on the theory of metals and alloys, the theory of dislocations, superconductors, and magnetic and electrical properties of non-crystalline solids, used in applications such as computers and solar energy devices.

Werner Karl Heisenberg (1901–1976; FMRS 1955) also carried out important research interpreting ferromagnetism in relation to electron behaviour. At Cambridge, Ralph Howard Fowler (later Sir Ralph, 1889–1944; FRS 1925) included in his theoretical research ferromagnetism and magnetostriction.

Edmund Clifton Stoner (1899–1968; FRS 1937) at the University of Leeds, where he became Professor of Theoretical Physics, made fundamental contributions to the theory of magnetism. An important feature of his work was that his approach to ferromagnetism was within the general framework of the electron theory of metals. Thus the magnetic properties of metals and alloys could be correlated with properties such as specific heat, electronic energy band structures and Fermi surfaces. Enrico Fermi (1901–1954; FMRS 1950) made important contributions to the electron theory of the solid state. Stoner's work included an interest in ternary alloys which have high remanence and high coercivity, as required for permament magnet applications.

Important research contributions on magnetism were made by Willie Sucksmith (1896–1981; FRS 1940) and Leslie Fleetwood Bates (1897–1978; FRS 1950). In the early 1920s at the University of Leeds Sucksmith collaborated with Arthur Prince Chattock (1860–1934; FRS 1920) and Bates in work on the gyromagnetic ratio (the ratio of the angular momentum and the magnetic moment of the elementary magnet gyrostat) in various ferromagnetic materials. He continued to work in the field of magnetism at Leeds and later at Sheffield University. Materials studied included the copper–aluminium–manganese Heusler alloy (which had previously been studied by Fleming and Hadfield), nickel and paramagnetic materials. He also applied measurements of magnetic properties of ferromagnetic alloys phases to the study of alloy constitution. His devising of the apparatus which became known as the Sucksmith ring balance for measuring susceptibility was an important experimental development. Measurements could be made in vacuum or controlled atmosphere, and eventually the balance was used in the temperature range

Charles Lees

William Tilden

George Stokes

Ezer Griffiths

from 1 to 2000 K. He produced a later version for measuring the saturation magnetisation of ferromagnetics. In the period after the 1939–45 war at Sheffield Sucksmith's research included the themes of hysteresis in ferromagnetic materials, magnetic analysis of iron carbon alloys – the tempering of martensite and retained austenite, and magnetic properties of dilute ferromagnetic alloys.

Bates, after a period at Leeds, moved to University College London in 1924 where he pursued research on ferromagnetism. He began a long series of experiments on the ferromagnetic properties of manganese compounds, including measurements of specific heats and electrical conductivities. To obtain the high purity manganese he required he developed a method of preparing an amalgam by electrolysis which led him to investigate the properties of amalgams of ferromagnetic metals, including iron, cobalt and nickel. Having moved to a chair of physics at Nottingham in 1936 his work on amalgams continued, and from around 1940 this was extended into the theme of thermomagnetic measurements to determine the temperature changes during the hysteresis cycle. In about 1950 he began to use the Bitter powder pattern technique to investigate magnetic domain structures studying various materials including single crystals (eg. of silicon–iron and cobalt) and permanent magnet materials. From around 1954 Bates's work in association with the Atomic Energy Research Establishment at Harwell included measurements of the electrical magnetic and thermal properties of uranium and thorium and some alloys based on these metals.

Albert James Bradley (1899–1972; FRS 1939) co-operated with Sucksmith in correlating magnetic properties and crystal structure. Bradley proposed ideas for the structure required to achieve high coercivity for permanent magnets. Arising from this he made detailed studies of the constitution of relevant ternary systems, such as that of iron–nickel–aluminium, using X-ray and metallographic methods.

Beginning in the 1920s the research of Charles Sykes (later Sir Charles, 1905–1982; FRS 1943) on alloy metallurgy included investigations of electrical and magnetic properties of some zirconium-based alloys (e.g. containing aluminium); also electrical resistivity measurements of iron–aluminium alloys were part of his studies of order–disorder phenomena.

Piotr (Peter) Leonidovich Kapitza (1894–1984; FRS 1929) was a major figure in magnetism and low-temperature physics over a period of more than half a century. Born in Russia, he came to England in 1921 and was accepted by Lord Rutherford (1871–1937; FRS 1903) to work in the Cavendish laboratory at Cambridge, where he remained for 13 years. In his work on alpha particles he had the novel idea of producing large magnetic fields lasting only a very short time (impulsive magnetic fields). This work led him into pioneering research during the 1920s in solid state and low temperature physics, inluding the development of large machines to achieve high magnetic fields. With his new equipment he made an extensive study of the increase of electrical resistance of metals with increases in magnetic field using a pulsed field at temperatures down to that of liquid nitrogen. In this work on magneto-resistance he realised that lower temperature experiments would probably be required for a better understanding of metals and he soon

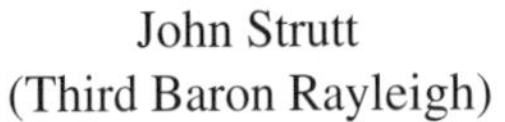

John Strutt
(Third Baron Rayleigh)

Percy Bridgman

began to work on hydrogen and helium liquefaction. He also carried out researches at high fields on the magnetisation of various substances and of magnetostriction, mainly of single crystal bismuth. The Royal Society Mond Laboratory was built to house the high field magnetic equipment and to provide cryogenic facilities to extend his work to lower temperatures. He developed equipment (completed in 1924) to produce liquid helium. In 1934 when he visited the Soviet Union, he was prevented by Stalin from returning to England. He became the director of a new institute in Moscow where he continued to make outstanding contributions to magnetism and low temperature physics; also research on plasma physics became his main personal interest.

Other Fellows whose research involved ferromagnetism were Lev Davydovich Landau (1908–1968; FMRS 1960) and Evgenii Mikhailovich Lifshitz (1915–1985; FMRS 1982); their interests also involved phase transitions in thermodynamic and crystallographic terms including second order changes. The research of Walter Charles Marshall (later Lord Marshall, 1932–1996; FRS 1971) in solid-state physics included magnetic properties, for example of the transition metals. In their work on low-temperature physics, Nicholas Kurti (1908–1998; FRS 1956) and David Shoenberg (1911–; FRS 1953) also carried out important research on magnetism. Metallic magnetism is the field of research of Gilbert Geroge Lonzarich (1945–; FRS 1989), including the electronic structure of metallic compounds.

Allan Roy Mackintosh (1936–1995: FRS 1991), following his investigation of the Fermi surface of metals using ultrasonic attentuation methods with Pippard at Cambridge, moved to Iowa State University. There, in the Ames laboratory, work had begun to make single crystals of the rare-earth metals. Mackintosh initiated research into the new area of the physical properties of these materials, particularly the magnetic behaviour, and became a leading expert. During his career, in which many years were spent in Denmark, he carried out extensive research on rare earth magnetism, including the use magneto-acoustic and magneto-resistance techniques as well as positron annihilation and inelastic neutron scattering.

In the field of electronic and magnetic properties of alloys, the theoretical research of Jacques Friedel (1921–; FMRS 1988), a contemporary Fellow of the Royal Society, has profoundly influenced understanding of the behaviour of electrons in solids. His work has included magnetism, electron transport, liquid and disordered metals. Rudolph Ernst Peierls (later Sir Rudolph, 1907–1995; FRS 1945) in his research on solid state physics included work on the theory of diamagnetism in metals. John Michael Ziman (1925–; FRS 1967) in his contributions to theoretical solid state physics has included the electronic properties of liquid metals.

The research of Bryan Randell Coles (1926–1997; FRS 1991) focused particularly on the electrical and magnetic properties of metals and alloys, involving extensive experimental work, linked with theoretical interpretation. Specific themes included the role of exchange effects in suppressing superconductivity, dilute alloys showing the Kondo effect, heavy fermion magnets and spin glasses; (he introduced the term spin glass relating to the state of frozen – in magnetic order.) The research of Michael Moore (1943–; FRS 1989) includes spin glasses.

Thermal and other properties

Physical properties, other than electrical and magnetic, were the subject of investigations by a number of Fellows of the Royal Society beginning in the 19th century. The work of Jean-Baptiste Biot (1774–1862; FMRS 1815) included measurements of the thermal conductivity of metals (1804) using a technique of maintaining one end of a bar at a known, high temperature, with thermometers in holes along the bar; he analysed the results quantitatively providing a basis for later work by Jean Baptiste Joseph Fourier (1768–1830; FMRS 1823) on heat conduction.

For Charles Herbert Lees (1864–1952; FRS 1906) the investigation of thermal conductivities was a major theme of his research; he carried out over a long period many measurements and showed both great experimental and mathematical ability. In his Bakerian Lecture (1908) he presented the results of extensive measurements of thermal (and also electrical) conductivities at temperatures down to that of liquid nitrogen. These measurements covered a number of metals of considerable degrees of purity, including high conductivity copper, silver (99.9 %), aluminium (99.9%), nickel (99.9%), high carbon steel and 70:30 brass. In discussing his data he made reference to electronic structure theory.

Pierre Dulong (1785–1838; FMRS 1826) collaborated with Alexis-Therèse Petit (1791–1820) beginning in 1815, in developing a reliable method of measuring specific heats and showed for a series of metals that the specific heats were inversely proportional to their respective atomic weights. From 1904–1913 Dewar devised and used a calorimeter to measure specific heats and latent heats at low-temperatures; he determined the atomic heats of elements and molecular heats of compounds between 80 K and 20 K; in 1913 he found that the atomic heats of the solid elements at a mean temperature of 50 K are a periodic function of atomic weights. Also in the field of low temperature studies Hermann Walther Hermann Nernst (1864–1941; FMRS 1932) and Frederick Alexander Lindemann (later Lord Cherwell, 1886–1957; FRS 1920) constructed a calorimeter and measured specific heats at very low temperatures. Other Fellows whose research included specific heats were: Albert Einstein (1879–1955; FMRS 1921); William Augustus Tilden (later Sir William, 1842–1926; FRS 1880); Peter Joseph Wilhelm Debye (1884–1966; FMRS 1933); Max Born (1882–1970; FRS 1939); Theodore von Kármán (1881–1963; FMRS 1946); Ralph Howard Fowler (later Sir Ralph, 1889–1944; FRS 1925) and Moses Blackman (1908–1983; FRS 1962).

The research career of Ezer Griffiths (1888–1962; FRS 1926) was predominantly concerned with various aspects of heat. His early work at Cardiff with Ernest Howard Griffiths (1851–1932; FRS 1895) involved measuring the specific heats of a number of metals at low temperatures to test the then current theories e.g of Nernst, Lindemann and Debye. From 1915 to 1953 Ezer Griffiths was at the National Physical Laboratory at Teddington and became a leading international authority on such subjects as heat insulation, heat transfer and evaporation. His work was of great importance to industry. It included measurements of the thermal conductivities of insulating materials and materials for furnace construction, metallic alloys (e.g. aluminium based and bronzes, including measurements of electrical conductivities, latent heats of fusion), physical properties of iron and a series of steels, pyrometry and furnace construction.

The velocity of sound in metals was the subject of interesting experiments by Biot. The laying of water mains in Paris in 1808 provided him with the opportunity of transmitting sound through a 951-metre length of cast iron pipe. From the time interval between the transmission and reception of the sound, in comparison with that through air he deduced that the velocity in the metal was approximately 10 times that in air; however, a factor limiting the accuracy of his experiment was the presence of lead used to join the individual pipes.

James Joule among his many scientific interests included extensive measurements of the specific gravity of a large number of substances, including metals.This work included measurements for liquid metals, some of these being of relatively high melting points. The procedure was later used by Robert Mallet (1810–1881; FRS 1854) to determine the density of molten cast iron, of silicate slags and of molten lead. Roberts-Austen included in his many research activities collaborative work to determine the densities of liquid metals.

Among Fellows who worked in the field of hydrodynamics and properties of fluids George Gabriel Stokes (later Sir George, 1819–1903; FRS 1851) made important contributions to the theory of viscous fluids in the period 1845–1850. Among his findings he derived an equation which came to be known as Stokes's law relating to the motion of small spherical particles through viscous fluids; this has become important in metallurgy, for example, concerning the flotation of non-metallic inclusions in metallic melts.

The research of Edward Neville da Costa Andrade (1887–1971; FRS 1935) on liquids in collaboration with Y. Schiong and Leonard Rotherham (1913–; FRS 1963) included the development of an oscillating sphere technique for measuring viscosities; in other collaborative work Andrade investigated the viscosities of liquid alkali metals as a function of temperature.

Osborne Reynolds (1842–1912; FRS 1877) in his work on fluids and heat included investigations on the condensation of steam on metal surfaces and of heat transfer between metal surfaces and liquids.

Diffusion

Pioneering work on the fundamental theme of diffusion was carried out by Thomas Graham (1805–1869; FRS 1836). He was particularly concerned with the constitution of matter and the manner in which atoms or molecules move. Notable among his researches was the discovery of the law of diffusion of gases: he showed the rate of diffusion to be inversely proportional to the square root of the density of the gas. His experiments on the passage of gases through small openings and through films greatly extended knowledge in this subject area. He showed the practicability of separating gases by diffusion, and oxygen was separated from the air using graphite. He also studied the manner in which liquids permeate membranes. Among other important research topics was the interaction of hydrogen with platinum and palladium (see Chapter 4). Striking features of his work were his originality and simplicity of his methods, nevertheless leading to important, fundamental results. In his later work he had the co-operation of his personal assistant, William Roberts-Austen.

Frederick Guthrie (1833–1886; FRS 1871) investigated the diffusion of zinc, lead, tin, sodium and potassium in liquid mercury. Also Roberts-Austen studied the diffusion of gold and platinum in liquid lead and bismuth, taking steps to minimise convection effects. He also did pioneering research on the diffusion of gold and lead in the solid state. Diffusion couples were made by fusing gold on to the end of cylinders of lead; the fusion process produced a layer of lead–gold alloy not more than 1 mm thick. The cylinders were heated for long periods at various temperatures: 250, 200, 165 and 100°C. The samples were sectioned and assayed to determine the distribution of gold, leading to a calculation of diffusivity using Fick's theory; diffusion was found to occur at a temperature as low as 100°C. Roberts-Austen presented his results in the Bakerian Lecture of 1896.

In ferrous metallurgy the cementation process in which wrought iron (or 'bar iron' as it was often known) was heated for several days at a bright red heat in contact with charcoal so that carbon diffused into the metal to produce 'blister steel' was an important industrial procedure, for example in Sheffield; blistering resulted from the reaction of carbon with slag particles to form carbon monoxide, and this, occurring sufficiently near the surface could lead to a gas pressure higher than the yield point of the steel. In the paper by Henry Clifton Sorby (1826–1908; FRS 1857) ‘On the microscopical structure of iron and steel’ (1887) he reported his observations on the microstructural features as a function of depth below the surface in a bar of Swedish iron partly converted into blister steel by cementation. In addition he referred to the structural gradient when white cast iron is decarburised by heating with iron oxide. John Arnold published an important paper in 1898 on ‘The microchemistry of cementation’ concerned with the carbon distribution in carburised Swedish iron; his results showed the relationship of time, temperature and the rate of diffusion of carbon. A paper in the following year by Arnold and A. McWilliam investigated the diffusion of other elements, reporting that sulphur, phosphorus and nickel diffused through iron at high temperatures.

Properties at high pressures

The properties of fluids at high pressures, including the compressibility of mercury, were studied by the French physicist Émile Hilaire Amagat (1841–1915; FMRS 1897). The imagination of Percy Williams Bridgman (1882–1961; FMRS 1949) at Harvard in the USA was aroused and he set about working at higher ranges of pressure at which new phenomena might be expected to occur. There were many practical difficulties; high tensile steels which were necessary for the construction of apparatus to withstand the stresses he had in mind were still in the early stages of development and were difficult to obtain commercially; there were no instruments available for measuring pressures of the required magnitude, and the existing methods of making joints, screwed closures and pistons were not sufficiently reliable. He tackled these problems with great energy and success, making much of his own apparatus. One of his earliest inventions was the self-tightening joint which allowed pressures to be developed and maintained far far in excess of those previously achieved. He next studied pressure-measuring devices and in two papers in 1909 he describe the construction and operation of a primary free-piston gauge designed for pressures up to 6,800 kg cm^{-2} (~670 MPa) and a secondary gauge based upon the variation of the electrical resistance of mercury with pressure. Two years later he reported an improved design of primary gauge which enabled him to measure pressures up to 13,000 kg cm^{-2} (~1270 MPa) and a new secondary gauge based upon the electrical properties of the alloy manganin. The resistance of manganin was found to be a linear function of pressure up to 12,000 kg cm^{-2} (~1180 MPa) and, by extrapolation, pressure measurements could be made up to 20,500 kg cm^{-2} (~2010 MPa).

With these new tools Bridgman began an extensive investigation of the thermodynamic properties of a wide range of liquids up to pressures of 12,000 kg cm^{-2} and at temperatures between 20 and 80°C. The results, published between 1911 and 1915, provided an huge amount of data on compressibilities, changes of state and melting curves. Bridgman’s next major work was concerned with the compressibilities of solids, melting phenomena and the occurrence of polymorphic transformations. Between 1912 and 1915 he published results on the melting curves of solids which showed that with the exception of bismuth, gallium and water, the curves rise with pressure. He then measured the measurement of the electrical resistance of metals and alloys and found a wide diversity of behaviour with increasing pressure. His final investigation published in 1959 was on the compression and alpha/beta transition in plutonium. In addition to being an inspired experimentalist Bridgman was fully aware of the importance of interpretation and his early speculation on the theory of the liquid and solid states demonstrate his keen interest in the scientific significance of his work. His achievements included the discovery of the polymorpism of many materials at high pressures and an early method of refining by zone-melting. Much of his work was done while solid state physics was in its infancy, and he provided a vast amount of data for the subsequent development of the field. He received the Nobel Prize for Physics in 1946.

Surface properties *

In the field of surface properties, Thomas Young (1773–1829; FRS 1794) made an important contribution, early in the 19th century when he published (1805) a theory of capillary action; this accounted for the angle of contact in surface-tension effects in liquids, which has proved of considerable interest in various fields, including heterogeneous nucleation during the solidification of metals.

The wide-ranging work of Lord Rayleigh (formerly John William Strutt, 1842–1919; FRS 1873) included a paper (1878) on the instability of cylinders of fluids with respect to periodic perturbations leading to the formation of spheres and relating to water jets; the parameters include surface area and hence surface energy. In recent years the theory has been applied to solid state phenomena such as spheroidisation of rod shaped particles.

The fundamental thermodynamic work of Josiah Willard Gibbs (1839–1903; FMRS 1897) included aspects concerning surfaces. As a rule solute elements which decrease the surface tension should concentrate at the surface, a feature that has assumed importance in metallurgy as applied to grain boundaries in solids as well as to surfaces.

Heinrich Gustav Magnus (1802–1870; FMRS 1863) carried out work on the condensation of gases on solid surfaces. Later, Michael Polanyi (1891–1976; FMRS 1944) in research in the chemical-physics of solids carried out studies of the adsorption of gases by solids, commencing in 1914. He suggested the existence of an attractive force between the atoms or molecules in the gas and the solid surface, and that multilayers form at adsorbed surfaces.

In late 1916 and in 1917 Irving Langmuir (1881–1957; FMRS 1935) at the research laboratory of the General Electric Company at Schenecdaty, USA, began studies of the constitution and fundamental properties of solids and liquids. He proceeded to investigate the properties of both solids and liquids in terms of chemical forces and more particularly the properties of liquid surfaces. Extending the earlier work of Lord Rayleigh and others on the behaviour of oil films on water surfaces, Langmuir made a revolution-

* Surface phenomena associated with corrosion, wear, erosion and oxidation are discussed in Chapter 14.

ary advance by measuring the spreading of pure substances on water surfaces. The Langmuir trough and balance became of great importance, for example, in deciding between alternative proposed configurations for complex molecules, and in the demonstration of gaseous, liquid, liquid-expanded and solid films on surfaces. He collaborated with Katherine Burr Blodgett (1898–1979) using a silk thread, suitably water-proofed, resting on the surface of the water as a barrier between film and water surface; the film could be placed under a constant pressure, and Langmuir and Blodgett showed how stearic acid monolayers could be built up into oriented layers on a metal surface by cycles of dipping the metal through the film and then withdrawing it again. Attention was given to monolayers on solid surfaces and this led to Langmuir's concept of gases on solid surfaces as a time lag between condensation and evaporation of the impinging gas molecules. Langmuir incorporated his ideas on monolayer adsorption at surfaces into a general theory of reactions (1921). On the new theory, which became the basic approach to surface kinetics, chemical reaction occurred between adjacently adsorbed reacting species on a surface.

Other research by Langmuir in Schenectady was concerned with the thermal conduction and convection in gases at high temperatures. Also working at Schenectady was William David Coolidge (1873–1975) who in 1913 produced a filament-heated cathode X-ray tube. Coolidge had previously (1908) developed a powder metallurgy route for preparing tungsten in the form of wires for metal filaments (see Chapter 8). Such filaments, notably of tungsten, provided the high temperature sources for Langmuir's research. From these researches early values for the heat of formation of hydrogen molecules were obtained. In 1913 studies of 'clean-up' of oxygen and nitrogen in a tungsten lamp were made. The reactions of hydrogen, oxygen and nitrogen with heated tungsten filaments were classic studies of reactions involving gases at very low pressures. For each gas studied, except for the special case of water vapour, Langmuir demonstrated that the blackening of a lamp was due solely to evaporation. He could 'conclude with certainty that the life of the lamp would not be appreciably improved even if we could produce a perfect vacuum'. The research on the rate of evaporation in high vacua was followed by studies of the rate in inert gases at different pressures; stemming from this work the nitrogen-filled lamp and, later, the argon-filled lamp became available. The work included measurements of the vapour pressure and melting points of tungsten, vapour pressures of platinum and molybdenum, and condensation pumps with an improved form of vacuum pump. Other work led to thoriated tungsten filaments becoming standard components for medium power tubes. The small thoria additions increased the electron emission from the filaments and aimed at preventing crystal growth of the tungsten.

In Britain a leader in the science of thermionic emission phenomena, Owen Willans Richardson (later Sir Owen, 1879–1959; FRS 1913) introduced the concept of an evaporation potential barrier for electrons. His work resulted in his receiving the Nobel Prize for Physics in 1928. The wide-ranging theoretical work at Cambridge by Ralph Howard Fowler included thermionic and photoelectric emission of electrons from metals.

The early research of George Ingle Finch (1888–1970; FRS 1938) involved studies of the effect of catalysis on the combustion of gases. His interest in platinum as a catalyst led him to work on surface effects, including the use of electron diffraction. Also the researches of Eric Keightley Rideal (later Sir Eric, 1890–1974; FRS 1930) included the fields of surface and colloid chemistry; his studies included bimetallic catalysis and kinetic investigations of the oxidation of metals. The research interests of David Tabor (1913–; FRS 1963) have included the adsorption of vapours and film formation.

The research of Sir Ronald Mason (1930–; FRS 1975) in surface science has included diffraction and spectroscopic studies of transition metal surfaces and their chemisorption of simple molecules, organometallic chemistry of transition metal surfaces and catalysis by transition metals. Also in surface science David Anthony King (1939–; FRS 1991) has carried out fundamental work on the mechanisms of the interaction of gases with metal surfaces. He has developed and enhanced spectroscopic and other physical techniques, which have led to greatly improved models of the behaviour of molecules on metal surfaces.

Cyril Clifford Addison (1913–1994; FRS 1970) in his research on surface chemistry made important investigations on the surface tension of liquid sodium and potassium-sodium alloys in the practical context of coolants in nuclear reactors; a significant feature was the use of argon as a protective atmosphere to avoid contamination by oxide films. Addison's publications included a book on *The chemistry of the liquid alkali metals.*

CHAPTER 12

Advances in Techniques for the Study of Materials' Structures

Introduction

In the field of physical metallurgy a huge leap forward occurred in the second decade of the 20th century when X-ray diffraction became a tool for elucidating the details of the crystalline structure of metals and alloys. As the century has progressed X-ray techniques have continued to play an indispensable role in research on the structure of both crystalline and non-crystalline materials of all types. Electron diffraction has played a similar role, including the development and application of transmission electron microscopy in the past half century. The resolution of structures at the atomic level by techniques such as field ion microscopy and scanning tunnelling microscopy has also become important for elucidating the structural and compositional features of materials. This chapter reviews the role of Fellows of the Royal Society in these fields, preceded by some comments on the history of crystallography.

Crystallography

In the subject area of crystals and their structure the idea developed in the late 18th century that a crystal could be built up from small units. René-Just Haüy (1743–1822; FRS 1818) is regarded as the founder of modern crystallography; his approach was mainly morphological, developing a geometrical classification of crystal forms. Torbern Olof Bergman (1735–1784; FRS 1765) in the 18th century had also contributed in this area. In 1812, William Hyde Wollaston (1766–1828; FRS 1793) expanded on earlier ideas of regular packing of spherical particles to explain the geometry of crystals.

Crystallography experienced rapid advances in the 19th century; the 32 crystal classes were demonstrated and descriptive knowledge of crystals and their behaviour was greatly extended. William Hallowes Miller (1801–1880; FRS 1838), building on earlier work by William Whewell (1794–1866; FRS 1820, Professor of Mineralogy at Cambridge University and Master of Trinity College) and Franz Ernst Neumann (1798–1895; FMRS 1862), introduced a new mathematical system, incorporating indices, which was far more simple, symmetrical and adapted to mathematical calculations than any which had previously been devised. As Maskelyne* said it 'gave expressions adapted for working all the problems that a crystal can present, and it gave them in a form that appealed at once

* Mervyn Herbert Nevil Story-Maskelyne, 1823–1911; FRS 1870; mineralogist.

to the sense of symmetry and appropriateness of the mathematician... he thus placed the keystone into the arch of the science of crystallography'. Miller's system, published in 1838, quickly found favour. Late in the century lattice structure theory was developed e.g by William Barlow (1845–1934; FRS 1908) who learned most of his formal knowledge of crystallography from Henry Alexander Miers (later Sir Henry; 1858–1942; FRS 1896) and William Jackson Pope (1870–1939; FRS 1902). Barlow's work included the derivation of the 230 space groups published in 1894. This work was independent of that of E.S. Federov in Russia, and of Arthur Moritz Schoenflies(1853–1928) in Germany, both of whom had independently published derivations in 1891. Barlow also considered crystal structures.in terms of close packing of spheres, showing that two closest–packed assemblages of spheres exist, one having full cubic symmetry and the other hexagonal symmetry. At that time the space lattices of the metals were not known and the internal structure of crystals had been a subject of much debate.

X-ray diffraction

X-rays had been discovered by William Konrad Röntgen (1845–1923) in 1895, who also searched unsuccessfully for diffraction. In February 1912 Max Theodore Felix von Laue (1879–1960; FMRS 1949) conceived the idea of X-ray diffraction by sending X-rays through crystals; at that time at the University of Munich he was writing a chapter of an encylopaedia on wave optics for Arnold Sommerfeld (1868–1951; FMRS 1926). Laue had a discussion with Paul Peter Ewald (1888–1985; FRS 1958) who had obtained his PhD in Munich with Sommerfeld. The thesis (submitted in February 1912) had been concerned with the explanation of the double diffraction of light produced by the anistropic arrangement of atoms in a crystal; before finishing it, Ewald had sought the opinion of Laue, who was very interested in Ewald's assumption of a regular lattice of atoms in crystals with an interatomic distance of about 0.4 A, which agreed well with the estimate of the wavelength of X-rays. Although the wave nature of X-rays was not yet clearly established Laue had been looking for a way to prove it.

In his Nobel lecture Laue told how, arising from discussions on the passage of light waves through a periodic, crystalline arrangement of particles, he had the idea that much shorter electromagnetic waves, such as X-rays were thought to be, should produce an interference or diffraction effect in such a medium. Proposed experimentation was discussed, but Sommerfeld and other eminent physicists concluded that the disturbing influence of the temperature would prevent success. However, Walter Friedrich (1883–1968), Sommerfeld's experimental assistant and Paul Knipping (1883–1935), a recently graduated doctor and, began experimenting in April 1912 with a quite primitive X-ray arrangement and made the first exposure, using a copper sulphate crystal. They placed the photographic plate between the crystal and the tube, expecting an effect similar to that of a reflection grating, but no result was obtained. However, in a second exposure, made with plates on the exit side, and to the sides, of the crystal, they obtained a regularly ordered pattern of strong, rather blurred spots around the trace of the direct beam.

Rene-Just Haüy

Henry Miers

William Pope

William Barlow

When Laue heard and thought about this result he quickly realised the mathematical formulation of the effect.

A more elaborate apparatus was constructed to enable an accurately oriented crystal to be exposed to a well-collimated X-ray beam. Röntgen came to see the experiment and while agreeing that the phenomenon was caused by the crystal structure he did not accept that the spots resulted from diffraction. Laue, Friedrich and Knipping, with the improved apparatus, aiming to show clearly that the expected diffraction effect had been obtained, used two zincblende plates, one normal to a fourfold axis of symmetry, the other to a threefold axis. The clarity and precision of resultant photographs removed doubt, particularly when Laue, using the mathematical formulation, showed that each spot could be accounted for as the diffraction of a certain order, described by three integers.

The fundamental discovery was published in the Bavarian Academy of Science in June and July 1912 and news spread rapidly; for example, at the British Association Meeting in Dundee in September 1912, Lord Rayleigh (formerly John William Strutt, 1842–1919; FRS 1873) commented on the discovery, and a series of letters to Nature by William Henry Bragg (later Sir William, 1862–1942; FRS 1907) appeared in the following month. This probably resulted from the reprints Laue sent out to those known to be working on X-rays. Both W. H. Bragg and his son, William Lawrence Bragg (later Sir Lawrence, 1890–1971; FRS 1921) were deeply interested and discussed the matter during their family holiday. In Cambridge, in 1912, W. L. Bragg, drawing on available knowledge, showed that the phenomenon of X-ray diffraction could be understood in terms of the reflection of the rays by planes of atoms in the crystal. Since the path difference between the waves of the reflected train is $2d \sin \theta$, where is the glancing angle at which the radiation falls on the planes and d is the interplanar spacing, it followed that the wavelengths of the different orders of reflection would be given by $n\lambda = 2d \sin \theta$, where n is an integer. On the basis of these ideas,he successfully explained the diffraction pattern from ZnS assuming a face-centred cubic structure. Thus, he showed that the Laue pictures resulted from a continuous range of X-ray wave lengths, and that X-ray diffraction could be used to obtain information about the atomic arrangement in crystals. This was the start of the X-ray analysis of crystals. William Pope, then professor of chemistry at Cambridge, suggested to W. L. Bragg a possible structure for NaCl, arising from the work of William Barlow, and that a study of the alkali halides would be advantageous. Bragg in 1913 confirmed by X-ray diffraction the atomic arrangements in sodium chloride and potassium chloride.

Combining W. H. Bragg's familiarity with characteristic radiation and W. L. Bragg's concept of considering reflection on a set of atomic planes rather than diffraction by a three-dimensional grating, father and son proceeded to found the subject of X-ray crystallography. The X-ray spectrometer, constructed by W. H. Bragg, enabled the planes of a crystal structure to be investigated and measured in terms of known wavelengths. In the hands of these two eminent physicists this instrument served to elucidate a series of crystal structures and to build up a school of physicists and chemists with a main interest

Max von Laue

Paul Ewald

William Henry Bragg

William Lawrence Bragg

in deciphering the atomic arrangement in crystals and hence obtaining increased fundamental understanding. After Laue had seen that his physical idea worked he wrote:

> Once I had conceived the idea of X-ray diffraction in crystals this appeared so evident to me that I never understood the astonishment it produced among scientists, nor, indeed, the doubts expressed in the course of the first few years. That Bragg's idea of reflexion on atomic planes was essentially equivalent to mine – this I saw at once, and I published it early in 1913. Now access to short electromagnetic waves had been found.

In the succeeding period over many years the researches of Laue, the Braggs and Ewald continued. Ewald, for example, discovered the concept of the reciprocal lattice and, in 1914, introduced the 'structure factor' concept. A few years later he conceived the dynamical theory of X-ray diffraction, a field which he continued to work in all his life. The theory, which received experimental proof from other workers in the 1930s, became of great importance in other fields; also other theories, including that of electron diffraction were built on it. Among the aspects in which W. H. Bragg was involved in later years was the phenomenon of diffuse spots observed by X-ray examination of single crystals. W. L. Bragg at Cambridge later promoted the development of X-ray microbeam methods with a view to studying the structural features of cold worked metals, and he also carried out extensive work on non-metallic materials.

Also in the early period following the discovery of X-ray diffraction Peter Joseph Wilhelm (Wilhelmus) Debye (1884–1966; FMRS 1933) published his first paper in the field in 1913. Having moved to Göttingen the following year he did important X-ray crystallographic research. Using his theory of specific heats he explained the effect of increasing temperature in decreasing temperature X-ray intensities, and also related the intensities of diffraction spots to wavelength and diffraction angle. A metal X-ray tube was built with a water cooled copper target and an aluminium foil for the escape of the X-rays. Debye had a young assistant Paul Scherrer who took a photograph of finely powdered lithium fluoride; they found that the diffraction produced sharp rings which were interpreted as originating from the intersection of conical beams from the randomly oriented crystals with the photographic plate. This important discovery (1916) of diffraction from randomly oriented particles became the basis of the powder (Debye–Scherrer) technique that provided an important addition to the use of single crystals. (The technique was independently discovered by Albert Wallace Hull (1880–1966) in the USA, and reported in 1917). Debye and Scherrer proceeded to determine the structure of some cubic crystals and also (1917) partially elucidated the more complex structure of graphite. Debye's later work included X-ray diffraction studies of liquids, for example mercury.

In 1913 Charles Galton Darwin (later Sir Charles, 1887–1962; FRS 1922) in Rutherford's department at Manchester published a paper with Henry Gwynn Jeffreys Moseley (1887–1915; killed in action in the Gallipoli campaign) on the diffraction of X-rays. Then in the following year Darwin published two papers under the title 'The theory of X-Ray Reflexion' which W. L. Bragg referred to as 'landmarks in the history of X-ray

Charles Galton Darwin

Albert Bradley

John Desmond Bernal

Charles Sykes

analysis of crystals'. Darwin deduced that crystals are not ideally perfect, but consist of a 'mosaic' of blocks of slightly differing orientations. Among associates of W. L Bragg at Manchester, Reginald William James (1891–1964; FRS 1955) carried out influential studies of a fundamental and quantitative nature, including diffraction intensities. In 1926 Bragg, Darwin and James presented an important paper 'The Intensity of X-Ray Reflexions by Crystals'. Other distinguished researchers, some associated with Bragg, made important contributions in X-ray crystallography, building on the earlier pioneering work, and were elected as Fellows of the the Royal Society: Albert James Bradley (1899–1972; FRS 1939); Charles Sykes (later Sir Charles, 1905–1982; FRS 1943); Henry Solomon Lipson (1910–1991; FRS 1957); Victor Moritz Goldschmidt (1888–1947; FRS 1943); Norman James Petch (1917–1992; FRS 1974); Michael Polanyi (1891–1976; FRS 1944) and Arthur James Cochran Wilson (1914–1995; FRS 1963). Their research included the study of metallic and other inorganic materials, while other Fellows were particularly associated with X-ray crystallography in the non-metallic field, e.g. Kathleen Lonsdale (later Dame Kathleen, 1903–1971; FRS 1945); Dorothy Crowfoot Hodgkin (1910–1994; FRS 1947, Nobel prize winner for chemistry, 1964, OM 1965); John Desmond Bernal (1901–1971; FRS 1937); John Turton Randall (later Sir John, 1905–1984; FRS 1946) and William Cochran (1922–; FRS 1962) working on organic and/or biological materials.

Albert Bradley became W. L. Bragg's first research student at Manchester University in the early 1920s. Bragg's main work lay in the determination of crystal structures from single crystal X-ray diffraction patterns but he realised that the powder method had possibilities and asked Bradley to explore them. Since the most exciting work with the method was being carried out in Sweden under Arne Westgren and G. Phragmen at the Institute for Metals Research at Stockholm, Bradley was sent there in 1926 to learn the techniques. Bradley was highly successful and soon was leading the way in his ability to solve complicated problems. During his year's stay in Sweden he solved the structure of alpha-manganese which had 58 atoms in the unit cell and this work made his reputation. Attracted by problems of unusual complexity he was intrigued by the diffraction patterns of the alloy Cu_5Zn_8, gamma-brass. Westgren, having obtained some single crystals and worked out the unit cell could not determine the atom positions and allowed Bradley, by then the world's expert in X-ray powder photography, to take his data back to Manchester in 1927. Bradley's greatest achievement was the determination of the structure of this copper–zinc compound from a powder photograph which contained over 50 lines, as compared with less than ten from a simple structure. He obtained an approximate arrangement of atoms which he then adapted to give the observed powder pattern. Although the alloy had little practical use, the result turned out to be of immense importance in the development of the electron-concentration theory of the metallic state as described by Nevill Mott (later Sir Nevill, 1905–1996; FRS 1936) and Harry Jones (1905–1986; FRS 1952).

Other work by Bradley included a redetermination of the crystal structure of gallium, and various studies of alloy systems containing intermetallic compounds (see later).

Henry Lipson

Arthur J.C. Wilson

Concerning experimental techniques he made an important contribution by the methods he developed for making precision measurements of lattice parameters; also he invented the demountable X-ray tube, which was then produced commercially by Metropolitan Vickers.

Charles Sykes in Manchester in 1930 investigated alloys from the iron–aluminium system, which were known to have good heat-resisting properties. He found a sudden increase in the coefficient of expansion of an alloy with 12 wt% aluminium (in the region of Fe_3Al) at temperatures above 400°C, together with changes in electrical resistivity, depending on the heat treatment; these changes could not be explained on the basis of existing phase diagram data. Sykes discussed these results with W. L. Bragg and with Bradley, who, with A. J. Jay, carried out X-ray diffraction of iron–aluminium alloys, and reported (in 1932) the change from a disordered to an ordered (superlattice) arrangement of atoms in the lattice below a critical temperature. Sykes with collaborators, during the 1930s made a number of important investigations relating to the ordering in iron–aluminium and in other alloys: copper–gold and copper–zinc, including X-ray diffraction and measurements of specific heat and electrical resistance. Bragg's interest in order–disorder phenomena, stemming from his association with Sykes, led to the development, with Evan James Williams (1903–1945; FRS 1939) of the first theory of this type of transformation. Hans Albrecht Bethe (1906–; FMRS 1957) also carried out theoretical studies of ordering.

Bradley, in addition to his work with Jay on the ordering of the intermetallic compounds Fe_3Al and FeAl, investigated aluminides in other systems. In work with A. Taylor, X-ray diffraction and precision measurements of density were used to investigate nickel–

aluminium alloys. Their results led them to the conclusion that on the aluminium-rich side of stoichiometry of NiAl lattice vacancies formed an essential feature of the structure; they also determined the crystal structures of Ni_2Al_3 and $NiAl_3$. Detailed investigations were made of the constitution of certain ternary systems, involving extensive use of X-ray diffraction: iron–nickel–aluminium (in relation to magnetic properties with A. Taylor), copper–nickel–aluminium (with Henry Lipson), iron–copper–aluminium, iron–copper–nickel and iron–nickel–chromium (with H.J. Goldschmidt). Also, work with J. W. Rodgers on the ferromagnetic ternary compound Cu_2MnAl (the 'Heusler' alloy) showed that the superlattice structure was similar to that of Fe_3Al. Bradley's work in these areas extended from the early 1930s over a period of around two decades.

Henry Lipson, as a research student in the early 1930s, was also influenced by W. L. Bragg at Manchester. He had been allocated to a research project to work with a recent graduate, Arnold Beevers, on solving some crystal structures using X-ray diffraction. The two research students, lacking proper equipment, constructed their own X-ray source and a camera based on a brass tube within which a film was wrapped against the inner surface. Eventually they obtained a diffraction pattern from nickel sulphate heptahydrate. They studied the literature but nobody in Liverpool had enough experience to help them get started with the process of interpreting the pattern. However, they approached W. L. Bragg who invited them to visit his laboratory at Manchester; some of his staff, particularly Dr W. H. Taylor, gave them guidance which enabled them to solve their first two crystal structures, which were of high symmetry. They then tackled a more difficult structure using the Fourier method (Jean Baptiste Joseph Fourier, 1768–1830; FMRS 1823) described by Bragg in 1929; having only tables of sines and cosines to aid computation it appeared that the Fourier summation work would take the best part of a year. However, Lipson successfully introduced a simplifying idea for the form of the summation. He also adopted a method of making a set of paper strips for different wave indices and different amplitudes that could be repeatedly used, instead of recording data in note books. With a good supply of strips available the time taken for the calculation was reduced to a few days. News of this advance quickly spread, and Bragg arranged for a loan from the University of Manchester so that the strips could be printed in quantity and sold. The Beever–Lipson strips were for many years the workhorse of crystallographic calculations.

Later, at Cambridge, Lipson was introduced by Ewald to the idea of the Fourier transform and to the reciprocal lattice. W. L. Bragg had introduced the idea of an X-ray microscope, which was really an optical analogue of X-ray diffraction; Lipson realised that the diffraction pattern of a crystal structure was essentially its Fourier transform. This idea formed the basis of his optical work related to X-ray crystallography which became a central theme for his future research.

Victor Goldschmidt in Oslo after the 1914–1918 war carried out investigations of rocks and minerals, including the use of the technique of chemical analysis by X-ray spectra. One investigation concerned the partition of chemical elements between systems of co-existing liquid phases-molten iron, liquid iron sulphide, and fused silicate;

this topic was important not only in metallurgy, but also in petrology and meteorite study, and the results contributed to the study of the chemistry of the earth's deep interior. Then from 1923 until 1929 at Göttingen, beginning at a time when crystal chemistry did not exist as a scientific discipline, Goldschmidt and his colleagues carried out a remarkable series of X-ray studies, involving the preparation and X-ray analysis of nearly 200 separate chemical compounds. Goldschmidt showed that the characteristics of atoms that determine their distribution in the crust of the earth are mainly ionic size, charge and polarisability. This work together with the knowledge of silicate structures from the Bragg school explained elegantly and simply the basis of ionic substitution in minerals. The theory was extended by Goldschmidt to explain the physical properties e.g. the hardness in relation to structures. His work was applied by crystallographers and crystal chemists, becoming the basis for all subsequent studies. In an important investigation of the metallic state Goldschmidt established sets of atomic radii corresponding, not to atomic charge, but to different coordination numbers, and thus illustrated the regularities of the periodic system of the elements.

Michael Polanyi in Berlin in 1920 began research on the X-ray analysis of fibrous structures, particularly of cellulose, but also of metals; his work included new methods of X-ray analysis in general and led on to his investigations of the structure and properties of crystals (see Chapter 13).

In 1939 Norman Petch joined the crystallography group in the Cavendish laboratory at Cambridge under the supervision of Harold Lipson. His first research, leading to a publication in 1940, was an X-ray investigation of the complex crystal structure of cementite; earlier workers had demonstrated the orthorhombic unit cell, found the space group and the positions of the iron atoms in this important constituent of steels. Petch used three-dimensional Fourier analysis which gave the electron density at all points in the unit cell directly from the X-ray intensity, to enable the positions of both iron and carbon atoms to be determined. To obtain the required accuracy a very large powder camera 35 cm in diameter was constructed, and precise lattice parameters were measured. In 1941 and 1942 he published two very significant papers. In one of these, using a high manganese steel, he deduced the positions of the interstitial carbon atoms in the face-centred cubic austenite. The other paper demonstrated the carbon atom positions in the body-centred tetragonal martensite in quenched steels. In 1944 he published another important paper on the structure of cementite.

Arthur Wilson made important contributions to the field of X-ray studies and crystallography. His early research at Cambridge included precision measurements of lattice parameters to determine accurately the thermal expansion of aluminium and magnesium as a function of temperature. From experimental and theoretical studies of diffuse lines in powder diffraction photographs he was led on to an interest in crystal imperfections; he studied stacking faults in cobalt (1941–42) and three-dimensional faults in Cu_3Au in 1943. X-ray statistics were an important area of his work and he established a really detailed theory of X-ray powder diffraction and all its related problems, especially for the acquisition of absolute intensities, and including modern powder diffractometry, in which technique he played a founding role. He also played a leading role in international activities concerning the compilation of crystallographic data.

John Bernal in his earlier X-ray work work on the nature of the crystalline state determined the structure of graphite (1924), using single crystals, a rotation method and the concept of the reciprocal lattice. Also he investigated around 1929 the complex crystal structure of three of the intermetallic compounds in the copper–tin system; for the delta phase he found a large cubic unit cell, closely related to gamma-brass. Bernal was also involved later in studies of the nature of the liquid state, including metals (Bakerian lecture 1962). Using evidence from X-ray diffraction on liquid metals of the distribution of nearest neighbour atoms he prepared various types of model to demonstrate geometrical features. He made several discoveries about the characteristics of homogeneous, random, close-packed assemblies.

Among other Fellows, with their research groups, beginning in the 1930s who made extensive use of X-ray diffraction in the study of alloy structures were William Hume-Rothery (1899–1968; FRS 1937) and Geoffrey Vincent Raynor (1913–1983; FRS 1959) (see Chapter 16).

The early research of John Randall during the period beginning in 1925 at the General Electric Company laboratories at Wembley included X-ray investigations of the structure of glasses and of liquid metals (sodium, potassium, rubidium and caesium).

In the USA Linus Carl Pauling (1901–1994; FMRS 1948) began his PhD research at the California Institute of Technology in 1922 on the X-ray determination of crystal structures, including the structure of an intermetallic Mg–Sn compound, published in 1923, the first determination of the structure of any such compound. Pauling during his career used X-ray crystallography and other techniques to establish an extensive collection of data on interatomic distances and angles and to calculate atomic radii; also he made outstanding contributions in the field of bonding and molecular structure in chemistry and biology.

Andrew Richard Lang (1924–; FRS 1975) has made important contributions in X-ray crystallography, particularly through the development of techniques for observing crystal defects such as disocations in relatively thick crystals; differences in diffraction intensity reveal the features and he investigated metals and semiconductors. Michael Hart (1938–; FRS 1982) in his work on X-ray optics has also been involved in this field.

Electron diffraction

The application of electrons, discovered in 1897 by Joseph John Thomson (later Sir Joseph, 1856–1940; FRS 1884), in diffraction provided an important experimental technique in metallurgy and other fields. Clinton Joseph Davisson (1881–1958; recipient of the Hughes Medal of the Royal Society in 1935) at the Western Electric Company in New York investigated the reflection of electrons from metal surfaces under electron bombardment. Some effects he obtained in 1925 using a nickel target provided the first experimental observations of the wave nature of electrons; in the following year he suggested that these effects were caused by diffraction of electron waves from the atomic planes in the nickel crystals. Davisson and his assistant Lester Halbert Germer (1896–1971) using a nickel single crystal demonstrated electron diffraction early in 1927. In the same year at Aberdeen University George Paget Thomson (later Sir George, 1892–1975; FRS 1930), the son of J. J. Thomson, carried out similar experiments with metal foils at higher voltages. Davisson and Thomson shared the Nobel Prize for Physics for this work in 1937. Thomson, later at Imperial College London, continued investigations on electron diffraction; in the late 1940s and early 1950s Donald William Pashley (1927–; FRS 1968) worked with Thomson on epitaxial phenomena. Another Fellow involved in this research on epitaxy was Ronald Charles Newman (1931–; FRS 1998). Moses Blackman (1908–1983; FRS 1962) was associated with theoretical aspects of this work.

Also George Ingle Finch (1888–1970; FRS 1938) at Imperial College, London applied electron diffraction innovatively. The initial scientific research of Finch and his students was concerned with the effect of catalysts on the combustion of gases. It was his interest in platinum as a catalyst that led him to the study of surface effects that destroy its efficiency. Here he collaborated initially with G. P. Thomson in the use of electron diffraction, which showed the presence of a contaminant on the surface. This led him into the field of electron diffraction, in order to link catalytic and other properties of surfaces and thin films with crystal structure and crystal texture. In the early 1930s he built a new electron diffraction unit (the 'Finch' camera) which represented a marked technical advance. This used a magnetic coil to focus the electrons, a vacuum of 10^{-5} mm or better, and a specimen holder, which gave flexibility of movement and tilting, and served for both reflexion and transmission patterns; another important feature was the provision of an evaporating device so that thin films could be produced in situ by vapour deposition. Finch initiated a very fruitful period of researches into surface structure and the properties of thin films, including epitaxial growth. Finch's appreciation of the technological aspects of his work showed very clearly in his study of 'Beilby layers'– surface features produced by polishing.

Finch also made important investigations of electrodeposition the main emphasis being the mode of growth of the deposit and the adhesion of the deposit to the base. The substrates used in electrolysis were found to divide into 'inert' and 'active' types. The former included mainly well polished metals (i.e. having an 'amorphous surface') the latter being crystalline metal surfaces. Two extreme types of growth were found, 'lateral' and 'outward'. This latter form was due to to essentially three-dimensional initial aggregates with relatively poor adhesion and was associated with 'inert' bases. In lateral growth, associated with 'active' bases, the initial growth started as a two-dimensional form with a well defined geometrical relationship between the deposit and the base, leading to good adhesion. Finch also investigated the effect of factors such as composition and temperature on the formation of thick electrolytic deposits.

Electron diffraction was also used by James Woodham Menter (later Sir James, 1921–; FRS 1966) working at Cambridge with Frank Philip Bowden (1903–1968; FRS 1948) to study the orientation of surface films.

Electron microscopy and other developments in microscopy

In the latter part of the 20th century the development and application of electron microscopy has had a profound influence in metallurgy. Following work by Ernst August Friedrich Ruska (1906–1988) in the early 1930s, many advances have occurred in this field, including most notably through the examination of thin foils by the transmission technique, giving high resolution, together with diffraction and compositional analysis facilities. Scanning electron microscopy of bulk specimens has also been of extraordinary importance. Among Fellows of the Royal Society whose work included electron microscopy Dennis Gabor (1900–1979; FRS 1956, Nobel Prize for Physics 1971, inventor of holography) made important advances in electron optics, including lens design. Karle Manne Georg Siegbahn (1886–1978; FMRS 1954), notable for his work on precision spectrometry in Uppsala, Sweden, in the mid 1940s, designed and developed an electron microscope which was used for electron diffraction.

Vernon Cosslett (1908–1990; FRS 1972) was prominent in many aspects of electron microscopy. Cosslett arrived at the Cavendish laboratory in September 1946 to run the RCA microscope and to promote electron microscopy in the University of Cambridge. His equipment was soon augmented by a 100 kV Siemens ' supermicroscope' from Krupps in Germany. Over the years many gifted research students and colleagues were attracted to the group that he built up. Many disciplines were represented, and Coslett brought in problems from these disciplines, often building the experimental analytical instruments needed to solve them.

With his students and collaborators, Coslett was responsible for the design and development of a number of highly successful electron optical instruments. The first of these, the X-ray microscope, which achieved sub-micrometre resolution, was from a PhD project by W.C. Nixon; the first book on this subject, *X-ray Microscopy*, was published by Coslett and Nixon in 1962. The second development, in collaboration with Peter Duncumb (1931–; FRS 1977) was the scanning electron probe X-ray micro-analyser. R. Castaing in Paris in the early 1950s had constructed a static electron probe instrument in which a metallurgical specimen on a mechanical stage was located with the aid of an optical microscope under an electron beam of approximately 1μm diameter; the X-ray emission spectra were used to obtain a chemical analysis. The instrument of Coslett and Duncumb was based on a modified scanning electron microscope for viewing the specimen, and was the prototype of many commercially produced instruments which became standard equipment in many metallurgy, geology and biology laboratories all over the world. Electron microscopes with accelerating voltages up to one million volts became important in the 1960s, and Coslett, with K.C.A. Smith, developed the Cambridge 750 kV electron microscope, and this formed the basis for the design of the AEI EM7 1 MeV instrument. A fourth development, with Nixon and others, was the Cambridge 650 kV ultra-high resolution microscope, which achieved a point to point resolution of 0.25 nanometre.

Alongside the activities of Coslett and others on developing electron microscopy at

George Paget Thompson

George Finch

Vernon Coslett

Charles Oatley

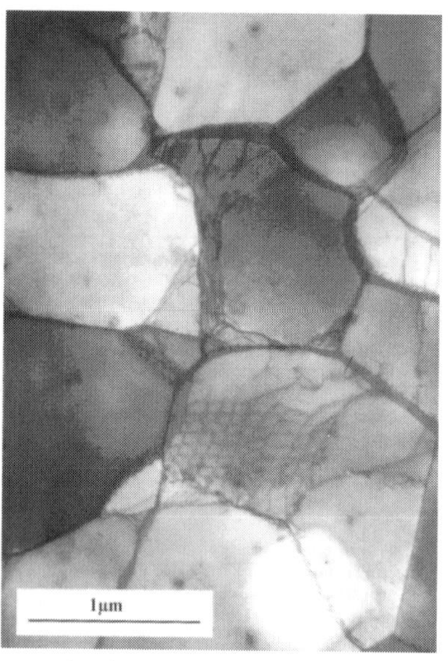

Transmission electron micrograph of heavily deformed aluminium, showing dislocations in sub-grain boundaries (M. Whelan – courtesy P.B. Hirsch)

Cambridge including the transmission technique, there took place pioneering development and design work on the scanning electron microscope (SEM), led by Charles William Oatley (later Sir Charles; 1904–1996; FRS 1969). Work in Germany and the USA in the 1930s, aiming to use a scanned beam of electrons and to obtain direct imaging of surfaces, had produced inconclusive results. Oatley believed that the new techniques available to him could be applied to the scanning concept and that, in spite of contrary views, the SEM would become an important laboratory instrument. With his first research student, Dennis McMullan, work was begun in 1948 leading to the production of a working instrument in 1951. With other students and technical support new instruments were developed and in 1965 the world's first production SEM, the Stereoscan became available, manufactured by the Cambridge Instrument Company. The scanning electron microscope, with its ability to produce three-dimensional images, and other advantages has become an essential and widespread technique not only in metallurgy and materials science, but in many other scientific fields.

Other groups in Cambridge University have also been also active in electron microscopy, particularly in the transmission technique involving other researchers who are now Fellows of the Royal Society. Their wide-ranging pioneering research has involved fundamental aspects, including the interaction of electron beams with metals leading to diffraction and image formation. It has also covered the application of electron microscopy to elucidate phase transformation phenomena and plastic deformation, including the role of dislocations and other crystal defects. Peter Bernhard Hirsch (later Sir Peter, 1925–; FRS 1963) and James Menter pioneered the transmission technique, including the revelation of dislocations in crystals. Archibald Howie (1934–; FRS 1978) and Michael John Whelan (1931–; FRS 1976) were among those who further developed the theory of electron diffraction to enable images to be obtained to be used to compute lattice defects. Other Fellows whose research using electron microscopy, including diffraction, to investigate metallurgical phenomena, involved a period at Cambridge, are Michael Farries Ashby (1935–; FRS 1979); Lawrence Michael Brown (1936–; FRS 1982); Robert William Honeycombe (later Sir Robert, 1921–; FRS 1981); Robin Buchanan Nicholson (later Sir Robin, 1934–; FRS 1978) and John Wickham Steeds (1940–; FRS 1988). At the Tube Investments laboratory at Hinxton Hall, the research of Donald William Pashley and Michael James Stowell (1935–; FRS 1984) using electron microscopy included the area of epitaxy. In 1963 there took place in Cambridge a summer school, from which stemmed the influential book *Electron Microscopy of Thin Crystals*, authored by Hirsch, Howie, Nicholson, Pashley and Whelan.

Electron microscopy, which flourished so strongly in Cambridge, has also seen important developments in other universities and research institutions in England and other countries. For example, Derek Hull (1931–; FRS 1989) and Raymond Edward Smallman (1929–; FRS 1986) have used the technique to make important contributions to metallurgical science, including studies of defects in crystals. Alan Kenneth Head (1925–; FRS 1988) in Australia has pioneered computer simulation of dislocation images in the electron microscope. In the earlier part of his career Sir Peter Michael Williams (1945–; FRS 1999), now Chairman of Oxford Instruments plc, carried out research in surface physics and electron microscopy. Among the contributions of David John Hugh Cockayne (FRS 1999) to electron microscopy have been the development of the theory, and joint responsibility for the experimental development of the 'weak beam' technique, that has improve the resolution for investigating lattice defects.

Also in the area of high resolution investigation of structure the development of field ion emission microscopy, with its associated techniques – stemming from the work of Erwin Wilhelm Mueller (1911–1977) allowed imaging of structures at the atomic level. In the field electron emission microscope which Mueller invented in 1936 a metal specimen with a fine tip is held in a vacuum near a phosphorescent screen. When a high negative voltage is applied to the tip electrons emitted from the tip move to the screen and form a high magnification image of the tip surface. In the field-ion microscope, which Mueller produced in 1951 the metal tip was held electrically positive in a gas at low pressure; positive ions from the gas adsorbed at the tip are repelled and produce the image on the screen. By cooling the tip cryogenically the resolution of the microscope was improved and Mueller obtained well resolved images of atoms. In Britain research began in the late 1950s. At Cambridge University during the period from 1958–1965 under the leadership of Alan Cottrell (later Sir Alan, 1919–; FRS 1955) a field ion instrument was built by David Brandon and was in use in 1959 and important research was carried out over a period of years by Brandon and others, including Brian Ralph and Michael Southon. Also during the period 1959–1963 when John Stuart Anderson (1908–1990; FRS 1953) was at the National Chemical Laboratory, a member of his group,

David Bassett, set up a field ion microscope. When Anderson moved to the chemistry department at Oxford University, field ion microscopy was used to investigate the chemisorption of monolayers of oxygen on nickel and other metals. George David William Smith (1943–; FRS 1996) carried out research with Anderson using the technique and then moved to the metallurgy department, headed by Peter Hirsch. There he played a major and continuing role in atom probe/field ion microscopy for the elemental analysis of materials on the atomic scale. He has applied the technique to problems in phase transformations and microstructural stability and has led the development of the position sensitive atom probe which gives three-dimensional images with near atomic resolution of individual elements, finding wide-ranging applications.

In a different area of high resolution microscopy now available to the metallurgist and materials scientist, for the study of surface features, the scanning tunnelling microscope was invented by Gerd Binnig (1947–) and Heinrich Rohrer (1933–) at IBM, Zurich; they received the Nobel Prize for Physics in 1986, together with Ernst Ruska, for his role in the invention of the electron microscope. The scanning tunnelling microscope has a very fine tip (made from tungsten or a platinum–iridium alloy) is scanned across the specimen at a very small distance from the surface by piezoelectric transducers. A constant distance of the tip from the surface is maintained. By 1981 Binning and Rohrer had reduced the tip size to a single atom to reveal very fine contour details of the surface, for example images of single atoms. Royal Society Fellows working with this type of microscope in Britain are David Tabor (1913–; FRS 1963) at Cambridge and Donald Pashley at Imperial College (on semiconductor materials). Another type of scanning probe instrument is the atomic force microscope of which Calvin Forrest Quate (1923–; FMRS 1995) was a co-inventor (1986). In this microscope the force experienced by the probe is used to form an image; applications have included modifying surfaces at atomic level for electron device fabrication. In the field of investigations on the atomic scale it was John Bernard Pethica (FRS 1999) who introduced the concept of force between tip and surface in scanning tunnelling microscopy, which led to the discovery of atomic force microscopy; he has recently modified the latter technique to permit mechanical characterisation of chemical bonds. Also he is the inventor of nanoindentation which has found important applications in the thin film and coating industries.

The interaction of neutron beams with materials has become an important technique for obtaining information on structural features. Colin George Windsor (1938–; FRS 1995) has made important contributions to industrial applications of neutron beams, including the use of pulsed beams, and the investigation of critical defects and internal stresses in engineering products; he has applied pattern recognition techniques and neural nets to areas such as weld defect classification. (See Chapter 17 for other examples of research on the use of neutron beams and synchroton radiation and also of spectroscopic techniques for the investigation of structural features of materials).

Alongside these advances, light microscopy has continued to be an important technique for investigating structural features of metals and other materials. The research of Samuel Tolansky (1907–1973; FRS 1952) included surface microtopography, in which field he became a leader. In 1943 he evaporated metal on to surfaces, producing highly reflective films which gave multiple beam interference (which provides extremely narrow and sharp fringes), and revealed surface steps of only 1.5 nm. He investigated a range of crystals, including diamonds, and his work covered features such as slip bands, etch pits and growth spirals. In hardness testing he used interferograms to characterise the indentations, revealing features such as the directional dependence of hardness of bismuth crystals. He studied the profiles of metal surfaces turned by diamond loads. The work on optics of James Dyson (1914–1990; FRS 1968) at the Associated Electrical Industries (AEI Ltd) research laboratory included the development of an interferometer microscope used for example for the study of the depths of defects in machined or worn surfaces and the effect of differential etching of constituents in alloys. The research of Cecil Reginald Burch (1901–1983; FRS 1944) included the construction of a reflecting microscope with aspherical reflecting surfaces and a very wide aperture; this microscope allows increased working distances between the specimen and the objective, which is advantageous in hot stage microscopy and for observations of fracture surfaces.

The study of surface features was a particular interest of Richard Edmund Reason(1903–1987; FRS 1971), whose career was spent with the firm of Taylor, Taylor and Hobson (Kapella) Ltd. (later Rank Taylor Hobson Ltd). Among Reason's many inventions were the 'Talysurf' and the 'Talystep' stylus-type instruments; the high sensitivity of the latter led to important industrial applications, for example for measuring the thickness of surface films on semiconductor devices.

In the technique of acoustic microscopy, invented by Calvin Quate, an ultrasonic wave is passed through the specimen and the transmitted sound is used to form an image on a computer screen. Eric Ash (later Sir Eric, 1928–; FRS 1977) has done important work in this field for example to reveal features not displayed by light microscopy.

CHAPTER 13

Fundamentals of Elastic and Plastic Deformation

Introduction

In the theory of elasticity, following the early milestone of Hooke's law in 1660 (Robert Hooke, 1635–1703; FRS 1663), the 19th century saw important developments. However, the theory of plastic deformation and fracture of metallic crystalline aggregates progressed only slowly during the century, awaiting the development of understanding of the crystalline nature and microstructure of metals. Many Fellows of the Royal Society in various countries carried out research on deformation phenomena. This chapter is concerned primarily with fundamental aspects, including the introduction of the concept of dislocations in the 1930s and its early application to deformation behaviour. The field of deformation is pursued in greater detail in Chapter 14 in the context of metallurgical challenges, involving topics such as creep, fatigue and fracture.

Elastic behaviour

As early as 1807 Thomas Young (1773–1829; FRS 1794) described the modulus of elasticity in tension and compression in his lecture on 'Passive Strength and Friction'. Although the original definition was obscurely stated it was later restated by others in essentially its present form;Young's modulus: the ratio between the stretching force and the resultant strain. From experiments in tension and compression of bars he noted that Hooke's law is only obeyed up to a certain limit of stress and that longitudinal deformation is accompanied by changes in lateral dimensions. Young introduced the term 'energy' to represent the product of the mass of a body and the square of the velocity. He also analysed stresses leading to the fracture of elastic bodies produced by impact, considering the kinetic energy.

Claude-Louis-Marie-Henri Navier (1785–1836) formulated the general equations of elasticity in 1821, while Augustin Louis Cauchy (1789–1857; FMRS 1832) had discovered by 1822 most of the pure theory of elasticity; among his conclusions he showed that stress is expressible by six component stresses, and that principal planes of stress exist. In 1828 Siméon Denis Poisson (1781–1840; FMRS 1818) obtained equations of equilibrium for isotropic elastic solids identical with Navier's, and these were applied by Gabriel Lamé (1795–1870) and Benoit-Pierre-Emile Clapeyron (1799–1864) to many problems of vibration and static elasticity. Also, Poisson, in 1829, having considered the ratio (which became known as Poisson's ratio) of lateral contraction to longitudinal extension of a bar under tension, deduced a value of 1/4.

Félix Savart (1791–1841; FMRS 1839) did experimental work using vibrations which demonstrated elastic anisotropy in a range of materials. He introduced a new way of obtaining information about the structure of metals, from the dependence of elasticity, deduced from the variation of acoustic nodal patterns from laminae (plates) cut from different planes of materials. He investigated various substances including, notably, metals and alloys, for example cast ingots of lead, tin, zinc and copper and in 1829 he reported some important observations. For example, he wrote

> Metals have hitherto been regarded as the solid substances which most closely approached conditions of homogeneity; they have been regarded as assemblages of an infinite number of little crystals united together without order and at random, and there was no suspicion that in any mass of metal whatever there could be differences of elasticity or cohesion as great as, or perhaps greater than, those observed in a fibrous body such as wood.

Tests on plates cut from various positions in a large cylinder of lead showed elastic axes which changed with the direction of excitation, and were thus not isotropic. He concluded that metals 'are not crystallised regularly'.

Savart examined a lead ingot whose crust had been pierced and drained of liquid before solidification was complete, and observed (what would now be termed dendrites) 'numerous octahedral crystals arranged in parallel rows, which form a great number of distinct systems...' and 'Seen with the magnifying glass, the little crystals which compose each of these systems seem to be grouped around three directions at right angles to each other, and they are arranged so that their axes are parallel to each other...' He also investigated the effect of different casting conditions, including vibration during solidification, which gave rise to a structure with a high uniformity of elasticity. For zinc, a rolled plate had a uniform preferred orientation. In studies of various metals he showed that annealing had little or no effect unless there was prior cold work, in which case the frequency and the orientation of the elastic nodes changed; the effects varied according to the metal and were sensitive to impurities.

The field of elasticity was of interest to a number of other Fellows of the Royal Society. The research of Adolph Theodor von Kupffer (1799–1865; FMRS 1846) included a long series of investigations on metals. The Russian government had founded an Institute of Weights and Measures to which was also entrusted the task of investigating factors which can affect the standards of measurements, namely temperature and elasticity. Kupffer carried out experiments on the torsional vibrations of metals in the form of wire: iron, copper, platinum, silver and gold including damping effects. He investigated after-strain effects, showing that deflections in steel bars do not disappear immediately after the removal of the load, but that a progressive decrease occurs over a period of several days. Measurements of the temperature dependence of the elastic modulus were also made.

Thomas Young

Siméon Poisson

James Clerk Maxwell

Ernst Neumann

In Britain, workers in elasticity included William Thomson (later Lord Kelvin, 1824–1907; FRS 1851) who did important work on both experimental and theoretical aspects. He introduced the concept of internal friction and in 1865 studied the elastic after-effect in quenched steel. Working with James Prescott Joule (1818–1889: FRS 1850) he made studies of the thermal effects of strains on solids and liquids; Thomson calculated the heating that should occur in deformed elastic materials and Joule measured the temperature rise in a compressed spring. William's brother James Thomson (1822–1892; FRS 1877) also worked in the field of strength of materials, including the elasticity and strength of spiral springs. James Clerk Maxwell (1831–1879; FRS 1861) presented a paper to the Royal Society of Edinburgh in 1850 'On the equilibrium of elastic solids'. At Cambridge, George Gabriel Stokes (later Sir George, 1819–1903; FRS 1851) and JohnWilliam Strutt (later Lord Rayleigh, 1842–1919; FRS 1873) included elasticity theory in their research. The work of Stokes presented in 1845 which dealt with viscous fluids included theories of the equilibrium and motion of elastic solids; a comparison was made of solids with viscous fluids, involving reference to the plastic deformation of metals.

Important work on the mathematical theory of elasticity by Franz Ernst Neumann (1798–1895; FMRS 1862) dealt with the elastic properties of crystals in relation to symmetry. He also studied residual stresses remaining in a body after removal of stresses which have produced plastic deformation. Woldemar Voigt (1850–1919; FMRS 1913) during his period at Königsberg was influenced by Neumann and chose elasticity as one of his main research interests. In the course of his research he carried out a classical series of experiments on the elasticity of crystals, e.g. of rock salt, including anisotropic behaviour of single crystals. His work on longitudinal impact of prismatic specimens included experiments on metal rods. Gustav Robert Kirchhoff (1824–1887; FMRS 1875), a student of Neumann's, worked on aspects of the theory of elasticity such as the bending of plates and the deformation of thin bars. Among others who made contributions in the field were Émile Hilaire Amagat (1841–1915; FMRS 1897); Anders Jöns Ångström (1814–1874; FMRS 1870; Wilhelm (William) Eduard Weber (1804–1891; FMRS 1850); Gustav Heinrich Wiedemann (1826–1899; FMRS 1884) and Jean Victor Poncelet (1788–1867; FMRS 1842).

The contributions of Poncelet in industrial mechanics included the publication of of *Introduction à la Mécanique Industrielle Physique et Expérimentale* first printed in 1841; this stemmed from a course of lectures given and published some ten years previously. This book incorporated a section on the strength of materials, dealing with mechanical test results for various materials, together with a discussion of their significance for the design engineer; he used graphs of 'traction vs stretch' i.e. tensile stress vs strain.

In the late 1870s, four years after Alexander Blackie William Kennedy (later Sir Alexander, 1847–1928; FRS 1887) had become Professor of Engineering at University College London he established a laboratory, the first of its type in England, where engineering students could obtain practical experience to supplement their theoretical knowledge. Kennedy's research included the strength and elasticity of materials and the strength of riveted joints; he also designed steel and concrete structures.

Jean Poncelet

Adolph Kupffer

Ludwig Prandtl

Horace Lamb

Augustus Love

The substantial book *A History of the Theory of Elasticity and of the Strength of Materials,* by Isaac Todhunter (1820–1884; FRS 1862), which was published posthumously, the first volume appearing in 1886, contains detailed information on the findings of these and many other researchers. The editing and extension of this book, leading to its publication, was carried out by Karl Pearson (1857–1936; FRS 1896), who graduated from Cambridge in mathematics and was appointed to the Chair of Applied Mathematics and mechanics at University College London, and did original research in elasticity.

Louis Napoleon George Filon (1875–1937; FRS 1910) succeded to Pearson's chair in 1904; as a mathematician he carried out important research in mechanics, including elasticity and the theory of 'generalised plane stress'. He also collaborated with Ernest George Coker (1869–1946; FRS 1916), an engineer, who was appointed to the chair of Civil and Mechanical Engineering at University College London in 1914. Coker had previously developed an interest in the stress-optical effect, and had started to experiment to develop the subject of photoelasticity, in which James Maxwell had previously developed the technique on photoelastic stress analysis using transparent model systems. The work of Coker and Filon included the publication of *A Treatise on Photoelasticity* in 1931, a classic book. Coker's research on photoelasticity made important contributions in relation to the determination of stress distributions in metals, materials and structures, including devising apparatus and models. Among his publications were papers on the effect of discontinuities such as holes, cracks and scratches, and on stress distributitions in fusion joints. Other important contributors to elasticity theory were Horace Lamb (1849–1934; FRS 1884) and Augustus Edward Hough Love (1863–1940; FRS 1894). Love's treatise on *The Mathematical Theory of Elasticity,* first published in 1892–93, had a wide scope including the equilibrium of solids, the propagation of waves in elastic solid media and torsional bending of rods, plates and shells. In Germany, Ludwig Prandtl (1875–1953; FMRS 1928) who was particularly noted for his work on aerodynamics, included elasticity and the strength of materials in his activities. For example, his doctoral thesis (1899) was concerned with lateral buckling of beams of narrow cross section; he devised a technique for the accurate determination of the critical load, which was subsequently widely used.

Stephen Prokofievitch Timoshenko (1878–1972; FMRS 1944) in Russia and subsequently in the USA, worked to advance mechanics as a science and to promote its application to practical engineering problems. In addition to his research, he published a number of influential textbooks including the themes of strength of materials and theory of elasticity. His book *History of Strength of Materials* (1953) gives a detailed account of the field.

Plastic deformation

Among work early in the 19th century, Louis Bernhard Guyton de Morveau (1737–1816; FRS 1788), in 1809, reported data for the tensile strength of platinum and various other metals (iron and copper) by determining the weight that 2 mm diameter wires would support without fracture. He commented 'the force of cohesion is not appreciably diminished while the ductility of a metal permits its parts to slide over each other without breaking'. During the 20th century many major contributions have led to extensive fundamental understanding of mechanical behaviour, involving studies of single crystals and polycrystalline aggregates.

James Alfred Ewing (later Sir James, 1855–1935; FRS 1887) and Walter Rosenhain (1875–1934; FRS 1913) in 1899 at Cambridge established the process of slip as a fundamental mechanism of plastic deformation. Rosenhain had commenced his doctoral work on steam jets, but arising from a consultation with Ewing he turned to metals for his research. Ewing suggested to Rosenhain that they should examine by optical microscopy how the grains of a metal behave during plastic deformation and Rosenhain visited metallurgical laboratories to learn the techniques of polishing samples for microscopical studies. In their experiment a strip of metal was placed on a microscope stage, fitted with a straining device, so that it could be observed, as it was stretched. Ewing recorded how during deformation, sharply defined parallel black lines appeared on the surface, within the area of a grain and changing their direction from one grain to another. The bands increased in number with increasing deformation, and sometimes new systems appeared, crossing the earlier lines. Both Ewing and Rosenhain had come to the conclusion when they met the morning after the experiment that the lines represented finite slips, occurring on parallel planes within the grain. This indicated the crystalline nature of the grains, and that deformation occurs by successive layers of the crystal slipping over one another like a pack of cards might do. By successive slips in different directions grains could change their shape and adapt to the change of the bulk metal. The black lines were thus steps, casting a shadow when illuminated. These 'slip-bands' provided the key to the plastic deformation of metals and the discovery was published in March 1899. The fact of slip was not universally accepted at once. However, Rosenhain, by depositing copper on a surface after slip and then cutting a section through the two metals, later showed the slip steps in profile, their height being only about 0.001 mm; it was proved by measurements that the blocks were not distorted. A paper (1902) of which Thomas Andrews (1847–1907; FRS 1888) was a co-author reported microscopical observations on the deformation of platinum, showing the formation of slip bands on the polished surface of a compressed sample. Rosenhain and Ewing also discovered recrystallisation and grain growth during the annealing of deformed metals such as lead and zinc.

George Thomas Beilby (later Sir George, 1850–1924; FRS 1906) investigated matter in the solid state, stemming mainly from his interest in the behaviour of metals when exposed to gases at high temperatures such as he observed in his manufacturing practice. This work, which could be carried out with comparatively simple resources in his own study, began with the investigation of the changes in structure that occurred when metals such as iron and copper were exposed to ammonia at temperatures from 700 to 800 °C, and which caused complete deterioration of their properties. Initially his aim was to find a metal or alloy which would be reasonably stable under such conditions. However, he proceeded to scientific investigations of the behaviour of metals subjected to various mechanical and thermal treatments. The outcome was his theory of the hard and soft states in metals, which formed the theme of the May Lecture of the Institute of Metals in 1911, and which profoundly influenced the development of physical metallurgy. Beilby was a skilful investigator, and his book on *Aggregation and Flow of Solids* represented 21 years of work. He spent much time on an investigation of cutting, filing, grinding and polishing the surface of solids, and concluded that in all cases a thin layer was formed which differed from the underlying material. He proposed that the action of the tool or polishing agent moving over the surface produced mobility in this thin layer such that it behaves like a liquid and is subject to the action of surface tension. Many of Beilby's early experiments were carried out with calcite, which was very suitable for microscopical studies and which showed cleavage fractures, producing flat surfaces on which no mechanical work had been done.

Beilby proceeded to a new field of study on the crystalline and vitreous (flowed) states of matter, and he carried out experiments which led him to introduce a new theory of the hardening of metals by plastic flow during cold working. He proposed that hardening of metals results from the formation, at all the internal surfaces of slip or shear, of mobile layers similar to those produced by polishing of the outer surfaces. These layers were assumed to remain mobile only for a very short time, and then 'solidified' in a vitreous state, thus forming a cementing material at all surfaces of slip and shear in the material. He concluded that the strength of the work-hardened metal depended on the 'texture' or type of structure developed and on the proportions in which the two states were present. The vitreous (work-hardened) state could be reconverted into the crystalline state by a suitable heat treatment. Beilby's hypothesis of the crystalline and vitreous states of metals was widely accepted by English and American workers, but not to the same extent by Continental workers. Rosenhain developed Beilby's theory and applied it in detail to the deformation of iron and steel. Some workers doubted whether the crystalline state was as mechanically unstable as Beilby supposed and later studies led to the modern interpretation of deformation mechanisms. For example, collaborative work in the 1930s by Herbert John Gough (1890–1965; FRS 1933) using X-rays to study the structural changes produced by plastic deformation of metals, showed the formation of small ' crystallites' 10^{-4}–10^{-5} cm in diameter.

The wide-ranging research of Willam Chandler Roberts-Austen (later Sir William, 1843–1902; FRS 1875) included a project, begun in the late 1880s, on the effects of small proportions of 'impurities' on the properties of metals. From experiments involving trace additions of a number of elements to high purity gold he discussed the embrittling tendency observed in relation to the atomic volumes of the elements present. The presence of lead or antimony gave rise to well developed intercrystalline fractures. In a paper published in the *Transactions of the Royal Society* in 1896 he speculated on the possible role of grain boundaries in determining the mechanical properties.

In France, studies of the tensile properties of non-ferrous metals were made by Henri Louis Le Chatelier (1850–1936; FMRS 1913), who observed the general decrease in strength and increase in ductility with increasing temperature. He also showed that the temperature at which blue brittleness appears in steel depends on the speed of testing.

Karl Pearson

George Beilby

Alexander Kennedy

Andrew Robertson

Gilbert Cook

The work of William Unwin (1838–1933; FRS 1886) included investigations of the yield point of iron and steel and the effect of repeated straining and annealing. Also in measurements of the tensile properties of various copper-based alloys up to about 450 °C he noted the marked effect of bismuth in reducing the strength.

Andrew Robertson (1883–1977; FRS 1940) and Gilbert Cook (1885–1951; FRS 1940) collaborated in research at Manchester University. In 1911 they published an important paper on the strength of thick, hollow cylinders under internal pressure; their aim of using a combined stress condition was to determine the relevance of the theoretical criteria for the breakdown of elasticity. For this work they devised a very accurate mirror extensometer. A second paper (1913) presented results on the transition from the elastic to the plastic state in mild steel, with reference to the yield drop. Subsequently Robertson did further research on the yielding of Armco iron and on the testing of mild steel beams; he established that the stress drop at yield is a property of the metal and not simply a consequence of the testing conditions. Robertson also carried out important work on struts using steel and aluminium alloy relevant to aircraft design. After World War I he continued his work on struts, including a thin tubular type using mild steel and nickel–chromium steels, investigating buckling behaviour. In this work he collaborated with Richard Vynne Southwell (later Sir Richard, 1888–1970, FRS 1925).

Concerning plastic deformation at cryogenic temperatures important collaborative work was carried out by Sir Robert Abbott Hadfield (1858–1940; FRS 1909). A paper presented in 1905 by Hadfield and Sir James Dewar (1842–1923; FRS 1877) reported results from over 500 specimens tested immersed in liquid air at the Royal Institution. A wide range of materials was investigated including iron, plain carbon steels, iron–nickel alloys and iron–nickel–chromium alloys and nickel. Later, extensive research with Wanda Johannes de Haas (1878–1960) at the Leyden cryogenic laboratory in 1934 involved testing in liquid hydrogen: pure iron, carbon steels, 39 alloy steels, copper, nickel and several non-ferrous alloys. These cryogenic studies demonstrated that iron had increased tensile strength with virtually zero ductility at –182 °C. Nickel, copper and aluminium also showed an increase in strength, and an increase in ductility. Plain carbon steels behaved similarly to iron. Certain alloy steels containing high nickel contents, and in some cases with additions of manganese, showed combinations of high tensile strength and excellent ductility. The formation of martensite from metastable austenite was found to occur in an iron–31.4% nickel–0.7% carbon during tensile straining.

The research on metals by Edward Neville da Costa Andrade (1887–1971; FRS 1935) with various co-workers, included fundamental studies of deformation mechanisms. He found that circular wires of mercury stretched plastically at –78 °C contracted elliptically into ribbon shapes along their lengths, and showed chisel-edge fractures; there were regular semicircular markings along each face of a ribbon. Andrade reported these observations in 1914, interpreting the mechanism of tensile deformation of a 'single crystal' by glide on slip bands. Returning to the subject in 1937 he developed a travelling furnace method for growing single crystals, and also began some careful studies to establish more precisely than hitherto, the crystallographic glide elements of body-centred cubic metals. After the 1939–45 war he interpreted the effects of surface condition on the plastic yield strength of cadmium and zinc crystals. In 1951 he developed a new method for growing single crystals of metals of higher melting point, and investigations of the stress-strain curves were made. The results, among which were the discovery of the 'easy glide' part of the stress-strain curves, subsequently played a key part in the development of the theory of strain hardening.

The production of single crystals of metals by heat treatment procedures marked an important development in the study of mechanical properties. Daniel Hanson (1892–1953) prepared large aluminium crystals at the National Physical Laboratory during the first world war. Subsequently the work of Henry Cort Harold Carpenter (later Sir Harold, 1875–1940; FRS 1918) from 1918 onwards included the growth and behaviour of metallic crystals. Studies of the recrystallisation of electrolytic iron in the form of thin strips, on being heated or cooled through the A_3 change point, showed that for certain critical ranges of thickness, extraordinarily large crystals were obtained. In 1920 the first of a number of papers with Miss Constance Elam (1894–1995; holder of the Royal Society's Armourers and Brasiers' Fellowship from 1924–29) was published on the production of single crystals of aluminium, by the strain–anneal method. These researches represented the beginning of a new era in physical metallurgy, involving the systematic determination of the conditions necessary to change the millions of crystals in a typical polycrystalline aggregate into a common orientation of single crystals. Carpenter and Elam made a careful study of the plastic deformation behaviour of single crystals of aluminium. The strain–anneal method of single crystal growth was later applied by other workers to various metals.

In Germany important research on the growth and deformation of crystals was carried out by Michael Polanyi (1891–1976; FRS 1944) and co-workers, which led Polanyi to his contribution to the concept of dislocations. Among other investigators who carried out important investigations of crystal deformation were Eric Schmid (a co-worker of Polanyi's), and Walter Boas; Schmid and Boas were the authors of *Kristallplastizitatät*, published in 1935, the same year as Constance Elam published *The Distortion of Metal Crystals*.

In the ferrous field, work on grain growth in iron and steel was carried out in Swansea by Charles Alfred Edwards (1882–1960; FRS 1930) and Leonard Bessemer Pfeil (1898–1969; FRS 1951) in relation to problems of deep pressing problems of sheet steel. It was found that local decarburization of the steel made it liable to excessive grain growth after critical strain. Single crystals of iron were prepared and mechanical property measurements made. Edwards and Pfeil also demonstrated the increase in strength accompanying a reduction in grain size.

The interest of Geoffrey Ingram Taylor (later Sir Geoffrey, 1886–1975; FRS 1919) in the strength of metals that had been stimulated by his work with Alan Arnold Griffith (1893–1963; FRS 1941) was resumed in the light of the research of Carpenter and Elam. During the mid-20s Taylor was mainly concerned with elucidating the mechanisms of plastic flow in metal single crystals. To produce the large crystals required to obtain homogeneous loading and accurate observations of dimensional changes he relied on the skill and experience of Miss Elam and on assistance from others in matters such as the design and operation of X-ray apparatus. Among the main principles established he showed that, at a macroscopic level of deformation, face-centred cubic crystals deform plastically by a specific type of crystallographic shear, which is activated when the corresponding component of applied stress reaches a critical value; body-centred cubic crystals deform plastically by 'pencil glide' in specific lattice directions but over non-specific planes. Subsequently Taylor carried out theoretical work to predict the bulk plastic behaviour of a random aggregate of crystals and he collaborated with colleagues in experiments on aggregates of crystals. William Scott Farren (later Sir William, 1892–1970; FRS 1945) collaborated with Taylor, designing apparatus for measuring the distortion of aluminium in compression and also for measuring the heat developed during the plastic extension of metals.

Rodney Hill (1921–; FRS 1961), in his wide-ranging research on the deformation of solids, has included the theme of the plasticity of crystalline aggregates, for example, of the face-centred cubic type.

Among the advances made in the period following the 1939–1945 war the work of Norman James Petch (1917–1992; FRS 1974) demonstrated and explained the relationship between the increase in yield strength and fracture strength of steels and a reduction in the grain size of the material. In 1961, Petch reported on the room temperature flow stress at constant strain for a number of polycrystalline metals and alloys. He showed that it followed the same relationship as that for the lower yield stress in iron; the grain size term arose from the resistance at a grain boundary to the formation of a slip band.

Fundamental investigations of deformation mechanisms of metallic single crystals of various metals e.g. in precipitation hardening aluminium alloys have been made by a number of contemporary fellows of the Royal Society, for example Anthony Kelly (1929–; FRS 1973) and Sir Robert William Kerr Honeycombe (1921–; FRS 1981).

Among important current themes is the increase in strength of certain intermetallic compounds with increasing temperature in contrast to the typical behaviour of metallic materials involving a decrease in strength as the temperature increases. The increase in strength with increasing temperature shown by Ni_3Al and nickel–based superalloys containing large proportions of this compound is of great importance in gas turbine engine technology and Sir Peter Bernhard Hirsch (1925–; FRS 1963) has contributed to the elucidation of this behaviour in terms of the complex dislocation effects involved. Robert Wolfgang Cahn (1924–; FRS 1991) has investigated the role of ordering in affecting yield stress and creep resistance in intermetallic compounds such as Fe_3Al.

A field which has become of considerable theoretical and industrial interest concerns the phenomenon of superplasticity, in which extremely high levels of ductility can be achieved in alloys depending on fine grain structures and appropriate conditions of temperature and deformation. Michael James Stowell (1935–; FRS 1984) and Donald William Pashley (1927–; FRS 1968) are among those who have been involved in fundamental investigations and in bringing superplastic forming into industrial practice. This included the development of alloys based on the aluminium–copper–zirconium system, using control of composition and thermomechanical treatment to achieve the fine grain size required for superplastic behaviour.

During World War II, Percy Williams Bridgman (1882–1961; FMRS 1949) carried out investigations for the Watertown Arsenal on the plastic properties of metals under high stress. From his results he suggested the possibility of constructing vessels to withstand pressures much greater than those determined by the elastic properties of simple or pre-stressed cylinders. In his Bakerian Lecture of 1950 he described apparatus for generating pressures up to approximately ~9,800 MPa and some of the results obtained by its use. In more recent years, Sir Bernard Crossland (1923–; FRS 1979) has included high pressure experimentation in his research, investigating aspects such as the strength of thick walled cylinders subjected to internal pressure.

Hardness testing in various forms has been a long established technique for obtaining information on mechanical properties of metallic materials. David Tabor (1913–; FRS 1963) has carried out important research in this field; for example, providing explanations of indentation hardness data in terms of yield properties, determining the effect of temperature and loading time on hardness, and correlation of hardness with creep properties. He is the author of the classic book, *The Hardness of Metals* (1951).

Among engineers, Fellows of the Royal Society, whose interests included the mechanical behaviour of metals in engineering construction, the work of Theodore (Todor) von Kármán (1881–1963; FMRS 1946) was focussed particularly in the aeronautical field, but also impacted on civil engineering. His early work at Budapest concerned the theory of buckling and compression tests on long slender columns. While at Göttingen his work included the strength of corrugated tubes and the deformation of thin walled tubes; also he made the first experiments on plastic flow and fracture of specimens under external hydraulic pressure thus contributing to the field of flow and fracture of materials under combined stresses. His theoretical findings from his research on vortices found many applications, for example, the resonance due to periodic vortices which led to the collapse of the bridge over the Tacoma Narrows in 1940. Later, collaborative work was concerned with the theory of propagation of plastic deformation.

In 1935 an important event in the sphere of civil engineering in Britain occurred with the setting up of the Steel Structures Research Committee. The code of design for the use of steel as a structural material for buildings was outdated, and the remit of the Committee involved reviewing existing methods for the design of structures, including bridges, investigating the application of modern theory to the design of steel structures, and making recommendations aiming to improve efficiency and economy in practice. Reginald Edward Stradling (later Sir Reginald, 1891–1952; FRS 1943) in his capacity as director of the Building Research Station of the Department of Scientific and Industrial Research had initiated the formation of this Committee, and served as its Executive Officer. In its five and a half years of existence, the Committee produced voluminous reports, including one of the subject of welding in steel structures; the work had far-reaching results.

In civil engineering design and construction the importance of the plastic deformation of metals, in particular steel, has assumed great importance since the 1939–1945 war. John Fleetwood Baker (Baron Baker of Windrush, 1901–1985; FRS 1956) was a leading civil engineer in this field. In 1930 he became the Technical Officer in the Steel Structures Research Committee. During the 1930s Baker began to work on the theme of plastic theory in engineering design, which he pursued throughout his career. His wartime work involved the application of this to the design of the Morrison air raid shelter of which more than a million were produced. By around 1948 the plastic theory could be used in practice to calculate the stages of collapse of a steel frame and hence lead to the economical and safe design of a structure.

Gilbert Roberts (1899–1978; FRS 1965) was a leader in the field of structural engineering, whose career included work on many major civil engineering projects such as bridges, including the Sydney harbour, Severn and Humber bridges. He was a pioneer in the use of high tensile steels and in the introduction of welding technology, for example, all-welded plates.

The civil engineering projects of Oleg Alexander Kerensky (1905–1984; FRS 1970), including particularly bridge construction, involved the pioneering use of high tensile steel rivets as well as bolts; he also utilised composite action between steel and concrete in construction.

Among a number of contemporary Fellows in the field of structural engineering Patrick Joseph Dowling (1939–; FRS 1996) has been particularly concerned with the design of steel-plated structures such as those used in bridge girder and ship hull design. Other aspects of his work are inelastic stability and damage resistance of major steel components in offshore platforms, and the strength of composite steel- concrete buildings.

Dislocations

The introduction and early development of the concept of dislocations is associated particularly with the names of three scientists: Geoffrey Taylor, Egon Orowan (1902–1989; FRS 1947) and Michael Polanyi. The account of their independent discoveries is an interesting one, and all three published papers on the dislocation mechanism of plastic deformation in crystals in 1934.

Taylor believed that the perfect lattice was very strong but that what he called a 'dislocation' could move under a negligibly small stress; this later became known as the 'Peierls force', named after Rudolf Ernst Peierls (later Sir Rudolf, 1907–1995; FRS 1945) and was not negligibly small in some materials. Comparison with experiment was given in Taylor's papers for both metals and rock salt and, for example, he showed that the density of dislocations deduced fitted reasonably with observations of the amount of stored energy which he had measured. His work was limited to cubic crystals and his analysis supposed that the dislocations in a cold-worked material are uniformly distributed in space. Taylor's contributions were of major importance, for example showing that the strength of a cold-worked material is due to the strains around dislocations, and that the flow stress depends on dislocation density and arrangement. Furthermore, his calorimet-

Geoffrey Taylor

Egon Orowan

Michael Polanyi

John Eshelby

ric technique for measuring stored energy was the forerunner of many similar studies carried out subsequently by others.

Michael Polanyi and his collaborators, working in Berlin grew crystals, principally metallic, investigated their properties under stress and found many interesting phenomena. In particular Polanyi was concerned with the shear and the rupture strength of crystals, which was less than a hundredth of what simple theoretical considerations would suggest to be the external force required to separate two crystallographic planes. According to the theory of Polanyi, the much lower rupture strength is caused by imperfections in the lattices around which the stress lines concentrate so that a wedge of increasing depth is formed. The situation is similar with respect to shearing and Polanyi's assumption required a lattice imperfection around every thousandth atom.

In 1928 Egon Orowan went to the Technical University of Berlin, where, after initially studying mechanical engineering and then electrical engineering, he transferred to physics, influenced by Richard Becker, the recently appointed Professor of Theoretical Physics. Alongside Orowan's studies in electrical engineering,which included the task of computing, designing and drawing a reversing rolling mill, he also spent some time weekly in the advanced physics laboratory course. He was asked by Becker whether he would be interested in checking experimentally a 'little theory of plasticity' he had worked out three years before. Orowan accepted this invitation and his task required the making of single crystals of zinc, tin etc. and determining whether they possessed any plasticity at liquid air temperature. However, Polanyi, Walther Meissner (1882–1974) and E. Schmid showed in 1930, before Orowan's equipment and crystals were ready, that the ductility of these metals was nearly as good in liquid air as at room temperature. Orowan was not able to complete his experiments before the work of Polanyi and his co-workers became known, but they formed the basis of his Diplomarbeit in February 1929. He recounted that one afternoon he was left with one zinc crystal which was bent when he accidentally dropped it. He straightened it, left it to anneal for some time, and tried a practice run; the crystal was found to extend with sharp jerks instead of flowing smoothly, and from this observation, often repeated, he made various deductions and was led on to the concept of dislocations. Orowan and Taylor independently developed the idea of dislocations and their role in plasticity. Also Orowan was in touch with Polanyi who (as referred to above) had recognised a similar concept.

Work by Orowan on the development of dislocation theory was delayed by the fact that his subsequent doctoral thesis was on a different subject, namely, the cleavage of mica, a highly anisotropic material. In 1937 Orowan moved to the physics department of the University of Birmingham, where his main interest was the theory of fatigue; during this period he introduced Rudolph Peierls to the problem of the structure of a dislocation core and the stress required to move a dislocation through the lattice, which Peierls solved.

Nevill Francis Mott (later Sir Nevill, 1905–1996; FRS 1936) as one of the outstanding theoretical physicists of the 20th century, included dislocation theory in his wide-ranging research interests. When Orowan went to the Cavendish laboratory, Cambridge in

Nevill Mott

Frederick Charles Frank

1939, his research included the theory of precipitation hardening and work in collaboration with Mott. The mechanism by which the formation of solid solutions and precipitates produce hardening had been a long standing topic of discussion. For example, Rosenhain had attributed the increase of hardness hardening on alloying to a warping of the lattice planes by the introduction of foreign atoms. In 1921 Rosenhain described a relation between the hardening effect of different metals in solid solution in a common solvent metal, the effect being approximately inversely proportional to their solubilities in the solid state. Mott and Orowan showed that when precipitate particles become large and widely spaced in the overaging period and the flow stress decreases, dislocations 'bulge' forward into the gaps between the particles, and eventually become separated from the particles, leaving them encircled by small dislocation loops. Also at Cambridge Sir William Lawrence Bragg (1890–1971; FRS 1921), following discussions with Orowan, invented the bubble model of a crystal structure which was used to illustrate features of plastic deformation and dislocation behaviour and to popularise the theory. James Dyson (1914–1990; FRS 1968) working at the Associated Electrical Industries (AEI Ltd) research laboratories used optical examination, including diffraction, to reveal details of the features of bubble rafts providing information on dislocations and other defects relevant to metallurgists.

Mott also worked with John Douglas Eshelby (1916–1981; FRS 1974) who became one of the chief authorities on the theory of dislocations and internal stresses in solids, commencing in the period following the 1939–1945 war when experimental evidence of the dislocation was largely lacking, but its theoretical implications were being explored in several centres world wide. Eshelby collaborated with Frederick Charles Frank (later Sir Charles, 1911–1998; FRS 1954), Frank Reginald Nunes Nabarro (1916–; FRS 1971) and Bruce Alexander Bilby (1922–; FRS 1977). Eshelby's work included the theory of mechanical damping in metals, and also solving the problem of determining the elastic field in a region that contains an 'inclusion' or has elastic constants different from the remainder (an inhomogeneity). He became interested in the newly discovered metallic 'whiskers' which displayed the enormous strength predicted for a perfect crystal, and he gave a tentative theory of whisker growth. Also he did important work in the field of point defects in solids, concerning the volume changes produced by these defects, and their energies and interactions.

The distinguished, wide-ranging research of Sir Charles Frank involved important investigations of dislocations including their sources and consequences in interfaces and crystal growth; he demonstrated that screw dislocations lead to growth spirals on the surfaces of crystals.

The role of contemporary Fellows of the Royal Society

Fundamental investigations of deformation behaviour, both experimental and theoretical, involving electron microscopy and dislocation theory have constituted a major area of work over some four decades, involving a substantial number of contemporary Fellows of the Royal Society.

The first experimental verification of the existence of dislocations was provided by Robert Cahn in 1947 through the polygonisation of bent metal crystals of zinc and aluminium. Soon after this John Wesley Mitchell (1913–; FRS 1956) made dislocations visible in silver chloride by a 'decoration' technique in which silver was precipitated using light (1953). The development and application of transmission electron microscopy to metal foils, for example the demonstration of dislocation movement by Peter Hirsch, opened up a new era in the experimental field.

The pioneering research of Alan Howard Cottrell (later Sir Alan, 1919–; FRS 1955) has spanned many aspects of mechanical behaviour, linking particularly with the role of dislocations. For example, he demonstrated that yield point phenomena in carbon steels derive from the interaction of carbon atoms with dislocations (Cottrell atmospheres). The research of Nabarro has made major contributions to the understanding of dislocations and deformation, including the theory of forces resisting dislocation movement (the Mott–Nabarro force). Bilby has also made distinguished contributions to the understanding of the plasticity of metals, particularly in relation to the interaction and movement of dislocations. The research of John Frederick Nye (1923–; FRS 1976) has included dislocation theory and the physical properties of crystals; he worked with Bragg on the bubble raft model. Other Fellows with research interest in dislocation mechanisms and mechanical properties are Peter Hirsch; Zbigniew Stanislaw Basinski (1928–; FRS 1980); Lawrence Michael Brown (1936–; FRS 1982); Derek Hull (1931–; FRS 1989) and Raymond Edward Smallman (1929–; FRS 1986). The mathematical research of Colin Atkinson (FRS 1998) has involved dislocation theory. In France the work of Jacques Friedel (1921–; FMRS 1988) has included a classic book published in 1956.

The pioneering mathematical work of John Raymond Willis (1940–; FRS 1992) has also been concerned with studies on dislocations. Anthony James Merrill Spencer (1929–;

FRS 1987) in the area of theoretical mechanics has worked on plasticity and impact loading. Michael Farries Ashby (1935–; FRS 1979) has also done fundamental research on deformation behaviour, identifying underlying mechanisms, and modelling theoretically a number of materials' properties of practical importance to engineers.

CHAPTER 14

Metallurgical Challenges of Engineering Applications

Introduction

From antiquity the choice of metals and alloys and devising processing routes have been primary tasks of those concerned with the production and working of metallic materials for specific applications. Examples of applications during the 17th and 18th century period of the Royal Society have been discussed in Chapter 2, including the use of brass and steel by the instrument makers and of copper–tin alloys required for mirrors in reflecting telescopes. In the 19th century the advances in the field of ferrous metallurgy, notably in the use of wrought iron and subsequently steel, and their impact on large scale engineering construction have been discussed in Chapter 9. Particularly from the latter part of the 19th century and increasingly throughout the 20th century, existing, established alloys have been improved and new alloys have been produced, together with the necessary processing techniques, to meet the challenges presented by new engineering developments. Experimental work to obtain data on properties such as mechanical, electrical and magnetic, as illustrated in Chapter 11, to provide a basis for selection of suitable materials to match the engineering service conditions has had an ongoing indispensible task. In parallel with this, advances in fundamental understanding of the structures of metals and alloys have enabled the role of empiricism to be reduced and scientifically based alloy design to become increasingly possible.

This chapter focuses on some of the main challenges confronted by metallurgists concerned with applications for service conditions where the responses of the material to the conditions of stress and environment are critical considerations. Specific aspects include the phenomenon of creep at elevated temperatures, leading to deformation or even fracture of components. The avoidance of fracture in service at room temperature or subzero temperatures is obviously also a crucial consideration. In addition to considerations of static stresses of various types, the phenomenon of fatigue failure resulting from cyclic stresses, often of a complex type, has posed major problems. The response of materials to corrosion, oxidation and wear is another critical field.

In the latter part of the 20th century there have been notable instances of engineering progress which have only been possible because of metallurgical advances to extend the range of properties of existing materials. One such example considered in this chapter is the development of materials for aircraft jet engines. In this area synergistic collaboration of design engineers and metallurgists has been a key feature. In the case of the nickel-based superalloys for turbine blades, over a period of around half a century the temperature capability has increased from around 800°C to 1100°C; this was achieved through changes in alloy composition and heat treatment and by processing techniques to produce single crystals by directional solidification incorporating channels for blade cooling, and also surface treatments.

The development of materials for jet engines illustrates the challenges of especially stringent service conditions. It is also important to refer to an important trend in the overall field of engineering applications, particularly in recent years, in which great emphasis has been shown in the selection of structural materials in relation to design and performance of engineering components. In this connection Michael Farries Ashby (1935–; FRS 1979) has pioneered a quantitative approach, devising maps which display key parameters for various specific properties, such as yield strength for metals, ceramics, polymers and composites, which form an important base for an analytical optimum choice.

Creep deformation at high temperatures

The time-dependent deformation behaviour of metals, which has become a major consideration in high temperature engineering, was first systematically investigated at room temperature by M. Vicat and reported in 1834. He investigated iron wires in the context of the long-term behaviour of suspension wires being used in bridge construction. Fairbairn at around the same period also explored the behaviour of cast irons under stress at room temperature for long periods (see chapter 9).

Walter Rosenhain (1875–1934; FRS 1913) became interested in the difference between the modes of deformation of metals at low and at high temperatures. He observed that whilst the fracture of a normal steel takes place through the crystals, at temperatures near the melting point the metal fractures preferentially between the grains. Rosenhain adopted and developed the concept proposed by George Thomas Beilby (later Sir George, 1850–1924; FRS 1906) of an amorphous intercrystalline cement and applied it over many years to explain a range of metallic properties. At low temperatures this amorphous material would be harder than the crystals, thus resisting fracture, but it would become softer on heating, and above a certain critical temperature failure would be intercrystalline. Moreover, the amorphous material would exhibit viscous flow so that on prolonged loading elongation and rupture would occur at stresses insufficient to cause fracture in short time tests. Although this hypothesis was valuable in relation to high temperature metallic creep it aroused considerable controversy and encountered strong theoretical objections.

Before the First World War engineers were becoming interested in the creep of metals under steady loads applied over long periods of time, and tests were carried out using constant load. This was the time when Edward Neville da Costa Andrade (1887–1971; FRS 1935) began to study time–dependent deformation of copper, lead and lead–tin alloys at room temperature. He also investigated lead wires in an oil bath at 162°C. This work is generally recognised as the first elevated temperature creep experiments (al-

Edward Andrade

Richard Bailey

Norman Allen

William Rankine

though he did not actually use the term creep to describe the phenomena). He published his first paper, a landmark in the field of mechanical properties in 1910 'On the viscous flow of metals, and allied phenomena' and another paper in 1914. He realised that a different approach from that used by engineers would be required if the physical processes involved were to be elucidated. First, it was necessary to investigate a material which readily undergoes creep – usually lead at room temperature – instead of, say, an engineering steel used under conditions in which it does not creep readily. Secondly, constant stress instead of constant load conditions should be used in the tests. In his first experiments Andrade stretched wires of lead, lead–tin, and copper, under constant stress, using a hyperbolically-shaped loading weight sinking into water. He distinguished three components in creep: (i) an immediate extension, (ii) a transient flow and (iii) a constant flow, currently generally referred to as 'steady-state creep'. During the 1930s Andrade's work included the creep behaviour of metallic single crystals. Andrade's students included Bruce Chalmers; in 1932 they published a paper concerned with changes of electrical resistivity during the creep of polycrystalline wires, aiming to relate these to the structural changes. After the second World War, Andrade became increasingly interested in the influence of the surface condition of a metal on creep behaviour. These later researches showed the complexity of the creep process, and identified features such as grain boundary migration and sliding, recrystallisation, the pattern of slip hands, and geometric freedom of surface grains.

Beginning in 1924 Richard William Bailey (1885–1957; FRS 1949), as an engineer at Metropolitan Vickers Electrical Company Ltd, Manchester, made important contributions concerning creep behaviour, with special reference to the development of materials for steam turbines as requirements of steam and thermal efficiency became increasingly demanding. In his laboratory, research testing was developed to obtain important new data. With increasing sensitivity of measuring equipment for creep strain it was found that creep was detected over continually widening conditions of stress, and that the concept of 'limiting creep stress', then widely adopted, was incorrect. In a paper published in 1926 Bailey considered the interpretation of features of plots of elongation versus time under constant load. Following the initial extension of a specimen on loading and a period of creep (primary) with diminishing creep rate, there is a period, when creep (secondary) occurs at a constant rate, followed by a period of increasing creep rate (tertiary). Engineers generally aimed to use the secondary range of 'minimum creep rate' for service conditions. Bailey introduced the idea that in this range the strain hardening resulting from the creep deformation is continuously counterbalanced by the annealing effects. Information on creep behaviour over periods of 100,000 hours was not available, and there was little data for 10,000 hours. Bailey played a role in developing procedures for extrapolation of data to the longer periods. He also studied changes in structure and properties associated with creep exposure; in joint work he showed that the spheroidisation of pearlitic carbides occurs in steel during creep at temperatures as low as 350°C. He was involved in work on the beneficial effect of molydenum in nickel–chromium steels and the discovery of the remarkable creep properties resulting from the presence of molydenum and vanadium.

Concerning creep under combined stresses Bailey was among early researchers; he tested lead pipes under internal pressure and under combined internal pressure and axial loading and he investigated the properties of steel pipes under axial tension and torsion at 482°C and 549°C. In other work he studied creep in thick walled cylinders under internal pressure, and in rotating discs. Bailey devoted much attention to the design of steam piping and pipe flanges. His laboratory was also involved in collaboration with the Royal Aircraft Establishment, beginning in 1938, on materials aspects for the experimental jet engine for aircraft propulsion.

The firm of Robert Abbott Hadfield (later Sir Robert, 1858–1940; FRS 1909) was active in developing steels for heat-resisting applications. Also William Herbert Hatfield (1882–1943; FRS 1935), as director of the Brown–Firth Research Laboratories, Sheffield, carried out important work on creep, including a significant practical contribution in the assessment of creep behaviour. Many laboratories had equipment for carrying out long term creep tests over periods of years to obtain relevant service data in the construction of steam plant such as turbines.However there was a need for some shortened form of tests which would at least make it possible to exclude unsatisfactory materials and to assist engineers in choosing between alternative types of steel. Hatfield devised, and widely used in his work, the 'time-yield' test. This consisted in determining, by static loading at a selected temperature, the stress within which dimensional stability is reached after 24 hours for a further 48 hours, the extension not exceeding the elastic deformation of 0.5%; stability was defined as a change of less than a millionth of an inch ($\sim 2.5. 10^{-5}$ mm) per hour. This stress was called the time-yield and Hatfield recommended that 2/3 of its value should be the limit of safe stress for design purposes. Hatfield's work included the first creep resistant steels containing 18 wt% nickel and 14 wt% chromium with small additions of other metals. Leonard Rotherham (1913–; FRS 1963) also working in the Brown–Firth laboratories, beginning in 1935 carried out research on creep and creep-resistant alloys. Some alloys were provided in the early days of the experiments made by Frank Whittle (later Sir Frank, 1907–1996; FRS 1947) in developing his jet engine. During the war years another member of the laboratories, John Anthony Hardinge Giffard, the Earl of Halsbury (1908–; FRS 1969) carried out creep studies. When Charles Sykes (later Sir Charles, 1905–1982; FRS 1943) became Director of the Laboratories in 1944, work was continued on the creep resistant steels under development for high temperature gas turbines e.g. alloys with 18 % chromium, 10 % nickel stabilised with niobium. The development of such special steels contributed significantly to the success of the first industrial gas turbine designed by Geoffrey Bertram Robert Feilden (1917–; FRS 1959), beginning in 1946, for Ruston and Hornsby Limited.

The National Physical Laboratory at Teddington has a long history of involvement in creep investigations. During the 1930s creep testing and microstructural studies were carried out in connection with power stations. Norman Percy Allen (1903–1972; FRS 1956), who became Superintendent of the Metallurgy Division in 1945 was associated with the continuation of this work. During the war NPL had done creep testing of the nickel–chromium based Nimonic alloys for the jet engine development programme, and after the war research was done on similar materials appropriate for use in industrial gas turbines. This work included investigations of samples cut from typical components of power station equipment e.g superheater tubes, turbines, blades. The NPL programme, under Allen, made important contributions in assisting turbine designers in achieving increase in size and operating efficiency of turbine units.

Among contemporary Fellows, Frank Reginald Nunes Nabarro (1916–; FRS 1971) in 1947 proposed the existence and mechanism of the phenomenon of diffusional creep, involving flows of atoms and vacancies, but no dislocation motion. Geoffrey Wilson Greenwood (1929–; FRS 1992) has made detailed analyses of diffusional creep behaviour. The work of Michael Ashby on deformation maps (plotting normalised stress and temperature) has included creep phenomena displaying the areas of dominance of specific flow mechanisms.

Fatigue and fracture

Engineers in the 19th century in their role as designers were familiar with failures of components and structures, for example under static tensile or compressive loads, associated with excessive deformation or fracture, and safety factors were used. Observations of fractures became common practice with the development of mechanical testing of materials by engineers a major activity during the century, but relatively little progress was made in achieving fundamental understanding. With engineering advances such as in steam engines and mechanical transport, fractures occurring after repeated loadings began to become common. Such fractures were found to occur at relatively low stresses and the term 'fatigue' eventually came into use, reflecting the view that the material had lost its ability to sustain the load as a result of repeated stress cycles. There was a popular view that metals would crystallise under repeated vibrations or sharp impact.

Many failures of wrought iron axles occurred during the early days of railway engineering, and there were also failures of some iron bridges. One of the distinguished investigators in this area was William John Macquorn Rankine (1820–1872; FRS 1853) whose book *Applied Mechanics* (1836) was an important event in the history of engineering design. Rankine presented a paper to the Institution of Civil Engineers in 1843 'On the causes of unexpected breakages of journals on railway axles...' He recommended journals with a large radius in the shoulder so that the fibres of iron could be continuous throughout, considering that with a sharp change of section there 'is an abrupt change in the extent of the oscillations of the molecules of the iron, these molecules must necessarily be more easily torn asunder'. He noted the characteristic brittle appearance of metals broken under repeated loading.

The great importance of the reliabiIity of cast and wrought iron was indicated by the appointment by Queen Victoria in 1848 of a Commission to examine the use of iron in railway applications (see Chapter 9). Eaton Hodgkinson (1789–1861; FRS 1841) carried out the first systematic studies of fatigue in extensive series of experiments on the behaviour of iron subject to oscillating loads and repeated impact; his findings were

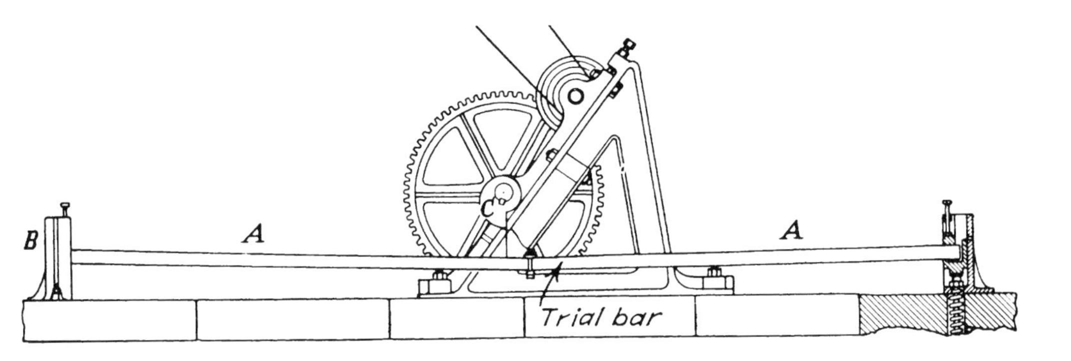

Fatigue testing machine (H. James and D. Galton c.1850) used for testing iron under cyclic loading. A rotating eccentric was used to deflect the bar and to release it suddenly at a frequency of 4–7 deflections per minute.

presented in an appendix to the Commission's report in 1849. Other results stemmed from work by Capt. Henry James (1803–1877; FRS 1848) and Capt. Douglas Strutt Galton (later Sir Douglas, 1822–1899; FRS 1859) who constructed a fatigue testing machine at Portsmouth and investigated the effect of as many as around half a million loading cycles on the strength of iron; it was concluded that 'iron bars will scarcely bear the reiterated application of one–third their breaking weight without injury'.

From the evidence of witnesses the Commission referred to the difference of opinion that existed as to whether the internal structure of iron was altered by being subjected to repeated light blows at a low temperature. They referred to the view of Isambard Kingdom Brunel (1806–1859; FRS 1830) which was that fracture appearance depends on the mode in which the iron is broken, and on the temperature. Brunel, Robert Stephenson (1803–1859; FRS 1849, who made microscopical observations) and Peter Barlow (1776–1862; FRS 1823) held an essentially correct viewpoint. At a meeting of the Institution of Mechanical Engineers (reported in 1850) concerning fatigue Stephenson remarked 'I therefore wish the members of the Institution,... to pause before you arrive at the conclusion that iron is a substance liable to crystalline or molecular change from vibration...' However, the prevailing, erroneous view that fatigue failure occurred as a result of crystallisation of the metal persisted for a long period. Among evidence counteracting this view was that reported in 1887 by Sorby; his microscopical observations on a bar of iron that had been fractured by repeated loading, showed similar crystallinity to that of the 'normal' material. In considering the question of fracture he stated that 'we must not look upon iron as a homogeneous substance, but simply as a mass of small crystals, which cohere less strongly than separate parts of individual crystals'. John Percy (1817–1889; FRS 1847) in the volume of his treatise concerned with iron and steel also presented his views on the crystalline behaviour of iron, together with comments on the matter of fracture. Thus the crystallinity of metals was recognised although a clear view of the nature of crystalline aggregates had yet to emerge.

An interesting large scale experimental investigation of repeated loading was made by William Fairbairn (later Sir William, 1789–1874; FRS 1850) in 1860–61 for the Board of Trade, and reported in 1864. He used a wrought iron beam 6.7 m in length by ~40 cm in width, made up from plates by ~5 mm thick. A mechanism activated by a water wheel was used to raise and lower a load (with impact) at the centre of the beam. The calculated breaking load was 12 tons (~120 kN) but Fairbairn found that 3 million applications of a 3 ton load (30 kN) produced failure. For higher loads fracture occurred at smaller numbers of reversals. Fairbairn concluded that there was a safe load for the structure which could either be sustained indefinitely, or would only cause failure after a number of loadings greater than that expected to be experienced during the normal life of a bridge.

The British Association meeting in Manchester in 1887 included a report from a Committee appointed to obtain information on the 'endurance of metals under repeated and varying stresses and the proper working stresses on railway bridges and other structures subjected to varying loads. Of the 11 members of the Committee, nine were at the time or subsequently became Fellows of the Royal Society namely: William Henry Barlow (1812–1902; FRS 1850); Sir Frederick Joseph Bramwell (1818–1903; FRS 1873); James Thomson (1822–1892; FRS 1877); Sir Douglas Strutt Galton; Benjamin Baker (later Sir Benjamin, 1840–1907; FRS 1890); William Cawthorne Unwin (1838–1933; FRS 1886); Alexander Blackie William Kennedy (later Sir Alexander, 1847–1928; FRS 1887); Henry Selby Hele-Shaw (1854–1941); FRS 1899) and William Chandler Roberts-Austen (later Sir William 1843–1902; FRS 1875). The report of the Committee reviewed earlier work in Britain and abroad, and described important tests being made by Baker associated with the construction of the Forth Bridge, in which the fatigue properties of the steel used in the bridge were compared with the properties of iron and a 'hard steel'. His experiments involved four procedures: spindles rotating with a weight at the free end; flat bars subjected to bending; small specimens tested with alternate direct tension and compression; and full sized girders subjected to many thousands of stress applications. It was noted by Baker from examination of a fracture that the failure involved 'the establishment of small but growing flaws'. The Committee's recommendations included statements of maximum stresses for wrought iron and steel in various service conditions.

During the period 1852–1861 important work on fatigue was done in Germany by August Wöhler (1819–1914), the chief locomotive engineer of the Royal Lower Silesian Railways. He made many tests to determine the cause of axle failures and constructed various types of fatigue machines, including the rotating bending type. Wöhler developed the first S-N (stress-number of stress cycles) curve, including the concept of a limiting stress for iron and steel below which fracture does not occur.

During the first half of the 20th century a number of distinguished engineers who became Fellows of the Royal Society contributed in the area of fatigue testing and in the elucidation of factors affecting, and mechanisms of, fatigue failure. This work involved various collaborators, and the National Physical Laboratory, Teddington and the Royal

Frederick Bramwell

Leonard Bairstow

Thomas Stanton

Charles Jenkin

Herbert Gough

Aircraft Establishment, Farnborough were important centres. At the NPL Leonard Bairstow (later Sir Leonard 1880–1963; FRS 1917) who became distinguished as an aeronautical engineer, began in 1904 research which included fatigue. In 1906 he published a joint paper with Thomas Edward Stanton (later Sir Thomas, 1865–1931; FRS 1914) and in 1909 a paper of which he was the sole author. Aspects covered concerned the effect of repeated stress cycles on the elastic behaviour of iron and steel, and stress–strain hysteresisloops in fatigue testing. Stanton and Bairstow devised direct stress fatigue testing equipment, including a machine imposing cycles of impact. Work included the effect of various types of discontinuity on fatigue properties under direct reversed stress for a series of steels.

Charles Frewen Jenkin (1865–1940; FRS 1931) who was Professor of Engineering Science at Oxford University from 1908–1929 worked at Farnborough during World War I. He was appointed head of a group responsible for all types of material used for aircraft and aircraft engines. A report was published in 1920. Jenkin was concerned with the effect of cracks and notches on the strength of machine parts and the influence of fatigue. He also worked on corrosion fatigue, using high-frequency testing.

Working at Cambridge University, Bertram Hopkinson (1874–1918; FRS 1910) constructed a high-frequency fatigue testing machine operating at 7,000 cycles/minute. He found no difference in the fatigue limit of a steel tested at this frequency compared with the behaviour for the same material found by Stanton at a lower frequency of testing.

Herbert John Gough (1890–1965; FRS 1933) worked at the National Physical Laboratory from 1914–1938, where his research included mechanical properties, such as fatigue. His book *Fatigue of Metals* (1924) included descriptions of testing machines and detailed discussion of theoretical and practical aspects of fatigue phenomena. For example, in relation to the theory of fatigue mechanisms he referred to Beilby and Rosenhain who had proposed an explanation in terms of the formation of an amorphous phase during repeated cycling of stress; Beilby suggested that lamellar regions of the amorphous phase formed until eventually cracks were initiated. Important features of Gough's research included the 'fatigue limit' beyond $\sim 10^7$ stress reversals, found in Wöhler tests on steels.

Gough's work also involved microscopical observations to elucidate the mechanisms of crack initiation and propagation. This followed earlier important work (1903) using microscopy by James Alfred Ewing (later Sir James, 1855–1935; FRS1887) and J. C. W. Humfrey who found slip lines after relatively few stress cycles, which broadened into bands in which cracks formed and grew; the higher the stress the greater the density of slip and the sooner cracks formed. Stanton and Bairstow also made microscopical studies. Gough found that slip lines appeared with stress ranges below and above the safe loading range. Cracks originated in regions of heavy slip and spread out through both slipped and unslipped regions. Among the materials investigated were Armco iron, steels and copper for a variety of stress conditions. When large single crystals became available (see Chapter 13), Gough made extensive investigations of their fatigue behaviour during the period 1922–1933, mainly using torsional loading. Crystals of aluminium and

Ben Lockspeiser

Charles Inglis

Richard Southwell

Stanley Dorey

of other metals, including iron, zinc and silver were used and some work on bicrystals was done. Gough's work included investigations under combined stress conditions, the influence of stress raisers (which enabled engineers to adopt more realistic safety factors) and corrosion fatigue. Another aspect of Gough's research was the use of an X-ray technique to investigate structural changes during cyclic and static stressing.

In 1921 Ben Lockspeiser (later Sir Ben; 1891–1990; FRS 1949) joined the Royal Aircraft Establishment where he worked in the Physics and Instruments Department, in the section headed by Alan Arnold Griffith (1893–1963; FRS 1941), and designated elasticity research. His work on mechanical properties, in collaboration with Griffith, included fatigue phenomena.

The many contributions Egon Orowan (1902–1989; FRS 1947) made to the field of mechanical properties, included work on the theory of fatigue during the period from 1937–39 when he was at the University of Birmingham. Richard Weck (1913–1986; FRS 1975) was, for a period, in charge of the fatigue laboratory of the British Welding Research Association. Alfred Grenvile Pugsley (later Sir Alfred, 1903–1998; FRS 1952) a leading structural engineer concerned particularly with the safety of engineering structures, included fatigue damage of aircraft in his work. Among contemporary Fellows the research of Bernard Crossland (later Sir Bernard, 1923–; FRS 1979) has included fatigue behaviour under complex stress conditions. Derek Hull (1931–; FRS 1989) has included investigations on fatigue, fracture and also creep in his research activities.

In addition to fracture arising from repeated stresses, the problems of failure under conditions involving factors such as low temperatures and impact stresses have constituted a major engineering challenge. Early work on fracture behaviour at low temperatures was done by James Prescott Joule (1818–1889; FRS 1850) who in 1871 published the results of experiments 'on the alleged action of cold in rendering iron and steel brittle'. In the very cold winter prior to this work there had been a large increase in the incidence of fracture of tyres of railway wheels. Joule considered the commonsense explanation of the problem was an increase in hardness of the ground. As regards the suggestion that the failures were due to change in properties of the materials, Joule considered that this was 'a pretence... that iron and steel become brittle at a low temperature' to excuse certain railway companies. Accordingly he carried out some very small scale tests using wires, darning needles and garden nails, with freezing mixtures of salt and snow and comparing the behaviour at 13 and at –11 °C. The wires were stretched by loading to fracture, and another procedure involving impact was also used. The tests indicated a 1% greater strength at the lower temperature, effectively an insignificant difference. Joule deduced that iron and steel were not rendered brittle by low temperatures, but, not surprisingly, he received criticism from engineers as to the relevance of such tests to the actual large-scale engineering situations.

Subsequently, also in the context of the properties of ferrous materials for the railway industry Thomas Andrews (1847–1907; FRS 1888) conducted a long and expensive series of experiments on the effect of temperature on the strength of axles. In these experiments, in which over 300 tonnes of snow were used in making freezing mixtures, 286

Alan Arnold Griffith

Norman Petch

George Irwin

Richard Weck

railway axles and forgings were tested to destruction. His test results at temperatures betwen approximately –18°C and 100°C demonstrated lower impact strengths at the lower temperatures. The detailed investigations of Robert Hadfield with James Dewar (later Sir James, 1842–1923; FRS 1877) and others at temperatures of liquid air and liquid hydrogen provided important data concerning reduction in ductility.

The work of Robert Mallet (1810–1881; FRS 1854) in relation to the construction of artillery, reported in 1856 was of special importance in showing the way in which fracture paths in cast ferrous artefacts depended on changes in section and planes of weakness where growth fronts of crystals met.

The role of flaws/cracks in reducing the fracture strength of materials came to be clearly recognised early in the 20th century. Charles Edward Inglis (1875–1952; FRS 1930) published an important paper in 1913 on the stresses in a plate due to the presence of cracks and sharp corners; the main work lay in the determination of the stresses around an elliptical hole in the plate, and he obtained general expressions so that the stresses could be found when the ellipse assumed its extreme limits from a circular hole to a straight crack. Thus a quantitative treatment of stress concentration effects was provided representing a far-reaching contribution which Alan Arnold Griffith later used in his calculations.

Geoffrey Ingram Taylor (later Sir Geoffrey, 1886–1975; FRS 1919) carried out an investigation with Griffith in 1915 on the stress distribution in cylindrical shafts under torsion. This was in the context of the finding that aircraft propellor shafts were weakened by keyways cut into them for the transmission of torque. Taylor and Griffith set up an analogue procedure using the displacement of a soap film, a technique previously used by Ludwig Prandtl (1875–1953; FMRS 1928). The film was stretched across a hole cut to the shape of the cross-section of a bar and when the film was distended by the introduction of a pressure difference across it the contours related to the stress distribution of the twisted bar. Taylor and Griffith formulated rough rules for the rounding required to prevent too large a stress concentration at the internal corners where cracks in the shaft would occur first.

In 1920 Griffith produced a famous paper 'Theory of rupture' which made a vital contribution to the science of the behaviour of materials, and applied the work of Inglis. In the context of the differences between ideal and observed strengths, Griffith proposed that materials contained invisibly small cracks or other lattice flaws that led to local stress concentrations that could cause rupture. The theory, which applied particularly to amorphous material, was the first to draw attention to the surface condition of the material. He treated the weakening effect of a crack as an equilibrium problem, in which the reduction of strain energy when the crack extends was equated to the increased surface energy due to the enlargement of the crack. He showed that very high strengths can be obtained from fine drawn filaments.

Egon Orowan in 1931 commenced a doctoral thesis, which he presented the following year on the cleavage of mica, a highly anisotropic materials. This work provided the first confirmation of the Griffith theory for a crystalline material. Orowan's technique was to stretch a sheet of mica in its plane, using grips much narrower than the sheet, so that the edges of the sheet were unstressed and cracks in the edge would not cause fracture. He found that the tensile strengths of these sheets with unstretched edges were up to ten times greater than those usually measured, showing that the usual tensile strength is controlled by defects in the edges of the sheets. Orowan concluded that dangerous defects are extremely rare in mica and he discussed in detail the fracture process. His interests in fracture continued into the early 1970s, including non- metallic materials.

During the early 1930s Richard Vynne Southwell (later Sir Richard; 1888–1970; FRS 1925) worked on impact testing while he was professor of engineering science at Oxford University. In the Izod and Charpy tests, which were widely used to obtain information on the energy required to fracture notched specimens one of the problems is that some of the energy of impact is transmitted into the foundations of the testing machine. Southwell, with collaborators designed and built an improved machine in which both the 'hammer' and the 'anvil' were in effect suspended on wires.

The problems of the brittle fracture of steel were brought to prominence in World War II by the catastrophic failures in ship's steels. Preparations for the final phases of the conflict involved the construction, largely in the USA, of a vast fleet of ships; eventually there were about 5,000 Liberty cargo carriers and T2 tankers alone, with displacements of 15,000 to 20,000 thousand tonnes each. In the revolution that occurred in fabrication methods the introduction of fusion-welded construction was the most significant. In Britain, as early as 1940, there were reservations concerning the integrity of large welded hulls, stemming from some previous brittle failures in welded bridges in Belgium and Berlin, and in a cargo carrier in transit from Liverpool. In the USA a significant number of fully-welded ships experienced serious cracking; two of these failures were particularly dramatic, since they occurred in moored vessels with relatively light external loading. The problem was found to stem from three causes. There were cracks in welds resulting from poor quality welding, and stress raisers e.g. at square deck hatch covers acted as initiating points for cracks. Also the steel used had a Charpy toughness acceptable for riveted construction, where propagating cracks could be arrested by the rivets but the steel was unsuitable for welded structures where cracks could propagate unhindered across the hull.

In the United Kingdom the wartime development of shipbuilding methods was placed in the hands of the Admiralty Ship Welding Committee set up in 1943. Eleven years later, with the Admiralty's ongoing concern with fracture, this committee became the Admiralty Advisory Committee on Structural Steels with Hugh Ford (later Sir Hugh, 1913–; FRS 1967) as Chairman. Among those working at Cambridge in World War II on the problems of brittle fracture in ships were John Fleetwood Baker (later Baron Baker of Windrush, 1901–1985; FRS1956), Egon Orowan and Richard Weck, who later became Fellows of the Royal Society. Another distinguished researcher at Cambridge was Constance Flig Tipper, née Elam (1894–1995). Beginning in 1943 Dr Tipper made important metallurgical investigations on the deformation and fracture of iron and steel; she developed an edge–notched tensile test and carried out an extensive test programme, including studies of fracture surfaces.

Orowan investigated notch brittleness effects, and his first research student at Cambridge was John Frederick Nye (1923–; FRS 1976). Weck joined the programme of research at the outset and was assigned the difficult problem of residual stresses in welded construction. His approach emphasized the role of plasticity and he explained the catastropic failure of some welded structures by drawing attention to the role of defects and ductility. His published work during the years 1945–52 on weld residual and reaction stresses involved firstly, estimation and verification of their magnitude and extent in relation to the characteristics of butt fusion welds in mild steels, and secondly, their influence on the initiation and propagation of notch brittle fractures, including the effect of test temperature. Weck devised methods for testing and carried out important experiments. He was invited to begin a programme of fatigue tests on rectangular panels of ship's steel provided with welded stiffeners and the testing of such large specimens presented a considerable challenge.

Stanley Fabes Dorey (1891–1972; FRS 1948), who from 1919 to 1956 worked with the Lloyd's Register of Shipping, becoming their Chief Engineer and Surveyor in 1932, was a leading figure in materials for marine engineering. The responsibilities of the Lloyd's surveyors included the examination of numerous failures of components and structures, and Dorey was much concerned with marine engineering defects – their causes and prevention; in this context he was interested in non-destructive testing e.g radiography. His work also included corrosion and corrosion fatigue, and a major activity in welded pressure vessels.

The postwar period at Cambridge saw important work on fracture carried out by Norman James Petch (1917–1992; FRS 1974) in the Cavendish Laboratory as a member of the metal physics research group headed by Egon Orowan. He investigated the yield and fracture stresses of steels at low temperatures, usually 77K, aiming to relate these properties to the structures of the materials. Eric Hall, a young physicist from New Zealand, who joined the group in 1948, investigated the yielding of mild steels, particularly the yield point and the development of Lüders bands. The researches of Petch and Hall continued in parallel. In the period 1946–48 Petch had found that, during testing of mild steel at 77K sharp upper and lower yield points occurred followed by limited plastic deformation, involving Luders band propagation, then brittle transgranular fracture. He found the relationship between the lower yield stress σ_i and the grain diameter, *d*, to be of the form: $\sigma_i = \sigma_0 + k_y d^{-1/2}$ where σ_0 and k_y are experimental constants. This result was also obtained by Hall carrying out tensile tests largely at room temperature. Petch also demonstrated that the cleavage strength followed a similar relationship with ferrite grain size.

The underlying relationship between yield stress and grain size was determined by both Petch and Hall in the period 1947–1950 while in the Cavendish Laboratory; this fundamental relationship became known as the Hall–Petch equation in which the s_0 term was the friction stress needed to move individual dislocations in pile-ups within the grains, while k_y measured the localised stress needed to propagate the deformation across grain boundaries. Their results subsequently influenced profoundly the industrial development of steels, for example in the work by the British Iron and Steel Association on low carbon steels combining strength with toughness at a reasonable cost. The work was also useful in the 1950s in relation to the problem of irradiation embrittlement of steels studied at Harwell by Alan Howard Cottrell (later Sir Alan, 1919–; FRS 1955) and co-workers. Later at Leeds much of Petch's research activity, with various co-workers, concentrated on the experimental constants in the Hall–Petch equation; aspects covered included the role of carbon and nitrogen in solution, and the role of manganese. The work broadened into a study of the ductile–brittle transition and the development of equations defining the criteria determining the transition temperature. In his later research at Newcastle upon Tyne, Petch included an investigation of the fracture behaviour of laminated steel cylinders, aiming to improve toughness.

Important research on the fracture of iron and iron alloys of high purity and including single crystals was carried out at the National Physical Laboratory in the post-war period, led by Normal Percy Allen.

In the field of theory of fracture John Douglas Eshelby (1916–1981; FRS 1974) made important contributions during his period at Sheffield, concerning the crack extension force in elastic materials; his approach used the concept of the elastic energy–momentum tensor that he had introduced in 1951.

A leading figure in research on fracture was George Rankin Irwin (1907–1998; FMRS 1987), who has been referred to as 'the father of fracture mechanics'. In addition to his work at the Naval Research Laboratory, Washington, DC for thirty years (1937–67) he continued his work at Lehigh University and at the University of Maryland until his recent death. He made major contributions on fracture behaviour, testing and control in the area of mechanics where the resulting fundamental concepts and procedures for testing and analysis were foundational for the modern field of fracture mechanics. Stemming from Irwin's work, methods of testing and analysis were used in many important problems of fracture; for example in: pressurised jet fuselages in the de Havilland Comets (1953–54): in heavy rotating components of large steam turbine generators (1955–56), and in ultra-high strength steel chambers of solid-propellant rockets (1957–60); welded structures were an important feature of his work. Other examples of Irwin's work concerned gas pipe-lines, steel bridges and adhesive joints. Among important concepts was his important contribution on the role of plasticity at crack tips. Irwin's work at the Naval Research Laboratory, beginning in his earlier years there, also involved important investigations of penetration ballistics and improvements in new types of armour materials. During the periods of his career in universities Irwin played a major role in the introduction of the subject of fracture mechanics into the teaching of engineering science.

A number of contemporary Fellows of the Royal Society have played leading roles in the field of fracture phenomena. The research of Sir Alan Cottrell has been influential in areas such as the behaviour of pressure vessels in the nuclear industry, a topic in which Sir Peter Bernhard Hirsch (1925–; FRS 1963) has also been involved. Edwin Smith

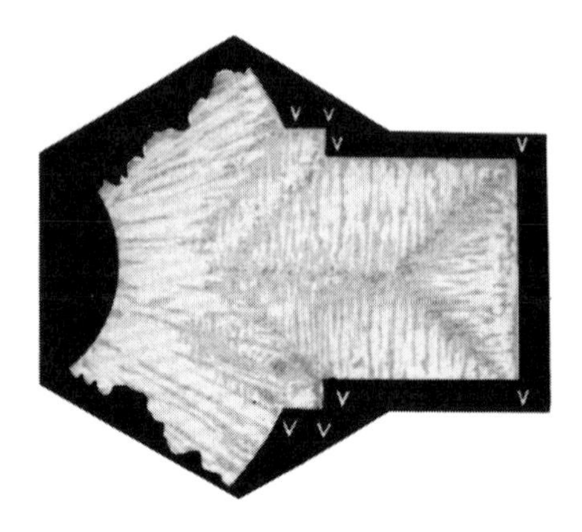

Top left: Result of brittle fracture tests on a steel pressure vessel, carried out by J. Thompson at the Welding Institute (1965) (*Int. Sci. Rev.*, 1997, 22, 4); Top right: Cast iron cannon: fracture showing grain structure as-solidified, with planes of weakness (Mallet 1885); Bottom centre: SS Schenectady, failure initiated in January 1943 in welded mild steel hull (*Int. Sci. Rev.*, 1997, 22, 4).

(1931–; FRS 1996) has also been involved in the field of safety in the nuclear industry, and has worked on the theory of dislocations, cracks and notches and on the effect of second phase particles on the cleavage fracture of steels.

Other Fellows who have made major contributions in the field of fracture and fracture mechanics are Alan Arthur Wells (1924–; FRS 1977), formerly Director General of the Welding Institute, and Frederick Michael Burdekin (1938–; FRS 1993) whose activities have included fractures in large welded structures.

The research of John Frederick Knott (1938–; FRS 1990) has contributed to the understanding of the fracture processes in metallic materials, particularly concerning the role of microstructure in crack initiation and propagation. Also the mathematical work of John Raymond Willis (1940–; FRS 1992) has advanced knowledge in ductile fracture and radiation damage. The mathematical contributions of Colin Atkinson (FRS 1998) to materials science and engineering have included pioneering research on dynamically expanding cracks.

Frank Philip Bowden (1903–1968; FRS 1948) applied high speed photographic techniques used for explosive studies to the investigation of impact, erosion by water droplets and fracture. More recently John Edwin Field (1936–; FRS 1994) has developed new methods of studying the behaviour of materials under impact, fracture and erosion. Fast moving cracks and explosive reactions have been studied using very high speed photo-graphy, up to 20 million frames per second. This work is of special importance in predicting and controlling the failure of metals subjected to sudden stress. Earlier work (1943) by Nevill Francis Mott (later Sir Nevill, 1905–1996; FRS 1936), as Superintendent of Theoretical Research in Armaments in the Government Laboratory at Fort Halstead made an important contribution to the theory of fragmentation of shell and bomb cases under the effect of their explosive charges. Charles Sykes (later Sir Charles, 1905–1982; FRS 1943), as Superintendent of Technical Ballistics at Fort Halstead was in charge of the development of the 'disintergrating sabot', which was a shot consisting of a tungsten carbide projectile carried in the barrel on an aluminium casing (the sabot), which broke away on leaving the barrel.

Also, during the 1939–45 war Frank Ewart Smith (later Sir Frank, 1897–1995; FRS 1957) who had graduated in mechanical engineering and carried out metallurgical postgraduate research, held the appointment of Chief, Engineering Armaments Design in the Ministry of Supply.

Corrosion, oxidation and wear

Problems of corrosion were recognised as of critical importance early in the 19th century, and attention was given to explaining corrosion phenomena. The idea that metallic corrosion was caused by electrochemical effects had been appreciated by Louis Jacques Thénard (1777–1857; FMRS 1824) as early as 1819, and he also worked on the oxidation of metals. Auguste Arthur de la Rive (1801–1873; FMRS 1846) in 1830 interpreted the rapid corrosion of impure zinc in terms of local electrolytic cells set up between the zinc and impurities; this led to the idea that corrosion might be prevented by applying a thin coating of metal that would corrode preferentially.

An important practical problem of national interest was the corrosion of copper sheathing on naval vessels. In 1823 the Navy Board asked Sir Humphry Davy (1778–1829; FRS 1803) to form a group of scientists to investigate the problems in the manufacturing procedure of copper sheets used for such sheathing, with the objective of obtain optimum quality, while maintaining the smoothest surface. Davy, in his younger days in Penzance, had become aware of corrosion phenomena, his attention having been drawn to the deterioration of the floodgates of the port of Hayle as a result of the contact of copper and iron and sea-water. Davy's distinguished group for the Navy Board consisted of Royal Society Fellows, namely William Thomas Brande (1788–1866; FRS 1809), Charles Hatchett (1765–1847; FRS 1797), John Frederick William Herschel (1792– 1871; FRS 1813) and William Hyde Wollaston (1766–1828; FRS 1793). It was known that when tin, zinc, lead or iron are attached to copper the latter becomes electropositive. Experiments were made using three model ships; for one of these ships copper was protected by bands of zinc, in the second, plates of iron were soldered on to the copper sheathing, while the third ship had no protection. The copper plates remained uncorroded (the principle later termed cathodic protection); however, when the technique was applied practically to ships, fouling by weeds and barnacles occurred, slowing the ships speed, so that the government in 1825 ordered the discontinuing of the procedure for sea-going ships, a decision which greatly disturbed Davy. Wollaston reported after small scale trials that proper control of the 'poling' process for copper production was important to remove impurities to a sufficient extent and hence to obtain good quality copper sheets; his recommendations were beneficially implemented.

In 1839 John Frederic Daniell (1790–1845; FRS 1814) carried out studies for the Admiralty of the rapid corrosion of copper sheathing of vessels employed on African stations, and found that the generation of hydrogen sulphide caused this problem. After 1832 Muntz metal, instead of copper, was used as sheathing on merchant navy ships.

Another worker in the field of marine corrosion was Edmund Davy (1785–1857; FRS 1826) who, in 1835, reported on the rapid corrosion of ironwork attached to new buoys in Kingstown harbour, Ireland. These buoys were made of wood, with copper sheathing at the top, and lead sheathing below, and had iron shackles. Davy reported that the corrosion was worst on the iron close to the lead, while it and the copper remained well-preserved. He found that separating the lead from the iron, and putting additional iron nails in contact with the lead, stopped the rapid corrosion of the other iron work. Robert Mallet also, by 1835, had begun an extensive series of investigations into the corrosion of iron, leading to a patent for protecting metals from corrosion and for preventing fouling of iron ships. The work of Thomas Andrews on the corrosion of metals included electrochemical properties and the action of tidal streams on iron and steel. James Prinsep (1799–1840; FRS 1828) in India made experiments on the protection of iron from rusting.

In the 20th century, Guy Dunstan Bengough (1876–1945; FRS 1938) began research in 1910 as investigator for a committee set up by the Institute of Metals to study the causes of failure of brass condenser tubes in ships and in land power stations. This work proceeded over a long period, adopting the view that corrosion was essentially an electrochemical process, and covering general and pitting corrosion. A series of reports was published and valuable work was done leading to materials for tubes with greater resistance to corrosion, so that tube failures became a less alarming problem. Bengough's contributions included the design of apparatus and of experimental methods which would allow of exact quantitative measurements under strictly controlled conditions. His work also extended to include the effects of marine organisms in the corrosion of ship plating.

Guy Bengough

Ulick Evans

Albert Portevin

John Forrest

In the early 1920s there was extensive empirical knowledge but the science of corrosion and oxidation of metals was in a primitive state. The dominant figure in the field who completely changed the situation was Ulick Richardson Evans (1889–1980; FRS 1949). In addition to believing that corrosion and oxidation behaviour could be explained by the available scientific knowledge, he made experiments which extended beyond the previous tradition of exposing specimens to various environments and attempting to correlate the observations with environmental conditions. Evans's approach was to alter the corroding system, for example by cutting the corroding specimens into pieces, connecting them into galvanometer circuits, measuring currents, changing their potentials, and bubbling oxygen around them. With simple 'string and sealing wax apparatus', and with good theory and brilliant design of experiments, important results were soon achieved.

Ulick Evans established, proved and then applied what he called the 'new' electrochemical theory of corrosion. It had been long known that the flow of electrical currents was involved in the corrosion of some metals, for example at the junctions of two dissimilar metals, where the contact between the poles provided a short-circuited cell. However, the fact that corrosion often occurred intensely on homogeneous metal, out of contact with any other metal could not be explained by the older electrochemical approach. Furthermore although the presence of oxygen was known to be necessary for the continuation of wet corrosion on many metals, the actual corrosive attack occurs at locations least accessible to the oxygen.

Many difficulties were removed by the Evans's idea of 'differential aeration' which proposed that different parts of the same homogeneous metal become differently polarised electrodes if they are in different environments. Thus, if there are local differences in oxygen content in the surroundings the higher oxygen region becomes cathodic and the metal is protected, while the lower oxygen region becomes the anode and is attacked. For a few years the theory attracted controversy including arguments with Bengough. However, Evans made further and more quantitative experiments, including work with T.P. Hoar; this work led to the important 'Evans diagram' which became a valuable guide in the field of corrosion control using applied currents as in cathodic protection or of inhibiting additives. Differential aeration proved capable of explaining many features of wet corrosion and a very useful principle in practice e.g. in intense, localised corrosion in crevices or in joints between riveted or bolted plates.

Among Evans's many contacts with scientists overseas was Marcel Pourbaix from Belgium. Late in 1939 Pourbaix gave Evans a paper concerning his graphical method of representing electrochemical equilibria, and when the end of the 1939–45 war made it possible, Evans gave Pourbaix very encouraging comments. Pourbaix diagrams, constructed using the equation for calculating electrode potential derived by Walther Hermann Nernst (1864–1941; FMRS 1932) have subsequently become of major importance.

An aspect of the work which Evans began in the 1930s involved a statistical approach to the probability and velocity of corrosion. Evans's interests had long included types of corrosion such as atmospheric corrosion, oxidation and tarnishing. In 1927 he isolated the air-formed oxide film from a metallic surface, and placed the oxidation theory of

Henry Hele-Shaw

William Hardy

John Charnley

Frank Bowden

passivation on a convincing basis. The pioneering discoveries of Evans and his co-workers led to to many practical advances in prevention and protection from corrosion, and contributed to progress in dealing with the problems of corrosion fatigue and stress-corrosion cracking. After the 1939–1945 war Evans was mainly interested in corrosion fatigue and the laws of growth of oxide films on metals; he also worked on the mechanism of atmospheric rusting. In 1960 he published his 1000-page treatise on *The corrosion and oxidation of metals*; eight years later he published a 500-page *First supplementary volume,* and then in 1976 a *Second supplementary volume.*

An interesting industrial investigation was carried out by John Walter Ryde (1898–1961; FRS 1948) at the General Electric Company Laboratories at Wembley, arising from an accident to a turbo-generator at one of the Central Electric Authority's power stations. Overspeeding of the machine had occurred rapidly, leading to overstressing of parts and subsequent disintegration. The small control pistons were found to show a black deposit, which were found, by X-ray diffraction and electron microscopy, to be almost pure magnetite (Fe_3O_4) of particle size ~0.05–0.5 micron. It was shown that the magnetite was formed by electrochemical reaction between dilute salt water and steel, resulting from a differential aeration effect. The magnetite was dispersed in the lubricating oil to the moving piston, where is was compacted and eventually led to failure.

In the development of corrosion-resistant ferrous alloys, research on various types of stainless steels was a major activity. The work of William Hatfield on corrosion, particularly concerned the 18:8 type of nickel–chromium steels (see Chapter 11) and included problems encountered in stainless steel weldments. As early as 1905 Albert Marcel Germain René Portevin (1880–1962; FMRS 1952) had noticed that chromium steels with chromium contents greater than 9–10 wt% were resistant to attack by common reagents, and he carried out a systematic corrosion study of these steels using many media. This led him into research into making the martensitic stainless steel usable and he established the heat treatment conditions. He had an ongoing interest in many aspects of corrosion. Much work on corrosion resistant steels was carried out by the laboratories of the Company of Robert Hadfield.

In the field of electricity transmission by the National Grid in Britain, John Samuel Forrest (1907–1992; FRS 1966) who became Director of the Laboratory of the Central Electricity Generating Board at Leatherhead, included in his work an investigation of the corrosion of steel-cored aluminium conductors. Another interesting example of corrosion science is the research of Cyril Clifford Addison (1913–1994; FRS 1970) on the attack of stainless steel by liquid N_2O_4, which is used as an oxidant in rocket propulsion.

In the field of friction and lubrication important results were presented in 1829 by George Rennie (1791–1866; FRS 1822) to the Royal Society on the friction of metals and other materials. Rennie used an apparatus in which the materials in surface contact under load were pulled along a surface under various experimental conditions e.g. of pressure and lubrication. Among the metals investigated were brass and iron. His conclusions included the finding that hard metals have less friction than soft metals. He noted that the ability of steel to be hardened and to resist abrasion made it the best material for reducing friction in delicate instruments such as pendulums and assay balances.

Another interesting illustration of the role of friction and lubrication in an engineering application is found in the experiments carried out in 1857 by Isambard Kingdom Brunel (1806–1859; FRS 1830) in connection with the launching of the steamship the *Great Eastern* on the River Thames. Brunel decided to use sliding surfaces of iron instead of greased wood, and the launching ways that were constructed consisted of iron rails, of the type used by the Great Western Railway, placed on a network of timbers, which rested on a thin layer of concrete. Experiments prior to the launching aimed to obtain an estimate of the force likely to be required in the launch. Two rails were laid at an inclination of 1 in 12; on these rails was placed an experimental cradle, which was loaded to about 8 tonnes and which had iron bars on the underside. The results showed that the friction became less as the velocity increased, in contrast to the generally held view that the friction between rubbing surfaces did not vary with velocity. Lubrication effects were included in the experimental work. Shortly after the commencement of the launch (1858), which was carried out sideways and slowly, the experimental rig was re-erected, incorporating modifications which provided quantitative data of the forces involved. William Froude (1810–1879; FRS 1870), who had previously worked as an assistant to Brunel, was associated with the investigations of friction.

The eminent engineer Osborne Reynolds (1842–1912; FRS 1877) at Owens College, Manchester, included lubrication in his many research interests. His 1886 paper 'On the Theory of Lubrication' was a classic on film lubrication; there stemmed from it bearings which could function at very high loads and at speeds exceeding those previously attained.

Another Fellow whose work in the latter part of the 19th century included the field of friction was Henry Selby Hele-Shaw (1854–1941; FRS 1899). William Bate Hardy (later Sir William, 1864–1934; FRS 1902), whose main field of research was in the life sciences, also had a significant interest in friction and lubrication and did pioneering work on the nature of surface forces and films. In his Bakerian Lecture in 1925, with a coauthor, he reported on the sliding of steel against plates of various metals and non–metallic materials in relation to Amonton's law.

In the 20th century Frank Bowden at Cambridge was a leading investigator in friction and lubrication, beginning around 1930. His work, with various collaborators, included the role of frictional heating and surface melting in the formation of Beilby polish layers, the measurement of the real area of contact between stationary surfaces and the action of long chain molecules as boundary lubricants. Some of his work included cooperation with David Tabor (1913–; FRS 1963) who, at Cambridge, has made seminal contributions to the basic study of friction and wear between solids; this work has been of considerable relevance to the design of machines involving mechanical and physical aspects of moving parts in contact. Also, at Cambridge,the work of John Field has included studies of erosion under surface impact conditions.

George Ingle Finch (1888–1970; FRS 1938) applied electron diffraction to important technological problems relating to surface phenomena. He investigated intensively the liquid-like 'Beilby layers' produced by polishing on a variety of crystals. He applied his experience to industrial problems associated with wear in internal combustion engines; this led on to research on the action of lubricants and to the influence of chemical compounds in lubricants on bearing surfaces. For example, it was shown that the success of cast iron as a bearing metal is associated with graphite flakes forming an oriented surface layer.

Biomedical materials constitute a rapidly expanding field which presents unusual problems in the choice of materials for clinical use. The temperature of the human body may be constant, but provides a saline environment with surprisingly high forces generated in the musculoskeletal system of many times body weight. John Charnley (later Sir John, 1911–1982; FRS 1975) was an orthopaedic surgeon who specialised in arthroplasty, or joint reconstruction. Prior to the work of Charnley, some joint replacements had been tried, inserting a gold foil between the surfaces and even using an ivory prothesis, but joint reconstruction was a rarely attempted operation. Most patients who suffered from arthritis were offered a fusion, or fixation of the joint, which although it removed the pain, reduced the patient's mobility and was not easy to perform successfully. The Judet brothers in France replaced the surface of the femoral head with an acrylic cup, but this either broke or wore excessively. Charnley saw a patient who had received a Judet replacement and complained about the squeak, indicating to Charnley the need for low friction in arthroplasties. His initial attempt was a polytetrafluorethylene (PTFE) ball-and-socket surface replacement, which failed by damage to the blood supply of the femoral head. This surface replacement was followed by a femoral prosthesis with a large head, made from a low-carbon austenitic steel, in a thin PTFE socket, where the socket loosened due to the torque, so the femoral head size was reduced enabling a thicker PTFE socket to be used. After three years of use high and accelerating wear was seen with the PTFE, producing large amounts of wear debris, leading to the formation of a granuloma.

Charnley was then shown high-molecular-weight polyethylene (HMWPE) which stood up well to wear testing and was used to replace PTFE in the acetabular cups; later the HMWPE was replaced with ultra-high-molecular-weight polyethylene (UHMWPE), which is currently used. The second significant material introduced into orthopaedics by Charnley was 'bone cement'. He realised that a solution to the loosening of protheses seen with all previous designs would be to grout them into the bone, rather than rely on either a biological response or a press fit. After discussions with Dennis C. Smith, who became a leader in the field of polymeric biomaterials in the late 1950s he decided to use self-curing acrylic cement to grout the implant in place. Once the femoral canal was prepared the polymethylmethacrylate (PMMA) normally used for the manufacture of dentures was pushed down the bone and the implant forced into position. As the PMMA polymerised the implant was stabilised. Over 30 years later the concept of a metal head articulating in a polymeric cup, both held in place with acrylic cement is still the basis for over 90% of all total hip replacements performed worldwide.

Throughout his career Charnley applied engineering principles to the practice of sur-

Leonard Pfeil

Frank Whittle

Stanley Hooker

Hayne Constant

gery. His work included analysing the forces used to fuse joints by measuring the deflection of fixation screws and comparing these with the deflections produced in identical screws by known loads. Also he measured the extremely low coefficient of friction in a normal joint (0.013). He set up the world's first biochemical testing laboratory in 1961, where various properties including mechanical, were investigated.

Duncan Dowson (1928–; FRS 1987) in the field of tribology has contributed to the theory of elastohydrodynamic lubrication; his work has provided greater insights into the operation of high performance of machine elements and of load-bearing human joints.

The phenomenon of fretting corrosion can occur at contact surfaces of closely fitting machine components under vibration. Herbert Gough showed that this is largely a mechanical effect associated with surface slip. Kenneth Langstreth Johnson (1925–; FRS 1982) has carried out original theoretical and experimental work in the field of contact mechanics. This concerns rolling contacts of metal surfaces, as for example in ball bearings and gears, relating to aspects such as plastic deformation, fatigue, internal stresses and wear.

A topic of critical importance in the development of alloys for high temperature service is that of minimising oxidation by achieving protective oxide layers, and among the most important areas in which this has been achieved is in the nickel-based superalloys, together with the application of protective coatings. Leonard Bessemer Pfeil (1898–1969; FRS 1951) in his earlier work (around 1921) made an important contribution to the understanding of the mechanism of high-temperature oxidation. He was engaged in producing single crystals of iron for mechanical property studies. This involved long heat treatments of hundreds of specimens, each having its unique history. He had little assistance and prepared his test pieces with his own hands. It was the failure of one of these heat treatments that led to a significant observation. Specimens nominally given prolonged heating in a reducing atmosphere were covered with a thick layer of shining iron oxide crystals. He broke away the scale, and looked for traces of the stamp marks on the exposed metal surface of the metal. The scale had a well marked layered structure, and, on separating these layers found the stamp marks, on the outer surface of one layer and on the inner surface of the next. This surface clearly represented the original surface of the bar. At that time dry oxidation was discussed entirely in terms of oxygen diffusion through the scale, but the growth of iron oxide crystals on the outside of the original surface indicated that diffusion of iron outwards must also be involved. Pfeil showed how a counter-current diffusion mechanism could explain both the sequence of structures and the special features introduced by variation of temperature and the presence of alloying elements. Portevin carried out collaborative work on the effect of additions of chromium, aluminium and silicon on the high-temperature oxidation behaviour of iron.

The research of Graham Charles Wood (1934–; FRS 1997) has ranged widely in the field of corrosion science and engineering. He has investigated the mechanism of formation of anodic oxide films on aluminium. Also his research on various binary and ternary alloys concerning transient and steady state oxidation has contributed to alloy development for high temperature applications in gas turbines and power plants. He has shown

Frank Whittle and his early gas turbine.

that transient oxides can act as vital lubricants in oxidative friction and wear processes.

Nickel-based superalloys and other alloys for aircraft gas turbine engines

The aircraft industry provides an outstanding illustration of technological advance stemming from the collaborative work of engineers and metallurgists – including a number who became Fellows of the Royal Society, in the application of gas turbines to aircraft propulsion. In the United Kingdom in 1929 Frank Whittle (later Sir Frank, 1907–1996, FRS 1947) conceived the idea of using a gas turbine for jet propulsion and a patent was filed in 1930, although Whittle did not receive official support. In 1926 A. A. Griffith wrote a report for the Royal Aircraft Establishment, *An Aerodynamical Theory of Turbine Design,* in which he demonstrated that an axial gas turbine was feasible as a power unit for aircraft with a compressor and a turbine, and a propellor instead of the jet which Whittle conceived. During the 1930s and early 40s an interesting period followed in which both Whittle and Griffith developed their concepts. Whittle experienced difficulties in obtaining support, but in 1936 the Power Jets Company was formed and in the following year experiments on the Whittle engine, which utilised a centrifugal type compressor, began at the British Thomson-Houston Company at Rugby. Griffith worked on the on the contra-flow turbine system which he had evolved in 1929. He was the first to suggest that turbine blades should be treated as aerofoils and to base their design on aerodynamic theory. During the war, as an engineer at Rolls-Royce, Griffith produced a contra-flow design comprising a multi-axial compressor and turbine combined with blades of improved aerodynamic design. However, tests showed difficulties in development and the urgent war-time requirement led to suspension of the tests, and the development of the Whittle jet engine proceeded to spectacular success.

Also in the area of engineering aspects of jet engine production among others involved during the 1939–45 war were Stanley George Hooker (1907–1984; FRS 1962) with Rolls-Royce and Geoffrey Feilden at Power Jets. Another engineer, Hayne Constant (1904–1968; FRS 1948) played an important role, particularly in the field of compressors. Working with Griffith at the RAE he designed an axial flow compressor as a supercharger for a piston engine and was involved in the collaborative project between RAE and Metro-Vickers Ltd on a turbine system for aircraft. Richard Bailey and his colleagues at Metro-Vickers played a role in recommending materials and dealing with development problems; Bailey devised a rig for thermal shock testing. In the early part of this work Constant in his interests in materials for the system, considering potential future requirements, discussed with representatives of the refractories industry the possibility of producing ceramic blade discs. Whittle had also made enquiries about ceramics. Constant met Whittle in 1937 and subsequently collaborated with him.

The metallurgical challenge that had to be faced in developing the jet engine was to provide alloys to withstand the conditions of stress and temperature, particularly in the turbine, necessary to yield a useful advantage over the piston engine; such alloys had not then been developed. In the compressor an aluminium alloy, known as RR 56, a product of High Duty Alloys, Slough, could be used. Firth Vickers at Sheffield had developed special steels for high temperature service and the first flight engine, W1, used Firth Vickers Rex 78 alloy for the turbine blades. However, improved turbine materials were an essential requirement.

Leonard Pfeil working at the Mond Nickel Company Laboratories at Birmingham played a key role in leading the alloy development work and Norman Allen was also an important participant. Research began in August 1939, mainly on age-hardenable nickel–chromium alloys. As war had become increasingly imminent, Dr William Griffiths (later Sir William), Head of the Laboratory, decided that a set of properties representing a sufficient advantage should be defined; these properties were somewhat arbitrarily defined as the ability to withstand a stress of 3 tons/in^2 (~46 MPa) at 750 °C under oxidising conditions, without the occurrence of more than 0.1% permanent deformation in 300 hours, the nominal life of a fighter plane in wartime. The laboratory had six creep machines, and Henry Wiggins's works were asked to provide samples of any experimental nickel–chromium alloys that had exhibited notable resistance to hot forging and rolling procedures. Six alloys that were selected and put on test gave extraordinarily variable results; however, the behaviour of one suggested the possibility of achieving success. This alloy was one that had been made at Pfeil's suggestion to meet a cracking problem encountered with the high-carbon nickel–chromium alloy used for coating the heads of aero-engine exhaust valves. It contained a small proportion of titanium, and after some studies of the combined effects of carbon and titanium, the alloy designated Nimonic 75 was produced; although it was never used as a turbine blading alloy it was widely applied for flame tubes and other sheet metal work in gas turbines. 1940 was spent in establishing the works production of this alloy; also, systematic explorations of alloys were begun with increased amounts of titanium, and since titanium could at that time only be obtained in the form of aluminothermic temper alloys, the alloy additions contained various proportions of aluminium and titanium. Very soon an alloy with the required performance at 3 tons/in^2 at 800 °C was available, and was designated as Nimonic 80. A key feature was the strengthening and resistance to creep provided by the precipitation of the intermetallic compound based on Ni_3Al (termed γ').

The temperature of the turbine blade in the Whittle engine was between 600 and 650 °C, but the blade stresses were high, around 12 tons/in^2 (~185 MPa). When tested under these conditions Nimonic 75 failed badly but Nimonic 80 was adequate, and could endure stresses high enough to ensure that the Whittle engine could be practicable as designed and would be capable of considerable development. Sufficient quantities of Nimonic 80 blading were then required to equip some experimental Whittle engines. Fabrication difficulties were overcome in time for Nimonic 80 blades to be adopted in the Whittle engine after an experimental E28 jet plane powered with a W2B engine had outpaced conventional fighter aircraft impressively in the presence of Winston Churchill in April 1942.

By 1943 it was possible to consider seriously the design of aircraft to fly at 1000 miles per hour. There ensued a huge programme of detailed development work in which the effects of every stage in the production process on the creep resistance of the alloy were examined, and more advanced alloys were systematically developed, necessitating the installation of great numbers of creep testing machines. Before the war ended the use of jet-propelled civil aircraft with long service lives was foreseen, and the testing periods were extended; possibilities of gas turbines in ship propulsion and electricity generation were also considered. The long task of measuring the acceptable stresses for the new alloys for periods up to 100,000 hours, and for service temperatures between 600 °C and 1000 °C was undertaken.

Since the earlier work, research and development in many countries have contributed to the evolution of nickel-based superalloys which can operate in turbine blades at temperatures up to around 1100 °C. Design engineers and metallurgists have worked in cooperation; on the metallurgical side control of alloy composition and the use of sophisticated processing techniques, including the production of single crystal turbine blades have led to the achievement of the property goals for complex service conditions. Beginning in 1972 at the INCO Europe research laboratory Robin Buchanan Nicholson (later Sir Robin, 1934–; FRS 1978) directed important investigations including nickel-based superalloys.

Among the engineering Fellows of the Royal Society whose careers have been prominent in the aircraft industry, including engines and materials, are Lionel Haworth (1912–; FRS 1971) and Sir Frederick William Page (1917–; FRS 1978).

In the compressor stages of jet engines the development of titanium alloys to operate at elevated temperatures continues to be a vital area in which the role of the metallurgist is complementary to that of the engineer. Distinguished contributions have been made to aero-engine design and development by Stewart Crichton Miller (1934–1999; FRS 1996) at Rolls-Royce, including directing the work on the hollow titanium fan blade which is now used in compressors on all Rolls-Royce by-pass engines. Philip Charles Ruffles (1939–; FRS 1998) in his distinguished career at Rolls-Royce in gas turbine engineering and technology, has been particularly associated with the RB211 and Trent engines, involving advanced applications of fluid dynamics and materials science.

In research laboratories, in relation to nickel-based alloys and all other important types of engineering alloys in wide-ranging applications, the role of electron microscopy in elucidating mechanisms of deformation has been enormous; fundamental knowledge of aspects such the role of dislocations, grain boundaries and precipitates has contributed to the current major emphasis on quantitative modelling of materials structure and properties.

CHAPTER 15

Processing for Engineering Applications

Introduction

Processing to achieve components of the required shape, dimensions and properties has been a vital aspect of metallurgical technology throughout the centuries, and has become of even greater importance in recent years. It serves to link the understanding provided by studies of the science of metals and other materials to the production of components required for engineering service. Major processing routes include casting, mechanical working, joining (particularly welding) and powder metallurgy. Some specific examples of the involvement of Fellows and Foreign Members of the Royal Society, have already been discussed, e.g. in Chapters 2 and 3, and other examples are presented here.

Melting and casting technology

When Henry Bessemer (later Sir Henry, 1813–1898; FRS 1879) came to London as a young man in 1830 he developed a successful technique for casting decorative objects of white metal – low-melting-point alloys – containing for example tin and antimony. He used ceramic moulds, and was able to produce fine detail in the castings. A further important aspect of his process was to coat the castings with copper. In the late 1830s Bessemer used water-cooled moulds for casting printing type; his process also used pressure to force molten metal into the mould and the subsequent application of a vacuum to remove air. He developed a new white metal (containing tin, zinc, copper and antimony) with a wide freezing range, allowing the production of fine detail on stamping dies and embossed rolls by applying pressure during the final stages of solidification. In 1846 Bessemer patented a process for the continuous casting of sheet of tin and lead, similar to that which he devised for producing sheet glass. Ten years later he took out a patent for continuous casting of steel strip (see Chapter 7).

Among the challenges in non-ferrous metal processing in the 19th century ongoing from the previous century was the need to produce large and high quality cast metal mirrors in speculum metal (copper–tin based alloys) for reflecting telescopes. This led to the involvement of several Fellows of the Royal Society concerned with astronomy, following earlier work. John Frederick William Herschel (later Sir John, 1792–1871; FRS 1813) took over from his father (Frederick William Herschel, 1738–1822; FRS 1781) various astronomical and instrumental techniques, and himself made the 18 inch (~0.5 m) mirror of his 20 ft (~6 m) telescope.

Another eminent astronomer, William Parsons, the Third Earl of Rosse, formerly Lord Oxmantown (1800–1867; FRS 1831) made a remarkable contribution to the production of large metal reflectors. He began experiments to improve the efficiency of the reflecting telescope in 1827, at his father's residence at Birr castle, Parsonstown, Kings County, Ireland. There was no established casting procedure since the processes of the Herschels had not been made public. He set up in the castle a foundry and the necessary associated equipment, with furnaces to allow slow cooling and solidification of the castings, and machinery for grinding and polishing the mirrors. A speculum alloy composition was selected of 'four equivalents of copper and one of tin'. Substantial practical difficulties were encountered in obtaining high quality castings in relation to the cracking tendency due to the brittleness of the alloy and to porosity resulting from gas evolution during solidification. Showing great perseverance and ingenuity in overcoming these difficulties he successfully cast and mounted a three ft (~0.9 m) speculum around 1839, and in 1842 and 1843 succeeded in casting two specula, each six ft (~1.8 m) in diameter, and about four tonnes in weight. Observations using the telescope were begun in 1845. The casting technology that Rosse developed was presented in great detail in two *Philosophical Transactions* papers (1840 and 1861) dealing respectively with the 0.9 and 1.8 m mirrors. Of particular interest was the novel casting technique involving a mould with an iron base, sides of sand and an open top, which gave directionality to the solidification process. Also he devised a venting system in the mould assembly to counter the problem of gas porosity. He recognised some of the advantages of directional solidification in achieving a high quality product. He wrote:

> Were it possible to satisfy the following conditions, viz that heat should be extracted rapidly and equally from the lower surface of a fluid disc of speculum metal, so that it should solidify from the bottom upwards in strata, or rather infinitely thin laminae, the surface being the last to solidify we should have a perfect casting: for the particles in that case being deposited not uniformly, indeed, owing to the unknown action of the forces of crystallisation, but in such a way as to fill up the interstices, there would be no flaws; and the temperature being uniform in a horizontal direction, and in the vertical varying in regular gradation from the lower surface to the upper there would be no strain.

Robert Mallet (1810–1881; FRS 1854) in his earlier research developed a method of minimising problems of oxidation and volatilisation, and hence of improving compositional control in the production of copper–zinc and copper–tin alloys; in this method, referred to in Percy's *Metallurgy* (John Percy, 1817–1889; FRS 1847) the alloys were melted in a vessel from which air was excluded. Mallet also carried out work aiming to correlate chemical composition with colour and certain physical properties of some copper-based alloys. Also, working in his father's company Mallet produced components for the railway industry in Ireland.

In the 1920s the porosity of copper and its alloys was the subject of research by Norman Percy Allen (1903–1972; FRS 1956). The phenomena associated with the poling process to produce tough pitch copper are complex and of considerable industrial impor-

William Parsons
(Third Earl of Rosse)

Robert Mallet

Benjamin Thompson
(Count Rumford)

William Armstrong
(Baron Armstrong)

tance. Allen's work involving, extremely difficult experimentation, showed the relationship between hydrogen and copper oxide in the liquid metal, and that blowhole formation resulted from the evolution of steam; it thus contributed greatly to the understanding of the process and control of porosity, by applying in detail the principles of physical chemistry and thermodynamics. Research of Albert Marcel Germain René Portevin (1880–1962; FMRS 1952) on casting concerned the complex effects of filling of moulds and of solidification.

One of the major advances in solidification processing in the 20th century was associated with the research of Henton Morrogh (1917–; FRS 1964) concerning cast irons. He produced the type of iron known as 'spheroidal' or 'nodular' graphite iron by the addition of small amounts of elements such as cerium to the melt. These irons show combinations of mechanical properties, which include superior ductility compared with irons containing flake graphite. Morrogh's extensive microstructural studies elucidated the mechanisms of solidification of these important engineering materials.

Experiments to melt materials using a large lens to focus the sun's rays had been carried out in the 18th century, and temperatures were achieved to melt iron. Henry Bessemer, in his later years at his home in Denmark Hill had an interest in the use of solar energy and constructed a large solar furnace. This consisted of a tower, (~9.2 m high), with an open front side. An angled mirror had a circular aperture in the middle, in which a lens (~0.8. m diameter) was mounted; the solar rays were reflected by the mirror upwards on to a concave mirror and thence downwards through the lens to focus on a crucible. The results were, however, disappointing; although copper was melted the efficiency of the furnace was not good.

Also concerning melting technology Charles Sykes (later Sir Charles, 1905–1982; FRS 1943) in the 1920s, following up work by Thomas Edward Allibone (1903–; FRS 1948) at the Metropolitan-Vickers Laboratory at Manchester, carried out investigations on the induction melting of zirconium alloys, for example, containing aluminium. He overcame experimental difficulties associated with segregation and vapourization during the vacuum heating of the materials and he also constructed molybdenum wire-wound furnaces operating up to 1750 °C for heat treatments and melting. The work opened up a new field of study of refractory alloy metallurgy. Stemming from observations of evaporation effects when certain metals were heated in vacuum, Sykes developed techniques for producing uniform metal films over large areas, for example for astronomical mirrors. Also in collaborative work at the Metropolitan-Vickers laboratory Cecil Reginald Burch (1901–1983; FRS 1944) developed the science of induction heating at very high frequencies; precious metals and steels were melted.

At the Associated Electrical Industries (AEI Ltd.) Research Laboratories, James Dyson (1914–1990; FRS 1968) through his research in optics contributed to the activities of the metallurgy section; this section used vacuum induction melting and a vacuum arc furnace for growing single crystals and making refractory alloys. Dyson devised a microscope with high magnification which was able to receive light from an object at high temperatures, and could be used, for example, to observe crystals during the growth process.

Bessemer's Solar Furnace

A contemporary Fellow, Henry Keith Moffatt (1935–; FRS 1986) in his mathematical research on fluid mechanics includes metallurgical magnetohydrodynamics, concerning the generation of magnetic fields in moving conducting fluids.

Armament production

Military technology has from antiquity been a major field of metallurgical activity, and the production of weapons such as swords, and defensive artefacts such as armour has involved the development of sophisticated processing techniques. The Tower of London has a fine collection of arms and armour and a Fellow of the Royal Society Robert Porrett (1783–1868; FRS 1848) had an interesting association with this. His father resided at the Tower, where he was ordnance store keeper. Robert became his assistant and later succeeded him to become head of the department. He was an authority in the field, as well as carrying out important research in several areas of chemistry.

Count Rumford (formerly Sir Benjamin Thompson; 1753–1814; FRS 1779), noted for his work on heat, included guns and other weapons in his interests. At the time when he was military commander for the Elector of Bavaria he was involved at the arsenal in Munich in the manufacture of cannon bored from blocks of iron. He noted in 1798 that the heating effect became greater as the drills became blunter and less effective in drilling. Rumford related the heat generated to the work done in the drilling process, and this work opened up a new approach to the theory of heat.

In the 19th century a number of Fellows of the Royal Society were involved, some to a major extent, in research and development relating to armaments. The research of Robert Mallet on casting technology concerned with cast iron and gun-metals is of particular interest in relation to the influence of macrostructure on fracture behaviour. His work on the production of cast-iron guns was reported in his book *Physical Conditions Involved in the Construction of Artillery* (1856). This research included the effect of the shape and size of castings on the as-solidified macrostructure and physical properties; the structures (showing columnar crystal growth) as clearly illustrated in the book, were revealed by fracturing castings e.g. of portions of guns, and observing the fracture surfaces. The castings included round and square bar sections, flat plates and parts of cannon. Mallet concluded that

> It is a law of the molecular aggregation of crystalline solids that when their particles consolidate under the influence of heat in motion, their crystals group and arrange themselves with their principal axes in lines perpendicular to the the cooling or heating surfaces of the solid; that is, in the lines of direction of heat wave in motion, which is the direction of least pressure within the mass.

Examples on fractured guns showed the dangers of salient and re-entrant angles:

> in producing an equally sudden change in the arrangement of the crystals of the metal, and that every such change is accompanied with one or more planes of weakness in the mass... Planes of fracture follow the tracks, with almost unerring precision, of all reentrant angles and of all sudden changes of scantling or dimension.

Mallet also discussed the change in fracture of wrought iron upon forging, but did not correctly appreciate the nature of the slag stringers. Other experiments were concerned with the variation in the density of large vertical cylinders of cast iron with the height of the feeder head.

In his book Mallet also gave an account of gun metals (typically copper–tin alloys with about 7–12 wt% tin) for casting of cannon. Discussion of the phenomena of major segregation was presented, and also comments on the microscopical examination of fractures, referring to crystal facets, and recognising the polycrystalline nature of the alloy.

Thomas Graham (1805–1869; FRS 1836) was appointed in 1847 by the Board of Ordnance to enquire into methods of casting guns. Also Henry Bessemer was closely associated with the development of armaments. He had shown that his proposal to use a grooved projectile in a smooth bore gun required stronger materials than those then available for the construction of guns; this led to work which culminated in the converter process for steel-making, (see Chapter 7).

William Armstrong (later Baron Armstrong of Cragside, 1810–1900; FRS 1846) having established a reputation in hydraulic machinery became involved in improvements in armaments which brought him an equally wide reputation. Just after the outbreak of the Crimean war in 1854, he received at Elswick his first commission from the war office; this was to devise submarine mines to blow up Russian ships that had been sunk in Sebastopol harbour. Having achieved success he turned his attention more especially to artillery, and developed a rifled bore gun which soon widely became known by his name. First, after thorough consideration of all possible materials, he selected shear steel and wrought iron. Then he proved experimentally that the ordinary procedure of making guns, in which the metal was forged into the form and and a hole was bored down it, was unsatisfactory. He introduced a construction method based on the fact that a cylinder offers the greatest resistance to bursting when the exterior layers are in tension and the internal layers are in compression. An internal cylinder of steel was enclosed in a jacket made by twisting a wrought-iron bar, and welding the turns to produce a cylinder of diameter slightly smaller than that of the steel component. The jacket was heated to cause expansion and slipped over the core; contraction during cooling produced the required stress distribution. Other rings as needed were in turn shrunk on this cylinder. He made other improvements in the gun construction; a very important feature was breech loading, with inventions for methods of closing the breech. Also, he determined the best shape,dimensions and charge for the projectile.

Armstrong's first 3-pounder was completed in 1855, but it was initially regarded by the artillery officers as a 'pop-gun'. He then constructed a 6-pounder, and continued experiments; a series of eleven patents was taken out commencing in 1857. At first the military authorities looked coldly upon the new gun, but in 1858 the committee on rifled cannon reported favourably. Armstrong gave the nation his valuable patents, and placed his talents at its service. In 1859 he accepted the appointment of engineer of rifled ordnance at Woolwich. The Elswick Ordnance Company was formed in 1859 specifically to make Armstrong guns for the British government under Armstrong's supervision; he had no financial interest in this new company. Three thousand guns were manufactured be-

Joseph Whitworth

William Anderson

Mallet's mortars in the Thomas Ironworks of C.J. Mare at Blackwell. From *Illustrated London News*, December, 1857.

tween 1859 and 1863, and at the latter date the British armanent was the best in existence. However, there was then a reaction in favour of the superiority of muzzle-loading guns. The government greatly reduced the orders they placed with the company, and returned to muzzle-loaders, and Armstrong resigned his official appointment in 1863. During the next fifteen years, England lost her supremacy in artillery.

The Elswick Ordnance Company was incorporated with Armstrong's engineering company and its works; blast furnaces were added and the company was largely employed by foreign governments. Armstrong improved the breech-action of his gun, and investigated the best method of rifling and the most advantageous calibre of the bore and structure of the cylinder, so as to obtain the greatest accuracy in shooting and the longest range with the minimum weight.

In 1877 Armstrong began trials on a gun construction technique in which a cylinder was wound with wire. His interest in this technique, in which wire in tension would replace hoops or jackets around the core, dated back to the mid-1850s and to his friendship with Isambard Kingdom Brunel (1806–1859; FRS 1830) who proposed that Armstrong should experiment with the technique to construct a gun. In 1855 it turned out that a patent for this technique was taken out independently by a Mr Longridge, although it appears that the patent was not put into practice. In 1877 when the patent had expired Armstrong returned to the subject. Following his initial trials he continued to work and finished the construction of a 6-inch (15 cm) breech-loading gun at the beginning of 1880. The results were so satisfactory that before the end of the year the ordnance authorities decided to abandon the manufacture of muzzle-loading guns and once more to adopt breech-loading guns with polygroove rifling.

Another development in ordnance was made by Robert Mallet who proposed, around 1850, the principle of ringed ordnance construction, which became widely used to increase the bursting strength of mortars; the thickness of the barrel was built up by using superimposed laminae or rings, with initial tension. At the time of the Crimean War he organised at the Thames Ironworks the construction of two very large mortars for projecting 36 inch (~0.9 m) shells. The contract was signed in 1855, but it was not until 1857 (after the war) that the mortars were tested at the Plumstead marshes; although the projectile was fired, fracture of a wrought iron ring occurred and the project was discontinued in 1858.

Isambard Kingdom Brunel, in his distinguished engineering career, worked on the improvement of large guns, and designed a floating gun-carriage for use in the Crimean war. Another important figure in the armaments industry was Joseph Whitworth (later Sir Joseph, 1803–1887; FRS 1857) whose inventions included a rifled cannon; however, despite its unrivalled ballistic power the weapon was rejected by the ordnance board in 1865. In the construction of guns from steel he was concerned with achieving adequate ductility, and he found that the difficulties of producing large steel castings of satisfactory soundness could be overcome by applying high pressure; he discovered that this could best be done using a hydraulic press. (Bessemer had taken out patents for casting under gaseous pressure and also under the pressure of a hydraulic plunger and had granted a licence to Whitworth). Whitworth steel, as it was called, was produced in this way around 1870, its special application being for large guns. Whitworth was the first to show the penetration of armour-plating up to four inches thick. In 1897 Whitworth's works at Openshaw, near Manchester for the manufacture of Whitworth guns was incorporated with Armstrong's industrial activity to form Sir W.G. Armstrong, Whitworth and Company Ltd.

The distinguished Victorian engineer William Anderson (later Sir William, 1835–1898; FRS 1891), worked as a young man at an ironworks in Dublin, where he was involved in general engineering and in design work. Later in his career his activities included the design of gun mountings and also work on explosives; in 1889 he was appointed as director-general of ordnance factories.

Other examples of Fellows in the 19th and also in the 20th century whose work included metallurgical aspects of armaments are referred to in Chapter 9.

Ingot production and mechanical working

In the sphere of mechanical working of metals an important development is associated with Robert Stirling Newall (1812–1889; FRS 1875). In 1840 he took out a patent for a method of producing wire ropes, and, with a partner, established a works at Gateshead-on-Tyne for manufacture. World-wide use quickly followed with an extensive industry of wiredrawing and Newall continued to improve his method. An important application was found in submarine cables. Newall studied the properties of the insulating material, gutta percha, and surrounded the insulation with strong wires; in 1853 he invented the 'brake drum' for laying cables in deep seas. Half of the first Atlantic cable was made at his works, and he himself directed much cable-laying work in the seas around England and elsewhere.

The researches of John Edward Stead (1851–1923; FRS 1903) covered a wide range of industrial problems, including especially ferrous process metallurgy. He had a particular interest in the influence of phosphorus on iron, and his investigations relating to this feature involved solidification phenomena and their implications in subsequent mechanical working. His researches on blast furnace 'bears' typify his methods of investigation; these 'bears' were large masses e.g. about 600 tonnes of metal which accumulate in the furnace hearth. Some time before the existence of eutectics in metallic alloys was widely recognised, Stead had determined the composition of the binary iron-phosphorus eutectic and the ternary iron-phosphorus-carbon eutectic by a novel method. Using hydraulic pressure he squeezed out of Cleveland iron the almost pure ternary eutectic and described its microstructure. His early experiments squeezed out only a small fraction of the total eutectic, since the temperature was not kept above that of the eutectic, 945 °C. Stead's investigations of blast furnace 'bears', however, showed what might have happened had that condition been fulfilled. During the exceedingly slow cooling, the 'bear' contracted circumferentially and compressed the central portion which was last to solidify, producing a vertical columnar arrangement, closely resembling on a small scale the basalt of the Giant's Causeway. These columns exhibited 'intercrystalline 'fracture' and Stead found by analysis that most of the eutectic had escaped vertically between the columns during the period of pressure.

Stead's investigations on 'ghosts' in steel forgings also relate to iron and phosphorus. He showed that in steels containing up to 0.45 %C, although carbon may be uniformly distributed in the steel at 1000 °C, yet on very slow cooling ferrite first appears in the parts richest in phosphorus; this is associated with carbon diffusing out of the phosphorus rich regions during cooling, the phosphorus-rich regions being called 'ghosts'. Stead concluded that if these features were not associated with a substantial amount of slag inclusions they were not dangerous or liable to lead to the failure of engineering structures. As early as 1906, Stead had concluded that sulphur, phosphorus and carbon are the main segregating elements in steel. The influence of Stead's ideas was apparent later when the Heterogeneity of Steel Ingots Committee produced a series of reports between 1926 and 1943.

In the technology of ingot casting, Robert Abbott Hadfield (later Sir Robert, 1858–1940: FRS 1909) developed a method for reducing shrinkage pipe; the liquid steel was covered with charcoal and a blast of air was applied to maintain the heat more completely. He also studied the mechanism of solidification by pouring molten copper into the mould after it had been filled with steel, and investigating the distribution of copper in the as-solidified ingot. Charles Algernon Parsons (later Sir Charles, 1854–1931; FRS 1898) set out procedures for ingot casting, involving chilling in the lower section and applying heat to the top. Also in the area of ingot casting, stemming from the work of Henri Etienne Saint-Claire Deville (1818–1881) and Henri Jules Debray (1827–1888) in furnace technology (1857), the achievement of George Matthey (1825–1913, FRS 1879) and collaborators in producing large ingots of platinum was of major industrial importance.

Non-metallic inclusions in steel play a crucial role in the final properties of the fabricated product and, among early workers, Thomas Andrews (1847–1907; FRS 1888) carried out microscopical studies of the distribution of manganese sulphide in steel forgings. Also Albert Portevin was concerned with inclusions and trace elements in ferrous alloys. Classic work was done by Andrew McCance (later Sir Andrew, 1889–1983; FRS 1943) arising from his concern with armour plate; two papers were published (1917 and 1918) which represented a great step forward in understanding inclusion formation. McCance showed that in addition to MnS formation inclusions of manganese silicate formed, and also that inclusions could be deoxidation products formed by the reaction of oxygen in the liquid steel with the deoxidisers added at the end of steelmaking. He also pointed out that to eliminate inclusions from liquid steel required aggregation of the smaller particles into larger ones and also time for these to rise to the surface. A great deal of experimental work was done, involving identification of the constitution and composition of the inclusions whose small size and their embedding in steel normally prevented analysis so that the method available for identification was mainly the examination under the microscope of colours produced by various etching agents. To relate constitution to com-

Barnes Wallis

Gilbert Roberts

position also required considerable work on the phase diagrams of the relevant oxide systems. Later (1930) McCance gave a quantitative treatment of the thermal problems of solidification in a mould.

William Herbert Hatfield (1882–1943; FRS 1935), who earlier in his career was concerned with castings for malleable iron and wrote a book on the chemistry of cast iron, later played an important role as chairman in committee work associated with important research programmes in the field of processing. Under the auspices of the committee dealing with the heterogeneity of steels ingots, up to 1943, 78 ingots were sectioned and examined, ranging in weight from 13 cwt (~ 0.65 tonnes) to ~170 tonnes. Two important subjects were the estimation of the quantity and character of the non-metallic inclusions and the determination of dissolved oxygen, nitrogen and hydrogen. The Alloy Steels Committee's work included the important subject of hair-line cracks, in large steel forgings and the role of hydrogen.

Charles Sykes also worked on the adverse effect of hydrogen on the ductility of low alloy steels and on the distribution of hydrogen in very large ingots in relation to the problem of the growth of hair line cracks. In other work on the complex problem of the development of internal stresses in large cylindrical forgings during cooling, he presented theory for the cooling of an infinite cylinder and included the effect of phase transformations. He described a procedure for determining residual stresses from the examination of trepanned samples. His work achieved success in developing treatments that reduced residual stress in large forgings.

Birmingham has long been an outstanding centre for many types of metallurgical processing. Arthur Neville Chamberlain (later Sir Neville, 1869–1940; FRS 1938), who became prime minister of Great Britain), received metallurgical instruction at Mason College. Neville's father Joseph Chamberlain (1836–1914; FRS 1882) had been associated over a long period with the screw manufacturing firm founded by John Nettleford. Neville became a director in 1897 in Elliott's Metal Company in Selly Oak. The company employed seven or eight hundred men and dealt mainly in copper and brass; Chamberlain's work included six months in the copper and metal sheet mill. Later in the same year, with family support, he bought Hoskins and Sons in Bordesley, a firm employing about 200 men, manufacturing ship's berths in metal; during the 1914–1918 war the firm built a steel mill. Chamberlain also became a director of the Birmingham Small Arms Company, whose activities during the war included the production of the first 'landships' (tanks).

Barnes Neville Wallis (later Sir Barnes, 1887–1979; FRS 1945) achieved particular fame during the 1939–1945 war for his work in developing novel bombs, including the bouncing type used in the RAF raids on the Ruhr dams. However, during his career he also made notable contributions as an engineer in the aeronautical industry, involving important metallurgical interest. He had special expertise in the use of the Duralumin type aluminium alloy. This was illustrated in his work in the 1920s on the design and construction of the R100 airship. For the construction he needed tubes, approximately 10 cm in diameter, which were too large for existing methods of manufacture. Wallis showed his skill as a mechanical engineer in developing a new method of fabrication, in which the tubes were made from an approximately 23 cm wide slab of Duralumin twisted spirally with edges overlapping to allow rivetting along the tube length. Later, including the wartime period, he also played an important role in aircraft design.

In the field of rolling Theodore von Kármán (1881–1963; FMRS 1946) published a paper in 1925 on the mechanics of the rolling of thin sheet metal. Egon Orowan (1902–1989; FRS 1947) during his period at Cambridge (1939–1950) also made important contributions to the theory of rolling. The applied mechanics research of Hugh Ford (later Sir Hugh, 1913–; FRS 1967) has included the theory of rolling of metals. Rodney Hill (1921–; FRS 1961) has worked on the theoretical mechanics of solids, especially in relation to plasticity; he was instrumental in developing the 'slip line field' theory and applied it to processes such as metal forming and machining. William Johnson (1922–; FRS 1982) has included in his research the mechanics of metal forming and pioneering investigations of manufacturing processes.

Joining

An interesting observation concerning welding was made early in the 19th century by Michael Faraday (1791–1867; FRS 1824) and James Stodart (1760–1823; FRS 1821) in their work on alloy steels; they found that bundles of wires of steel and platinum could be welded together by heating in the solid state. In the latter stages of the 19th century welding began to have a major influence on the processes of metal fabrication. Among the interesting developments during the century was the work of James Prescott Joule (1818–1889; FRS 1850) when in 1855 he suggested the use of electricity in welding.The first experiments were made in the laboratory of William Thomson (later Lord Kelvin, 1824–1907; FRS 1851) in Glasgow; bundles of iron wire were welded by surrounding them with charcoal and passing a powerful electric current through them. Joule followed this by welding several steel wires into one, and by joining iron with brass and with platinum. Gas torch and electric arc welding were developed during the century. For example, Matthey developed (1861) the use of an oxy-hydrogen blowpipe for the fusion welding of platinum, while Henri Louis Le Chatelier (1850–1936; FMRS 1913), arising from his work on the combustion of acetylene, developed the oxy-acetylene torch and its use in welding. In the 20th century, Irving Langmuir (1881–1957; FMRS 1935) at the General Electric Research Laboratory, Schenectady, USA, in the 1930s was associated with the development of the hydrogen welding torch. This stemmed from observations of one of his colleagues that, in a low-pressure hydrogen discharge, a platinum wire in a part of the tube remote from the discharge could become hot. Langmuir concluded that this effect resulted from hydrogen atom recombination, and that it could be used with an arc discharge in hydrogen, propelled against metal parts to be welded. Sebastian Ziana de Ferranti (1864–1930; FRS 1927) was involved in the electric welding of turbine blades.

Interesting observations on pressure welding were reported by James Alfred Ewing (later Sir James, 1855–1935; FRS 1887) and Walter Rosenhain (1875–1934; FRS 1913) in a Royal Society paper in 1900. Lead discs, with clean surfaces were pressed together in a testing machine, and were sometimes subsequently annealed. The welds were sectioned and examined microscopically to determine the nature of the interface. The effect was also examined of interposing a lead–bismuth eutectic between the surfaces to be welded and of annealing above the eutectic melting point after welding.

An area which became critical in welding technology was the occurrence of cracking in carbon steels, relating to the formation of brittle martensite above critical cooling rates. Portevin was active in welding and introduced the concept of weldability. Finding solutions for the problems of fracture in welded structures, such as ships, became a major field of investigation during the 1939–1945 war; Richard Weck (1913–1986; FRS 1975) played a major role in this area (see Chapter 14).

Stanley Fabes Dorey (1891–1972; FRS 1948)at the Lloyd's Register of Shipping carried out important work on welded joints in pressure vessels, beginning around 1930. He visited works in Britain and abroad and collaborated with boiler manufacturers in examining their problems. In relation to Class I pressure vessels recommendations were made for highly controlled welding process conditions and extensive radiographic examination. Dorey also worked with other welding applications.

Gilbert Roberts (1899–1978; FRS 1965), a leading civil engineer associated with many major projects such as bridges, was a pioneer in the introduction of welding in construction, beginning in 1931. He drafted rules for the design and making of butt welds and he recognised the need to use scientific knowledge in the control of materials. His work included the use of welded plate girders and the construction of all welded landing craft in World War II.

In the latter part of the 20th century among leading figures in welding technology, including fracture mechanics, are Alan Arthur Wells (1924–; FRS 1977), and Frederick Michael Burdekin (1938–; FRS 1993) whose work on large welded structures has led to the development of standard fracture tests and is important in North Sea oil and gas platforms and equipment. The work of John Frederick Knott (1938–; FRS 1990) has been important in the establishment of codes of practice for welded joints and pressure vessels. Also the research of Bernard Crossland (later Sir Bernard, 1923–; FRS 1979) has included friction welding and explosive welding.

Joining techniques for metals involving the use of adhesives have played an increasingly important role, particularly in the aircraft industry, during the 20th century. A notable inventor in this field, Norman Adrian de Bruyne (1904–1997; FRS 1967) founded his company, The Cambridge Aeroplane Company, in 1932, where he produced a revolutionary design from which he built the *Snark* monoplane (The name of the company became Aero Research Limited in 1934). In 1937 de Bruyne and his team developed a new synthetic glue – Aerolite – which was particularly effective for wood bonding. Five years later he developed the important adhesive – Redux – and the concomitant Redux bonding process, which was, and still is, used for structural metal-to-wood and, most importantly, metal-to-metal bonding of a vast range of military and civil airplanes (from the de Havilland *Hornet* to the BAe-146)

De Bruyne's work on adhesive bonding was crucial during the 1939–45 war for the bonding of clutch plates in tanks and for aircraft construction (e.g the de Havilland *Mosquito* and *Hornet/Sea Hornet*). The production of bonded aluminium honeycomb structures, combining light weight with high strength and stiffness, was also an important contribution to the aircraft industry, for example, in the construction of the de Havilland *Comet* series of jet airliners and their military version, the *Nimrod*.

Powder metallurgy

Powder metallurgy became a major industrial route for the production of a wide range of components from various types of alloys during the 20th century. It had previously assumed importance in the development of platinum metallurgy (see Chapter 4) and another interesting example of the production of metal powders is found in the early career of Henry Bessemer.

In around 1840, Bessemer turned his attention to the production of bronze powder and gold paint. This was an industry that had been known in China and Japan for many centuries, and had been imitated very successfully in Germany. Bessemer found that the German product was made from various copper alloys beaten into thin ' leaves' which were then ground by hand to a powder on a marble slab. His objective was to produce powder from a solid sample of brass at a low cost. His first approach was based on the turning lathe, but incorporating a novel procedure of using grooved rollers to form very small truncated pyramids on the periphery of a rotating brass disc. The turning tool machined off a very thin film of the metal from the apex of the pyramids, and large quantities of powder could be rapidy produced. However, Bessemer found that this powder did not have the same appearance as that of the commercially available material. He examined his powder and the commercial powder with a microscope, finding that the small particles from his process had a morphology he described as 'curled-up pieces, one side being bright and the other rough and corrugated and destitute of any brilliancy'; when these particles were placed on an adhesive surface, their random array reflected very little light. In the case of the commercial product Bessemer saw the thin flake morphology, which led to the particles lying flat on a surface giving bright reflections.

He then had the idea of a new manufacturing process, which he decided to develop, involving great effort in terms of equipment and extreme secrecy as to the techniques involved. He succeeded in his venture and in one room had 30 pieces of brass being operated on simultaneously, each piece yielding 2,000 or 3,000 'fine, needle like filaments' per minute, ~9 mm in length; the shape was attributed to the intense vibration of the machine. The flakes were subjected to further processing, involving a series of passes through rollers, which reduced the section of the particles; the product was a 'leafy, flaky powder of varying degrees of fineness'. The powder was subjected to a 'polishing' process and then separated into various grades of size. The control of colour of the powder was achieved by an oxidation treatment, the colour depending on alloy composition and on temperature and time of heating; careful process control was achieved. For control of colour Bessemer made a series of copper-based alloys e.g. one containing silver gave a brilliant purple colour when oxidised. For the production of a 'white bronze' Bessemer developed a process in which brass powder and small spherical shot of pure tin were subjected to a churning action in a solution of sodium carbonate, by which tin was transferred to the surface of the brass. For many years Bessemer supplied very fine powder of pure copper to Messrs. Elkington of Birmingham for metallising the surface of non-metallic moulds used to produce works of art by electrodeposition of metals.

In the 20th century in an important application of powder metallurgy, namely the production of cutting tools, Charles Sykes carried out work during the 1930s at the Metropolitan Vickers Electrical Company. This work used metallic carbides, for example of titanium, molybdenum and tungsten; mixed carbides were included in the work, and these were used for armour piercing shot in the 1939–45 war.

Consolidation of metal powders is an important and extensive field of work in which Geoffrey Wilson Greenwood (1929–; FRS 1992) has investigated the effects of hydrostatic pressure, residual gases and microstructure on the removal of small pores during sintering.

CHAPTER 16

Alloy Structure and Constitution

Introduction

The 20th century has seen remarkable advances in the fundamental understanding and development of many types of alloy, which have been accompanied by a greatly extended range of metallic materials available to industry. These advances have built on the scientific foundations and experimental data from the 19th century discussed in earlier chapters. Also they have only been made possible by the use of X-ray diffraction, electron microscopy and other modern experimental techniques. This chapter reviews the role of Fellows and Foreign Members of the Royal Society in foundational work on phase equilibria and phase diagrams, and phase transformations in steels and other alloys, leading on to the theory of alloy constitution.

In addition to the work of Fellows described here, research by other distinguished scientists proceeded very strongly in various countries beginning in the 19th century. Individuals such as Dmitri K. Tschernoff (1832–1921), Floris Osmond (1849–1912), Albert Sauveur (1863–1939), Henry Marion Howe (1848–1922) and Gustav Heinrich Johann Apollon Tammann (1861–1938) with their associates,made important contributions in physical metallurgy, for example the determination of phase diagrams and the study of phase transformations in steels and other alloys.

Phase equilibria and phase diagrams

A milestone in fundamental understanding of alloys was established by the thermodynamic work of Josiah Willard Gibbs (1839–1903; FMRS 1897) on heterogeneous equilibria, carried out in the USA in the latter part of the 19th century and published rather obscurely there. Although he attained a world-wide reputation Gibbs did not draw many graduate students to Yale University, New Haven. His influence came chiefly from his writings and he devoted himself to the development and presentation of his theory of thermodynamics. There was already a general basis of this science, through the first and second laws of thermodynamics, and these laws had been worked out mathematically and applied to homogeneous substances. The first two of Gibbs's scientific papers presented in 1873 an exhaustive study of geometrical methods of representing diagrammatically the thermodynamic properties of homogeneous substances. The second of these papers attracted the attention of James Clerk Maxwell (1831–1879; FRS 1861), who understood its importance; Maxwell constructed a model illustrating part of this work, and sent a plaster cast of the model to Gibbs. In 1876 Gibbs published the first half of his great memoir *On the Equilibrium of Heterogeneous Substances* in the *Transactions of the Connecticut Academy of Arts and Sciences;* the second half followed in 1878. This work developed and virtually completed the theory of chemical thermodynamics and provided basic theory for the development of physical chemistry, which was to be created by Friedrich Wilhelm Ostwald (1867–1932) and others in the 1880s, although it was some years before many of Gibbs's theoretical developments were experimentally verified. The memoir contains the invaluable Gibbs Phase Rule and his name is also associated with the Gibbs free energy. Henri Louis Le Chatelier (1850–1936; FMRS 1913), having become aware of the memoir of Gibbs and having discovered that his own main results had been anticipated, considered that the purely mathematical methods of Gibbs had prevented most chemists from understanding its chemical applications; by Le Chatelier's translation (published in 1899) of the original memoir, the pioneer work of Gibbs became known in France. (A translation into German had been published around 10 years previously). In the metallurgical field, Gibbs's work provided an essential basis and guiding principle for the extensive activity in phase diagram determination and application that occurred as a major feature in the development of physical metallurgy from around the turn of the century. Gibbs also carried out other important work on themes including the electromagnetic theory of light, and statistical mechanics, introducing in the latter the fundamental concept of Gibbsian ensembles.

The work of Francois Marie Raoult (1830–1901) and Jacobus Hendrik Van't Hoff (1852–1911; FMRS 1897) on solutions became important in relation to phase diagram work. (Although Raoult was not elected to the Royal Society he was awarded the Davy Medal in 1892.)

There was extensive and important research in Britain. William Chandler Roberts (later Sir William Chandler Roberts-Austen,1843–1902; FRS 1875) published a paper in the *Proceedings of the Royal Society* on the copper–silver system, reporting liquidus temperatures across the whole range of compositions. His technique was to immerse an iron sphere in a liquid alloy and then to transfer the sphere to a calorimeter when the alloy began to solidify. The temperature rise in the calorimeter was measured and then, using a calibration procedure, the temperature of the sphere which corresponded to the liquidus of the alloy was calculated. Although the temperatures determined were not of high accuracy, the work was a significant achievement.

Frederick Guthrie (1833–1886; FRS 1871) investigated the freezing points of solutions of salts, of alloys and of fused salts. In 1884 he published *On Eutexia,* using the term eutectic to represent the alloy of lowest melting point; the term was also used later as an adjective to describe the characteristic structure. Guthrie, using bismuth to form binary alloys with zinc, lead, tin and cadmium respectively, determined by thermal studies the compositions and temperatures of the eutectics, finding that the compositions did not correspond to simple atomic proportions. Data were also obtained for the eutectic temperature (71 °C) and composition in the quaternary system: bismuth–lead–cadmium–tin.

In the period approximately 1889–1894 Charles Romley Alder Wright (1844–1894: FRS 1881) conducted experiments with C. Thompson reported in a series of papers in

Josiah Willard Gibbs

Frederick Guthrie

Charles Wright

Frederick Abel

the *Proceedings of the Royal Society* on an extensive series of binary and ternary alloys with reference to liquid solutions and the existence of miscibility gaps. Binary alloys were prepared from combinations of nine elements: lead, bismuth, aluminium, zinc, tin, silver, antimony, cadmium and arsenic. In some of the alloys miscibility gaps were found, with the separation by gravity of two liquid alloys layers of different densities. Composition limits for these gaps were determined. e.g. for the lead–zinc system at 800 and 650 °C. Ternary alloys from the same series of elements were also investigated. Gibbs had reported a method of displaying ternary alloy compositions using an equilateral triangle; the method was rediscovered by Sir George Gabriel Stokes (1819–1903; FRS 1851) and used to display Wright's results. The concepts of 'critical curve' and 'tie-line' were introduced, and the boundaries of the miscibility gaps were determined for various cases where the gap derived from one of the binary systems e.g. the lead–zinc–tin and lead–zinc–silver systems. In the latter system, relevant to the Parke's process of desilverisation of lead bullion, evidence was obtained of the presence of zinc–silver compounds. The situation where the gaps derived from two binary systems was also investigated. Reference was made to quaternary alloys whose compositions were represented graphically within a tetrahedron. These experiments made an important early contribution to the study of binary and higher-order phase diagrams, the earliest of the papers being the first published work on ternary alloy phase relationships.

Particularly notable was the accurate experimental work of Charles Heycock (1858–1931; FRS 1895) and Francis Neville (1847–1915; FRS 1897) which extended into the 20th century and represented an important advance in phase diagram studies. The first of the series of metallurgical papers from this partnership, published in 1889, dealt with the depression of the freezing points of metals by others metals in solution; tin was used as the solvent metal in the initial work. In this, and later papers, the addition of small amounts of a second metal was reported to depress the freezing point to an extent directly proportional to the weight of metal added, and in approximately inverse proportion to the atomic weight. Raoult's law for ordinary solutions was thus extended to alloys, and a method indicated for calculating the latent heat of fusion of a metal by the application of the equation of Van't Hoff to the freezing point depressions.

It is interesting to note that also in 1889 William Ramsay (later Sir William, 1852–1916; FRS 1888, at University College London, and subsequently renowned for his investigations of the inert gases) published a paper of metallurgical interest. Raoult had shown the effect of dissolved solutes on the vapour pressure of a solution, raising the boiling point; Ramsay reported on the determinations of the increase in boiling point of mercury resulting from the addition of other metals and his paper was discussed alongisde that of Heycock and Neville.

Heycock and Neville initially used mercury thermometers for their temperature measurements and only alloys of low melting point could be studied; the introduction by Hugh Longbourne Callendar (1863–1930; FRS 1894) and Ernest Howard Griffiths (1851–1932; FRS 1895) of the platinum resistance resistance thermometer made it possible for Heycock and Neville to extend the scope of the investigations to metals of high melting

Charles Heycock

Frances Neville

The laboratory of Heycock and Neville at Sidney Sussex College, Cambridge.

point. At that time the melting points of silver, gold and and copper were not known accurately, partly because of the difficulty of making the measurements, and partly because the need to use metals of high purity and to protect them from contamination during melting had not been realised. A number of fixed points on the platinum resistance pyrometer had to be established, and these were determined by Heycock and Neville with very high accuracy, with the aid of Griffiths.

The study of dilute solutions led to the determination of the complete liquidus curve of many binary metallic systems, such as those of silver or copper with a second metal; in many cases the cooling curves showed arrest points below the solidus temperature. John Edward Stead (1851–1923; FRS 1903) and William Chandler Roberts-Austen (later Sir William, 1843–1902; FRS 1875) had done much to elucidate this subject, but the causes of these heat evolutions were not properly understood. Accordingly Heycock and Neville turned their attention to the examination of solid alloys and in about 1897 began work on the gold–aluminium system, probably choosing this because of its complexity. They developed a transmission technique for taking photographs through an alloy sample by using Röntgen X-rays (William Konrad Röntgen, 1845–1923). Although this method gave some valuable results it was soon abandoned because the microscopical examination of etched surfaces was simpler and more effective. In 1891, working on ternary alloys of gold, cadmium and tin, they established the existence of an intermetallic compound of gold and cadmium. Also work was done to confirm the existence of the $AuAl_2$ compound found by Roberts-Austen, who in 1891 had exhibited at the Royal Society a new gold–aluminium alloy, remarkable for its intense purple colour. (In modern times this compound has become of interest in the electronic industry as the 'purple plague' problem). Thus Heycock and Neville contributed to acceptance of the fact that stoichiometric intermetallic compounds can exist in alloy systems.

Their subsequent research on the bronzes was the basis of the Bakerian lecture presented in 1903 'On the Constitution of the Copper–Tin Alloys', which can be regarded as the foundation stone of modern metallographic study of phase equilibria and microstructure. It was the first substantially complete and accurate description of a complex series of alloys; also it attracted great interest and encouraged many others to undertake similar work. When the work was completed Heycock and Neville returned to the investigation of the gold–aluminium system, but progress was slow because of inherent difficulties of the problems encountered and because earlier records were lost in a fire in their laboratory; in 1914 however, they published a classical piece of work on the gold-rich alloys.

The phase diagram of the most important metallurgical system, that of iron and carbon, was the subject of many investigations. Roberts-Austen in 1897 determined freezing points; Hendrik Willem Backhius Roozeboom (1854–1907) using the data of Roberts-Austen and of Osmond (thermal studies) published an iron–carbon phase diagram in 1900, which made an important contribution. Later the first metallurgical publication of Henry Cort Harold Carpenter (later Sir Harold,1875–1940; FRS 1918), with B. Keeling, concerned the freezing point range and other transformations in the iron-carbon system.

They not only attained the necessary temperatures, which was a significant achievement in a laboratory in those days, but also successfully guarded against errors from thermocouple contamination and other difficulties. Their data gave some indication of the changes corresponding to the delta range of the system and the decrease with decreasing temperature of the solubility of carbon in ferrite. Carpenter later investigated high speed tool steels, with special reference to the influence of tungsten, molybdenum and chromium on the thermal critical points of steel.

In the 1930s and 1940s important investigations of alloy constitution were made by Albert James Bradley (1899–1972; FRS 1939) and co- workers; these involved the structures of ordered aluminide compounds and extensive determinations of a number of important ternary phase diagrams. The iron–nickel–aluminium system, significant regarding magnetic properties, was among those studied, and X-ray diffraction played a vital role as a technique for phase diagram determination. Also in this period another feature of interest is found in a collaborative investigation by Henry Solomon Lipson (1910–1991; FRS 1957) on the copper–nickel–aluminium system, which included an example of what came to be identified later as a solid-state spinodal decomposition.

Phase transformations and the heat treatment of steels

The enormous industrial importance of steels, both plain carbon and alloy, naturally led to intense activity, relating to structural features relevant to processing and heat treatment. There were long-running controversies centred around the phase transformations stemming from the allotropy of iron; the nature of the various structures observed in heat treated steels, particularly martensite, was one of the great and controversial metallurgical topics, leading up to and around the turn of the century. It attracted many researchers, including Fellows of the Royal Society. Techniques of investigation were primarily microscopical observations – following the work of Henry Clifton Sorby (1826–1908; FRS 1857) – thermal analysis, dilatometry and physical property measurements. Some of the phases investigated were named after distinguished workers: austenite, after Roberts-Austen; martensite, after Adolf Martens (1850–1914); also one of the constituents formed during the relatively rapid cooling of austenite, later found to be a fine, lamellar pearlite, was termed sorbite, after Sorby.

George Gore (1826–1908; FRS 1865) made an interesting observation (1868) when a strained iron wire was heated to redness by an electric current. During cooling the wire suddenly showed an increase in length and then gradually contracted; this effect was accompanied by a change in magnetic permeability. A length change (contraction) on heating was demonstrated by William Fletcher Barrett (1844–1925; FRS 1899), and he also observed a recalescence (increase in temperature) during cooling in the range ~700 –800 °C. Thus an insight was obtained into what came to be known as property changes associated with the allotropy of iron. Earlier in the 19th century William Hasledine Pepys (1775–1856; FRS 1808) had used the heating effect of an electric current in an interesting carburising experiment; he showed that steel was produced when a pure iron wire was heated in close contact with diamond.

Following earlier work, such as that of Torbern Olof Bergman (1735–1784; FRS 1765), on the extraction and analysis of carbides in steels, the discovery of iron carbide, cementite Fe_3C, in steels by Sir Frederick Augustus Abel (1827–1902; FRS 1860) in 1885 was a significant finding. John Oliver Arnold (1858–1930; FRS 1912) in 1884 published the first of a series of papers on carbides in steels, describing investigations he had made with A.A. Read who was at Cardiff. They isolated cementite by electrolytic dissolution and deduced that the bright lamellae in pearlite were identical with cementite; they extended their work to steels containing alloy elements.

Concerning allotropy, three forms of iron were considered to exist: alpha, beta and gamma. In the controversies, one theory, held by the 'allotropists', took the view that beta iron was a hard constituent, which was retained by quenching. In contrast, the 'carbonists' suggested that a compound 'hardenite' was formed between iron and carbon at high temperatures, and that this hard form could be retained by quenching. Le Chatelier concluded that martensite was a solid solution of carbon in the low temperature allotropic form of iron, and Andrew McCance (later Sir Andrew, 1889–1983; FRS 1943) considered that carbon in martensite exists in a distorted ferrite lattice. The first paper published by McCance 'The constitution of troostite and the tempering of steel', appeared in 1910, the results of undergraduate work. There followed in 1914 a major paper, 'A contribution to the theory of hardening', based on studies of magnetic, electrical and volume changes. This paper also contained calculations and measurements on the cooling rates produced by quenching, and considered quench-cracking. Distinguished researchers including Walter Rosenhain, (1875–1934; FRS 1913), Sir Robert Abbott Hadfield (1858–1940; FRS 1909), William Herbert Hatfield (1882–1943; FRS 1935), Le Chatelier and Stead commented favourably on the paper. Eventually the beta-form of iron was proved to be the paramagnetic form of alpha-iron.

In France, Le Chatelier and his colleagues carried out important work on various aspects of ferrous physical metallurgy, including alloy steels. Albert Marcel Germain René Portevin (1880–1962; FMRS 1952)with co-workers, made substantial contributions relevant to steel heat treatment. He extended (1919) the work of Osmond through studies of steels covering a range of carbon contents, determining the transformation temperatures at different cooling rates obtained by varying the size of the specimens. The 'split transformation' was reported in which both pearlite and martensite form, and the critical cooling velocity at which pearlite formation is just avoided. Measurements were made of the depth of hardening by cooling round bars of different diameters at a constant cooling rate. The effect of carbon and manganese content in decreasing the critical cooling velocity was also shown. Among later findings relevant to quench-hardening was that of Harold Carpenter and J. M. Robertson, who showed that the temperature of martensite formation is not influenced by the cooling velocity.

Investigations of isothermal decomposition of metastable austenite were made in the USA (1929) by Edgar Collins Bain (1891–1971) and co-workers, particularly E. S. Davenport, using dilatometry and quenching of a series of isothermally treated samples for microscopical study. The plotting of the progress of transformation in the form of Time–

Temperature–Transformation (TTT) diagrams was a major advance, and illuminated many features relevant to heat treatment phenomena; these included the nature of the constituent that came to be called bainite. Investigations of the isothermal transformation kinetics of steels became a major research theme and, in the UK, an important centre of activity was the Mond Nickel Laboratory at Birmingham. Here, in the latter part of the 1930s the head of laboratories was Leonard Bessemer Pfeil (1898–1969; FRS 1951) with Norman Percy Allen (1903–1972; FRS 1956) as second in command. Allen was concerned with major projects on the transformation characteristics of low alloy constructional steels, using dilatometry to study the isothermal transformation of metastable austenite. Investigations were also made of the effect of alloy elements on critical cooling rates in relation to quench hardening of large diameter sections, in order to obtain maximum economy in the use of strategic elements. This work made a major contribution to a field which, subsequently in many countries expanded into topics such as hardenability and Continuous-Cooling Transformation (CCT) diagrams.

In the contemporary list of Fellows of the Royal Society, Robert William Kerr Honeycombe (later Sir Robert 1921–; FRS 1981) in his wide-ranging research on the structures and properties of alloys, has included major contributions in ferrous physical metallurgy. At Cambridge with his research group he made detailed studies of phase transformations associated with austenite decomposition in steels, for example, concerning the control of alloy carbide precipitation in microalloyed steels e.g. interphase precipitation, by thermomechanical treatments. Harshad Kumar Dharamshi Hansraj Bhadeshia (1953–; FRS 1998) has carried out extensive research on solid state phase transformations, involving thermodynamics, kinetics, mechanisms and structure-property relationships; the work includes modelling of microstructures in multicomponent steels and the design of steels with bainitic structures with good resistance to wear and impact, which has assisted in the large scale production of rail steels. The field of phase transformations has also been included in the mathematical research of Colin Atkinson (FRS 1998).The research of Kenneth Henderson Jack (1918–; FRS 1980) has included novel treatments of nitriding to produce nitride precipitate dispersions in steels.

Advances in fundamental understanding of metallic materials in the 20th century

In the second decade of the 20th century, fundamental experimental research on the nature of the grain structure of metallic aggregates was carried out by Cecil Henry Desch (1874–1958; FRS 1923) using alloys such as brasses which can be separated into their constituent grains. Desch showed that there was a close analogy between the distribution of faces in soap foam 'grains', whose form had been determined by surface tension, and metallic grains where surface tension is secondary in its effect; both cases in fact depend on the same general principle, studied by Lord Kelvin (formerly Sir William Thomson, 1824–1907; FRS 1851) namely the division of space into 'cells' without the presence of voids (see Chapter 11).

Grains and grain boundaries continued to be a vital theme in the 20th century, including associated phenomena such as recrystallisation and grain growth. The term recrystallisation had been used by Sorby when he observed (1887) the elongation of the grains when iron was cold hammered,and the production of equiaxed grains on subsequent annealing at red heat. Walter Rosenhain and James Alfred Ewing (later Sir James, 1855–1935; FRS 1887) investigated recrystallisation and grain growth during the annealing of deformed metals such as lead and zinc, noting the relationship between the recrystallisation temperature and the melting point. They proposed that grain growth occurred by grain boundary migration in contrast to an earlier proposal by John Stead that the mechanism was grain rotation and coalescence. Subsequently Carpenter and Constance Elam (1894–1995) established that strain energy was the driving force for recrystallisation and grain boundary energy that for grain growth. They determined quantitatively the phenomenon of critical strain for recrystallisation and the relationship between grain size and the extent of prior strain; they also found that grain boundary migration was the mechanism of grain growth.

A number of contemporary Fellows of the Royal Society have worked in this field. The work of Robert Wolfgang Cahn (1924–; FRS 1991) on the annealing of deformed metals has included mechanisms of recovery (e.g. polygonisation) and recrystallisation, while that of Michael Farries Ashby (1935–; FRS 1979) and Geoffrey Wilson Greenwood (1929–; FRS 1992) has included grain boundary movement. Ernest Demetrius Hondros (1930–; FRS 1984) at the National Physical Laboratory carried out research on the structure, chemical composition and energy of surfaces and interfaces of solids.

Phase transformation phenomena have formed a major theme, both in steels and nonferrous alloys, linked to processing–structure–property relationships. Topics covered have included precipitation- (age-)hardening, martensitic transformations, ordering and radiation-induced effects. The understanding derived has been utilised creatively in alloy development, together with advances in processing procedures such as powder metallurgy, surface treatments, continuous casting, rapid solidification techniques(including metallic glasses), joining technology, and nanotechnology. Extensive research in industrial and government laboratories and in universities has been carried out involving many scientists who became Fellows of the Royal Society. In addition to those whose research commenced in the 19th century there is an impressive list of past Fellows whose research was carried out mainly, or entirely, in the period since the turn of the century: Desch, Carpenter, Rosenhain, Portevin, Hatfield, Pfeil, Allen and Charles Alfred Edwards (1882–1960; FRS 1930). The activities of these men involved many themes in physical metallurgy, especially concerning alloys and including mechanical behaviour.

The research of a number of contemporary Fellows closely relates to microstructure and its control. In the work of John Wyrill Christian (1926–; FRS 1975) there has been a particular focus on phase equilibria and on phase transformations, covering crystallographic, microstructural and kinetic aspects including the key area of martensitic transformations. His book *Transformations in Solids* first published in 1965 is a classic in the field; the second part of this major work was delivered to the publishers in 1998.

One of the most significant fields of solid state phase transformations has been that concerned with the precipitation of fine dispersions of particles, typically of intermetallic phases, from supersaturated solid solutions. Beginning in the early years of the 20th century (see Chapter 12) numerous studies of the associated phenomena have been made seeking to develop high performance alloys and/or to understand the fundamental mechanisms. In the latter area, the X-ray investigations in the late 1930s on aluminium alloys by André Guinier and George Dawson Preston (1896–1972) made an important advance; the concept of metastable precipitate phases (including Guinier–Preston Zones) was established. With the advent of transmission electron microscopy of metal samples in the 1950s detailed understanding of the precipitation process opened up. Among Fellows who have made notable contributions by electron microscopy Sir Robin Buchanan Nicholson (later Sir Robin, 1934–; FRS 1978) has carried out fundamental studies on aluminium alloys; he has also studied Ni_3Al precipitation in nickel-based alloys, another example where properties depend vitally on the precipitate dispersion. The phenonemon of 'rafting' of particles in nickel-based alloys, important in relation to creep properties, has been studied by Frank Reginald Nunes Nabarro (1916–; FRS 1971). Alongside his work on mechanical properties Michael Ashby has investigated various important microstructural phenomena. Also the contributions of Geoffrey Greenwood have included phase transformation phenomena, such as coarsening of precipitate particles.

Robert Cahn, in addition to his distinguished investigations including the deformation, recovery and recrystallisation of metals has also played a major role as an author and editor in metallurgy and materials science, e.g, through the book *Physical Metallurgy* first published in 1965, and subsequently reissued in updated and expanded versions.

The names of William Hume-Rothery (1899–1968; FRS 1937) and Geoffrey Vincent Raynor (1913–1983; FRS 1959), who collaborated over many years are particularly associated with advances in the fundamental theory of metallic alloys, based on highly accurate experimental studies of phase diagrams. Hume-Rothery's PhD thesis (1925) at the Royal School of Mines, Imperial College, London was concerned with the structure and properties of intermetallic compounds. This thesis and the publication which stemmed from it contained the first mention of 'electron compounds' – intermediate phases in certain alloy systems whose structure and composition appeared to be controlled by the number of valency electrons per atom in the phase. Working at Oxford from 1925 and from 1932–55 holding a Warren Research Fellowship, he made a brilliant contribution to the science of metals and alloys. Together with his research activities Hume-Rothery was the author, or co-author of seven books; of these, his monograph on *The Structure of Metals and Alloys* (1936) gave to metallurgists for the first time an account of the theory of atomic structure, an outline of the theory of the metallic state, and also considered the factors influencing alloy constitution.

From early experimental work (1934) on the determination of liquidus and solidus curves and solid solubility limits for various alloys of copper and silver, the significance of the relative atomic size of the components, and factors of an electronic origin were clearly demonstrated for the first time. This work was the origin of the 'Hume-Rothery' rules which quickly attracted much attention among metallurgists and solid state physi-

William Hume-Rothery

Geoffrey Raynor

Harry Jones

John Lennard-Jones

cists and led to interaction with Nevill Francis Mott (later Sir Nevill, 1905–1996; FRS 1936) and Harry Jones (1905–1986; FRS 1952). Other research was concerned with intermediate phases forming at particular values of electron/atom ratio, magnesium alloys, high temperature alloys (mainly alloys of the transition metals) and iron alloys.

In the experimental aspects, Hume-Rothery installed X-ray equipment in about 1933 and carried out very accurate lattice parameter measurements. Later, high-temperature and low-temperature diffraction cameras were developed. For phase diagram determinations he made extensive use of metallographic methods, and elegant techniques were developed for thermal analysis and heat treatment of high temperature materials.The interests of the research group widened to include the theory of the deformation of metals and solid-state transformations.

In 1936 when Raynor commenced his postgraduate research at Oxford in Hume-Rothery's group the basic principles of alloy formation, the 'Hume-Rothery rules', had proved successful in their applications to alloys of many metals with copper, silver or gold. Raynor's work as a research student was mainly concerned with the exploration of alloys based on a metal of higher valency such as magnesium, and the application of the newly developed quantum-mechanical theory of metallic crystals. With Hume-Rothery's support he made a wide-ranging investigation of Brillouin zone effects in alloys (Marcel Léon Brillouin, 1854–1948). The intellectual and experimental contributions made by Raynor were such that, within a few years the partnership of 'Hume-Rothery and Raynor' had become the most famous in the science of alloy constitution, certainly since that of Heycock and Neville.

During the 1939–1945 war the Oxford research group carried out technological alloy researches for the Ministries of Supply and of Aircraft Production with Raynor playing a leading role. In particular this work was concerned with the constitution of multicomponent alloys such as those from the aluminium–magnesium–manganese–zinc system for the development of high-strength aircraft materials, and also with carbides and nitrides in titanium-containing steels. One discovery was that of ultra-light alloys, more ductile than magnesium, obtained by adding about 30 at% of lithium to magnesium, which produces a homogeneous body-centred cubic phase. Raynor moved to Birmingham University in 1945 where he built up and maintained for many years a large and brilliant research group on alloy constitution concerned with transition metals in aluminium alloys, including electron compound formation, copper- and silver-rich ternary phase diagrams, and rare earth metals in alloys.

At Oxford in the late 1940s John Christian and Bryan Randell Coles (1926–1997; FRS 1991) were co-workers with Hume-Rothery. Bryan Coles developed an on-going interest in the physics of transition metals and alloys; he began his career as a metallurgist and went on to become a solid-state physicist. He made important contributions in the field of electrical and magnetic properties and to the fundamental understanding of metals and alloys. Also Zbigniew Stanislaw Basinski (1928–; FRS 1980) worked in the Oxford laboratory.

The National Physical Laboratory was another important centre for the accurate determination of equilibrium phase diagrams, and during Norman Allen's period as superintendent of the metallurgy division, beginning in 1945, a section was set up to derive such diagrams from first principles using thermodynamic data: heats of formation, free energies etc. Methods of measurement for the determination of the data were critically reviewed, and experimental work was carried out, particularly on the iron–nickel and iron–chromium systems. The NPL is now a leading centre for the thermodynamic calculation of phase diagrams (CALPHAD).

In the field of fundamental understanding of the electronic structure of metals and alloys, including its link with properties, a number of Fellows of the Royal Society have played outstanding roles. Among these Enrico Fermi (1901–1954; FMRS 1950) made important contributions, including for example the concept of the Fermi level. Nevill Mott, in the course of a long career, made outstanding contributions to the theory of metals and alloys, including work on superconductors and other materials.

In 1930 Frederick Alexander Lindemann (later Lord Cherwell, 1886–1957; FRS 1920) a member of the government's Advisory Council for Scientific and Industrial Research, was among those who were concerned at Britain's neglect of fundamental work on the behaviour of electrons in metals, in view of the importance of the metallurgical industry and the new insights given by quantum mechanics. Lindemann persuaded the Advisory Council for Scientific and Industrial Research, of which he was a member, that this situation should be remedied. He told John Edward Lennard-Jones (later Sir John, 1894–1954; FRS 1933), then professor of theoretical physics at Bristol, that, if he could establish a group on this subject, funds would be provided. Lennard-Jones accepted this offer, and Harry Jones as a theoretical physicist was appointed as a research assistant in 1930. He began to study the recently published papers, notably the work of Felix Bloch (1905–1983), Rudolf Peierls (later Sir Rudolph, 1907–1995: FRS 1945), Brillouin and Alan Harries Wilson (later Sir Alan, 1906–1995; FRS 1942). Mott, when he moved to Bristol in 1933, wanted to support Jones in work on metallic alloys. Early in 1933 Jones attended a lecture given by William Lawrence Bragg (later Sir William, 1890–1971; FRS 1921) on the structure of alloys, and in particular on the rules found experimentally by Hume-Rothery, showing that the structure of the brasses and bronzes depended on the ratio of electrons to atoms. Jones showed that an explanation in terms of quantum mechanics was possible. During the period 1933–1937 Jones cooperated with Mott and the influential book *The theory of the properties of metals and alloys* was published jointly by them in 1936. They also published a paper jointly with Herbert Skinner (1900–1960; FRS 1942) at Liverpool who was carrying out research on X-ray emission bands in solids, experimenting with light metals. Skinner made an important advance in soft X-ray spectroscopy (1940); using targets coated with liquid nitrogen he examined the effect of temperature on the electron energy of metals.

Alan Harries Wilson was distinguished for his research on the electronic theory of metals, insulators and semiconductors. He made important advances in the theory of the optical properties of metals, of diamagnetism, and of thermal conductivity and thermoelectric effects at low temperatures. Rudolf Peierls included in his research work the movement of electrons in solids, particularly the effect of magnetic fields.In the USA, the research of Philip Warren Anderson (1923–; FMRS 1980) has included the electronic behaviour especially of magnetic and disordered materials. Also in the USA at the end of the 1939–1945 war, John Bardeen (1908–1991; FMRS 1973) who during the 1930s had worked on the theory of the alkali metals joined the research laboratories of the Bell Telephone Company at Murray Hill, New Jersey, where he began his research on semi-conductors, that led to the invention of the transistor.

Volker Heine (1930–; FRS 1974) is distinguished for his work on the electron theory of solids; he has been closely involved in the development of the pseudopotential method for explaining bonding and structure of many metals and alloys, and has extended his work to applications in insulators and minerals.

Particularly during the latter half of the century, the fundamental understanding achieved in the field of atomic structure and thermodynamics has led to major advances in the prediction and modelling of phase equilibria and alloy behaviour, underpinning alloy design. The research of David Godfrey Pettifor (1945–; FRS 1994) uses fundamental theories of atomic structure to develop highly predictive models of cohesion and structure in alloys, with important practical applications. Also his collaborative work on the behaviour of iron under pressure and temperature represents a major advance in understanding metallic magnetism.

CHAPTER 17

Non-Metallic Materials and the Royal Society

Introduction

The previous chapters include consideration of the role of metals and alloys in a range of engineering applications; brief mention is also made of non-metallic materials such as glasses, ceramics and cements (for example in Chapter 3). The present chapter outlines some of the advances made in the 20th century in the now greatly extended range of non-metallic materials developed for numerous engineering applications. In these advances many distinguished chemists, physicists and engineers, elected to the Royal Society have been prominent.

The latter half of the 20th century has seen the evolution of materials science, technology and engineering, by the converging of advances in metallurgical knowledge with developments in various other streams of interdisciplinary activity, particularly from chemistry and physics, and involving semiconductors, superconductors, ceramics, glasses, polymers and rubbers, biomaterials and composites. The term 'advanced materials' has come into use, referring to products where progress has achieved novel, sometimes unique properties opening up new fields of technology, for example in the electronics industry, and the development of smart materials and products for biomedical applications. At the same time progressive improvements have been crucial, including importantly advances in processing of long-established existing materials. In parallel with the advances in scientific understanding, technological advances have spanned the whole spectrum of materials.

An interesting overview of advanced materials as perceived in 1987 and looking ahead into the 1990s was presented in the proceedings of a discussion meeting held at the Royal Society.* The topics ranged widely covering aspects such as aerospace materials, electronic materials, rapidly solidified materials, advanced thermoplastics, ceramics and composites; other features included the importance of market requirements and the transfer of new inventions into industrial usage. An interesting historical perspective of the evolution of materials for mechanical and civil engineering is presented schematically overleaf. It illustrates how the relative importance of metals as compared with non-metallic

*Technology in the 1990s: the promise of advanced materials. Discussion organised by E. D. Hondros FRS, A. Kelly FRS, and Sir Robin Nicholson FRS, edited by E. D. Hondros FRS and A. Kelly FRS, *Phil. Trans.*, 1987, **A322**, 307–492.

EVOLUTION OF ENGINEERING MATERIALS

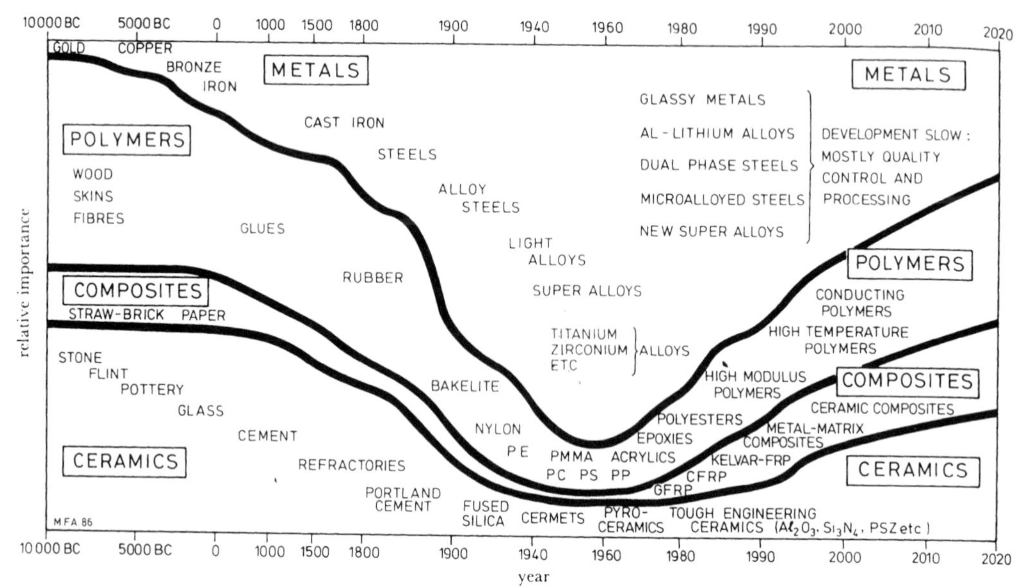

M. F. Ashby: 'Technology in the 1990s: advanced materials and predictive design', *Phil. Trans.*, 1987, **A322**, 393–403.

materials increased from the centuries of early civilisation to a maximum at around the mid-20th century; in the latter half of the century, and projecting into the 21st century, the increased importance of non-metallic materials, notably those regarded as of the advanced type is shown.

In recent years the terms 'structural materials' and 'functional materials' have become widely used. The former term refers to materials whose engineering applications depend mainly on their mechanical behaviour. The latter type describes materials whose specific applications depend essentially on physical properties such as semiconducting or optical behaviour.

As with attempts to define a 'metallurgist' it is difficult to draw a boundary for materials science, technology and engineering. The research of individuals can involve projects in which much or all of the spectrum of processing–structure–properties–applications relationships is strongly involved. Other researchers may focus on particular aspects such as fundamental understanding of physical or chemical phenomena, crystallography, instruments for materials characterisation, the analysis of mechanical behaviour in service, or the development of materials systems such as electronic devices. Taking a broad view of materials, as mainly relating to the production of artefacts for the various fields of modern life, but including scientists associated with fundamental work in solid state physics and chemistry, and also work by some engineers, the current list of Fellows and Foreign Members can be said to include approximately 150 names. About a fifth of them are particularly concerned with metallic materials. The contributions of those concerned with non-metallic materials are briefly reviewed here. The total for materials, representing more than 10% of the complete Fellowship (currently around 1300 including Foreign Members) clearly illustrates the importance of the field.

Semiconductors, liquid crystals and solid-state science

John Bardeen (1908–1991; FMRS 1973) joined the Bell Telephone Company laboratories, New Jersey at the end of World War II and began his research on semiconductors; this led to the invention of the transistor in 1947 for which he jointly with Walter Hauser Brattain (1902–1987) and William Bradford Shockley (1910–1989) received a Nobel Prize for Physics in 1956. The period from the 1950s has seen remarkable advances in electronic materials and electronic devices, including the area of optoelectronics. Of the many Fellows and Foreign Members of the Royal Society associated with electronic materials some have been mainly concerned with the engineering approach in the development of devices, while also being interested in the materials science aspects of fundamental electronic structure and properties; other Fellows are mainly interested in relevant theoretical areas of solid state physics. The list of Fellows (mainly contemporary) given here, with brief statements of their interests, does not completely cover all those who have contributed.

Alan Harries Wilson (later Sir Alan, 1906–1995; FRS 1942) made fundamental advances in the understanding of semiconductors and insulators as well as metals. The research of Evgenii Mikhailovich Lifshitz (1915–1985; FMRS 1982) included photo-electromagnetic effects in semiconductors. Peter George LeComber (1941–1992; FRS 1992) worked on amorphous semiconductors, particularly silicon; his investigations on 'doping' led to application in large scale junctions for solar energy conversion. Robert Allan Smith (1909–1980; FRS 1962) included in his research the investigation of impurity bands in semiconductors.

Among contemporary Fellows, the research of Sir Roger James Elliott (1928–; FRS 1976) is concerned with the theory of the solid state with particular reference to magnetism (rare earth metals), semiconductors and disordered materials. The research of Sir Peter Bernhard Hirsch (1925–; FRS 1963) involves a major interest in semiconductors, including dislocation effects. Brian Kidd Ridley (1931–; FRS 1994) has investigated quantum processes in semiconductors and electrons and phonons in semiconductor multilayers. Peter Neville Robson (1930–; FRS 1987) has worked on semiconductor devices and compounds and on magnetic theory. The research field of Walter Eric Spear (1921–; FRS 1980) has involved electronic and transport properties in crystalline solids, liquids and amorphous semiconductors. In his research in solid state theory, materials science and computational physics, the contribution of Dennis Lawrence Weaire (1942–; FRS 1999) include laying many of the foundations of the theory amorphous semiconductors. The research of Cyril Hilsum (1925–; FRS 1979) has included the science of compound semiconductors and the development of semiconducting devices. In the electronics industry, Derek Harry Roberts (later Sir Derek, 1932–; FRS 1980) has played a leading role in silicon integrated circuits, optoelectronics and memory devices. In optoelectronics the research of William Alexander Gambling (1926–; FRS 1983) and David Neil Payne (1944–; FRS 1992) includes optical fibre technology. Electron microscopy, electron beam lithography and integrated circuit fabrication come within the sphere of work of Sir Alec Nigel Broers (1938–; FRS 1986).

Edward George Sydney Paige (1930–; FRS 1983) has included in his research electron transport in semiconductors. The research of Sir Gareth Gwynn Roberts (1940–; FRS 1984) on the physics of semiconductors has included insulating films. Arthur Marshall Stoneham (1940–; FRS 1989) has been concerned with the theory of defects in semiconductors and insulators, particularly those which control optical, electronic and chemical properties. Ronald Charles Newman (1931–; FRS 1998) has made important contributions on epitaxial growth of thin films of metals and also of semiconductors. Also, his semiconductor research has contributed to understanding of impurity behaviour; he was the first to use infra-red methods to identify and characterise the vibrational spectrum of carbon as a substitutional impurity for silicon, and to demonstrate directly the segregation of elements such as carbon and oxygen to dislocations. Alan Herbert Cowley (1934–; FRS 1988) in his distinguished chemical research has investigated the synthesis of semiconductors, for example, compounds of phosphorus, arsenic and antimony. Pioneering work has been done by Donald Charlton Bradley (1924–; FRS 1980) in producing precursor materials and in their conversion to metal oxides and nitrides for chemical vapour deposition in the field of micro-electronics.

Pioneering work on the application of high pressure techniques to investigations of semiconducting materials has been carried out by Alfred Rodney Adams (1939–; FRS 1996). Robert Hughes (Robin) Williams (1941–; FRS 1990) has applied surface science techniques to study semiconducting solids and interfaces in solid state electronic devices. Low-dimensional semiconductors and semiconductor physics for devices are among the interests of Michael Joseph Kelly (1949–; FRS 1993). Laurence Eaves (1948–; FRS 1997) is concerned with the understanding of the electronic properties of semiconductors; his work on low-dimensional structures has shown the importance of space charges. Michael Pepper (1942–; FRS 1983) is concerned with semiconductors and the physics and chemistry of the solid state. The research of Alan Frank Gibson (1923–1998; FRS 1978) in solid state physics has dealt particularly with semiconductors and devices, including investigations of optical properties using lasers. Optical properties of semiconductors are the subject of research carried out by David Andrew Barclay Miller (1954–; FRS 1995). X-ray studies of semiconductors have formed part of the research of Michael Hart (1938–; FRS 1982).

The investigations of Richard Henry Friend (1953–; FRS 1993) cover the electronic properties of novel semiconductors, conducting polymers and of metals, including some superconducting materials; practical applications include transistors and light-emitting diodes. Recent research of John Brian Pendry (1943–; FRS 1984) is concerned with a technique for producing photonic materials by coating very small metal spheres with one of two types of DNA, leading to bonding when the spheres are brought together.

Louis Eugène Felix Néel (1904–; FMRS 1966, Nobel Prizewinner for Physics 1970) in his work on magnetism predicted in 1936 the type of ordering called anti-ferromagnetism; this prediction was subsequently confirmed experimentally. Néel also suggested (1947) the effect known as ferrimagnetism, observed, for example, in ferrites, important in the electronics industry.

In the field of theoretical physics and quantum mechanics, properties of the solid state at low temperatures have been among the interests of Maurice Henry Lecorney Pryce (1913–; FRS 1951). Edgar William John Mitchell (later Sir William, 1925–; FRS 1986) has studied electrons and defects in solids by optical and neutron scattering techniques.

Liquid crystals, which have been investigated extensively since the 1960s and 70s, have led to a major field of technology in electronic display systems. Consisting in essence of aggregates of long molecules, they are intermediate in nature between crystals and liquids and when subjected to small electric signals alignment of certain structural features occurs. It is interesting to note that as early as 1933 Sir William Henry Bragg (1862–1942; FRS 1907) turned his attention to liquid crystals, and contributed to the understanding of the smectic type of structure; in November 1933 he gave a lecture on the subject at the Royal Institution. Pierre-Gilles de Gennes (1932–; FMRS 1984) was awarded the Nobel Prize for Physics in 1991 for his research on liquid crystals and polymers; his contribution in theoretical physics had also previously involved magnetism and superconductivity. The wide-ranging research of Sir Frederick Charles Frank (1911–1998; FRS 1954) included liquid crystals in which he elucidated the elasticity of the nematics and considered the defects termed 'disclinations'. John Douglas Eshelby (1916–1981; FRS 1974) also worked on defects in liquid crystals. Sivaramakrishna Chandrasekhar (1930–; FRS 1983) has predicted and discovered liquid crystals of disc-like molecules. George William Gray (1926–; FRS 1983) and Cyril Hilsum have made pioneering investigations on the molecular structure and properties of liquid crystals and materials for liquid crystal electronic displays. Also Edward Peter Raynes (1945–; FRS 1987) has worked on the physics and chemistry of liquid crystals and on their applications, while the research of Ian Alexander Shanks (1948–; FRS 1984) has included liquid crystal displays. The mathematical work of Frank Matthews Leslie (1935–; FRS 1995) has included fundamental studies providing models of the alignment and motion of liquid crystals in magnetic fields. Pioneering investigations have been made by Athene Margaret Donald (FRS 1999) on thermotropic liquid crystalline polymers, using transmission electron microscopy.

Ceramics and solid-state science

Considering research in the important field of inorganic materials with a strong chemical approach John Stuart Anderson (1908–1990; FRS 1953) with co-workers carried out extensive research over a long period, particularly on the constitution and behaviour of non-stoichiometric compounds and solid solutions. Part of his career was spent in the Chemistry Division of the Atomic Energy Research Establishment at Harwell. His work included the semiconducting properties of oxides at high temperatures, uranium oxides and hydrides, ternary oxide systems, rare earth carbides, and (from 1987) high temperature superconductors. Electron microscopy was used as an important technique.

In the areas of surface science, membranes, porous solids and diffusion, Richard Maling Barrer (1910–1996; FRS 1956) made outstanding contributions. Among his achievements was the laying of the foundations of research on zeolites and the industrial applications of these microporous materials, which possess among other characteristics the abilty to act as 'molecular sieves'. His work included aspects such as the synthesis of aluminosilicates and other zeolites. Barrer's research career also involved extensive investigations on molecular transport in microporous media and polymer membranes; he also carried out theoretical work on rate processes across gas/metal interfaces in metallic membranes.

Also in solid-state chemistry, the electronic and magnetic properties of inorganic compounds, including superconductors, are the research themes of Peter Day (1938–; FRS 1986). The research of Sir John Meurig Thomas (1932–; FRS 1977) in solid state and surface chemistry includes catalysis and the influence of crystalline imperfections. Theoretical aspects of surfaces and interfaces are part of the research interests of Michael Ellis Fisher (1931–; FRS 1971).

James Derek Birchall (1930–1995; FRS 1982), much of whose career was spent with Imperial Chemical Industries, carried out important and extensive research on inorganic materials. Concerning crystallisation of common materials his investigations included the determination of the best processing conditions to achieve the growth of dendritic crystals of sodium chloride. One of his major interests was the extinction of fire and he developed the product 'Monnax' – a composite consisting of potassium carbonate dispersed finely in urea as a non-toxic solid. During the 1960s he began research aiming to produce a cheap fibrous filling agent for composites. This work focussed on refractory oxides such as zirconia and alumina, and he developed techniques involving sol-gel processing for growing fibres by a spinning process, extrusion and heat treatment. An alumina fibre, termed 'Saffil' was produced having high strength and great thermal resistance; it has found important applications such as for thermal insulation and for heat resistant tiles for the space shuttle, and as fibres for a reinforcement phase in metal-matrix composites based on aluminium and on ceramics and glass. The improvement of Portland type cements was a major area of research for Birchall, aiming for example to achieve maximum particle packing, reduction of porosity and the prevention of macrovoid formation. He and his group developed processing techniques to achieve a defect-free condition with particle dispersion and lubrication by adding water soluble organic polymers, and by using high shear mixing. Various types of cements of improved mechanical properties were produced; for example, springs produced from cements were demonstrated. Work on 'green body' ceramics led to patents on what is now referred to as viscous plastic processing, which Birchall applied to ceramic high-temperature superconductors.

Kevin Kendall (1943–; FRS 1993) whose wide-ranging interests include particularly mechanical properties such as crack creation and propagation, played a major role working with Derek Birchall on the research on cements and glasses described above; also his work includes themes such as adhesion and delamination of composites.

William Watt (1912–1985; FRS 1976) at the Royal Aircraft Establisment, Farnborough, beginning in 1944, worked on the development of non-metallic materials, such as sintered oxides (for example alumina) and interstitial carbides (for example zirconium carbide) for high-temperature applications. Kenneth Henderson Jack (1918–; FRS 1980) has investigated interstitial structures important in metallurgy, and oxides and oxynitrides such as the SIALONS, for high temperature applications. Also in research on ceramics the contributions of Frank Philip Bowden (1903–1968; FRS 1948) have involved studies of the mechanical and structural properties of high temperatures solids such as carbides and borides of transition metals. Sir Ronald Mason (1930–; FRS 1975) has been concerned with the processing of ceramics, involving innovative manufacture.

In spectroscopy Lawrence Michael Brown (1936–; FRS 1982) is investigating the chemical state of embedded ions in materials.The interaction of energetic ion beams with materials, including ion implantation is an area in which Geoffrey Dearnaley (1930–; FRS 1993) is a leader. In the field of crystallography, the work of Alan Lindsay Mackay (1926–; FRS 1988) has correctly predicted the occurrence of five-fold symmetry (quasi-crystallinity) in nature, and he is a leading authority on the geometry and symmetry of crystalline and quasi-crystalline materials. The applied mathematical research of John Macleod Ball (1948–; FRS 1989) includes phase transformations and crystal microstructure.

Outstanding contributions to the solid state chemistry of ionic substances and to chemical spectroscopy have been made by Chintamani Nagesa Ramachandra Rao (1934–; FRS 1982). Sir Alan Walsh (1916–; FRS 1969) has also made distinguished contributions in spectroscopy, including emission and infrared spectroscopy and the atomic absorbtion method of quantitative analysis. Anthony Kevin Cheetham (1946–; FRS 1994) has developed and exploited neutron beams and synchroton radiation, especially powder profile methods, to analyse the structure of solid-state materials, not amenable to single crystal and spectroscopic techniques; he has interpreted the behaviour of microporous catalysts and absorptents of great potential industrial value. John William White (1937–; FRS 1993) has used neutron scattering techniques to study the structure of crystals and liquids and the structure and dynamics of adsorbed monolayers. Robert Kemeys Thomas (1941–; FRS 1998) has applied neutron and X-ray scattering methods to the study of the structures of wet interfaces including the liquid–solid type.

The research of David Keith Bowen (1940–; FRS 1998) has made unique contributions in applied physics on instrumentation and to nanoscience and metrology. He has instrumented precision techniques, particularly for surface metrology and characterisation and has been the prime mover in setting up the X-ray topography line at the Daresbury laboratory.

The research of several Fellows of the Royal Society in the subject area of earth science has considerable relevance to materials science. For example, James Desmond Caldwell (1933–; FRS 1987) is concerned with silicate, oxide and sulphide minerals, including structural transformations, and Colin Trevor Pillinger (1943–; FRS 1993) includes in his interests investigations of meteorites and the mineral physics of diamonds. The research of Alfred Edward Ringwood (1930–1993; FRS 1977), a distinguished Australian geochemist, included investigations of polymorphic transformations in a number of minerals at high pressures and temperatures in the context of his special interest in the origins and constitution of the Earth. Among his other activities he invented a new cutting tool material, consisting of diamond crystals bonded with silicon carbide.

Carbon

Carbon, with its various structural forms, is a material that has attracted a number of Fellows of the Royal Society in their research. For example, John Desmond Bernal (1901–1971; FRS 1937) determind the structure of graphite in 1924. Alfred René Jean Paul Ubbelohde (1907–1988; FRS 1951) whose research covered various important themes in solid state science, carried out extensive collaborative work on carbons and graphite, including compounds of graphite. He coined the term 'synthetic metals' to refer to the production of materials showing metallic conduction, but consisting only of non-metallic atoms such as carbon, nitrogen, hydrogen, the halogens and oxygen. Among the achievements of the work carried out at Imperial College during the 1960s was the development and characterisation of highly oriented, dense pyrolitic graphite. William Watt in the 1950s also worked collaboratively on commercial graphites and pyrolitic graphite, and developed a technique for producing impermeable graphite fuel cans for the nuclear industry. He also produced a polycrystalline carbon fibre with the graphite basal planes parallel to the fibre axis; his approach was based on a polyacrilonitrile (PAN) textile fibre and a process was patented in 1962 for a high strength, high modulus fibre. Watt's research covered both processing and the determination of properties. Anthony Kelly (1929–; FRS 1973) who has a major interest in strong solids and composites has included in his investigations the deformation of graphite and the elastic constants of graphite fibres.

Research on diamond has a long history (see for example the work of Henry Moissan (1852–1907; FMRS 1905), attempting to produce 'artificial' diamonds by high temperature/high pressure treatments). In recent years research on the physical properties of diamond, involving elegant and difficult experiments, has been carried out by Trevor Evans (1927–; FRS 1988); he has reproduced in synthetic diamonds many of the processes that occur over geological times in natural diamonds. Sir Harold Walter Kroto (1939–; FRS 1990) in his work has pioneered the exciting field of new carbon structures – the fullerenes. As part of his wide-ranging, innovative research Denis Henry Desty (1923–1994; FRS 1984) devised a plasma reactor process for producing buckmaster fullerene.

Polymers and rubbers

Polymer science and technology constitute a field of immense practical importance, originating from research in the 19th century; a brief survey is given here with some historical background, before considering the role of 20th century Fellows in both the chemical and physical approaches. More than a century ago Alexander Parkes (1813–1890), who made important inventions in metallurgy developed the first commercial plastic, celluloid, which he exhibited in 1862. Soon after the turn of the century James Swinburne (later Sir James, 1858–1958; FRS 1906) worked to develop a resin product from the reaction between phenol and formaldehyde, but his application for a patent in 1907 was anticipated by Leo Hendrik Baekeland (1863–1944), with the product bakelite (see Chapter 8). Swinburne continued to work as an influential figure in the plastics industry, and has been referred to as the 'father of British plastics'.

Polymer science really grew from the research of Hermann Staudinger (1881–1965, awarded the Nobel Prize for Chemistry in 1953) who fought off the opposition of the chemistry hierarchy to establish the concept of chain molecules and thus the science of polymeric materials. Another chemistry Nobel Prize (1974) in the field of polymers was that awarded to Paul John Flory (1910–1985). His signal contributions to the science stemmed from the applications of the principles of statistical thermodynamics to the behaviour of chain molecules in thermal and chemical equilibrium.

Major contributions have been made to polymer science by Sir Samuel Frederick Edwards (1928–; FRS 1966), who redirected the equations of quantum theory to the study of chain molecules, and Pierre Gilles de Gennes, whose application of scaling laws added a layer of mathematical elegance to the science and positioned it firmly within the broader discipline of 'Soft Solids'. The introduction of carefully designed and executed experiments to parallel the theoretical development owes much to Sir Geoffrey Allen (1928–; FRS 1976) whose studies of rubber deformation coupled with advanced techniques for the study of chain dynamics provided key insights. Julia Stretton Higgins (1942–; FRS 1995) has applied neutron scattering to the study of polymer systems and advanced the understanding of the thermodynamics and processing of polymer blends in many systems of central industrial importance. The research field of Peter Nicholas Pusey (1942–; FRS 1996) includes the structure, dynamics and phase behaviour of colloidal and polymeric systems, pioneering the technique of dynamic light scattering. Another Fellow concerned with polymer physics through his interest in statistical mechanics is Michael Arthur Moore (1943–; FRS 1989).

The structural and geometric approach to polymer science was pioneered by Sir Charles Frank as the 'Bristol School' which in the hands of Andrew Keller (1925–1999; FRS 1972) spanned the elucidation of the mechanisms of polymer crystallisation, the generation of ultra-strong fibres and the application of basic thermodynamics to the understanding of polymer systems in general and the development of order in particular. As a distinguished crystallographer Charles William Bunn (1905–1990; FRS 1967) carried out wide-ranging X-ray structural determinations including long molecular chain polymers such as polyethyelene and nylons, and also inorganic materials. Ian Macmillan Ward (1928–; FRS 1983) has pioneered the study of the relationship between polymer structure and chain orientation on the one hand and the achievable mechanical properties on the other, and has used these insights to invent a wide range of plastic products with quite outstanding mechanical properties. The research of Alan Hardwick Windle (1942–; FRS 1997) in polymer science has included studies of degrees of structural order between fully crystalline and amorphous materials. Diffraction analysis has been combined with computer molecular modelling in investigations of glassy polymers. Also he has developed new methods of measuring and understanding molecular orientation in deformed polymers and has investigated diffusion in polymer-solvent systems. Relationships between the structure and mechanical properties of polymers are among the research interests of Athene Donald, concerning the understanding of brittleness and ductility. Of importance in biology and food science, she has developed X-ray methods for characterising starch.

In the field of engineering properties of polymers James Gordon Williams (1938–; FRS 1994) has studied the way in which polymeric materials develop cracks and fractures, and react under deformation, fatigue and impact loading; this work has been used extensively in machinery design and in the applications of polymers in many industries. The work of Frank Brian Mercer (1927–1998; FRS 1984) has been concerned with industrial applications of polymeric materials, including the invention of the Netlon process of forming, by a novel extrusion technique, integral net and mesh structures; these structures have found important and widespread industrial applications, for example in stabilising embankments. The numerous inventions of Denis Desty included a large device for clearing oil spills in the sea; it consisted of a large glass fibre hull containing approximately 320 m of folded boom made from polypropylene synthetic rubber; he also developed other light weight structures of fibre glass.

The UK polymer industry has also maintained a close identification with the Royal Society. Alexander Fleck (later Baron Fleck of Saltcoats, 1889–1968; FRS 1955) who worked with Imperial Chemical Industries Ltd from its establishment in 1926, and became chairman in 1953, included in his responsibilities the development of the manufacture of polymers such as polythene and nylon. Christopher Frank Kearton (later Lord Kearton, 1911–1992; FRS 1961) in his distinguished career at Courtaulds was associated with research on a range of materials including nylon, polyesters, elastomerics and also carbon fibres. The research of Anthony Ledwith (1933–; FRS 1995) has included the mechanisms of polymerisation of monomers into plastics and rubbers and the development of methods of creating polymers whose chemical and physical properties are affected by light. He has devised new ways of initiating the 'curing' of polymers using photochemical means that are used in printing (xerography and lithography), surface coatings and dental fillings and has also worked on glass-polymer composites. Also he has designed polymers for xerography and lithography.

Among Fellows concerned with fundamental chemical aspects are Sir Harry Work Melville (1908–; FRS 1941) who investigated chain reactions, including the kinetics of polymerisation, and Michael Szwarc (1909–; FRS 1966) whose work has involved polymerisation. Current UK advances in polymer synthesis are represented by William James Feast (1938–; FRS 1996) who has developed novel synthetic routes for producing conjugated polymers, in morphologically controlled form. Polymer chemistry has thrived under the leadership of Cecil Edwin Henry Bawn (1908–; FRS 1952) and Clement Henry Bamford (1912–; FRS 1964), including polymerisation processes.

William Lionel Wilkinson (1931–; FRS 1990) has carried out analysis of polymer processing operations as a basis for the design of plant, while the contributions made by Raghunath Anant Mashelkar (1943–; FRS 1998) are concerned with polymer engineering particularly in the modelling of polymerisation reactors, diffusion in polymer media and transport studies on swelling polymers as well as non-Newtonian fluids. The research of Kenneth Walters (1934–; FRS 1991) in theorectical and experimental rheology has been concerned in particular with the measurement and analysis of properties of non-Newtonian fluids, such as elastic liquids, relevant to manufacturing industry.

In the field of the chemistry and physics of rubber and rubber-like substances two Fellows of the Royal Society, Geoffrey Gee (1910–1996; FRS 1951) and Leslie Clifford Bateman (1915–; FRS 1968) have made major contributions.

Research at the Road Research Laboratory of the Department of Scientific and Industrial Research under the leadership of William Henry Glanville (later Sir William, 1900–1976; FRS 1958) included investigations of vehicle skidding on wet roads. Among the relevant features, collaborative work was carried out with David Tabor (1913–; FRS 1963) on the influence of the properties of rubber tyres. Other important work by Granville involved investigations of road materials such as tar and bituminous materials, including visco-elastic behaviour.

Composites

Composite materials, particularly those based on polymers and incorporating fibres for strengthening, have now become established as important engineering materials, combining, for example, enhanced strength and stiffness. The development of this field has involved the pioneering research of Anthony Kelly. The research of Rodney Hill (1921–; FRS 1961) on the mechanical behaviour of materials has included investigations of composites, especially of the fibre type. Derek Hull (1931–; FRS 1989) has also carried out important work on composites, including mechanical properties. Also the theoretical mechanics research of Anthony James Merrill Spencer (1929–; FRS 1987) has included the properties of fibre reinforced solids.

Retrospect and Prospect

Looking back one to one and a half centuries various roots of modern materials science, technology and engineering can readily be traced. Metallurgical knowledge, especially of physical metallurgy has contributed much in the nucleation and growth of the broader field of materials particularly through the developments in understanding of structure from the atomic to the microstructural levels. The role of chemistry and physics has continued, as from the early years of the Royal Society, to have enormous influence, while the continuum mathematical approach involving modelling, powerfully used in engineering has also advanced the field of materials.

The wide recognition of materials as a broad, interdisciplinary field of endeavour is less than half a century old; this was greatly aided by the industrial and economic impact of advances at the atomic level of understanding, leading to the development of materials and devices dependant on semiconducting and optical behaviour. The analogy of a composite is perhaps appropriate as a description of the contemporary field, with its components represented by the respective contributory disciplines, blended together with diffuse rather than discrete interfaces. The whole field provides characteristic features in terms of its achievements, superior to those of the individual components. There is the promise of continued benefits based on quantitative modelling to enable artefacts to be designed and economically produced to meet the challenges of new developments in all fields of engineering in the 21st century.

CHAPTER 18

Metals and the Atomic Age

Introduction

The unravelling of the structure of the atom, and the associated practical developments stemming from this, including the generation of electricity through atomic energy, have been among the most spectacular achievements of 20th century science and engineering. Just as the 19th century may be called 'The Electrical Age' (see Chapter 8), which in England was initiated by the letter from Allessandra Volta (1745–1827; FRS 1791) to the Royal Society in 1800, concerning his 'pile' or 'battery', so the 20th century could be called 'The Atomic Century' or 'The Atomic Age'. The critical experiments by Luigi Galvani (1737–1798) and by Volta preceeding Volta's letter had taken place near the end of the 18th century. In a comparable way the Atomic Century had its origins in the dying years of the 1800s, through the experiments of William Konrad Röntgen (1845–1923) and Antoine-Henri Becquerel (1852–1908; FMRS 1908).

This chapter is concerned with the role of metals in the evolution of the Atomic Age. It begins with a review of the exciting discoveries concerning radioactivity and atomic structure. It then proceeds to describe briefly the history of atomic energy in the military and civilian fields, in which radioactive metals such as uranium and plutonium were fundamental; in particular, attention is given to the post-war atomic programme for electricity generation in Britain. Immense challenges have been confronted and mastered in the relevant metallurgical aspects, involving metal extraction and other processing techniques, and also necessitating fundamental and applied investigations concerning the structure and properties of the new materials required for the exacting service conditions in nuclear reactors. The contributions of many Fellows and Foreign Members of the Royal Society have been crucial in the dramatic developments during the 20th century, and the Royal Society collectively has been influential through its advisory activities.

In the section of this chapter dealing with the British atomic energy programme the scientific and technological features are presented in some detail to exemplify the magnitude of the achievements; also, reference is made to many members of the teams who made important contributions, in addition to those, relatively few, who were elected to the Royal Society.

The Atomic Century

In 1895 Röntgen discovered X-rays and a few months later, though in 1896, while attempting to repeat Röntgen's experiments, Antoine-Henri Becquerel (1852–1908; FMRS

1908) found that some uranium salts produced emissions which blackened photographic plates. Becquerel investigated the phosphorescence of uranium compounds after 'insolation' (i.e. exposure to the sun's rays); the compounds, in glass cells, were inverted on photographic plates enclosed in black paper. He made an experiment in which, between the plate and the uranium salt, he placed a sheet of black paper and a small cross of thin copper. On bringing the apparatus into the daylight the sun had gone in, so it was put back into a dark cupboard and left there for another opportunity of solar exposure. Following several cloudy days Becquerel developed the plate but was surprised to find that it had darkened under the uranium salt, the image of the copper cross appearing white against the black background. This was the foundation of a long series of experiments which led to the remarkable discoveries of radioactivity which made 'Becquerel Rays' a standard expression in science at that time. Having found that the radiation behaved like X-rays in penetrating matter and ionizing air, he showed that it was associated with uranium in the crystals and that uranium metal is radioactive.

In 1897 Marie Curie (1867–1935) and Pierre Curie (1859–1906) isolated two new metals, polonium and radium, both of which were far more radioactive than uranium. (Although neither Marie nor Pierre were elected to the Royal Society, they were recipients of the Davy Medal in 1903, and shared the Nobel Prize for Physics with Becquerel for the discovery of radioactivity; Marie also received the Nobel Prize for Chemistry in 1911).

Also in 1897 two English physicists, Joseph John Thomson (later Sir Joseph, 1856–1940; FRS 1884) and John Sealy Edward Townsend (later Sir John, 1868–1957; FRS 1903) discovered that emanations from cathodes in evacuated chambers were streams of negatively charged particles which were far lighter than atoms and which are now called electrons. By subjecting the radiation from radium to magnetic fields Becquerel proved from the extent of the deflection that the radiation consisted of electrons.

The discovery of the electron has obvious implications for the development of the theory of electricity and how it is transported, but this fundamental particle is of equal importance in deriving an atomic model of the atom. However, an even more immediately significant inheritance from the 19th century was the astonishing theoretical work of James Clerk Maxwell (1831–1879; FRS 1861). Starting with the observations of Michael Faraday (1791–1867; FRS 1824) and ideas of lines of electrical and magnetic force, Maxwell introduced the concept of electromagnetic radiation and he derived his field equations; this paved the way for the special theory of relativity proposed by Albert Einstein (1879–1955; FMRS 1921), which established the equivalence of mass and energy.

Maxwell's ideas also stimulated, though rather indirectly, the other major innovation of 20th century physics, the quantum theory. His description of electromagnetic radiation led to the development of a law of heat radiation which, when compared to experiment, turned out to be unsatisfactory. To reconcile theory with experiment Max Karl Ludwig Planck (1858–1947; FMRS 1926) conceived the idea that radiant heat energy is emitted, not continuously, but in discrete packages or 'quanta'.

Joseph John Thomson

John Townsend

Ernest Rutherford

Frederick Soddy

As support for Planck's idea grew it became apparent that in the world of atomic dimensions classical mechanics was inapplicable – a profound revolution in physical theory was in the making. Appropriately, in the first year of the new century (October 19th, 1900) Planck in a lecture to the Berlin Physical Society, announced his new quantum theory. This concept was to have as profound an effect on the development of physics in the 20th century as Volta's letter to the Royal Society in March 1800 had had on the physics of the 19th century. These developments in physics, revealing as they did the structure of the atom and the basics of the metallic state, form the base rock for modern metallurgy and hence are further discussed in this chapter. A fuller account in terms of the Periodic Table and of the electronic structure of the elements, with particular reference to bonding and the nature of metals is given in Chapter 11.

It was ironic that it was Albert Einstein, who grew to dislike so intensely so many of the manifestations of the quantum theory, who was the first to provide the strongest support for Planck's ideas. In 1905 he used the quantum theory to account for the photoelectric effect, that is to say the ability of light to knock free a limited number of electrons from certain metals. An hitherto unexplained phenomenon was that the energy of these liberated electrons did not depend, as common sense would predict, on the intensity of the light but instead on the colour of the light, that is to say, on its frequency.

In order to explain this it was necessary for Einstein to account for not only the quantum theory but also to make the epoch-shattering assumption that light travelled, not as waves, but as extremely small particles, which he labelled 'energy quanta' and which today are called 'photons'. He attributed a distinct energy to each photon, the value of which would dictate the energy of the released electron. It followed that a brighter light would certainly release more electrons, but none of them would have a higher energy if the frequency of the incoming light remained the same.

Quite independently of the developing ideas of Planck and Einstein, Ernest Rutherford (later Lord Rutherford, 1871–1937; FRS 1903) had been adopting an experimental approach to studying radioactivity. After working with J.J. Thomson at the Cavendish Laboratory, mostly on radio waves, Rutherford was appointed Professor of Physics at McGill University in Montreal in 1898 at the early age of 28. In 1902 Rutherford, with Frederick Soddy (1877–1956; FRS 1910) proposed that radioactive elements are undergoing spontaneous transformation into new elements. It was Soddy who, a decade later (1913) proposed the important concept of isotopes of elements, which have the same atomic number and chemical and physical properties as the elements, while having different atomic masses. The development of the mass spectrograph by Francis William Aston (1877–1945; FRS 1921; Nobel Laureate 1922) led to vital advances, and Aston investigated isotopes in more than fifty elements.

Rutherford, while at McGill found that even a very thin film of mica could deflect alpha particles, albeit by only up to two degrees. As he realised, this was an astonishing result; for a particle as massive as an alpha particle moving at such a high velocity, a deflection of two degrees was enormous. In 1908, by then a Professor of Physics at Manchester, Rutherford instructed his student, Ernest Marsden (1889–1970; FRS 1946)

Niels Bohr

James Chadwick

John Cockcroft

Harold Urey

to fire alpha particles at a gold foil and it was found that very occasionally a particle experienced such a marked change in direction it almost bounced back along its incident path. Clearly a few particles had encountered immensely powerful, highly concentrated, forces inside the atoms and from this observation it was immediately clear that J.J. Thomson's 'plum pudding' model of the atom (that it consisted of negatively-electrified corpuscules immersed in a 'sea' of uniform positive electrification, like raisins in a pudding) must be incorrect.

After pondering Marsden's result for many months Rutherford proposed his new model of the atom as consisting of an extremely small positively charged nucleus which contains most of the mass of the atom and surrounded by 'a uniform spherical distribution of opposite electricity charge quantitatively equal in amount'. In the light of knowledge of electrons it became necessary to become more specific about their distribution and gradually the idea emerged of the electrons circulating the nucleus rather like the sun's planetary system. The model, incidentally, bore a striking resemblance to the 'Saturnian' model put forward some five years earlier by the Japanese theoretical physicist, Hantaro Nagaoka (1865–1950) which involved electrons revolving like Saturn's rings around a positively charged nucleus.

The main drawback with both Rutherford's and Nagaoka's models was that no new physics was proposed, in spite of the fact that according to the laws of classical mechanics, the atom would clearly be unstable. For example, the branch of classical physics known as electrodynamics would require that the electrons by following a curved path would radiate energy continuously and hence would soon spiral into the nucleus. Some new physics was required with someone bold enough to apply it who was familiar not only with the thinking of Planck and Einstein but who also had a detailed knowledge of the current state of experimental atomic science. The new physics was the quantum theory and the man to meet the challenge was Niels Henrik David Bohr (1885–1962; FRS 1926).

Bohr recognised the strength of the experimental evidence which had led Rutherford to put forward his model of the atom and he was attracted to the idea of a central positive nucleus surrounded by rotating electrons. If classical mechanics was inadequate to describe such a system then Bohr reasoned that classical mechanics must be abandoned at the atomic level and in 1913 he proposed simply that there must exist 'stationary states' within which electrons could 'orbit' for ever without loss of energy. Electrons could jump from one orbit to another but could not occupy any intermediate energy level, in other words they could only make quantum jumps during which they emitted or absorbed set quantities of energy. On this basis he was able to derive a mathematical formula which accounted exactly for certain of the lines in the hydrogen spectrum. When Einstein heard of this success he exclaimed 'This is one of the greatest discoveries'.

When Bohr's, and subsequent, ideas were applied to the complete table of the elements, the reason for the periodicity of properties became clear, and it also became possible to understand how it came about that although about 80% of the elements were metals, the vast majority of them managed to arrive at a configuration where only one or two electrons occupied their outermost shell, (See Chapter 11). In addition to this, the ability of these 'valency' electrons to become 'free' from any individual nucleus and wander at large, and also their role in chemical reactions, went a considerable way to explaining why it was that metals were such good conductors of heat and electricity, were ductile and reflective, and took part readily in so many chemical reactions. It would not be too much of an exaggeration to claim that Bohr's revision of Rutherford's model of the atom was the initiating event for modern physical metallurgy.

Rutherford was pre-eminently an experimentalist and he held a deep suspicion of theoreticians. Nevertheless, he had been very helpful to the young Bohr and raised only one objection to his new model and this was incorporated in his question: 'How does an electron decide what frequency it is going to vibrate at when it passes from one stationary state to another?' This question was eventually answered by Einstein in 1917 when he pointed out that any frequency is possible and the one that is arrived at simply had the best odds of being selected. In other words it was all a matter of statistics and probabilities, concepts which continued to give Rutherford difficulties throughout his life.

In retrospect Rutherford's model of the atom was so manifestly wrong, clinging as it did to the principles of classical mechanics, that it is puzzling to understand why it was ever given any credence. It was, however, the model's very shortcomings which inspired Bohr to develop his mould-breaking ideas. Coincidentally, it was another deeply-flawed paper, by another British-based scientist, John William Nicholson (1881–1955; FRS 1917) of King's College London, which stimulated Bohr to apply his theory to the generation of spectral lines.

Nicholson had attempted to explain the unusual spectrum of the corona of the sun on the basis of a quantized Saturnian model, a concept which was riddled with fallacies and inconsistencies which Bohr immediately recognised. However, while attempting to sort out the errors made by Nicholson, Bohr saw the way clear to apply his theory to explain spectral lines, at least for the hydrogen atom. Evidentally, scientists see so far not so much because they stand on the shoulders of giants but rather because they stand on top of the decaying remains of earlier scientists' misconceptions, erroneous data and discredited theories.

In turn Bohr's model too was soon found to have important shortcomings. While it did account successfully for the spectral lines of hydrogen, it broke down almost completely when it was applied to other atoms, that is atoms with more than one electron spinning around the nucleus. Also, as Bohr himself frequently pointed out, his model was based on some serious inconsistencies. In particular Bohr had rejected classical electrodynamics so that he could account for perpetual rotation of the electrons but had reverted to classical mechanics to calculate the energy of an electron in its orbit.

It took the genius of Werner Karl Heisenberg (1901–1976; FMRS 1955) and the development of a new science, quantum mechanics, to put atomic theory on a firm basis. This was achieved by Heisenberg and his two collaborators, Max Born (1882–1970; FRS 1939) and Ernst Pascual Jordan (1902–1980) towards the end of 1925. Wolfgang Pauli (1900–1958, FMRS 1953) applied the new science to the hydrogen atom and derived the same result that Bohr had achieved much earlier, but unlike Bohr's model quantum mechanics could also be applied to more complex atoms.

Quite independently of this work on quantum mechanics being carried out in Germany, Louis-Victor Pierre Raymond duc de Prinze de Broglie (1892–1987; FMRS 1953) in Paris had a brilliant idea. Planck and Einstein had shown that light, hitherto regarded as a kind of wave, could sometimes act as though it were a particle associated with a certain parcel of energy (a quantum). De Broglie reasoned that the reverse could also be true, that is to say the electron, although thought of traditionally as a particle, might also behave as waves, in other words it might also demonstrate wave–particle duality.

The Austrian, Erwin Schrödinger (1887–1961; FMRS 1949), picked up this idea and reasoned that for an electron to be continuous the length of its orbit must be equal to a whole number of electron wavelengths. It would also follow that the distance of the orbit from the nucleus would be governed by this 'whole-number' requirement. Using these ideas and de Broglie's formula for the wavelength, Schrödinger found his permissible electron paths for the hydrogen atom were exactly those determined from Bohr's analysis. As an important bonus Schrödinger's theory successfully predicted the properties of other atoms. Subsequently, Schrödinger demonstrated that the quantum mechanics and the Schrödinger–de Broglie formulations were in fact identical.

As if all this was not enough, Heisenberg then introduced an idea that so changed our view of the physical world that it is hard to overemphasise its importance in the long history of man's intellectual development. This was of course his Uncertainty Principle, one manifestation of which is that the precise velocity and actual position of an electron could not be determined simultaneously. At a stroke causality was abolished and our understanding of the physical had to come to terms with a completely new paradigm.

Everything after the Uncertainty Principle runs the danger of being an anticlimax, but an account of the astonishing theoretical achievements of the second half of the 1920s cannot be concluded without a mention of Pauli's Exclusion Principle which propounded that when one electron occupied a particular orbital it could not be joined by any identical electron. This explained how it was that all the electrons of an atom were prevented from collapsing towards the lowest energy orbital. (A more general statement of the Exclusion Principle is that no two fermions in a system can exist in identical quantum states, i.e. have the same set of quantum numbers. As protons and neutrons are also fermions, nuclear structure too is also bound by the requirements of the Exclusion Principle.)

The only Englishman to take part in this astonishing explosion of theoretical atomic physics in the 1920s was Paul Adrien Maurice Dirac (1902–1984; FRS 1930) who provided its crowning achievement. In 1928 he combined the ideas of de Broglie, Schrödinger, Heisenberg and Born with the relativity theory of Einstein and produced a remarkable synthesis. One of the outcomes of his analysis was to provide an explanation of why it had proved necessary to postulate that all electrons as they rotate around the atomic nucleus also spin on their own axes.

If Dirac's achievements are set on one side then Britain's major contributions in this period continued to be the carrying out of brilliant experiments, mostly under Rutherford's direction at the Cavendish. Rutherford had moved to Cambridge from Manchester in 1919, the year he produced the first artificial transmutation by bombarding nitrogen with alpha particles and producing oxygen.

During the following five years Rutherford and his now close collaborator, James Chadwick (later Sir James, 1891–1974; FRS 1927) achieved transmutations with ten further elements. The source of their alpha particles were radioactive elements, for example radium, so the experimentalists had no control over the energy of their projectiles. To speed up particles in an accelerator seemed an obvious development but when Rutherford and his associates, which now included John Douglas Cockcroft (later Sir John, 1897–1967; FRS 1936) and Ernest Thomas Sinton Walton (1903–1995) calculated the electrical potentials needed to get useful results the prediction came out to be millions of volts and hence totally impracticable at that time.

By a stroke of good fortune George Gamow (1904–1968) visited the Cavendish Laboratory during 1929 and pointed out that particles could enter the nucleus without passing over the top of the energy barrier, in other words he told them about quantum mechanical 'tunnelling'. This meant that the building of an effective accelerator became feasible and having received the green light from Rutherford the efforts of Cockcroft and Walton culminated, in April 1932, in the breaking up of lithium elements by accelerated protons. For the first time a transmutation had been achieved by purely artificial means, and as an added bonus the result confirmed the validity of the quantum theory and the assumptions on which the experiment had been based.

1932 was the *annus mirabilis* for atomic research. As well as Cockcroft and Walton's production of artificially-induced transformation, the positron had been discovered in that year by Carl David Anderson (1905–1991) in a shower of cosmic rays (its existence had been predicted by Dirac in 1928), and Harold Clayton Urey (1893–1981; FMRS 1947) and his collaborators isolated deuterium. Most portentous of all these 1932 achievements was James Chadwick's discovery that the penetrating irradiation emitted by beryllium on being bombarded with alpha particles from polonium, was in fact a stream of neutrons. The existence of a fundamental particle as heavy as the positron but with no electrical charge had been predicted by Rutherford in 1920 so it was fitting that the discovery of the neutron should be achieved by his close associate in his laboratory at the Cavendish.

The following year, 1933, in a public lecture, Rutherford uncompromisingly stated that the idea that nuclear energy would ever have practical application was pure 'Moonshine'. Of more import, Hitler came to power in Germany and the Nazi anti-semitism laws accelerated the departure of Jewish scientists from Germany itself, Italy and other occupied and threatened European countries. The Nazis, like the French Revolutionaries in the 1790s when they executed Lavoisier, evidently felt their state 'had no need of savants'.

In 1934 Irene Curie (1897–1956), Madame Curie's daughter, and her husband Jean-Frédéric Joliot (1900–1958; FMRS 1946) discovered that when they irradiated aluminium with alpha particles from polonium the aluminium continued to irradiate particles after the polonium source had been removed – they had discovered artificial radioactivity. This important discovery compensated them for having obtained in their earlier experiments evidence for the existence of both neutrons and positrons, but failing to realise the significance of their observations. The following year they were awarded the Nobel Prize for Physics for their discoveries.

Returning to the neutron, it was its discovery which was the dominating event of the start of the thirties; indeed Hans Albrecht Bethe (1906–; FMRS 1957) has suggested that Chadwick's discovery was a watershed separating prehistory from history as far as the development of nuclear science was concerned. Neutrons have two major advantages as atomic missiles. Firstly, being electrically neutral, they could penetrate relatively easily the negative field of the electron cloud and also the positive field of the nucleus. Secondly, having a mass almost two thousand times as great as electrons and positrons they have the ability when they impact to cause profound disturbances within the nucleus of the atom.

No one saw the potential of neutron bombardment more clearly than Enrico Fermi (1901–1954; FMRS 1950), the youthful Professor of Physics at the University of Rome, and news of the Joliot–Curies' discovery of artificial radioactivity in early 1934 spurred him into a feverish period of activity. He and his collaborators set about experimenting with as many elements in the Periodic Table as they could acquire. They succeeded in irradiating 60 elements, 40 of which became radioactive after neutron irradiation. Right at the start of these studies Fermi realised that polonium was too weak a source of alpha particles to stimulate the beryllium target into emitting a very powerful stream of neutrons. Fortunately, within his physics building the Rome health department held a gram–source of radium which of course produced a continuous supply of radon – a far more energetic source of alpha particles than polonium. It did not take Fermi long to produce a powerful radon/beryllium neutron source.

To their astonishment, the Italian investigators found that their neutron source appeared to operate more efficiently, ie induced a greater degree of activation, when the experiment was carried out on a wooden table rather than on a steel or marble table (where else but in Italy would you find a marble table in a laboratory?). It did not take the brilliant Fermi long to work out what was happening. The neutrons were colliding with the hydrogen atoms (and perhaps the carbon atoms too) in the wood and thus being slowed down and in some way these slower neutrons carried a greater propensity to interact with the nuclei of the target atoms.

Fermi went on to reason that as a neutron does not carry an electrical charge it does not need a high speed to enable it to penetrate the electrical barriers of the target atom. Moreover, on arrival in the vicinity of the target atom's nucleus, the slower the neutron is moving the longer the time available for its capture. Fermi had not forgotten his classical mechanics and he recognised that elements with nuclei of similar mass to the neutron would be most effective in slowing the neutron down – materials with a high concentration of the hydrogen atoms being ideal. The process of 'moderation' has of course proved to be of profound importance in the development of nuclear technology and Fermi has rated its discovery as one of his greatest achievements.

With his radon/beryllium neutron source and using paraffin wax as a moderator, Fermi and his collaborators should have been in a powerful position to discover nuclear fission, and they did indeed include uranium in their experiments. In fact they amost certainly did induce fission in their uranium samples but they completely misinterpreted their results by concluding that their irradiations had simply produced transuranics, including a completely new man-made element. These erroneous results were published in a paper which received Rutherford's imprimatur and was published in the *Proceedings of the Royal Society.* (The Joliot–Curies made a similar misinterpretation with their experiments with uranium – they must hold the record for missed opportunities having been the first to obtain evidence for the existence of positrons, neutrons and fission products but each time failing to capitalise on their good fortune due to mistakes in the interpretation of their observations.)

In 1938 Otto Hahn (1879–1968 ; FMRS 1957) and Fritz Strassman (1902–1980) discovered, as a result of superbly skilful chemical analysis of neutron-irradiated uranium samples that the radioactive species in their specimens was not in fact a transuranic element but the much lighter element, barium. Lisa Meitner (1878–1968; FMRS 1955) and Otto Robert Frisch (1904–1979; FRS 1948) were the first to point out that the only possible explanation of these results was that the neutrons had caused the uranium nucleus to divide, or 'fission' thereby forming elements of much lower atomic number, subsequently termed 'fission products'. In arriving at this conclusion Meitner and Frisch had found Bohr's liquid-drop model of the atomic nucleus very helpful. They suggested that the fission process would be expected to release large amounts of energy, a prediction subsequently confirmed by Frisch himself.

Meitner and Frisch published their findings in *Nature* in February 1939 and within a matter of weeks the Juliot–Curie group demonstrated that as a result of the fission process more neutrons were released than had been adsorbed in the process and this pointed to the possibility of the occurrence of a chain reaction and an associated explosive release of energy. Bohr calculated that fission was much more likely to occur in the lighter isotope of uranium, U-235, than in the much more abundant U-238 isotope; (natural uranium consists of 99.3% U-238 and only 0.7% U-235).

Bohr went on to produce with J.A. Wheeler a definitive analysis of the fission process and this was contained in a paper published just two days before the outbreak of the Second World War. Already Einstein and Leo Szilard (1898–1964) had written to President Roosevelt drawing his attention to the possibility of producing an atom bomb and arguing that the USA should do more research in this area if only to keep abreast of possible developments in Germany; (fissioning of uranium had after all been discovered in Germany and, in spite of the exodus of its Jewish scientists, that country contained many nuclear scientists of world renown).

In the physics department at Birmingham University during the early months of 1940, the two German–Jewish expatriates, Rudolph Ernst Peierls (later Sir Rudolph, 1907–

DEBYE AND HIS CONTEMPORARIES AT THE SOLVAY CONFERENCE IN BRUSSELS, 1927.

A. PICCARD E. HENRIOT P. EHRENFEST Ed. HERZEN Th. DE DONDER E. SCHRÖDINGER E. VERSCHAFFELT W. PAULI W. HEISENBERG R.H. FOWLER L. BRILLOUIN

P. DEBYE M. KNUDSEN W.L. BRAGG H.A. KRAMERS P.A.M. DIRAC A.H. COMPTON L. de BROGLIE M. BORN N. BOHR

I. LANGMUIR M. PLANCK Mme CURIE H.A. LORENTZ A. EINSTEIN P. LANGEVIN Ch.E. GUYE C.T.R. WILSON O.W. RICHARDSON

Absents : Sir W.H. BRAGG, H. DESLANDRES et E. VAN AUBEL

Photograph by courtesy of Instituts Internationaux de Physique et de Chimie Solvay.

Acknowledgements to Fonds des Instituts Internationaux de Physique et Chimie Solvay, 112, Archives de la Université Libre de Bruxelles.

1995; FRS 1945) and Otto Robert Frisch (1904–1979; FRS 1948) calculated that as little as a kilogram or two of the uranium isotope U-235 would constitute a 'critical mass'. That is to say a relatively small volume of this naturally-occurring isotope would not lose from its surface sufficient neutrons to prevent a chain reaction occurring with a concomitant catastrophic release of energy. If a means could be devised for separating the 0.7% U-235 from the major isotope, U-238, the construction of an atomic bomb would be feasible.

This was a shattering result and, on being informed of its implications, the UK government agreed to form the MAUD Committee to oversee developments in the nuclear field. Sir John Anderson (later Viscount Waverley, 1882–1958; FRS 1945), as Lord Privy Seal was involved at Cabinet level in this secret project; he had taken a science degree at Edinburgh University in 1913 and had then done research on the chemistry of uranium in Germany. The membership of the MAUD Committee included George Paget Thomson (later Sir George, 1892–1975; FRS 1930), James Chadwick and Walter Norman Haworth (later Sir Walter,1883–1950; FRS 1928). Among matters within the ambit of the Committee was the production of uranium metal, and Haworth, who was Professor of Chemistry at Birmingham University, used an electrolytic method; however, there were problems in achieving a suitable product in substantial quantities, and the task was taken over

Enrico Fermi

Jean-Frédéric Joliot

Otto Frisch and Rudolf Peierls

by Imperial Chemical Industries. (It is of interest to note that nearly 40 years earlier another Fellow, Eric Keightley Rideal (later Sir Eric, 1890–1974; FRS 1930) did electrochemical work on uranium for his PhD thesis; this included the preparation of the tetrachloride and the pure metal , and he constructed a 100 atmosphere hydrogen furnace to prepare uranium from its oxide.). Another vital issue for the MAUD Committee was the separation of the U-235 from the U-238 isotope and diffusion techniques were explored with a view to achieving this (see later).

In 1941 the MAUD Committee was replaced by an organisation under the Department of Scientific and Industrial Research: this was designated the Directorate of Tube Alloys, an ambiguous, metallurgical 'cover' title for the highly-secret project. It was headed by Wallace Alan Akers (later Sir Wallace, 1888–1954; FRS 1952) who was seconded from his post as research director at ICI.

Subsequently it was decided that Britain simply did not have the resources to develop a nuclear weapon on its own, and in any case was too vulnerable to aerial inspection and attack, and the nuclear secrets and many of the nuclear scientists moved to the USA to form part of the core of the Manhattan Project. Important developments were already taking place in the USA (where many of the European Jewish scientists had already settled). Notably Glenn Theodore Seaborg (1912–1999; FMRS 1985) and his collaborators discovered plutonium in December 1940; subsequently they produced other transuranic elements. An isotope of plutonium, Pu-239, like U-235, was found to be fissile and its production in nuclear reactors provided an alternative route to the atomic bomb. During the rest of the war-years atomic weapon research was concentrated in America and Canada and the first major achievement of the programme took place on December 2nd, 1942 when Fermi's graphite/uranium atomic pile, constructed in a squash court at the University of Chicago, went critical. This was the first man-controlled large-scale fissile experiment. The history of the Manhattan Project is well documented so only the highlights are mentioned here.

The Manhattan Project

The production of the atomic bomb was called the Manhattan Project because much of the earlier preliminary work had been carried out by Fermi, Szilard and Urey and others, at Columbia University, which of course is in Manhattan. By the time General Leslie Groves had been put in operational charge of the Project it had already been decided that parallel development programmes would be carried out aimed at manufacturing both a uranium-235 bomb and one using as its fissile heart plutonium-239.

Three possible ways of separating U-235 from U-238 were being actively studied. At Columbia University Harold Clayton Urey who had been awarded the Nobel Prize for Physics for his discovery of heavy water, was pursuing gaseous diffusion; at the University of California at Berkeley, Ernest Orlando Lawrence (1901–1958) the inventor of the cyclotron, and recipient of the Nobel Prize for Physics in 1939, was working on an electromagnetic technique for separating U-235 and at the Westinghouse Research Labo-

John Anderson

Walter Haworth

Wallace Akers

ratory at Pittsburgh, Eger Murphree was developing a separation process utilising a centrifuge. Among the many scientists who came from Britain to participate in the atomic energy research in the USA, Mark (Marcus) Laurence Elwin Oliphant (later Sir Mark, 1901–; FRS 1937) worked with Lawrence on the electromagnetic separation programme at Berkeley and was his deputy.

The work on producing plutonium was being carried out at the University of Chicago under the direction of Arthur Holly Compton (1892–1962) and Enrico Fermi; indeed the world's first atomic pile was close to completion in the University's squash court. A scaled-up version of the Chicago air-cooled graphite pile was subsequently built and commissioned at the Clinton Engineering Works site in Tennessee (subsequently named the Oak Ridge National Laboratory).

The gaseous diffusion process for separating U-235 eventually proved to be the one most capable of being developed on a 'commercial' scale (i.e. capable of producing kilogram quantities of highly enriched, weapons-grade U-235). The huge diffusion plant was also constructed at Oak Ridge, and the whole site occupied an area of 70 square miles. When the plant was fully operational it consumed as much electrical power as the whole of the United Kingdom.

For plutonium production it was decided to stay with graphite-moderated reactors, but to take advantage of the higher thermal capacity of water cooling. Water-cooled graphite reactors have highly undesirable reactor-physics characteristics (which subsequently was to cause the accident at Chernobyl) so a remote 1000 square mile desert site was chosen for them alongside the Colorado river at Hanford in Washington State.

Julius Robert Oppenheimer (1904–1967) was appointed head of the scientific programme at Los Alamos, the purpose-built laboratory in a remote part of New Mexico where the first atomic bombs were researched, developed, constructed and tested. By the summer of 1945, a few months after the end of the European war, enough U-235 was available to manufacture a single bomb and sufficient Pu-239 positioned at either end of a cylinder, and a nuclear explosion was achieved simply by firing one portion into the other to obtain a supercritical mass. This was such an uncomplicated arrangement that it was felt there was no need to carry out a test explosion (which is just as well because such a test would have used up all the available U-235).

This simple 'gun barrel' arrangement could not however be adopted for the plutonium weapon because adventitous neutron release would have led to premature detonation and a low yield. A more complicated arrangement was necessary involving the packing of explosives around a hollow sphere of the fissile isotope. A full-scale trial of such a complicated design was thought essential – hence the 'Trinity' test at Alamogordon in New Mexico on July 16th 1945.

The test was a complete success and, despite the opposition of a number of nuclear scientists led by Leo Szilard, the uranium bomb was dropped on Hiroshima on August 6th, and the second plutonium bomb on Nagasaki three days later. The issue of whether or not the Americans were justified in using the atomic bombs against a nation which was already on the point of defeat, remains to this day a subject of heated debate.

Japan capitulated shortly thereafter and immediately was brought into focus the whole subject of how the existence of nuclear weapons might change the world scene. The United Nations Charter had been signed in June 1945 i.e. before the destruction of Hiroshima and Nagasaki and hence it became the responsibility of the first meeting of the UN General Assembly in January 1946 to take on board the entirely new circumstances; its very first resolution established an Atomic Energy Commission mandated to 'deal with the problems raised by the discovery of atomic energy and other related matters'.

At the first meeting of the Commission, on 14 June 1946, the US representative, Bernard Baruch presented the US proposal for international control. This became known as the Baruch Plan and proposed that an International Atomic Development Authority be set up which would work towards the creation of a nuclear weapon-free world with power to punish severely any transgressors. The main drawback with the scheme, from the Russian standpoint, was that until such a world order was set up the US would retain its nuclear weapons and hence have a monopoly of them. Another serious disadvantage as far as Russia was concerned was that there would be no veto available to them in a West-dominated Authority.

Although the Security Council approved the Baruch Plan in 1948 no measures could be taken to implement it while Russia remained intransigent. The plan in fact was doomed to failure, as should have been clear to its proposer. The final meeting of the Atomic Energy Commission took place on July 29th 1949; a month later Russia exploded its first atomic bomb. The gap between East and West atomic capability was far narrower than the most pessimistic of the US observers had realised. America retaliated by developing the hydrogen bomb, carrying out its first thermonuclear test explosion in May 1951, but the Russians were not far behind, they exploded their first hydrogen bomb in August 1953. The gap was getting even narrower; the nuclear arms race was well and truly underway, the Cold War had started.

Metallurgy and the British post-war atom programme

Early developments

About fifty British scientists had moved to America to take part in the Manhattan Project. It was tacitly assumed that at the end of the war the UK would re-establish and expand its own nuclear research programme, certainly with the objective of applying the new source of power for peaceful purposes, such as generating electricity and diagnosing and treating diseases, and possibly also for developing an independent nuclear weapon. It was believed that in this endeavour Britain would benefit from substantial help from the USA; indeed an obligation to provide such assistance was implied by the Quebec Tripartite Agreement signed on behalf of the USA, the UK and Canada by Roosevelt, Churchill and Mackenzie King in August 1943.

However, in August 1946 Britain's plans for an independent nuclear research programme received an unexpected setback by the passing by the US Congress of the McMahon Bill which prohibited the transfer of any nuclear information to any foreign country, including, Britain, Canada, France and of course Russia. Both Britain and Canada

Harrie Massey

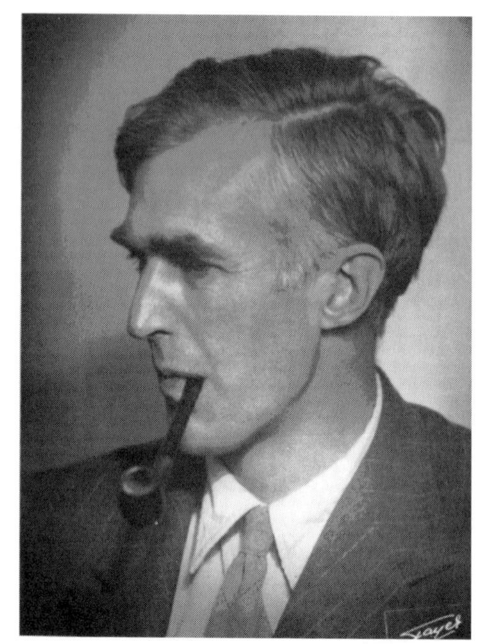

Herbert Skinner

William Penney

Christopher Hinton

felt a sense of outrage as the Bill appeared to violate the terms of the 1943 Quebec Tripartite Agreement.

The intention of the supporters of the McMahon Bill, however, was not to renege on former agreements with the allies but primarily to wrest control of nuclear weapons from the US Military and place it firmly under the judistriction of the US Civil authorities. The initiator of the Bill, Brien McMahon, had no idea that Britain had played such an important role in the conception and development of 'America's' atom bomb. In fact the passing of the Bill tended to strengthen Britain's resolve to develop an 'independent' nuclear weapon system, and a decision to follow such a course was finally taken by Prime Minister Attlee and just a few members of his cabinet in January 1947. France did much the same, while Canada decided, to her credit many would argue, not to develop a weapon's system but to concentrate on the peaceful applications of the new source of power.

The first major step in planning Britain's post-war nuclear research programme was taken some years earlier, during a chance meeting of three Fellows of the Royal Society – John Cockcroft, James Chadwick and Mark Oliphant – in a Washington hotel in November 1943. As well as discussing in broad outline what Britain's objective and research policy would be, it was at this meeting that Cockcroft was persuaded to take charge of the Anglo–Canadian Atomic Energy Project at Montreal, it being already perceived that this would be a valuable prelude to his assuming responsibility subsequently for Britain's post-war nuclear research programme.

A year later Cockcroft, Chadwick and Oliphant were joined by Rudolf Peierls, Harrie Stewart Wilson Massey (later Sir Harrie, 1908–1983; FRS 1940, and future Secretary of the Royal Society) and Herbert Skinner (1900–1960; FRS 1942) for an important meeting where they defined more precisely the sites and facilities for a possible British post-war nuclear programme. The first requirement was to select a suitable site for a central laboratory, and to provide an estimate of its manning levels and major equipment such as research reactors, accelerators and 'hot cell' facilities. Subsequently, a number of potential sites for the laboratory near Oxford and Cambridge were inspected and an RAF station and aerodrome at Harwell on the Berkshire Downs was eventually chosen.

In July 1945 Cockcroft was offered the post of Director of the Harwell Laboratory (which of course had yet to be established) by Sir John Anderson who promised him that he would have 'the utmost measure of freedom in the control of the Establishment'. This was a degree of flexibility which Cockcroft was to exploit to the full in the following years. Cockcroft's appointment as Director of the Atomic Energy Research Establishment (AERE) at Harwell was announced in January 1946, and almost at the same time William George Penney (later Lord Penney, 1909–1991; FRS 1946) was appointed Chief Superintendent, Armament Research in the Ministry of Supply (it being generally recognised that this appointment was merely a prelude to him becoming responsible for the design, manufacture and testing of an atomic bomb). Somewhat earlier, in November 1945, Christopher Hinton (later Lord Hinton, 1901–1983; FRS 1954) was invited to take charge of the industrial organisation to be created inside the Ministry of Supply, and which would be responsible for planning and constructing major industrial-scale plants in support of the nuclear programme.

The types of industrial plant needed would of course depend on major decisions yet to be taken – on whether the UK programme would be for purely peaceful applications or would include the developments of nuclear weapons and whether these were to be powered by U-235 or Pu-239. In any event it seemed likely that Hinton's organisation would be required to construct nuclear fuel manufacturing facilites, enrichment plant, nuclear reactors and plant designed to reprocess spent fuel – altogether a formidable undertaking.

This is how it came about that at the time of the fateful announcement, in January 1947, of the country's intention to construct an atomic bomb, the famous triumvirate – Cockcroft, Hinton and Penney – were in post within the Ministry of Supply with Marshal of the Royal Air Force, Lord Portal, in overall control. Cockcroft's main base was to be Harwell, Hinton's Risley and Penney's Atomic Weapon's Research Establishment (AWRE), Aldermaston.

The task facing Portal and his three directors, Cockcroft, Hinton and Penney was formidable. The atomic development work for the previous four years had been almost entirely concentrated in North America so there were no production facilities in the UK to manufacture enriched fuel or heavy water. Recalling that the uranium enrichment plant at Oak Ridge, with its 2 mile perimeter fence, had at its peak consumed as much electricity as that used by the whole of the United Kingdom, and then only produced one bomb's-worth of U-235 by the summer of 1945, it was obvious that Britain would have to set its eyes initially on developing a plutonium bomb. This would require the construction of nuclear reactors (or 'piles') of appreciable size to 'breed' the plutonium. These reactors would have to be charged with fuel elements manufactured in a special factory and at the end of their lives the 'spent' fuel would have to be discharged and chemically treated in a 'reprocessing' plant in order that its plutonium content could be extracted.

Having decided on a plutonium route to manufacture an atom bomb then there might have appeared little need for Britain to construct an enrichment plant (enrichment is the process by which the fissile isotope of uranium, U-235, is artificially increased in concentration above its naturally-occurring level of 0.7%; to make weapons-grade uranium a U-235 concentration above about 92% is required). In fact Hinton was ordered to give the construction of an enrichment plant (at Capenhurst in Cheshire) a high priority. The main reason for this was that there was concern over uranium supplies. When spent fuel is reprocessed and its plutonium extracted there remains relatively large quantities of 'depleted' uranium (ie having U-235 contents lower than 0.7%). This is virtually useless unless it can be enriched back at least to its 'natural level' and recycled through the core. The net result is that having an ability to enrich uranium reduces the demand for fresh uranium ore.

Perhaps also, as was the case with the Manhattan Project, it was felt that it was prudent to have an alternative route to a fissile weapon in case something went badly wrong with the plutonium route. There were also real fears that the plutonium-producing reactors might not have enough reactivity to achieve criticality if they were loaded with only natural uranium. (In fact, subsequent experience demonstrated that there were grounds for such concerns – on attempting to commission the first Windscale reactor it was found there was insufficient reactivity in the core for criticality to be achieved – see later.) Finally, there might already have been a perceived need to manufacture enriched uranium fuel to power nuclear submarines

As enrichment played an important part in the British post–war nuclear programme it is appropriate to discuss it here, since the development of the equipment for reprocessing has strong links with the country's metallurgical industry, and a number of the pioneers in this field were elected to the Royal Society.*

Uranium enrichment in war and peace

In 1940, when Peierls and Frisch first calculated that the critical mass of U-235 was only a few kilograms, plutonium had yet to be discovered. Hence the practicability of developing an atom bomb then rested entirely on the feasibility or otherwise of separating U-235 from U-238, that is to say, the viability of enrichment procedures. Peierls and Frisch at first recommended thermal diffusion as an enrichment process in their original memorandum though they later abandoned this approach in favour of gaseous diffusion.

In this process a gaseous compound of uranium (actually the gas uranium hexafluoride was chosen) is forced under pressure to diffuse through minute pores in a membrane. The gas molecules containing U-235 diffuse through the pores at a slightly faster rate than the molecules containing U-238. By repeating the process through a huge number of stages the desired enrichment can be achieved. It was clear that if such an enrichment process was successful then the manufacture of an atomic bomb would be feasible.

In the summer of 1940 in the Clarendon Laboratory at Oxford University, Professor Francis Eugen Simon (later Sir Francis, 1893–1956; FRS 1941) and his team consisting of H.S. Arms, H.G. Kuhn and Nicholas Kurti (1908–1998; FRS 1956) started to do research to develop an efficient membrane. They enlisted the help of Mr S.S. Smith, the Research Manager of ICI Metals who had previously collaborated in the early attempts to produce membranes by rolling out metal gauzes. Smith went on to work on a method which depended on photograving and lithography but the process did not find practical application. At around this time Christopher Frank Kearton (later Lord Kearton, 1911–1992; FRS 1961) of ICI became involved in the gaseous enrichment project and in due course became a leading light in enrichment technology.

ICI also worked on methods of production of the gas, uranium hexafluoride, which was to be the diffusing species in the enrichment process. Unfortunately this was highly reactive and ICI had a significant programme of research studying its propensity to corrode structural materials with which it would come in contact during operation of the

*It is interesting to note that the Royal Society had a Committee on Nuclear Physics from 1952–57, comprised of a number of Fellows, including some prominent from the wartime period. Chadwick was the first chairman, followed by G.P. Thomson.

Francis Simon

Nicholas Kurti

Christopher Kearton

Hans Kronberger

Eugen Glueckauf

Robert Spence

Heinz London

Norman Franklin

enrichment plant. Professor Walter Haworth at Birmingham University carried out similar studies and liaised closely with Peierls and Frisch who were both still located at the University, though Frisch was shortly to join Chadwick at Liverpool. As the work progressed a contract to build prototype diffusion units was placed with Metropolitan Vickers who constructed these at a disused Ministry of Supply factory at Valley in North Wales.

The development of the diffusion membrane was a most fascinating metallurgical exercise. The holes through which the hexafluoride was to diffuse had to be extremely small and close together; (Smith specified a hole size of 0.0025 mm and spacing of 0.05 mm in a sheet which was itself only 0.05 mm thick!). For such a component to retain mechanical stability and not have its pores blocked during the transport of a highly corrosive gas, was no small order and pushed the metallurgy of the day to its limit. The gas had to be forced through the membranes under pressure and Peierls had consulted the great G.I. Taylor (Geoffrey Ingram Taylor, later Sir Geoffrey, 1886–1975; FRS 1919) on their design.

More than any aspect of nuclear technology, enrichment had been taken to an advanced stage in wartime Britain, even though the process did not find a practical application during the war years. Towards the end of 1942 a British team of experts visited America to discuss membrane technology and were amazed at the resources the Americans were putting into enrichment processes. They found that the Americans had achieved much success with membranes manufactured from nickel powder.

That these developments have been reported as causing much surprise to the British visitors is something of a mystery for the distinguished British metallurgist Leonard Bessemer Pfeil (1898–1969; FRS 1951) of Mond Nickel in Birmingham, had been liaising with the Americans on the manufacture of nickel powder membranes. Moreover, to produce the membranes, the Americans used a type of nickel powder which was only made by the Mond Nickel Company at their factory in Wales. The factory was expanded to meet the new demand and by the end of June 1945, approximately 3000 tonnes of powder had been supplied for the American enrichment programme.

Although the comprehensive work by the British on enrichment technology was impressive, it did not in the event prove to be of much help to the Americans. They did however appreciate the assistance Pieirls and Klaus Fuchs were able to provide on the theoretical aspects of their manufacturing operations. When the British mission returned home in January 1944, Peierls and Fuchs stayed behind in New York to do theoretical work on enrichment for the Americans. Kurti and Kearton also spent some time in New York to establish an office for liaising with the American enrichment programme.

Britain's wartime research and involvement with the American enrichment organisation meant that we had a strong pool of expertise for Hinton to call upon when he was planning the gaseous diffusion plant at Capenhurst. Four Fellows (or future Fellows)of the Royal Society – Simon, Peierls, Kurti and Kearton – joined the Capenhurst design committee and the advice they gave was of inestimable value to the Risley design engineers. Even so there were immense difficulties to be overcome and Hinton was disappointed he could not persuade Kearton to take charge of the whole operation.

Recognising the problem, Cockcroft then stepped in with what Hinton referred to as a characteristically generous action by seconding from Harwell two of his best enterpreneurial scientists to the Capenhurst project – H.S. Arms and Hans Kronberger (1920–1970; FRS 1965). Arms was an American who started life as an engineer but became involved in gaseous diffusion at a very early stage – at Oxford in 1940. Kronberger was a brilliant physicist and an even more brilliant organiser who was to become a pillar of the nuclear programme until his tragic and untimely death in 1970. A great deal of the success of Capenhurst was due to his technical leadership and organisational flare. It has been suggested that Harwell's greatest contribution to the enrichment programme was the seconding of Arms and Kronberger; the laboratory itself concentrated its research on the alternatives to gaseous diffusion, magnetic separation and centrifugal enrichment. Neither of these were important during the first phase of the British nuclear programme, though centrifuges became important at a later stage.

An equal debt was owed to S.S. Smith, as Research Director of ICI Metals who, together with his colleague, W. Lake, continued development work on the diffusion membrane, armed now with the knowledge of the success achieved in America with the use of nickel powder. They brilliantly solved the problem inherent in developing a large-scale production facility from experimental-scale laboratory components. Smith also resolved a major metallurgical problem facing the designers – whether to employ plated steel or aluminium to manufacture the miles of piping needed to be incorporated into the full-scale plant. The Americans had used steel but Hinton, against the advice of his design team, had a preference for aluminium. The main objection to aluminium was its allegedly poor welding properties but Smith carried out a crash development programme and gradually evolved an entirely satisfactory welding procedure. This success won the day and aluminium tubing was selected.

Another future Fellow of the Royal Society, Eugen Glueckauf (1906–1981; FRS 1969) who had joined the Chemistry Division at Harwell in 1947, made many important contributions to the atomic energy programme, including work on membranes. During 1952 and 1953 , when the gaseous diffusion plant was being constructed at Capenhurst, Glueckhauf represented Harwell on the design work. He assisted with theoretical and experimental investigations of various aspects such as the corrosion of membrane materials by uranium hexafluoride. Heinz London (1907–1970; FRS 1961), a distinguished physicist, as part of his work at Harwell was also involved in isotope separation, and designed the trapping system for the Capenhurst diffusion plant.

Choice of gas–cooled reactors

The selection of reactor system was something of a Hobson's choice. With only natural uranium available as fuel then light water could not be used as a moderator and the absence of an indigenous supply of heavy water left graphite as the only alternative. In principle there remained the possibility of cooling the pile with light water but the highly undesirable safety characteristics of a water-cooled graphite pile effectively restricted the choice of reactor coolant to air or some reasonably-inert gas (the devastating accident at Chernobyl some forty years later in a water-cooled graphite-moderated RBMK reactor, confirmed the wisdom of the decision to opt for gas cooling). As a prelude to constructing large plutonium-producing reactors it was decided to build two air-cooled, graphite-moderated uranium-fueled experimental reactors. The first of these, GLEEP (Graphite Low Energy Experimental Pile), was constructed and went critical in August 1947 and the much larger, but still experimental, BEPO, became operational in July 1948.

Harwell's Metallurgy Division

With the possible exception of the US space programme in the 60s and 70s, no greater challenge has faced the metallurgists and other scientists and engineers than that presented by the need to support the development of nuclear power. Apart from having to explore the almost completely unknown terrain of the influence of irradiation on the physical, chemical and mechanical properties of materials, there was the requirement to study the conventional properties of exotic metals, such as uranium, plutonium, thorium, zirconium, beryllium and liquid sodium, which hitherto had been thought of as little more than metallic curiosities. In addition, conventional materials, such as aluminium, magnesium, graphite and steel, would have to be alloyed or otherwise treated to withstand temperature, chemical environments and stress regimes quite outside the range of common practice.

As well as difficulties arising from the clampdown on exchange of information with the Americans following the passing of the McMahon Act, very few British metallurgists had been seconded to the Manhattan Project so they lacked first-hand experience of nuclear technology. The Division suffered yet another disadvantage in that the staff were initially divided between Harwell and the Royal Aircraft Establishment (RAE) at Farnborough. The Division Head, Dr Bruce Chalmers, and the majority of his staff were located at RAE.

In spite of all these drawbacks remarkable progress was made by the Division especially during its earliest days. Sadly, in spite of these great achievements, none of the true pioneering hands-on scientists and technologists of the Metallurgy Division were elected to the Royal Society. There are a number of reasons for this, an important one being that in the very early days of Harwell there was great pressure to support the technology of nuclear weapon production rather than to attempt to obtain a fundamental understanding of the physical processes underlying the various problems which had arisen.

There were also cultural differences between industry-based laboratories such as Harwell and the universities, which were much more geared towards cultivating and nominating candidates from metallurgy for election to the Royal Society and conforming to its requirements. (At a later stage in the evolution of nuclear power the Central Electricity Generating Board (CEGB) did even worse; during its 30 year existence and with a graduate research staff rising to over a 1000, only two of its researchers were elected FRS for work carried out entirely within the organisation). However, from about the mid-1950s, when the nuclear programme was more established and reactor-construction in progress, there was more time for fundamental metallurgical studies and Harwell's crop of FRSs grew to a respectable number. It was still the case though that officers who achieved high administrative positions within the Atomic Energy Authority were far more likely to be elected than those who had made individual scientific discoveries.

It is very fortunate that Stanley Pugh, himself one of the Harwell pioneers and the final Division Head, has written a full account of the early days of the Metallurgy Division*. Pugh has described how the Harwell staff were very aware that coincident with those early days of setting up the Division something of a renaissance was taking place in physical metallurgy and this was spearheaded by academic scientists in Departments of Physics and Metallurgy particularly in the Universities of Bristol, Birmingham and Oxford.

There were from the start fruitful interactions between Harwell staff and leading academics and Pugh has illustrated this by describing how the work on the theory of alloy constitution, pioneered by William Hume-Rothery (1899–1968; FRS 1937) at nearby Oxford University provided useful pointers for the choice of alloying elements for uranium and plutonium. The division also received more direct assistance from Nevill Francis Mott (later Sir Nevill, 1905–1996; FRS 1936), Frederick Charles Frank (later Sir Charles, 1911–1998; FRS 1954), and Frank Reginald Nunes Nabarro (1916–; FRS 1971) and also Alan Howard Cottrell (later Sir Alan, 1919–; FRS 1955) who later joined the Metallurgy Division.

Plutonium

Having decided to manufacture a plutonium bomb then two immediate tasks faced the Metallurgy Division: one was to study the metallurgy of plutonium itself (it was known to be complicated, having numerous allotropic forms) and to solve the multitude of materials problems generated by the need to construct and operate a number of experimental and plutonium-producing reactors. In order to achieve the first objective – to study the properties of plutonium – it was obviously necessary to obtain a quantity of the metal. This was to prove extremely difficult, at least in the short term.

It was hoped to obtain some samples of plutonium from America, or at least receive from them information on its physical, chemical and nuclear properties. Expectations were raised during discussions with the Americans in late 1947 when it seemed to be possible that a deal would be struck whereby in exchange for samples of plutonium to study, Britain would agree that the total 1948 supply of uranium ore from the Belgium Congo would be allocated to America. These and other discussions led to a general arrangement on exchange of nuclear information known as the Anglo–American *Modus Vivendi*, verbally agreed on 7th January, 1948. It was hoped that during his subsequent visit to England, the British-born Cyril Stanley Smith (1903–1991) the ex-Chief Metallurgist of Los Alamos, would supply this badly-needed information about plutonium, but it turned out he was under strict orders to make no such revelations. In the end the British

*'The Harwell Metallurgy Division 1946 to 1954', *Report No. AERE R 12109*, March 1986.

metallurgists had to wait until Britain's separation pilot plant was working before receiving significant quantities of the fissile metal; the first gram-quantities of plutonium were eventually supplied to Harwell in February 1951.

The fact that even this delayed supply became available was largely the result of the efforts of the British chemist Robert Spence (1905–1976; FRS 1959). At the end of 1946 Spence was head of a team of twelve chemists and chemical engineers at the Chalk River Laboratory in Canada and he was requested to develop by the following September a plutonium-separation process which would be applicable on an industrial scale. Their only source of plutonium to perform their experiments on was that contained in four irradiated rods from an American reactor. From these they extracted just 12 milligrams of plutonium and with this minute amount performed sufficient experiments to establish an industrial separation process which was eventually adopted by the UKAEA – a very considerable achievement. In 1947 Spence left Chalk River to take up an appointment as Head of the Chemistry Division at Harwell, where he played a leading role in many aspects of the Harwell programme.

Research on the properties of plutonium was shared mainly between Harwell and AWRE Aldermaston, although some work was carried out at Dounreay and in British industry. As far as construction of the separation (later known as reprocessing) plant is concerned, ICI played its usual important part. The strength of Britain's physical metallurgy effort was well illustrated by the progress made on unravelling the properties of this strange element.

Quite soon restrictions on information transference between national groups were relaxed and considerable data was released at a conference in Chicago in 1957 and at the 1958 Geneva Conference on Peaceful Applications of Nuclear Energy. At the Second International Conference on Plutonium Metallurgy held in Grenoble, France in April 1960, a third of the papers presented were by British authors. The Introductory Address was given by Montague Harold Finniston (later Sir Monty, 1912–1991; FRS 1969), and the Rapporteurs (chosen for their authority and distinction) included G.L. Hopkin (Aldermaston), M.B. Waldron (Harwell), P. Murray (Harwell), J.G. Ball (Imperial College but ex-Harwell) and D.S. Oliver (Dounreay). As far as plutonium metallurgy was concerned, Britain had well and truly arrived on the international scene.

Materials for GLEEP, BEPO and the Windscale reactors

Of equal importance for the Metallurgy Division in the late 40s was to develop fuel elements for the GLEEP and BEPO reactors. It soon became clear that the method of manufacture of the uranium fuel bar could critically influence the performance of the fuel element in the reactor. The crystal structure of the low temperature form of uranium was so anisotropic that irradiation, or simple thermal cycling, could cause it to 'grow', that is to say the individual crystals would increase in length in one crystallographic direction, decrease in another and remain unaltered in a third. As a consequence any macroscopic texture in a bar, induced by, say, rolling or extrusion processes, could result in an uncontainable increase in length when irradiated or thermally cycled in a reactor.

Partly as a consequence of these effects it was decided to use cast bars of uranium. By cooling the bars very rapidly during casting, a process which became known as 'beta quenching', a fine and random alpha grain structure could be achieved and this minimized 'wrinkling', that is the formation of a very rough surface on coarse–grained bars due to the differential growth of the individual grains. It was also found that naturally-occurring impurities in the bar (particularly iron, silicon and carbon) assisted in the process of producing a fine and random grain size and by 'adjusting' the levels of these naturally-occurring impurities even better quality bars could be produced. It was from this process that the adjective 'adjusted' was applied to the uranium used in UK reactors. Coincidentally, it was found that these modified impurity levels also impeded the 'swelling' of the uranium due to the formation of fission gas bubbles in the microstructure at high irradiation levels. In fact British 'adjusted' uranium achieved international acclaim for its excellent in-core properties.

The uranium bars had to be protected from oxidation by the air used as a coolant in the reactors. Simply spraying aluminium onto their surfaces was first attempted for the fuel used in the GLEEP reactors, but subsequently aluminium cans were used (pure aluminium was known to be a suitable canning material because it had been used successfully in the air–cooled 'Clinton' graphite pile constructed at Oak Ridge as part of the Manhattan Project).

GLEEP and BEPO, as well as being valuable experimental reactors were forerunners of the huge air–cooled plutonium–producing piles planned to be constructed at Windscale. Initially it was intended that the core of the Windscale piles, consisting of aluminium-clad natural uranium fuel elements located in horizontal channels in a graphite moderator, would be pressurised. However, it was discovered that when the BEPO fuel element cans incorporated fins on their external surfaces to increase their surface area, the heat transfer characteristics were enormously improved. It became clear that providing the fuel cans for the Windscale piles also had fins, their heat transfer properties would be so improved that it would no longer be necessary to pressurise the core.

The fact that plans to construct a large steel pressure vessel around each of the Windscale cores could be set aside enormously reduced the complexity and cost of the reactors and enabled a very tight construction schedule to be readily achieved. The fuel element fins did though contribute to one unexpected problem – at a late stage, after the fuel elements had been loaded, the first Windscale reactor could not be made to go critical because there was too much neutron–absorbing material in the core. The solution was to reduce the amount of aluminium in the core by decreasing the height of the fuel element fins. This mammoth task was accomplished by the Windscale staff unloading all 70,000 fuel elements and clipping off by hand a narrow strip from each fin.

The evolution of the Magnox reactor

It was then realised that if both finned fuel elements and a pressurised core were incorporated into a design then a reactor capable of generating electricity might be achieved. What was conceived was to become the world's first 'commercial scale' nuclear power station – the 'Magnox' reactors at Calder Hall. The Calder reactors also used natural 'adjusted' uranium as fuel loaded into channels in a graphite moderator, only this time the fuel channels were vertical and each core was contained within a steel pressure vessel through which circulated pressurised carbon dioxide. The heat in the fuel produced by the fission process was transferred to the carbon dioxide which circulated through a boiler which generated steam which powered the electrical generators. The reactors were designed to produce military grade plutonium as well as to generate electricity and their original name, PIPPA (Power and Plutonium Production Assembly) reflected this dual role.

Another major difference with the Calder Hall reactors was that a magnesium alloy was used, in place of pure aluminium, to contain the fuel bars and to protect them from chemical attack by the reactor coolant. The main problem with aluminium cans was found to be that at higher operating temperatures, characteristic of the Windscale piles and Calder reactors, a chemical reaction took place between the aluminium and the uranium bar forming the intermetallic compound UAl_3, with an associated large volume increase. This process, known as 'pimpling', was found to be very damaging and occasionally resulted in penetration of the can wall. Diffusion barriers inserted at the interface between the aluminium and the uranium, such as graphite, were developed but were not completely successful at higher operating temperatures.

At that time, ie the early days of the Harwell work, it was known that magnesium did not interact chemically with uranium and had other attractive properties, but it was not considered for canning purposes because early American data had indicated that magnesium had an unacceptably high neutron absorption cross section. However, subsequent measurements by Fred Colmer and Derek Littler using the GLEEP oscillator revealed that the earlier results were in error and that magnesium did in fact have an acceptably low neutron cross section.

As soon as it was realised that magnesium was a possible canning alloy Roy Huddle and Les Wyatt and others started to investigate the possibility of improving its corrosion resistance by alloying it with other low-neutron-cross-section metals. Eventually they produced a promising material – a magnesium/0.8%aluminium alloy to which they added trace quantities of beryllium to improve its resistance to ignition. This alloy they named Magnox AL80 (MAGnox No OXidation) and when it was adopted for the fuel elements it lent its name to the complex reactor system, a fitting tribute to the efforts of the Harwell pioneers. Even today, fifty years on, the fuel elements employed in the still-operating Magnox reactors still consist of 'adjusted' uranium encased in Magnox AL80 cladding, another tribute to the skill of the pioneering metallurgists.

In 1955, the year he was elected to the Royal Society, Alan Cottrell resigned his important position as Professor of Metallurgy at Birmingham University to become Deputy-Head of Harwell's Metallurgy Division. He was already recognised as one of the world's leading metallurgists and his move from academia surprised many people. His motives were though entirely idealistic; he reasoned that the country was about to invest so much of its resources into developing nuclear electricity, with its frightening range of potential materials' problems, that it was incumbent on the metallurgical profession to put its weight to the wheel. The industry did not have to wait long before his presence was felt.

'Cottrell creep'

The year Cottrell moved to Harwell saw the first 'Atoms for Peace' conference at Geneva, where a Russian scientist, S.T. Konobeevsky, reported that under neutron irradiation the creep rate of uranium was enhanced by up to a factor of two. Cottrell wondered if this effect might be related to the well-known phenomenon of irradiation-induced 'growth' of uranium, that is to say the irradiation-induced increase in length in some crystallographic directions in a crystal, and a corresponding reduction in others. He also asked himself whether or not Konobeevsky's observations had any implications for the performance of the Calder Hall fuel elements, which at that time were already being loaded into the reactor.

Unlike the Windscale reactor, which had horizontal fuel channels, the fuel elements in Calder Hall, which were stacked vertically, had to withstand the weight of those in higher positions in the channel. The Calder engineers had satisfied themselves that, as far as conventional thermal creep was concerned, there was no possibility of creep strains developing. On the other hand, irradiation-induced growth in the individual 'grains' or crystals of the polycrystalline uranium bar would generate internal stresses at the grain boundaries which would soon exceed the yield stress. The critical question was would even a small externally applied stress, such as might arise from the weight of the fuel elements in higher channel positions, have the ability to 'guide' the local deformations and produce an appreciable macroscopic strain? Such a strain might manifest itself as a 'buckling' or 'bowing' of the lower fuel elements, making them impossible to discharge at the end of their useful lives in the core.

Cottrell calculated the magnitude of his irradiation creep and, with the assistance of A.C. Roberts, tested his ideas by irradiating in the BEPO reactor a stressed spring specimen of uranium. In fact the creep strains observed corresponded with those predicted by Cottrell's model. By this time Calder Hall's first charge had been loaded and the reactor was operating, so a carefully selected group of fuel elements were discharged prematurely and it was found that they indeed exhibited bowing strains which Cottrell had predicted. The reactor was closed down and a proportion of the fuel elements were replaced by ones which were strengthened by incorporating into their design an anti-bowing device.

There can be few, if any, comparable examples in the history of technology of a theoretician conceiving an entirely new deformation mode which might give rise to a practical difficulty, confirming this was so by experiment, and suggesting solutions even before the plant operators had any evidence of the existence of the problem.

Finniston and the status of metallurgy

In spite of such successes, in those early post-war years metallurgy was hardly recognised as a science, particularly by nuclear physicists (Rutherford thought that science was divided into physics and stamp collecting!). Indeed many years later* Cottrell himself speculated that it was this disdain from the physicists which drove Bruce Chalmers to abandon Harwell and return to academia. Chalmer's replacement, Monty Finniston, was made of sterner stuff and was in fact one of the most talented metallurgist/managers of his generation. He was determined that the Metallurgy Division would become an internationally-recognised centre of excellence for materials research, would have a separate existence outside any General Physics organisation, and of course assist in development of UK's nuclear technology. He was successful in all these objectives; when Finniston started at Harwell, in 1948, nuclear electricity was just a distant dream, but when he left ten years later Calder Hall was powerfully supplying the National Grid.

It is true that pressure of events did not always make it immediately possible to obtain a detailed understanding of the underlying atomic mechanisms of processes which were causing problems. In fact responding to such demands, Roy Huddle, the inventor of the Magnox alloy, once declared that his job was not to research corrosion, but to prevent it happening. Such an approach might have been acceptable as a transient response to a crisis but as a general policy it was completely at variance with Finniston's philosophy. For the new Division Head, being successful was not enough, it was of equal importance to find out, by fundamental study, why one had been successful.

A contemporary Harwell scientist, Robert Wolfgang Cahn (1924–; FRS 1991) needed no encouragement to study metallurgical phenomena at the deepest level. Already well-known for his innovative work on polygonisation and recrystallisation, he studied the deformation mechanisms in uranium and he added significantly to the knowledge of twinning in this and other metals. Other distinguished scientists in the Division included Alan LeClaire, Michael Thompson, Robert Barnes and Kingsley Williamson, in spite of their major contributions to nuclear metallurgy, and subsequent distinguished careers, none of them has won election to the Royal Society. LeClaire's field was diffusion and he addressed the question of whether or not insoluble inert gases, formed inside metals by nuclear transmutations, could diffuse. He found that they could, and escape at free surfaces etc. Thompson did distinguished work on radiation damage and became well known for his work on 'channelling' of knock-on collisions; he was knighted and became Vice-Chancellor of Birmingham University. Barnes's interests lay mainly in the field of mechanical properties of irradiated materials, particularly how they were influenced by the formation of inert gas bubbles. He went on to become Director of Research for the British Steel Corporation.

As a Harwell research fellow Williamson was allowed a wide choice of research topics but to his colleagues surprise chose to work on plutonium, the most difficult of metals with its high toxicity, instability and bewildering number of allotropic forms. Building on the work of Waldron and Ball and many other pioneers both at Harwell and Aldermaston, Williamson soon became an international authorities on the properties of this enigmatic element which lies right at the heart of nuclear technology.

In 1953 Raymond Edward Smallman (1929–; FRS 1986), Williamson's former stu-

Interdisciplinary Science Reviews*, 1992, **2(17), 120.

dent at Birmingham joined the Division and the two friends renewed their scientific collaboration. Initially Smallman worked on liquid metals and studied textures developed during deformation of metals. Later, in collaboration with Kenneth Westmacott, Smallman was one of the first to study the dislocation structures of quenched and irradiated metal samples in the transmission electron microscope. They were, for example, the first to observe prismatic dislocation loops in quenched aluminium and they published this work in a paper with Peter Bernhard Hirsch (later Sir Peter, 1925–; FRS 1963) and P. Silcox in a paper which became an ISI 'citation classic'.

No less than fourteen members of the Harwell Metallurgy Division went on to become university professors, and a number of these became vice-chancellors or pro-vice-chancellors. Prominent among this select band of FRS's were Derek Hull (1931–; FRS 1989) and Geoffrey Wilson Greenwood (1929–; FRS 1992). Hull carried out important studies of the role twinning plays in initiating brittle fracture in body centered cubic metals, but he is equally well known for his now classic work with Dennis Rimmer on a quantitative model for creep fracture due to cavitation. This work was not only of considerable theoretical interest, it also had great practical significance because the reactor fuel cans were failing by this mode.

Finniston thought of Greenwood as Cahn's replacement, but the two metallurgists worked in quite different fields. Trained as a physicist Greenwood has turned his hand to the study of many processes and phenomena in physical metallurgy in general and in nuclear metallurgy in particular. In 1956 he derived the basic law of particle coarsening, also known as Ostwald Ripening, a process of fundamental importance in metallurgy, as it is in many other disciplines. He developed these studies with practical and theoretical investigations of the coarsening of inert gas bubbles in irradiated solids. In this case, intriguingly, the driving force is not the reduction in interfacial energy (it actually remains constant) but an increase in entropy. This work was relevant to the swelling of uranium due to the formation of bubbles of fission gas.

Greenwood had always been fascinated by creep cavitational failure and developed and strengthened Hull's pioneering work, and at a later stage in his career became interested in diffusional creep processes. On leaving Harwell in 1959 he joined Trevor Churchman, Kingsley Williamson and John Buddery to form the nucleus of the Materials Division of the Berkeley Nuclear Laboratories of the CEGB (see below).

Properties of graphite

Williamson also became a pioneer in the use of the transmission electron microscope to study the properties of materials, concentrating initially on investigating irradiation damage of graphite. On being seconded to the Department of Metallurgy at Cambridge he worked closely with a distinguished electron microscopist, Clive Baker, and published extensively on the microstructure of graphite. Baker continued these studies at Cambridge, working in collaboration with Anthony Kelly (1929–; FRS 1973). Graphite moderators were of course common to all the UK gas-cooled reactors, from GLEEP to the Advanced Gas Cooled Reactor (AGR), so it was not surprising that the UK nuclear industry was

Thomas Marsham

Walter Marshall

prepared to sponsor research which would elucidate the response of the material to irradiation and corroding gases.

A particularly worrying characteristic of graphite is its propensity, when irradiated at relatively low temperatures, to store what has become known as 'Wigner' energy. This was discovered by Eugene Paul Wigner (1902–1995; FMRS 1970, Nobel Laureate in physics 1963) during the Manhattan Project and it resulted from carbon atoms being displaced from their equilibrium positions and occupying high-energy interstitial sites. The danger here is that this energy can readily anneal out at higher temperatures with an associated uncontrolled release of energy in the form of heat.

Wigner energy storage can be a greater hazard in water-cooled graphite piles, as operated by the Americans at Hanford during the Manhattan project, so the UK should have been alerted of the danger from Amercian experience. In fact, while the interchange of information with the American nuclear scientists was greatly inhibited by the McMahon Act there was operating an Anglo–American *modus vivendi* which did allow exchange of information in certain areas, including matters relating to safety. In January 1949 Edward Teller (the 'father' of the H-bomb), while on a visit to Harwell for a *modus vivendi* exchange of information, predicted that the energy stored in a graphite pile might lead to fuel rods catching on fire

There were two ways of avoiding a dangerous build-up of Wigner energy – either operate the reactor continuously above the annealing temperature for the graphite, or periodically carry out a controlled annealing of the irradiated core. The former course was not possible in the case of the two Windscale reactors and it was during a so-called 'controlled' annealing of the Windscale No 1 reactor in October 1957 that the uranium fuel elements caught on fire and gave rise to the worst nuclear reactor accident the world had seen to that time. The accident sequence was very much along the lines of that predicted by Edward Teller almost nine years earlier; it is perhaps unfortunate that no one from the Northern Group was at the Harwell meeting so the full import of the message was not conveyed to those designing and operating the Windscale reactors.

The accident resulted in both the Windscale reactors being permanently closed. By that time, however, they had almost become redundant because weapons-grade plutonium could be produced in the Calder Magnox reactors, the first of which had already been commissioned. Fortunately the graphite moderators of the Magnox reactors operate at a sufficiently high temperature that Wigner damage did not accumulate.

During the planning and construction stages of the Windscale reactors the designers were far more concerned with another phenomenon in graphite – the anisotropic dimensional changes of the graphite which was known to occur under prolonged neutron irradiation. This phenomenon was also named after Wigner; it was in fact known as 'Wigner growth'. These dimensional changes, of unknown magnitude, obviously made the job of the reactor designers much more difficult and numerous ingenious design solutions and devices were proposed and incorporated into the reactor cores to accommodate the anticipated stresses and strains. Sir Christopher Hinton was far from convinced that the Harwell researchers were paying sufficient attention to Wigner growth and a crash research programe was at one stage instituted under the overall supervision of Alan Cottrell.

In this instance American experience was not very relevant because it was found that British graphite had different characteristics – an exaggerated growth along the extrusion direction, for example. There were also fears that radiolytic corrosion of graphite due to its exposure to the reactor coolant gas (ie either air or carbon dioxide) could result in such loss of carbon that the reactors would become insufficiently moderated. Were this to happen the life of the core would be the life–limiting feature for the reactor and accordingly, dimensional changes and radiolytic corrosion has been extensively studied and the progress of both processes are still being monitored in both Magnox and Advanced Gas-Cooled Reactors.

For a while graphite became probably the most thoroughly investigated material in the UK, work being carried out in most branches of the Atomic Energy Authority. The studies were spearheaded at Harwell by J.H.W. Simmons and W.M. Lomer at Harwell and J.C. Bell, H. Bridge and G.B. Greenough of the Northern Group.

Harwell's Physics Division

Of course not all the UK's pioneering work in nuclear metallurgy was performed in Harwell's Metallurgy Division; much ground-breaking studies relating to irradiation damage was carried out in their Physics Division. What was needed in the early 50s was a theoretical model for what had become known as displacement cascades which form as a result of primary knock-ons in irradiated fissile material, and such was produced in a paper which has become a classic of the genre written by G.H. Kinchin and Rendel Sebastion Pease (1922–; FRS 1977). Pease later became Programme Director for the UKAEA fusion project and then Director of the Culham Laboratories.

What can broadly be classified as Pease's extra-mural activities which have made him one of the more significant figures in the country's scientific and nuclear establishment. For example, he has served on the Royal Society Working Group on 'Management of Separated Plutonium' which reported in February 1998. In 1995, when Pease was chairman of British Pugwash and member of its International Council, the Pugwash organisation shared the Nobel Peace Prize with its then President, Joseph Rotblat (later Sir Joseph, 1908–; FRS 1995). Pease is an outstanding example of the importance of 'crossbenchers' in the nuclear debate – that is those who were distinguished advocates of the peaceful application of nuclear power but at the same time actively opposed the proliferation of the use of this unique source of energy in the manufacture of weapons of mass destruction.

Another distinguished member of the Physics Division, who specialised in the solid state, was Ronald Bullough (1931–; FRS 1985). He graduated in physics at Sheffield and started his working career at the AEI Fundamental Research Laboratory at Aldermaston Court under the general direction of one of the pioneers of nuclear physics, Thomas Edward Allibone (1903–; FRS 1948) (who still thinks of Bullough as one of his 'boys'). Bullough is a theoretician with special knowledge of dislocation theory and the interactions of point defects with dislocations. The canon of his works is astonishingly diverse and his researches have led to new insights into the mechanisms of a large number of irradiation-induced phenomena in fuels, cladding and structural alloys. A great deal of his success arises from his ability to communicate easily with experimentalists and other theoreticians with whom he co-authored many papers. His major collaborators include John Raymond Willis (1940–; FRS 1992), R.S. Nelson, A.B. Lidiard, M.R. Hayns, D.R. Harries, J. Gittus, G.K. Kulcinski, A.D. Brailsford, D.J. Mazey, R.G. Anderson, K. Malen, M.J. Makin, R.C. Perrin and B. Eyre.

By the mid 60s many of the phenomena associated with thermal reactors had been studied in some detail both experimentally and theoretically and the research scientists at Harwell and elsewhere were looking for new problems to solve. Until about 1967 the most detrimental effect expected by irradiation of stainless steel cladding of liquid metal fast breeder reactors (such as that at Dounreay) was hardening of the steel at low temperatures and helium embrittlement at high temperatures. Then came the bombshell: C. Cawthorne and E.J. Fulton discovered that under fast neutron irradiation voids formed in the microstructure of stainless steel, giving rise to significant 'swelling' with associated unacceptable macroscopic dimensional changes.

From the characteristics of the swelling (ie its dependance on neutron flux and temperature) a qualitative model for the formation of the voids could readily be formulated. Irradiation with fast neutrons displaces atoms from their sites giving rise to large numbers of vacancy-interstitial pairs. Most of these point defects are annihilated by recombination or migrate to other sinks such as dislocations and grain boundaries. However, the dynamic balance between the point defect creation and destruction during irradiation sustain, within certain temperature and flux regimes, a point defect concentration far in excess of the thermal equilibrium value. In such circumstances two dimensional assemblies of interstitials, and similar (collapsed) assemblies of vacancies, form dislocation loops, and other vacancies form three dimensional agglomerations which act as nuclei for voids which grow by continued vacancy condensation.

It was found that the formation of voids could be suppressed in the presence of a fine dispersion of precipitate particles, especially where the precipitates were coherent with the matrix. It was thought that such interfaces acted as recombination sites, that is to say regions where vacancy-interstitial recombination could occur thereby inhibiting swelling; this was an assumption incorporated into one of the several models proposed by Brailsford and Bullough. Occasionally the voids form in a periodic fashion forming a sort of super–lattice, an early example being observed in irradiated high-purity molybdenum. Arthur Marshall Stoneham (1940–; FRS 1989) another brilliant product of Harwell's Physics Division, has published an explanation and theoretical analysis of the phenomenon. Towards the end of his career at Harwell, Stoneham became Chief Scientist of AEA Technology.

Of at least equal importance to voidage and associated swelling is the occurrence of irradiation creep under fast neutron bombardment. A clue that swelling and creep may be related phenomena is provided by the fact that they tend to occur within the same temperature range (350–500 °C for stainless steel). Pioneering work on irradiation creep was carried out by R. V. Hesketh in the early 60s when, as a founder-member of the Berkeley Nuclear Laboratories of the Central Electricity Generating Board (CEGB), he had been seconded to the Dounreay Laboratories in the North of Scotland.

Two types of irradiation creep had been identified, a transient creep (due it was thought to climb of pinned segments of the dislocation network) and a steady-state creep. The latter, according to Hesketh, is a consequence of the applied stress causing the preferential nucleation of interstitial loops on suitably orientated crystallographic planes. Two colleagues of Hesketh's at the Berkeley Laboratories, P. Heald and M.V. Speight broke new ground when they developed what has become known as a SIPA (Stress-Induced Preferential Adsorption) model to explain the occurrence of irradiation creep, and to quantify its magnitude. As its name implies, the model relies on the fact that an applied tensile stress will produce a drift of intersititials towards those dislocations whose climb contributes to creep in the direction of the stress, an effect which was first pointed out in 1957 by John Douglas Eshelby (1916–1981; FRS 1974). Bullough and Willis later produced an analysis of irradiation creep based on a SIPA concept.

It should be emphasised that this discussion is largely restricted to research carried out within the UK, no attempt has been made to achieve balance and give due credit to the huge amount of research in foreign laboratories, particularly in the USA and Europe.

Three FRSs and the evolution of the Magnox fuel element

Harwell invented 'adjusted uranium' and the canning alloy Magnox AL80, but it was still necessary to put these two materials together in the form of hundreds of thousands of fuel elements, first to charge Calder Hall and Chapel Cross and then the nine Civil stations. In this tremendous technological endeavour three non-Harwell future Fellows of the Royal Society, Leonard Rotherham (1913–; FRS 1963), Thomas Nelson Marsham (1923–1989; FRS 1986) and Norman (Ned) Laurence Franklin (1924–1986; FRS 1981) played major roles.

Of these three only Rotherham would be regarded as a fully-fledged metallurgist, but the other two, by training a physicist and chemical engineer respectively, became so embroiled with problems related to materials that they could well be designated 'honorary metallurgists'. Rotherham had been Head of the Metallurgy Department at RAE Farnborough for four years when in 1950 Hinton invited him to become Director of R&D for the UKAEA Industrial Group, a position he held until becoming Board Member for Research in the Central Electricity Generating Board in 1958. Throughout this time the development of the Magnox, and then the Advanced Gas Cooled, fuel elements was seen as the prime objective, though in the event it was other components of the reactors which gave rise to most operational difficulties.

Marsham joined the United Kingdom Atomic Energy Authority (UKAEA) in 1953, working first as a physicist on uranium enrichment and diffusion plant problems in the Technical Policy Branch at Risley, and in 1955 he was appointed Reactor Manager of Calder Hall station. He took full part in commissioning the Calder Hall 'A' reactor which became the model for all subsequent Magnox reactors. The reactor became operational in July 1956, just two years and eleven months from first digging a hole in the ground, a very remarkable achievement. It was officially opened by the Queen three months later, the first nuclear power station in the world to supply power to a national grid.

As well as producing electricity, and generating plutonium for military purposes (a task which acquired greater importance with the fire and closedown of the Windscale plutonium-producing piles) the Calder reactors were test beds for the advanced fuel element designs for the civil stations. Marsham played an active part in these civil fuel element developments, although the Calder elements themselves were capable of producing surprises – for example, bowing under the stack load due to 'Cottrell Creep', a hitherto unknown deformation mode discussed earlier. It was found that the helical fuel elements designed for the civil stations tended to vibrate in the Calder gas stream leading to fatigue failures of the cladding. This problem was solved by fitting a device known as a 'spring arm spider' which braced the fuel element against the channel wall and thereby minimised rattling.

With the higher burn-ups planned for the civil Magnox stations a greater proportion (typically 40%) of the heat generated by the fuel elements at the end of life would be produced by fission of Pu-239 which had been 'bred' in the fuel itself by transmutation of the more abundant but non-fissile isotope, U-238. This alters the physics of the core and affects the characteristics of methods of controlling the core's reactivity. Marsham was particularly interested in this important effect and initiated experiments to quantify the expected reactivity changes.

After playing an important part in the design, commissioning and operation of the Windscale AGR, Marsham became responsible for the UK's fast reactor programme and in due course was universally recognised as a world authority in this field. In 1978 his achievements were recognised by the Royal Society by the award to him and four colleagues of the Esso Energy Gold Medal. Soon after his election as Fellow of the Royal Society he started, with colleagues, to organise a major discussion meeting 'The Fast Neutron-Breeder Fission Reactor' which was held in May 1986. His contribution was a paper with a characteristic optimistic title – 'Energy for a 1000 Years'. It was a great disappointment for him that his ambitious plans for a large fast reactor programme were not supported, and indeed development in this field has now virtually ceased in Britain. There is though little doubt that the attraction of this reactor system will one day again come back into fashion, but perhaps not until the middle of the next century.

Franklin was recruited into the nuclear industry in 1955 by John McGregor Hill (later Sir John, 1921–; FRS 1981), who became Chairman of UKAEA and joined a small group at Risley whose task was to formulate a policy for the Production Group of the AEA. More of an academic researcher by temperament, Franklin's brief period as special assistant to the Managing Director, Sir Leonard Owen, greatly broadened his perspective, and this experience facilitated his subsequent rapid promotion eventually to the most senior positions in the organisation.

Again it was thought that it was the performance of the fuel elements which would be the key to making or breaking nuclear power as a reliable source of electricity so Franklin, being one of the brightest people available, was given the task of overseeing fuel element design and evaluation. He performed these duties with great distinction for many years. Perhaps his greatest achievement though was recognising the potential of the ultra–centrifuge for uranium enrichment. His enthusiasm, even passion, for this process eventually led to the establishment of the Urenco company, a successful tripartite centrifuge organisation involving British, German and Dutch interests. Franklin went on to become Managing Director of British Nuclear Fuels and a Member of the Board of the Atomic Energy Authority. Against his better judgement he was persuaded to leave British Nuclear Fuels Limited (BNFL) to become Chief Executive of the National Nuclear Corporation, the name of the merged nuclear construction companies. This was not a happy time for him and he left the industry in 1984 to return to his first love, academia – he became Professor of Nuclear Engineering at Imperial College.

Of course there were many others in the Industrial Group who made distinguished contributions to materials aspects of the nuclear programme. Among the names that spring to mind are A. B. McIntosh, Chief Metallurgist at Risley for many years. V. W. Eldred directed with great distinction teams of mostly metallurgist who monitored the performance of the Magnox and AGR fuel elements and G.B. Greenough was generally regarded as perhaps the cleverest man in the fuel technology field. Others working in a senior capacity within the materials discipline include T. Edge, T. Heal, E. Hyam, R. Stacy, R. Shaw and J. Gittus.

The Berkeley Nuclear Laboratories

It is clear from the above discussion of the achievements of Hesketh and Heald and Speight that from the mid-sixties onwards Harwell, and other laboratories of the UKAEA, were no longer the only UK centres of excellence for research on nuclear materials. The Materials Division in the CEGB's Berkeley Laboratories, may have been newly-born and a progeny of Harwell, but it was already clamouring for attention and challenging the supremacy of the older organisation. The Berkeley Laboratories came into being directly as a result of Sir Christopher Hinton's insistence, on being invited to become chairman of the newly-formed CEGB in 1957, that he must have 'his own laboratory'.

The establishment of the Berkeley Laboratories was oppossed on the grounds that they duplicated expensive facilities and effort which already existed within the UKAEA. In fact as the reactors became operational the problems, many of which were related to shortfalls in the performance of materials, multiplied. Dealing with these and carrying out the related research not only exhausted Berkeley Laboratories effort but also that of the CEGB's other two national laboratories, Marchwood and Leatherhead, and its Regional laboratories. The magnitude of the task facing the CEGB should not be underestimated – at one point the Board was not only the owner of the largest concentration of nuclear reactors in the world, it was generating more nuclear electricity than the combined total output of the rest of the 'free' (ie non-communist) world.

As mentioned earlier, Leonard Rotherham was the CEGB Board Member for Research and although he had never worked at Harwell he was in complete sympathy with Finniston's policy of allowing his researchers to study practical materials problems at a fundamental level, a sort of curiosity-driven progress towards a practical solution. As if to reinforce the Harwell tradition for the small management team in Berkeley's Materials Division, Rotherham chose three Harwell scientists Kingsley Williamson, Geoffrey Greenwood and John Buddery as Section Heads, each of whom reported to the Division Head, A.T. (Trevor) Churchman who was, in effect, another Harwell scientist, having been seconded there from the AEI Laboratories at Aldermaston Court.

To make sure that the CEGB research programme was carried out at the highest level, Rotherham sought help from a number of Fellows (or future Fellows) of the Royal Society. Alan Cottrell acted as an advisor and took an especial interest in the work of Berkeley's Materials Division, Churchman having been one of his first research students at Birmingham University. Later Cottrell chaired the Electricity Council Research Committee which monitored, amongst others, the performance of the Materials Division. Christopher Kearton and Alan Brian Pippard (later Sir Brian, 1920–; FRS 1956) played an important part in the working of the Committee and in advising on the research programme. During this period Rotherham became President of both the Institution of Metallurgists and the Institute of Metals and served on the Council of the Royal Society, so creative links with the learned societies were maintained. Publication of research papers was strongly encouraged.

The initially quite small team of metallurgists in the Materials Division at Berkeley was faced with a formidable task right at the start of the 60s. The first of the civil Magnox stations, Berkeley and Bradwell, were soon to be commissioned and no less than four, later five, consortia were constructing, or competing to construct, seven more stations in England and Wales (the bailiwick of the CEGB) and another in Scotland. Although again much attention was given to monitoring fuel element performance it was also realised that at least these components were routinely discharged and improved designs could readily be introduced should they be necessary. In contrast, the steel components, in particular the pressure vessels, could not be replaced, and repaired only with great difficulty, so problems with these could be life-limiting, as could shortfalls with the graphite moderator.

G.K. Williamson had then every encouragement to continue his interest in radiation damage of graphite and in this work he was ably assisted by G.M. Jenkins and W.T. Eeles, and his old collaborator Clive Baker. At the same time Buddery formed a powerful team of chemists to study such processes as the radiolitic corrosion of graphite and E. Welch organised a programme of monitoring the in–pile corrosion of graphite specimens placed inside the core during the last stages of construction of the various reactors. When Buddery left to form his own company he was replaced by John Antill, another immigrant from Harwell. He, and later Derek Dominey, recruited a very talented team of chemists, including Kay Simpson and Geoffrey Allan.

Williamson also extended his experience of electron microscopy by obtaining funds to purchase a 1MeV Hitachi high voltage electron microscope (HVEM). Such an instrument, it was thought, would be invaluable for carrying out in situ radiation damage studies and for doing electron transmission studies on high atomic number materials, such as uranium. J. Buswell, S. Fisher, J. Harbottle and D. Norris demonstrated the usefulness of the HVEM by irradiating samples of Nimonic PE16 and type 316 stainless steel and measuring their propensity to swell (this work was carried out with the benefit of close collaboration with Harwell).

Brian Edmondson utilised the greater penetrating power of the HVEM to study by transmission dislocation arrangements in uranium and, at a later stage in his career, he turned his attention to the integrity of pressure vessel steels. In order to study fracture in scaled-down models of pressure vessels Edmondson commissioned the construction of fracture-testing facilities in a near-by disused quarry. He subsequently became a recognised authority on the fracture of pressure vessel steels and this expertise proved to be invaluable when the CEGB eventually decided to purchase a Pressurised Water Reactor (PWR). Edmondson subsequently became Director of the Laboratories and finally was appointed Safety Officer for the whole of the Utility. B.C. Masters, another scientist who joined the Berkeley Laboratories as a research scientist specialising in electron microscopy, also became Director of the Laboratories.

Greenwood was responsible for setting up a fuel element monitoring programme but also found the time to collaborate in a number of successful research activities. He developed a theory of irradiation swelling in uranium in collaboration with M.V. Speight, carried out studies of the structure of irradiated uranium with E.C. Sykes, and developed a theory with R.H. Johnson to account for enhanced creep, leading to superplastic flow, during cycling through an allotropic boundary in uranium. He developed a model for the possible role of an externally–applied hydrostatic pressure on high temperature creep fracture and supporting evidence for his ideas were obtained by R.T. Ratcliffe, a member of his team.

John Edwin Harris (1932–; FRS 1988), another member of Greenwood's group, collaborated with R.B. Jones to analyse the wealth of creep data on the Magnox AL80 canning alloy (Mg/0.8%Al) and themselves carried out creep tests on the material. They concluded that Herring–Nabarro (diffusion) creep might well be the dominant deformation mode in some reactor magnesium components (this was subsequently observed by V. Haddrell, the first time this fundamental deformation mode had been proved to have occurred in an engineering component). Harris and Jones subsequently found that, surprisingly, precipitate particles in the microstructure inhibited diffusion creep and they collaborated with Greenwood in proposing a mechanism for this inhibition. Jones subsequently produced the first experimental evidence for the occurrence of Coble creep, that is to say deformation achieved by movement of vacancies along the grain boundaries. I. Crossland extended Jones's work on Coble creep.

At a somewhat later stage Greenwood recruited a young, but already outstanding material scientist, G.B. Gibbs who, among other topics studied the interactions between oxidation and mechanical properties. Greenwood left Berkeley in 1965 to become a Research Manager at the Electricity Council Research Centre at Chester though he left there a year later to become Professor of Metallurgy at Sheffield University. His first two students, W. Beere and B. Burton turned out to be outstanding scientists and both eventually gravitated towards the Berkeley Laboratories to continue the tradition started by Greenwood.

When the nuclear reactors swung into full operation, problems arose with such frequency that much of the laboratory had to put to one side their longterm studies and work on more immediate concerns. An early problem was the occurrence of 'fast bursts' that is a fuel element suddenly failing because it suffered a leak which although below the limit of detection of the burst cartridge detection gear nevertheless was large enough to cause the almost complete oxidation of the uranium fuel bar. In such circumstances the resultant volume increase due to oxidation could suddenly lead to a rapid failure sometimes requiring the closedown of the reactor. The solution was higher standards of fuel element manufacture and improved detection and discharge procedures.

Curiously enough the same phenomenon which caused the fast bursts, the expansion that takes place on oxidation or corrosion of a metal, was the primary cause of two other major problems – the occurrence with the fuel elements of 'fin waving' and the 'oxide jacking' of steel bolts and other components. 'Fin waving' was the name given to the severe distortion of some of the magnesium alloy fins incorporated into the design of the Magnox fuel elements to improve heat transfer. This was found to be due to the stresses generated by the expansive force of corrosion deforming the metal substrate. To compensate for the loss in fuel element heat transfer, two of the reactors had to be de-rated.

The corrosion and fracture (again due to oxide expansion) of steel bolts in some of the reactors was a far more serious matter. The operating temperatures in all the civil stations (except Berkeley which already operated at a low temperature) had to be reduced with an associated overall loss in power of some 1000 MW (e), equal to a quarter of the combined total design output of the Magnox system. This was a permanent restriction and made this particular example of oxide jacking the most serious materials problem experienced with the Magnox reactors.

Following the identification of this problem a Working Group under the chairmanship of Alan Cottrell recommended the installation of additional emergency shut-down systems for those reactors having steel pressure vessels (that is all the reactors except those at Oldbury and Wylfa which had concrete pressure vessels). The shut down devices involved the ability to inject into the core during an emergency large quantities of small boron balls, or boron dust, to damp-down the reactivity.

It is ironic that with all the high quality materials research effort being devoted in the early days to the study of esoteric, irradiation-induced, potentially damaging processes, it was conventional rusting which proved to be the most troublesome to the reactor operators.

The move to AGRs – technical aspects

When the decision was taken in 1965 for the CEGB to order an Advanced Gas Cooled Reactor (AGR) for Dungeness B, the laboratories were already doing research on the radically-different AGR fuel element, which consists of enriched uranium oxide pellets protected from attack by the coolant by being enclosed in a stainless steel can or 'pin' (plans to use beryllium as canning alloy had had to be abandoned due to the element's inherent brittleness and toxicity). These reactors were again carbon dioxide cooled and graphite moderated though were more compact and operated at a higher temperature than the Magnox stations.

A single stainless steel pin in an AGR can be identified with a single fuel element in the Magnox system, and this fact can be used to illustrate an important drawback with the AGR. When a fuel element failure is indicated in a Magnox channel (which generally contains eight elements) the elements can be removed individually and the failed one identified during the discharging process. Subsequently, if it is considered desirable, the discharged sound elements can be reloaded. In contrast, when a failed-pin-signal occurs with an AGR channel, a whole stringer containing typically 288 fuel pins must be discharged and all must be discarded. This clearly demands a much higher degree of fuel pin integrity.

Another cause for concern was that the radiolytic corrosion of the graphite moderator by the carbon dioxide coolant could dictate and restrict the operating life of the reactor. Fortunately, although the operating conditions were less severe in the Magnox reactors, a strong research team had been built up over the years who were well capable of taking on board the AGR graphite problems.

A problem the AGRs did *not* have was the possibility of rapid failure of its pressure vessel. As already mentioned, in common with the last two Magnox stations, the steel pressure vessels of the earlier Magnox design had been replaced by vessels constructed from prestressed concrete, the catastrophic failure of which could be deemed inconceivable. This took some of the pressure off Edmondson and his collaborators who were working on steel pressure vessel fracture, but a considerable programme was still required bearing in mind the number of pressure-vessel reactors still in operation and the cataclysmic consequences should rapid fracture occur. Of particular concern was the possibility that under prolonged irradiation mild steels are known to become progressively embrittled.

A great deal of nuclear research was also carried out at the CEGB's other two national laboratories – Marchwood and Leatherhead. It was at Leatherhead during the 1960s that two young material scientists, John Frederick Knott (1938–; FRS 1990) and Edwin Smith (1931–; FRS 1996) established themselves as two of the rising stars of their generation in the general field of deformation and fracture of metals and alloys. Knott had graduated in metallurgy at Sheffield and obtained his PhD at Cambridge where he came under the inspiring influence of Alan Cottrell. One of Knott's notable achievements was that he had elucidated the role of microstructure in the initation and propagation of cracks and showed how this understanding can permit the design of structures which are tolerant to defects in service. His work on stress relief cracking, on ductile crack initiation in pressure vessels, and on fracture processes near weld joints has been generally recognised as being of high importance. Following his time at Leatherhead, Knott returned to Cambridge and then moved to Birmingham University to become Feeney Professor of Metallurgy.

Smith graduated in mathematics at Nottingham but then moved on to the home of metallurgy, Sheffield, and with Bruce Alexander Bilby (1922–; FRS 1977) and others became well known for his work on continuous distribution of dislocations. After early fracture work using the concept of crack opening displacement, studies of wedge cracking led to his quantitative account of the effect of second phase particles on the cleavage fracture of steels. This understanding has led to his work in the USA and Canada as well as this country, on the safety of nuclear reactor pressure vessels and pipework. He has also become interested in pellet/clad interactions in nuclear fuel and stress corrosion cracking and crack arrest and crack growth in nuclear materials. He is widely consulted on the general subject of brittleness and fracture in metal stuctures and is generally regarded as an international expert in the field. In 1968 Smith left Leatherhead to become Professor of Metallurgy at Manchester University (UMIST).

Choice of reactor system – the big mistake

Many scientists, when they observe Britain's relative decline as an industrial nation, console themselves with the thought that if only scientists were in a position to make the major industrial/commercial decisions, all would be well and the UK would prosper. In fact the history of decision-making in the nuclear field in this country during the past half century should explode this comfortable belief. Almost without exception the leaders of the nuclear industry have been scientists of world class, and yet the annals of decision-making with respect to civil nuclear power is a catalogue of missed opportunities and major errors of judgement. Perhaps the biggest mistake can be traced to the decision in 1965 to base this country's nuclear future on the AGR, which was a design unique to this country and for which the only operating experience related to that of a completely inappropriate so-called prototype – the Windscale AGR.

It is necessary to go back a stage or two to be able to put the choice of the AGR in perspective. As discussed earlier, the decision in 1947 to manufacture a British atomic bomb, coupled with the absence of nuclear facilities and materials (no means of enriching uranium or producing heavy water for example) almost inevitably demanded the construction of a graphite moderated, gas-cooled, natural uranium reactor for plutonium production. It was the next step, taken in March 1953, which was controversial – the evolution from the Windscale air-cooled pile to the dual-purpose, plutonium producing plus electricity generating, Calder Magnox reactors. According to Margaret Gowing, many Harwell scientists opposed the move to the Magnox system because they could see no long-term development potential for gas-cooled reactors (the Americans had already come to this conclusion). The same critics strongly opposed the decision to construct a prototype AGR in 1957, but this 'negative view' was not allowed to surface in case it punctured the euphoria surrounding Britain's 'world lead' in nuclear technology.

Even when every benefit of hindsight is taken into account, the events surrounding the choice of an AGR for Dungeness B in 1965, and the decision to commission Atomic Power Constructions to build it, remain bizarre. When in 1964 the CEGB invited tenders for a second nuclear station at Dungeness, to be known as Dungeness B, it represented the start of the UK's second nuclear programme and was to consist of the construction of reactors more advanced than the Magnox design. The AEA strongly supported their own system, the AGR, but the CEGB insisted that consideration also be given to American Light Water Reactors (LWRs) that is to say the Boiling Water Reactor (BWR) and the Pressurised Water Reactor (PWR). Two of the three consortia offered the generating boards a choice of an AGR or an American LWR, but the third, Atomic Powers Construction Ltd (APC), restricted its bid to an AGR design. The AEA, fearing that LWRs might displace the AGR for the whole of the CEGB's 'Second Nuclear Phase', collaborated with APC to put forward a more advanced AGR design than that originally tendered or requested by the generating boards.

By comparing the predicted performance of this futuristic, entirely unproven, AGR, with an older version of a BWR, APC won the contract. In reality, the claims for the proposed 'commercial' 600 MW (e) AGR with a concrete pressure vessel were based on extrapolation of limited operational experience with the minute 30 MW (e) 'protoype', the Windscale AGR (WAGR). This reactor had the additional disadvantage, as far as being a pre-cursor is concerned, of having a steel pressure vessel. The claim that the shift to concrete vessels for the AGRs could be justified on the basis of experience with the two concrete-vessel Magnox stations at Oldbury and Wylfa, were premature and unrealistic because neither station was even close to operating at the time the Dungeness B decision was taken.

Soon after the start of construction of Dungeness B the shortcomings in the design of the station began to manifest themselves in mistakes and delays. In due course APC folded and over twenty years were to elapse before the station was finally completed. Delays with later AGR stations, starting with Hinkley Point B and Hunterston B, although not as serious as at Dungeness, compounded the loss in confidence with the reactor system and confirmed that serious mistakes in policy had been made. Potential foreign customers noted these difficulties and shortcomings and, not surprisingly, the British consortia bids to sell AGRs in competition with LWRs were unsuccessful in Belgium, Germany, Greece, Spain and Italy. Indeed no export order for an AGR was achieved. There is no doubt that the Dungeness B fiasco ended Britain's hopes of becoming a major force in the then-expanding international trade in nuclear reactor.

To add to Britain's problems the collaborative OECD programme of work on developing a Gas-Cooled High Temperature Reactor (GCHTR), also known as 'Dragon', fell well behind competing programmes on similar systems in Germany and America and in 1976 the project was abandoned. The AEA's Fast Breeder Reactor programme too was losing ground to the French nuclear industry and in any case, by the mid 70s, it was becoming abundantly clear that the fast reactor would not become viable as an electricity producer at the very least until decades into the future.

If the AGR was proving, to say the least, disappointing and the GCHTR removed from consideration then the only other thermal reactor system Britain had under development in the early 70s was the Steam Generating Heavy Water Reactor (SGHWR). Although it was yet another indigenous reactor of a unique design which had only operated as a small (100 MW (e)) prototype it had the political advantages of being British and a water reactor and these two facts carried great weight with a Labour government which was embarrassed by a previous Labour adminstration's unquestioned enthusiasm for British technology in the shape of the AGR and the new consensus in favour of water systems.

In 1974 the CEGB, whatever reservations they themselves might have, were required to order SGHWRs up to a combined capacity of 4000 MW (e) in the period to 1978. Even the AEA, who had developed the system had their doubts. In July 1976 it became known that the Chairman of the AEA, John McGregor Hill (later Sir John, 1921–; FRS 1981), had written to the Minister in charge, Anthony Wedgwood Benn, that the SGHWR gave no clear-cut advantages over other systems and the programme should be abandoned and the CEGB should choose the PWR, or failing that the AGR.

In accordance with this advice, in January 1978 the Government authorised the CEGB and the South of Scotland Electricity Board to order one AGR station each; the CEGB station, to be at Heysham in Lancashire was planned to be commissioned in 1987 and the SSEB station at Torness in Lothian in 1986. In addition, and this was seen as the ground–breaking decision, the CEGB was, at long last, given permission to order a PWR, providing the necessary clearances were obtained. It was decided that the PWR would be the B station alongside the old Magnox station at Sizewell. The Sizewell B Enquiry started at Aldeburgh in January 1983, and the station commissioned some twelve years later. The station has been operating successfully, but plans for it to be the first of a tranche of similar stations have been aborted, partly as a result of the harsher economic criteria following privatisation of the industry.

Before discussing the controversy which surrounded the choice of the PWR, which itself was dominated by metallurgical factors, it is instructive to examine to what extent the problems with the various earlier reactor system were due to shortcomings in their materials of construction. One common thread is immediately apparent and that is that a material's ability to resist corrosive attack by the reactor coolant is of dominant importance. It has also become very evident that the behaviour of materials in small prototype reactors is a very poor guide to the performance of similar materials in the more highly rated cores of commercial systems where temperatures and pressures are so much higher.

To illustrate these points, two of the most severe operating problems with the commercial Magnox reactors were both due to vital components being corroded in the gas stream. The first to be observed was the distortion of the fuel element fins (known as 'fin waving') and the second was the expansive rusting and consequent fracture of in-core steel components. Neither of these 'oxide jacking' phenomena occurred in the much lower-rated Calder and Chapel Cross prototype reactors.

In addition to this, the radiolytic corrosion of the graphite moderator at Calder and Chapel Cross reactors proved a poor guide to the extent of these process in the commercial stations. Similarly, corrosion of the moderators in the AGR prototype at Windscale did not alert the scientists to the fact that corrosion of the moderator in the commercial AGRs could well be a life-limiting feature. As already mentioned, the Windscale prototype did not have a concrete pressure vessel so none of the problems associated with the insulation of the AGR concrete pressure vessel came to light during its period of operation. Neither of course did the Windscale prototype have its boiler located inside the pressure vessel so the difficulties due to undue corrosion of its 9% chromium-steel boiler-tubes did not come to light during the development stage. Having the boiler inside the pressure vessel did of course make maintenance and repair of the boiler enormously more complicated.

Walter Marshall and the PWR

Walter Charles Marshall (later Lord Marshall, 1932–1996; FRS 1971), was one of the cleverest and most remarkable people to achieve high office in the British nuclear industry. He came from a modest home in South Wales and left his grammar school in Cardiff to study mathematical physics at Birmingham University. After graduating he was fortunate enough to work with Rudolph Peierls and gained his PhD for work on 'Antiferromagnetism and neutron scattering from ferromagnets'. He was recruited into the Theoretical Physics Division at Harwell by Brian Hilton Flowers (later Lord Flowers, (1924–; FRS 1961) in 1954. After a period in the USA, but still only 28, he became Head of Theoretical Physics at Harwell. Unusually, as he ascended the promotion ladder to high office, he kept up his interest in, and contributions to, fundamental research, mostly related to the magnetic properties of the solid state. He was appointed Director of Harwell in 1968, and was the youngest FRS at the time of his election.

During this period Harwell's role was changing – a number of other laboratories were competing for diminishing funds for nuclear work (including of course the Berkeley Nuclear Laboratories). Throughout his life Marshall was a visionary and wanted to establish Harwell as a national centre for materials research. (It might have been a better idea for it to have evolved under his leadership into a postgraduate science university along the lines of MIT). He also proposed the construction at Harwell of a high magnetic field laboratory and a high flux neutron beam reactor, but was frustrated with both these ambitions. Instead he was required to direct his formidable abilities towards making Harwell less involved in fundamental work and more geared towards solving the practical problems thrown up by British industry. He became Deputy-Chairman of the UKAEA in 1975 and Chairman in 1981.

In 1974 Marshall was appointed Chief Scientist in the Department of Energy and became interested in such subjects as combined heat and power, energy conservation and offshore energy technology. He retained his enthusiasm for nuclear power and became even more convinced that the right course for Britain was to construct PWRs. However, the Minister at the Department of Energy, Anthony Wedgwood Benn, from a starting point of enthusiasm for nuclear power had switched to being opposed to its continued expansion, particularly if this involved buying American technology. The result was inevitable – in 1977, Marshall was dismissed, and, deeply hurt, returned to his full-time duties at Harwell.

By the mid 70s the leading authorities in the CEGB, like Marshall, were convinced that the PWR was the best system for Britain, though the Scottish Board remained strangely attached to the AGR. However, the Government Chief Scientific Adviser, Sir Alan Cottrell, had some worries about the safety of the PWR pressure vessel. It was true that the UK had much experience of manufacturing and operating nuclear reactors with steel vessels (ie the early Magnox reactors and Windscale AGR) but the ambient pressures within these gas-cooled reactors were much lower than those of the PWR and hence required much thinner pressure vessels.

To appreciate why thinner vessels are in most cases safer than thicker ones it is necessary to understand what is meant by the term 'critical length' of a pre-existing crack. It is the length above which the crack will spontaneously propagate rapidly when subject to a design or fault stress. If this critical length is greater than the thickness of the vessel then it will leak, and thus give a warning of the fault, before catastrophic failure can occur (known as the 'leak before break' safeguard). Contrariwise, if the critical length is less than the wall thickness a rapid failure without prior warning is possible. The fact that irradiation embrittles pressure vessel steels is an added complication. For the Magnox vessels 'leak before break' will occur, but for the PWR pressure vessel the critical crack length (~2.5 cm) is much less than the wall thickness. It follows that unless inspection techniques during manufacture can guarantee that no such defect exists, catastrophic failure during service is in principle possible.

Sir John Hill, as chairman of the UKAEA, invited Marshall to set up a study group to review the integrity of PWR pressure vessels. Marshall set about forming his group and began calculating the probabilities of failure under PWR operating, fault and emergency conditions taking into account the estimated distribution of crack sizes in vessels about to enter service, the distribution of material properties and the possibility of crack growth during service. His group published a review for external assessment in 1976 and another in 1982 which contained no less than 57 recommendations. Cottrell was finally convinced that, subject to a very intensive ultrasonic inspection during manufacture, Sizewell B's pressure vessel would be safe. (Incidentally Marshall's Study Group has been extended and is still operating under the chairmanship of Sir Peter Bernhard Hirsch.)

Marshall was appointed Chairman of the CEGB in 1982 so was centrally placed to push ahead with preparations with Sizewell B. He found that the industry, backed by the Thatcher government were preoccupied with other things, mainly preparing what was seen to be an inevitable national coal strike within the next few years. By stockpiling coal beforehand, making nuclear work flat out and burning £2b of oil in the much more expensive oil fired plants, Marshall and his team kept the lights burning and the miners were defeated.

By 1985 Marshall was riding high as head of one of the country's leading industries, and all seemed set fair for him to realise his ambition for the construction of a whole series of PWRs beyond Sizewell B. However, the government had determined to privatise virtually all of the great nationally–owned industries and utilities. Marshall may not have been philosophically opposed to privatisation *per se*, providing the CEGB was kept as a single unit. Marshall was convinced that only a very large, albeit monopolistic, electrical utility could accommodate and possibly expand the nuclear component.

Unfortunately for Marshall's point of view, British Telecom and British Gas had already been privatised as single entities and this had resulted in a storm of protests from a large number of Conservative MPs. The backbenchers argued (reasonably) that the point of privatisation was to create competition and this could only be achieved by breaking up the monopolistic nationalised industries and creating two or more independent and competing companies. It was one of those unfortunate accidents of history that had the electricity industry been privatised before British Telecom, or even before British Gas, then it might well have been able to preserve its monolithic organisational structure. As it was it became inevitable that the CEGB would be broken up – the National Grid would be hived off and become the property of the 12 Area Boards and the main part of the CEGB split up into two or more separate generating companies.

From Mrs Thatcher's autobiography *The Downing Street Years* it is evident that there were disagreements amongst the Ministers who had an involvement in energy matters. One view was probably content to accept Marshall's single-company approach, another involved a determination to introduce 'competition' by breaking the CEGB up into two or more companies, while a third view favoured the creation of four or five generating companies. Mrs Thatcher finally agreed to splitting the generating part of the CEGB into two unequal parts. The larger company, referred to as 'Big G' later to become National Power, would become responsible for 70% of the generating capacity, and the smaller company ('Little G' and then PowerGen) would have the remaining 30%, all conventional, generating capacity. Marshall, very reluctantly agreed to become chairman of Big G.

Things did not work out as planned. During the run-up to privatisation a reassessment was made of the costs of decommissioning the ageing Magnox plant and this was about twice hitherto supposed. This made the privatisation unattractive to the City so the Magnox reactors were withdrawn from the sale. Later, somewhat similar escalations in the estimates of the decommissioning liabilities of the AGRs led to the withdrawal of all the nuclear plant from the sale.

Marshall had no wish to be head of a non-nuclear generating company and resigned. It was a sad ending to an outstanding career, though he continued to play a useful role internationally as Executive Chairman of a new nuclear organisation, formed as a result of the Chernobyl accident, called the World Association of Nuclear Operators (WANO). Its function was to monitor and assist nuclear reactor operators throughout the world to make sure high standards of safety are universally obtained.

To take over the nuclear stations two new publicly–owned companies were formed, Nuclear Electric and Scottish Nuclear. An old friend of Marshall, a colleague from his Harwell days, John Gordon Collier (1935–1995; FRS 1990) became Chairman of Nuclear Electric and so successful has been the performance of Sizewell B and the AGRs that they were subsequently privatised, though their total selling price was less than the cost of constructing Sizewell B alone. In effect the new company, British Energy plc, was presented with the seven AGR stations at zero cost. The reactors are currently performing well.

The Royal Society and privatisation

Perhaps the main attraction of privatisation to the government was the money it raised for the Treasury. By selling to the public assets it already owned, sufficient cash could be generated to make it unnecessary to raise taxes to compensate for a general failure to curb public expenditure. This was electorally very advantageous especially as the public assets were sold off so cheaply that handsome profits were made by millions of small investors.

The other motivation is the deeply held conviction held by right-wing politicians that private enterprise businesses operating in an environment where market forces dominate will *always* be more efficient and effective than the same business carried out by a state-owned monopoly. The idea has wide appeal, – a number of other countries around the world are following Britain's example and privatisating their State-owned industries. In some cases, notably in Russia and other former Soviet Union countries, ill-considered and badly executed privatisation programmes have had disastrous consequences.

When discussing the advantages or otherwise of privatisation it is important to recall how the industries came to be under public control and ownership in the first place. In the case of electricity in Britain, as discussed in Chapter 8, it was an almost inevitable consequence of shortfalls in previous private and municipal ownerships. At the outset of both World Wars Britain's electrical industries, both for manufacturing and supply, were a poor shadow of their German counterparts, and only by urgent reform did Britain manage to avoid a disastrous shortfall in munition production due to electricity shortages. At the end of the Second World War there were in Britain no less than 560 separate supply undertakings operating with 300 generating stations half of which had outputs of less than 10 MW (e). The passing of the 1947 Electricity Act, authorising full scale nationalisation of the generation, transmission and distribution of electricity for public supply, seemed at the time entirely sensible, indeed inevitable.

Even if it accepted that privatisation does indeed lead to a harsher more competitive environment and hence, perhaps, to a more efficient industry, there are what could be called second-order effects that can have unforseen and important implications. The most significant of these consequential factors is what has become known as 'short-termism', that is to say industries have to respond to City pressure for a rapid return on their capital investments. Such pressures have destroyed at a stroke the possiblity of constructing more nuclear stations to replace the ageing Magnox stations as they are progressively retired. This is ironic as traditionally the Conservative Party in general, and Mrs Thatcher in particular, are strong supporters, in principle, of nuclear power.

Planning permission had been obtained to construct a PWR at Hinkley Point and the industry had hopes of building another PWR at Sizewell. These ambitions had to be abandoned. In his valedictory address to the British Nuclear Energy Society, Marshall pointed out that whereas the CEGB had assumed an 8% cost of capital over a 40-year life for its PWR programme, the banks wished to recover their capital in 20 or even in 10 years. In free market economics no credit could be given for ensuring diversity of supply by maintaining a nuclear construction industry or for any lessening of our generation of greenhouse gases.

The scientific community in general and Fellows of the Royal Society in particular are broadly supportive of maintaining the nuclear component of our generating capacity at about its present level, instead of the gradual run-down, and loss in morale which is an inevitable consequence of following strict market economics. Another casualty of short-termism has been the research function of the CEGB which had universally been regarded as world class and indeed the most comprehensive in support of any country's electrical system.

The CEGB had three 'national' laboratories, Marchwood, Leatherhead and Berkeley Nuclear Laboratories. Marchwood was the first casualty; almost 80% of its work had been related to the nuclear programme; of especial note was its fundamental work on welding technology and remote inspection and repair of nuclear reactors. It was nevertheless allocated to PowerGen which had no nuclear stations! When asked for an explanation it was pointed out that 'Marchwood was the right size' (Leatherhead was too large for the smaller of the two companies and Berkeley Laboratories were entirely nuclear so had to stay with the nuclear reactors). This was a good illustration of the cavalier fashion major decisions were made in the self-inflicted frenetic circumstances which prevailed during the privatization exercise. Almost inevitably, Marchwood was soon closed down and the country lost a national asset.

The Leatherhead Laboratories formed the jewel in the CEGB's crown. It was the largest and oldest of the three laboratories being noted for its work on all aspects of electrical generation, distribution and supply. Its Materials Division contained one of the strongest teams of metallurgists and materials scientists in the country and its group working on atmospheric pollution was building up a world-wide reputation. Although the environmental and operating problems were the same before as after privatisation, the Leatherhead Laboratories were closed down. The Berkeley Laboratories still exist as buildings but one can no longer refer to themselves as 'Laboratories'; they have transmogrified into the 'Berkeley Centre' and accommodate most of the administration staff of the new state-owned company, Magnox Electric plc, charged with operating the ageing Magnox stations. More recently Magnox Electric has become a Division of British Nuclear Fuels.

Most of the major scientific disciplines were affected severely by the run down of research following electrical privatisation but metallurgy and materials science was hit particularly hard. The CEGB had built up strong links with the universities with numerous contracts, studentships and fellowships. Of more importance though was the encouragement of long-term research in its own laboratories; in materials science it was by far the largest concentration in Britain of research effort working entirely within an industrial context. This dominance was enhanced with the run-down of the Materials Division at Harwell, and the restrictions on research activity in the steel companies due to the decline in profitability of that industry. Other privatisations, for example of British Gas had also resulted in run-downs of industry-based research. To add to these difficulties there has been the closedown of the National Engineering Laboratories and the changes at the National Physical Laboratory.

An incidental consequence of privatisation could be that the UK loses its international position as one of the leading countries in the field of materials research and development. An upsetting aspect of this is that in none of the biographies of those involved in the decisions to privatise electricity, nor in any of the books written about this development (including the book *The British Electricity Experiment Privatisation* edited by John Surrey) has *research* received a single mention.

Another casualty of electricity privatisation was the coal industry. A continued gradual decline in the demand for coal was inevitable and accepted but the largely unnecessary acceleration in the *rate* of decline resulting from electricity privatisation has caused much suffering in the mining communities. During the 1970s and 1980s the CEGB had been closing coal fired plant almost every year to make room for the brand new coal stations which had been ordered during the boom years of the 1960s. Accordingly National Power and PowerGen had inherited efficient coal stations of the latest design. Then, almost coincident with privatisation, came the liberalisation of gas, that is to say the European-wide ban on using natural gas to generate electricity was set aside.

Even though Combined Cycle Gas Turbine (CCGT) stations were quick and easy to build and to operate, were extremely efficient and the fuel relatively cheap, when their capital cost was taken into account they could not compete with an existing ex-CEGB coal station where, in effect, the capital cost had been written off. Nevertheless, because many of the senior staff of the 12 new Regional Electricity Companies had spent much of their former working lives under the shadow of the all-powerful CEGB, they were determined to forge a degree of independence from the two new generating companies, National Power and PowerGen. Hence even though there was no economic case for the creation of new gas stations, they nevertheless did invest in such constructions, usually in collaboration with one or other of the newcomers to the industry, the Private Generators. The 'Big Two' (National Power and PowerGen) fearing that they would be left behind in some new enterprise also began to invest in gas.

This was 'The Rush for Gas' which as well as signalling a virtual elimination of the indigenous coal industry caused great concern that Britain's own resources of natural gas would soon be exhausted. In addition a huge over-capacity in generating plant was signalled – the National Grid forecast a 'plant margin' of nearly 60% by the late 1990s compared to the usual 'safe' level of 20%.

In December 1991 a director of the National Grid Company, John Harold Horlock (later Sir John, 1928–; FRS 1976), drew attention to these highly disturbing developments in the electrical industry in a letter to the President of the Royal Society. He pointed out that during the next three or four years it was planned that the National Grid Company would be connecting some 15 GW (e) of new generating plant (mostly gas) representing about a third of maximum in the UK. He suggested a group of Fellows of the Society consider the government's fuel policy in preparation for the 1994 review.

A meeting of those Fellows who were knowledgeable about energy was convened and the gravity of the situation was confirmed, but even assuming all the wisdom of the Society was brought to bear and communicated to the government, little could be done at this late stage to assist the situation. A reading of the autobiographies of the Ministers involved in the early stages of electricity privatisation – Thatcher, Parkinson and Lawson – makes clear that in the frenetic circumstances of the times, and the imminence of the general election, there was no consultation with the Royal Society about the wisdom of their policy.

The decision–making procedures were well summarised by the author Andrew Holmes in his outstanding Financial Times book *Privatising British Electricity*,

> Now that the system is up and running, there is a temptation to look for underlying logic in the way events have moved and to invent retrospectively a degree of calculation which was never there. Starting from a position of unproven beliefs, the privatisation and restructuring which went with it, proceeded at a breakneck speed through a series of improvisations, compromises and responses to emergencies which arose along the way.

Nuclear waste

In 1976 the public's worry about the nuclear waste problem was crystallised and apparently supported by the findings of a Royal Commission Report entitled *Nuclear Power and the Environment* produced under the Chairmanship of Professor Brian Flowers. It has become a landmark document and is universally known as *The Flowers Report*. It contains the following major conclusion:

> There should be no commitment to a large programme of nuclear fission until it has been demonstrated beyond reasonable doubt that a method exists to ensure the safe containment of long-lived, highly radioactive waste for the indefinite future.

At first sight this appears to be an entirely reasonable recommendation, but in reality it is open to criticism. Even at that time there existed sufficient knowledge of geology and the migration rates of active species to define a satisfactory disposal route for all forms of nuclear waste. Where research was still needed was to provide information on the *best* choice of disposal route amongst many competing and acceptable alternatives; in no sense was the future of civil nuclear energy *dependent* upon new discoveries in the field of waste management, as implied by the Flowers' recommendation. There are two other important points that can be made.

Firstly, the final definition of the preferred disposal route is not a matter of extreme urgency, as may be implied by the Flowers recommendation. By definition, HLW is so radioactive it produces considerable amounts of heat which have to be removed and it is simpler to do this on the surface than say a kilometre underground. Hence, as was always recognised by the cognoscenti, there is much advantage in delaying final burial of HLW for at least fifty, and more probably a hundred, years. Secondly, the choice of a site for a geological depository will certainly generate intense opposition from those who live nearby so inevitably the government will be involved in the final decision, and it will be a political rather than a technical decision.

As far as finding a final resting place for high level waste is concerned, the industry, taking up Flowers' recommendation, planned a series of test drillings at various sites throughout the United Kingdom but these plans generated such a storm of local objections to the planning applications that the government of the day ordered it to be abandoned. In announcing the change in policy to Parliament in December 1981, the Minister for Local Government and Environmental Services quoted from the Second Report of the Radioactive Waste Management Advisory Committee (RAWMAC), which advocated that serious consideration be given to storing high-level waste at the surface in solid form for 50 years or possibly much longer.

One positive outcome of the Flowers' Report was the eventual formation by the nuclear industry and government of a separate organisation whose sole objective was to design and locate a site for the disposal of Intermediate Level Nuclear Waste (ILW) together with a small quantity of Low Level Waste (LLW) which had alpha radiation levels too high for the Drigg near-surface disposal site. This organisation was named

Nirex (Nuclear Industry Radioactive Waste Executive) and set about looking for suitable sites and attempting to carry out test drillings. These activities generated a huge backlash from the general public living in areas which were potential sites, and all planning applications were bitterly opposed.

In the period 1983–87 a number of locations for shallow disposition were investigated for short-lived Low Level Waste (LLW) and Intermediate Level Waste (ILW), and by the end of this period four sites were under active consideration. However, stemming from political factors associated with the 1987 General Election all four sites were withdrawn from considersation. The government also decided that both ILW and LLW should be buried together in deep depositories – a strange decision based on the assumptions that the marginal costs of burying LLW deep underground were no higher than the cost of near-surface disposal.

In 1988 Nirex, continuing its search for a LLW/HLW depository embarked on a national consultation with about 50,000 questionnaires circulated. Thousands of replies were received and the importance of retrievability was emphasised. The next stage was a programme of site-selection, Starting with 500 potential sites these were whittled down by 1989 to 12 deep sites. Since each of these 12 met the government guidelines on operational and post closure safety, selection was based on a wide range of safety and non–safety parameters: ease of transport, site accessibility, land ownership, national parks' regulations, special rules covering areas of outstanding national beauty, environmental impact, public acceptability etc.. Attention became focused on two sites, Dounreay and Sellafield, and in 1991 Sellafield was chosen; although it was not ideal, it was good enough and met all the government criteria. At the time the initial decision to focus on Sellafield was made Nirex had the support of the responsible Minister and that of the independent Radioactive Waste Management Advisory Committee (RAWMAC).

In June 1994 Nirex applied for permission to build an underground laboratory to investigate further the Sellafield site; this was to be known as the Rock Characterisation Facility (RCF). In December Cumbria County Council rejected the planning application, even though the laboratory was just an exploratory stage in the process. Nirex appealed and a public enquiry was initiated which started in September 1995; in late 1996 the Inspector reported, supporting the rejection. In March 1997, the Environmental Minister announced that he supported the rejection of Nirex's application. These events led to substantial costs; Nirex and the industry had spent something like £400 million, a not inconsiderable sum; the Planning Enquiry alone cost Nirex £20 million. Such enforced expenditures in Britain and in other countries have contributed to an ending of the development of nuclear electricity in the Western democracies.

In 1994 Nirex invited the Royal Society to comment on its research programme relating to the Rock Characteristic Facility and a study group was formed under the chairmanship of Sir Alan Muir Wood (1921–; FRS 1980). For Nirex itself to initiate such a study was generally welcomed as a sign that the organisation was willing to seek external inputs to its scientific strategy. The working group made helpful recommendations and was impressed by Nirex's scientific programme in support of their main objectives.

The group did reserve their position on whether or not Sellafield has the right potential as a site for a deep depository but this hesitance may reflect a lack of appreciation of the socio/political difficulties surrounding such a choice.

Reprocessing and the plutonium 'mountain'

It was German researchers in the late 30s who obtained the first hints of the existence of a chemical element of atomic number 93 or 94 but had not obtained sufficient material to determine its nature. Working with the physicist Edwin Mattison McMillan (1907–1991) in 1940 Glenn Seaborg discovered again element 94 and named it plutonium (appropriately *pluto* is a synonym of Hades!). A year later they prepared and identified plutonium 239, an isotope which fissions with slow neutrons. In 1942 Seaborg moved to the metallurgical laboratory at Chicago University to devise a chemical process for separating plutonium from uranium and in September 1942 he produced the first significant fraction of a gram of pure plutonium. Seaborg was then the first 'reprocessor' and for his discovery of plutonium he and McMillan were awarded the Nobel Prize for Chemistry in 1951. Over a period of 20 years he was involved in the creation and identification of synthetic transuranic elements from 95 (americium) to 106 (seaborgium). Having discovered more metals than anyone else Seaborg deserves alongside Humphry Davy, to have a place in the pantheon of metallurgical heroes.

In Britain reprocessing irradiated fuel, firstly from the Windscale piles and then from Calder Hall and Chapel Cross, was an essential part of the production of plutonium for Britain's first atom bombs. Other products of reprocessing – depleted uranium and fission products went into storage. This is how it came about that at the outset of the civil nuclear programme a fully developed reprocessing programme was in full swing. It seemed common sense to continue this activity and reprocess the spent civil fuel, particularly as some of the fuel from the initial charges of the Magnox reactors would be lightly irradiated and provide plutonium suitable for military use.

On the other hand, reprocessing spent *civil* fuel to provide plutonium for atom bombs was seen by many to be reprehensible and inconsistent with the spirit of the Atoms for Peace movement. Another, more acceptable, motive was needed to justify reprocessing, and such an objective was conveniently to hand; it was to provide the plutonium for the initial charges of an unspecified number of future UK fast reactors. In the mid 50s and early 60s it was thought that nuclear power would burgeon in many countries and there would hence arise a global shortage of high quality uranium ore to fuel the reactors. As is well known, the drawback with thermal nuclear reactors is that they can only consume (initially at least) the minor fissile isotope U-235, which constitutes just 0.7% of natural uranium, so this type of reactor uses the earth's uranium resources quite prodigally.

It is at this point that the advantages of a mix of thermal and fast reactors become apparent. The plutonium generated in thermal reactors can be extracted by reprocessing their spent fuel and this plutonium can then be manufactured into fuel elements to form the initial charges of fast reactors. Around the core of these reactors could be placed a blanket of natural (or depleted) uranium whose U-238 atoms would absorb fast neutrons and be converted to fissile plutonium, by amounts in fact that could exceed the quantity consumed in the core itself. Alternatively a 'breeding blanket' of thorium could be used which would partially transmute to U-233, another fissile isotope of uranium which can also be used to fuel fast reactors.

By manipulating such a combination of reactor systems and concurrent reprocessing, useful heat would be derived, not just from 0.7% of the atoms in natural uranium, but from close to all the atoms. By so doing it would increase, at a stroke, by almost two orders of magnitude, the amount of energy potentially available from the world's reserves of uranium ores. To this very considerable figure could be added the earth's enormous reserves of thorium ores. In short, the successful development of the fast reactor would mean that sufficient uranium and thorium ores would be available to supply mankind with close to all his energy requirements for many millennia and, in effect, its adoption would solve the energy crisis.

The fast reactor sounds too good to be true; closer inspection uncovers serious shortcomings. The widespread adoption of this system would require the construction of numerous reprocessing plants throughout the world and yet still involve extensive transport of plutonium to and from plants and nuclear stations, and this would represent a severe security risk. Indeed in 1977 the newly-elected US President, Jimmy Carter, let it be known that in his view the proliferation risks associated with fast reactors outweighed their potential benefits. He set in motion stopping the construction of their Clinch River fast reactor, and at the same time announced that he would order the ending of reprocessing of all spent civil fuel in the US. Indeed he extended this ban on the reprocessing to all spent nuclear fuel of American origin even if irradiated in foreign-owned reactors. Other countries which also ban reprocessing are Canada, Sweden and Finland.

An even greater disadvantage with the fast reactor was that although several prototypes had been constructed and operated they were far more complex, and hence far more expensive, than thermal nuclear reactors. Moreover, the anticipated near-exhaustion of uranium ore reserves did not take place so nuclear fuel remains relatively cheap. These trends were apparent as long ago as the second half of the 1970s when it became clear that a viable fast reactor lay many decades into the future. The system looks even more uneconomic today. Indeed some authorities argue that it would be cheaper to tap the vast reserves of uranium in sea water than to construct fast reactors. If this be the case, and current research into extracting uranium from sea water is very encouraging, then the construction of fast reactors will be delayed at least until the 22nd century.

In spite of these unpromising auguries BNFL went ahead with the construction of its Thermal Oxide Reprocessing Plant (THORP) at Sellafield and actively sought foreign customers. These were not hard to find. In Germany and Japan in particular internal political pressures within each country, albeit with varying motives, were forcing the nuclear operators to do *something* with their spent fuel. Sending the fuel to Britain or France bought *time* and to achieve this breathing space the two countries were prepared to pay high reprocessing charges and also, in effect, pay an appreciable proportion of the capital costs of the THORP plant.

An implied condition, as far as BNFL gaining foreign reprocessing orders were concerned, was that the CEGB would also have to reprocess its Magnox and Advanced Gas Cooled Reactor spent fuel. In spite of disquiet amongst some CEGB staff, and the fact that reprocessing charges would greatly exceed the cost of dry storage and that the fast reactor was already proving unviable (and in any case the reprocessing programme would eventually produce an order of magnitude more plutonium than was needed to launch any conceivable fast reactor programme) BNFL managed to get government backing for its policy and the CEGB was 'persuaded' to conform with the reprocessing policy.

A consequence of this largely-unnecessary reprocessing is that BNFL has now accumulated at Sellafield almost 50 tonnes of separated plutonium of British origin and another 35 tonnes separated from or incorporated in the spent fuel of foreign customers. The agreement with overseas customers is that the plutonium and highly active waste extracted from the imported spent fuel must be returned to the country of origin. Such transport of radioactive materials, and particularly plutonium, is of course highly undesirable.

The distinguished nuclear analyst, Professor Frank von Hippel (*Nature*, 1998, **394**, 415) suggested that instead of fulfilling its contracts to reprocess the spent fuel for its foreign customers, BNFL should simply store the fuel, ie *not* reprocess it, but instead in due course send to these customers the agreed amount of plutonium, but taken from its existing plutonium store. It should do the same with the high level waste and thereby fulfil its contractual obligations to the satisfaction of its customers and at the same time keeping its indigenous critics contented. This is certainly an ingenious idea but any scheme which might accelerate movement of plutonium from its very secure store at Sellafield to foreign parts would have to be viewed with a great deal of caution.

In December 1996 the Council of the Royal Society set up a working group under the chairmanship of Sir Ronald Mason (1930–; FRS 1975) to study the science needed for management of this civil plutonium stockpile and they completed their studies in February 1998. The group has pointed out that at a time when the USA and Former Soviet Union countries are taking strenuous efforts to convert their military plutonium to the 'spent fuel standard' it is inappropriate for BNFL to add to its already large stockpile of separated 'civil' plutonium. They further point out that if reprocessing continues as presently planned the inventory of civil plutonium will reach 150 tonnes by 2010 and then constitute *two thirds of the predicted global separated civil inventory*. The Royal Society have recommended that the British government reviews the strategy and options for stabilising and then reducing the stockpile.

The group list the options in the medium term as (a) adopting the once-through cycle for standard fuel (i.e. stop reprocessing), (b) recycling MOX fuel in British and foreign reactors (MOX, Mixed OXide is a mixture of plutonium oxide and uranium oxide in such proportions that it can be used as fuel in thermal reactors) and (c) constructing a new British LWR specifically designed to burn MOX.

The Royal Society Report (*Management of Separated Plutonium*, 1998) has been widely welcomed for being a well-written and thorough account drawing attention to a most important issue. The report though does have some shortcomings as has been pointed out by von Hippel *(loc cit)* and more comprehensively by Dr Frank Barnaby *(Medicine Conflict and Survival* 1998, **14**, p197). It is difficult in the text and tables to know whether or not the authors are referring to plutonium oxide (as it is stored at Sellafield) or to metallic plutonium. This confusion can lead to misleading comparisons. For example the Royal Society report suggests that the critical mass of reactor grade plutonium may be an *order of magnitude* greater than that of weapons-grade plutonium. In fact the only way the Royal Society group could have arrived at this conclusion is by comparing the critical mass of reactor plutonium *oxide* with that of *metallic* weapons-grade plutonium. Comparing like with like, the critical mass of reactor grade plutonium (13 kg) is almost the same as that of weapons grade plutonium (11 kg). In the present context this is not of course a trivial matter. Incidentally, it is a relatively simple matter to reduce plutonium oxide to the metallic state. To some extent the study redeems itself by going on to say that reactor-grade plutonium is 'a plausible target for determined terrorist groups or states wanting to make nuclear weapons.'

It is noteworthy that at no point in the report is BNFL criticised for following its aggressive policy of advocating commercial reprocessing. In fact it would have been difficult for the group to make such criticisms bearing in mind that a number of its members had formerly supported the policy on which they were now expected to pass judgement. This highlights a very fundamental difficulty for the Royal Society. When it is required to judge a controversial scientific/technological issue it can either choose from among its Fellows those with the greatest expertise, but who almost inevitably have some previous involvement in the issue, or scientists with lesser expertise but who bring to bear judgements unencumbered by previous involvements or commitments.

It can be argued that it is almost invariably better to opt for conspicuous independents. In the case of the reprocessing issue almost any random group of FRSs could quite readily take on board and understand the issues involved and be capable of forming a wise judgement. This lesson seems to have been learned. The most recent study on nuclear power is a joint Royal Society/Royal Academy of Engineering effort with the Royal Society's Treasurer, Sir Eric Ash as chairman. Sir Eric, wisely, had chosen a talented study group with experience extending over a wide range, most having an interest in nuclear matters but few having previous executive responsibilities in the area. They have in fact produced a balanced, and very-well received report, (*Nuclear Energy: The Future Climate, Royal Society/Royal Academy of Engineering*, June 1999). The group were very concerned about the consequences of projected increases in emissions of greenhouse gases and do not believe that these can be contained to reasonable levels by increased generation efficiency, conservation and use of renewables alone, but will also require the construction of new nuclear plant within the next two decades. They further urge that the planning of new nuclear plant be carried out early enough to allow nuclear power to play its full, long-term role in national energy policy. They were very critical of the government's plan to levy a tax on all energy generation indiscriminately. Any such tax should, in their view, be based on net carbon dioxide emitted.

CHAPTER 19

The Royal Society and the Industrial Revolutions

Introduction

A typical member of the international community of research scientists living in almost any part of the world will have heard of the Royal Society, will know something of its long and distinguished history, and most probably regard it as a benevolent and valuable organisation with an important role in contemporary affairs. In contrast, at home, amongst the British public at large, the Society is much less well known. In conversation, should a scientist happen to mention that he or she is a Fellow of the Royal Society, the most likely reaction will be confusion and the question 'Fellow of the Royal Society of what?' Should it be explained that the Society is the oldest and one of the most distinguished of the world's National Academies of Science, then the most likely response will be 'What is the Royal Society for?' In this final chapter an attempt is made to answer this reasonable question, particulary in respect to the power the Society has to influence UK science and technological policy, making reference where appropriate to the subjects of metallurgy and materials.

Throughout this volume the history of the Royal Society has been linked to the evolution of metallurgy – the science and technology of metals and alloys. In fact of course the beginning of man's use of metals pre-dates the origin of the Society by several millennia; it extends as far back as the dawn of civilisation. The lustre, formability and strength of metals filled primitive man with awe and stimulated his curiosity. The extraction of metals from their ores created an interest in chemical reactions as did the observed increase in weight on calcination of metals. This led to the concept of the existence of phlogiston which in due course gave way to ideas on the composition of air. The technology of mining ores and their treatment and the assaying of the product were other activities which eventually stimulated scientific investigation.

The desire to turn base metal into gold was the foundation of alchemy, the forerunner of chemistry. The two English philosophers, William Gilbert and Francis Bacon, who around the beginning of the 17th century created the intellectual framework which nurtured in due course the fledgling Royal Society, held artisans who worked with metals in the highest regard. In Bacon's Solomon's House in his 'The New Atlantis', a special area was set aside for metallurgical experiments.

Note: Dates of Fellows and Foreign Members of the Royal Society, and for some other individuals referred to in this chapter are listed in the Index.

Bacon classified experiments into those which simply satisfied curiosity by providing new knowledge or 'light', *experimenta lucifera*, and those which produced technological information of immediate usefulness, *experimenta fructifera*. The widely held view that the Royal Society concentrated exclusively on the former cannot survive examination, certainly during the 17th century this was far from being the case. A survey by the distinguished historian and writer, Robert Merton, of 300 problems studied by the Royal Society during four of its earliest years revealed that almost two thirds were topics with some potential practical application, many of which related to mining and navigation.

There were attempts to set up sub-groups of Society members with the Baconian aims of applying science to the improvement of techniques in arts and manufacture. These efforts were though largely unsuccessful as the active members with relative experience were too few and too overworked with little contact with the practitioners in the various applied fields of endeavour. This inability to penetrate and influence industrial activities persisted certainly to recent times, as will be discussed.

While many of the ambitions and efforts of the fledgling Society were worthy and on the whole sensible, it is difficult to take seriously many of the investigations listed in Thomas Sprat's *History of the Royal Society* published in 1667:

> 'Experiments of destroying Mites by several fumes: of the equivocal Generation of Insects: of feeding a Carp in the Air: of making insects with Cheese, and Sack: of killing Water-Newts, touching their skin, with Vinegar, Pitch, or Mercury: of Spiders not being inchanted by a Circle of Unicorns-horn, of Irish Earth, laid round about it.'

John Ziman (*Reliable Knowledge*, 1978) though strenuously defends these studies, arguing that every old wives' tale had to be tested, for until there was a consensus on such matters, there was no basis for biological science.

Taking the broader view it is undeniable that in promoting scientific discovery the Society succeeded brilliantly during its early years and made England the focal point for the scientific revolution. How could it have been otherwise having as Fellows such luminaries as Wren, Boyle, Hooke and Halley and, above all, Newton, the Shakespeare of science. Much has been made of the support given by Charles II who throughout his life was fascinated by science and thoroughly approved of the Society's astronomical studies because of the value of such work in the improvement of ship navigation. He was though scornful of the Fellows' attempts to 'weigh air', thereby revealing his own inadequate understanding of the new science.

Charles granted the Society a Royal Charter and presented it with a Mace and a Charter Book, but it is significant that no Royal cash was forthcoming and this created a separation between government and science which has existed right up to the present time. Almost the only source of money for the Society was the subscriptions of its Fellows, many of whom were poor payers, with the result that the organisation was frequently chronically short of funds. This was in marked contrast to King Louis XIV's gift of a salary and pension to the Academicians of the French Academy of Science established in 1666. This generosity attracted to Paris distinguished foreign scientists to work in the Academy, the first being Christiaan Huygens, the Dutch physicist and astronomer. Among his many achievements were a widely regarded wave theory of light and his invention of the pendulum clock. His assistant Denis Papin was a pioneer of the atmospheric steam engine (see Chapter 2 and below).

The absence of a state subsidy did mean that in theory the Royal Society had more freedom in its choice of subjects to study, or as it was quaintly described 'it could follow the whims of the afternoon'. In addition, the Society could in principle enjoy more independence and the power to challenge government decisions and policy relating to scientific issues. There is though little evidence that it has taken advantage of this freedom (the dispute with George III over the design of lightning conductors being an exception). On the other hand, the French government has a long record of creating excellent educational establishments specialising in training students for careers in science and technology and providing economic conditions favourable for the development of science-based industries. These enlightened initiatives may well be a consequence of the economic ties and closer relationships between French academicians and their government.

Scientific Academies in the 18th century

The century started and ended with strong Royal Society Presidents. Newton was elected PRS in 1703 and carried out his presidential duties with considerable skill and dedication. He remained in office until his death in 1727, when he was succeeded by the redoubtable Sir Hans Sloane who served until 1741. Sir Joseph Banks was elected in 1778 and ruled the Society with a rod of iron for 42 years. Sadly though, between Sloane and Banks, from 1741 to 1778, there were elected a succession of undistinguished men. In fact the complete century was a lean period for the Society, whose performance was characterised by a lack of professionalism. It became more of a social club than a scientific body, the non-scientist 'social' Fellows exceeding the number involved in scientific endeavours by a factor two or more. The absence of government financial sponsorship played a part in this decline; Fellows with scientific interests needed private funds to be able to pursue their scientific studies.

In 1957 the author D.S.L. Cardwell (*The Organisation of Science in England*) calculated that of the 106 scientists active in Britain during the 18th century, about half were amateurs. In contrast in France, from the beginning of the century to the Revolution of 1789, scientific and technological activity enjoyed strong government support, often working through the National Academy of Science. Prizes were offered for discoveries such as the power propulsion of boats, institutes were set up to study such subjects as hydraulics and a major survey was initiated to record the then current states of French Arts and Crafts.

Scientific activity was complemented in France by other intellectual and cultural intercourse. It was in Paris that Jean-Jacques Rousseau met Denis Diderot and was persuaded by him to compete for the prize offered by the Academy of Dijon for an essay on the question of whether the restoration of the sciences and the arts had tended to purify morals. Rousseau won the competition and became famous as a result of this success. In that year (1750) Diderot published the first volume of *The Encyclopedia;* the 17th and last volume together with the books of plates were published by 1772. For this mammoth task Diderot had recruited the help of numerous scientists, philosophers and litterateurs with the ultimate objective of bringing out the essential principles and applications of every art and science. Diderot's accounts and drawings of various metal-producing furnaces have been of enormous assistance to metallurgical historians, as demonstrated in the books of R.F. Tylecote (eg *A History of Metallurgy*, 1976). In French society the gap between science and the arts was more effectively bridged and evidence for this was the success of Voltaire's popular account of Newton's achievements in his book *Elements de la Philosophie de Newton.*

The Royal Society may have been passing through an inactive period but this did not stop it, and the French Academy, being adopted as models for the creation of other Scientific Academies throughout Europe during the 18th century. The most significant of these creations was the Berlin Academy, the brainchild of Gottfried Wilhelm von Leibniz. He envisaged that the Academy would act as a catalyst promoting the unity of Germany which would become strong through the industrial application of 'useful' science. He was supported in these endeavours by the Elector of the Palatinate, later King Frederick I of Brandenburg-Prussia, who reigned from 1701 to 1713.

As Elector during the 1690s Frederick I had embarked on a programme of reforms of the cultural life which included the establishment of the University of Halle and various other colleges. Among the reforms were the revision of the Prussian calendar and Liebniz proposed that the new Academy would produce accurate almanacs and calendars and receive the income from their sale, and thus become economically self-sufficient. Unfortunately this source of income did not materialise, the cash was pocketed by the state, and it was not until the accession of Frederick II the Great to the throne in 1740 that the Berlin Academy was placed on a firm financial basis.

The 'revised' Academy was to be modelled more on the French Academy rather than the Royal Society. In particular it was to receive financial support from the government making it possible to attract foreign scientists and mathematicians. Prominent amongst these distinguished immigrants was Leonhard Euler, the remarkable Swiss mathematician who moved to Berlin at the request of Frederick II in 1741 and became Director of the Academy in 1744. He remained in Berlin until 1766 when he was recalled to Russia by Empress Catherine the Great to become Director of the St Petersburg Academy of Science. Frederick II was also successful in persuading the French mathematician and Newtonian, Pierre Maupertuis, to become President of the Academy in 1746. Maupertuis, famous for formulating the principle in mechanics of 'least action', was a successful administrator and established the scientific eminence of the Berlin Academy to rival that of the French Academy and the Royal Society.

Back in England, the shortcomings of the Royal Society were not in any way made good by the activities of the Universities of Oxford and Cambridge which were still steeped in studying and debating the theology of the Middle Ages. According to Patrick

Blackett (President of the Royal Society, 1965–70); during this period, the time of the Industrial Revolution, universities slept in Britain, and remained asleep for at least another century. It is true that Edinburgh was in a cultural ferment as the Scottish Enlightenment unfolded – Edinburgh's Medical Society evolved into its Philosophical Society and in 1783 into the Royal Society of Edinburgh.

If British science was in such poor shape then why did the Industrial Revolution take place in Britain and not in France, Germany or Italy? Reference must again be made to Blackett and his address on 12th October 1966 in Birmingham to mark the bicentenary of the founding of the Lunar Society. As President of the Royal Society he considered that the explanation must lie mainly in political and social factors rather than to the quality of British scientific activity at that time. He argued that contemporaneously the Italian state was too fragmented to stage an industrial revolution. Germany too was divided politically into many parts and this was particularly unfortunate because the strength of its mining and metallurgy industries and technologies would have made it a favourable candidate to host the Revolution.

France was not fragmented and its government supported science to a much greater degree than Britain, and as early as the 1720s Réaumur had been unlocking the secrets of the role of carbon in steel. Blackett concluded that what held France back was simply the absence of a strong middle class with an entrepreneurial spirit and money to invest in fledgling industries.

There were essentially two technological developments in Britain at the end of the 17th century and the beginning of the 18th which were of fundamental importance in this context. The first was metallurgical – the discovery by Abraham Darby I in about 1709 that the sulphur and phosphorus content of iron could be kept to an acceptably low level if coke were used as the fuel during smelting of iron in place of the expensive and increasingly scarce charcoal. There are those who would regard Darby's discovery, and its industrial exploitation at Coalbrookdale by him and his descendants, as the initiating event and initial fruition of the Industrial Revolution, yet none of the Darbys were elected FRS.

The second seminal development was the invention of the steam engine, and in this case, during the early stages of its development, the Royal Society *did* play an important part. As has been described in Chapter 2, the story starts with Denis Papin, a Frenchman who graduated in medicine at Angers University in 1669 and then became an assistant to the Dutch scientist, Christiaan Huygens at the French Academy of Science in Paris. Huygens had had the rather bizarre idea of developing an engine driven by multiple explosions of gunpowder. The power was to come from the expansive forces of the discharge followed by implosion and atmospheric pressure moving a piston to destroy the consequential vacuum.

Huygens's idea turned out to be both impracticable and dangerous but it caused Papin to think in terms of power being generated from atmospheric pressure reacting to the vacuum created by condensing steam – the principle of the first reciprocating steam engine. He put forward his idea in the following elegant and now historic paragraph:

> 'Since it is a property of water that a small quantity of it turned into vapour by heat has an elastic force like that of air, but upon cold supervening is again resolved into water, so that no trace of the said elastic force remains, I concluded that machines could be constructed wherein water, by the help of no very intense heat, and at little cost, could produce that perfect vacuum which could by no means be obtained by gunpowder.'

In fact one volume of water yields 1,300 volumes of steam so that a very high vacuum could indeed be achieved by condensing the steam back into water. Papin made a small device which although not having any possibility of practical application, did illustrate the principle of the atmospheric engine.

Subsequently, Thomas Savery invented a more practical atmospheric engine which he demonstrated to the Royal Society in 1699. The first practical atmospheric engine though was developed by a blacksmith from Devon, Thomas Newcomen and his assistant, the plumber John Calley; the first being installed at a colliery near Dudley Castle in Worcestershire in 1712. This engine was enormously successful; within just four years of its invention its use had spread to eight countries. Although Newcomen's engine may have owed little to Savery's discoveries, patents taken out by Savery were so comprehensive that royalties had to be paid to him on every Newcomen engine manufactured.

The first thorough scientific study of the Newcomen engine was carried out by John Smeaton in 1767. By measuring the quantity of water which it could raise one foot for each bushel of coal burned Smeaton derived a method of determining engine efficiency and used this measurement to assess the effect of modifications to the design. The true breakthrough occurred though when a model of a Newcomen engine was sent for repair to James Watt, the young instrument maker for Glasgow University. He soon realised that what was needed to increase efficiency was a separate 'condensing' chamber into which the steam could be released from the main cylinder and condensed thereby creating a vacuum in both the cylinder and condensing chamber. Subsequently Watt introduced other major innovations – making the engine double-acting for example, thereby doubling the engine's power. Other improvements included a 'parallel motion' mechanism, a governor to regulate the inflow of steam and a 'sun and planet' gear to convert the machine into a rotative engine and a separate vacuum pump for the condenser.

Watt's first partner was John Roebuck who manufactured his engines at his famous Carron Ironworks. The engines were considerably improved by having their cylinders bored by a refined technique previously developed by John Wilkinson for the production of cannons. When Roebuck ran into severe financial difficulties, Watt teamed up in 1774 with Matthew Boulton who manufactured his engines at the Soho factory in Birmingham. By the end of the century Boulton and Watt had sold about five hundred of their engines.

Technology, like science evolves. Watt constructed a successful steam engine because, like Newton, 'he stood on the shoulders of giants', in his case Papin, Savery, Newcomen and Smeaton. With the exception of Newcomen each of these innovators were nurtured, encouraged or rewarded by the Royal Society so it is quite wrong to imply, as some authorities have done, that the Society played no part in the early stages of the First Industrial Revolution. The Royal Society was also not slow to recognise the importance of the work of the members of the Lunar Society (see Chapter 2). 11 of its 14 members were elected FRS, in spite of the fact that only Joseph Priestley had made significant contributions to fundamental science.

Special mention should also be made of the metallurgical inventors and entrepreneurs, particularly the Darbys and John Roebuck and not forgetting the ironmaster John Wilkinson who was throughout his life obsessed by the metallurgy of iron and was buried in an iron coffin. Benjamin Huntsman's crucible process of 1742 and Henry Cort's puddling and rolling process of 1884 were both significant developments that owed nothing to science (what 'science' which existed was due to Réaumur who in 1722 was perhaps the first to realise that steel is iron with not too little, nor too much, carbon). The universities too were not entirely inactive, particularly the Scottish Universities. Watt gained much benefit from his position as instrument maker at Glasgow University and learned a great deal from Joseph Black, the university don who was the first to measure the latent heat of evaporation of water. Roebuck (whose contributions are frequently overlooked) had his interest in chemistry stimulated while a medical student at Edinburgh and was one of the many who benefited from Scotland's link with Holland, in particular with the University of Leyden.

It is also important to remember that the mechanical ingenuity of the British which became apparent at the beginning of the Industrial Revolution, did not spring suddenly into being in the 18th century but evolved from a great British tradition of manufacturing astronomical and navigation instruments, microscopes and clocks and watches (see Chapter 2). In all these activities the Royal Society was involved during the 17th and early 18th centuries. In fact high standards of mechanical ingenuity existed long before this; some medieval cathedral clocks involved the gyration of figures, for example. As early as 1596 Queen Elizabeth I sent as a present to the Sultan of Turkey an automatic organ built by a Lancashire clock-maker, Thomas Dallam, in which birds flapped their wings and sang, bells chimed, trumpets sounded and figures gyrated and which played a whole sequence of madrigals.

The 19th century

One of the major disadvantages of the absence of state funding for the Royal Society was its inability to finance its own laboratories. It is not then surprising that when Sir Benjamin Thompson (Count Rumford of the Holy Roman Empire) proposed setting up such a laboratory in London in 1799 as an integral part of what was to be the Royal Institution, he received the enthusiastic support of Sir Joseph Banks, then President of the Royal Society. Thompson was an American who had been elected to the Royal Society in 1781 for his work on gunpowder and endowed the Society's Rumford Medal. He had become wealthy by marrying a rich American widow (his second wife was Antoine Lavoisier's widow) and was handsomely rewarded during his appointment as Grand Chamberlain to the Elector of Bavaria. He had a great affection for England and clearly recognised that the Industrial Revolution here could not be a continued success without some means of training practical-minded and mechanical scientists and technicians. In his appeal for funds for the Institution he described its objectives as follows:

> '..diffusing the Knowledge and facilitating the general Introduction of useful mechanical Inventions and Improvements, and for teaching by Courses of Philosophical Lectures and Experiments the applications of Science to the common Purposes of Life.'

The huge success of the experiments of Humphry Davy and Michael Faraday which among other things launched the Electrical Age (see Chapter 8) speaks volumes for what the Royal Society lost in the 18th century by not having its own laboratory. Throughout the first few decades of the new century there were strong interactions between the Royal Institution and with industry, and applied research was encouraged. Davy invented the miners' safety lamp and carbon arc light and during the time he was President of the Royal Society assisted the Navy to improve the performance of copper sheathing on ships. Faraday developed improved glasses and investigated alloy steels and toured a number of foundries and other steel plant to explore ways in which the Institution could be of assistance to the metallurgical industry. Of course it was Faraday's discovery of electromagnetic induction leading to the development of practical dynamos and electric motors which had the most widespread industrial application: electric lighting for general use and for lighthouses, electroforming, electroplating, telegraphic communication and electric transport, to name but a few.

In 1800 Sir Joseph Banks was about halfway through his uniquely long Presidency, having being elected to that office in 1778 and eventually retiring in 1820. He was a well-respected botanist, having been made famous by his sea journeys with Cook. He was rich and well-connected and a friend of George III (who had by that time recovered from the unpleasantness of the dispute with the Society over the design of lightning conductors). Banks had been made a baronet in 1781 and a Privy Councillor in 1797 and was comfortably installed in spacious accommodation in Somerset House, to which the Society had moved in 1780. All seemed set fair and every prospect was pleasing.

Banks though was an autocratic President and tended to favour for election to the Society candidates who worked in the biological rather than the physical sciences. He considered that the Society should consist of two social classes; the working men of science, and those who from their position in society or fortune would become patrons of science, and of the Royal Society. In defence of this deferential policy it should be mentioned that his courtship of the rich and influential is but another manifestation of the poverty which lack of state funding brought to the Society's affairs. Nevertheless the election of rich but scientifically-illiterate gentry was to continue for many years and to cause acute irritation to many professional scientists, above all to the irascible, Charles Babbage.

Babbage was one of the most remarkable men of the 19th century. He was born in Devonshire in 1792 and in 1810 went to Cambridge to study mathematics where he became a close friend of John Frederick William Herschel who, like his father Frederick William Herschel, was to become a distinguished astronomer. The year after he left Cambridge Babbage published three papers on 'The Calculus of Functions', and on the strength of these he was elected FRS in 1816. He was obsessed by the possibility of constructing a super calculating machine. With the backing of the Royal Society he was able to persuade the government to invest £20,000 to construct a difference engine, but before it was completed he had a better idea – to manufacture an analytical engine; this was not completed before his death in 1871, but it is generally regarded as the forerunner of the modern computer.

When Banks announced his retirement in May 1820 due to failing health there were a number of potential contestants for the Presidency, including two Royal candidates. As soon as he heard of Bank's retirement Davy rushed home from Paris to throw his cap into the ring. Although he was by now both rich and a baronet, Davy's background was too humble to make him a suitable candidate for the Presidency in the view of many Fellows. To add to this, he was arrogant, intolerant and a snob (his general unpleasantness was manifest in 1824 when he tried to block the election to the Fellowship of his own protege, Michael Faraday). The youthful trio from Cambridge, Babbage, Herschel and George Peacock, were strongly opposed to Davy's election and supported the candidacy of William Hyde Wollaston, a Cambridge physician who had made original contributions to physiology, optics, chemistry and crystallography, and had become wealthy through his discovery of a method for making platinum malleable. His qualifications for the office were extremely impressive having been Secretary of the Society from 1804–16 and Copley Medallist in 1802.

Unfortunately he was a man of retiring disposition and did not really want to become President though he agreed to serve in that capacity during a short interregnum during 1820. He had no desire for social junketing but was instead fascinated by Oersted's discovery that year that there is a relationship between electricity and magnetism and simply wanted to get on and take the revelation further. In short Wollaston was a true scientist. He himself, somewhat reluctantly, supported Davy who was duly elected on 30th November 1820.

Davy suffered a stroke in late 1826 and was advised to travel abroad to a drier climate. In charge of the Society in his absence was Davis Gilbert, the Vice President and Treasurer of the Society; Gilbert was an authority on bridge design and had worked on Cornish steam engines and was a competent, but not brilliant, mathematical engineer. He had helped Babbage in obtaining government finance for his difference machine. After Davy officially resigned in July 1827 Gilbert was involved in an apparent attempt to having the then ex-MP, Robert Peel, elevated to the Presidency. Eventually Gilbert himself took on the office and remained President till 1830.

The election of 1830 was a close run thing. Augustus Frederick, Duke of Sussex and son of George III, received 119 votes and John Herschel, 111 votes. So Sussex was elected. Herschel was a friend of Babbage and no doubt his defeat by the noble Lord confirmed Babbage's poor opinion of the workings of the Society.

It was in that year, 1830, that Babbage published his highly controversial book *Reflections of the Decline of Science in England and some of its Causes* in which he attacked what he perceived as the corrupt state of the Royal Society. More perceptively though he drew attention to the lack of recognition paid to scientists by the British community at large. He contrasted this with the attitude to scientists in France and Germany where science was properly taught in the universities and where scientists could become Ministers of State. With the support of Sir David Brewster, physicist and critic of government neglect of science, Babbage and others formed the British Association for the Advancement of Science in 1831. This was supposed to compensate for the elitism and conservatism of the Royal Society. This started well and raised money for research grants and equipment, but soon acquired the British disease of respectability.

As far as Babbage's call for reform of the Royal Society is concerned, the Duke of Sussex turned out to be a much more liberal and reforming President than was expected, and when he resigned in 1838 the Society was far more open and more 'modern' than it had been in 1830. It was well on the way to responding positively to some of Babbage's strictures. Sussex was succeeded by the Marquess of Northampton as President, another nobleman and even less well qualified scientifically, but during his term of office, in February 1847, much of Babbage's and other critics wishes were fulfilled and only those with scientific accomplishments to their credit would in future be elected to the Society. It would take time to change but the age of the (almost) exclusively scientific Royal Society was launched.

Nevertheless the next President, elected in 1848, was William Parsons, Earl of Rosse, yet another nobleman, and this fact was regretted in 'The Atheneum'. However, the Earl was also an outstanding astronomer and during his term of office continued his observational studies using his own telescope with a 6ft (~1.8m) mirror, his major achievement being the discovery of spiral nebulae. When Rosse resigned in 1854 he was succeeded by another noble astronomer, Lord John Wrottesley, but when he resigned in 1858 the Council tried to persuade the country's most famous commoner-scientist, Michael Faraday, to become President. The 67 year old scientist, then in poor health, declined and Sir Benjamin Collins Brodie assumed the office, the first surgeon to become President. From 1861–71 a military man, General Sir Edward Sabine reigned. He wanted a noble successor and he himself approached the Duke of Argyll and Lord Salisbury but the Astronomer Royal, George Bidell Airy was elected, the only President of the century to be a man of modest means. He was succeeded in 1873 by Sir Joseph Dalton Hooker. He was the director of Kew and the first botanist to be elected President since Banks. The mathematician William Spottiswoode served from 1878 to 1883 and he was followed, rather reluctantly, by Thomas Huxley, 'Darwin's Bulldog'. As though to compensate for Huxley's vigorous atheism, the deeply religious George Gabriel Stokes succeeded him and served from 1885 to 1890. Stokes was a Cambridge mathematician and Lucasian Professor and although less well known than Huxley, was regarded as the more distinguished scientist. The century ended with more of a bang than a whimper with the presidencies of Lord Kelvin (1890–1895) and Lord Lister (1895–1900).

Summing up the 19th century

Considerable space has been allocated in this chapter to the history of the Royal Society during the 19th century, particularly the characters, beliefs and policies of its Presidents. The reason for this attention is that during this hundred year period the Society had been transformed from one where social graces and aristocratic connections were as important as scientific achievement, to one where the only important criterion for election was that the candidate was a practising and successful scientist. Some Presidents have resisted this change in philosophy but most have embraced it. Curiously enough, the most radical changes towards a more open, democratic and scientific Society came about under the Presidencies of courtly and noble Presidents.

It is in fact difficult to overestimate the importance of the 19th century in the evolution of the Society. In 1984 Dr Marie Boas Hall published her important book *All Scientists Now: The Royal Society in the Nineteenth Century* in the Foreword to which the then President of the Society, Sir Andrew Huxley, suggested that the history of the Society during the 19th century was almost as dramatic as the much better-known beginnings of the Society in the 17th century.

During this century the Society had also become, *de facto*, the country's National Academy of Science, responsible for advising the government on scientific issues. Reciprocally, Society members were beginning to have access to government ministers and might aspire to influencing government decisions on science policy. An event of pivotal importance bearing on the relationship between the Society and the government was the creation of the Government Grant in 1849, though the Society was quick to point out that any government money it might receive would be used for the purpose specified, and not be employed to defray the general running expenses of the Society itself. In other words the Society still cherished its freedom of action which came with financial independence.

In spite of many organisational vicissitudes, throughout the 19th century the Society's Fellows produced world-class science; from Davy, Dalton and Faraday to Darwin, Wallace, Lyell and Huxley, and then on to Maxwell, and Kelvin. Then, as a fitting climax, at a time when Lord Kelvin feared there was little more to discover in physics, J.J. Thomson in 1897 identified the electron, the fundamental particle the movement of which constituted the flow of electricity. The discovery of the electron could then be thought an appropriate end to what has sometimes been called 'the Electrical Century'. Thomson's discovery was also an important step in launching the 'new physics', which was eventually to lead to nuclear power and the Atomic Age. Was there anything left for the Royal Society to worry about in this best of all possible worlds?

As a matter of fact there was. Establishing a route along which advice can flow from Society to Government is only of use if the initiator knows what counsel to offer, and the receiver the wit to understand and act on the guidance given. Neither of these criteria seem to have been present as the 19th century drew to a close, as Britain failed to play a dominant part in what has become known as the science-based 'Second Industrial Revolution'. Being able to make recommendations to the King on the construction of clocks and interpreting astronomical data to help navigation, as was done by the Society in the 17th century and early 18th century, is one thing, making recommendations on the governance of a new, burgeoning, highly-technological, internationally competitive, industrial society, is quite another. It required knowledge in the political, social and economic fields as much as in the physical sciences and in these former areas the Society's Fellows possessed no particular expertise (and apparently neither did the various governments).

Apart from providing some encouragement for the evolution of the steam engine the Royal Society had not been much involved in the Industrial Revolution during the whole of the 18th century, and neither had the Universities of Oxford and Cambridge. The springboard of the industrial revolution was not the golden triangle of Oxford, Cambridge and London but Coalbrookdale, Leeds, Manchester, Birmingham and Glasgow and many other, mostly Northern, centres. (More often than not these were locations where deposits of coal and iron ore happened to co-exist.) This gulf between the Society and the practical Northerners who were busy transforming Britain into the world's most powerful industrial nation, perisisted throughout most of both the 18th and 19th centuries.

By 1800 the widespread adoption of the Watt steam engines in mining, metallurgy and textiles increasingly emphasised the need for improved transport of coal, ores, metals and manufactured goods. It seemed only logical to create steam-driven locomotives and the early stages of such a development took place in the mining industry, including the adoption of a 'permanent way' of metal rails. A major problem was that most steam engines were far too heavy to be capable of being employed in a locomotive. As early as 1801 Richard Trevithick demonstrated that a high-pressure engine would be needed and that further weight savings could be achieved by dispensing with the condenser and releasing the exhaust steam directly to the atmosphere. In 1804 he had produced the first railway locomotive, able to haul 10 tonnes and 70 people for 15 km on rails used by horse-trains at Penydarren Mines near Merthyr Tydfil in South Wales.

Many problems in the development of a commercially viable steam locomotive remained but most of these difficulties were eventually overcome by the remarkable George Stephenson, the self-taught mechanical engineer and son of a colliery fireman. His locomotive, the *Rocket*, constructed under the supervision of his son Robert, won the prize at the Rainhill trials of 1829, and the rest is history. During the next couple of decades, the locomotive evolved and a network of railways spread all over Britain, and during the remainder of the century over much of the rest of the world.

George Stephenson, like many of the 18/19th century pioneers came from a humble background and was poorly educated, but then science was not involved in his discoveries in any inspirational or creative sense. Neither Richard Trevithick nor George Stephenson were elected FRS, though George's son, Robert, was so honoured in 1849, the year following George's death. Astonishingly, the Institution of Civil Engineers (founded 1818) also refused to elect George as a member, and to find a home for this greatest of 19th century engineers the Institution of Mechanical Engineers was formed in 1847. These two Institutions can still be found occupying adjacent 'mausoleums' in

Westminster, symbols of the rejection of Monty Finniston's recommendation to unite the fifty or so engineering institutions into one coherent body (though the formation of the Council of Engineering Institutes in 1965 and the Fellowship of Engineering in 1976 have been tentative steps in the right direction).

Other pioneers of the Industrial Revolution who came from modest backgrounds and were unencumbered by a university education, include Joseph Bramah (carpenter), his one-time apprentice, Henry Maudslay (mechanic whose firm made the engine for I.K. Brunel's *Great Western*) and Joseph Whitworth, another mechanic and brilliant precision engineer who in later life became rich and philanthropic. The great Isambard Kingdom Brunel did not attend university but he did enjoy the advantage of having been educated in France. His election as FRS in 1830, at the early age of 24, was a recognition of his precocious genius.

During this period the wealth created by an expanding Empire was making Britain even richer (or so it was thought – see later). In addition, the absence of competition for the products of British industrial activity meant that during the first half of the 19th century there seemed no need for Britain's industrialists to worry overmuch about such matters as efficiency and high productivity and the potential threat from foreign competitors. That in large measure was the cause of Britain's undoing.

Apart from being London-based there were other aspects of the lives of Fellows which separated them from the ordinary working people. As the above account of the background and position in society of the Presidents has illustrated, many of the Fellows in the 19th century were rich, privileged and well connected at a time when the gap between rich and poor was, by present day standards, almost unimagineably large. To many Fellows the bustling industrial parts of Britain would seem stranger and more foreign than many of the exotic, far-flung foreign lands visited by the various Society-supported scientific expeditions.

Moreover, those Fellows who *were* scientifically active would be preoccupied by matters which would appear esoteric to the industrial worker and of little relevance to his everyday life and the welfare of his company. The Fellows would be excited by news of hitherto unknown flora and fauna discovered during the various voyages of exploration. They would be arguing about the age of the earth and the solar system and attempting to reconcile the latest findings of palaeontologists, geologists and cosmologists with biblical accounts of the earth's and man's creation. They would be fascinated by the observation of spiral nebulae, intrigued by the electromagnetic theory of light and puzzled by the news seeping through from the continent of the mysterious rays discovered by Röntgen and Becquerel.

In short, to use once again Newton's imagery, the 19th century FRSs could be thought of as strollers along the sea shore wondering how long it had taken the oceans to grind the stones into smooth pebbles. Occasionally they would find unrecorded flora and the odd fossil and this would prompt discussions on biological evolution and the age of the earth; the new cosmology would also be a subject for debate. While they were thus preoccupied the great ocean of technology lay undiscovered before them and within which a great storm was brewing which threatened to overwhelm the Society and the country, and change the world.

Britain's electrical industry to 1914

The above representation of the 19th century FRS as an impractical, ivory-towered, aesthete is fair in many instances, but there were exceptions where such a description would be unjust, even Swiftian. In the case of electricity, a number of the pioneering researchers *did* take an interest in the practical application of their discoveries. The carbon arc lamp had been invented by Davy but its use only became possible after the magneto had been developed following Faraday's discovery of electromagnetic induction. Faraday himself took much interest in the installation of magnetos and carbon arc lights in lighthouses and supervised such an installation at Blackwell in 1857. In the related field of telegraphy, the magnificent achievement in 1865/66 when Brunel's immense steamship the *Great Eastern* was used to lay the Atlantic cable, the whole enterprise would not have been technically possible without the active participation, and brilliance, of William Thomson (Lord Kelvin).

In the late 1870s Joseph Swan, an almost solitary researcher in his laboratory in the best British tradition, developed a carbon filament lamp which was as successful as those produced by the blitzkrieg attack on the problem by Thomas Edison and his many assistants in his Menlo Park 'industrial' laboratory. Swan went on to display his entrepreneurial skills by taking part in the creation of the world's first central, electrical power station at Godalming. Ferranti with his design of the huge Deptford station which was years ahead of his its time, and Parsons, the inventor of the steam turbine, had practical objectives from the outset and their contribution to electrical technology is unquestioned.

To illustrate Britain's contribution and potential strength in the general field of electricity all that is necessary is to list some of the names of Britain's 19th century electrical pioneers: Davy, Faraday, Sturgeon, Daniell, Joule, Wheatstone, Kelvin, Maxwell, Swan, Armstrong, Crompton, Hopkinson, Parsons, Rayleigh, Heaviside, Ferranti, Thompson and Merz. It constitutes a formidable assembly; their combined contributions should have established Britain as a world leader in electrical technology at the start of the 20th century. It was not to be.

One way to judge a country's industrial progress is to compare its performance with its foreign competitors – a tedious exercise often only of interest to other economic experts. In wartime, however, interest quickens, relative industrial performances are no longer the subject of academic study but are of vital and immediate importance upon which may depend the future existence of the state. There is nothing like a sentence of death for focusing the mind. Such a testing time came for Great Britain at the outbreak of war with Germany in 1914. How did the strength of Britain's electrical industry compare with Germany's and other participating countries at this time?

Astonishingly, Britain's electrical manufacturing industry in 1914 was only half the size of Germany's. Moreover, as discussed in Chapter 8, in the immediate pre-war period three quarters of Britain's electrical machine and electric lamp imports came from Germany and Britain also relied on that country for imports of large generator sets. In the period 1880–1914 nearly all the major developments in electrical engineering were made abroad. Britain's electrical supply side was equally archaic; in 1914 500 electrical power stations were operating in the UK but their combined output was only 75% of that supplying Chicago from only four units. In comparison with Chicago's four power stations London had 70 stations with a bewildering number of systems and control authorities. New York too was supplied by only four stations and similar concentrations had taken place in Detroit, Boston, Hamburg, Berlin and Paris.

It is clear that Britain had completely failed to take advantage of the very favourable position Britain's success in electrical discoveries and pioneering technological developments had provided during the first eighty years of the 19th century. This was of course very disturbing but electricity was one of the new 'high-tech' science-based industries; maybe applying the '1914 test' to a more traditional industry, such as iron and steel, would produce a more favourable result. This was, after all, a technology in which Britain had reigned supreme for a century and a half. This subject is now explored.

Britain's steel industry to 1914

J.D. Bernal has argued in his book *Science in History, Volume 2*, that the ready availability of iron and steel, and *the revolution in metallurgical technique* that this entailed, were factors in the Industrial Revolution *of comparable importance to the invention of textile machinery and and to the steam-engine*. As was the case with other developments which led to the Industrial Revolution, there was little science involved until right at the end of the 19th century when large-scale production of steel became necessary to feed the burgeoning needs of the expansion of railways and other manifestations of the later stages of the First Revolution.

One of the most important developments of the second half of the 19th century was the discovery in 1856 by Henry Bessemer that excess carbon in an iron/carbon melt could be burnt out by passing a stream of air through the melt itself (see Chapter 7). This oxidation process had the additional advantage of generating useful amounts of heat which raised the melt temperature and made it less viscous and easier to pour into moulds. This process was so obvious, and at the same time so ingenious, that inevitably it was copied or concurrently discovered in other countries. (The American metallurgist, William Kelly, independently, made the same discovery during experiments he started as early as 1846, but he has not received the credit he deserves. To rectify this the procedure is now sometimes referred to as the Bessemer–Kelly or Kelly–Bessemer process, depending on which side of the Atlantic the commentator happens to be!)

Many of Bessemer's early casts from his converter were of poor quality, largely as a consequence of his lack of understanding of the chemical processes involved in the steel-making process. He was considerably helped by the metallurgical expertise of Robert Mushet, the discoverer of tungsten steel, who introduced the idea of de-oxidation by means of a manganese-containing alloy. The Bessemer converter's main competitor for large-scale steel production was the ubiquitous Siemens' open hearth furnace, designed by William Siemens, the German-born British inventor. Siemens knew a little more chemistry than Bessemer but neither of them could sort out why it was that they could not produce good steel from pig iron derived from ore that had a high phosphorus content. This ruled out major iron ore deposits, particularly in the USA, Germany, France and Belgium.

A method of removing phosphorus from Bessemer (and Siemens) steel was developed not by the steel companies, which did little research and development, but by a talented amateur with a passion for science, Sydney Gilchrist Thomas. He was a police clerk who in 1870 attended evening classes in chemistry at Birkbeck Institution and it was here that he heard about the phosphorus problem with the Bessemer converter. During the conversion process the stream of air oxidises the phosphorus impurity in the metal charge to form phosphorus pentoxide but this does not react with either the acidic flux or the siliceous furnace lining. At the high temperatures achieved (up to 1600 °C) the pentoxide is reduced and the phosphorus returns to the melt and becomes a highly undesirable impurity in the final product.

It was already known that if lime is added to the iron melt a basic flux is formed and this reacts chemically with, and thereby removes, the acidic phosphorus oxides. Unfortunately, at that time the only known furnace lining capable of withstanding the high temperatures involved was pure sandstone ie silica and, being acidic, could quickly be destroyed by reacting with a basic flux. Clearly what was needed was a basic lining and as a result of carrying out experiments with his cousin, Percy Gilchrist, over a period of nine months in 1877–8 at Blaenavon steelworks in South Wales, a successful basic lining containing rammed dolomite was produced.

Adopting basic fluxes and furnace linings was as applicable to the Siemens open hearth process as it was to the Bessemer converter. Indeed the open hearth furnaces became much more widely used than Bessemer converters because the melt in the latter picked up nitrogen from the injected air. A high nitrogen content in the steel turned out to be particularly embrittling and deleterious in welded structures and the problem was only finally resolved as recently as the 1950s with the substitution of oxygen for air as the gas forced through the melt.

Considerable attention has been given here to the Gilchrist–Thomas discovery for the following reasons. Firstly it illustrates very vividly how successful the scientific method can be in finding a solution to an apparently intractable industrial problem. Secondly, it is hard to overestimate the importance of the Gilchrist–Thomas discovery – it in fact inaugurated the age of steel. As good quality steel became abundantly available it rapidly replaced wood and cast iron for structural purposes (replacement of wrought iron took a little longer and was not always progressive, for example while the Forth Bridge, completed in 1890, was constructed of Siemens open hearth steel, the Eiffel Tower, erected nine years later, was fabricated from wrought iron).

Thirdly, and most importantly in the context of this chapter, it turned out that, paradoxically, the Gilchrist–Thomas innovation, although developed in Britain with a British furnace in mind, actually harmed Britain's *relative* position in the league of steel producing countries. The reason for this is that Britain, unlike Belgium, France and Germany, had quite good reserves of low phosphorus ores, so the new process had less impact, relatively speaking, on the volume of production of our steel industry. It was Britain's main rival, Germany, which benefited most; in 1871 it had annexed from France a large part of the province of Lorraine with its huge reserves of high phosphorus ores, known as *minette*. The integration of these ores with the coal of the Ruhr led to the creation of one of the largest steel-producing regions in the world. As steel is so important in armaments production, these developments were to have a profound influence on the relative strengths of Britain and Germany during the two World wars of the 20th century.

A consequence of these developments, and shortcomings in the organisation of technology in Britain, was that at the outbreak of war in 1914 Britain's steel industry was only half the size of Germany's. This could have had devastating consequences on the outcome of the hostilities. Were it not for huge imports of steel from America Britain would have lost the war by 1916, according to the official *History of the Ministry of Munitions*. The magnitude of this relative decline in Britain's steel industry in the two or three decades at the end of the 19th century can be gauged from the fact that in 1870 Britain produced 50% of the world's pig iron and 43% of the steel *and accounted for 75% of the world's iron and steel exports*. As late as 1885 Britain's pig iron production was still twice that of Germany's and almost twice that of the USA.

Britain's failures during the Second Industrial Revolution

In 1914 Britain's electrical industry and steel industry were only half the size of the corresponding industry in Germany. This was amazing in view of Britain's previous ascendancy in electrical science and iron and steel technology. Moreover these two shortfalls were manifestations of a wider malaise – Britain's catastrophic failure to establish a dominant position in what has become known as the Second Industrial Revolution. This started during the second half of the 19th century and was well underway from about 1870. The revolution was spearheaded by the USA and Germany and consisted of applying high quality management techniques to the 'older' industries such as steel, shipbuilding and mining, and the creation and expansion of new science-based industries such as electricity, communications and chemicals.

In his outstanding book *The Lost Victory* the Cambridge political historian, Corelli Barnett, summed up Britain's appalling industrial and scientific performance in the period which straddled the two centuries in the following words:

> 'Leadership of the second, science based, industrial revolution from the 1870s onwards – complex precision machine-tools, electrical equipment, the large scale manufacture of dyes and drugs – belonged not to Britain but to America and the continent of Europe, and, within Europe, to newly united Germany above all. In the early 1900s only 5% of Britain's indus-

trial labour-force were working in the new technologies, which accounted for a mere 7% of her exports; 25% of her industrial labour force was locked up in older technologies such as coal, iron and steel and textiles, which still accounted for 70% of her exports."

Britain's fall from its position of power and influence has also been vividly described in the following words by the forthright historian, Max Nicholson, in his book *The System: The Misgovernment of Modern Britain*:

> 'Up to then (1850) the genius of a smallish number of inventors, engineers like George and Robert Stephenson, Brunel and Paxton, and other technologists and scientists had unfolded vast possibilities of rapid economic expansion. Some of these had been vigorously and ably taken up by the early contractors and industrialists, who had a fairly free hand operationally and financially until the politicians, administrators and bankers began to catch up with them after the mid-point of the century. From then onwards, for reasons which are complex and obscure, the inventors and pioneers ceased to be able to command the necessary resources and power to keep Britain on the move into the newer industries....It would seem more correct to regard those who ran Britain between 1850 and 1900 as chronic bunglers who inherited the most commanding advantages ever bequeathed to any generation in Britain and who lost no time in throwing them to the winds.'

Corelli ascribes Britain's decline to a failure, from the end of the 18th century onwards, to give scientists and technologists a high status and neglecting to provide institutions for advanced instruction in science and engineering. He contrasts the paucity of scientific and technological institutes in Britain with the 19th century achievements on the Continent and in the USA. France created the world's first technological university, the *École Polytechnique* as early as 1794 and the Berlin Technical Institute opened its doors in 1821 and technical high schools started at Karlsruhe, Dresden and Stuttgart in the period 1825 to 1829. The German language became the *lingua franca* of science and German science exerted a considerable influence on the science of most of continental Europe, Russia, the United States and Japan. During the last two decades of the 19th century Prussia's education budget was an order of magnitude larger than Britain's. In the United States the Massachusetts Institute of Technology was inaugurated in 1865, decades before anything even vaguely comparable was set up in Britain.

The Royal Society and education for industrial decline

Throughout the 19th century, as has been discussed, the Royal Society, through the aristocratic connections of its Presidents and other members, had at least the possibility of influencing government policy in relation to science and technology. The Society was, as is the case today, the country's leading scientific organisation and the *de facto* National Academy of Science and could have assumed responsibility for keeping government abreast of scientific developments and advising on science policy. If Britain's abject failure in the Second (Scientific) Industrial Revolution was the result of a neglect of science teaching in our schools and colleges, coupled to a failure to create and support science-based industries, can the Society escape at least some censure for our country's inadequacies?

The Society might be culpable but it is not known to what extent the various elected Councils of the Royal Society recognised these absolutely basic shortcomings in the application of science to industrial health and national defence and privately warned the government of the dangers of this policy, or rather the absence of a policy. If they gave such warnings, then clearly they were not heeded. More probably, the leading Fellows came from similar backgrounds to the politicians they were advising and were not that interested in trade, and had not recognised the dominant part science and technology would take in future conflicts.

On the other hand, it has to be recognised that there were very deep religious, cultural, social and imperial elements affecting British society in the 19th century which the Royal Society was powerless to influence. The most influential educationalist of the first half of the 19th century was Dr Thomas Arnold, headmaster of Rugby School from 1827 to his death in 1841. Religion for Arnold was the most important element in education and he resigned from the governing body of London University because religion was not to be a compulsory examination subject. In a letter written in May 1836 he emphasised how Christian morality was very much more important to him than scientific knowledge:

> 'rather than have it (science) the principal thing in my son's mind, I would gladly have him think that the sun went round the earth, and that the stars were so many spangles set in the the bright blue firmanent. Surely the one thing needed for a Christian and an Englishman to study is a Christian and moral and political philosophy...'

Partly as a result of Thomas Arnold's influence Britain's public schools in the 19th century steered their most intelligent students towards becoming Christian gentlemen, steeped in the classics, and fit for careers in the Civil Service and in the British Raj. Thomas's son, the distinguished poet, Matthew Arnold, also became involved in education; he was a lay inspector of schools from 1857 to 1867. Matthew was frequently sent by the government to inquire into the state of education on the Continent, particularly, France, Holland and Germany and he drew attention to Britain's shorcomings in numerous forthright reports. In 1871–2 he pointed out the connection between the excellence of the Prussian education system and its recent victory over France, but the lesson was not taken. However, Matthew never could quite escape the overbearing influence of his father and in 1882 he argued that 'knowledge' that cannot be directly related to our sense of conduct and our sense of beauty are mere 'instrument knowledges', whence followed his view of the overriding importance of classical greek as part of our culture.

Cardinal Newman too followed the Thomas Arnold tradition and was a strong influence, particularly on the Roman Catholic public schools. While Newman recognised that mechanical knowledge, ie science and technology, had its practical value, he rated it far below the humanities. In 1852 he spoke of the proper purpose of education in the following terms:

> '...there are two methods of Education; the one aspires to be philosophical, the other mechanical; the one rises towards ideas the other is exhausted upon what is particular and external.'

John Stuart Mill, like Newman, conceded that a knowledge of engineering could be useful to society and instruction in this discipline might be given to a limited extent in institutions linked to establishments where 'proper' subjects (ie the humanities) are taught. Mill did not think that engineering knowledge was needed by many people, and these should be encouraged to acquire such knowledge by their own efforts. However, such knowledge was 'no part of what every generation owes to the next, as that on which its civilization will principally depend'. In 1867 Mill summed up his views in his inaugural address as Rector of St Andrews University:

> 'The proper function of an University in national education is tolerably well understood. At least there is a tolerably wide agreement about what an University is not. It is not a place of professional education. Universities are not intended to teach the knowledge required to fit men for some special mode of gaining their livelihood. Their object is not to make skilful lawyers, or physicians, or engineers, but capable and cultivated human beings.'

Three reforming FRSs

In 1853 Lyon Playfair, the Scottish Member of Parliament and Professor of Chemistry, in his book *Industrial Education on the Continent* predicted that unless Britain changed her whole industrial outlook and methods she would be overtaken by Continental countries. His doleful prophesy seemed confirmed when the Royal Commission on Technical Instruction reported in 1886 that the Europeans had passed Britain in the application of science to industry and in the efficiency of their general organisation. The Commission also praised the German polytechnic system:

> 'To the multiplication of these polytechnics may be ascribed the general diffusion of a high scientific knowledge in Germany, its appreciation by all classes of persons, and the adequate supply of men compentent, so far as theory is concerned, to take the place of managers and superintendents of industrial works. In England there is still a great want of this last class of person.'

The teaching of science in the 19th century found its most forceful advocate in Thomas Huxley, who served both as Secretary and as President of the Royal Society. It was already clear that the inadequacy of science teaching was not a result of passive neglect but rather to the active antipathy towards science of the Christian theologians who had a stranglehold on our educational establishments. The science versus religion debate reached a crescendo with the publication in 1859 of Charles Darwin's *The Origin of Species*, perhaps the greatest intellectual landmark of the 19th century. Huxley was the theory's most enthusiastic supporter and as a result of his constant advocacy of it he became known as 'Darwin's Bulldog'. At a celebrated British Association meeting in Oxford in 1860, Huxley responded to an offensive question from Bishop Wilberforce about his simian ancestry with the assertion that he would rather be descended from an ape than a bishop! This exchange, which stimulated much notoriety, was significant because it was a notable and public proclamation of science's freedom from theology.

In 1870 Huxley became an active member of the first London School Board and helped set up general principles which influenced elementary education for the following three quarters of a century. He persuaded the Guilds and Livery Companies of the City of London to use some of their vast wealth to promote technical education and he persuaded Eton College to start serious attempts to teach science. While doing all this he continued to teach full-time at the Royal School of Mines, London.

Huxley was no narrow-minded opponent of instruction in the classics; indeed in later life he taught himself Greek in order to study the philosophy of Aristotle in its original language. However, to countermand some of Mill's influence, he suggested that science should be an alternative to the classics as a means of cultivating the minds of British students. This excellent idea though received little backing except from his friend and supporter, Herbert Spencer.

Spencer, who was not elected FRS and hence is not one of the troika referred to in the title of this section, was a teacher, railway engineer and journalist, before becoming one of the country's outstanding savant and social philosophers. It is interesting that he had published his idea of the evolution of biological species before the views of Darwin and Wallace were known, but he had attributed evolution to the inheritance of acquired characteristics. However, he readily accepted the ideas of natural selection and coined the phrase 'survival of the fittest'. He had a lifelong interest in science and the teaching of science and in his book *Education: Intellectual, Moral and Physical* contrasted the stained-glass and white-marble ideals of British education and the iron foundries, and cotton mills and gas works that sustained British prosperity. He also made the following perceptive comments on our education system:

> 'That which our school-courses leave almost entirely out, we thus find to be that which most nearly concerns the business of life. Our industries would cease, were it not for the information which men begin to acquire, as best they may, after their education is said to be finished.'

Corelli Barnett illustrates in his excellent book *The Audit of War* that he is a great admirer of Spencer's ideas, particularly his views on the inadequacies of Victorian education.

The third and last FRS referred to here, Sir William Huggins, also warned the country, perceptively but fruitlessly, of the mortal danger of inadequate provisions for science teaching in our education institutes. In 1903, while he was President of the Royal Society, he speculated on how it could come about that the government could be so foolish as to neglect scientific education:

> 'the absence in the leaders of public opinion, and indeed throughout the more influential classes of society, of a sufficiently intelligent appreciation of the supreme importance of scientific knowledge and scientific methods in all industrial enterprises and indeed all national undertakings.... In my opinion the scientific deadness of the nation is mainly due to the too exclusively medieval and classical methods of our higher public schools.'

In spite of these warnings there is no evidence of the British government taking any effective action to correct these deficiencies in scientific education, or in scientific research, from the Great Exhibition of 1851 to the outbreak of the Second World War in 1914. A consequence of this neglect could easily have been that Britain would have been rapidly defeated by Germany; in fact this would have come about were it not for speedy and effective assistance, particularly in the form of steel imports from the United States. It is true that during the course of the war Britain's soldiers, industrialists and their workers rose to the level of events, resulting in eventual victory, but at terrible cost, and it had indeed been a 'close run thing'.

'The sciences are never at war'

In 1919, just after the end of the First World War, Ernest Rutherford was late for a meeting in Paris of an inter-allied committee on anti-submarine warfare. He subsequently wrote to the chairman, Karl Compton, to apologise for his lateness and said the reason for it was he had been observing the first artificial nuclear transformation! – a scientific result which if confirmed, he added, would be of far more importance than the war. This opinion, which contained much truth, was perhaps the remnant of a belief held by many Fellows throughout the lifetime of the Society – that science floats above, and is unaffected by, such mundane matters as conflicts between nations.

This conviction had found expression in a letter written by Edward Jenner in which he made his famous statement that 'the sciences are never at war'. The Fellows usually were forgiven this eccentricity, though it is true that in 1667 Henry Oldenburg was arrested and imprisoned in the Tower 'for dangerous designs and practices'. Samuel Pepys assumed his crime was writing to a scientist in a hostile country, but more probably he had criticised Charles II in a letter, and this had been spotted by an official censor.

This separation of science and international conflict certainly had held sway in Britain's relationship with France, its nearest foreign country and traditional adversary. From the 1660s until the downfall of Napoleon, Britain and France were at war, or close to war, for an aggregate of almost 60 years, yet throughout this time scientists in the two countries never ceased to communicate freely. They gave each other prizes, and elected one another to membership of their respective academies. Moreover, throughout the Napoleonic Wars, Britain and France exchanged astronomical tables in spite of the fact that these assisted navigation and might well have been regarded as information helpful to the enemy.

While the two countries were at war Napoleon had no difficulty in granting Davy and his wife and Faraday free access to France to enable them in an eighteen month period in

1813–15 to travel in France and have discussions with many of its leading scientists. Incidentally, while in Paris Davy determined that a substance that a French scientist, Bernard Courtois, had extracted from seaweed, was not, as the French believed, a compound but a new element which he named iodine. He communicated this finding in a paper which he sent from Paris to the Royal Society and was freely allowed to do so.

Scientific advice to the Armed Forces

The fact that scientists feel they have a right to communicate with other scientists across barriers created by conflict, did not mean that they would not assist their own military. Indeed at times of national crisis scientists are usually only too keen to fulfil what they see as their patriotic duty and offer every assistance to their own Armed Forces. Newton often advised the Admiralty on astronomical calculations and even recommended that the Navy should employ its own mathematicians to sail with the ships and interpret the astronomical data as soon as it was obtained. In the 19th century both Michael Faraday and Sir Frederick Abel acted as official advisers to the Services, but their influence seems to have been slight.

The employment of science in more modern conflicts has been brilliantly surveyed in R. W. Clark's *The Rise of the Boffins*, with a foreword by Sir Solly Zuckerman, the 'boffin' with perhaps the highest profile. Clark describes how at the start of the present century a new tool became available to aid navigation – the wireless. To cope with this innovation the Admiralty in 1908 appointed its first full-time scientist, H.A. Madge, who came from the Marconi organization with the responsibility of developing wireless communication for ships and sea.

During the first couple of decades of this century the Services had much assistance from three national scientific organisations: the Imperial Institute (founded 1887), the National Physical Laboratory (1902) and the Imperial College of Science and Technology (1909). During its early years the National Physical Laboratory was under the supervision of the Royal Society. The Society's main link with government was via the Liberal peer, R.B. (Lord) Haldane, Secretary of State for War, 1905–12 (R.B. was a member of the very remarkable Haldane family, his brother was the physiologist J.S. Haldane, and his nephew the biologist J.B.S. Haldane). R.B. Haldane saw very clearly that Britain's perfunctory attitude towards science, compared to Germany's counterparts, could put Britain at a very severe disadvantage should war break out between these two countries. He created the organisational basis for a scientific defence structure.

R.B. Haldane forced through an association between the newly-created National Physical Laboratory and the organisation that had started life as the War Office's Balloon Equipment Store at Woolwich in 1878. Research on balloons passed to Farnborough Common on the Surrey–Hampshire border in 1905 and it was here in 1909 that a further major division took place. One wing became in succession the Farnborough Air Battalion, the Royal Flying Corps and the Royal Air Force. The other wing, the H.M. Balloon Factory, developed at Haldane's initiative into the Army Aircraft Factory and later into the Farnborough Royal Aircraft Establishment – described by Clark as 'the brainbox of the great technological revolutions which were to help to give Britain victory in the air in 1940'.

The Royal Society and the First World War

An official of the Royal Society has written that the First World War had little effect on the Society as an Institution; indeed the war was rarely mentioned in the official Record of the Society for the war years. On the other hand, while the structure of the Society itself may have been relatively unaffected by the conflict, the Society's contribution to the war effort, particularly in guiding research and development in support of the Armed Services, was very significant.

Even the relatively reactionary Matthew Arnold had recognised in the early 1870s that the superior education and technical knowledge of the Prussian troops had played an important part in gaining victory over the French. However, the British High Command, in spite of R.B. Haldane's efforts, enjoyed no such insight at the outset of hostilities in 1914. When talented (and extremely rare) young scientists and technicians volunteered to serve in the Forces, in spite of protests from the Royal Society, they were promptly sent to the front line, where many were killed. Included amongst these casualties was the 27 year old Henry Moseley, Rutherford's former assistant, who lost his life at Gallipoli in August 1915. (Moseley was perhaps the most brilliant of Britain's young experimental physicists; he had been the first to establish the atomic numbers of the elements by studying their X-ray spectra, leading to a complete classification of the elements and an experimental basis for an understanding of the structure of the atom.)

Rutherford had made strenuous efforts to secure for Moseley a non-active-service role within the army but obviously was unsuccessful. The man responsible for many of these placements, Lord Derby, admitted the extreme folly of not discriminating between those whose skills and training would be of great value in a prolonged and increasingly technological conflict, and those who possessed no such qualifications. These mistakes reflect perfectly the ignorance of the importance of science by those who governed the affairs of the nation.

With the introduction of conscription in 1916, and under much pressure from the Royal Society, a limited degree of exemption for students of technology in their third and fourth years was introduced. However, science graduates received no such special treatment and the best of these were sometimes given falsified medical exemption to allow them to continue with their vital warwork. Another aspect of the lack of appreciation of the importance of technological developments was that scientific discoveries which were important for the war effort were slow to be applied. For example, Chaim Weizmann's discovery in 1911 of a new method of producing acetone from the bacteria in fermenting grain, was only finally applied to manufacturing badly-needed cordite in 1916. (Incidentally, Weizmann was a Zionist and gratitude for his discovery led to the Balfour Declaration of 1917, which proposed the establishment of a Jewish National home in Palestine.

When the State of Israel was created in 1948, Weizmann became its first President, Einstein having turned down an offer of the post.)

As it became progressively clear that the war was to last an appreciable time, and that it could be won, not by bravery in hand-to-hand fighting on the battlefield, but by the superiority of one's technology, the Royal Society redoubled its efforts to stress the importance of science. In May 1915 the Society sent a deputation to the Presidents of the Boards of Trade and of Education urging government assistance for scientific research. In July 1915 a committee of the Privy Council was set up for this purpose with an advisory body, the chairman being Sir William McCormick, who was already the Chairman of the Advisory Committee on University Grants, the predecessor of the University Grants Committee (UGC).

The Committee began work in 1915 with a fund of one million pounds to try to encourage firms to group together to form industrial research associations and to support 'research of special timeliness and promise'. This organisation became the Department of Science and Industrial Research (DSIR) in 1916. In this period much important war research was carried out at the National Physical Laboratory, and during this period this laboratory was controlled by the Royal Society. At long last scientific research aimed at improved technology was being taken seriously. As Sir James Dewar put it, 'technology became a cult in Great Britain between 1915 and 1919'. (A very good account of these developments appears in the book *The Universities and British Industry* by Michael Sanderson, 1972). The question which hung in the air was whether or not these initiatives would be built upon in the post-war years.

Many initiatives and projects started during the First World War bore fruit in the Second. A young physiologist, A.V. Hill (Nobel Laureate 1922) investigated the accuracy of anti-aircraft fire and was head of the Anti-Aircraft Experimental Section of the Ministry of the Munitions; working with him were four other soldier-scientists who were to become FRSs, including Rutherford's son-in-law R.H. Fowler. This Section was to evolve into the Operational Research Group in the Second World War. Other projects which were resuscitated in the Second World War were the development of sound-ranging, a method of locating enemy batteries developed by William Lawrence Bragg, the development of magnetic mines by the Naval Mine Design Department and the development of ASDIC for locating submarines, on which worked Rutherford and Lawrence Bragg's father, William Henry Bragg.

There was also a significant concentration of precocious scientific talent working on the science of flight at Farnborough, most of whom were to make their larger impact, not in the First, but in the Second World War. Among this galaxy were F.A. Lindeman, (later Lord Cherwell); G.P. Thomson (later Sir George), E.D. Adrian (later Lord Adrian), William Farren and perhaps cleverest of all, G.I. Taylor (later Sir Geoffrey). At Martlesham Heath in Suffold H.T. Tizard (later Sir Henry) was put in charge of all experimental flying, in spite of having been turned down by the RAF because of bad eyesight.

To sum up: there seemed to be a gathering recognition by the military of the contribution scientists can make to the war effort, and also an appreciation by the Armed Forces of the wealth of scientific talent in the country. It is though salutary to recall Ronald Clark's conclusion from his extended study of the relative contributions of British and German scientists to their respective war efforts:

> '....it was realised that (the reasons for) the length of the war, and the narrowness of the margins which had separated victory from defeat, might well lie in the Germans' more realistic utilization of science and in their keener appreciation of the part which it could play in any future struggle. The Germans had, after all, applied more scientifically than their enemies the new power to be gained from four of the six latest technological advances – railways, smokeless powder, the machine-gun, and the petrol engine. Only in wireless and powered flight could the British claim the lead, and even there it was hardly overwhelming.'

Nevertheless, shortcomings in application of science to the war effort were much less serious than structural failings in basic industries such as electricity, steel, glass, some chemicals, dyestuffs, scientific instruments and explosives. These problems were to plague the country for the rest of the century, as has been discussed in *Science and Society* by Hilary and Steven Rose.

The interwar period

In 1918 as soon as the euphoria of victory began to wear off the full enormity of Britain's industrial weakness began to be realised, and shortcomings in medicine and agriculture also became painfully apparent. A far-reaching report of the R.B. Haldane Committee on the Machinery of Government was published and hardly an aspect of Britain's public and academic life did not receive attention, and criticism in many cases. Haldane recommended that the Medical Research Committee should be reconstituted as the Medical Research Council under a Privy Council Committee with a status equivalent to that of the DSIR (which covered industrial science and the physical sciences related to it, i.e. largely chemistry and physics). Eminent scientists on the MRC, like those in the DSIR would be approved by the President of the Royal Society. There was also an Agricultural Research Council. In this immediate post-war period the DSIR concluded that one of its main functions was to try and ensure that civil research also benefitted the nation. To this end they created four research boards – dealing with chemistry, physics, engineering and radio, with Henry Tizard in overall charge. Unfortunately, these boards were not altogether successful largely due to lack of co-operation from the Services. The Army was an exception, partly due to the fact that it had had no option for many years but to accept the importance of physics in improving gunners' accuracy and effectiveness. The Military College of Science gradually evolved from the Royal Arsenal at Woolwich, and in 1920 the distinguished scientist, E.N. da Costa Andrade was appointed to run a physics course which although biased towards military matters achieved university standard of attainment.

In order to promote technological research in industry the DSIR assisted in the creation of a number of Co-operative Trade Research Associations, in which joint research projects supported by groups of manufacturers would receive financial assistance from the government. By 1920, twenty research associations were established covering about half the country's manufacturing activities. The traditional industries, particularly coal and shipbuilding, did not take part in these arrangements and continued along their empirical paths.

This expansion in DSIR activities and its stimulation of science and industry was of course very welcome. Its budget in 1939 at just under one million pounds was three times its level in 1919, and its permanent staff had risen from 900 to 2000. These seem very satisfactory until it is realised that in 1939 it only represented a spending on research and development of 0.1% of GNP, compared with about 0.6% in the USA. Hilary and Steven Rose point out that these figures would not have satisfied J.D. Bernal, who published his *The Social Function of Science* in 1939, in which he was highly critical of Britain's support of science.

Britain's industries and the Second World War

Aircraft

At the outbreak of the First World War a British aircraft industry scarcely existed in spite of the fact that it was perfectly clear to all military experts that the aeroplane would play a dominant part in any future international conflict. From 1914 to 1916 Britain had to import French aeroengines in large numbers while it created its own aeroplane construction industry. This it did quite successfully and by the end of hostilities Britain was manufacturing planes at a rate of 30,000 per year. There are two lessons here. Firstly, that the British seem quite incapable of re-organising their industrial base in response to rapidly changing circumstances or likely future eventualities, and secondly, that where situations arise which threaten the very future of the country – such as the possibility of losing a war – then such organisational changes are almost miraculously brought about.

During the course of the 1930s, as it gradually became clear that yet again war with Germany was highly probable, great emphasis was placed by the British government on the importance of air power, though it was more talk than action. Excellent military planes had been designed, including Sir Sydney Camm's *Hurricane* and Reginald Mitchell's *Spitfire*, but there was simply not the industrial capacity to produce them in large numbers. Additionally there was an acute shortage of machine tools needed for manufacturing the engines (many were imported from Germany which commanded nearly half the world trade in these products). Supporting industries, such as manufacturers of instruments, armaments and radios, were not equipped for any sudden expansion in demand from the aircraft industry and trained technicians of all types were in short supply. When messages got through of these severe difficulties with aircraft production it pushed the government towards adopting its policy of appeasement at Munich, principally to buy a few more years for Britain to get its act together and make progress with its now near-panic rearmament programme.

The upgrading of the *Merlin* engine and the adoption of all-metal stressed fuselages made great demands on the metallurgical industry but generally speaking their research and development departments were up to the task and made important contributions to aircraft evolution. (On a visit to Germany in 1938 Air Marshall Sir Hugh Dowding concluded that the failure to keep up with German aircraft development lay primarily with aeroengine designers rather than with any shortfalls in light alloy metallurgy.)

In spite of the importance of aircraft development to Britain's national security it is little short of astonishing that the private aircraft manufacturing companies during the whole of the interwar period carried out no central research of their own; it relied entirely on state research laboratories such as the National Physical Laboratory and the Royal Aircraft Establishment at Farnborough. It is not surprising that it was in the general area of research and development that Britain fell behind its international competitors.

These shortcomings led Clement Attlee, Leader of the Opposition, to ask the Prime Minister in the House in January 1938 why it was that so many developments in aircraft technology seemed to originate from abroad. The examples he listed include the retractable undercarriage, the variable pitch screw, blind flying apparatus including artificial horizon devices, enclosed cockpits, power-driven turrets and so on. His *coup de grâse* was to ask why did Britain continue with the development of the biplane while it was universally being abandoned for the monoplane in countries with advanced aircraft industries?

At the start of the Second World War Britain was as reliant on assistance from America as was the case 25 years earlier in 1914. There seemed to be an inability to learn from experience; to quote once again Corelli Barnett:

> 'The bleak historical truth is that those great symbols of British myth, the Battle of Britain *Spitfire* and *Hurricane* and their *Merlin* engine, were largely fabricated on foreign machine tools; more, their armaments and much of their instrumentation too were foreign in design, and, in the case of earlier production batches, foreign in manufacture as well'.

Electricity Supply

As discussed in Chapter 8, and above, in spite of many past achievements in electrical science Britain's electrical industry was only half the size of Germany's in 1914. In 1917 R.B. Haldane and C. Merz presented to Lloyd George the findings of their committee which had concluded that the *six hundred* electrical undertakings in the country operating at a multiplicity of frequencies had developed almost entirely along the wrong lines and were hugely uneconomic. They advocated the construction of fewer stations, but each of a modern design and generating at a uniform frequency and feeding a national grid. Little action was taken but when in 1925 similar recommendations were made by the Weir Committee a national grid was indeed established and was operating successfully in 1939 at the outbreak of the war. Little progress had been made on increasing the size and reducing the number of power stations, but this turned out to be an advantage because it is more difficult to disrupt by bombing a country's electrical supply system if it has many small power stations rather than fewer larger ones (such as that at Battersea). The outcome was that although generating efficiency was low, adequate electric power levels were maintained throughout the Second World War.

The Steel Industry

As has been discussed above, an official history acknowledged that Britain would have lost the First World War by 1916 had it not been for huge imports of steel from the USA. During the Second World War it is doubtful if Britain could have survived for even a shorter period of time without similar massive imports from the USA. Braving the hazardous Atlantic crossing, and at great loss to life and shipping, in the period 1940–44 Britain had been forced to import no less than ~14 million tonnes of steel from America, representing a quarter of British total production during this period.

This was in spite of a massive government-assisted modernisation programme for the steel industry from the mid 1930s onwards resulting in major constructions in Lincolnshire (United Steels), Clydeside (Colvilles), Shotton (Richard Thomas), Corby (John Summers) and Ebbw Vale (Richard Thomas). These were welcome co-operative initiatives by the larger and more advanced companies, but they only represented less than half the Britain's total steel output; thus inefficiency was still the characteristic of the industry. The net result failed to satisfy Britain's steel requirements even during the rearmament period. Steel shortages affected modernisation and expansion in other vital industries in the late 1930s, including shipbuilding, mining, transport and the manufacture of tanks and other military requirements.

Some causes of Britain's industrial decline

Britain's weaknesses in aircraft manufacture, electrical power supply and steel production were mirrored by shortcomings in other major industries, notably shipbuilding, transport and mining, and in many other manufacturing activities. These deficiencies were at first highlighted by Britain's complete inability to counteract and match German industrial power in two world wars, and was subsequently disguised by the almost unbelievable scale of American aid. One result of this American largesse was that for the majority of the war years and beyond the inadequacies were hidden, even from the British themselves.

There are as many explanations of Britain's disastrous industrial performances during the past century and a half as there are commentators. There is though general agreement that Britain attached too much importance to its role as a world power, and leader of a large Empire, long after the benefit and strength, which might once have characterised such an alliance, had disappeared. The influence of the 'Empire factor' on the education system persisted long after other prejudices had disappeared. Even today there is a belief in many circles that a knowledge of the humanities is of first importance for developing leadership qualities compared to burdening young people with a technical education.

The education system in Britain lies at the heart of what remains a divided society. (Even today about half of the students at Oxford and Cambridge, the pinnacle of educational excellence, come from the 7% of the young people whose parents can afford to send them to public schools.) Privilege can lead to resentment amongst the mass of the people and this feeling of being two nations was pervasive in the '20s and '30s leading up to the Second World War. On the outbreak of hostilities there was a drawing together of the various social classes but old enmities in the industrial arena were not slow to reappear.

When one considers the horrors a Nazi victory would have brought in its wake, it now seems incredible that during the 1939–45 War 'wildcat' strikes were not uncommon, particularly in some parts of the aircraft industry and in the 'older' industries such as mining and shipbuilding. In a statement to the House in July 1944 the Minister of Fuel and Power, Gwilym Lloyd George said: ‘We had a case recently where a 1,000 tons of coal was lost and the reason for the strike was that the men wanted the dismissal of the lady in charge of the canteen’. No more vivid example could be imagined of the gap between the industrial leader and those they were supposed to have led.

Class divisions in Britain are though only one factor amongst many. What became increasingly clear, at least from about 1880 onwards, was that being competitive internationally required keeping up-to-date with modern industrial equipment and practice and increasing significantly the size of Britain’s individual companies. This contrasted with how the industrial structure in Britain had evolved in the unbridled free enterprise culture of the Victorian era. In each sector there was a tendency for numerous small companies to be created to feed expanding demand both at home and abroad. As foreign countries developed their own manufacturing and engineering industries British companies became less profitable but still provided adequate cash to keep contented their already wealthy and small ownership base.

Modernisation and amalgamations may have been desirable from the point of view of the national economy, but in general such policies were opposed by the owners. One way of forcing both amalgamations and modernisation would have been widescale nationalisations but this was anathema to Conservative administrations which tended to dominate the political scene. Widespread nationalisation was possible and did occur under the post-war Labour governments (1945-51) but so much money was spent on welfare reform that little was left for modernising the new industries or providing an improved industrial infrastructure. Taking the Second World War as a reference point then it could be said that from the point of view of industrial health Britain suffered from an excess of Conservatism before the war and an excess of Socialism afterwards.

Creativity, the Royal Society and the Second World War

Traditionally, the typical FRS is thought of as a rather unworldly university don with his head (it could only be *his* head in those days) full of science and not too interested in politics or sociology. He, or indeed his Society, could hardly be blamed for the weaknesses of Britain's industries – the causes were too complicated and too political and sociological for his understanding. What was clear, as international tension mounted during the late 1930s, was that any future war might be won or lost depending on the relative strength of a nation's creativity in engineering design and in science. Thinking along these lines, in July 1938, two scientists, G.T.R Hill and C.F. Goodeve sent to the Secretaries of the Royal Society a 'Memorandum on Peacetime Organization for Voluntary Training of Scientific Workers for Service in the Event of a National Emergency'. The general idea here was that scientists should be given training relating to tasks they might be required to carry out should war be declared.

The suggestion in the Memorandum fell on fertile ground because one of the Royal Society's Secretaries was A.V. Hill who had introduced operational research for anti-aircraft performance during the First World War (the other Secretary was A. Egerton). Moreover the Society's President at that time was W.H. Bragg who was also experienced in military matters. Bragg almost immediately took up the suggestion and proposed something along these lines in a letter to the Prime Minister.

Although this correspondence did not lead to an adoption of the original training idea it did result in the creation of a Central Register of scientists and other technical experts who could be called upon in times of war to perform especial duties for the country. The Royal Society agreed not only to organise the Register but also to pay from its own funds the operating expenses of its creation and maintenance throughout the period of hostilities. The compilation started in February 1939 and by October, a month after outbreak of war, the Register contained 6484 names.

The Central Register was an essential tool in ensuring that the scientific manpower of the country was available on a planned basis. (This of course was in marked contrast to the position at the outbreak of the First World War when scientists were sent to fight as ordinary soldiers in the battle fields of France.) The Register was invaluable for identifying suitably qualified people for urgent war work but this was only half the problem – the individual had then to be persuaded to change direction and perhaps interrupt his or her own research and academic career. The person to spot this problem was W.L. Bragg, Head of the Cavendish Laboratory, who, after consultation with the Heads of the Services Research Organisation, wrote to Egerton, as the joint Secretary of the Royal Society, outlining a scheme for paying seconded scientists, and guaranteeing a return to their former position when their war work was over. Such a scheme was eventually adopted and worked successfully throughout the war.

During 1938 W.L. Bragg had alerted his colleagues at the Cavendish to the importance of radar but H. Tizard realised that further work in this area was vital and would require the efforts of an appreciable proportion of the output of physicists from British universities. He invited for lunch at the Athenaeum, J. Cockcroft, who in addition to his distinguished work in nuclear physics also had an interest in radar, and asked him to think of some scheme whereby recently qualified and training physicists could be introduced to the complexities and potential of radar. With the help of other Fellows of the

Royal Society Cockcroft arranged for almost a hundred physicists from the Cavendish and Clarendon Laboratories and from the universities of Birmingham, Bristol and Manchester to spend some time at the radar stations. Many of these were subsequently recruited into radar work.

Another contribution by the Royal Society to the development of radar was their granting of permission for J.T. Randall to work on radar transmitters, which related to an Admiralty contract, rather than on the work for which he had been commissioned by the Society. As a result of this relaxation, Randall who was working in Oliphant's Department at Birmingham University, collaborated with H.A.H. Boot and developed the cavity magnetron. This was one of the major discoveries of the war and made possible airborn radar equipment which played a significant part in Britain's success in the 'Battle of Britain'.

From its very inception in 1660, physics has been the dominant discipline in the Royal Society, and throughout the first four decades of this century, nuclear science has dominated physics. It is not then surprising that the Society, from the end of 1938 onwards, was pre-occupied with the prospect of developing a nuclear weapon and fearful that such a development might first take place in Germany. As war approached the Society realised that the agreement with Kaiser Wilhelm Gesellschaft for an exchange of lectures must come to an end, but not before Otto Hahn had lectured in London on 23rd June on 'The fission of uranium nuclei by neutrons'.

For many of the interwar years Germany had maintained a lead in theoretical physics and it was also the country where nuclear fission had first been demonstrated, by Hahn and Strassmann at the end of 1938. Germany's dominance of nuclear physics would have been much greater were it not for the fact that the numerous Jewish scientists were forced to flee Germany, and also from neighbouring countries which were likely to be invaded. Many of these Jewish scientists settled in Britain, and were encouraged to do so by the Royal Society, who assisted them by finding grants and other funding.

When war was declared most of the immigrant Jewish scientists were German nationals and because of this many were interned. The Royal Society argued for their release for humanitarian reasons and also because they had much to contribute to Britain's wartime science programme, and indeed had just as much reason to work for the downfall of Nazi Germany than the indigenous scientists. Amongst these Jewish/German expatriates were Peierls and Frisch who worked on atomic physics in Oliphant's Physics Department at Birmingham University (the same department in which Randall and Boot were developing the cavity magnetron – work kept secret from the two expatriates). As is well known, Peierls and Frisch demonstrated that the critical mass of U-235 was quite small so the construction of an atomic bomb was a theoretical and practical possibility. This led to the formation of the MAUD Committee (initially consisting of Thomson, Chadwick, Cockcroft, Oliphant and Moon) which, according to the historian Margaret Gowing 'met almost invariably at the Royal Society'. Once again the Society, simply by acting as host and facilitator, was smoothing the path for scientific progress at a vital point in the nation's history.

Atomic research and radar tended to absorb a high proportion of the country's available supply of outstanding physicists, but progress was also made on other 'physics' fronts, including the development of the proximity fuse and anti-magnetic mine defences. Biology and medicine were also vital subjects with relevance to warfare; it was important to recall that in past wars quite frequently more soldiers died through disease and incompetent treatment of wounds than were killed on the battlefields.

Alexander Fleming had accidentally discovered penicillin as early as 1928, but isolating, purifying and manufacturing useful quantities of this anti-bacterial enzyme represented quite a different order of complexity. However, success in this area was achieved in 1941 by yet another German expatriate, Ernst Chain, who together with Howard Florey at Oxford, developed techniques which were capable of being converted into large scale manufacturing processes. (For this work Chain and Florey shared the Nobel Prize for Medicine with Fleming in 1945.)

To sum up, during the couple of years preceding and following the declaration of war in 1939, Britain's engineers and scientists, and their immigrant colleagues, exhibited remarkable creativity. It is only necessary to mention the names of some of the pioneers to support this contention: de Havilland, Handley Page, R. Chadwick, Camm and Mitchell (aircraft designers); Lovesey, Rubbra and Hooker (aeroengine engineers); Whittle and Barnes Wallis (aero inventors); Watson, Watt, Lovell, Oliphant, Randall and Boot (radar); Blackett, Thomson, Cockcroft, Chadwick, Tizard, Peierls, Frisch, Haworth, Simon and Kurti (atomic energy); R.V. Jones and Frank (secret-Science war); Florey and Chain (penicillin).

Somewhat unfortunately, in the three areas where most progress had been made in Britain: in radar, atomic energy and penicillin, the industrial base was insufficient to see the projects through to completion. There proved to be no alternative but to transfer the activities to the USA. In the case of radar, if full advantage were to be taken of the discovery of the cavity magnetron, it was clear that Britain's electronics industry was too small, and too incompetent, to cope with the large expansion in development and manufacture needed. (During the early stage of radar, British manufacturers, who carried out no research, could not even supply sufficient thermionic valves of a suitable quality for the planned installations, and the radio industry was little more than a collection of assembly plants for imported components).

In the case of developing a nuclear weapon, Britain simply did not have the industrial capacity to tackle such a vast project, and the construction of suitable plant would reveal to the Germans our intention and also would be vulnerable to aerial attack. With penicillin, manufacture would have been possible, but not on the scale needed to take advantage of such an important discovery, so again the correct decision, to transfer the project, lock, stock and barrel to the USA, was taken.

In all these activities the role of the Royal Society was the same as it has always been, since 1660; it acted as a focal point, a gatherer and disseminator of information, and above all as a brilliant *facilitator*. Its *modus operandi* was subtle and sagacious, it encouraged scientific excellence here, made creative introductions there, and now and then touched the tiller of government with a few discrete words in the ear of a government Minister.

All this may seem irrelevant and inconsequential against the background of the turmoil which was Europe in 1939, as country after country collapsed under the German *blitzkrieg*, but such a view would underestimate the vital contributions of the Society in what was to become the world's first war which was dominated by science. To support this view, this section ends with a quotation from one of the best books written about the science aspects of World War II *The Rise of the Boffins*. It relates to the compiling of the Central Registrar:

> 'As Europe slid slowly down the slope to war, Britain thus had ready a muster roll of the scientists in university and research station who could be called upon to help the Services. Credit for this rested more with the Royal Society and its Fellows than with the Government, which had only reluctantly agreed that it would be useful...'

From the Second World War to the present

From the above it is clear that without enormous assistance from the USA Britain would have quite rapidly lost the 1914–18 war. One of the major sources of weakness was a complete failure to modernise basic industries and invest adequately in the newer science-based companies. It might have been thought that a realistic assessment of these shortcomings would have stimulated in the post-war years a determination to strengthen and modernise Britain's industrial base and improve its international standing in technology.

In fact, in 1939, with the outbreak of yet another war with Germany, it was already clear that industrial performance had not improved enough and history was about to be repeated, that is to say that once again Britain only avoided defeat because of massive aid received from America.

In both conflicts Britain was severely disabled by inadequacies in its steel industry, and this was a particularly important disadvantage in view of steel's central importance in the manufacture of armaments and the hazards involved in having to ship large quantities of ferrous metals and alloys across a submarine-infested North Atlantic. The question now arises, if Britain completely failed to learn a vitally important and obvious lesson in the 25 year period from 1914 to 1939, could it be equally wayward and irresponsible in the 60 years from 1939 to 1999?

Although having accumulated huge debts in the period 1945–50, in a devastated world Britain should have been in a commanding position as far as modernising its industries was concerned. Indeed during the period 1950–60, industrial productivity per man did increase by 25%, which seems satisfactory, until it is compared with a 59% increase in Germany and a 77% increase in France. Germany and France might have been starting from a lower base but other statistics force a relentlessly depressing message.

In 1957 both Germany and Japan overtook Britain in shipbuilding – in fact in that year British shipyards produced less that they had done in 1913. Traditionally Britain has been famous for its motorcycles but, also in 1957, it became a net importer. By January 1994, when British Aerospace sold Rover plc to the German company BMW, 99% of the motor industry became owned by foreign countries; most of the recent purchases having been made by German and Japanese companies.

Britain's poor industrial performances in 1914–18 and 1939–45 and shortly thereafter have been mostly brilliantly, even cruelly, revealed in the writings of Corelli Barnett, particularly in his books *Audit of War* and *The Lost Victory*. A book of comparable importance summing up the state of Britain half way through the present decade is Will Hutton's *The State We're In*. In the following he compares industrial training in Britain and Germany:

> 'Despite the imperative to develop what economists call 'human capital', 64 per cent of the British workforce have no vocational qualifications, compared with 26 per cent in Germany. Britain in 1993 had 250,000 people on apprentice schemes compared with more than 2 million in Germany. Efforts at improvement are hamstrung by lack of resources and an inability to build a national consensus on training.'

on Britain's investment in R&D he comments as follows:

> 'While the world's top 200 companies spent three times more on R&D than on dividends, Britain's 'top innovators' research spend was a miserable two-thirds of what they hand out in dividends.'

and he sums up the outcome of the massive denationalisation exercise of the 1980s and the state of Britain's education as follows:

> 'If the public sector was bloated in the 1970s, it had now become the twilight zone of the second-best. Its employees are poorly rewarded and its services under-resourced. Education has become a creator of class division, with opted-out schools in the state sector and private schools offering access to qualifications, prestigious universities and, for some, lucrative careers.'

Hutton is not very reassuring. How does Britain stand in the international league of industrial nations? The traditional index of relative wealth is the Gross Domestic Product (GDP), but in 1993 the IMF recommended that instead of making comparisons between GDPs by using commercial exchange rate tables, account should also be taken of the much lower costs of non-traded labour-intensive services in the Third World. The new conversion factor is known as Purchasing Power Parity (ppp) and the adjusted Gross Domestic Product as GDP(ppp). In an important article in *The Economist* (October 1994), its economics editor, Pam Woodall, produced a league table comparing various countries according to their GDP(ppp) per person rating. Britain had slipped into eighth position having just been overtaken by Italy. Our rating was about half that of Germany's and a little more than a third that of Japan's. Woodall further predicted that on past trends there would be a further decline during the next quarter of a century to something like fourteenth position at the end of the period.

The steel industry has had a particularly chequered history. It was nationalised in 1949, denationalised in 1953 and in April 1966 the then Prime Minister, Harold Wilson, invited Lord Melchett, a merchant banker, to chair an organising committee for the renationalisation of the British steel industry; he accepted and from 1967 to his tragically early death in 1973 he was Chairman of what became British Steel Corporation. Before accepting he had consulted the heads of fourteen major iron and steel companies. His task was to weld all these into a single financially-viable corporation; it was the biggest amalgamation ever conceived in British industry. He proved to be an enlightened Chairman, establishing a two year notice of any major plant closure so that workers could be trained for alternative duties. He also introduced worker directorships. In 1972 he won approval from Prime Minister Edward Heath and his Cabinet for a large modernisation and expansion programme to involve an expenditure of three billion pounds over a ten year period.

Following Lord Melchett's sudden death, Monty Finniston, his Chief Executive, became Chairman and responsible for the ten-year expansion programme. A production target of ~35 million tonnes of steel per annum proved to be over-ambitious, but many essential modernisation schemes were successfully carried through. The cost of transporting vast quantities of ore and coke was a major factor and as the use of rich imported ores gradually replaced indigenous supplies this forced a movement of activity from traditional locations near ore fields to coastal sites. Expansion took place at Port Talbot and docking facilities capable of accommodating 100,000 tonne vessels were constructed.

Of the Chairmen of the British Steel Corporation and later British Steel plc, Robert Scholey and Brian Moffat (the current Chairman) have exhibited interesting differences in priorities and style. Scholey has been responsible for the tough, but inevitable, pruning and plant closures, turning the company into more of a specialised, high value-added organisation. From a workforce in the late 60s of 270,000 the number has gradually been decreased to the current level of about 40,000. It has been said of Scholey that he started a renaissance in British Steel and made it saleable. Moffat is by training an accountant and this was important in the run-up to flotation. He changed the divisional structure and was responsible for a cultural change, freeing managers and encouraging them to use their own initiative.

Determined efforts have been made to turn British Steel into an international company; it owns 51 per cent of Avesta Sheffield AB, a quoted Swedish company, which manufactures stainless steel products. It also owns a company in the US and has been bidding for Polish steel companies and has interests in the Pacific Rim. It recently announced (June 1999) a new and major merger with a Dutch manufacturer of steel and aluminium products, Koninklijke Hoogovens. Following the approval of this merger by the shareholders (BSKN) has become the largest steel group in Europe and one of the World's top steel companies. Perhaps, at long last, the British steel industry is providing a model for the future, and one that other British industries might seek to emulate.

In the meantime the Royal Society continues its serene and virtuous course. Some of its important activities have long and distinguished histories. For example, *Philosophical Transactions* (the oldest established scientific journal still current) and the *Proceedings* continue to publish papers of the highest standard, including themes of metallurgy and materials. In addition research appointments are made and medals awarded for excellence in science, of the Prize Lectures, the Bakerian is especially noteworthy for its inclusion of materials topics.

The President of the Society and other office holders continue to represent the country as scientific ambassadors on many international occasions. During the past couple of decades the Society has recognised that sensible scientific and industrial policies can only be forged by an electorate and a Parliament which is well versed in science and technology. Accordingly, and prompted by a report on the Public Understanding of Science organised by Sir Walter Bodmer in 1985, it set up in 1986 COPUS (Committee on the Public Understanding of Science) in collaboration with the British Association for the Advancement of Science and the Royal Institution. This provides a focus for a broad-ranging programme of education in science, and the Committee organises, in association with the Science Museum, the prestigious Rhone-Poulenc Prizes for science books. Another of its promotions is the awarding of bursaries to assist scientists and engineers learn skills in communicating the results of their researches.

Other activities in this general area include the organisation of the Royal Society Michael Faraday Award presented annually to the scientist or engineer who has done most to promote the public understanding of science. It has also, in association with the British Association for the Advancement of Science, been responsible for producing the magazine *Science and Public Affairs* which publishes papers discussing topical issues in science and technology. Yet another initiative is the annual summer science exhibition 'New Frontiers In Science' which typically attracts up to a thousand visitors consisting of schoolchildren, the media, professional scientists and interested members of the general public.

The Society sees as one of its especial responsibilities to put forward an independent view on scientific issues which are controversial and have entered the public arena. Debate is either stimulated by holding Press Conferences in collaboration with the Association of Science Writers, sponsoring study groups or issuing publications. Subjects covered in recent years include: Mad Cows Disease Syndrome, Cloning, Genetically Modified Food, Science and Social Responsibility and Teaching of Science in Schools. Its activities in the field of nuclear power have been discussed in Chapter 18.

In 1992 the Society established the National Academies Policy Advisory Group (NAPAG) which enables its four constituent Academies (The Royal Society, The British Academy, The Royal Academy of Engineering and The Academy of Medical Royal Colleges) to tackle multidisciplinary issues of public interest. It has carried out studies on bio-diversity, long term energy policy, intellectual property rights and research, the research capability of the university system, and intelligent medical devices.

The impact of all this on government policy is hard to quantify and may be marginal.

With changes that are currently taking place in the structure of the House of Lords there exists the opportunity of appointing a number of FRSs as life peers and making similar elevations to Fellows of other National Academies. Such appointments could also be made to members of the Institute of Directors, Confederation of British Industry and the Trades Union Congress. It might then be possible to raise the level of debate on such subjects as science policy, technology and the restructuring of Britain's industrial base for the 21st century.

Biographies of Fellows and Foreign Members

The biographies presented here refer predominantly to those past Fellows and Foreign Members, discussed in the previous chapters, whose work relates significantly to the metallurgical themes of the book; biographies are not included for those Fellows, who were mentioned in the previous chapters solely in the context of aspects other than metallurgical.

For contemporary Fellows and Foreign Members the scope has been extended to cover the field of materials science, technology and engineering: in this context a very wide interpretation has been adopted, to include a number of researchers in areas such as solid state chemistry, crystallography, solid state physics and various branches of engineering. Some important names, especially from theoretical areas, may have been omitted by oversight, or by the difficulty in deciding the limit of the scope of the materials field as a whole.

The biographies vary considerably in length and detail, reflecting the relevance of the individuals' work to the emphases of the book, rather than to their overall achievements and distinction. For contemporary Fellows and Foreign Members, the information* is highly selective and is presented briefly in annotated form: it refers to school and university education and main career appointments, but not generally to visiting professorships, emeritus titles, directorships, appointments on committees, honorary degrees, fellowships, prizes and awards, and positions relating to Learned Societies (except for some aspects relating to the Royal Society).

A number of Fellows of the Royal Society have also been elected to the Royal Academy of Engineering, and this important information is included in the biographies. As from June 1st 1999 the designatory letter of Fellows of the Royal Academy of Engineering became FREng; this is used throughout this chapter in place of the previous designation FEng.

SIR FREDERICK AUGUSTUS ABEL 1827–1902; FRS 1860
Born at Woolwich, Frederick Abel was the son of Leopold Abel and Louisa née Hopkins. He entered the Royal College of Science, London, in 1845, one of the original twenty six students. From 1846 to 1851 he worked at St Bartholomew's Hospital, including being a demonstrator for John Stenhouse FRS. He was appointed in 1852 as Lecturer in Chemistry at the Royal Military Academy, Woolwich, in succession to Faraday. In 1854, he became Ordnance Chemist at Woolwich, and two years later Chemist to the War Department. From 1854, until his retirement from Woolwich in 1888 he was the chief official authority in the field of explosives on which he did extensive research. He also carried out much research on the composition of alloys in relation to their properties, and discovered iron carbide (cementite) in steels in 1885. Following the death

*For contemporary Fellows and Foreign Members the information presented is based, with acknowledgements, on sources available at the time of going to press.
Details of reference sources for each biography, in this chapter, are given in the Bibliography (Appendix 8).

of his wife Sarah Selena Blanch in 1888 he married Giulietta de la Feuillade; there were no children. Abel was the recipient of the Royal Medal in 1877, and also received other honours: CB (1877), knighted in 1883, KCB in 1891, made Baronet in 1893, and GCVO (1901).
DNB 2nd Supp. VI, 5, 6. RS.

ALFRED RODNEY ADAMS 1939–; FRS 1996
Rayleigh Technical School; Westfield High School. Univ. of Leicester, BSc, PhD, DSc; Research Fellow, 1964. Univ. of Karlsruhe 1965. Univ. of Surrey: Lecturer, 1967; Reader 1984; Professor of Physics 1987–. Royal Society/Japanese Society for the Promotion of Science Fellow 1980; Hitachi Professor 1992.
Semiconductors.

CYRIL CLIFFORD ADDISON 1913–1994; FRS 1970
Cyril Addison was born at Plumpton, near Penrith, Cumbria; he was the elder of two children of Edward Thomas Addison and Olive née Clifford. His secondary education was at Workington Grammar School, which he entered in 1926, and at Millom Grammar School, from 1928–1931. He went to Durham University with a scholarship and graduated BSc in 1934 in chemistry with subsidiary mathematics and physics. There followed a postgraduate course in physical chemistry and a PhD in 1937 in the field of surface chemistry. In 1939 he married Marjorie Whineray Thompson and they had one daughter and one son. When Addison left Durham in 1936 he worked until 1938 at the British Launderers Research Association on various surface chemistry topics. His next post was for a period of a year as a lecturer at the Harris Institute in Preston. With the coming of war in 1939 he moved to the Ministry of Supply Chemical Inspection Department and Chemical Defence Research Establishment. In 1946 he became a Lecturer in Inorganic Chemistry at the University College of Nottingham, which later became the University of Nottingham; promotion to reader occurred in 1952 and to the first Chair of Inorganic Chemistry in the University in 1960; he held this appointment until his retirement in 1978. Addison's research included important investigations of the surface tension of liquid alkali metals, and of the corrosion chemistry of stainless steel. His surface chemistry work led to the award of the DSc degree in 1946.
BMFRS 1997, 43,1–12. Lord Lewis of Newnham, FRS and B.F.G. Johnson, FRS.

SIR WALLACE ALAN AKERS 1888–1954; FRS 1952
Wallace Akers was born in Walthamstow, the second of five children of Charles Akers and Mary Ethelreda née Brown. He was educated at Bexhill-on-Sea and at Aldenham School. At Christ Church, Oxford he was awarded a first class honours in chemistry in 1909. He joined Brunner Mond and Co. in Cheshire in 1911, and after a short time in the research laboratory he moved into process work. In 1924 he joined the Borneo Company and worked as General Manager to the Far East. He returned to London in 1928 to join Imperial Chemical Industries (in which Brunner Mond had been merged), working with the Technical Director in London. His career with ICI involved responsibility for the Billingham activities (1931), an appointment at headquarters (1937), Executive Manager (1939), a Director (1941), and Director for Research (1944–1953). In 1941 he was seconded to the special section of the Department of Scientific and Industrial Research to be Director (under Sir John Anderson, later Lord Waverley) for the British programme for the development of atomic energy. In 1953 he married Bernadette la Marre. He was awarded a CBE in 1944 and was knighted in 1946.
BMFRS 1955, 1, 1–4. Waverley, incorporating an addendum by Alexander Fleck. DNB 1951–60, 11–12, Fleck.

SIR GEOFFREY ALLEN 1928–; FRS 1976

Clay Cross, Tipton Grammar School. Univ. of Leeds: BSc, PhD. National Research Council of Canada: Postdoctoral Fellow, 1952–54. Univ. of Manchester: Lecturer 1955–65; Professor of Chemical Physics, 1965–75. Imperial College, London: Professor of Polymer Science, 1975–76; Professor of Chemical Technology, 1976–81. Science Research Council: Chairman 1977–81. Royal Society: Vice-President, 1991–93. Unilever plc: Head of Research, 1981–90. Director, Unilever Plc and NV 1982–90: Director, Kobe Steel Ltd. 1990: Executive Adviser, 1990–. Univ. of East Anglia: Chancellor 1994–. Royal Society Philips Lecturer, 1983. FREng 1993. Knighted 1979.

Chemical physics of polymers.

NORMAN PERCY ALLEN 1903–1972; FRS 1956

Norman Allen was born at Wrexham, North Wales, the fifth in a family of 10 children of Sidney Edward Allen and Emily née Davis. His father was then an accountant in the Borough Treasurer's Department at Wrexham and later moved to Sheffield to a similar post before becoming Borough Treasurer of Burton-on-Trent. Allen was educated at Ranmoor Elementary School, Sheffield and then at the Burton-on-Trent Boys' Grammar School from 1913 to 1920. He obtained an open scholarship at Shefffield University where he studied metallurgy under Professor Cecil Desch graduating with a second-class honours degree and the associateship in metallurgy.

Allen then worked in a research group at the university concerned with low-melting-point die-casting alloys. He investigated the flow behaviour of the alloys under stress and also twinning and intergranular corrosion. In 1925 he moved to Swansea University College as an investigator for the British Non-Ferrous Metals Research Association working on the porosity of copper and copper alloys. In 1928 he became an assistant lecturer at Birmingham University; a year later he married Olive Williams whom he had met in Swansea, and they had two sons and a daughter.

Allen continued at Birmingham the work he had started at Swansea on the complex problem of porosity in cast copper and its alloys, and made an important contribution to the control of the industrial process of producing 'tough pitch' copper. This led to the renewal of his appointment in 1927, and to the award of the DSc by Birmingham University in 1934.

Early in 1933 Allen joined the Mond Nickel Research Laboratory in Birmingham where he remained until 1945. New laboratories were built in 1934–36, with the task of developing additional uses for nickel and thus expanding the demand for the metal. Allen was second in command to Leonard Pfeil and was concerned with two major research projects. One dealt with the transformation characteristics of austenite in various alloy steels. The second subject, from 1939 onwards, was concerned with the development of highly alloyed nickel base materials (the Nimonic series of alloys) required for the Whittle jet engine.

In 1945 Allen was appointed as Superintendent of the Metallurgy Division of the National Physical Laboratory where he re-orientated long-term programmes in the light of the many new metallurgical problems and ideas which had arisen during the war period. There was great expansion in staff and space; his policy was to keep a research balance ranging from the academic to the very practical industrial areas. The activities included: creep, the effect of alloying elements on the properties of pure iron, superconductivity and calculations of phase diagrams from thermodynamic data.

In 1966 Allen was appointed Deputy Director of the NPL and until he retired in 1969 he was concerned with the changes needed in the formation of a materials group. He travelled widely and gave numerous lectures abroad and spent much time and effort in promoting the welfare of

others. Allen received many honours, including that of CB in 1966. His career was marked by outstanding work on the casting of non-ferrous alloys, on Nimonic alloys and on the mechanical properties of iron and contributions to metallurgical progress as a whole.
BMFRS 1973, 19, 1–18. Sir Charles Sykes FRS. DNB 1971–1980, 11, 12. P.B. Hirsch FRS.

WILLIAM ALLEN 1770–1843; FRS 1807
William Allen was the son of a Quaker silk manufacturer. He went to school at Rochester, Kent, and was employed in his father's business for a period; however, his interest in chemistry led him to enter the Plough Court Pharmacy in Lombard Street, London where in 1797 he took sole charge of the business. He carried out scientific experiments, and associated with friends of similar interests. In 1796 he married Mary Hamilton, who died when their daughter was born; Allen was twice remarried.

In 1796 with William Haseldine Pepys he took the initiative in forming the Askesian Society; the activities of which led to the foundation of the British Mineralogical Society, whose members would be able to chemically analyse minerals. It was intended that this collaborative action in applying science to mining and metallurgy would remedy the absence in England of a college concerned with scientific mineralogy such as the École des Mines in France. Allen was appointed lecturer at Guy's Hospital in 1802, and he also lectured at the Royal Institution at the request of his friend Humphry Davy. His many interests included work beginning in 1805 on the refining and fabrication of platinum. Allen became active in the philanthropic movements of the time; for example, as a quaker, he worked for the abolition of the slave trade, which involved him travelling widely in Europe meeting statesmen and monarchs.
PRS 1844, 5, 532, 533. DNB. Ref. Ch. 4.1.

THOMAS EDWARD ALLIBONE 1903–; FRS 1948
Central school, Sheffield. Univ. of Sheffield: PhD. Univ. of Cambridge: Gonville and Caius College, PhD. 1851 Exhibition Student, Cavendish Laboratory 1926–30, PhD. Metropolitan Vickers Electrical Co., Manchester: 1936–46. Member, British Mission on Atomic Energy, Berkeley California and Oakridge, Tennessee 1944–45. AEI Aldermaston: Director of Research, 1946–63. AEI (Woolwich) Ltd.: 1948–63. AEI: Scientific Adviser, 1963. Central Electricity Generating Board: Chief Scientist, 1963–70. FREng 1976. CBE, 1960.
High voltage and transient electric phenomena; fission and fusion.

ÉMILE HILAIRE AMAGAT 1841–1915; FMRS 1897
Émile Amagat, who was born at Berri (Saint-Satur Cher), France, studied physics at the Sorbonne. His career included professorships of physics at Fribourg, Switzerland and at the Faculté Libre des Sciences of Lyons. From 1869–1872 for his doctoral thesis Amagat studied the effect of elevated temperatures on the compressibility and expansion of gases. A brilliant experimentalist Amagat constructed apparatus for high pressure experiments and for measuring pressures above 3,000 atmospheres. Between 1886 and 1893 his research included the compressibility of gases and the behaviour of liquids. Also his work was broadened to include the elasticity of solids.
PRS 1914–15, A 91, lxv–lxviii. SY. DSB I, 128, 129. J. Payen.

ANDRÉ-MARIE AMPÈRE 1775–1836; FMRS 1827
André Ampere was born in Lyons, France, the son of Jean-Jacques Ampere, a wealthy merchant and Jeanne née Desutieres-Sarcay. The family moved to the nearby village of Poleymieux where

André grew up. He showed great mathematical ability and was educated largely from his own private studies. After teaching mathematics at a school in Lyons (1796–1801) he was appointed Professor of Physics and Chemistry at a school in Bourg in 1802. He became an Assistant Lecturer in Mathematics and in 1809 Professor at the École Polytéchnique in Paris. In 1808 he was appointed as Inspector-General of the newly formed university system in France, a post which he held, except for a few years in the 1820s, throughout his life. He was appointed at the Collège de France as Assistant Professor of Astronomy in 1820 and Professor of Experimental Physics in 1824. In his family life Ampère experienced tragedy through the death of his father, executed in 1793 during the Revolution, and the death of his wife Julie Carron, whom he married in 1799, and by whom he had a son; in 1806 he married Jeanne Potot, and they had a daughter. In his research from 1800–1814, Ampère's main work was mathematical and he was outstanding in this field. From 1808–1815 he also carried out chemical research, and in the period 1820–1827 he founded and developed the science of electrodynamics.
DSB 1, 139–147. L. Pearce Williams.

SIR JOHN ANDERSON (FIRST VISCOUNT WAVERLEY) 1882–1958; FRS 1945
John Anderson was born in Edinburgh, the son of David Alexander Pearson Anderson and Janet Kilgour née Briglemen. He was educated at George Watson's College and at Edinburgh University, where he graduated BSc in 1903 with special distinctions in mathematics, natural philosphy and chemistry, and MA with first class in mathematics and natural philosophy. He then spent a period in Leipzig studying the chemistry of uranium. In 1905 he was placed first in the Civil Service examination for that year and entered the Colonial Office. His career then involved a series of senior administrative appointments up to 1922 when he became Permanent Secretary at the Home Office, a post which he held for 10 years. In 1932 Anderson went to India as the Governor of Bengal, at a difficult and dangerous time.

Anderson was elected to the House of Commons as an independent nationalist for the Scottish Universities in 1938, serving until 1950. He became Lord Privy Seal in 1938 with special responsibilities for manpower and civil defence, and invited Sir William Paterson to design what became known later, with some modifications as the 'Anderson shelter'. In 1940 he entered the War Cabinet as Lord President of Council with overall responsibility for civilian and economic resources. He took charge of much of the secret work, aided by his scientific background, including the atomic bomb and atomic energy. From 1943–1945 he was Chancellor of the Exchequer. His work on atomic energy research led to his election as FRS in 1945, and after the war he was Chairman of the Advisory Committee on Atomic Energy (1945–48) and of the committee which led to the establishment of the Atomic Energy Authority in 1954.

In 1907 Anderson married Christine Mackenzie and they had one son and one daughter; his wife died in 1920 and in 1941 he married Ava Bodley. Anderson's many honours included CB (1918), KCB (1919) , GCB (1923), PC (1938), OM (1958) and elevation to the peerage in 1952.
BMFRS 1958, 4, 307–325. Lord Bridges. DNB 1951–60, 21–24. Salter.

JOHN STUART ANDERSON 1908–1990; FRS 1953
John Anderson was born in Islington, London, the only son and younger child of John Anderson, a master cabinet maker, and Emma Sarah née Pitt. There was a time of poverty for the family, particularly when his father died in 1916. His schooling included periods at the Hugh Myddleton LCC school at Clerkenwell (1917–19), and with a scholarship, at Highbury County School (1919–24). He left school in 1924, deciding to study chemistry and with a scholarship he took a BSc

course at the Northern Polytechnic (1924–26). In 1926 he obtained a Royal Scholarship, placed at the top of the list and entered the Royal College of Science. Two years later he was at the top of the first class list of BScs. The degree was awarded in 1929 after completion of the required research year. He continued on postgraduate research involving nickel carbonyl and obtained a PhD in 1931. Anderson then spent two semesters in Heidelberg with a University of London scholarship, and then returned to Imperial College as a Demonstrator in 1932. In 1935 he married Joan Habershon Taylor and they had three daughters and one son.

Anderson's career involved periods in Australia and Britain. In 1937 he became a Senior Lecturer in the University of Melbourne. He returned to England in 1946 as Deputy Chief Scientific Officer in the Chemistry Division at the Atomic Energy Research Establishment at Harwell. A further period in Melbourne began in 1954 in a Chair of Chemistry followed by return to Britain in 1959 to become Director of the National Chemical Laboratory. Anderson held the chair of Inorganic Chemistry at Oxford from 1963 until retirement in 1975. There followed a period of association with the University College of Wales at Aberystwyth until he returned to Australia in Canberra in 1979.

Anderson's main research activity was in the chemistry of the solid state, notably concerning the constitution and behaviour of non-stoichiometric compounds. He was the recipient of the Davy medal in 1973.

BMFRS 1992, 38, 3–26. B.G. Hyde and P. Day. DNB 1986–90, 7, 8. P. Day.

PHILIP WARREN ANDERSON 1923–; FMRS 1980

Harvard University: BS, 1943, MA, 1947, PhD, 1949. Naval Research Laboratory: Chief Petty Officer, US Navy, 1943–45. Bell Telephone Laboratories: member technical staff, 1949–76; Director, 1976–84. Princeton University, NJ: Joseph Henry Professor of Physics, 1975–1996. Bell Laboratories: Consulting Director of Physical Research, 1980–. Nobel Prize for Physics, jointly, 1977.

Solid state physics – magnetism, superconductivity, the disordered state.

SIR WILLIAM ANDERSON 1835–1898; FRS 1891

William Armstrong was the fourth son of John Anderson, a banker and merchant, and Frances née Simpson. He was born in St Petersburgh, Russia, where he attended the High Commercial School. By the time he came to England in 1849 he was proficient in English, Russian, German and French. He entered the Applied Science Department at Kings College London as a student and became an associate. He served a three year period of pupelage at Fairbairn's works in Manchester. In 1855 he joined the firm of Courtney, Stephens and Company of Blackwall Place Ironworks, Dublin where his work involved general engineering and the design of cranes. He married Emma Eliza Brown in 1856. In 1864 he went to the firm of Easton and Amos of Southend, which had built a new works at Erith; he later become a partner and head (Easton and Anderson). His activities included the construction of pumping machinery and the design of gun mountings. In 1888 the Explosives Committee of the War Office asked him to design machinery for the production of cordite; in the following year he was appointed Director-General of the Ordnance Factories. He was the first Chairman of the Alloys Research Committee established in 1889 at the instigation of the Institution of Mechanical Engineers. Also he was a member of the Iron and Steel Institute. Anderson was appointed CB in 1895 and was knighted(KCB) in 1897.

DNB Supplement, i, 47–48. EIC.

EDWARD NEVILLE DA COSTA ANDRADE 1887–1971; FRS 1935

Edward Neville da Costa Andrade was born in London, the second son of Henry da Costa Andrade and Amy Eliza née Davis. The family of Andrade had come from Portugal to England during Napoleonic times. In 1897 he went to St Dunstan's College, Catford in south-east London from whence he entered University College London with a scholarship in 1905, obtaining first class honours in physics in 1907. He then proceeded to do research on the flow of solid metals such as lead under constant stress and published an important paper in 1910.

In the same year Andrade went to Heidelberg with an 1851 Exhibition Scholarship to work on the electric properties of flames. He was awarded the PhD degree *summa cum laude* at the end of 1911 and returned to England where he did research at the Cavendish Laboratory and then at University College London. In 1913 he went to the University of Manchester as John Harling Fellow to work with Rutherford, measuring the wavelength of gamma-rays from radium.

With the coming of war in 1914 Andrade was commissioned and served as an artillery officer until 1920, becoming a captain. On the French front from 1915 to 1917 he served first with a battery of 60-pounders; later, with a group of counter-batteries on the Arras salient, he was one of several young scientists, including Lawrence Bragg, who worked near the enemy guns to locate their exact position. Andrade was mentioned in dispatches and also suffered injuries. Returning to England in 1917 he worked for the Ministry of Munitions on explosives. Also, in the same year he married Katherine Barbara Evans, by whom he had two sons; in the 1930s the marriage broke up and he married Mona Wilkinson in 1938. At the end of the war, Andrade was awarded the degree of DSc at London University and in 1920 was appointed Professor of Physics at the Ordnance College, Woolwich (later the Military College of Science), where he studied the viscosity of liquids.

Andrade was appointed as Quain Professor of Physics at University College in 1928, a post which he held for 22 years. The strong, wide-ranging research school he built up and sustained covered the fields of the physics of metals, viscosity, acoustics, brittle fracture, deposition and sputtering of thin films, radiation damage, and sensitive flames. This period saw the publication of many of his most important scientific papers, especially on the plasticity of single metal crystals, and on surface cracks in glass. Andrade wrote various general scientific books, presented in elegant English, showing his clarity of thought and his effectiveness as a teacher.

In the 1930s, Andrade's activities included writing poems and building up his famous collection of old scientific books. The history of science, especially of physics, and of the early days of the Royal Society, became one of his main subjects.

During the Second World War Andrade became increasingly occupied with scientific advisory work for the Ministry of Supply, including scrutinizing the ideas for inventions submitted to the Government. In 1941 he lost all his personal effects at the College – books, prints, engravings, research papers, apparatus – when his laboratory was destroyed in an air raid. However, he continued writing and lecturing and he also served on the council of the Royal Society from 1942–1944. After the war he soon returned to investigations with new research students on the creep of metals and viscosity of liquids. In 1950 he was appointed Director in the Royal Institution and the Davy–Faraday Laboratory; unfortunately however, disputes led to his resignation in 1952. He continued for a few years as a research consultant. Then in 1957, he became a Senior Research Fellow in the Department of Metallurgy at Imperial College and, with Leonard Walden, who had been his laboratory steward since 1919, he continued research on the creep of metals until 1971.

Andrade's many honours included the Hughes Medal of the Royal Society in 1958. Over a period of 60 years his wide ranging research as a great experimental physicist made important contributions in metallurgy and materials science.
BMFRS 1972, 18, 1–20. A.H. Cottrell. DNB 1971–80, 14, 15. A.H. Cottrell.

THOMAS ANDREWS 1847–1907; FRS 1888
Thomas Andrews was born in Sheffield, the only son of Thomas Andrews, proprietor of the Wortley ironworks, and Mary née Bolsover. He was educated at Broombank School, Sheffield, later studying chemistry and early showing a deep interest in scientific research. Having also obtained training in practical metallurgy and engineering with his father, he became proprietor of the Wortley works when his father died in 1871. In 1870 he married Mary Hannah Stanley, and they had three sons and one daughter.

He began investigations of scientific metallurgy and came to be recognised as a metallurgical expert; he was consulted by the the Board of Trade, the Admiralty, and many leading railway and naval companies, on matters relating to iron and steel and this work involved examining and reporting on many serious accidents caused through fracture of steel. Andrews was among the first to apply the technique of microscopical examination of metals, particularly to ferrous materials. A few years before his death he gave a course of lectures to engineering students at Cambridge University on the microscopical examination of metals and its relation to engineering. In the field of mechanical properties he made a substantial series of experiments on the effect of sub-zero temperatures on the strength of railway axles and forgings. His research also included investigations of corrosion and of magnetic properties.
PRS 1908, A, 81, lxxxii–lxxxiv. J.E. Stead FRS. DNB 1901–11, 43, 44. TEJ.

ANDERS JÖNS ÅNGSTRÖM 1814–1874; FMRS 1870
Anders Ångström was was born at Logdo, Sweden and entered Uppsala University in 1833, where having obtained a doctorate in physics in 1839 he was appointed as a lecturer. In 1843 he became an Observer at the Uppsala Observatory. From 1858 until his death he occupied the Chair of Physics at the University. Ångstrom was one of the pioneers in the development of spectroscopy, for example in the sun's spectrum; in 1868 he published *Researches on the Solar System*. His measurements of the wavelengths of Fraunhofer lines were made to six significant figures in units of 10^{-8} cm and the unit, the Ångström was officially adopted by 1907. His earlier work included studies of electrical and thermal conductivity. He was awarded the Rumford Medal in 1872.
PRS 1876–1877, 25, xviii–xxii. DSB 1, 166, 167. C.L. Maier.

DOMINIQUE FRANCOIS JEAN ARAGO 1786–1853; FMRS 1818
Dominique Arago was the eldest son of François Bonaventure Arago and Marie née Roig, and was born in Estagel where his father became mayor. The family moved to Perpignon in 1795 where he completed his classical education, and having prepared by study he obtained entry to the École Polytechnique in 1803. He became Secretary of the Bureau des Longitudes and was sent to Spain on a geodetic expedition with Biot. In 1809 he became Professor of Descriptive Geometry at the École Polytechnique. Two years later he married and had three sons. His most important scientific work was done before 1830; his research related to electricity and magnetism, and included the influence of non-magnetic metals on a magnetised needle. He received the Copley medal in 1825 and the Rumford Medal in 1850. After 1830 he was involved in politics,

entering the Chamber of Deputies and reaching high office in government.
PRS 1853, 6, 359–365. DSB 1, 200–203. R. Hahn.

SIR WILLIAM GEORGE ARMSTRONG (BARON ARMSTRONG OF CRAGSIDE) 1810–1900; FRS 1846

William Armstrong was born in Newcastle upon Tyne. He was the son of William Armstrong and Ann née Potter, who also had a daughter. His father who achieved success in business in Newcastle, became mayor of the city in 1850. From an early age William showed an interest in machines, and was educated at private schools and at the Grammar School at Bishop Auckland. On leaving school he became an articled clerk in a solicitor's office in Newcastle; he completed his preparation for the legal profession in London, and then returned to Newcastle in 1833 to become a partner in the legal firm to which he had been articled; the firm flourished and he was entrusted with the interests of many important families, estates and companies. In 1834 he married Margaret Ramshaw, whose father had an engineering works.

After his return to Newcastle he regularly visited Watson's High Bridge Engineering Works, and began to publish and invent in the field of water-driven machinery. He constructed a hydro-electric machine, which formed the most powerful means then known of generating electricity at high tension; electricity was generated by steam escaping through nozzles from an insulated boiler. The uses and applications of water at the time chiefly occupied him, and he was heavily involved in the scheme to construct a reservoir for providing water to Newcastle. Armstrong's scientific achievements became widely recognised, and on his election as FRS he was described as 'a gentleman well known as an earnest investigator of physical science, especially with reference to the electricity of steam and the hydro-electric machine'. Faraday was among those who supported his election. Much interest was shown in his cranes, and this led to the founding of the great engineering works at Elswick-on-Tyne, with Armstrong as Manager. The business initially consisted in the manufacture of his newly devised hydraulic machinery; in 1848 the firm received the order for cranes for the Liverpool docks. Armstrong's activities included the continued improvements of his hydraulic machinery, and developing many applications. His hydraulic accumulators were large vertical cylinders with pistons carrying heavy iron ingots; the pistons were raised by water pumped by a steam engine. This source of hydraulic power was widely used for machines such as cranes and for the operation of dock gates. The ingenuity and utility of his inventions brought him wide recognition.

In middle life Armstrong began a major involvement in armament manufacture in which he also achieved widespread recognition and established a lage industrial enterprise. He developed an improved form of gun, using a new manufacturing technique and devoted much attention to naval matters in ship design and in advances in gunnery. For a period he served as Engineer of Rifled Ordnance at Woolwich.

Armstrong also carried out important work in connection with the economical use of fuel; in 1866 he served on a Royal Commission to inquire into the duration of British coalfields and in 1883 he suggested that a thermo-electric engine might be used to utilise the direct heating action of the sun's rays.

Armstrong's varied activities brought him great wealth which he put to enlightened uses and he was a notable benefactor of the City of Newcastle. He served as High Sheriff of Northumberland and also as a County Councillor. In 1863 he purchased some land on the east of Rothbury, and built a stately home, 'Cragside' where he transformed the bleak hillside into a magnificent park. In later life he carried out electrical research in his laboratory at Cragside. His honours

stemming from his public services included CB and a knighthood in 1859; in 1887 he was raised to the peerage.
PRS 1905, 75, 217–227. AN. DNB.

JOHN OLIVER ARNOLD 1858–1930; FRS 1912
John Arnold, the son of David Nelson Arnold, General Manager of the Midland Railway Carriage and Wagon Company, was born at Peterborough. Having been educated at King Edward VI's Grammar School, Birmingham, he became a naval cadet on *HMS Conway* in which capacity he travelled to India. Although he soon left the navy he retained a love for the sea, and was a keen yachstman throughout his life. When he was 20 years of age he entered the Engineering Department of Messrs Brown, Bayley and Dixon, and subsequently held appointments with the Farnley Iron Company, Leeds, and then in the Spanish Steel Works, Sheffield; the latter appointment was combined with a metallurgical consulting practice. When W.H. Greenwood, the Professor of Engineering and Metallurgy at Firth College, Sheffield, returned to industry, Arnold was invited to give the lectures in metallurgy, while still continuing his consulting work. He was soon appointed full-time Professor. In his teaching Arnold brought science into direct touch with the industry from which his students came; he was a clear and fluent lecturer, and starting from the one student who had attended his first lecture, there was a rapid increase in numbers. By 1890 his success enabled him to install a small-scale steel works, whose equipment included a 25 cwt (1. 25 tonne) Siemens furnace, to provide students with an insight into manufacturing processes. In 1897 the Firth College became a founding part of University College, Sheffield, which in 1905 attained the rank of an independent university. In 1917 when a Faculty of Metallurgy was established, Arnold was its first Dean. At the end of 1919, after ill-health he resigned his chair and lived in retirement first in London and then at Bowness-on-Windermere. He had married in 1883 a daughter of the Rev. James England, and they had two sons and a daughter.

Arnold applied scientific knowledge to the manufacture and use of steel, and also devoted much attention to chemical analysis. As a friend of Sorby he applied his methods of microscopical examination and persuaded industrialists that the technique could be valuable for the control of manufacture and the detection of causes of failure. He carried out important investigations of carbides in steels and was involved in the introduction of vanadium and molybdenum as alloying elements in steels. Arnold was very influential as a teacher and was devoted to the interests of his students.
PRS 1930–1931, 130, xxiii–xxvii. C.H. Desch.

SVANTE AUGUST ARRHENIUS 1859–1927; FMRS 1910.
Svante Arrhenius was born at Wijk, near Uppsala, Sweden, the second son of Svante Gustav Arrhenius, a surveyor and estate manager, and Carolina Christina née Thunberg. He attended school in Wijk and then entered Uppsala University at the age of 17, where he studied chemistry, physics and mathematics. After graduating in 1878 he continued at Uppsala, where in 1884 he presented a doctoral thesis concerned with the electrolytic theory of dissociation and the conductivity of electrolytes. Although the thesis was not highly rated by the examiners, leading chemists were impressed by the work and Arrhenius was appointed as a Lecturer in Physical Chemistry later in the same year. In 1887 he published further classic work in a paper 'On the dissociation of substances in water'. His contributions included the influence of temperature on reaction velocity and the concept of activation energy. In 1891 he became a Lecturer at the Technical Institute, Stockholm, and Professor in 1905. Also in 1905 he became Director of the Nobel Institute

of Physical Chemistry in Stockholm. Arrhenius married Sofia Ruhbeck in 1894 and they had one son; from his second marriage in 1905 to Maria Johansson there were two daughters. Arrhenius was awarded the Davy medal in 1902 and the Nobel Prize for Chemistry in 1903.
PRS 1928 A, 119, ix–xix. JW. DSB I, 296–302. H.A.M. Snelders.

PETER ASCANIUS ?–?; FRS 1755
Peter Ascanius was a Swedish scientist who for a long period was the Inspector of Mines in Norway and he made a special study of mineralogy. He visited London in 1755 and published papers in *Philosophical Transactions*, including studies of an iron ore deposit in Sweden.
Bulloch's Roll.

SIR ERIC ALBERT ASH 1928–; FRS 1977
University College School. Imperial College, London: BSc(Eng), PhD, DSc. Univ. of Stanford, California: Research Fellow, 1952–55. Queen Mary College: 1954–55. Standard Telecommunications Laboratories Ltd.: 1955–63. University College London: Senior Lecturer, Electronic and Electrical Engineering, 1963–65; Reader, 1965–67; Professor, 1967–85; Pender Professor and Head of Department of Electronic and Electrical Engineering, 1980–85; Professor, Department of Physics, 1993–. Imperial College: Rector, 1985–93. FREng 1978. CBE 1983. Knighted 1990. Royal Medal 1986. Royal Society: Treasurer and Vice-President, 1997–.
Electronics, acoustic microscopy.

MICHAEL FARRIES ASHBY 1935–; FRS 1979
Campbell College, Belfast. Univ. of Cambridge: Queen's College; BA, MA, PhD; post-doctoral work, 1960–62. Univ. of Göttingen: Assistant, 1962–65. Harvard Univ.: Assistant Professor, 1965–69; Professor of Metallurgy, 1969–73. Univ. of Cambridge: Department of Engineering, Professor of Engineering Materials, 1973–89; Royal Society Research Professor, 1989–. *Acta Metallurgica*: Editor 1974–95. FREng 1993. CBE 1997.
Mechanical properties of metals and alloys; modelling of properties; deformation and fracture maps, selection of materials and design; microstructural phenomena.

FRANCIS WILLIAM ASTON 1877–1945; FRS 1921
Francis Aston was born at Harborne, Birmingham, the third of seven children of William Aston, a merchant and farmer, and Fanny Charlotte nee Hollis. He attended Harborne Vicarage School and then spent four years at Malvern College. In 1893 he entered Mason College, Birmingham where he studied for the London Intermediate Science Examination. From 1898–1900 he studied chemistry at Mason College and from 1900–1903 he worked as a Brewery Chemist at Wolverhampton; while in this post he passed the examinations for the Associateship of the Institute of Chemistry. In 1903 he returned Birmingham University (formed from Mason College) as a Physics Research Student, investigating gaseous discharges, and spent a short period as an Assistant Lecturer. Aston moved in 1910 to Cambridge to work with J.J. Thomson at the Cavendish Laboratory on the effects of electric and magnetic field on positive rays, showing the deflection of the rays. He was awarded the Cambridge BA by research in 1912, and became a Clerk Maxwell Scholar in 1913, but his research was interrupted by the 1914–18 war, when he worked at the Royal Aircraft Establishment, Farnborough as a Technical Assistant; this work included the treatment of aeroplane fabrics. In 1919, after his return to Cambridge he became a Fellow of Trinity College. He improved his earlier equipment to produce in 1919 a mass spectrograph, which could separate atoms of similar mass and measure the masses separately. He went on to develop

improved versions of the equipment, and to investigate isotopes in more than 50 elements. In 1926 Aston became a Research Fellow of Trinity College. He remained unmarried.

Aston received the Hughes Medal in 1922, and was Bakerian Lecturer in 1927. He was the recipient of the Nobel Prize for Chemistry in 1922.

ONFRS 1945–48, 5, 635–51. G. Hevesy. DSB 1,320–322. W.H. Brock. DNB, 1941–50, 24–26. N. Feather.

COLIN ATKINSON; FRS 1998

Imperial College, London: Professor of Applied Mathematics.

Mechanical behaviour; dislocations, crack propagation; transformations in solids.

CHARLES BABBAGE 1791–1871; FRS 1816

Charles Babbage was born at Teignmouth in Devonshire, the son of a banker who left him a large inheritance. As a young boy he developed an interest in mathematics and went to Cambridge in 1810 to study the subject, graduating in 1814. Subsequently he proceeded to develop machines to produce mathematical tables, e.g. for navigation by mechanical computation and Government support was obtained. Babbage lived in London, while from 1827–1835 he held the Lucasian Chair of Mathematics at Cambridge. He was also interested in railway engineering and corresponded with William Fairbairn concerning the model experiments for tubular bridge construction.

DSB 1, 354–356. N.T. Gridgeman. DNB.

FRANCIS THOMAS BACON 1904–1992; FRS 1973

Francis Bacon was born at Ramsden Hall, Billericay, the son of Thomas Walter Bacon, an electrical engineer. He attended St Peter's Court Preparatory School in Broadstairs, Kent. His hope for a career in the Navy having been frustrated because of failure in an eyesight test, he attended Eton from 1918 to 1922, before entering Trinity College, Cambridge and completing the Mechanical Science Tripos in 1925. Bacon then became an apprentice at C.A. Parsons and Co. Ltd at Newcastle, where he came to know Charles Parsons. Among various aspects of his work, he was involved in the improvement of reflectors for searchlights and was in charge of the production of silvered glass reflectors from 1935–39. In 1934 he married Barbara Papillon, a marriage from which there were two sons and a daughter.

In the early 1930s Bacon became interested in the field of fuel cells and in 1937 proposed to the directors of Parsons that a workable fuel cell might be developed. Although the proposal was rejected he began experiments, which initially repeated Grove's early work but moved on to using activated nickel electrodes instead of precious metal catalysts. He left Parsons in 1940, and obtained financial support to enable him to carry out further electrochemical work; he was also provided with space in a laboratory at King's College, London. It became apparent that fuel cells could not be developed for use in submarines in the short term, and in 1941 he went to Fairlie in Scotland to work for the Admiralty on ASDIC, the underwater submarine detection system. After the war he sought industrial support for fuel cell research; eventually the Electrical Research Association agreed to give support and he went to the Department of Colloid Science at Cambridge in 1946 to work under the direction of Eric Rideal.

Bacon and his collaborators carried out extensive work which led to the development of a cell. The Pratt and Whitney Division of United Aircraft in the USA took out a licence on the patent, and developed the system which was used in the Apollo moonshot. Following the success of Apollo a new firm, Energy Conversion Ltd, was formed in Britain to which Bacon acted as

consultant. Bacon received the OBE in 1967 and the S.G. Brown Award of the Royal Society in 1965.
BMFRS 1994, 39, 3–18. K.R. Williams.

RICHARD WILLIAM BAILEY 1885–1957; FRS 1949
William Bailey was born in Romford Essex, the son of James William Bailey, a railway official, and Ann née Durley; he was the second in a family of seven children. He became an apprentice at the locomotive works of the Great Eastern Railway at Stratford and during his time there he obtained a Whitworth exhibition and a Whitworth Scholarship, which enabled him to spent periods of study at the East London Technical College, (which became Queen Mary College). Subsequently he spent a year on an electrical engineering course at the City and Guilds Central Technical College at South Kensington, and a year's college apprenticeship with the British Westinghouse Electric and Manufacturing Company at Trafford Park, Manchester. For four years, beginning in 1908 Bailey was a Lecturer in Mechanical Engineering at Battersea Polytechnic. In 1909 he married Mary Florence Dormer Alderman and they had two daughters. After an appointment from 1912–1919 as the first principal of the Crewe Mechanics' Institute he returned to British Westinghouse as a senior member of the research staff in charge of the section concerned with mechanical, metallurgical and chemical aspects. He remained with the company, whose name soon changed to Metropolitan Vickers Company Ltd, for the rest of his career. In 1945 he became Consulting Engineer to the Company, until he retired in 1957. He served on the Council of the Royal Society from 1951–52.

Bailey's research, including collaboration with colleagues, made outstanding contributions to in the field of high temperature creep, with particular relevance to the development of materials for steam turbine technology covering fundamental and applied themes.
BMFRS 1958, 4, 15–25. D.M. Smith.

SIR LEONARD BAIRSTOW 1880–1963; FRS 1917
Leonard Bairstow was born in Halifax, the son of Uriah Bairstow, a commercial clerk, and Elizabeth née Lister. Having received his elementary and secondary education in Halifax he obtained a scholarship at the Royal College of Science, London in 1898, where four years later he became a Whitworth Scholar. In 1904 he joined the Engineering Department at the National Physical Laboratory, where he worked on metal fatigue and aerodynamics. He moved to the new Section of Aerodynamics in 1909 and carried out pioneering research, which included wind tunnel design and aircraft stability. Bairstow was appointed to the Air Board in 1917, where his responsibilities included work on structural strength. In 1920 he was appointed Zaharoff Professor of Aerodynamics and later served as Head of the Department of Aeronautics at Imperial College until 1945. Among his activities he served on two government enquiries which investigated accidents to the R38 and R101 airships. Bairstow married Eleanor Mary Hamer in 1926 and they had one son and one daughter; following her death he married Florence Katherine Stephens. He was awarded the CBE in 1917 and was knighted in 1952.
BMFRS 1965, 11, 23–40. G.Temple (with notes from A. Fage, J.L. Naylor and E.F. Relf). DNB 1961–70, 65, 66. G. Temple.

SIR BENJAMIN BAKER 1840–1907; FRS 1890
Benjamin Baker was born at Keyford, Frome, Somerset, the son of Benjamin and Sarah Hollis; his father became principal assistant at the Tondu Ironworks in Glamorganshire. Benjamin was

educated at Cheltenham Grammar School and then obtained four years experience (1856–1860) as an apprentice in the Neath Abbey Ironworks. Having moved to London in 1860, where he obtained civil engineering experience, he then joined the staff of John Fowler, later becoming a partner with him from 1875, including involvement in the construction of the Forth Bridge. Baker achieved a high reputation as a consulting engineer in Britain and abroad, involving work on railways and dam construction. From the early stages of his career he had a deep interest in the theory of construction and the strength of materials. He was active in many government enquiries including a committee appointed by the Board of Trade in 1900 concerned with the strength of steel rails. In the Royal Society he was a Member of Council (1892–1893) and a Vice-President from 1896 until his death. His honours included KCMG in 1890 and KCB in 1902.
PRS 1910–1911 A, 84, i–xi. WCU. DNB 2nd Suppl. 1901–1911, 85–87. WFS.

HENRY BAKER 1698–1774; FRS 1740

Henry Baker was born in London, the son of William Baker, a clerk in chancery, and Mary née Pengry. At the age of 15 he was apprenticed to a bookseller and when he completed this period he went to stay with a relative, whose daughter was born deaf. Having taught the girl to read and speak, Baker went on to become in great demand as a teacher for the deaf and dumb. He had wide-ranging scientific interests including the microscopical examination of water creatures and fossils. Among his writings on microscopy was *Employment for the Microscope* (1753), in which he included observations on the growth of dendritic crystals of various metals. For this work he received the Copley Medal in 1744. He bequeathed £100 to the Royal Society to establish an oration – the Bakerian Lecture.
DSB 1, 410–412. L'E. Turner. DNB. Ref. Ch. 2.8.

SIR JOHN FLEETWOOD BAKER
(BARON BAKER OF WINDRUSH) 1901–1985; FRS 1956

John Baker was born at Liscard, Cheshire, the son of Joseph William Baker and Emily Caroline née Fleetwood. In 1915 he entered Rossall School from whence he won an open mathematical scholarship to Clare College, Cambridge in 1919. He took the course in mechanical sciences commencing in 1920 and obtained firsts in all three years, showing special talent for the theory of structures. The employment situation when he graduated in 1923 was difficult and he obtained a temporary post funded by the Air Ministry at University College, Cardiff. The research was on structural problems of airships and involved the design and building of a braced tubular framework, and the study of its behaviour under load. Good progress was made and Baker transferred to the Royal Aircraft Works at Cardington in 1925 as a member of the design staff, with particular responsibility for stress calculations of complex structures. In 1926 he became an Assistant Lecturer at Cardiff. Two years later he married Fiona Mary Macdonald Walker and they had two daughters. He fell ill which led to a considerable period of absence away from work. However, in 1930 he was able to take up the responsibilities of Technical Officer in the Steel Structures Research Committee concerned with rationalising the numerous building regulations on steel-framed structures. Baker made notable contributions and was appointed to the Chair of Civil Engineering at Bristol in 1933 continuing to work for the Committee for another three years. During the 1930s Baker began to work on the plastic theory which he pursued throughout his career. With the outbreak of the war Baker became Scientific Advisor to the Ministry of Home Security at Princes Risborough. His wartime work included the design of the Morrison air raid shelter.

In 1943 Baker returned to Cambridge University as Professor of Mechanical sciences and Head of the Department of Engineering, which experienced great expansion during his 25 year period of headship. Baker received support for his work from the British Welding Research Association at Abington. He received a Royal Medal in 1970 and was elected FREng in 1976. He was awarded the OBE in 1941, received a knighthood in 1961 and in 1977 was made a life peer.
BMFRS 1987, 33, 3–20. J. Heyman. DNB 1981–85, 24, 25. J. Heyman.

STANLEY BALDWIN (EARL BALDWIN OF BEWDLEY) 1867–1947; FRS 1927
Stanley Baldwin was born at Lower Park, Bewdley, Worcestershire, the only son of Alfred and Louisa née Macdonald. His father was a third generation ironmaster who became wealthy and over the years extended his business into tinplate in Monmouthshire, carried through some amalgamations, and turned it into a public company in 1902. In his childhood the family lived at Wilden; then after attending a preparatory school near Slough Stanley went to Harrow in 1881, from whence he entered Trinity College, Cambridge in 1885 where having obtained a first class in the first year he finished in the final year with a third class honours in the Historical Tripos in 1888. He then entered the family business, where he worked for 20 years and was second-in-command to his father, maintaining the family tradition of making the employees his friends. Apart from attending in his early business career a brief lecture course on metallurgy given by Turner at Birmingham he had no scientific training. In 1892 he married Lucy Ridsdale and there were two sons and four daughters from the marriage.

Baldwin became a parish and county councillor, and when his father died in 1908 he entered Parliament as member for Bewdley, continuing to hold this seat until he went to the House of Lords in 1937. His business interests included holding the post of Vice-Chairman of Baldwins. In his political career the offices that he held included Chancellor of the Exchequer (1922–23) and three periods as Prime Minister and First Lord of the Treasury (1923 –24; 1924–29; and 1935–37). When he retired from public life in 1937 he was created Earl Baldwin of Bewdley and was invested with the Order of the Garter. His admission to Fellowship of the Royal Society in 1927 was under the statute which allows the election of 'those who have rendered conspicuous service to the cause of science or are such that their election would be of signal benefit to the Society'.
ONFRS 1948–49, 6, 3–5. J.C. Irvine. DNB 1941–50, 43–51. T. Jones. Roy Jenkins, *Baldwin,* Collins 1987.

JOHN MACLEOD BALL 1948–; FRS 1989
Mill Hill School. Univ. of Cambridge: St John's College; BA (Mathematics) 1969. Univ. of Sussex: DPhil. 1972. SRC Postdoctoral Fellow, Heriot-Watt Univ.; Department of Mathematics and Lefschetz Centre for Dynamical Systems. Brown Univ. USA, 1972–74. Heriot-Watt Univ.: Lecturer in Mathematics, 1974–78; Reader, 1978–82, SERC Research Fellow, 1980–85, Professor of Applied Analysis, 1982–96. Univ. of Oxford: Sedleian Professor of Natural Philosophy, 1996–.
Applied analysis; phase transformations, crystal microstructure.

CLEMENT HENRY BAMFORD 1912–; FRS 1964
St Patrick's and King Edward VI Schools, Stafford. Univ. of Cambridge: Trinity College; Senior Scholar, 1931; Fellow, 1937. MA, PhD, ScD. Joined Special Operations Executive, 1941. Messrs. Courtaulds Ltd: Fundamental Research Laboratory, Maidenhead, 1948; Head of Laboratory, 1947–62. Univ. of Liverpool: Campbell Brown Professor of Industrial Chemistry, 1962–80; Dean Fac-

ulty of Science, 1965–82; Pro-Vice-Chancellor, 1972–75; Head of Department of Inorganic, Physical and Industrial Chemistry, 1973–78; Hon. Senior Fellow, Department of Clinical Engineering, 1980–96.
Polymer science and biomaterials science.

SIR JOSEPH BANKS 1743–1820; FRS 1766
Joseph Banks, the only son of William and Sarah Banks, came from a family established among the landed gentry. He was born in London, but his early years were spent at Revesby Abbey, Lincolnshire. He attended Harrow School from 1752 and then Eton from 1756. Botany became his goal in life, a subject in which he obtained formal training at Christ Church, Oxford, beginning in 1760. Having inherited wealth in 1764 he came to London, and travelled widely in various parts of the world, becoming a famous traveller and botanist. He had a long tenure of the Presidency of the Royal Society (1778–1820). Ensuing from his receiving a supply of Indian wootz steel, he arranged for investigations of its nature and properties to be caried out by George Pearson and James Stodart, which subsequently led to the involvement of Faraday in work on steels. In 1779 Banks married Dorothea Weston-Hugessen. He was made a Baronet in 1781, knighted (KCB) in 1795 and appointed a Privy Councillor in 1797.
DSB 1, 433–437. G.A. Foote. DNB.

JOHN BARDEEN 1908–1991; FMRS 1973
John Bardeen was born in Madison, Wisconsin, the second of five children of Charles Russell Bardeen, Dean of the University of Wisconsin Medical School, and Althea née Harmer. After schooling in Madison he entered the University of Madison at the age of 15; he majored in electrical engineering and having spent a year working in Chicago he graduated with a bachelor's degree in 1928, followed by a master's degree a year later. In view of the employment situation in the Depression he took a post in the oil industry with the Gulf Research and Development Corporation in Pittsburgh, where he devised a method of using a magnetic survey to study rock conformation.

In 1933 Bardeen enrolled as a graduate student at Princeton University, receiving a PhD in 1936; he worked on the theory of alkali metals and calculated a value for the work function of sodium. In the autumn of 1935 he had been elected a member of the Society of Fellows at Harvard and he extended his work on alkali metals. In 1938 Bardeen moved to the University of Minnesota as an Assistant Professor, where his interest in the theory of superconductivity emerged. However, in 1941 he went to the Naval Ordnance Laboratory in Washington, where his work was connected with magnetic mines. In 1945 he joined the Research Laboratories of the Bell Telephone Company at Murray Hill, New Jersey. He began the research on semiconductors that led to the invention of the transistor, for which he together with Walter Brattain and William Shockley received the Nobel Prize for Physics in 1956.

In 1949 Bardeen began an association with the Xerox Corporation, in which he provided the main theoretical support of the experimental work that led to the commercial development of xerography. Bardeen accepted an invitation to a chair at the University of Illinois in Urbana in 1951 where he stayed for the rest of his life. With his team he worked on the development of the theory of superconductivity. His collaboration in this activity with Leon Cooper and Robert Schrieffer, beginning in 1957, led in 1972 to the joint award of a Nobel Prize for Physics.
BMFRS 1994, 39, 21–34. Sir Brian Pippard FRS.

PETER BARLOW 1776–1862; FRS 1823

Peter Barlow who was born in Norwich had an ordinary school education but by his own efforts gained considerable scientific knowledge. He obtained by a competitive examination the post of Assistant Mathematical Master in the Royal Military Academy at Woolwich; subsequently he was promoted to professor, a post which he held until he resigned in 1847. He published his first book in 1811 on the theory of numbers and in 1814 a mathematical and philosophical dictionary, followed by his *New Mathematical Tables,* which represented a vast amount of accurate and highly useful information. Barlow's high ability became conspicuous by his *Essay on the Strength of Timber and other Materials,* published in 1817; this was based on the results of numerous experiments he made in the Woolwich dockyards and provided important data for engineering calculations. The experiments on the 'resistance' of iron relating to the design for the Menai suspension bridge were printed as an appendix to the third edition of his *Essay* (1826).

Magnetism was another area in which Barlow made important contributions, including work on the magnetic behaviour of iron, and he was awarded the Copley Medal in 1825. His work on optics included innovation in lens construction. He was also involved in experiments designed to provide practical data for steam locomotion, and on several occasions in the 1830s and 1840s served as a Commisioner of Railroads. He came to know leading engineers and architects and was much consulted on important works of construction.

Peter Barlow had two sons, both of whom were engineers and were elected Fellows of the Royal Society: Peter William Barlow (1809–1885; FRS 1845, a civil engineeer whose work involved railway projects and the construction of the Tower subway in London using cylinders of cast iron) and William Henry Barlow (see below, 1812–1902: FRS 1850). They revised their father's *Treatise on the Strength of Materials* which was published as a new edition in 1867.

PRS 1862–63, 12, xxxiii–xxxiv. DSB 1, 459, 460. H.L. Sharlin. DNB.

WILLIAM BARLOW 1845–1934; FRS 1908

William Barlow, who was born in Islington, inherited from his father an estate and building property business, through which he obtained financial independence and was able to devote himself to the study of crystal structure in his early 30s. He had an interest in physical science and marked mathematical talent, but did not have a formal scientific education. Most of his knowledge of formal crystallography was acquired through his contact with Henry Alexander Miers and William Jackson Pope. Then in Germany he studied the geometry of crystal structure. On his return to England he carried out theoretical crystallographic work throughout his life, including studies of space groups (published 1894); his theory of the properties of crystals was based on a model of the close packing of spheres representing atoms. His skills as a cabinet-maker were helpful in the construction of complex models of crystal structures.

ONFRS 1935, 4, 367–370. W.J. Pope. DSB 2, 460–463. W.T. Holser.

WILLIAM HENRY BARLOW 1812–1902; FRS 1850

William Barlow was born at Woolwich, the younger son of Peter Barlow, the engineer. Following education at home by his father he obtained practical training at Woolwich Dockyard and at the London docks. At the age of 20 he began a period of six years in Constantinople, Turkey, on engineeering work for a British company. Back in England in 1838 he entered railway engineering in which he was involved for many years. He also carried out scientific research which included the field of the theory of structures. Barlow was a member of a committee of engineers set up in 1868 to investigate the application of steel to structures. He recommended the advantages

of steel and in 1873 he was appointed to a Board of Trade enquiry; he was also a member of the Court of Enquiry into the Tay Bridge disaster; he was consulted in relation to the bridging of the Forth and was appointed (1881) to the Ordnance Committee. He married Selina Crawford Caffin, and they had four sons and two daughters. In the Royal Society he served as Vice-President from 1880–81.
DNB 2nd Suppl., 98, 99. WFS.

RICHARD MALING BARRER 1910–1996; FRS 1956
Robert Barrer was born in Wellington, New Zealand, the third of four children of Thomas Robert Barrer and Nina Agatha Rosamond née Greensill. The early period of his life was spent on his parents' farm, where his mother provided his schooling. When he was nine years old the family moved to Masterton, where he attended the Cole Street School, prior to his secondary education at Wairarapa High School from 1923–27. Barrer entered Canterbury University College in 1928 with a scholarship, and graduated with a BSc degree in mathematics, physics and chemistry in 1931, and an MSc, first class in chemistry in the following year. He then came to Clare College, Cambridge with an 1851 Exhibition Scholarship. He carried out research in the Colloid Science Laboratory, headed by Eric Rideal, and was awarded a PhD degree in 1935; he was elected by Clare College to a Senior Research Scholarship, and two years later he became a Research Fellow. He was awarded a full blue for athletics as a runner, and a half blue for tennis. In 1939 he married Helen Francis Yule and they had one son and three daughters.

From 1939–46 Barrer was Head of the Chemistry Department at the Technical College, Bradford. There followed a period from 1946–49 when he held an appointment as a Reader in Chemistry in London University. In 1948 he was awarded the ScD of Cambridge University. He was appointed Professor of Physical Chemistry at Imperial College in 1954, and served as Head of the Department from 1955–76; also he was Dean of the Royal College of Science from 1964–66. He became a Senior Research Fellow in 1977 and continued his research actively.

In his research at Cambridge Barrer developed what became a life-long interest in surfaces and membranes; in particular he carried out research on zeolites, and continued to do outstanding work and to become a world authority on these important materials. His industrial interests included acting as a consultant for the Union Carbide Corporation, USA. His research also covered extensive work on diffusion through porous membranes and inorganic porous materials.
BMFRS 1998, 44, 37–49. Lovatt V.C. Rees FRSE. *Who's Who*.

SIR WILLIAM FLETCHER BARRETT 1844–1925; FRS 1899
William Barrett was the son of the Rev. William Garland Barrett and was educated in Manchester. In 1869 he became a lecturer in physics at the Royal School of Naval Architecture. He also worked with Tyndall on sound at the Royal Institution. From 1873–1910 he was Professor of Physics at the Royal College of Science, Dublin. He was knighted in 1912. In 1916 he married Dr Florence Willey. His research in the 1880s included investigations of the electrical and magnetic properties of manganese steel and other steel alloys prepared by Hadfield; also he made observations on the allotropic changes in iron, reported in 1873.
Bulloch's Roll. *Nature* 1925, 2901, 115, 880, 881. Ref. Ch. 11.5.

ZBIGNIEW STANISLAW BASINSKI 1928–; FRS 1980
Lyceum of Krzemieniec, Poland. Polish Army Cadet School, Camp Barbara, Palestine. Univ. of Oxford: BSc, MA, DPhil, DSc; Research Assistant, 1951–54. MIT: Department of Mechanical Engineeering (Cryogenic Laboratory); staff member. National Research Council of Canada: Prin-

cipal Research Officer; Head of Materials Physics, 1956–87. McMaster Univ., Hamilton: Institute for Materials Research; Department of Mechanical Engineering, 1987–92; Department of Materials Science and Engineering, Research Professor, 1987–92.
Crystal defects and mechanical properties of metals; plasticity of metals, particularly in relation to the interaction and movement of dislocations.

LESLIE CLIFFORD BATEMAN 1915–; FRS 1968
Bishopshalt School, Uxbridge. University College, London: BSc Hons Chem., 1935; PhD, 1938; DSc 1955. Univ. of Oxford: Oriel College, 1940–41. National Rubber Producers Research Association: 1941–53; Director of Research, 1953–62. Controller of Rubber Research Malaysia: 1962–74. International Study Group; Secretary General; 1976–83. CMG 1965.
Chemistry and physics of rubber-like substances; technical-economic status of natural rubber and its developments.

LESLIE FLEETWOOD BATES 1897–1978; FRS 1950
Leslie Bates was the eldest of the six children of William Fleetwood Bates and Henrietta Anne née Pearce; he was born in Bristol where his father was a clerk. He was educated at the Hanham Road School before entering the Merchant Venturers Secondary School with a scholarship. In 1913 a city scholarship took him to Bristol University to study physics. Because of the war he took a BSc pass degree in physics and mathematics, combining his studies with work in the X-ray Department in a military hospital. In 1916 he qualified as a radiographer in the Royal Army Medical Corps; he was commissioned as a Lieutenant and served in India until 1920. Back in Bristol, and having been unsuccessful in obtaining a grant to study medicine, he became a Student Demonstrator in the Physics Department, where he collaborated with A.P. Chattock and W. Sucksmith in research on measurements of the gyromagnetic ratio.

In 1922 Bates obtained a grant to carry out research on radioactivity in the Cavendish laboratory in Cambridge under Rutherford leading to a PhD. An appointment as Assistant Lecturer in the Physics Department at University College London followed in 1924. In the following year he married Francis Furze Ridler, and they had a son and a daughter. In his career promotion took place to Lecturer (1928), Senior Lecturer (1929) and Reader (1930). At University College he pursued research on ferromagnetism; he continued in this field when in 1936 he became Lancashire-Spencer Professor of Physics at the University College of Nottingham where he worked until his retirement in 1965. During the 1939–45 war Bates was involved with the degaussing of ships to counteract magnetic mines; from 1946–76 he consulted for the Admiralty Compass Laboratory at Slough. From around 1954 he carried out some work in association with the Atomic Energy Research Establishment at Harwell. He served as Deputy Vice-Chancellor of Nottingham University from 1953–56, and received the CBE in 1966.
BMFRS 1983, 29, 1–25. N. Kurti FRS. DSB 17, 50, 51. T.I. Williams.

CECIL EDWIN HENRY BAWN 1908–; FRS 1952
Cotham Grammar School. Univ. of Bristol: graduated 1929; PhD, 1932. Univ. of Manchester: Assistant Lecturer in Chemistry, 1931–34; Lecturer, 1934–38. Univ. of Bristol: Lecturer in Physical Chemistry, 1938–45; Reader in Physical Chemistry, 1945– 48. Armament Research Department, Ministry of Supply, in charge of Physico-chemical Section, 1939–45. Univ. of Liverpool: Grant-Brunner Professor of Inorganic and Physical Chemistry, 1948–69; Brunner Professor of Physical Chemistry, 1969–73. CBE 1956.
Chemistry of high polymers.

ALEXANDRE EDMOND BECQUEREL 1820–1891; FMRS 1888

Alexandre Becquerel was the second son of Antoine-César Becquerel, Professor at the Muséum d'Histoire Naturelle. He joined his father at the Muséum at the age of eighteen to assist him. Subsequently he held several academic appointments prior to succeeding his father as Director at the Museum in 1878. His scientific work included contributions in electricity, magnetism, electrochemistry and luminescence; he identified many new phosphorescent substances.
PRS 1892, 51, xxi–xxiv. WC. DSB 1,555–556 J.B. Gough.

ANTOINE-CÉSAR BECQUEREL 1788–1878; FMRS 1837

Antoine-César Becquerel was born in Chatillon-sur-Loiret, France, where his father was the Royal Lieutenant. Having graduated from the École Polytéchnique, which he had entered in 1806, he commenced a military career in the Corps of Engineers, attaining the rank of captain in 1812. He was appointed Sous Inspecteur of the École Polytéchnique in 1813, in which year he married Aimée-Cecile Darlui. The following year he returned to active service, but in 1815 devoted himself to science. In 1838 he became the first holder of the Chair of Physics at the Muséum d'Histoire Naturelle. His important researches on minerals included the discovery of the piezo-electric effect and investigations of thermoelectric effects. Becquerel also worked on the voltaic cell and established that all chemical reactions can generate electricity. He developed a battery that could supply current at a reasonably constant electromotive force, with which he synthesised mineral substances. He wrote many papers in collaboration with his second son Alexandre Edmond who joined his father at the Museum at the age of 18 to assist him. He received the Copley Medal in 1837.
PRS 1843, 4, 22, 23. DSB 1,557–558. D.M. Knight.

ANTOINE-HENRI BECQUEREL 1852–1908; FMRS 1908

Antoine-Henri Becquerel was the son of Alexandre Edmond Becquerel and the grandson of Antoine-César Becquerel. He was born in Paris in the professor's house in the Museum, and received his early education at the Lycée Louis le Grand. He entered the École Polytechnique in 1872 and two years later he went as a pupil to the École des Ponts et Chaussées, (1874–77) to receive engineering training; from here he entered the administration of bridges and highways where he attained high rank. His teaching career began with an appointment as Demonstrator at the Polytéchnique in 1876. In 1878 he was appointed Aide-Naturaliste to the Museum under his father. From this time his professional activities were shared between the Museum, the Polytéchnique, and the Ponts et Chaussées. He succeeded his father in the Chair of Physics in l'École Polytechnique. Then in 1892 Henri was appointed Professor of Physics at the Muséum d'Histoire Naturelle, a chair previously occupied by his father and grandfather; he was also appointed in succession to his father to the Chair of Physics at the Conservatoire National des Arts et Metiers. He married Lucie-Zoë-Marie Jamin, who died a few weeks after the birth of their son; the following year he married E. Lorieux.

In Becquerel's earlier years his researches which led to his doctorate were mainly optical, including making observations by means of the light from certain phosphorescent crystals under infra-red illumination. Following Röntgen's discovery of X-rays he investigated the phosphorescence of uranium compounds after exposure to the sun's rays, and this led him to a series of experiments which revealed the phemomenon of radioactivity of uranium (Becquerel rays). Later, he showed that the radiation from radium consists of electrons.

Becquerel's genius was widely recognised, and his awards included the Rumford Medal in 1900, honorary membership of learned societies, and a Nobel Prize for Physics (with the Curies)

in 1903.
PRS 1909–1910, 83, xx–xxiii. WC. DSB 1, 1970, 557–561. A. Romer.

SIR GEORGE THOMAS BEILBY 1850–1924; FRS 1906
George Beilby was born in Edinburgh, the youngest son of George Thomas Beilby MD and Rachel née Watson and was educated at private schools and at Edinburgh University. Although in his career Beilby was primarily a chemical manufacturer and chemical engineer, he also had an important role in the fields of extraction and physical metallurgy. During the first 20 years of his professional life (1870–1890) his work centred on the manufacture of ammonia and oils from the Scottish shales. He invented a process for the manufacture of cyanide which was used in gold extraction by the Cassel Gold Extracting Company in Glasgow. In 1900 he joined the Board of the Castner–Kellner Alkali Company, who manufactured pure caustic soda; this was used as the raw material for the electrometallurgical extraction of sodium, an activity with which he continued.

Through his experiences as a manufacturer and his ensuing studies he became the leading authority on the scientific utilisation of fuel in Britain. He was appointed as the first Chairman and Director of the Fuel Research Board in the Department of Scientific and Industrial Research in 1917. In the war years his intense activity and great service to the nation included membership of Lord Fisher's Central Committee of the Board of Invention and Research. He was knighted in 1916.

One of Beilby's interests was the field of natural philosophy although the demands of his professional work did not allow him much spare time for this pursuit. He chose for his main work the study of matter in the solid state. His interests in manufacturing practice led him to an interest in the behaviour of metallic materials subjected to various thermal and mechanical treatments. He produced a theory on the hard and soft state in metals, and published a book *Aggregation and Flow of Solids*. In the field of surface behaviour of solids subjected to cutting, polishing etc he proposed the formation of a surface layer that behaves in a liquid-like manner. Concerning plastic deformation he proposed that a mobile layer formed at internal surfaces of slip.

Beilby rendered great services to education and he became a Member of the Governing Body of the Royal Technical College in Glasgow in 1900, and chairman in 1907. He married Emma Clarke Newman in 1877 and they had one son and one daughter; he and his wife were influential in assisting in the admission of women to the professions, especially to medicine. He served for several years on the Council of the Royal Society. In his career Beilby achieved success as a manufacturer and scientist and was also a great public servant.
PRS 1925 A, 109, i–xvii. HCHC. DNB 1922–30, 71, 72. C.H. Desch.

BERNARD FOREST DE BÉLIDOR 1693?–1761; FRS 1726
Bernard Bélidor was born in Catalonia, Spain, where his father, a French officer of dragoons, was serving; his mother was Marie née Herbert, and when she and her husband died within five months of the boy's birth, Bernard was brought up by his godfather's widow. He had a flair for practical mathematics, and his career included appointment as Professor of Mathematics at the Artillery School at La Fere; there he wrote text books and a technical manual. He saw military service in the war of Austrian succession, and eventually settled in Paris having attained the rank of brigadier. Although not a great pioneer in structural technology, he wrote important books on architecture and engineering, bringing together in a critical manner theories and practices of the time.
DSB 2, 581–582. C.C. Gillispie. Gen. Ref. 26.

SIR ISAAC LOWTHIAN BELL 1816–1904; FRS 1874

Isaac Lowthian Bell was born in Newcastle upon Tyne; he was the eldest son in a family of four sons and three daughters, of Thomas Bell, a native of Cumberland, and Catherine née Lowthian. In 1808 his father had entered the services of Messrs Losh and Co., merchants, who were embarking on the manufacture of both alkali and iron. Realising the importance of physical science in relation to industrial problems, Thomas gave his son a training in physics and chemistry. After completing his school education at Bruce's Academy, Newcastle, Bell spent some time in various centres: in Germany, in Denmark, at Edinburgh University, and at the Sorbonne in Paris; finally at Marseilles he studied a new process for the manufacture of alkali.

In 1835 at the age of 19 Bell entered the office of Messrs Losh, Wilson and Bell in Newcastle and a year later joined his father at the firm's ironworks at Walker to manage the rolling mills. In 1842 he married Margaret, second daughter of Hugh Lee Pattinson, the chemical manufacturer. Bell, in collaboration with others, including his brothers, built up a huge industrial enterprise, particularly in the field of iron and steel production, and also including mines and collieries. He also carried out important scientific researches on blast furnace practice.

Bell held many offices in public life. He was twice Mayor of Newcastle upon Tyne, and Deputy Lieutenant and High Sheriff for the County of Durham. He sat in Parliament for the Hartlepools constituency from 1875 to 1880. In 1885 he received a Baronetcy.

PRS 1906–07, 78, xiv–xviii. DNB 2nd Suppl. v.1, 132–135. WAB.

GUY DUNSTAN BENGOUGH 1876–1945; FRS 1938

Guy Bengough, the son of Major E.B. Bengough of Chislehurst, Kent, was educated at Malvern and at Selwyn College, Cambridge. Having studied metallurgy at the Royal School of Mines from 1899 to 1902 under Roberts-Austen, with whom he also obtained research experience for several months at the Royal Mint, he then became Lecturer in Metallurgy and Demonstrator in Chemistry at the Sir John Cass Institute, London. Two years later he took up a position as Metallurgist and Assayer with a tin syndicate in Burma. When he returned to England he lectured for a time in Birmingham University before taking an appointment in 1907 as a Lecturer in Metallurgy in Liverpool University.

Bengough's earliest work was on brass and other alloys of copper, for example concerning the relationships betwen microstructure and mechanical properties. His name, however, is chiefly associated with metallic corrosion. He became official investigator in 1911 to the committee set up by the Institute of Metals concerned with the failure of brass condenser tubes in ships and in land power stations; in 1911 he became official investigator to that committee and carried out a systematic study of the corrosion of marine condenser tubes.

When the war began in 1914 Bengough was gazetted to the Royal Garrison Artillery, and from 1915 to 1916 he was Captain and Adjutant to the Commandant, RA Mersey Defences; in 1916 he was seconded for research work at the request of the Admiralty and did similar work for the Royal Flying Corps. He returned in civilian life to his work for the Corrosion Committee at Imperial College. The seventh report of the Institute of Metals Committee was the last with which Bengough was connected. Bengough was transferred in 1928 to the newly established Chemical Research Laboratory of the Department of Scientific and Industrial Research at Teddington. Here he equipped a laboratory for corrosion studies, brought together a group of research workers and directed the department until his retirement in 1936, continuing then as a consultant. Bengough was also actively involved in the work of a committee established in 1928 by the Iron and Steel Institute. Then in 1938, he became a member and later chairman of a committee set up to deal with marine corrosion involving marine organisms.

Bengough's interests included mountaineering, rowing and swimming. In his later years he and his wife Constance Helen née Jelf-Sharp lived in Highgate Village, where he built a house and spent much time gardening.

Bengough was devoted to the field of corrosion and although he did not escape the controversies that are part of the history of the subject his opinions were always received with respect and attention.

ONFRS 1945–48, 5, 14, 169–178. C.H. Desch.

TORBERN OLOF BERGMAN 1735–1784; FRS 1765

Torbern Bergman was the son of Barthold Bergman and Sarah née Hägg; his father was Sheriff on the Royal estate at Katrineberg , Sweden. Having been educated in classical subjects at Skara, and also having received private instruction in natural history, he entered Uppsala University in 1756 where he studied mathematics, philosophy, physics and astronomy, graduating in 1756. He gained a doctorate in 1758 and became a Physics Lecturer at the University. Promotion to Associate Professor of Mathematics occurred in 1761 followed in 1767 by appointment as Professor of Chemistry, a subject that was new to him.

Bergman's interests were wide ranging; for example, while a student he made important contributions to natural history, and his early research in physical science included studies of the rainbow, twilight and the aurora borealis. In 1766 he published an important treatise on physical geography, which included a long account of minerals, and he soon became actively interested in mineralogy. In his teaching he believed in applying chemistry to mining and industry; he taught not only theoretical chemistry, but also new experimental methods, especially in mineral analysis, a field in which he worked and published extensively.

Bergman's important research in metallurgy included the phenomena associated with the solution of metals in acids involving quantitative studies of carbon in various ferrous materials. His published work included the treatise *De Praecipitatis Metallicis.* Bergman corresponded with scientists throughout Europe, and became a leading 18th century chemist.

DSB 2, 4–8. W.A. Smeaton. Gen. Ref. 21.

JOHN DESMOND BERNAL 1901–1971; FRS 1937

John Bernal was born at Nenash, county Tipperary, Ireland, the eldest of three sons and two daughters of Samuel Grange Bernal, a farmer, and Elizabeth née Miller. After attending a local school he came to England for education at Stonybrook and Bedford Schools. Having entered Cambridge University in 1919 he took the Mathematical Tripos Part I in 1920 obtaining a II i; the natural science tripos Part I, first class in 1922, and physics Part II, class II in 1923. While an undergraduate he wrote a thesis on crystallography dealing with the analytical theory of point group systems. Having moved to the Royal Institution, London to work with W.H. Bragg, he determined the crystal structure of graphite (1924) using single crystals, the rotation method, and the concept of the reciprocal lattice. In 1922 he married Agnes Eileen Sprague and they had two sons.

Bernal returned to Cambridge in 1927 to the newly created Lectureship in Structural Crystallography, where he worked on the determination of structures of biochemical compounds such as vitamin B1 and pepsin. He also carried out some work on the determination of complex metallic intermediate phases, including the delta-phase in the copper–tin system. In 1931 he was appointed an Assistant Director of Research. Other work was concerned with the structure of

liquids. In 1937 Bernal moved to Birkbeck College, London as Professor of Physics. With the outbreak of war in 1939, Bernal worked with the Ministry of Home Security; he carried out important war work, including serving as scientific adviser to Lord Mountbatten, the Chief of Combined Operations.
BMFRS 1980, 26, 17–84. D.M.C. Hodgkin. DNB 1971–1980, 53, 54. J.C. Kendrew.

DANIEL BERNOULLI 1700–1782; FRS 1750
Daniel Bernouilli was born in Groningen, Switzerland, the son of Jean Bernouilli and the nephew of Jacques Bernouilli, both of whom were mathematicians. He was educated in Basel, first studying philosophy and logic and obtaining a master's degree by the age of 16. In 1717 he began to study medicine and received a doctorate in 1721. His interest in mathematics, in which he had been given some instruction by his father and uncle, led to a publication in 1724 and the following year he was appointed to the chair of mathematics at the St Petersburg Academy. In 1732 Bernouilli became Professor of Anatomy and Botany in the University of Basel and in 1750 Professor of Natural Philosophy, a post which he held until 1777. His wide-ranging research interests included physics and mathematics. His publication *Hydrodynamica* (1738) was an outstanding contribution, representing a theoretical and practical study of fluids in terms of equilibrium, pressure and velocity.
DSB 2, 36–46. H. Straub.

PIERRE EUGÈNE MARCELIN BERTHELOT 1827–1907; FMRS 1877
Marcelin Berthelot was born in Paris, the second of three children of Jacques Martin Berthelot, a doctor of medicine, and Ernestine Sophie Claudine née Biard. At the age of 11 he entered the Collège Henri IV in Paris, where he graduated bachelier es lettres in 1847. He then attended courses in the Faculty of Medicine and the Faculty of Science, graduating in 1849. After working in a private laboratory he became a Demonstrator at the Collège de France in 1851, obtaining a doctorate in 1854. In 1859 he was appointed to a Chair in Organic Chemistry at the École de Pharmacie. He married Sophie Caroline Nicaudet in 1861 and they had six children. He was appointed to a Chair of Organic Chemistry at the Collège de France in 1865. During the Franco-Prussian war he was made President of the Committee for the defence of Paris. He entered the Senate in 1871, and had a career in politics which included appointments as Minister of Education and Foreign Minister. His research activities extended beyond organic chemistry and included thermochemistry (e.g. the invention of the bomb calorimeter). He made an outstanding contribution to the history of chemistry through his research on alchemical texts and his investigations of metallic objects from Mesopotamia and ancient Egypt. In the Royal Society Berthelot received the Davy medal in 1883 and the Copley Medal in 1900.
PRS 1908 A, 80, iii–x. DSB 2, 63–72. M.P. Crosland.

CLAUDE-LOUIS BERTHOLLET 1748–1822; FRS 1789
Claude-Louis Berthollet was born in Talloire, in the Italian Savoy, the son of French parents. He obtained the MD degree from the University of Turin in 1768 and lived in Piedmont until in 1772 he moved to Paris where he studied chemistry and also obtained an MD at the University of Paris in 1778. He joined the staff of the École Polytechnique in 1795. As a friend of Napoleon he visited Italy and Egypt on missions for him, and became a senator in 1804. His career included the promotion of various chemical and metallurgical manufactures, and work on chemical reactions and iron and steel production.
DSB 2, 73–82, S.C. Kapoor.

FERDINAND BERTHOUD 1727–1807; FRS 1764

Ferdinand Berthoud was born in Plancecourt, Switzerland, where his brother, to whom he became apprenticed at the age of 14, was a clock maker. His father Jean was an architect. Later he studied in Paris where he became involved in solving the problems of finding accurate longitude values at sea. In 1764 he was appointed as Horologer de la Marine. In the period mainly 1760–1768 his research led him to develop an accurate and practical marine chronometer with improvements from earlier instruments of John Harrison and Pierre le Roy. He continued his work for around 30 years and brought the marine chronometer close to its modern version. He was an outstanding chronologist and author of treatises on timekeeping.

Gen. Ref. 20, v. IV.

JÖNS JACOB BERZELIUS 1779–1848; FMRS 1813

Jöns Berzelius was born in the village of Vaversunda in East Gothland, Sweden. His father Samuel was a teacher who died when his son was four years old; his mother, Elizabeth Dorothea, married Anders Elmark, a pastor, and Berzelius and his sister were raised with the Elmark children. When Elizabeth died in 1788 the two Berzelius children were sent to the home of their maternal uncle. Berzelius spent a period in the Linkoping Gymnasium, and then in 1794 took a position as a tutor on a nearby farm where he developed a strong interest in flowers and insects. He had originally intended to become a clergyman, but instead, at the age of 17 he began to study medicine in the University of Uppsala, where in 1798 he received a scholarship to support his studies.

Having been introduced to chemistry by his oldest step-brother he began to study and to carry out experiments. He worked for a while in a pharmacy at Vadstena where he learnt glass blowing. Then in 1800 he became Assistant to the Chief Physician at the Medevi mineral springs, where a chemical analysis of these springs formed the subject of his first scientific publication. Having read about the voltaic pile he built one, and, in 1802 he used it in medical research; after graduating as doctor of medicine in 1804 he established himself in practice at Stockholm. The reputation that he had acquired through his researches led to his appointment as Assistant to the Professor of Medicine, Botany and Pharmaceutical Chemistry in Stockholm University; in 1807, when the chair became vacant, he was elected to it and occupied it until 1832. He now had an increased salary and access to a laboratory, and from 1810, the year in which he became President of the Swedish Academy of Science, he was able to devote most of his time to chemistry. He travelled widely including a visit to England in 1812, where he met Humphry Davy, and he became acquainted with most of the leading chemists of his day; while in Paris in 1819 he collaborated with Pierre Dulong in determining the gravimetric composition of water. He created such a strong impression that a number of younger chemists, including Friedrich Wohler came to work with him in Sweden. In 1818, he was elected Secretary of the Swedish Academy, which provided him with increased income and an excellent laboratory, and enabled him to concentrate substantially on research and voluminous writing.

The research carried out by Berzelius ranged widely, including electrochemistry and analytical chemistry. He was a brilliant experimentalist and his contributions included included a system of symbols for elements and compounds, determination of atomic weights, new methods of analysis and the discovery (or co-discovery) of several elements. In mineralogy he established a chemically based classification for which he received the Copley Medal in 1836.

Berzelius suffered from poor health for much of his life, and in later years this interfered with his scientific work. His marriage to Elizabeth Poppius took place when he was 56 years old, and there were no children. During his life, titles and decorations came to him in recognition of his achievements, including the conferment of the title of Baron by the the King of Sweden, and

election as a fellow of almost every scientific society in the world. His research had spanned a period of 50 years.
PRS 1849, 5, 872–876. DSB 2, 1970, 90–97. H.M. Leicester.

SIR HENRY BESSEMER 1813–1898; FRS 1879

Henry Bessemer was born in the village of Charlton, near Hitchin, Hertfordshire, of French Huguenot stock. His father, Anthony Bessemer, also a notable inventor and engineer, was born in the city of London, but, with his parents, moved to Holland in early childhood, and was apprenticed to an engineer. When he was 21 the elder Bessemer went to Paris and soon gained scientific distinction. Later he was appointed to a leading position in the Paris Mint, where his artistic skill in die-sinking, and his invention of a copying machine, brought him reputation and financial rewards. However, in danger in the the French Revolution, Anthony Bessemer lost most of his fortune and escaped to England; in Charlton he established a small factory for making gold chains, which later gave place to a type-founding enterprise.

Henry inherited from his father the qualities of energy, inventive talent and artistic feeling; apart from the time devoted to an elementary education his early years were spent in his father's workshop. At the age of 17 he came to London to seek his fortune. Using his knowledge of easily fusible alloys, and in casting them, he traded in art work of white metal, and afterwards in copper-coating such castings, the earliest practical application of electro-plating. A natural step was to embossing metal, card, and fabrics, greatly assisted by his skill as a draughtsman, and his ability as a die-sinker.

Bessemer's first great invention stemmed from his connection with Somerset House (through the annual art exhibitions) and the attention he was then paying to stamping and embossing. At that time (about 1833) fraud on the government by the repeated use of stamps fixed to deeds was a serious problem involving an annual loss to the revenue of £100,000. Bessemer invented perforated dies so that a date could be indelibly impressed on each stamp, thus making the fraud impossible. The promise of a permanent official appointment as a result of his gift of the invention to the government was not kept. In 1833 Bessemer married Anne Miller by whom he had two sons and a daughter.

Following technical ventures involving the processing of plumbago and, the invention of a type-composing machine, Bessemer in about 1840, turned to the manufacture of bronze powder and gold paint, an ancient industry from China and Japan and very successfully imitated in Germany. After many trials and failures he produced material at least equal to the German product at a price to defy all competition. His process was operated by a few of his relatives for nearly 40 years, and from the substantial revenues that Bessemer derived he was able to fund the development of his third great invention concerning the production of steel.

The most important stage of Bessemer's career, namely his invention of a steelmaking process, dates back to the Crimean war, and the need for improvements in British artillery. He began work to produce a stronger material, and developed a process in which molten pig iron from the blast furnace or cupola was run into a large tipping vessel – the Bessemer converter – with an air blast being introduced through tuyeres to pass up through the charge; associated patents were established in the period December 1855 – March 1856. His extensive experimentation led, after initial problems, to the successful and world wide introduction of his process on an industrial scale.

When Bessemer's more active and financial interest in steel manufacture had ceased he turned his attention to other matters. Among these the invention which most attracted public attention was his swinging saloon for sea-going vessels; however, there were disappointing results and the venture was abandoned. Over a period of around 60 years his inventions led him to 120 patents. His later years were years of busy leisure. He erected a fine observatory at Denmark Hill, and devoted much time to the construction of a telescope and to a mechanism for grinding and polishing lenses. From this he was led to experiments on the application of solar heat to achieve high temperatures.

Bessemer was the recipient of many honours, including a knighthood in 1879 in recognition of his services rendered to the Inland Revenue Office 40 years before. He received the Freedom of the City of London in 1880; in the United States six manufacturing towns and one county were named after him.

Ref. Ch. 7.1, 2.

HANS ALBRECHT BETHE 1906–; FMRS 1957

Goethe Gymnasium, Frankfurt/Main. Universities of Frankfurt and Munich: PhD, 1928. Univ. of Frankfurt: Instructor in Physics, 1928–29, then at Stuttgart. Universities of Munich and Tübingen: 1930–33. Univ. of Manchester: 1933–34. Univ. of Bristol: Fellow, 1934–35; Assistant Professor, 1935; Professor, 1937–75. Cornell University: 1935–37; John Wendell Anderson Professor of Physics, 1935–1975. Los Alamos Scientific Laboratory: Chief of Theoretical Physics Division, 1943–46. Bakerian Lecturer 1993. Nobel Prize for Physics, 1967.

Atomic and nuclear processes. Electron densities.

HARSHAD KUMAR DHARAMSHI HANSRAJ BHADESHIA 1953–; FRS 1998

City of London Polytechnic, 1973–75. BSc, Hons Metallurgy. Univ. of Cambridge: Department of Metallurgy and Materials Science, Research Student, 1976–79, PhD; Science Research Council Research Fellow, 1979–81; University Demonstrator, 1981–86; Department of Materials Science and Metallurgy, University Lecturer, 1986–94; Reader in Physical Metallurgy, 1994–; Royal Society Leverhulme Trust Senior Research Fellow, 1996–1997. Armourer and Brasiers' Company Award, 1997.

Physical metallurgy; phase transformations in steels; microstructural modelling.

BRUCE ALEXANDER BILBY 1922–; FRS 1977

Dover Grammar School. Univ. of Cambridge: Peterhouse College, BA. University of Birmingham: PhD. Admiralty: 1943–46. Univ. of Birmingham, 1946–51. Univ. of Sheffield: Royal Society Sorby Research Fellow, 1951 57; J.A.Andrew Research Fellow, 1957–62; Reader in Theoretical Metallurgy, 1958–62; Professor 1962–66; Consultant; Professor of Theory of Materials, 1966–84.

Theory of dislocations and its applications to the deformation, transformation and fracture of metallic crystals.

ROBERT BINGLEY ?–1847?; FRS 1809

Robert Bingley was the King's Assay Master at the Mint and resided at the Tower of London.

Bulloch's Roll.

JEAN-BAPTISTE BIOT 1774–1862; FMRS 1815

Jean-Baptiste Biot was born in Paris, France, the son of Joseph, of peasant stock who had risen socially to hold a position in the treasury. Jean-Baptiste attended the Collège Louis-le-Grand,

Paris, where he distinguished himself in the classical curriculum; on leaving the school around 1791 he took private lessons in mathematics. His father intended him to have a career in commerce, but he took the opportunity of the Revolutionary War to join the army as a gunner in 1792. However he returned to mathematics, entering the Écoles des Ponts et Chaussées in 1794, but soon after this when the École Polytechnique was founded he transferred there and showed his outstanding ability. He married the daughter of Antoine Francois Brisson in 1797, and after graduating in the same year he was appointed Professor of Mathematics at the École Central of the Oise department at Beauvais. There followed appointments as Professor of Mathematics at the Collège de France (1800) and Assistant Astronomer at the Bureau des Longitudes (1806). He was appointed Professor of Astronomy at the University of France at the Paris Faculté des Sciences in 1808, where he also taught physics related to his research and gave courses on light, sound and magnetism; he retired in 1849.

Of metallurgical interest was his work to determine the velocity of sound in metals and thermal conductivities of metals. He was also concerned with the chemistry of the voltaic pile. Among his other researches his important contributions on the polarisation of light led to the award of the Rumford Medal of the Royal Society in 1840.

PRS 1862–63, 12, xxxv–xlii. DSB 2, 133–140. M.P. Crosland.

JAMES DEREK BIRCHALL 1930–1995; FRS 1982

James Birchall, who was born in Leigh, Lancashire was the son of David Birchall and Valetta née Marsh, who died in childbirth. He went to the Sacred Heart School at Leigh in 1935 and left school at the age of 14. He went to work at an industrial laboratory, Sutcliffe Speakman, as a General Industrial Assistant and was promoted to Assistant in the research laboratory. When he was 17 he worked under Dr Oliver Coligny de Champfleur Ellis on activated carbon and flame. In 1950 he left the firm and spent a year as personal assistant to Dr Ellis. The following year he joined John Kerr and Co.(Mc) Ltd. and obtained a Higher National Certificate in chemistry at Wigan and District Mining and Technical College. From 1952–54 he did National Service with the Royal Corps of Signals. While at John Kerr his work had included the combustibility of various substances and he had filed his first patent relating to fire extinguishers. When he had completed his army service he worked on developing several inventions. In 1956 he married Pauline Mary Jones and they had two sons.

In 1957 he joined ICI Salt Division as an Assistant Technical Officer and when the Salt Division was taken over by the Alkali Division in 1961 he was promoted to Technical Officer. His activities including work on crystallisation relating to the manufacture of 'dendritic' salt and also on the extinction of fire. Birchall became a Research Leader in 1968 and extended his interests in inorganic chemistry. In 1974 he was promoted to Senior Research Associate.

As a materials scientist Birchall made many original and important contributions in the field of inorganic materials. These included investigations of the mechanical properties of silicates and carbonates in nature, e.g. shells, the development of fibres of refractory oxides such as alumina for use in metal matrix composites and for insulation purposes, and the structure and properties of cements.

Birchall was appointed as Professor of Inorganic Chemistry at Keele University in 1992 and a Non-executive Director of Ceramic Research in 1994. He held a number of visiting and associate professorships e.g. at the University of Surrey and at MIT; from 1977–79 he was a visiting Research Fellow at Wolfson College, Oxford. He was awarded the OBE in 1990 and the Royal Society Armourers and Brasiers' Company award in 1993. The output of his work is evidenced

by his numerous patents and he received the Queen's Award for Industry in 1983.
BMFRS 1997, 43, 87–104. Anthony Kelly, CBE. FRS.

GOLDING BIRD 1814–1854; FRS 1846
Golding Bird was born in Downham, Norfolk. Following education at a private school he was apprenticed in 1829 to an apothecary in London. In 1833 he became a student at Guy's Hospital, and was licensed to practise in 1836. He worked in London in general practice, and in 1836 he was appointed Lecturer in Natural Philosophy at Guy's Hospital. In 1838 he took the MD degree from St Andrews (without residence) and MA in 1840. He became LRCP in 1840 and FRCP in 1845. In 1843 he became Assistant Physician at Guy's Hospital. He was married in 1842 and had five children. Bird was keenly interested in natural sciences and his work included electrodeposition.
DNB.

MOSES BLACKMAN 1908–1983; FRS 1962
Moses Blackman, or Morris Blackman, as he came to be known by his friends, was born in South Africa, the son of Joseph and Esther Blackman. He spent his early childhood in Capetown, where his father who had emigrated from Russia had established a dairy. In about 1915 the family moved to Upington in Cape Province where his father became a rabbi to the Jewish community and where Moses attended a small school. In 1922 a move to Grahamstown occurred where he attended the Victoria Boys' High School. A scholarship took him to Rhodes University College in 1925 to take a BSc degree and to continue on an MSc course in physics. When he had spent a year at Rhodes as a demonstrator he went to Göttingen in 1931. After about a year he began to work with Max Born on crystal physics and was awarded a DPhil in 1933. Blackman obtained a Beit Fellowship at Imperial College, London in the Mathematics Department where he worked on the vibrational spectrum of crystals and heat capacity, and obtained a PhD degree. There followed a period in Cambridge with a senior research award from the Department of Scientific and Industrial Research; his research led to his third PhD degree.

In 1937 Blackman joined the Physics Department at Imperial College, headed by G.P. Thomson, as an Assistant Lecturer in the electron diffraction group. He became recognised as the theoretical physicist in the department and worked in association with Thomson on various projects including the application of dynamical theory to electron diffraction. From 1942–45 he was involved in projects for the Ministry of Home Security. After the war he continued his work on crystal physics and electron diffraction and became Head of the Diffraction Group in 1952. Promotion to Reader was followed by the award of a Personal Chair in 1959; in the same year he married Anne Olivia Court. Projects in the electron diffraction group with which he was involved included epitaxy, electron microscopical studies of thin metal films and the effect of particle size on the melting of metal droplets,for example, of tin.
BMFRS 1987, 33, 49–64. D.W. Pashley, FRS.

BREBIS BLEANEY 1915–; FRS 1950
Westminster City School. Univ. of Oxford: St.John's College; Lecturer in Physics; Balliol College. Harvard Univ. and MIT 1949. Univ. of Oxford: University Demonstrator and Lecturer in Physics, 1945–57; Dr Lees Professor of Experimental Physics, 1957–77; Warren Research Fellow of the Royal Society 1977–80. Hughes Medal 1962. CBE 1965.
Electrical and magnetic properties; electron paramagnetic resonance.

NIELS HENRIK DAVID BOHR 1885–1962; FMRS 1926

Niels Bohr was the second of three children of Christian Bohr, Professor of Physiology at the University of Copenhagen, and Ellen née Adler. He was born in Copenhagen and studied at the University, carrying out his first research in 1906, while still a student. Having presented his doctoral thesis concerning the electron theory of metals in 1911 he went to Cambridge and studied with J.J. Thomson. In the following year he spent several months with Ernest Rutherford in Manchester and developed a model of the atom with electrons in rings around the nucleus. In 1912 he returned to Copenhagen as a Lecturer and also married Margrethe Norlund, a marriage from which they had six sons. In 1913 he advanced his work on atomic structure and went back to Manchester for two years before returning to Denmark where he was appointed to a chair; in 1920 he became Director of the Institute of Theoretical Physics established for him by the authorities. Bohr's interests turned to nuclear reactions in the 1930s. When Denmark was occupied the Germans in 1940 he was active in the Resistance; in 1943 he escaped to Sweden and then came to England and to the USA, where he was involved in the development of the atomic bomb. Later he became active in efforts to control nuclear weapons.

Bohr's outstanding contributions to the understanding of the atom were recognised by the award of the Nobel Prize for Physics in 1922. In the Royal Society he was awarded the Copley Medal in 1938.

BMFRS 1963, 9, 37–54. Sir John Cockcroft, OM. DSB 2, 239–254. Leon Rosenfeld.

WILLIAM ARTHUR BONE 1871–1938 ; FRS 1905

William Bone was born in Stockton-on-Tees, the eldest son of Christoper Bone and Mary Elizabeth née Hutchison. He was educated at Middlesbrough High School, the Friends' School at Ackworth, and at Stockton High School. He then attended the Leys School at Cambridge for a year. His uncle was the Manager of the Skinningrove Ironworks, and William spent spare time there, which was influential in his choice of career. He entered Owens College, Manchester in 1888 and graduated in chemistry three years later; then he continued research at the College. He married Kate Hind in 1893 and they had one son and two daughters; following his wife's death he married Mabel Isabel Liddiard in 1916. In 1896 he won a scholarship and went to study for a year in Heidelberg. On his return to England Bone was appointed Head of the Chemistry Department at Battersea College, London. In 1898 he returned to Owens College as Lecturer in Chemistry and Metallurgy. In 1906 he moved to the University of Leeds to organise a Department of Fuel Technology; he was appointed as the first Livesey Professor of Coal, Gas and Fuel Industries in 1910. Bone moved to the Royal College of Science at Imperial College, London, in 1912 to establish a Department of Fuel Technology, as Professor of Chemical Technology.

Bone's principal research field concerned fuels and their combustion, and his work included organic combustion and catalytic combustion; also he worked on iron smelting, including the chemical reactions in the blast furnace. He was awarded the Davy Medal in 1936 and was the Bakerian Lecturer in 1932.

ONFRS 1936–38, 2, 587–611. G.I. Finch and A.G. Egerton. DNB 1931–40, 85–86. A.G. Egerton.

MAX BORN 1882–1970; FRS 1939

Max Born was the son of Gustav Born, a professor at the University of Breslau, and Margaret née Kaufmann. Max attended the König-Wilhelm Gymnasium in Breslau and in 1901 he entered the University, where he studied a wide range of subjects; he also studied for summer semesters in 1902 and 1903 at Heidelberg and Zurich respectively. In 1904 he went to the University of

Göttingen, where he was awarded a doctorate in physics and astronomy in 1907. He also spent a period of six months at Caius College at the University of Cambridge. Returning to Breslau he then moved to Göttingen, where in 1912 he obtained an appointment as Privatdozent. In 1913 he married Hedwig Ehrenberg and they had two daughters and a son.

Born moved in 1915 to a professorial appointment at the University of Berlin; military service involved commissioned rank for research duties concerning artillery sound ranging. He moved to Frankfurt-am-Main in 1919 as Professor Ordinarius and Director of the Institute of Theoretical Physics. In 1921 he was appointed to the post of Director of the Physical Institute of the University of Göttingen. In 1933, with the rise of Nazism, Born was deprived of his chair, and came to England to Cambridge, where he held the Stokes Lectureship, He was appointed to the Tait Chair of Natural Philosophy at the University of Edinburgh in 1936 and held this post until his retirement in 1953, when he returned to Germany.

Born's research in solid state physics made important contributions on the structure of matter; this included lattice dynamics, a subject in which he published jointly with von Kármán a seminal paper in 1912. His achievements brought him may honours including the award of the Nobel Prize for Physics in 1954.

BMFRS 1971, 17, 17–52 N. Kemmer FRS and R. Schlapp. DSB Suppl., 1, 39–44. A. Hermann.

ROGER JOSEPH BOSCOVICH (RUGGIERE GUISEPPE BOSKOVIC) 1711–1787; FRS 1761

Roger Boscovich was born in Dubrovnik in 1711, the son of Nikola Boskovich, a merchant, and Paula née Bettera. He began his education in the Jesuit College at Dubrovnik and then studied in Rome, first at the novitiate of Sant' Andra, and then at the Collegium Romanum, where he was an outstanding student. In 1735 he began to study Newton's Optics and Principia, and five years later was appointed Professor of Mathematics at the Collegium Romanum. He became a Jesuit and his career and his scientific work were particularly wide-ranging, involving astronomy, optics, mathematics, mechanics and natural philosophy; among specific topics were cohesion and extension of solids and various physical and chemical properties of matter. In 1759 and 1760 he spent periods in Paris and London and then travelled widely for several years in various countries in Europe, before returning to Rome in 1763. Boscovich was apppointed Professor of Mathematics at Pavia University in 1764, and played a leading role in organising and directing the Jesuit Observatory near Milan. From 1770–72 he was in Milan; he then returned to Paris where he became Director of Optics for the Navy. Having returned to Italy around 10 years later he settled in Bassano.

Bulloch's Roll. DSB 2326–2332, Zeljko Markovic.

JAMES THOMSON BOTTOMLEY 1845–1926; FRS 1888

James Bottomley was born in Belfast, the eldest son of William Bottomley, a propsperous merchant, and Anna née Thomson, sister of William Thomson (later Lord Kelvin). He studied at Queen's College, Belfast and at Trinity College, Dublin, and became assistant to Thomas Andrews of Belfast. Having been a Demonstrator in Chemistry and then in Physics at King's College, London, he went to Glasgow as private assistant to his uncle, then Sir William Thomson. He was appointed Demonstrator and carried out varied research, including topics in the field of electrical engineering. He was the first to make electrical measurements on a specimen of manganese steel made under the Hadfield patents; his observations were reported to the British Association in 1885. Bottomley was twice married; to Annie Heap, by whom he had two sons, and to Eliza Blandy.

PRS 1927, A 113, xii–xiii.

MATTHEW BOULTON 1728–1809; FRS 1785

Matthew Boulton was born in Birmingham, where his father Matthew, had long been in trade as a silver stamper and piercer. He worked with his father and by the time his father died in 1759 he had begun to extend the business. In 1760 he married Anna Robinson whose dowry enabled him to found the famous Soho Works, which opened in 1762 and which acquired a reputation for high quality work. Boulton set out to improve workmanship and also the artistic merit of the products. James Watt having visited Soho was keen to obtain Boulton's help and to use the resources of the works in perfecting his engine; the formation of a partnership ensued. In 1788, using coining presses in the works at Soho, Boulton made coins for the East India Company, foreign governments and colonies. He also undertook, beginning in 1797, the production of new copper coinage for Great Britain and supplied machinery for the new Mint. He had extensive metallurgical interests, including assaying and materials for engines. Boulton was a leading member of the Lunar Society, and was a prominent figure in the science of his time.
DNB. Ref. Ch. 2.7.

FRANK PHILIP BOWDEN 1903–1968; FRS 1948

Philip Bowden was born in Hobart, Tasmania, the fifth of six children of Frank Prosser Bowden, telegraph and telephone manager for Tasmania, and Grace Elizabeth née Hill. Bowden attended the Hutchins School, Hobart. He took a post as a Junior Laboratory Assistant with the Electrolytic Zinc Company and entered the University of Tasmania in 1921 to study science. Because of illness in his second year he spent six months on an Australian inland station in New South Wales and then, fully recovered, he returned to the University and obtained a BSc degree with distinctions in 1924. He was co-author of papers on electrochemistry in 1925 and 1926 and was awarded the MSc degree with first-class honours. Bowden was the first recipient of the Electrolytic Zinc Company's Scholarship which enabled him to continue his work for a year. In 1926 with an 1851 Scholarship he left for England to work in Cambridge.

Bowden entered Gonville and Caius College in 1927 as a Research Student and began investigations of electrode potentials with Eric Rideal in the Physical Chemistry Department. He became friends and a research collaborator with C.P. Snow. In 1931 Bowden married Margaret Hutchinson from Tasmania, by whom he had three sons and one daughter. Bowden became a University Demonstrator and then a Lecturer (1937). He also received College appointments as a Fellow and Director of Natural Sciences and was awarded the Cambridge DSc in 1938.

Bowden's work in the first dozen years at Cambridge achieved considerable recognition, first in electrochemistry, and secondly in surface phenomena related to friction and lubrication. In the latter field he attracted the attention of industrial companies in England and abroad. In 1937 Shell established under his supervision a small research unit on wear and lubrication. The Air Ministry, the Fuel Research Board and the War Office showed increasing interest. His research with various collaborators included the formation of Beilby polish layers and the measurement of the real area of contact between stationary and sliding surfaces; he also investigated the friction of ice and snow and developed his ideas of the physics of skiing.

In 1939, after a lecture tour in America, Bowden decided to return via Australia. When the war began Bowden was apointed to the staff of the Council for Scientific and Industrial Research (forerunner of the Commonwealth Scientific and Industrial Research Organization, CSIRO) in Melbourne in charge of a section called 'Lubricants and Bearings'. By the end of the war the team had grown to nearly 20 members; (later it became known as the Division of Tribophysics). Work included the initiation and growth of explosions, the evaluation of special lubricants for machine tools and for aircraft, and the development of satisfactory casting techniques for the production of aircraft bearings.

Bowden returned to Cambridge in 1945 and re-built his research group, with support from the Ministry of Supply (Air). He became a Reader in Physical Chemistry in 1946. Work continued on explosives, friction and lubrication; other work was concerned with impact, erosion by water droplets, fracture, the study of surface forces and the mechanical and structural properties of high-temperature solids. When the department began to move into new accommodation Bowden became a Reader in Physics (1957). His laboratory, which acquired the title of the physics and chemistry of solids, became a sub-department of the Cavendish laboratory, and he was appointed to a Personal Chair in Surface Physics in 1966.

Bowden had a keen interest in industrial co-operation and in 1953 he was appointed adviser to Tube Investments Ltd. In the following year he established, for them, a research laboratory at Hinxton Hall, near Cambridge, aiming to contribute to fundamental scientific knowledge and to the commercial requirements of the Company.

As an experimental scientist Bowden showed great originality in his elegant researches. His interests included art, literature, the theatre and also mountaineering and skiing. Among the awards that he received were the CBE (1956) and in the same year the Rumford Medal.

BMFRS 1969, 15, 1–38. D. Tabor FRS. DNB 1961–1970, 123–125. David Tabor.

DAVID KEITH BOWEN 1940–; FRS 1998

Christ's Hospital. Univ. of Oxford, St. Edmund Hall: MA 1966, DPhil Metallurgy 1967, SRC Research Fellow 1966–68. University of Warwick, Department of Engineering: Lecturer, 1968–78; Senior Lecturer, 1978–85; Reader, 1985–89; Professor of Engineering, 1989–97. Bede Scientific Inc 1996–. FREng 1997.

Microscopy. High resolution X-ray diffractometry and topography.

ROBERT BOYLE 1627–1691; FRS 1663

Robert Boyle was the seventh son and fourteenth child of Richard Boyle, the first Earl of Cork, by his second wife, Catherine née Fenton. He was born to affluent circumstances at Lismore Castle, in the province of Munster, Ireland. His mother died when he was three years old, and at the age of eight, having had a private tutor he was sent to Eton, whence after nearly four years he went to his father's recently purchased estate of Stalbridge in Dorsetshire. In 1638 with an elder brother, and a French tutor, he travelled via Paris and Lyons to Geneva, for a period of 21 months, where he received tutoring including practical mathematics. During subsequent travels, including a stay in Florence, he mastered Italian, and was introduced to the 'new science'. With the outbreak of the Anglo-Irish war and the Civil War in England they set off to return to England, where they eventually arrived in 1664 after various adventures. After his father had died, Robert went to the manor of Stalbridge, which he had inherited, and where he set up a laboratory.

His interests having turned to medicine, agriculture, and then to chemistry, Boyle soon became a skilful experimenter and original chemical thinker. He became part of the 'Philosophical College' in London, one of the groups which were later incorporated as the Royal Society, and he rose to a leading role. In 1654, having returned from visits to his estates in Ireland, he settled in Oxford. There he linked up with other leaders of the English scientific movement. His lodging was one of the two regular meeting places of the group; he erected a laboratory and engaged Robert Hooke as his chemical assistant. Using the air pump constructed by Hooke he carried out work leading to 'Boyle's law', reported in 1662.

Boyle settled in London in 1668 in Pall Mall, having been one of the founders of the Royal Society; he was elected President in 1680 but declined to serve, for reasons of conscience in relation to the oaths. He had undergone a conversion experience at the age of 14, and this led to a devout religious commitment throughout his life. By the age of 21 he had already written on ethical and religious subjects; he supported projects for the dissemination of the Scriptures and his activities included being Governor of the Society for the Propagation of the Gospel in New England. He suffered considerably from ill health for much of his life and he never married.

His prolific writings over the period from 1660 to 1691 brought Boyle an immense reputation. His book the *Sceptical Chymist or Chymico-Physical Doubts and Paradoxes* (1661) has been seen as the first truly scientific work, insisting that theories required validation by experiment. His work based on experimental results contributed greatly to the foundations of modern science.
DSB 2, 377–382. DNB. Gen. Ref. 20, v. III.

SIR CHARLES VERNON BOYS 1855–1944; FRS 1888

Charles Boys was born at Wing, Rutlandshire, the eighth child of the rector, the Rev. Charles Boys, and Caroline Goodrich née Dobbie. He entered Marlborough College in 1869 and then, from 1873–76, he was a student at the Royal School of Mines in London; his studies included physics, taught by Frederick Guthrie, and chemistry, but he was largely self-taught in mathematics. He graduated in mining and metallurgy and then, having spent a short period of work in a colliery, he became an assistant to Guthrie. In 1881 he married Marion Amelia Pollock and they had one son and one daughter. Having been appointed a Demonstrator in Physics in 1881 he became an Assistant Professor of Physics in 1889. Opportunities for further promotion were rare, and in 1897 he took an appointment with the Metropolitan Gas Board as a Referee. Three years later he became one of the three Gas Referees for the whole of Britain. He retired formally in 1939.

Boys was a highly skilled experimenter and an inventor, particularly in the design and construction of instruments for making highly accurate physical measurements. He invented a 'microradiometer' (1890) for the detection of infrared radiation; in this connection he developed a technique for producing very fine fused quartz fibres, which were superior for use as galvanometer suspensions, as compared with silk, metals or glass. He produced these fibres by firing a sample of molten quartz from a crossbow. Subsequently he constructed a sensitive torsion balance, incorporating a quartz fibre suspension, and made an important measurement of Newton's gravitational constant. Quartz fibres came into wide use in instruments. Boys had wide-ranging professional interests; he served on the Ordnance Committee at Woolwich, acted as an expert witness in patent cases, and also took out numerous patents. In the Royal Society he was awarded a Royal Medal (1896), and the Rumford Medal (1924). He was knighted in 1935.
ONFRS 1944, 4, 771–788. Rayleigh. DSB 15, Suppl. I, 59–61. R.V. Jones. DNB 1941–50, 96–97. A.H. Cooke.

ALBERT JAMES BRADLEY 1899–1972; FRS 1939

Albert Bradley was born in Chesterfield one of two sons of Henry Bradley who had taken a degree in pharmacy and became a retail chemist, and Amy née Bingham. Bradley commenced studies in chemistry at Manchester University in 1916, but at the end of the first year he entered the Officers Training Corps at Scarborough and Oxford. In the spring of 1918 he went to France with the Sherwood Foresters and was seconded to the 9th Welsh Regiment, which had suffered many casualties; this experience greatly affected him. In 1919 he returned to his university studies and after graduating in 1921 he stayed on to do research obtaining an MSc in the following year. He then approached W.L. Bragg, the newly appointed Professor of Physics, and was offered a grant to work on crystal structures, thus becoming Bragg's first research student. He had some background in crystallography from lectures from Henry Miers and was awarded a PhD in 1924 for a thesis entitled 'On the investigation of crystal structure'.

Bragg asked Bradley to explore the potentialities of the X-ray powder technique. In 1926 when Bradley was appointed an 1851 Exhibitioner he went to the Institute for Metals Research in Stockholm, where Westgren and Phragmen were working on the technique. While in Sweden Bradley solved the alpha–manganese structure and when he returned to Manchester in 1927 he proceeded to solve the gamma–brass structure by the powder method. He became the world's expert in X-ray powder photography. In 1929 Bradley married Marjorie Dinnis and they adopted a son in 1936. Bradley became Royal Society Warren Research Fellow in 1932 holding this post until 1938; during the last year of this post he was in the Metallurgy Section of the National Physical Laboratory to which Bragg had moved. In 1938 he became Assistant Director of Research at the Cavendish Laboratory, when Bragg moved to Cambridge.

Bradley's work included the redetermination of the structure of gallium and the determination of the $NiAl_3$ structure. Also he developed methods for the precision measurement of lattice parameters. He became involved in practical applications of his methods, and he collaborated with Sucksmith in correlating magnetic structures and crystal structure. This work led him into detailed studies of alloy constitution of various binary and ternary systems e.g. iron–nickel–aluminium. His collaborative work included projects with research students and he was unmatched in his ability to extract information from powder photographs.

Stemming from his experiences in the first world war Bradley was keenly affected by the 1939–45 conflict and finally had to relinquish his charge of the crystallography section in the Cavendish Laboratory. Arrangements were made for him to have an appointment at the BSA Research Centre at Greystokes Hall, Sheffield. Although the start was promising, the improvement did not continue and Bradley retired but was able after medical treatment to live a normal and happy life.

BMFRS 1973, 19, 117–128. H. Lipson.

DONALD CHARLTON BRADLEY 1924–; FRS 1980

Hove County School for Boys. Univ. of London: Birkbeck College; BSc (Hons Chemistry), PhD, DSc. British Electrical Allied Industries Research Association: Research Assistant, 1941–47. Birkbeck College: Assistant Lecturer in Chemistry, 1949–52; Lecturer in Chemistry, 1952–59. University of Western Ontario, Canada: Professor of Chemistry, 1959–64. Univ. of London: Queen Mary College: Professor of Inorganic Chemistry, 1965–87, Head of Chemistry Department, 1978–92.

Synthesis and structure of metallic assistant compounds for chemical vapour deposition for microelectronics.

SIR WILLIAM HENRY BRAGG 1862–1942; FRS 1907

William Bragg was born at Westward, near Wigton, in Cumberland. He was the eldest of the three sons of Robert John Bragg, who after serving as an officer in the Merchant Navy had purchased a farm at Westward. William's mother, Mary Wood, the daughter of the vicar of Westward, died when he was barely seven, and he went to live with a bachelor uncle in Market Harborough, where he attended a school. Then in 1875 his father sent him to King William's College, Isle of Man. In 1881 he entered Trinity College, Cambridge as a scholar. He was placed

Third Wrangler in the Mathematical Tripos in 1884, and continued for a further year taking Part III of the Tripos and attending lectures, including those being given by J.J. Thomson. He obtained first class honours in 1885, and was appointed to the Elder Professorship in Mathematics and Physics at Adelaide, South Australia; he spent time during the voyage in studying physics. In 1889 he married Gwendoline Todd and two sons and a daughter were born in Australia.

Bragg became interested in experimenting and following Röntgen's discovery in 1895 he set up the first X-ray tube to operate in Adelaide, although he did not carry out any original investigation. However, he developed an interest in the new field of radioactivity and carried out fundamental researches in Adelaide between 1904 and 1908 on the range of the alpha particle and the related questions of ionisation produced and of the stopping power of the substances. Through this research he soon became recognised internationally as an original investigator of the first rank.

In 1906 Bragg returned to England as Cavendish Professor of Physics at Leeds University where he worked on X-rays and mastered the difficult technique of studying them by their ionisation. Following the discovery by Laue, Friedrich and Knipping, in 1912, that X-rays could be diffracted by passing through crystals Bragg proceeded to make two great improvements in the experimental technique by using ionisation and by studying glancing angle reflections at the crystal face. In all this work he co-operated closely with his son William Lawrence and their joint work in 1913 and 1914 founded the field of the X-ray analysis of crystal structure (X-ray crystallography). They published their book *X-rays and Crystal Structure* in 1915, the same year that they shared the Nobel Prize for Physics.

In 1915 Bragg was appointed Quain Professor of Physics at University College, London and was involved in war work for the Admiralty, becoming Resident Director of Research at the Admiralty experimental station at Hawkcraig. The hydrophone or underwater receiver, developed by Bragg and his team, was important into anti-submarine warfare. After the war Bragg returned to University College where he founded a research school and actively carried out experimental work himself. He began work on the structure of organic crystals. His reputation as a popular, original lecturer was established by the Christmas lectures he gave in 1919 at the Royal Institution.

In 1923 Bragg succeeded Sir James Dewar as Fullerian Professor of Chemistry at the Royal Institution, a post which also included residence in the building and being director of the Davy–Faraday laboratory. He directed the work of others, mainly on X-rays, now applied to the study of biological materials, while Lawrence, now at Manchester, concentrated on inorganic crystal structures. Bragg had an extraordinary ability to take up and contribute to a new subject; a notable example was his work on liquid crystals in 1933.

An essential part of Bragg's family life was the country cottage at Chiddingfold, Surrey. In his personal life he had a range of interests, and religion was a strong influence. He received the Copley medal in 1930, having previously received the Rumford Medal in 1916, and he was the Bakerian Lecturer in 1915. His honours included membership of the leading foreign societies. In 1917 he was appointed CBE, in 1920 KBE and in 1931 he was admitted to the Order of Merit. From 1935 until 1940 he served as President of the Royal Society. With the coming of World War II, he served as chairman of the Cabinet Committee on Scientific Policy and on other bodies; at the age of 77 he performed a range of responsible tasks continuing through the worst period of the war.

Bragg had an extraordinary career as a great teacher and as an outstanding representative of an important period of English physics.

ONFRS 1942–44, 4, 277–300. E.N. da C. Andrade FRS. DSB 2, 397–400. P. Forman. DNB 1941–1950, 99–101. C.G. Darwin. FRS.

SIR WILLIAM LAWRENCE BRAGG 1890–1971; FRS 1921

William Lawrence Bragg, the eldest son of William Henry Bragg and Gwendoline née Todd was born in North Adelaide, Australia. After attending a convent school he went to Queen's Preparatory School until he was about 11 years old; he then went to St Peter's College where he was in the sixth form at 14 and left school at 15 to enter his father's department in Adelaide University. He read mathematics with chemistry and physics as subsidiary courses, graduating with first class honours in mathematics at the age of 18. In 1908 when his father returned to England W.L. entered Trinity College in 1909, where he began by reading mathematics. In the following year he was awarded a major scholarship. He read Part I of the Mathematics Tripos, (first class 1910), and then transferred to the Natural Sciences Tripos (physics) for Part II, obtaining first class honours in 1911; his third year mainly involved research into the velocity of ions in various gases, suggested by J.J. Thomson, and he graduated with first class honours in 1912. Early in his time at Cambridge he had enlisted as a trooper, and served for four years in King Edward's Horse, a Cambridge unit of the Special Reserve. They were mounted infantry and trained during the year and at summer camps involving marksmanship, riding and the care of horses.

By the autumn of 1912 Bragg had shown that the phenomenon of X-ray diffraction can be understood in terms of the reflection of X-ray waves by planes of atoms in the crystal. He developed the relationship between the interplanar spacing of crystals and the wavelength of the radiation and the angle of reflection. He demonstrated the effectiveness of this relationship and working with his father they proceeded to further research e.g. establishing the structure of diamond. In the summer of 1914 Bragg was elected to a Lectureship and Fellowship at Trinity College, Cambridge and then in August he was commissioned as a 2nd Lieutenant in the Leicestershire Royal Horse Artillery. Following training in Norfolk he went to France in September 1915 where he worked successfully to develop a method of locating enemy guns from the sound of their firing. From the beginning of 1917 to the end of the war he supervised the successful application of sound ranging and a great expansion of the number of units. Bragg was awarded the OBE and MC, was mentioned in dispatches three times and rose to the rank of Major.

Early in 1919 Bragg returned to Trinity College but in the same year he was appointed Langworthy Professor of Physics at Manchester University; there he created the leading school of crystallography in which he and his co-workers set out to make X-ray analysis a quantitative science, concentrating on metals and minerals, and crystal physics more generally.

Bragg introduced the idea of atomic radii, experimentally verified the theory of X-ray diffraction from perfect and imperfect crystals and developed methods for analysing increasingly complex structures, particularly the Fourier method of calculating the electron-energy density in a crystal. From 1925 he concentrated on an intensive programme of research into the structure of a wide range of silicate minerals, which led to a detailed understanding of silicate chemistry. Important work was also done in the field of metals and alloys in association with Albert Bradley and Charles Sykes.

In 1921 Bragg had married Alice Hopkinson, whom he had met at Cambridge; there were four children from the marriage. In 1930 Bragg experienced a crisis resulting from various factors including the strains of his office. However, he recovered quite quickly with family support and having spent the spring of 1931 in Sommerfeld's Laboratory in Munich. Everything improved in 1933 when he moved his family to a house at Alderley Edge.

In 1937 Bragg became Director of the National Physical Laboratory, but in October of the following year he moved back to Cambridge as Cavendish Professor of Experimental Physics. He was also elected to a Professorial Fellowship at Trinity College. He made further advances in X-ray optics. Metals research was continued with significant emphasis on practical problems. Ensuing from discussions with Egon Orowan he invented the bubble model of a metal structure which did much to popularise the theory of dislocations. During the 1939–1945 war Bragg acted as a consultant for the Sound Ranging Section in the Army and for the Admiralty on the development of Asdic (sonar), and he served on important Government Committees. In 1941 for six months he was Scientific Liaison Officer in Ottawa, Canada.

After the war Bragg reorganised the Cavendish laboratory. His most direct interest was in crystallography and metal physics. The latter activity, particularly studying mechanical properties and relating them to structure was led by Orowan, and Bragg included further development of the bubble-raft model. He also promoted the development of X-ray microbeam methods for the study of cold worked metals.

Appointment as Fullerian Resident Professor and Director of the Davy-Faraday Laboratory at the Royal Institution, London occurred in 1954 and Bragg remained there until he retired at the age of 76. He achieved closer collaboration with industry, and built up a research team studying protein structures. Among his other activities he gave popular science lectures. He was a lucid writer whose books included *The Development of X-Ray Analysis.* The Braggs had a family house at Waldringfield in Suffolk. He enjoyed domestic activities and a country life by the sea with bird watching, gardening, sailing and painting.

Bragg served on the Council of the Royal Society from 1931–1933. He was the recipient of many honours, including, in addition to the Nobel Prize for Physics (shared with his father in 1915), the Hughes Medal (1931), a Royal Medal (1946), the Copley Medal (1966), a knighthood in 1941, and appointment as Companion of Honour in 1967.

Bragg, essentially a classical physicist, was a great creative scientist. His fundamental idea concerning X-ray diffraction and his continued research in X-ray crystallography were of outstanding significance in the advances that have occurred in physics, chemistry, metallurgy and materials science, mineralogy and molecular biology.

BMFRS 1979, 25, 75–145. Sir David Phillips, Sec. R.S. DSB 15, 61–64. John Law. DNB 1971–1980, 77–79. D.C. Phillips FRS.

SIR FREDERICK JOSEPH BRAMWELL 1818–1903; FRS 1873

Joseph Bramwell was born in London, the younger son of George Bramwell, a partner in a firm of bankers, and Elizabeth née Frith. Having attended the Palace School at Enfield he was apprenticed to a mechanical engineer in 1834. When he completed his apprenticeship he became chief draughtsman and then manager in the organisation and was involved in railway work. Subsequently he became manager of an engineering factory in the Isle of Dogs. He married Harriet Leonora Frith in 1847 and they had three daughters. In 1853 Bramwell set up in business for himself and came to concentrate on legal and consulting areas e.g. in patent cases. He was a liveryman of the Goldsmith's Company and served as Prime Warden (1877–88). In 1881 he became a Member of the Ordnance Committee. In the Royal Society he served on the Council from 1877–78. In 1889 Bramwell was made a Baronet.

DNB 20th century, 213–215. WLW.

WILLIAM THOMAS BRANDE 1788–1866; FRS 1809

William Brande was born in London where his father was an apothecary in Arlington Street. After attending a private school in Kensington he spent eight years at Westminster School. His father had hoped that he would enter the ministry of the church but William preferred the medical profession and in 1802 he was apprenticed to his brother, who was a member of the Society of Apothecaries. About this time the family moved to Chiswick, where he came to know Charles Hatchett who was keenly interested in chemistry and mineralogy, and one of whose daughters, Anna Frederica, he subsequently married. Hatchett allowed him to assist in his laboratory and taught him chemistry and mineralogy, encouraging him to study the classification of rocks and ores. In 1802 Brande visited his uncle in Hanover and in the following year was in Brunswick and Göttingen. Returning to England in 1804 he resumed his apprenticeship and became a pupil at a school of anatomy, and also studied chemistry at St George's Hospital under George Pearson. He carried out some chemical experiments and made his first contribution to the scientific literature when he was about 16 years of age. He received the Copley Medal in 1813 for work concerning alcohol in fermented liquors.

In about 1801 he had been introduced to Humphry Davy at the Royal Institution, and, when he returned from Germany he renewed his acquaintance and attended Davy's lectures. In 1808 he began lecturing on chemistry and pharmacy at London medical schools. He became Professor of Chemistry and Superintendent of Chemical Operations at Apothecaries' Hall in 1812, soon afterwards becoming Professor of Materia Medica. He succeeded Davy as Professor of Chemistry at the Royal Institution in 1813 where he remained in this post until 1852. When Faraday returned from his travels in Europe in 1815, he assisted Brande, so that the two men were associated for many years both in chemical investigations and in teaching. Brande's career included work at the Royal Mint, where in 1854 he became Chief Officer of the Coinage Department. Among his many writings his *Manual of Chemistry* was an important text book. From 1816–1826 Brande was one of the Secretaries of the Royal Society.

PRS 1867–68, 16, ii–vii. AST. DSB 2, 420. E.L. Scott. DNB.

PERCY WILLIAMS BRIDGMAN 1882–1961; FMRS 1949

Percy Bridgman was born in Cambridge, Massachusetts the only son of Raymond Landon Bridgman, a journalist, and Ann Maria née Williams. He entered Harvard College in 1900, graduated with an AB *summa cum laude* in 1904 obtained an AM the following year and a PhD in 1908. His experimental ability, including in the handling of machine tools and in the manipulation of glass, together with capacity for intensive work were apparent early in his career. His main interest was in physics and he carried out research research in the Jefferson Physical Laboratory of the College. Bridgman set out to extend studies of other workers on the effects of high pressure on the properties of fluids. Showing great experimental skills he constructed apparatus to withstand the pressures envisaged and also measuring instruments. He published extensive data on the thermodynamic properties of liquids up to pressures of ~11,800 MPa and then went on to study the compressibilities of solids, melting phenomena and polymorphic transformations. The massive output of the years 1909 to 1927 resulted mainly from Bridgman's own personal efforts and his avoidance of all external commitments. He had been elected Assistant Professor in 1913, Professor in 1919 and Hollis Professor of Mathematics and Natural Philosophy in 1926, but he remained dedicated to the laboratory. In 1950 he was appointed Higgins University Professor and in 1954 Emeritus Professor, continuing his work after retirement.

In 1912 Bridgman had married Olive Ware and they had a son and a daughter. He had purchased an old farm in New Hampshire and for many years this provided external interests including gardening, mountainéering, photography and music. In later years his custom was to spend the winter months in his Harvard laboratory and the summer tending his garden, writing scientific papers and writing.

Bridgman wrote *The logic of modern physics* published in 1927 and *The intelligent individual and society* just before the Second World war; in the latter he attempted an analysis of social conditions hoping to find a parallel between the simple situations presented by physics and the more complex situations of human life.

In Bridgman's lifetime there was a great revival of interest in high pressure phenomena. Developments leading to increases in pressures, both static and also by dynamic shock wave techniques producing extended the boundaries of solid state physics, geology and metallurgy. His pioneering work was of great importance, recognised by the conferment of many honours and distinctions, including the award of the the Nobel Prize for Physics in 1946.

BMFRS 1962, 8, 27–40. D.M.Newitt. DSB 2, 457–461 E.C. Kemble, F. Birch and G. Holton.

SIR ALEC NIGEL BROERS 1938–; FRS 1986

Geelong Grammar School, Australia. Melbourne Univ.: BSc, Physics, 1958, Electronics, 1959. Univ. of Cambridge, Caius College: BA, Mechanical Sciences,1962, PhD, Mechanical Sciences, 1966, ScD 1991. IBM Thomas Watson Research Centre: Research Staff, 1965–67; Manager Electron Beam Tech., 1967–72; Manager Photon and Electron Optics, 1972–80. IBM East Fishkill Lab.: Manager Lithography Systems and Technology Tools, 1981–82; Manager Semiconductor Lithography and Process Development, 1982–83; Manager Adv. Develp., 1983–84. Univ. of Cambridge: Head of Electrical Division, 1984–92 and Department of Engineering, 1992–96; Professor of Electrical Engineering, 1984–96; Master, Churchill College, 1990–96. FREng 1985. Univ. of Cambridge; Vice-Chancellor, 1996–. Clifford Paterson Lecturer, 1986. Knighted 1998.

Electron microscopy; electron beam lithography; integrated circuit fabrication.

LOUIS-VICTOR PIERRE RAYMOND,
DUC DE, PRINZ, DE BROGLIE 1892–1987; FMRS 1953

Louis de Broglie was born in Dieppe, France, into a distinguished French family. He was the second son in a family of five children; his father was Louis-Amede-Victor-Albert de Broglie and his mother was Pauline née d'Armaille. He had an elder brother Louis-Cesar-Victor-Maurice (1875–1960; FMRS 1940). Maurice had a career in the French Navy and also had a private laboratory in the family home, and carried out research including important investigations of X-ray spectra. Louis entered the Sorbonne in Paris in 1909 to study history, but a year later began to study physics. In 1913, for his military service he entered the Engineers as a private soldier and was seconded during the war to wireless communications work at the Eiffel Tower; at the time of his demobilisation in 1918 he held non-commissioned officer rank. After the war he carried out research in association with his brother, and published work on the absorption of X-rays by metals (1920) and on X-ray spectra. Louis obtained a PhD at the Sorbonne in 1924 and continued there until 1928. He was appointed Professor of Theoretical Physics at the Henri Poincare Institute in 1933, a post which he held until 1962. He developed the principle of the wave nature of particles and was awarded the Nobel Prize for Physics in 1929. He remained unmarried.

BMFRS 1988, 34, 23–41. A. Abragam FMRS.

LAWRENCE MICHAEL BROWN 1936–; FRS 1982

Univ. of Toronto: BASc. Univ. of Birmingham: PhD; Athlone Fellow, 1957. Univ. of Cambridge: Gonville and Caius College; Tapp Research Fellow, 1963; University Demonstrator, Cavendish Laboratory, 1965; Lecturer, 1970–83; Reader, 1983–90; Professor of Physics, 1990–; Lecturer, Robinson College, 1977–90. ScD, 1992.

Structure and properties of materials; electron microscopy; spectroscopy.

WILLIAM BROWNRIGG 1711–1800; FRS 1742

William Brownrigg was born at High Close Hall, Cumberland. After studying medicine in London for two years he completed his medical education at Leyden, graduating MD in 1737, and publishing a thesis. He entered practice in Whitehaven where he began to investigate the gaseous emissions from the neighbouring coal mines. He communicated several papers on this subject to the Royal Society in 1741. He had a laboratory erected and continued his research on firedamp in mines. He was also the first to give any detailed accounts of platina, as brought by his relative, Charles Wood from the West Indies in 1741; he showed that no known material approached nearer to gold. He was the recipient of the Copley Medal in 1766. He lived all his life in his native district, including the period of his retirement to Ormathwaite Hall near Keswick. He was an original experimental philosopher with broad interests beyond his science.

DSB 2, 1124–1126. E.L. Scott. DNB. *Platinum Metals Review*, 1965, **9**, 1, 20–25. Donald McDonald. Ref. Ch. 4.1.

ISAMBARD KINGDOM BRUNEL 1806–1859; FRS 1830

Isambard Brunel, the only son of Sir Marc Isambard Brunel and Sophie née Kingdom, was born at Portsmouth. His education included attending school at Hove and a period in Paris at the Collège Henri-Quatre. He entered his father's engineering office at the age of 17, and took part in the work on the Thames tunnel, where afterwards he was appointed resident engineer. His engineering projects included the Clifton suspension bridge, and serving as engineer to the Bristol docks and the Great Western Railway; he was an expert in the use of iron, masonry and timber. In his shipbuilding activities the *Great Western,* the *Great Britain* and the *Great Eastern* were major achievements. Brunel also devoted much attention to the improvement of large guns. He was an active promoter of the Great Exhibition of 1851. Brunel married Mary Horsley in 1836 and they had two sons and a daughter. He became a member of most of the leading scientific societies in London and of many abroad but he did not accept a knighthood. Brunel possessed high mathematical knowledge, with the ability to apply it and he also had great mechanical skill.

PRS 1860, 10, vii. DNB. Refs Ch. 9.6, 16.

SIR MARC ISAMBARD BRUNEL 1769–1849; FRS 1814

Marc Brunel was born at Hacqueville, Normandy where members of his family had farmed for generations. He was the son of Jean Charles Brunel and Marie Victoire née Lefèvre and the father of Isambard Kingdom Brunel. At the age of eight he was sent to the college of Gisors to begin classical studies, his parents intending him to enter the ministry of the church. However, he showed an interest in mechanical pursuits and drawing, and at the age of 11 he went to a seminary at Rouen with the aim of qualifying for the Navy. In due course, having obtained a nomination to a corvette, he served for six years before returning to Paris in 1793. In the same year he travelled to America, where he adopted the profession of civil engineer and architect. He became the Chief Engineer of New York, in which post he erected an arsenal and cannon foundry. Back in

England in 1800, he married Sophia Kingdom. Brunel's engineering projects included bridges and a tunnel under the Thames. He was knighted in 1841, served as Vice-President of the Royal Society and received honours from other countries also.
DNB. Ref. Ch. 9.6.

COMTE DE GEORGES-LOUIS LECLERC BUFFON 1707–1788; FRS 1739
Georges-Loius Buffon was born in Montbard, France, the eldest of five children of Benjamin-François Leclerc and Anne Cristine née Marlin. In 1717 the family moved to Dijon where they occupied a prominent place in society, and where he was educated at the Collège des Jesuits and showed mathematical ability. He spent a period at Angers from 1728, but having been forced to leave in 1730 because of a duel he travelled in southern France and Italy until returning to France in 1732; following his mother's death he received a substantial legacy. In Paris he began to become well known in political and scientific circles. For six years he divided his time between finance, and research in botany, forestry and mathematics. From 1740 onwards he spent each spring at Montbard administering his estate, continuing his research and writing, then returning to Paris in the autumn. His memoirs presented to the Académie Royale des Sciences dealt with his wide ranging research in mathematics, astronomy, physics, forestry, physiology and pyrotechnics. He was renowned as a naturalist, author of a multi-volume *Histoire Naturelle* and also superintendent of the Jardin du Roi from 1739. Of interest in the context of metallurgy was Buffon's work in the field of strength of materials, in which he carried out mechanical tests on timber and on iron. He was also involved in ironmaking.

In 1752 Buffon married Françoise de Saint-Belin Malain, aged 20, who died in 1769 leaving a five year old son. During his lifetime he developed his estate and became rich; he also became famous throughout Europe and in America, and was made Comte de Buffon by Louis XV.
DSB 2, 576–582. J. Roger. Gen. Ref. 26. Ref. Ch. 4.1.

RONALD BULLOUGH 1931–; FRS 1985
Univ. of Sheffield: BSc, PhD, DSc. AEI Fundamental Research Laboratory, Aldermaston: Research Scientist, 1956–63. Harwell Research Laboratory, Didcot: Theoretical Physicist and Group Leader, 1963–84; Head of Materials and Development Division, Harwell, 1984–88; Director for Underlying Research, Harwell, 1988–90. UK Atomic Energy Authority: Chief Scientist, 1988–93 and Director for Corporate Research, Harwell, 1990–93. Scientific Adviser to the National Physical Laboratory and Consultant in the UK and USA.
Defect properties in crystalline solids, particularly in relation to irradiation and mechanical response of materials.

CHARLES WILLIAM BUNN 1905–1990; FRS 1967
Charles Bunn was the second of five children of Charles John Bunn and Mary Grace née Murray. He was born in Camberwell and after attending Denmark Hill London County Council Elementary School he went to Wilson's Grammar School in 1916, with a scholarship. In 1923 he entered Exeter College, Oxford with an open scholarship to study chemistry. Having obtained first class honours in 1927 he took an appointment in the Research Department of the ICI Alkali Division at Winnington, Cheshire. In 1931 he married Elizabeth Mold and they had one son and one daughter. Bunn was transferred to the Plastics Division of ICI at Welwyn Garden City in 1946 where he later became Leader of the Molecular Structures Division. Following his early retirement in 1963 he joined W.L. Bragg at the the Davy–Faraday Laboratory of the Royal Institution as Dewar Research Fellow, an appointment from which he retired in 1972.

Bunn was an eminent chemical crystallographer. His research covered methods of identification of crystalline materials, work on inorganic crystals of industrial importance, crystal growth, structure determination of long chain polymers and the relationship of physical properties to molecular structure and investigations of the structures of pennicilin and rennin. His work on the identification of crystalline materials included a technique for determining the lattice parameters of hexagonal and tetragonal lattices using a graphical procedure. Concerning long chain polymers, examples of his research were polyethylene, rubber and nylons.
BMFRS 1991, 37, 69–84. U.W. Arndt FRS.

ROBERT WILHELM EBERHARD BUNSEN 1811–1899; FMRS 1858
Robert Bunsen was the youngest of four sons; his father Christian Bunsen was Professor of Modern Languages at the University of Göttingen and his mother was the daughter of a British-Hanoverian officer called Quensel. Following schooling at Göttingen, where he was born, he attended the Gymnasium at Holzminder from which he graduated in 1828, and then studied chemistry, physics, mineralogy and mathematics at Göttingen University. Having obtained a doctorate in 1830 he travelled until 1833 in Europe, visiting factories, laboratories and places of geological interest; while in Paris in 1832 he worked in the laboratory of Gay-Lussac. A series of academic appointments followed: Privatdozent at Göttingen (1833); successor to Friedrich Wohler at Cassel (1836); Professor of Chemistry at Marburg (1838); and at Heidelberg (1852), where he worked until his retirement in 1889 at the age of 78. From 1855 in a laboratory built for him by the government of Baden he carried out research and guided the work of many young men who later became well known scientists.

Bunsen's scientific interests in chemistry were wide ranging, and he also was interested in geology. After early research in the organic field of chemistry, he concentrated on the inorganic field, where his most important research was the development of analytical techniques. Much of his work involved metallurgical topics, and he was also concerned with applying experimental science to industrial problems including work relating to the gases evolved from iron smelting furnaces. Also, in the field of gases Bunsen's research covered analysis, diffusion and combustion. He invited the Bunsen burner that became important in chemistry. Bunsen made various improvements in the galvanic battery and and isolated a number of metals, in some cases with co-workers and he also carried out collaborative research in photochemistry and spectroscopy.

Bunsen excelled as a laboratory teacher and possessed great manipulative skill. He never married but devoted most of his time to teaching, research and travel. His wide-ranging achievements were recognised by the Royal Society by the award of the Copley Medal in 1860, and, in 1877 with Kirchhoff, the first Davy Medal.
PRS 1905, 75, 46–49, HER. DSB 2, 586–590. S.S. Schacer.

CECIL REGINALD BURCH 1901–1983; FRS 1944
Cecil Burch was one of the sons of George James Burch FRS and Constance Emily née Jeffries. He was born in Oxford and was educated at the Dragon School, Oxford and at Oundle. In 1919 he was awarded scholarships to enter Gonville and Caius College, Cambridge, where he obtained second class honours in both parts (1921 and 1922) of the Natural Sciences Tripos. He became a college apprentice at the Metropolitan–Vickers Electrical Company at Trafford Park, Manchester, where he learnt mechanical skills, invented complex electronic circuits and did radio transmission work. In collaboration with a colleague he developed the science of induction heating and used the technique for metal melting. Another important area of his work involved vacuum

techniques e.g. evaporation distillation. Burch had a long standing interest in optics and mastered the technique of grinding large areas of cast iron and speculum metal to high degrees of flatness. In 1933 he went with a Leverhulme Scholarship to Imperial College, London, where in the optics group he produced lenses and mirrors of spherical and aspherical shapes of high accuracy.

Burch was awarded a PhD degree and in 1936 moved to Bristol University as a Research Associate. In 1937 he married Enid Grace Morice and they had one daughter. He was a research fellow at Bristol from 1944–66. His work included the construction of a microscope with an aspherical reflecting system with a very high aperture. In 1953 he experienced ill health and during convalescence in Cornwall he became interested in tin mining, learning to 'pan' for tin and gold. Also he invented a novel mineral classifier. Burch was awarded the Rumford Medal in 1954 and the CBE in 1958.

BMFRS 1984, 30, 3–42. T.E. Allibone FRS. DNB 1981–85, 58, 59. T.E. Allibone.

FREDERICK MICHAEL BURDEKIN 1938–; FRS 1993

Kings School, Chester. Univ. of Cambridge: Trinity Hall, MA, PhD. Univ. of Manchester: MSc. Sandberg Consulting Engineers: Associate, 1968–77. UMIST: Professor of Civil and Structural Engineering, 1977–. FREng 1987.

Fracture, particularly large welded structures.

ROBERT WOLFGANG CAHN 1924–; FRS 1991

Univ. of Cambridge: BA, 1945, PhD,1950, ScD 1963. AERE, Harwell: Scientific Officer, Senior Scientific Officer, 1947–1951. Univ. of Birmingham, Department of Physical Metallurgy: Lecturer, Senior Lecturer, Reader, 1951–62. Univ. College of North Wales, Bangor: Professor of Materials Science, 1962–64. Univ. of Sussex: 1965–1981; Dean of Engineering, 1973–78. Universite de Paris-Soud: Professor of Physical Metallurgy, 1981–83. California Inst. Tech.: Fairchild Distinguished Scholar, 1985–86. Univ. of Cambridge: Department of Materials Science and Metallurgy; Distinguished Research Fellow (formally Senior Associate), 1986–.

Deformation, recovery and recrystallisation of metals: intermetallic compounds, ordering, mechanical properties.

HUGH LONGBOURNE CALLENDAR 1863–1930; FRS 1894

Hugh Callendar was the eldest son of the Reverend Lee Callendar, rector of Hatherop, Gloucestershire, and Anne Cecilia née Longbourne. He was educated at Marlborough College and then went to Trinity College, Cambridge in 1882 where he obtained a first in the Classical Tripos at the end of the second year, and was sixteenth Wrangler in 1885. He entered the Cavendish Laboratory with J.J. Thomson and was given a project to measure accurately the electric resistance of platinum as a function of temperature aiming to devising a technique for temperature measurement. His thesis on platinum thermometry led to his appointment to a Fellowship of Trinity College in 1886. He also started to study medicine and then law for a period. He married Victoria née Stewart in 1894 and they had three sons and a daughter.

During his career he held Chairs in Physics (1888) at Royal Holloway College, Egham; (1893) McGill University; University College, London, and (from 1898 until his death) at the Royal College of Science, London (later incorporated into the Imperial College of Science and Technology). He carried out important research on steam and in 1902 he published his treatise *The properties of Steam and Thermodynamic Theory of Turbines*. He was awarded the Rumford Medal in 1924 and was appointed CBE in 1920. He was a talented engineer as well as an experimental

physicist and made important scientific contributions through his patient and highly accurate research.
PRS 1932 A, 134, , xviii–xxvi. SWJS. DNB 1922–1930, 152–154. Ref. Ch. 4.1.

SIR HENRY CORT HAROLD CARPENTER 1875–1940; FRS 1918
Harold Carpenter was born at Bristol, the second son of William Lant Carpenter, an engineer, and Annie Grace née Viret; he was the great-great-grandson of Henry Cort, the inventor of the puddling process for making wrought iron, and the introducer of grooved rolls. Following his education at St Paul's School and at Eastbourne College he gained the Science Postmastership at Merton College, Oxford, in 1893.

Whilst at Eastbourne he became a close friend of Frederick Soddy who also went to Oxford. Carpenter was awarded a first-class honours degree in 1896 and at the end of the summer term, he and Soddy went to Germany to study organic chemistry for two years; he was the first foreigner at Leipzig or at any German university to obtain the PhD degree *summa cum laude*.

After Leipzig Carpenter went to Owens College, Manchester to work with W.H. Perkin but his interests soon turned to metallurgy and in 1901 he was appointed as Head of the two new Departments of Chemistry and Metallurgy at the National Physical Laboratory (NPL), Teddington. By 1905 his interest had become entirely centred around metallurgical research. He left the NPL in the autumn of 1906, having in these few years been responsible for a number of original researches. In 1905 he married Ethel M. Lomas; they had no children.

Following important research on transformations in the iron–carbon system he investigated high speed tool steels. Most of Carpenter's other researches and work, from the early part of 1901 up to the time he left Teddington were concerned with two extensive investigations which were incorporated in the volumes of the Seventh and Eighth Reports of the Alloys Research Committee of the Institution of Mechanical Engineers. The first of these was on steel and the second on the properties of copper–aluminium alloys; Carpenter had the ability to marshal clearly the enormous amount of data.

In 1906 he was appointed to the newly created Chair of Metallurgy at the Victoria University at Manchester and built up the department quickly into a flourishing research centre. In spite of a heavy load of committee work he helped to direct the researches of students and to do a considerable amount of original work himself. The research topics included the growth of cast iron as a result of heating, and the production of copper–aluminium alloy castings to withstand hydraulic pressure.

In the late summer of 1913 he was appointed to the Chair of Metallurgy at the Royal School of Mines, London. Before starting this new work he made a six months' tour of the metallurgical works and research centres in the USA and in Canada.During the war, Carpenter served on the Admiralty board of invention and research. After 1918 Carpenter's work was divided between teaching, research and service on important government committees; also over different periods he served as President of each of the three main metallurgical Institutes in Britain: Metals; Iron and Steel; and Mining and Metallurgy. On the outbreak of war in 1939 the Metallurgy Department of the Royal School of Mines was transferred to Swansea.

From 1918 onwards Carpenter's numerous publications covered a range of subjects such as the growth and behaviour of metallic crystals, including a series of papers, beginning in 1920 with Miss C. Elam on the production and mechanical properties of aluminium single crystals. In 1939, in collaboration with J.M. Robertson, he published a major book *Metals*. Carpenter's distinguished services were recognized by many honours including the conferment of a knighthood in 1929.
ONFRS 1939–41, 3, 611–625. C.A. Edwards. DNB 1931–1940, 145, 146. C.H. Desch.

AUGUSTIN-LOUIS CAUCHY 1789–1857; FMRS 1832

Augustin-Louis Cauchy, born in Paris was the eldest of a family of four sons and two daughters. His father was Louis-Francois Cauchy, who achieved a high administrative position, and his mother was Marie-Madeleine née Desestre. During the Terror the family went to the village of Arcueil, near Paris. Having received his early education from his father, he entered the École Centrale de Pantheon. In 1805 he entered the École Polytechnique in Paris and became a military engineer. He wrote a classical mathematical paper on polyhedra in 1811. His career included lecturing (1815) and holding the Chair of Mechanics (1816) at the École Polytechnique; he also held other professorial appointments. His marriage to Aloise de Bure took place in 1818 and they had two daughters. In 1830 having gone into exile with Charles X, he held a professorship at Turin, assisted the education of the Crown Prince in Prague (1833) and was made a Baron by the ex-king. He returned to Paris in 1838; then 10 years later during the second republic he occupied a chair at the Sorbonne. Cauchy was the author of more than 700 papers including areas of mathematics such as calculus and complex functions. In mechanics, especially elasticity, he created fundamental theory relating to elastic and anisotropic media.

PRS 1857–59, 9, 45–49. DSB 3, 131–148. H. Freudenthal.

TIBERIUS CAVALLO 1749–1809; FRS 1779

Tiberius Cavallo who was born in Naples, the son of a physician, carried out his scientific work in England where he settled in 1771 intending to acquire some experience in business. However, he turned to experimental philosophy through acquaintance with some English physicists. His research covered the physics of the atmosphere, and also magnetism; he also described a new type of pyrometer. Cavallo gave the Bakerian Lecture 13 times in the period between 1780 and 1792.

Journal Book Roy. Soc., xxxxiv, 208. DSB 3, 153–154. J.L. Heilbron.

HENRY CAVENDISH 1731–1810; FRS 1760

Henry Cavendish was born in Nice, France. He was descended from dukes on both sides of his family; his father was Lord Charles Cavendish and his mother was formerly Lady Anne Grey. After attending school in Hackney, near London, he studied at Peterhouse College, Cambridge as a fellow commoner from 1749–1753. He left Cambridge without taking a degree and after spending some time in Paris studying physical science and mathematics he lived in and near London. He was of independent financial means, particularly after receiving a fortune bequeathed to him. Although he did not have a title, he was often addressed by the title 'Honourable'. His research was carried out in a laboratory which he fitted out and he became well known in scientific circles. In his scientific activities which covered topics in chemistry and physics he made outstanding contributions. Cavendish demonstrated the existence of hydrogen as a distinct substance, for which work he received the Copley Medal in 1766. Of metallurgical interest was his work on heat and on the action of acids on metals.

DSB 3, 155 - 162. A.V. Carozzi. DNB.

SIR WILLIAM CAVENDISH (SEVENTH DUKE OF DEVONSHIRE) 1808–1891; FRS 1829

William Cavendish was born in London, the eldest son of William Cavendish and Louisa née O'Callaghan. He was educated at Eton and at Trinity College, Cambridge where he graduated BA in 1829 as Second Wrangler, first Smith's Prizeman and Eighth Classsic. In the same year he married Blanche Georgina Howard and they had three sons and one daughter. He was elected by Cambridge University as Member of Parliament and subsequently represented another constituency. When he entered the House of Lords in 1858 he left politics and devoted himself to scientific and industrial concerns of the country. His most significant industrial enterprise was the establishment of a major steel making activity in the Barrow-in-Furness based on mining of the haematite ore deposits on his estates in Lancashire and which led to the expansion of the town. Cavendish was an enlightened employer and also promoted the application of science to industry. As first President of the Iron and Steel Institute, he referred in his presidential address to the importance of scientific themes relevant to the iron and steel industry.

Cavendish was the founder of the laboratory at Cambridge University that bears his name; also he was Chancellor of London University (1836–1856), of Cambridge (1861–1891), Victoria University, Manchester (1880) and chairman of the Royal Commission on Scientific Instruction and the Advancement of Science. He was made KG in 1858 and a Privy Councillor in 1876.

PRS 1892, 51, xxxviii–xl. GDL. DNB Suppl. v I 400, 401 EIC.

SIR JAMES CHADWICK 1891–1974; FRS 1927

James Chadwick was born in Bollington near Manchester, the eldest of three sons and one daughter, of John Joseph and Anne Mary Chadwick; his father was a cotton spinner and his mother a domestic servant. James attended the local primary school and then the Manchester Municipal School. At the age of 16 he entered Manchester University with a scholarship and studied physics; he carried out his third year project under Rutherford and graduated in 1911 with first class honours. He carried out research leading to the award of an MSc in 1913. With the award of an 1851 Senior Research Studentship he went to Berlin, but with the outbreak of war, English people were interned and he spent four year in Ruhleben, where it proved possible to form a scientific society and to carry out some experimental work. In 1918 he returned to a post with Rutherford at Manchester, but when Rutherford moved to Cambridge in the following year Chadwick was awarded a studentship by Gonville and Caius College. He worked at the Cavendish Laboratory, being awarded a research fellowship by his College in 1921 and later became Assistant Director of Research of the laboratory. In 1925 Chadwick married Aileen Stewart-Brown and they had twin daughters. 1932 saw Chadwick's discovery of the neutron. From 1932–39 Chadwick was the Lyon Professor of Physics at Liverpool University and also served as Head of Department.

Chadwick was a member of the MAUD Committee concerned with the war-time atomic energy project; he coordinated experimental work at Liverpool, Cambridge and Bristol and was involved in the organisation of the supply of uranium and its hexafluoride. He played a crucial role in the atomic bomb project and from 1943 led the British team in the USA. In 1946 he returned to his post at Liverpool; then from 1948–1958 he was Master of Gonville and Caius College at Cambridge. In the Royal Society Chadwick was Bakerian lecturer in 1933, and was awarded the Hughes medal in 1932 and the Copley Medal in 1952; he served as Vice-President in 1949. His honours included the Nobel Prize for Physics in 1935, a knighthood (1945), and CH (1970).

BMFRS 1976, 22, 11–70. Sir Harrie Massie Sec RS and N. Feather FRS. DSB 17, 143–148. John Hendry. DNB 1971–1980, 131, 132. Denys Wilkinson. Ref. Ch. 18.2.

ARTHUR NEVILLE CHAMBERLAIN 1869–1940; FRS 1938

Neville Chamberlain was born in Edgbaston, Birmingham the only son of Joseph Chamberlain and his second wife Florence née Kenrick. Neville's grandfather, also Joseph, had taken a share in the screw-making business of John Nettlefold, and the success of Nettlefold and Chamberlain brought substantial wealth to the family. Neville attended Rugby School and then in 1888 took courses in metallurgy and engineering design at Mason College, Birmingham (which later became the University of the City). Later in his life Chamberlain referred to the metallurgical laboratory and the teaching by Professor Turner at the College on the properties of brass and cast iron. Chamberlain entered a firm of accountants and then in 1890 went to take charge of a large estate in the island of Andros in the Bermudas, which his father had bought to grow sisal. He spent seven years of hard work and socially isolated conditions, and although the venture did not succeed because of the soil conditions, he read in history, biography and science. Back in Birmingham in 1897 he began a business career in the engineering industry as a Director of Elliott's Metal Company dealing mainly in copper and brass, and then as the owner of Hoskins and Sons, a firm making ships berths in metal. He also later became a Director of the Birmingham Small Arms Company. He was a good employer and maintained his industrial activities over many years, becoming one of the outstanding figures in the industrial life of the city.

Chamberlain's political career began with participation in local public affairs. He was active in the Chamber of Commerce and Chairman of the General Hospital. A long and active association with the University of Birmingham was another sphere of his service; for example, in 1908 he was involved in an appeal to raise finance, showing particular interest in the Mining Department. In 1911 he married Annie Vera Cole and there was one daughter and one son of the marriage. Also in 1911 he was elected to the City Council, subsequently becoming Mayor. He first entered Parliament in 1918 as the member for the Ladywood Constituency of Birmingham. During the period 1922–24 he held several ministerial points in succession, including the Chancellorship of the Exchequer. He was a major figure in the pre-war politics of Great Britain and became Prime Minister in 1937, in which capacity he was prominent in the events preceding the outbreak of war in 1939 and in the first year of the war prior to his resignation in 1940.

ONFRS 1939–1941, 3, 731–734 W.W. Hadley. DNB 1931–40, 155–163 W.W. Hadley. D. Dilks, *Neville Chamberlain vol I 1869–1929*, Cambridge University Press 1984.

JOSEPH CHAMBERLAIN 1836–1914; FRS 1882

Joseph Chamberlain was born in Camberwell; he was the eldest child of Joseph Chamberlain and Caroline née Harben. In 1850 he began studies at University College School, and when he was 16 he entered the family business and also attended science lectures. Two years later the opportunity arose for the expansion of the screw manufacturing business of his father's brother-in-law, John Nettlefold in Birmingham. Joseph went to Birmingham to represent his father's interest in the firm, which expanded and with which he was actively associated for 20 years. Chamberlain entered local politics and became an influential figure. In 1861 Chamberlain married Harriet Kenrick and they had a son; following his wife's death he married Florence Kenrick in 1868 and had a son Arthur Neville, who became Prime Minister; after the death of his second wife he married Mary Endicott. Chamberlain entered Parliament in 1876 and had a distinguished career in national politics, becoming a cabinet minister ; his posts included the Presidency of the Board of Trade (1880–85).

DNB 1912–21, 102–118. HEE.

SIVARAMAKRISHNA CHANDRASEKHAR 1930–; FRS 1983

Nagpur University: MSc, DSc. Univ. of Cambridge: Pembroke College, PhD, 1958, ScD 1987. Raman Research Institute, Bangalore, India: Research Scholar. Univ. of Cambridge: Exhibition Scholar, 1954–57. University College London, Department of Crystallography: DSIR Fellow, 1957–59. Royal Institution, London, Davy Faraday Research Laboratory, 1959–61. Univ. of

Mysore: Professor and Head of Department of Physics, 1961–71. Raman Research Institute, Bangalore: Professor, 1971–90; Centre for Liquid Crystals Research, Bangalore, Director, 1991–. Royal Medal 1994.
Physics of liquid crystals; disc-like molecules.

SIR JOHN CHARNLEY 1911–1982; FRS 1975
John Charnley was born in Bury, the elder of two children of Arthur Walker Charnley, a pharmacist, and Lily née Hodgson, a nurse. He was educated at Bury Grammar School, and entered Manchester University as a medical student, where he had a distinguished career, graduating BSc in 1972 and MB, BCh in 1935, while winning scholarships and prizes. He became a Fellow of the Royal College of Surgeons in 1936 and after house surgeon posts in Manchester and Salford he spent six months at King's College, London, before returning to Manchester. In 1940 Charnley joined the Royal Army Medical Corps as an orthopaedic specialist, reaching the rank of Major and serving in the North African campaign. After demobilisation he returned to Manchester and was appointed Consultant Orthopaedic Surgeon to the Manchester Royal Infirmary and Wrightington Hospital. In 1957 he married Jill Margaret Heaver and they had a son and a daughter. Charnley was awarded the DSc of Manchester University in 1964 and was appointed Professor of Orthopaedic Surgery at Manchester in 1976 .

Charnley became a distinguished surgeon and researcher in the field of arthroplasty – the reconstruction of joints rendered stiff and painful by disease. Throughout his career he invented and developed various surgical tools to aid the surgeon. In the 1950s he started investigating joints and methods of replacing them. All the versions of hip replacements he developed were carefully tested, particularly in the friction rigs he devised. Charnley had superb manual dexterity and learned to use a lathe on which he manufactured all the cups for his early patients. He also used it to build prototype tools for use in joint replacement and other orthopaedic applications. The high clinical success rates with his protheses were also due to the care with which he trained surgeons before they were allowed to use one of his devices. He later realised that problems were produced by infections implanted along with the protheses and he developed the 'clean air tent', which when used by surgeons decreased the rate of infection to <1%. All his clinical results were carefully recorded, enabling good follow-up and analysis. In 1961 Charnley moved his clinical work entirely to Wrightington and set up the world's first biomechanical laboratory.

Charnley's pionéering work, involving collaboration with others, has led to the benefit of improved quality of life for numerous people. He was the first practising orthopaedic surgeon to be elected FRS. The honours that he received included the CBE in 1970 and a knighthood in 1977.
BMFRS 1984, 30, 119–137. N.W. Nisbet and Sir Michael Woodruff FRS. DNB 1981–85, 83, 84. R.B. Duthie. Information from W. Bonfield and E. Tanner.

ARTHUR PRINCE CHATTOCK 1860–1934; FRS 1920
Arthur Chattock was born in Solihull, Warwickshire, the eldest of seven children. He attended University College School and then, when he was 17 he spent two years abroad at Dunkerque and Stuttgart, mainly to study languages. He then entered the engineering course at University College, London and later studied physics there. A period was then spent with Siemens Brothers before he was appointed a Demonstrator in Physics at University College, Bristol (1885–1887); there followed a two year period as assistant to Oliver Lodge at Liverpool University. In 1889 he was appointed as a Lecturer in Physics at Bristol becoming a Professor in 1893 with responsibili-

ties for teaching electrotechnology. Chattock married Mary Frances Cooke in 1906. He retired from his post at the age of 50 but in 1919 he received an independent appointment for research at what had become the University of Bristol. Research in the field of magnetism was one of his interests and he was associated with L.F. Bates in the determination of the gyromagnetic ratio for iron.
ONFRS 1934, 3, 293–298.

ANTHONY KEVIN CHEETHAM 1946–; FRS 1994
Stockport Grammar School. Univ. of Oxford: St.Catherine's College; Hon. Scholar; Wadham College, Senior Scholar, BA, Chemistry, 1968; DPhil, 1971; Lincoln College, Cephalosporin Fellow, 1971–74; St. Hilda's College, Lecturer Inorganic Chemistry, 1971–85; Lecturer in Chemical Crystallography, 1974–90; Christ Church, Tutor Inorganic Chemistry, 1974–91; Reader in Inorganic Materials, 1990–91. Univ. at Santa Barbara: Materials Research Laboratory, Director, Professor of Materials,1991–. Professor of Chemistry, 1992–. Royal Institution, London: Professor of Solid State Chemistry, 1986–.
Chemical structure of materials, especially inorganic. Analysis of solid state materials by neutron beams and neutron radiation; microporous catalysts and absorptents.

RICHARD CHENEVIX c.1774–1830; FRS 1801
Richard Chenevix was the son of Colonel Chenevix, of Huguenot extraction; his birthplace was near Dublin, Ireland. He studied at Glasgow and began to contribute papers to French chemistry journals and lived in Paris for a period. He established a reputation as an analyst, and worked on platinum and palladium. He was awarded the Copley Medal in 1803. When William Wollaston reported the discovery of a new noble metal (palladium) in 1802, Chenenvix experimented on a sample and erroneously reached the view, and reported it in a Royal Society paper, that it was not a new element. He moved to live in France in 1804.
Bulloch's Roll. DSB, 3,232–233, H.A.M. Snelders. DNB. Ref. Ch. 4.1.

JOHN HUGH CHESTERS 1906–1994; FRS 1969
John (Jack) Chesters was the second son of a methodist minister, the Rev. George Chesters. He was educated at High Pavement School, Nottingham and then at King Edward VII School in Sheffield. The family having settled in Sheffield, he studied at the University taking honours physics, with fuel technology as a subsidiary subject, and graduated BSc in 1928. After graduation, with the award of a scholarship from Metropolitan–Vickers, he carried out work at the university on the linings of induction furnaces, leading to the award of the PhD degree in 1931. There followed periods of research abroad, at the Kaiser Wilhelm Institute für Silikatforschung in Berlin, with the award of a Robert Blair Fellowship (1931–32) and at the University of Illinois, with a Commonwealth Fund Fellowship (1932–34). While in the USA he met Nell Knight from Minnesota; they were married in 1936, and had three sons and one daughter.

In 1934 Chesters was invited to form a refractories section in the new research laboratory for the United Steel Companies in Sheffield. There, important work was done that included a major contribution to the war effort in the period 1939–45. In 1945 Chesters was awarded the DSc Tech degree. He held the appointments of Assistant Director of Research (1945–62); Deputy Director of Research, United Steel Companies Ltd (1962–67); and Midland Group, British Steel Corporation (1967–70). From the late 1940s to the early 60s he led research on furnace design and fuel efficiency. He travelled and lectured on refractories and played a leading role in national institutions. He also became interested, and wrote and lectured, in what he called the human factor in industrial research.

In 1970, only 18 months before his retirement, he became Director of what had been the British Iron and Steel Research Association (BISRA) . When he retired he took on various consultancies and commissions and continued lecturing. In 1976 Chesters became Chairman of the Watt Committee on Energy, a role which included raising funds and recruiting experts in various fields to give definitive analyses on energy-related subjects.

Chester's publications included several books; his definitive *Steelplant Refractories* was first published in 1944. He participated actively in many learned and professional societies and was the recipient of many awards in Britain and from abroad, including the OBE in 1970 and FREng in 1969. In his distinguished career the contributions he made in his chosen fields were of outstanding importance metallurgically particularly in the iron and steel industry.

Materials World, 1995, 3, 3. J. Mackenzie. *Who's Who*.

JOHN GEORGE CHILDREN 1777–1852; FRS 1807

John Children, born at Ferox Hall, Tonbridge, Kent, was the only son of George Children (1742–1818); his mother died a few days after his birth. His father was a bencher of the Inner Temple who owned much property near Tonbridge and was a successful banker, who spent his leisure in scientific pursuits. George Children did not marry again, but devoted himself to the care of John, who was educated initially at the Grammar School at Tonbridge and then at Eton, followed by entry to Queen's College, Cambridge in 1794. He left College in 1798 to marry a Miss Howell. She died two years later and Children travelled abroad to Lisbon, and then, in 1802 to North America; there with a cousin he travelled into uncleared backwoods. He suffered from a severe fever, but was fully recovered by the time he had returned to England. He then entered the West Kent militia as a Captain at a time when national defences were being reorganised, and he held this post until 1805. From this period his time was principally devoted to science in which he had been greatly interested since his youth; his special interests were mineralogy, chemistry and electricity, and he soon came to know leading men of science.

Following the discovery by Volta of the galvanic pile Children constructed new and large galvanic batteries. He built a laboratory at Tonbridge in which many experiments were carried out, including some by Humphry Davy. The household was broken up in 1816 by the bankruptcy of George Children in paying debts incurred by the bank in which he was a partner. John took a small house for him in Chelsea, and obtained work as a librarian in the Department of Antiquities in the British Museum. He married his second wife in 1809 but she died about a year later, and he married again in 1819. He was transferred by Davy's influence to the department of zoology in 1824. He served the Royal Society as one of the Secretaries in 1826–7 and again from 1830 to 1837. When his wife died in 1839 he resigned from his British Museum post and went to reside with his daughter at Halstead Place, Kent; he then took up the science of astronomy.

PRS 1850–54, 6, 245–247. DNB. *Platinum Metals Review*, 1987, **31**(1), 32–41. L.B. Hunt.

JOHN WYRILL CHRISTIAN 1926–; FRS 1975

Scarborough Boys High School. Univ. of Oxford: The Queen's College, BA 1946, DPhil 1949, MA 1950. Pressed Steel Company Fellow, 1951–55; Lecturer in Metallurgy, 1955–58; George Kelley Reader in Metallurgy, 1958–67; Professor of Physical Metallurgy, 1967–88; Fellow of St.Edmund Hall, 1967–93.

Physical metallurgy. e.g. phase equilibria; theory of transformations.

SAMUEL HUNTER CHRISTIE 1784–1865; FRS 1826

Samuel Christie was the son of James Christie the auctionéer, who founded the world famous 'Christies'. He was born in Pall Mall and educated at Walworth School, Surrey, where he showed great mathematical ability. In 1801 he entered Trinity College, Cambridge as a sizar, graduating in 1805 as BA, Second Wrangler, and joint Smith's Prizeman; during his time at Cambridge he was an officer in the grenadier company of the University Volunteers. In 1806 he became Third Mathematical Assistant at the Royal Military Academy at Woolwich where he was promoted to Second Mathematical Assistant, then becoming Professor in 1838, a post which he held intil 1854. He was married twice, and had a son by his second wife Margaret Malcom.

The research carried out by Christie extended beyond the field of mathematics, including the magneto-electrical properties of metals, which led to his giving a Bakerian Lecture in 1833; it is likely that he used some of Faraday's alloys for his research, since Faraday mentioned the preparation of a nickel steel alloy for 'Mr Christie'. Christie was the actual inventor of the 'Wheatstone Bridge' which facilitated quantitative electrical and magnetic measurements. In the Royal Society Christie was Secretary from 1837 to 1854.

PRS 1866–1867, 15 xi–xiii. DSB 3, 259–261. E.W. Morgan. Ref. Ch. 11.6.

JOSIAH LATIMER CLARK 1822–1898; FRS 1889

Josiah Clark, who was born at Great Marlow, Bucks, began his career as a chemist with a chemical manufacturer in Dublin. In 1847 he worked on railway surveying, and in the following year he became assistant engineer to his brother, Edwin, who had been appointed by Robert Stephenson as Superintending Engineer for the Menai bridge project. In 1850 he was again appointed Assistant to his brother at the Electric and International Telegraph Company; some 10 years later he succeeded his brother as Chief Engineer serving until 1870. Clark made improvements to the telegraph system and in 1860 he served on a government committee concerned with submarine telegraphy. His career included industrial partnerships. He was the joint author of a paper concerning standards of electrical quantities and resistance and the inventor of a battery. Clark was twice married and had children.

DSB 3, 288, 289. B.S. Finn. DNB Suppl. up to 1901, 451–452. AFP.

WILLIAM TIERNEY CLARK 1783–1852; FRS 1837

William Clark was born in Bristol; his father died when he was young and he was apprenticed at an early age to a millwright. When his apprenticeship was completed he went to work at the Coalbrookdale Ironworks of Messrs Darby and Co. in Shropshire where he gained considerable experience in cast and wrought iron. In 1808 John Rennie who was working in construction with cast iron offered him a post at his works in Blackfriars, London; this gave Clark the opportunity of superintending important projects. He was appointed Engineer to the West Middlesex Waterworks at Hammersmith in 1811 and held this post until his death. He also practised as a consultant civil engineer, particularly in bridge building.

PRS 1850–54, 6, 244, 245. DNB.

WILLIAM COCHRAN 1922–; FRS 1962

Boroughmuir School, Edinburgh. Univ. of Edinburgh: Assistant Lecturer, 1943–46. Univ. of Cambridge: Demonstrator and Lecturer, 1948–62; Reader in Physics, 1962–64. Univ. of Edinburgh: Professor of Physics, 1964–75; Professor of Natural Philosophy,1975–87; Dean, Faculty of Science, 1978–81; Vice–Principal, 1984–87. Hughes Medal, 1978.

X–ray crystallography; dynamics of atoms in crystals.

DAVID JOHN HUGH COCKAYNE; FRS 1999
Univ. of Sydney: Professor of Physics. Univ. of Oxford: Professor-elect in the Physical Examination of Materials.
Electron microscopy and its application to materials science.

SIR JOHN DOUGLAS COCKCROFT 1897–1967; FRS 1936
John Cockcroft was born in Todmorden, Yorkshire, the eldest of five sons of John Arthur Cockcroft and Annie Maud née Fielden; his father inherited the family cotton mill business at a time of financial difficulties and his mother played a major role in bringing up the children. John was educated at Walsden Church of England School from 1901–1908 and at Todmorden Elementary School the following year, before attending the Todmorden Secondary School from 1909–1914. In 1914 he obtained a County Major Scholarship and began to study mathematics and physics at Manchester University. He joined the Royal Field Artillery as a signaller in 1915, saw active service, was twice mentioned in dispatches and was commissioned as a Second Lieutenant shortly before the end of the war. In 1918 he returned to Manchester to study electrical engineering at the College of Technology of the University and in 1920 he became a College Apprentice with the Metropolitan Vickers Electrical Company; this involved working in the research department of the Company and he obtained an MScTech degree in 1922. In the same year he went to St John's College, Cambridge with a Studentship from the Manchester College of Technology and distinguished himself in the Mathematical Tripos Part II in 1924. Then, with the award of scholarships he joined Rutherford's group in the Cavendish Laboratory. In 1925 Cockcroft married Eunice Elizabeth Crabtree and they had two sons, one of whom died in childhood, and four daughters.

In the Cavendish Laboratory Cockcroft used his engineering skills to collaborate with Peter Kapitza, for example on the design of electromagnets. In 1924 he made experiments using molecular beams to deposit thin films on surfaces cooled to low temperatures. Subsequently he worked with E.T.S. Walton to develop high-voltage equipment for accelerating protons and in 1931 they made the first observations of a nuclear transformation – of lithium into helium. Cockcroft went on to investigate transformations produced by protons in other light elements such as boron and beryllium. In 1935 Cockcroft took over the direction of research in the Mond Laboratory. From 1939–1946 Cockcroft was Jacksonian Professor of Natural Philosophy at Cambridge. Also in the wartime period he played a leading role in work on radar and served as Chief Superintendent of the Air Defence Research and Development Establishment at Christchurch from 1940–1944. In addition he had a major involvement with the wartime atomic energy project, spending a period in the USA and then being appointed to direct the construction of the first nuclear reactor in Canada at Chalk River. When he returned to Britain in 1946 he was appointed as the first Director of the Atomic Energy Research Establishment at Harwell. When the Atomic Energy Authority was established Cockcroft became one of the full-time Technical Members, a post which he resigned when he was elected as Master of Churchill College, Cambridge in 1959; he continued, however, as a part-time Member of the Authority.

Cockroft was the recipient of many honours including the CBE (1944), knight bachelor (1948), KCB (1953) and OM (1957). In the Royal Society he was awarded the Hughes Medal (jointly with E.T.S. Walton) in 1938 and a Royal Medal in 1954. He received the Nobel Prize for Physics, jointly with Walton in 1951.
BMFRS 1968, 14, 139-188. Sir Mark Oliphant and Lord Penney. DSB, 3, 328–331. Robert Spence. DNB 196–70, 224–226. Penney.

JAMES COCKSHUTT c.1742–1819; FRS 1804

James Cockshutt was a civil engineer, who was associated with the Wortley Ironworks at Sheffield, and was often employed by John Smeaton. In ironmaking he did early work using multiple tuyeres.
Bulloch's Roll.

ERNEST GEORGE COKER 1869–1946; FRS 1916

Ernest Coker was born at Wolverton, Buckinghamshire, the son of George Coker and Sarah née Tompkins. When he left school he entered the London and North Western Railway Carriage Works at Wolverton in 1883, where he worked as an apprentice for three years, followed by two years in the drawing office and laboratory. By evening study he obtained a scholarship, which enabled him to go to the Normal School (later Royal College) of Science, London. He was awarded a Whitworth Exhibition at the end of the second year, and a year later (1890) he obtained an Associateship of the College, with first class in mechanics. Also he was awarded a Whitworth Scholarship, with which he went to Edinburgh University and obtained a BSc in engineering. He was successful in the open examination for the Patent Office, and was appointed as an Assistant Examiner in 1892. However, in 1893 he entered Peterhouse at Cambridge where he studied for the Mechanical Engineering Tripos, and was awarded a BA with first class honours in 1896. During the vacations he spent periods in the railway works at Wolverton and did experiments that probably stimulated his interest in materials.

In 1898 he moved to McGill University Canada, initially as an assistant professor and later as an Associate Professor in Civil Engineering. He married Alice Mary Ming in 1899. His next appointment, beginning in 1905, involved returning to England as Professor of Mechanical Engineering and Applied Mathematics, and Head of the Department, at the City and Guilds Technical College at Finsbury. In this appointment he began his interest in the stress-optical effect and its applications in studying stress distributions in engineering components and structures. In 1914, when he was appointed to the Kennedy Chair of Civil and Mechanical Engineering at University College London, he began to develop the subject of photelasticity and became involved in collaborative work with Louis Filon. Over the period until 1934, when he retired, Coker's investigations covered a range of stress problems, and devising apparatus and techniques relevant to engineering applications. He was awarded the Rumford Medal in 1936.
ONFRS 1952–53, 8, 389–395. H.T. Jessop. DNB 1941–5, 165, 166. Alfred R. Stock.

BRYAN RANDELL COLES 1926–1997; FRS 1991

Bryan Coles was born in Cardiff, the son of Charles Frederick and Olive Irene Coles. He was educated at Canton High School and entered University College, Cardiff as a student in the Metallurgy Department in 1944. He graduated with first class honours in 1947 and having developed a particular interest in the important and advancing field of the theory of alloys he went to Jesus College, Oxford and became a research student with William Hume-Rothery. His research leading to the DPhil degree included important experimental work on phase diagram investigations. In 1950 he went to Imperial College, London as a Lecturer in Metal Physics in the Physics Department, appointed by Sir George Thomson.

From 1954–56 Coles spent two years, on leave from Imperial College, as a research fellow at the Carnegie Institute of Technology, Pittsburgh. In 1955 he married Merivan Robinson of St Paul, Minnesota and they had two sons. At Imperial College, he was appointed Senior Lecturer in Physics in 1959, and Reader in 1962. His teaching included lectures to students of metallurgy as well as to physics students. During the summers of 1962 and 1969 he was a visiting professor at the University of California in San Diego. In 1966 he was appointed as Professor of Solid State Physics at Imperial College, where he continued to build up his research group and to carry out important investigations in metal physics. Coles played a major role in the major task of planning and equipping the new physics building, the Blackett laboratory. In 1983 he spent a period as a visiting professor at the University of Minnesota. From 1984–87 he served as Dean of the Royal College of Science and then became the Pro-Rector of Imperial College, serving until 1991. During these periods of major administrative responsibilities and during the period beginning 1991 when he became an Emeritus Professor and Senior Research Fellow his research activities continued.

Coles's research work involved important experimental investigations of electronic and magnetic structures of alloys, including superconductivity, accompanied by the interpretation of the results in terms of quantum mechanics. He also made distinguished contributions to studies of spin glasses. He published many research papers and was co-author of the fifth revised reprint (1969) of the book *Atomic Theory for Students of Metallurgy* (W. Hume-Rothery and B. R. Coles).

Beyond his work at Imperial College, Coles had numerous commitments, including service on international committees, and travelling extensively. He became a member of the Council of the Royal Society in 1996. His wide-ranging interests included natural history, mediaeval architecture, literature, and music; he was a gifted conversationalist and public speaker. A notable feature of his career was the unusual transition that he made from metallurgy graduate to eminent solid state physicist. His early background in which he acquired experimental and theoretical expertise in physical metallurgy served him well in his distinguished research career in which he contributed substantially to the fundamental understanding of alloys.

Who's Who. *The Times* 18.3.97. *The Independent* 15.3.97, D. Edwards.

JOHN GORDON COLLIER 1935–1995; FRS 1990

John Collier was born in London, the son of John Collier and Edith Georgina née de Ville. He was educated at St Paul's School, which he left at the age of 16 to take a Student Apprenticeship in Mechanical and Chemical Engineering at the Atomic Energy Research Establishment at Harwell. With a scholarship he entered University College London where he was awarded a first class honours BSc(Eng) degree in chemical engineering in 1956 before returning to Harwell to the Chemical Engineering Division, as a Scientific Officer. In 1956 he married Ellen Mitchell, and they had one son and one daughter.

At Harwell Collier carried out research on heat transfer and fluid flow, which was significant in relation to improving the design and operation of boilers in nuclear power stations. From 1962–64 he was seconded as Head of Experimental Engineering for Atomic Energy of Canada Ltd (AECL). Promotion to Senior Scientific Officer and then to Principal Scientific Officer at Harwell occurred. From 1966–70 Collier was Head of the Engineering Division of Atomic Power Construction Ltd at Heston. In the 1970s and 80s he was given wide-ranging management roles with the UKAEA, involving the technology and safety of light-water reactors, and important policy issues. From 1970–75 he was Head of the Enginering and Sciences Group at Harwell and from 1975–77 he headed the Chemical Engineering Division. Subsequent UKAEA appointments included: Member, Atomic Energy Technical Unit (1979–81); Director of Technical Studies (1981–82); Deputy Chairman (1986–87); Chairman (1987–90). He was Director General, Generation Development and Construction Division, Central Electricity Generating Board, Barnwood (1983–86); Deputy Chairman (1986), Chairman (1987–90). From 1990–95 Collier was Chairman of Nuclear Electric plc. He was elected FREng.

The Independent 23.11.95, Brian Eyre. *The Guardian* 28.11.95, Ian Fells.

JOHN CONDUITT 1688–1737; FRS 1718
John Conduitt, who came from Cranbury Park, Hampshire, attended Westminster School, London (1701) and was elected to Trinity College Cambridge in 1705. He was Judge Advocate to the British forces in Portugal in 1711, and the following year became a Captain in a Regiment of Dragoons serving there. He represented Whitchurch in Parliament from 1715 to 1734, when he became Member for Southampton. In 1717 he married Isaac's Newton's niece, Mrs Katherine Barton, and they had one daughter. On Newton's death in 1727 he was appointed Master of the Mint. He wrote an essay in 1730 on 'Observations on the present state of our gold and silver coins', showing a great knowledge of the history of the currency and much care in experimental assaying.
DNB.

HAYNE CONSTANT 1904–1968; FRS 1948
Hayne Constant was born at Gravesend in Kent, the second of a family of six children. He spent his early childhood in Folkestone, where his father was a dental surgeon. He was educated at several schools: King's College Choir School, Cambridge; King's School, Canterbury; the Technical Institute, Folkestone and Sir Roger Manwood's School, Sandwich. In 1924 he entered Queen's College, Cambridge, having obtained an open exhibition and a state scholarship. Having taken Part I of the Mathematical Tripos he proceeded into the Mechanical Science Tripos, obtaining a first class degree and becoming a scholar of his college. He did reseach at Cambridge on the torsional vibration of crankshafts before joining the Royal Aircraft Establishment at Farnborough in 1928 where for the next six years he worked in the Engine Department on various projects. From 1934–36 he was a Lecturer at Imperial College. Returning to RAE he was in charge of the Supercharger Section of the Engine Department, which was headed by A. A. Griffith; Constant designed an axial flow compressor on the basis of Griffith's ideas and became involved in a project relating to the use of turbines for aircraft propulsion. Constant met Frank Whittle in 1937 and subsequently worked in collaboration with him making a particular contribution in relation to compressor design. In 1946 he was appointed Deputy Director, when the large complex at Pyestock became the National Gas Turbine Establishment; two years later he became Director until in 1958 he was appointed as Scientific Adviser to the Air Ministry and Chief Scientist (RAF) Ministry of Defence. Constant remained a bachelor.
BMFRS 1973, 19, 269–279. Sir William Hawthorne, FRS, H. Cohen and A. Howell.

GILBERT COOK 1885–1951; FRS 1940
Gilbert Cook was born in Liverpool, the son of William Cook; he attended Roomfield School, Todmorden. At the University of Manchester he was awarded an honours degree in engineering. From 1906–1920 he was an Assistant Engineer with the Lancashire and Yorkshire Railway. In 1910 he became the Vulcan Research Fellow at the University of Manchester. From 1914–17 he was a Lieutenant in the Royal Garrison Artillery and then transferred to the RNVR for service in the technical branch of mining, minesweepers and antisubmarine services. He was Senior Lecturer in engineering at Manchester from 1919–1921 and then was appointed Professor of Mechanical Engineering at King's College, London. He married Florence Davies in 1922. Cook was appointed to the Regius Chair of Civil Engineering and Mechanics at the University of Glasgow in 1938 and during the 1939–45 war he served on many national committees. Cook's research included the strength and elasticity of materials.
ONFRS 1952–53, 81, 109–117. A. Robertson.

VERNON ELLIS COSSLETT 1908–1990; FRS 1972

Ellis Cosslett was the eighth of eleven children of Edgar William Coslett and his wife Annie née Williams – a Welsh family. Coslett was born in Cirencester, where his father had become a builder's foreman. Because of illnesses he did not have regular schooling until around seven years of age and when he was 12 he won a scholarship to Cirencester Grammar School. He entered Bristol University in 1926 with a county scholarship and a teacher's training grant and studied chemistry, with physics and mathematics as subsidiaries. In 1929 he was awarded an ICI Research Studentship for two years to work on gas kinetics. Also he received further financial support from the Kaiser Wilhelm Institute (KWI) for Physical Chemistry in Berlin–Dahmen to enable him to work there for a year. He attended lectures by Planck, Nernst and Einstein and also he was drawn into anti-Nazi politics.

Back at Bristol in the summer of 1931, ICI arranged for Coslett to transfer to University College London, where he did research on electron diffraction in gases. He came into contact with a group of researchers of left wing views; a reformer at heart, he became a secret member of the Communist Party. He was awarded the PhD degree by Bristol University in 1932 and obtained a temporary post in which he constructed an electron diffraction camera for the study of *in situ* deposited films. His studies of recently published papers by M. Knoll and E.A.F. Ruska, who had constructed the world's first two-stage transmission electron microscope, led to his life long commitment to electron microscopy. While in Bristol he married Rosemary Wilson in 1936.

Coslett was unsuccessful in obtaining a research post in the time of slump, and eventually accepted a post in London, teaching elementary physics and chemistry. However, within a year he took up part-time research at Birkbeck College where he designed a large magnetic lens for focusing high-energy beta rays; circumstances led to the work leading to an MSc in 1939 rather than a PhD. In 1937 Coslett discussed with Bernal the possibility of designing an improved X-ray tube with increased brightness. With the outbreak of war, Bernal's group was evacuated to Oxford, and Coslett was given space in the electrical laboratory at Oxford, where he set up an electron-optical bench.

During the period in London Coslett and his wife had played a part in helping refugees fleeing from Hitler's regime. The marriage ended in divorce in 1939, and, in 1940, he married Anna Wischin, a physicist and light microscopist; there were two children from this marriage. The scientific allocation panel decided that Coslett should stay at Oxford to teach officer cadets applied physics and wireless telegraphy, but he also maintained his interest in electron microscopy. He became a Lecturer and Demonstrator at Oxford.

In 1946 W. L. Bragg at Cambridge obtained an ICI fellowship for Coslett to go to the Cavendish laboratory to run the RCA microscope and to promote the subject of electron microscopy. Here Coslett applied widely for research grants and created within his group an information centre. He had wide ranging interests involving physics, metallurgy, materials science, biology and medicine. He built up a major research group and was appointed a University Lecturer in Physics in 1949. With colleagues he designed and developed highly successful electron optical instruments: the X-ray microscope; the scanning electron probe analyser; the Cambridge 750 kV electron microscope, and the Cambridge 650 kV ultra-high resolution microscope.

Coslett was awarded the Cambridge ScD, was appointed University Reader in 1965 and retired in 1975. With his political views he believed that reform of the Cambridge establishment was needed, but in 1963 Corpus Christi College offered him a fellowship, which he accepted, although with misgivings. He received a Royal Medal in 1979.

Coslett played an important role in building up national and international activities in electron

microscopy; for example, he was the first Secretary and President (1972–75) of the International Federation of Electron Microscope Societies. These activities have had an enormous influence in many fields of science and Cosslett held a greatly respected position in the world-wide electron microscopy community.
BMFRS 1994, 40, 63–84. T. Mulvey.

SIR ALAN HOWARD COTTRELL 1919–; FRS 1955
Moseley Grammar School. Univ. of Birmingham: BSc 1939, PhD 1942; Lecturer in Metallurgy, 1943–49; Professor of Physical Metallurgy, 1949–55. AERE, Harwell: Deputy Head of Metallurgy Division, 1955–58. Univ. of Cambridge: Goldsmith's Professor of Metallurgy, 1958–65; Fellow of Christ's College, 1958–70; ScD (Cantab) 1976. Ministry of Defence: Deputy Chief Scientific Adviser (studies), 1965–67; Chief Adviser, 1967. Deputy Chief Adviser to HM Government, 1968–71; Chief Scientific Adviser, 1971–74. Univ. of Cambridge: Master, Jesus College, 1974–86; Vice–Chancellor, 1977–79. FREng, 1979. Royal Society: Vice-President 1964, 76, 77. Hughes Medal, 1961, Rumford Medal, 1974. Knighted 1971.
Physical metallurgy: dislocations in crystals; plastic deformation and mechanical properties (e.g. the yield point of steels in relation to the interaction of carbon atoms with dislocations; the effect of radiation on crystals).

CHARLES ALFRED COULSON 1910–1974; FRS 1950
Charles Coulson was born in Dudley, Yorkshire, one of twin sons. When his family moved to Bristol in 1920 he was educated at Clifton College, whence he won an entrance scholarship in mathematics to Trinity College, Cambridge in 1928. After graduating with a first class in mathematics in 1931 he stayed for a further year to take Part II, in which he obtained a first, before carrying out research with Lennard-Jones, and obtaining a PhD in 1936. In 1934 he had been elected to a Prize Fellowship at Trinity. In 1938, the year of his marriage to Eileen Burrett, Coulson took up a post as Senior Lecturer in Mathematics in the University of Dundee. He moved to Oxford in 1945 to the Physical Chemistry Department as an ICI fellow, and two years later he was appointed to the Chair of Theoretical Physics at King's College, London. In 1952 he was appointed to the Rouse Ball Chair of Mathematics at Oxford; in 1972 he became the first Professor of Theoretical Chemistry at Oxford. He was awarded the Davy Medal in 1970. Coulson made important contributions in theoretical chemistry in the field of molecular theory and the solid state, including metals.
BMFRS 1974, 20, 75–134. S.L. Altmann and E.J. Bowen FRS. DNB 1971–89, 182, 183. W.H. March.

ALAN HERBERT COWLEY 1934–; FRS 1988
Univ. of Manchester: BSc., MSc. Dalton Chemical School, 1956–58, PhD 1958. ICI: Technical Officer: 1960–61. Univ. of Texas at Austin: Assistant Professor of Chemistry, 1962–67; Associate Professor, 1967–70; Professor, 1970–84: George W. Watt Centennial Professor, 1984–88; Richard J.V. Johnson Regents Professor of Chemistry, 1989–91. Imperial College, London: Sir Edward Frankland Professor of Inorganic Chemistry, 1988–89. Univ of Texas at Austin: Robert Welch Professor of Chemistry, 1991–.
Organo-metallic chemistry; inorganic syntheses of semiconductors.

LORENZ FLORENZ FRIEDRICH VON CRELL 1745–1816; FRS 1788
Lorenz Crell was the son of Johann Friedrich Crell, Professor of Medicine at the Duchy of Bruns-

wick's University at Helmstadt. His early education was supervised by his maternal grandfather, also a Professor of Medicine and a leading surgeon. Crell entered the local university in 1759, and graduated MD in 1768, having spent two and a half years studying studied medicine and chemistry in Strasbourg, Paris, Edinburgh and London. In 1771 he was appointed to a Chair in Chemistry in Germany; there followed in 1774 appointment as Professor of Medical Theory and Materia Medica in his native city of Helmstadt. He carried out some unsuccessful attempts to melt platinum. He was an active correspondent with scientists in other countries and founded the first journal to be devoted to chemistry, *Chemische Annalen.*
DSB 3, 464–466. K. Hufbauer. Ref. Ch. 4.1.

ROOKES EVELYN BELL CROMPTON 1845–1940; FRS 1933
Rookes Crompton was born near Thirsk, Scotland, the fourth son of Joshua Samuel Crompton and Mary née Alexander. During the Crimean War he was enrolled as a naval cadet and allowed to accompany his mother's cousin, and visited the trenches at the age of 11. His education included a period at Harrow (1858–1860). He served four years in India as an Ensign in the Rifle Brigade; during this period, having earlier built a workshop at home, he had a travelling workshop, and introduced steam road haulage. In 1871 he married Elizabeth Gertrude Clarke and they had two sons and three daughters. On returning to England in 1875 he became a partner in a Chelmsford engineering firm; also, as advisor at an ironworks in Stanton belonging to a branch of his family, he bought dynamos to illuminate the foundry. From about 1878 his work focussed on electricity and engineering, and he began to make electric light plant. During the period 1890–1899 he revisited India to advise the Government on electrical projects. Then in 1900 he saw war service in South Africa on electrical engineering work. He received the CB in 1900.
ONFRS 1941, 9, 395–403. A. Russell, DNB 1940, 203, 204. W.L. Randell.

SIR WILLIAM CROOKES 1832–1919; FRS 1863
William Crookes was born in London, the eldest son of 16 children of Joseph Crookes, a prosperous tailor, and Mary née Scott. He entered the Royal College of Chemistry in London in 1848 where he became personal assistant to A.W. Hoffmann from 1850–1854 and then came to the attention of Michael Faraday. Following other appointments he worked as a free lance chemical consultant from 1856 based in London. He married Ellen Humphry in 1856 and they had ten children.

Crookes's research included spectroscopy, leading to his discovery of the element thallium, and investigations of the rare earth elements. His success in producing a vacuum of the order of 10^{-6} atmosphere made possible the discovery of X-rays and electrons. During the 1880s he worked on the technology of electric lighting with evacuated tubes. Of metallurgical interest was his work on gold and silver amalgams and his translation and editing, with a co-worker, of a substantial German book on metallurgy.

In the Royal Society Crookes held many important posts: Foreign Secretary (1908–1912); Vice-President (1895–1896, 1907–1908, 1915–1916) and the Presidency from 1913–1915. He was the recipient of medals; Royal (1875), Davy (1888) and Copley (1904), and was Bakerian Lecturer in 1878, 1879 and 1883. He was awarded a knighthood in 1897, and the Order of Merit in 1910.
PRS 1919–20, 96, i–ix. WAT. DSB 3, 474–482 W.H. Brock. DNB 1912–21, 136–137. CNH.

WILLIAM CROONE 1633–1684; FRS 1663

William Croone, who was the fourth child of a London merchant, was educated at Merchant Taylors' School and at Emmanuel College, Cambridge where he took his first degree in arts in 1650, and was elected to a Fellowship. He was created MD by Royal mandate 12 years later, and then in 1674 he was made a Fellow of the Royal College of Physicians. In 1659, having been appointed Professor of Rhetoric at Gresham College, he became part of the group from whose meetings the Royal Society grew. As an active experimenter he became involved in planning the early programme of the Society in 1660; he became one of the first Fellows, often serving on its Council. Among his discoveries in physics was the fact that water has its maximum density above its freezing point, and he was associated with Robert Boyle (who was possibly his cousin) in his work on aerial physics. He is reported to have carried out tests of strength on silver wires at a meeting of the Royal Society in 1662. His main research was in the field of muscle physiology. Croone, in 1670, married Mary Lorymer, daughter of an alderman of London, and having had to give up his professorship at Gresham College, he became a Lecturer in Anatomy at Surgeon's Hall. On his death he left a scheme for the endowment of two lectureships, one of these being at the Royal Society; the Croonian Lecture became the Society's premier lecture in the biological sciences.

Notes and Records Roy. Soc. 1960, 15, 158–178. DSB 3, 482, 483. L.M. Payne. DNB. Gen. Ref. 20. v. III.

SIR BERNARD CROSSLAND 1923–; FRS 1979

Simon Langton's school, Canterbury. Rolls Royce, Ltd: Apprentice, 1940–41. Univ. of Nottingham: 1941–43. Rolls Royce: Technical Assistant, 1943–45. Univ. of Bristol: Assistant Lecturer; Lecturer; Senior Lecturer in Mechanical Engineering, 1946– 59. The Queen's University, Belfast: 1984–; Professor and Head of Department of Mechanical and Industrial Engineering, 1959–82; Research Professor 1982–84; Dean 1964–67; Pro-Vice-Chancellor 1978–82. Royal Society: Vice-President 1984–86.FREng 1979. CBE 1980. Knighted 1990.

Fatigue of metals; effect of very high fluid pressures on properties of materials; strength of thick walled vessels; explosive welding, friction welding; design and testing of engines.

SIR WILLIAM CUBITT 1785–1861; FRS 1830

William Cubitt was born at Bacton Wood, near Dilham, Norfolk, the son of Joseph Cubitt, a miller, and his wife née Miss Lubbock. Later his father moved to South Repps, where William was employed at an early age in his mill, before being apprenticed to a cabinet maker at Stulham for four years. At Bacton in 1804 he again worked with his father, and then joined an agricultural machinery maker at Swanton, where he became well known for the accuracy and finish of the patterns he made for iron casting.He then settled in Horning as a millwright, prior to obtaining employment in 1812 in the works of Messrs Ransome, Ipswich; he soon became Chief Engineer, then after 9 years, a partner until he moved to London in 1826. He became involved in most of the important civil engineering ventures of the time: canals, river improvements, docks, railway tunnels, and the erection of the Great Exhibition building. Cubitt was knighted in 1851.

PRS 1862–63, 12, iv–v. DNB.

WILLIAM ERNEST DALBY 1862–1936; FRS 1913

William Dalby, born in London, was eight years old when his father died and his mother was left to maintain and educate her three sons. In view of the financial circumstances, Dalby entered the Stratford Works of the Great Eastern Railway at the age of 14, where he worked from 6 a.m. to

5.30 p.m. but also attended evening classes. He was awarded a scholarship at the Science School at South Kensington at the age of 20, followed the next year by a Whitworth Scholarship. In 1883 after finishing his apprenticeship he entered the Crewe Works of the London and North-Western Railway, where by spare time study he obtained a London BSc degree. After six years at Crewe he received promotion, but in 1891 he accepted an invitation from James Ewing at Cambridge University to join his teaching staff. For the next six years he made an important contribution to the development of the University Engineering Department and in 1894 received an honorary MA and became a member of Trinity College.

In 1896, having become Professor of Mechanical Engineering and Applied Mathematics at the Finsbury Technical College, he did research and acquired an international reputation in the balancing of engines. Another move came in 1904, when he was appointed to the Chair of Civil and Mechanical Engineering in the University of London, at what was then the Technical College of the City and Guilds which later became the City and Guilds Engineering College of the Imperial College of Science and Technology. This appointment with its heavy administrative responsibilities he held until his retirement in 1931; he was involved also in important committee work in the sphere of government and professional societies.

Dalby made many contributions to engineering science. His research included collaboration with Hugh Callendar on the measurement of temperatures in the cylinder of a gas engine and he also did important work on mechanical testing.

ONFRS 1936, 5, 2, 145–149. C.E. Inglis FRS.

JOHN DALTON 1766–1844; FRS 1822

John Dalton was born at Eaglesfield, near Cockermouth in Cumberland. His father Joseph Dalton, a weaver, married Mary Deborah Greenup, who like himself was member of the Society of Friends. John was the youngest of three children who reached maturity from their family of six. Having attended a Quaker school, at the age of ten he entered the service of a Quaker gentleman, who taught him mathematics, and encouraged him to improve himself. In 1778 Dalton set up a school for boys and girls of all ages, but after two years the school closed and he did field work to earn a living. However, in 1781 he joined his brother as an assistant in a school in Kendal, which they carried on independently when the master retired. John was much occupied with private study on subjects including mathematics, mechanics, optics and astronomy; over the 12 years he spent in Kendal he read extensively, acquiring the basis of knowledge for his future research. He also kept an extensive meteorological journal, constructed some apparatus and was involved in botany and zoology.

In 1793 he accepted a Professorship of Mathematics and Natural Philosophy in New College, Manchester. By the discoveries that he made over a considerable period meteorology was established as a science. Moving into chemistry he presented his important theory based on the concept that matter is composed of minute particles, or atoms. When New College moved to York in 1799, Dalton supported himself by giving private lessons in mathematics, and carrying out chemical analyses and other chemical work at low charges. His first visit to London had been in 1792, to attend the yearly meeting of the Society of Friends, but he had no scientific acquaintances there at that time. However, he returned in 1803 to give a course of lectures at the Royal Institution. He came to know William Wollaston and Sir Joseph Banks and foreign scientists came to visit him at Manchester. Humphry Davy offered him a nomination to the Royal Society in 1810, which he refused, probably on the grounds of expense; however, he was elected in 1822 with no consent asked and paid the usual fees. His work included investigations of metallic oxides and alloys. Among his distinctions were honorary degrees and the award of a Royal Medal in 1826.

Dalton declined the offer made by a gentleman in Derby of a home and laboratory, with an annual salary, because of his wish for independence. However, Charles Babbage and others succeeded in 1833 in obtaining a pension for him so that he could give up the burden of teaching; a year later an inheritance from his brother placed him in a position of comparative wealth. Throughout his life Dalton never married, but lived for many years with a family in Manchester, retaining his Quaker traditions and dress. In 1837 he suffered from paralytic attacks, leaving him a semi-invalid. From 1817 to his death he was President of the Manchester Philosophical Society, before which Society he had read 117 papers, many of them of outstanding importance. In his old age he enjoyed widespread recognition; when he died he was given a civic funeral, recognising not only his intellectual achievements, but also the increasing importance of men of science in national life.
DSB 3, 537–547. A. Thackray. DNB. Gen. Ref. 20. v. IV.

JOHN FREDERIC DANIELL 1790–1845; FRS 1814

John Daniell was born in Essex Street, Strand, the son of George Daniell, Bencher of the Inner Temple who provided him with a good classical education at home. Having shown an interest in science Daniell was placed in a sugar-refining establishment of a relative, where he introduced important manufacturing improvements; however, not finding business life attractive he did not remain in it long. In 1817 he married Charlotte Rule; there were two sons and three daughters of the marriage. He became a friend of William Brande and they worked together and made various journeys in Britain and on the Continent. He soon commenced his important research on meteorology which included the invention of a hygrometer (1820) giving precise measurements. Daniell, in 1816 and 1817 reported important work on the acid attack on metals and etched steels. His work on the development of a pyrometer based on the expansion of a platinum rod led to his receiving the Rumford Medal in 1830.

When King's College, London was established in 1831 Daniell was appointed Professor of Chemistry, becoming a very successful teacher. He was a great friend and admirer of Michael Faraday and along with his original contributions to chemistry he also worked principally in electricity; his work included the invention of the 'constant battery'.

In 1839 he was appointed a member of a commission constituted by the Admiralty to inquire into the best method of protecting Royal Navy ships from lightning; also, for the Admiralty, he carried out corrosion studies. In the Royal Society he was Foreign Secretary, a post which he held from 1839 until 1845. He was awarded the Rumford Medal in 1832, the Copley Medal in 1837 and a Royal Medal in 1842.
PRS 1845, 5, 577–580. DSB 3, 556–558. Arnold Thackray. DNB. Gen. Ref. 20. Ref. Ch. 4.1.

SIR CHARLES GALTON DARWIN 1887–1962; FRS 1922

Charles Galton Darwin was the eldest son of Sir George Darwin, FRS (Plumerian Professor of Astronomy at Cambridge) and Maud née du Puy. He was born at Newnham Grange, Cambridge and attended St Faith's Grammar School. He went to Marlborough College in 1901 where he specialised in mathematics in the sixth form. In 1906 he entered Trinity College, Cambridge with a major scholarship; he took the Mathematical Tripos, becoming Fourth Wrangler in 1909 in Part I and obtaining a first class in Part II in the following year. Darwin then went to Manchester as Schuster Reader in Mathematical Physics under Rutherford. After a period of research on the upper atmosphere and on radioactivity he began work on X-ray diffraction. His publications in this field included two highly significant papers in 1914.

Darwin was in the Officers Training Corps at Manchester, and was sent to France when war broke out. Subsequently he was attached to the Royal Engineers to serve with the units that W.L. Bragg was organising to detect enemy guns by sound ranging; he commanded a section and was awarded an MC. Late in 1917 he was attached to the Royal Flying Corps at Orfordness for work on aircraft noise.

In 1919 Darwin was appointed a Fellow and Lecturer of Christ's College, Cambridge and carried out research in collaboration with R.H. Fowler on classical atomic statistics in relation to thermodynamics. He was appointed Tait Professor of Natural Philosophy in the University of Edinburgh in 1924, where his research included magneto-optics. In 1925 he married Katharine Pember, and they had one daughter and four sons. Darwin returned to Cambridge in 1936 to become Master of Christ's College. In 1938 he was appointed as Director of the National Physical Laboratory at Teddington, and carried out a reorganisation when war began in 1939. He was seconded to Washington for a year in 1941 as Director of the British office concerned with Anglo-American scientific co-operation for the war effort. When he returned to England he became Scientific Adviser to the War Office. He retired from NPL in 1949. Darwin received the KBE in 1942, and in the Royal Society he served as a Vice-President in 1939 and received a Royal Medal in 1935.

BMFRS 1963, 9, 69–85. Sir George Thomson. DNB 1961–70, 272–274. Margaret Gowing.

EDMUND DAVY 1785–1857; FRS 1826

Edmund Davy, the son of William Davy, was born in Penzance and received his early education there. He moved to London in 1804, where he was appointed assistant to his older cousin Humphry, then Professor of Chemistry in the Laboratory of the Royal Institution. In 1813 he was elected to the Chair of Chemistry at the Royal Cork Institute in Ireland and in 1821 he became Professor of Chemistry of the Royal Dublin Society. While still at the Royal Institution Davy had made several studies on the chemistry of platinum. Then at Cork he found the oxidising effect of finely divided platinum on alcohol vapour at room temperature, with considerable heat generation, which was pursued later by Döbereiner, playing a part in the field of catalysis. Davy was active in advocating the extension of scientific knowledge, and through his influence popular courses of lectures to which he contributed were instituted in many parts of Ireland.

DNB. Ref. Ch. 4.1.

SIR HUMPHRY DAVY 1778–1829; FRS 1803

Humphry Davy was born in Penzance, Cornwall, the eldest of five children. His father was Robert Davy, a wood carver, who came from an old family with a modest inheritance, and his mother was Grace née Millett. When Humphry was a child the family moved to Varfell, their estate in Ludgvan. After preparatory education he went to Penzance Grammar School, where he showed a remarkable memory and rapidly acquired knowledge from books. He enjoyed story telling to a circle of friends and developed a love of poetry, including the writing of verses and ballads. He acquired a taste for experimental science, mainly by instruction from a local saddler, a Quaker, who had constructed apparatus. In 1793 he went to Truro to finish his education privately. During his boyhood he gained a knowledge of classics, a love of books, and a devotion to his native countryside.

After the death of his father in 1794, Davy was apprenticed the following year to a surgeon in Penzance, where he became a chemist in the apothecary's dispensary. A garret in the house where he lived was the scene of his earliest chemical experiments and his interest turned to science. He came to know a local doctor who was a lecturer in chemistry at St Bartholomew's Hospital and who allowed him to use his apparatus and appears to have directed his attention to the corrosion of the floodgates at Hayle. A personal contact led Davy to take a post in Bristol in 1798 at the 'Pneumatic Institution' where patients came to be treated with therapeutic gases; here he was an assistant involved in experiments on medical aspects of air and gases. He intended to return to medical study but soon became committed to his involvement in the laboratory. Among his acquaintances were the poet Coleridge and when Davy visited London for the first time in 1799, his circle of friends was much extended. In the same year he published results from his research, including the subject of 'nitrous oxide and its respiration'; he continued this work the next year, running considerable health risks and demonstrating that the 'laughing gas' could be used as an anaesthetic.

An important event occurred in 1799 when an 'Institution for Diffusing Knowledge' (i.e the Royal Institution) was founded in Albermarle Street, London, stemming from a suggestion from Count Rumford. In 1801 the managers decided to appoint Davy as Assistant Lecturer, Director of the Chemical Laboratory and Assistant Editor of the Journals of the Institution, with a salary of £100 per annum and a room in the house. Rumford presented to Davy the prospect of his becoming Professor of Chemistry in two or three years time, and agreed that Davy should have every facility for his private philosophical investigations. There followed 12 years of full-time appointment at the Institution, during which Davy established a laboratory which was amongst the finest in the world. He became Professor of Chemistry in 1802, Director in 1804 (with an annual salary of £400) and held an Honorary Professorship up to the time of his death. When he joined the Institution in 1801 Davy gave courses of lectures, including the subjects of 'Galvanism' and of 'Pneumatic Chemistry'. In the following year he also gave lectures on 'chemistry' and 'chemistry applied to the arts', and continued to give courses of lectures over many years. He was a brilliant lecturer, who used demonstrations so that lay audiences were greatly attracted, and the lectures became important social functions. Coleridge said he went to listen 'to renew my stock of metaphors' and claimed 'that had he (Davy) not been the first chemist, he would have been the first poet of the age'. Davy became friends with members of high society and was an honoured guest at the great country houses.

In research Davy made early experiments on galvanism in 1800 and in 1801 his first communication to the Royal Society was in this field. In 1807 he reported his use of the voltaic battery to produce from alkalis 'two new inflammable substitutes' (potassium and sodium). Davy presented his results in this field to the Royal Society in a Bakerian Lecture; in total he gave seven Bakerian Lectures over a period of years covering various subjects. He became famous in Europe and he was awarded the Napoleon prize, a substantial monetary award of the Institut de France in 1807. At the end of 1807 Davy had a severe illness, and was not able to resume his electrochemical research for several months. In the succeeding years there ensued a range of research and other activities, including agriculture.

The year 1812 saw Davy knighted by the Prince Regent and a few days later, his marriage to Mrs A. Preece, a wealthy widow. In 1813 he resigned from his Directorship at the Institution, having appointed Michael Faraday as an assistant in the laboratory. Sir Humphry and Lady Davy planned an extended tour of the Continent and they set off in the autumn of 1813 for France, Italy and Switzerland; Faraday went with them as secretary and scientific assistant and they took some scientific equipment for use en route. Davy's reputation was such that he was allowed to enter France during a time of war. In Paris they experimented with iodine, a newly discovered element, and met scientists such as Gay-Lussac and Count Rumford. At Florence, Davy experimented, using a great burning glass on the combustion of diamond, showing that it was composed of pure carbon. Also in Italy he met Volta and examined papyri and pigments from ancient sites. Back in London in 1813 Davy returned to the laboratory in the Royal Institution, where he became involved in solving the problem of explosions in coal mines; his invention of a safety lamp in 1815 was an outstanding achievement. In 1823 he commenced an enquiry into the corrosion of copper sheathing on naval ships.

In 1818 Davy was created a Baronet and two years later he was elected as President of the Royal Society; he had served as Secretary and Member of Council of the Society in 1807, and had received the Copley and Rumford Medals in 1805 and 1816 respectively. During his Presidency he tried to improve the quality of the Fellowship, in terms of the level of scientific attainment for election. Failing health, including a stroke, occurred in 1826 and he travelled to winter in Italy on medical advice. He resigned from the Presidency of the Royal Society in 1827 and made another journey on the Continent in the following year, where he died in Geneva, a man of genius, and described as 'the greatest chemist of his age'.
DSB 3, 598–604. D.M. Knight. DNB. Refs. Ch. 8.7, 9.

PETER DAY 1938–; FRS 1986

Maidstone Grammar School. Univ. of Oxford: Wadham College, BA 1961, MA, DPhil 1965. Cyanamid European Research Institute, Geneva, 1962. Univ. of Oxford, St John's College: Junior Research Fellow, 1967–65; Official Fellow 1965–91; Departmental Demonstrator, 1965–67; Lecturer Inorganic Chemistry,1967–89; Ad Hominem Professor of Solid State Chemistry, 1989–91. Inst. Laue–Langevin, Grenoble: Director, 1988–91 (on secondment). Royal Institute and the Davy–Faraday Research Laboratory: Director, 1991–; Fullerian Professor of Chemistry 1994–. Royal Institution Professorial Research Fellow, Univ. College London, 1995–.
Electronic and magnetic behaviour of inorganic compounds.

NORMAN ADRIAN DE BRUYNE 1904–1997; FRS 1967

Norman de Bruyne was the son of a Dutchman, Pieter Adriaan de Bruyne and Maud née Mattock from England. He was born in Chile, but the family moved to England in 1906; he was educated at Lancing College, Sussex before entering Trinity College, Cambridge in 1923 to study physics. He graduated in natural sciences and carried out atomic research in the Cavendish Laboratory with Lord Rutherford obtaining a PhD. He became a fellow of Trinity College in 1928 and was awarded an MA and a PhD in 1930. He developed an interest in flying, obtained a pilot's licence, and in 1930 bought a light aeroplane.

In 1932 de Bruyne founded The Cambridge Aeroplane Construction Company; in 1934 the company name was changed to Aero Research Limited and in the following year a move to a site at Duxford occurred. His initial concern was the complacency shown by the, then, current generation of aero-engineers; he wanted to use his own ideas to 'revolutionise' aircraft design and hence designed, built and flew his own aircraft – *The Snark*. Research led to the launch of the Aerolite range of adhesives which made the name of the company in 1937. Five years later the adhesive Redux, and the Redux (standing for REsearch at DUXford) bonding process. The system, which was used to bond wood-to-metal and metal-to-metal components, came to be widely applied in aircraft construction, e.g. in the de Havilland Comet and in the BAe–146. The availability of a high-strength adhesive allowed de Bruyne's other, immediate post-war development – aluminium honeycomb – to realise its full potential with the possibility of producing light weight, stiff, bonded honeycomb sandwich panels for wing, tail and fuselage structures.

De Bruyne had married Elma Lilian Marsh in 1940 and they had one son and one daughter. In 1948, when the Ciba-Geigy Company bought his controlling interest in his company he continued as Managing Director until 1960, while starting a new company, Techne (Cambridge) Ltd to design and produce laboratory instruments. In 1956 he started an American Techne Company in New Jersey and moved to Princeton where he took American citizenship. He returned to Britain in 1974 and was elected FREng in 1976. His technological innovations in adhesives and light weight structures and in aircraft design were highly influential in the aeronautical industry.
The Times, 31 March 1997. *Who's Who*. John A. Bishopp, 'The History of Redux and the Redux Bonding Process', *The International Journal of Adhesion and Adhesives*, 1997, **17**(4), 287–301.

ARTHUR-AUGUSTE DE LA RIVE 1801–1873; FMRS 1846
Arthur-Auguste de la Rive was the eldest son of Charles-Gaspard de la Rive and Marguerite Adelaide née Boissier. He was born in Geneva (then French) where he attended the Collège Publique and the Académie (where his father was a professor). From 1816 to 1823 he studied successively letters, philosophy and law. His interests were primarily scientific and in 1823 he was appointed Professor of General Physics at the Académie; two years later he became Professor of Experimental Physics. In 1826 he married Jeanne-Mathilde Duppa and they had two sons and three daughters. His research included electricity and the chemical pile; de la Rive was known as the leader of the Swiss scientific community.
PRS 1875–76, 24, xxxvii–xl. DSB 8, 35–39. K.L. Canera.

GEOFFREY DEARNALEY 1930–; FRS 1993
Univ. of Cambridge: MA 1955, PhD 1956. Pembroke College; Research Fellow,1955–58. AERE, Harwell, Nuclear Physics Division: 1958; individual merit promotion, 1975; Chief Scientist, Surface Technologies, 1991– 93. Southwest Res. Institute, Texas: Inst. Scientist, 1993–94; Materials and Structures Division, Vice-President, 1994.
Interaction of energetic ion beams with materials, ion implantation.

PETER JOSEPH WILHELM DEBYE 1884–1966; FMRS 1933
Petrus (Peter) Debye was born in Maastricht, the Netherlands, the son of Wilhelmus and Maria Reumenkens Debye. After attending school in Maastricht he studied at the Technische Hochschule in Aachen, Germany, where he graduated in electrical engineering in 1905. When Sommerfeld moved from Aachen to Munich, Debye accompanied him, and obtained a PhD in physics in 1908 and became Privatdozent. Subsequently Debye held professorial posts in physics at a number of universities: Zurich, 1911, and again in 1920, Utrecht, 1912, Göttingen, 1914, Leipzig, 1927 and Berlin, 1935. In 1913 he had married Matilda Alberer, and they had one son and one daughter. In 1935 he became Director of the Kaiser Wilhelm Institute of Physics in Berlin. He left Germany just before World War II to go to Cornell University, USA, where he remained, taking USA citizenship and formally retiring from his appointment in 1952.

Debye carried out notable work on the theory of electrolytes and electrolytic dissocation in solution; the important theory, which became known as the Debye–Huckel theory was published in 1923. Debye also studied the specific heats of solids at low temperatures. While he was at Göttingen his important work on X-ray diffraction included the development with P. Scherrer of the technique which became widely used involving the use of a powdered sample for the study of crystal structures; this work was published in 1916. He also did some research on polymers including viscosity and diffusion, and on synthetic rubber. Debye was awarded the the Rumford

Medal in 1930 and the Nobel Prize for Chemistry in 1936.
BMFRS 1970, 16, 175–232, M. Davies. DSB 3, 617–621. C.P. Smyth.

JOHN THEOPHILUS DESAGULIERS 1683–1744; FRS 1714

John Desaguliers was born in La Rochelle, France, of Huguenot parents who settled in Islington, London in 1694. He entered Christ Church, Oxford in 1705 and graduated BA in 1709. He held a lectureship at Hart Hall and became MA in 1712. In the same year he married Joanna Pudsey and they had several children. Also in 1712 he moved to Westminster, London where he gave scientific lectures and also took up orders, being appointed to a living of a parish near Edgware. His practical abilities attracted the interest of the Royal Society and in the winter of 1713–1714 at Newton's suggestion he was invited to repeat some of Newton's experiments on heat; he became, *de facto,* Curator of Experiments. In recognition of his experimental ingenuity he was three times the recipient of the Copley Medal (1734, 1736 and 1741). He presented around 50 papers in *Philosophical Transactions* on optics, mechanics and electricity.
DSB 4, 43–46 A. Rupert Hall.

CECIL HENRY DESCH 1874–1958; FRS 1923

Cecil Desch was born in Tottenham, Middlesex one of three children of Henry Thomas Desch and Harriet née Ingerson. His father was a surveyor who held senior positions with W. Cubitt and Co. and who often visited France, taking Cecil with him. His mother and visiting teachers gave Cecil his early education. When nine years old he went to St John's School, Tottenham and three years later he was transferred to Birkbeck School, Kingsland. Having decided to study chemistry, he entered at the age of 15 the City and Guilds Technical College at Finsbury. After two years as a student in organic chemistry Desch was placed first in the College examinations and continued as a research student for another year. He was advised to spend a period at one of the German universities but since his finances did not make this feasible he obtained a position as an Assistant Chemist with F. Kendall & Son in Stratford-on-Avon, a firm that produced chemicals for the brewing industry. Desch was involved in routine testing and control of the manufacturing processes, but also did some research. He became interested in photo-micrographs and microscopes. Desch's work at Stratford also included a project on the rare earths.

In his spare time Desch worked by private study for an external degree from London University and took geology as his principal subject in the final examination. He was awarded a scholarship of £40, tenable for two years and in 1900 he was able to go to Wurzburg University where he obtained a PhD degree *summa cum laude* in 1902 for research on organic compounds of ferric iron. When he returned to England Sir William Ramsay, then Professor of Chemistry at University College, London offered him an appointment during which he carried out work on tautomerism, leading to the award of the London DSc also in 1902. Desch became became interested in archaeology and carried out analyses of ancient copper and bronze objects, using ordinary chemical methods and spectroscopic analysis. The scope of his work increased in this field as curators of various museums provided various other copper and bronze objects.

At King's College, London, the Professor of Metallurgy, A.K. Huntington, required an assistant with experience of a metallographic microscope and Desch was appointed. His interest in metallography grew alongside his interest in organic chemistry. Working on metallurgical analysis in the laboratory was another assistant, Miss Elison Macadam; in 1909 they were married.

Desch was appointed to a Lectureship in Metallurgical Chemistry at Glasgow University where there was good metallographic equipment and time for research, and work on corrosion was started. Also he started to collect the information that was used for his *Textbook of Metallography* which immediately won recognition when it was published in 1910; the sixth edition was issued in 1944. Desch taught the engineering students who came from many parts of the world.

A close neighbour at the University was Sir George Beilby with whom he became friends. Beilby financed an enquiry under the aegis of an Institute of Metals committee on solidification from the liquid state; Desch was appointed official investigator and proceeded to carry out work which did much to establish his scientific reputation. In the first report Desch summarised existing knowledge, while in the second report important experimental results on grain formation and shape were presented and discussed.

Desch accepted an invitation to occupy the Chair of Metallurgy at the Royal Technical College, Glasgow in 1918; following his recommendation that the University course should be merged with the College course a unified metallurgical centre was formed. In 1920 Desch was appointed to the Chair of Metallurgy at Sheffield University where he was involved in a major reorganisation; laboratories were extended and equipment provided through the generosity of Sir Robert Hadfield and other Sheffield industrialists.

Desch was appointed Superintendent of the Metallurgy Department at the National Physical Laboratory, Teddington in 1932. Before taking up the appointment he gave the George Fisher Baker Lectures at Cornell University on 'The chemistry of solids'.

On his retirement in 1939 Desch was appointed as a Scientific Adviser to the Iron and Steel Research Council. In 1943 he joined the board of Messrs Richard Thomas & Co. Ltd, until a merger led to the formation of Richard Thomas and Baldwins Ltd; he then served as Scientific Adviser to the Whitehead Iron and Steel Co. Ltd for 13 years.

Desch was honoured by many awards and appointments. He served on the Council of the Royal Society from 1932 to 1934. His career, in which he is best known for his achievements as a metallurgist, showed him to be a man of an unusually wide range of interests.
BMFRS 1959, 5, 49–68. A. McCance. DNB Missing Persons, Oxford Univ. Press 1993, 182, J. Nutting.

DENIS HENRY DESTY 1923–1994; FRS 1984

Denis Desty was born in Southampton, the son of Ernest James and Alice Desty. He was educated at Taunton's School, Southampton and at Unversity College, Southampton, where he studied chemistry. His studies were interrupted by wartime service as a signals officer with the RAF from 1942–46, including a period in India. Having resumed his studies at Southampton he was awarded a BSc (London) honours degree in 1948. He then joined BP as a Technologist, later becoming successively Group Leader (1952), Senior Chemist(1962) and Senior Research Associate; he retired in 1981as a result of ill health. In 1978 Desty was appointed a Visiting Professor in the Engineering Department at the University of Surrey; from 1984–86 he was a Visiting Professor in the Department of Chemical Engineering and Chemical Technology at Imperial College, London. Desty was not only a distinguished scientist but also a prolific inventor; his activities ranged widely including the fields of gas chromatography and combustion technology. Concerning materials his work included the use of polymers and he devised a plasma reactor process for the production of buckmaster fullerene.
BMFRS, 1995, 41, 137–143. C.S.G. Philips.

SIR JAMES DEWAR 1842–1923; FRS 1877

James Dewar was born at Kincardine-on-Firth, Scotland, the youngest son of Thomas Dewar, a vintner, and Ann Eadie Dewar. His education at local schools was interrupted for two years when he had an attack of rheumatic fever at the age of 10. He attended Dollar Academy before entering Edinburgh University in 1858, initially as an arts student, but his interest was directed to physical science. Dewar's career involved appointments as assistant to Playfair (1867–1868) and to his successor, a Lectureship in Chemistry in the Royal Veterinary College at Edinburgh (1869) and Assistant Chemist to the Highland and Agricultural Society (1873). In 1875 he moved to Cambridge to the Jacksonian Chair of Natural Experimental Philosophy and two years later he was elected Fullerian Professor of Chemistry at the Royal Institution; both of these chairs he held for nearly 50 years. Before leaving Edinburgh he married Helen Rose Banks in 1871; they had no children.

Dewar's earlier research (1867–1877) ranged widely in the field of chemistry, physics and physiology. The Royal Institution, with its superior experimental facilities, became the centre of Dewar's experimental activities. His research on cryogenics included the liquifaction of hydrogen, the invention of the vacuum-jacketed flask (1872), low temperature calorimetry, and measurements of electrical and magnetic properties of metals and alloys at temperatures down to 73 K. He also carried out investigations on spectroscopy jointly with Liveing at Cambridge.

Dewar served as a consultant to government and industry and was the recipient of many honours, including a knighthood in 1904. He served as Vice-President of the Royal Society from 1899–1900, received three medals; Rumford (1894), Davy (1909), Copley (1916), and gave the Bakerian Lecture in 1901. His range of interests included literature and music. As an outstanding experimentalist he made a major contribution to science.
PRS 1926 A, 8, xiii–xxiii. DSB 4, 78–81. A.B. Costa. DNB 1921–1930, 255–257.

LEWIS WESTON DILLWYN 1778–1855; FRS 1804
Lewis Dillwyn was the son of William Dillwyn and was born at Ipswich. His early education was in Tottenham. Beginning in 1798 he spent a period in Dover where he began his study of plants, marking the beginning of his scientific career as a naturalist. He settled in Swansea in 1803 in charge of a pottery which his father had bought; in the porcelain that was manufactured he introduced botanical decorations. In around 1800 Dillwyn married a daughter of John Llewellyn and they had two sons and two daughters. He served as Member of Parliament for Glamorgan from 1832–1841.
DNB.

PAUL ADRIEN MAURICE DIRAC 1902–1984; FRS 1930
Paul Dirac was born in Bristol, the second of three children of Charles Adrian Ladislas Dirac, an émigré from Switzerland, and Florence Hannah née Holten. Having attended the Bishop Road Primary School he entered the Merchant Venturers' Technical College at the age of 12. He then studied electrical engineering at Bristol University from 1918–21 and obtained first class honours. He spent a further two years at Bristol studying mathematics and obtained a first class in the examinations. In 1923 he went to Cambridge with a Research Studentship from the Department of Scientific and Industrial Research, followed in 1925 by the award of a Senior Studentship of the 1851 Exhibition. He was awarded a PhD in 1926 for a thesis entitled 'Quantum Mechanics', and then travelled, visiting various centres of theoretical physics and working with leading figures in quantum mechanics. Dirac was elected a Fellow of St John's College in 1927, was appointed a University Lecturer in 1929 and became Lucasian Professor of Mathematics at Cambridge in 1932. In 1937 he married Margit Wigner and they had two daughters.

Dirac made a major contribution to the field of quantum mechanics and predicted the exist-

ence of the positron. During World War II he assisted the Theoretical Group in Birmingham on matters relating to the atomic energy project, for example on uranium isotope separation, a subject on which he had previously done some experimental research. Following retirement from his chair in 1969 he became Professor of Physics at Florida State University in 1971. Dirac's honours included a Royal Medal (1939), the Copley Medal (1952) and the Nobel Prize for Physics, shared with Schrödinger, (1933). In 1973 he was awarded the OM.
BMFRS 1986, 32, 137–185. R.H. Dolitz FRS and Sir Rudolf Peierls FRS. DSB 17, 224–233. O. Darrigol.

GEORGE DOLLOND 1774–1852; FRS 1819
George Dollond was the grandson of John Dollond. In 1787 he began to learn the trade of mathematical instrument maker and the following year commenced his apprenticeship with his uncle Peter Dollond. He suffered a serious illness in 1792, but in 1805 entered partnership with his uncle, and after his uncle's retirement was in sole charge of the business. He was a highly skilled optician and mechanician and made instruments for applications in astronomy, geodesy and navigation.
DNB.

JOHN DOLLOND 1706–1761; FRS 1761
John Dollond was born at Spitalfields, London of Huguenot parents who had come from Normandy, and was brought up to the hereditary trade of silk-weaving. His father died while he was young and he studied widely to improve himself, including algebra and geometry. He married at an early age and had two sons and three daughters, one of whom married his apprentice Jesse Ramsden. When Dollond's eldest son, Peter, set up as an optician he joined him in 1752, abandoning his silk-weaving, and rapidly acquired practical skills to accompany his theoretical knowledge. He developed an achromatic telescope which led to his receiving the Copley Medal in 1758 and he served on the Council of the Royal Society.
DSB 4, 148, 149. G.L'E. Turner. DNB.

ATHENE MARGARET DONALD; FRS 1999
Univ. of Cambridge: Professor of Physics.
Polymers: structure and mechanical properties. X-ray characterisation of starch.

BRYAN DONKIN 1768–1855; FRS 1838
Bryan Donkin was born in Sandoe, Northumberland; his father was an agent for the Errington Estates. When he left home he worked for two years as a land agent to the Duke of Dorset at Knole Park, Sevenoaks, Kent. Subsequently he served an apprenticeship at a works in Dartford and became involved in the production of paper-making machinery. In 1804 he installed a machine at Frogmore in Kent, and he went on to manufacture many machines for paper-making. He also invented a screw-cutting machine. His later career included about 40 years of civil engineering work. In the Royal Society he served on the Council a number of times. He had a son by his wife Mary.
PRS 1855, 7, 586–589. DNB.

STANLEY FABES DOREY 1891–1972; FRS 1948
Stanley Dorey was the second son of William Dorey of London and Worthing. He was educated at Owen's School Islington and at the Royal School of HM Dockyard at Chatham, where he

served an apprenticeship. With a Whitworth Scholarship and a Lloyd's Register Scholarship in marine engineering he went to Armstrong College, Newcastle upon Tyne in 1912, obtaining a BSc in 1914. In the same year he was commissioned in the Durham Light Infantry, but was soon transferred to the Royal Navy where he served as an Engineering Officer throughout the war, reaching the rank of Lieutenant. Dorey worked for a short period with Sir W.G. Armstrong, Whitworth and Co. Ltd at Elswick in 1919 and then joined the Lloyd's Register of Shipping as a Ship and Engineer Surveyor at the headquarters. In 1932, the year in which he was awarded a DSc by Durham University, he became Chief Engineer Surveyor. He had married Dorothy Ellen Midworth in 1920 and had one son and two daughters; following his wife's death he married Evelyn Josephine Moffatt in 1956.

Dorey in his career was a manager and user of research and established an Engineering Research Department at Lloyds in 1935. He was an authority on materials in marine engineering, including areas such as defects and their prevention in marine machinery; another field was the welding of pressure vessels. His honours included the award of the CBE in 1946.

BMFRS 1973, 19, 305–316. D.M. Smith FRS.

PATRICK JOSEPH DOWLING 1939–; FRS 1996

Christian Brothers School, Dublin. University College, Dublin: BE NUI, 1960. Imperial College, London: PhD 1968. Dublin: Demonstrator in Civil Engineering, 1960–61. Imperial College: postgraduate studies, 1961–65. British Constructional Steelwork Association, 1965–68. Imperial College: Research Fellow 1968–74; Reader in Structural Steelwork, 1974–79; British Steel Professor of Structures, 1979–94; Head of Civil Engineering Department, 1985– 94. Univ. of Surrey: Vice-Chancellor and Chief Executive, 1994–. FREng. 1981.

Steel-plated structures; damage resistance of steel components in offshore applications; steel concrete structures.

DUNCAN DOWSON 1928–; FRS 1987

Lady Lumley's Grammar School, Pickering, Yorks. Univ. of Leeds: BSc Mech. Eng. 1950, PhD 1952, DSc 1971. Sir W.G. Armstrong Whitworth Aircraft Co.: Research Engineer, 1953–54. Univ. of Leeds: Lecturer in Mechanical Engineering, 1954; Senior Lecturer, 1963; Reader, 1965; Director Institute of Tribology, 1967–87; Professor of Engineering Fluid Mechanics, 1966–93; Head of Department Mechanical Engineering, 1987–92; Pro-Vice-Chancellor, 1983–85; Research Professor, 1995–; Head of Department Mechanical Engineering. FREng. 1982. CBE 1989.

Tribology; elastohydrodynamic lubrication; biomechanics.

PIERRE LOUIS DULONG 1785–1838; FMRS 1826

Pierre Dulong was born in Rouen, France; after he was orphaned at the age of four, an aunt took him into her home at Auxerre. Having shown mathematical ability he entered the École Polytechnique in Paris in 1801. For health reasons he did not complete the course and later, since formal qualifications were not required in post-revolutionary France, he turned to medicine and practiced for a while in a poor district of the city. Leaving medicine he turned to botany and then to chemistry, obtaining a post as an assistant in Thénard's laboratory, and subsequently a place in Berthollet's laboratory. He married Emelie Augustine Riviere in 1803 and they had four children. Dulong's career included appointments as Professor of Chemistry at the Faculté des Sciences in Paris, and Professor of Physics at the École Polytechnique. His research included important work on the relationship between specific heats and atomic weights.

PRS 1838, 35, 97–98. DSB, 4, 238–242. M.P. Crosland.

JEAN-BAPTISTE-ANDRÉ DUMAS 1800–1884; FMRS 1840

Jean-Baptiste-André Dumas was born in Alès, France, the son of the town clerk. Having received a classical education he was apprenticed to an apothecary. In 1816 he moved to Geneva where he studied pharmacy together with chemistry, physics and botany, and was also allowed to carry out experiments in the chemical laboratory of a pharmaceutical firm. His earliest researches were in medicine and physiology. He returned to France in 1823 where he obtained an appointment at the École Polytéchnique, prior to being appointed to a Chemistry Chair at the Athenaeum. In 1826 he married Hermine Brongniart, daughter of the director of the royal porcelain works at Sèvres. He was a co-founder of the École Centrale des Arts et Manufactures in 1829. His academic appointments included a chair at the École Polytéchnique (from 1835) and the Professorship of Organic Chemistry at the École de Médicine (1839). Dumas was actively involved in politics after the February Revolution of 1848, including holding the office of Minister of Agriculture (1850–1851) and being made a Senator. He was involved in the modernisation of Paris, supervising the installation of modern drainage, water supply and electrical systems.

In addition to being a brilliant teacher, Dumas's work ranged widely, including some metallurgical topics. He also developed a new method for directly measuring vapour densities, and by calculation, the relative molecular weights of substances in the gaseous state. He was awarded the Copley Medal in 1843.

PRS 1884, 37, vx–xxvii AWH. DSB 4, 242–248. S.C. Kapoor.

PETER DUNCUMB 1931–; FRS 1977

Oundle School. Univ. of Cambridge: Clare College, BA, 1953, MA 1956, PhD 1957; DSIR Research Fellow, 1957–59. Tube Investments, subsequently TI Group Research Laboratories, Hinxton Hall: Research Scientist and Group Leader, 1959–67; Head, Physics Department, 1967–72; Assistant Director, 1972–79; Director and General Manager, 1979–87. Univ. of Cambridge: Research Centre in Superconductivity, Director : 1988–89. Warwick University: Honorary Professor, 1990.

Electron microscopy and analysis.

JAMES DYSON 1914–1990; FRS 1968

James Dyson was born in Bradford, the son of George Dyson and Mary Grace née Bateson; he had an older sister. He attended the Queen Elizabeth Grammar School in Kirkby Lonsdale, Lancashire. In 1933 he entered Christ's College, Cambridge with a scholarship; having studied mathematics in his first year he transferred to mechanical engineering and graduated with a first in the mathematical tripos. From 1936–1939 Dyson was an Apprentice at the British Thomson Houston Co. of Rugby (BTH); he then continued at BTH as a Research Engineer. In 1940 he married Ena Lilian Turner and they had a daughter; following the dissolution of the marriage in 1948 he married Marie Florence Lamb and following her death he married Rosamund Pearl Greville Shuter in 1975.

In 1946 Dyson moved to the Associated Electrical Industries Laboratory (AEI) at Aldermaston, where he used his special expertise in optics on various projects relevant to metallurgy, including studies of bubble rafts as models for dislocations, the construction of a long working distance microscope and an interferometer microscope. In 1946 Dyson moved to the National Physical Laboratory as Superintendent of the Division of Mechanical and Optical Metrology. He was promoted to Deputy Chief Scientific Officer in 1975 and retired in 1976.

BMFRS 1991, 37, 151–174. T.E. Allibone FRS. FREng.

LAURENCE EAVES 1948–; FRS 1997

Rhonnda Co. Grammar School. Univ. of Oxford: Corpus Christi College, BA Physics, 1969, MA, 1973, DPhil, 1973; Research Lecturer, Christ Church, and Research Fellow Clarendon Laboratory, 1972–74. Univ. of California, Berkeley: Miller Fellow; 1974–75. Univ. of Nottingham: Lecturer, 1976–84; Reader, 1984–86; Royal Society Leverhulme Senior Research Fellow, 1992–94; EPSRC Senior Research Fellow, 1994–; Professor of Physics; 1986–.

Electronic properties of semi-conductors, low dimensional structures.

CHARLES ALFRED EDWARDS 1882–1960; FRS 1930

Charles Edwards was born in Kitchener near Toronto, Ontario, Canada, one of a family four sons and two daughters of Samuel and Elizabeth Edwards. His parents moved to England when Charles was two years of age. In 1898 he was apprenticed to the Lancashire and Yorkshire Railway Works at Horwich, Lancashire where he developed an interest in metals and alloys and spent some years in the chemical laboratory. In 1905 he became assistant to Harold Carpenter at the National Physical Laboratory and when Carpenter was appointed to the Chair of Metallurgy at Manchester University Edwards joined him there as Lecturer and carried out research on steel and the theory of hardening by heat treatment. In 1908 he married Florence Edith Roberts and they had one son.

Edwards returned to industry in 1910 as a metallurgist with Bolckow, Vaughan and Company and Dorman, Long and Company of Middlesbrough. In 1914 when Carpenter left Manchester, Edwards was appointed to the chair, having obtained the Manchester DSc degree in 1913. He carried out research on the hardening of steel, and wrote a book on *The Physico-Chemical Properties of Steel.*

In 1920 Edwards became Professor of Metallurgy and Vice-Principal at the newly formed University College of Swansea where he soon developed a deep interest in the local steel industry of tin plate and black plate manufacture, and industrially relevant researches flourished. In 1926 he was appointed as Principal of the College whilst retaining the post of Professor of Metallurgy. He continued his research and teaching and on his retirement he became consultant to Messrs Guest, Keen and Nettlefolds.

BMFRS 1960, 6, 33–38. L.B. Pfeil.

SIR SAMUEL FREDERICK EDWARDS 1928–; FRS 1966

Swansea Grammar School. Univ. of Cambridge: Caius College, MA, PhD. Harvard Univ., Institute for Advanced Study, Princeton: 1952. Univ. of Birmingham, 1953. Univ. of Manchester, 1958; Professor of Theoretical Physics, 1963–72. Department of Energy: Chief Scientist, 1983–88. Science Research Council: Chairman, 1973–77. Univ. of Cambridge: John Humphrey Plummer Professor, 1972–84; Cavendish Professor of Physics, 1984– 95; Pro-Vice-Chancellor, 1992–95. Caius College, Fellow 1972–; President, 1992–. Royal Society, Vice-President 1982–83. Davy Medal 1984. Knighted 1975.

Polymer chemistry; thermodynamic aspects; viscoelasticity of polymer melts; transport properties of polymer molecules.

ALBERT EINSTEIN 1879–1955; FMRS 1921

Albert Einstein was born in Ulm, Wurttemberg, Germany, the only son of Hermann and Pauline Einstein. He graduated from the Eidgenössische Technische Hochschule, Zurich in 1900. He obtained technical employment in the Swiss Patent Office in Berne, where, in his spare time, he

worked on theoretical physics and developed ideas that later led to a revolution in physics. As his discoveries became known and their importance appreciated he received professorial appointments at Prague and then at Zurich before becoming Director of the Institute of Physics at the Kaiser Wilhelm Institute in Berlin in 1914. In 1915 he published his general theory of relativity. In 1933 he accepted a position at the Princeton Institute for Advanced Study, where he spent the rest of his life, becoming a US citizen in 1940. His earlier work included the study of specific heats. He was the recipient of the Nobel Prize for his work in theoretical physics in 1921, particularly relating to the photo-electric effect.
BMFRS 1955, 1, 37–67. Sir Edward Whittaker. DSB 4, 312–338. M.J. Klein and N.L. Balazs.

FRANCIS ELGAR 1845–1909; FRS 1896
Francis Elgar was born in Portsmouth, the eldest of nine children of Francis Ancell Elgar and Susannah née Chalkley. He became an apprentice at Portsmouth Dockyard at the age of 14, and then, in 1864, when the Royal School of Naval Architecture and Marine Engineering at South Kensington was opened, he entered with a scholarship. Having graduated with he highest class of diploma in 1867 he became a Junior Officer in the Ship Building Department of the Royal Navy until 1871. During the period 1871–79 he was the Chief Technical Assistant to Sir Ernest Reed in Reed's consulting practice based in London and from 1874–76 he was General Manager to Earls Shipbuilding Co at Hull. Elgar was an advisor on naval construction in Japan to the Japanese government from 1879–81. There followed a period of private practice in England until 1886, while he also held the appointment of Professor of Naval Architecture at Glasgow University from1883–86. Elgar was Director of Dockyards from 1886–92. He married Ethel Colls in 1889; there were no children from the marriage. Elgar continued in consulting work until 1907 and was also Director of Fairfields Shipbuilding Company in Glasgow. Although he retired in 1908 he agreed to act as Chairman for the reorganisation of Cammell Laird and Co. of Maryport in Cumberland, who were concerned with steel and armour plate manufacture as well as shipbuilding. During his career Elgar had an interest in scientific work; he gave much attention to investigations of the structural strength of ships, and, as a manager of the Royal Institution, he helped to promote the low temperature investigations of James Dewar.
PRS 1910, 83A, viii–xi. WHW. DNB Second Suppl. v. 1, 614–616. WHW.

JOHN ELLICOTT c.1706–1772; FRS 1738
John Ellicott was the son of John Ellicott a clockmaker and his wife Mary. He became a clockmaker and man of science, working in London; he became clockmaker to George III. His work included measurements of the linear expansion of solids and in 1736 he presented to the Royal Society an improved pyrometer for this work. From his marriage he had eight sons and three daughters.
DNB.

SIR ROGER JAMES ELLIOTT 1928–; FRS 1976
Swanwick Hall School, Derbyshire. Univ. of Oxford: New College, MA, DPhil. Univ. of Berkeley, California: Research Fellow 1952–53. UKAEA, Harwell: 1953–55. Univ. of Reading: Lecturer 1955–57. Univ. of Oxford: St John's College, Fellow, 1957–74; Honorary Fellow, 1988; University Reader, 1964–74; Wykeham Professor of Physics, 1974–89; Professor of Physics, 1989– 96. Royal Society, Physical Secretary and Vice-President, 1984– 88. Knighted 1987.
Theory of the solid state, with particular reference to magnetism (e.g. of rare earth metals), semiconductors and disordered materials.

JOHN DOUGLAS ESHELBY 1916–1981; FRS 1974

John Eshelby was born at Puddington in Cheshire, the eldest of four children of Alan Douglas Eshelby, a regular army officer, and Phoebe Doris Mason Eshelby. He attended St Cyprian's School, Eastbourne, but he became seriously ill with rheumatic fever when he was about 13. In 1930 the family moved to Farrington Gurney in Somerset where his education continued privately at home through tutors in science and classics.

Eshelby graduated from the University of Bristol with first class honours in physics in 1937. He began some research work in experimental physics on the soft X-ray spectra of solids and on magnetism, working under H.W.B. Skinner and W. Sucksmith. With the outbreak of war in 1939, after an initial period in the Admiralty dealing with the degaussing of ships, he joined the technical branch of the Royal Air Force the following year. During 1941 and 1942 he was associated with the Coastal Command Development Unit (CCDU) and served at several centres. The work concerned performance trials and a wide range of operational devices. Eshelby flew in a variety of aircraft and was often allowed by the pilots to do some flying. After leaving the CCDU he had a number of postings concerned with radar duties before being transferred to disarmament duties. Prior to his release from the RAF in September 1945, with the rank of Squadron Leader, he had also served in the Air Historical Branch.

Returning to Bristol he resumed his research in the Physics Deparment and obtained a PhD degree in 1950. His original intention had been to become a theoretical physicist and during the war when off duty he had studied applied mathematics and theoretical physics; with his great interest in languages, he did this partly by learning Russian and reading Russian texts. Following a suggestion by Nevill Mott he began some calculations on the theory of dielectric loss and this led on to his considering the theory of mechanical damping in metals in terms of dislocation theory. He studied to learn the theory of elasticity and after a period of work he ' came up with the whole of dislocation theory'.

The academic year 1952–53 was spent as a Research Associate in the Physics Department of the University of Illinois at Urbana on the newly discovered metallic 'whiskers' and on point defects in solids. Having returned to England he accepted an appointment as a Lecturer in the Physical Metallurgy Department at the University of Birmingham, where he remained for 10 years apart from a period in 1963 when he was a visiting professor at the Technische Hochschule and Max Planck Institute at Stuttgart. He continued to work and to publish many research papers, some with various colleagues, on point defects and dislocations; also there were reviews incorporating new ideas and methods of approach. Two academic years (1964–1966) were spent in Cambridge, first as Senior Visiting Fellow in the Metal Physics Group at the Cavendish Laboratory and then as a Fellow and Lecturer at Churchill College. Eshelby then joined the Department of the Theory of Materials at Sheffield as a Reader, and was elected to a Personal Chair in 1971. Much of his work in Sheffield was on the theory of fracture and he was also concerned with defects in liquid crystals. Among those with whom Eshelby during his career collaborated were Frank, Nabarro and Bilby. He always regarded himself as primarily a theoretical physicist, but he had much to do with engineers and materials scientists and was very practical in outlook.

Eshelby's wide scholarly interests included archaeology and the ancient world, an encyclopaedic general knowledge and a special interest in philology. His linguistic attainments included many European languages and Latin and Greek. He was knowledgeable on Sanskrit and Chinese and other ancient languages. Eshelby, 'Jock' as he came to be known to many scientists and engineers was consulted by colleagues from all over the world and had a major influence on many areas of science and technology.

BMFRS 1990, 36, 127–150. B.A. Bilby FRS.

LEONHARD EULER 1707–1783; FRS 1747

Leonhard Euler was born in Basle, the son of Paul Euler, a protestant minister, and Margarete née Brucker. At the University of Basel he obtained a master's degree in 1723. He was invited by Daniel Bernouilli four years later to join him at St Petersburg where he spent three years at the Naval College and then became Professor of Physics at the Academy of Sciences in 1731. In 1733 he succeeded Daniel Bernouilli as Professor of Mathematics. In the same year he married Katharina Gsell and they had two sons. In 1741 he went to Berlin, and in 1744 he became Director of the Berlin Academy of Sciences. In 1766 he became Director of the Academy of Sciences at St Petersburg, and although soon after this he lost his sight he continued effectively in this post. Euler carried out important and wide ranging mathematical research, including the fields of astronomy, fluid mechanics and mechanics; in the last mentioned field he introduced the 'buckling formula' to determine the strength of columns.

DSB 4, 467–484. A.P. Youschevitch.

TREVOR EVANS 1927–; FRS 1988

Bridgend Grammar School. Univ. of Bristol: BSc,PhD, DSc. British Nylon Spinners: 1955–56. Tube Investments Research Laboratory: 1956–58. Univ. of Reading: Physics Department; Research Physicist, 1958–61; successively Lecturer, Reader, Personal Professor, 1961–92; Head of Department 1984–88.

Physical properties of diamond; synthetic diamonds.

ULICK RICHARDSON EVANS 1889–1980; FRS 1949

Ulick Evans was born in Wimbledon, the son of Richardson Evans and Amy née Feeney. His father, after a period in the Indian Civil Service, became a journalist and leader writer in London. Ulick was the fifth child in the family with six sisters. His education included a period at a Grammar School at Wimbledon, followed by attending Marlborough College from 1902 to 1907. Having begun to be interested in science at an early stage, he readily accepted the chance to specialise in science, half-way up Marlborough College. He entered King's College, Cambridge in 1907 where he remained until 1911; he became an Exhibitioner in 1908, took Part I of the Natural Sciences Tripos in 1909 and Part II in chemistry obtaining a BA in 1910 an MA in 1914. He received supervisions from Charles Heycock, which helped guide him towards the electrochemistry of metals. Financial support, which came through his mother's family enabled him to follow his own choice of research. After leaving Cambridge, Ulick Evans spent a year at Wiesbaden, studying electrochemistry before working in London with Samuel Rideal and his son Eric Rideal.

From August 1914 until 1919 Evans served with the Army, first in the East Surrey Regiment and then in the Royal Engineers (Signals) based mainly in the Middle East. Back in Cambridge he began in 1921 the research on corrosion which became his life's scientific work. Heycock and T.M. Lowry provided facilities for him in the Goldsmiths' Laboratory of Metallurgy and in the Laboratory of Physical Chemistry respectively. Following the establishment of the Department of Metallurgy successive holders of the Goldsmith's Chair provided accommodation and facilities.

Evans completed a book which he had started before the war, entitled *Metals and Metallic Compounds;* this was published in 1923 as was his first paper on corrosion. In 1924 he began lecturing on corrosion to students of metallurgy and engineering. As his reputation grew rapidly research students and collaborators came to work with him and many papers were published.

William Hatfield provided some industrial support without 'strings'. After obtaining his Cambridge ScD in 1932, and having built up an internationally famous corrosion school the university appointed him to an Assistant Directorship of Research in 1933, a post that he held until 1945. In the early days of his work, some pure scientists viewed corrosion as a 'dirty' subject meriting little attention but the quality of his work gradually compelled attention. His research transformed understanding in the field of corrosion and oxidation of metals contributing new ideas in the electrochemical field such as that of 'differential aeration'.

The Royal Society awarded Evans an Armourers and Brasiers' Fellowship from 1932 to 1936, and the University of Cambridge finally made him a Reader in the Science of Metallic Corrosion in 1945. He had many invitations to make overseas visits and to give lectures; he could lecture with ease in French and German. He built up a wide circle of overseas collaborators and friends.

Evans had wide-ranging interests – cross-country running, climbing, swimming (including in the River Cam in the winter) music, and the preservation of the countryside. He remained unmarried.

Evans continued his work during the 1939–1945 war, including service on committees, and after the war his interest included corrosion fatigue, and oxide growth on metals. Another interest which lasted throughout his scientific life was in fuel cells. He continued to give lectures to scientific audiences in England and abroad; he enjoyed writing books, having published four between 1923 and 1948, and he continued his writings until a very advanced age. He received many awards, including that of CBE in 1980. The outstanding contributions of Evans's research over nearly 60 years led to his being regarded as the 'Father of the modern science of corrosion and protection of metals'.

BMFRS 1981, 27, 235–253. Sir Alan Cottrell FRS and J.E.O. Mayne.

PAUL PETER EWALD 1888–1985; FRS 1958

Paul Ewald was born in Berlin where his father, also Paul, was a historian, Privatdozent at the University of Berlin. His mother Clara Philippson Ewald was a painter, who had set up a professional studio in Berlin in 1885. Paul's father died in 1887, three months before Paul Peter was born, and Clara augmented the income from her inheritance by painting. She took Paul on many travels as a young child, including periods in Paris and Cambridge.

At the age of five, having returned to Berlin, Paul attended the Real Gymnasium, and later the Wilhelm Gymnasium. When the family moved to Potsdam he attended the Victoria Gymnasium; here he showed an interest in chemistry and physics, which was stimulated by a visit to Cambridge in 1899. In 1905 Paul, having passed the final school examination, Abitur, spent the winter of 1905–6 at Caius College. In the spring of 1906 he moved to Göttingen, where he decided to make mathematics his main subject.

In the autumn of 1907 Ewald moved to Munich and having attended a course on hydrodynamics by Sommerfeld, he came to consider himself a theoretical physicist. He obtained his PhD with Sommerfeld on a thesis dealing with the double diffraction of light in crystals. Discussions that he had with von Laue played a role in the events that led on to Laue's ideas on X-ray diffraction and thence to the experimental demonstration of the phenomenon. In February 1912 he returned to Göttingen as the physics assistant to David Hilbert. Ewing extended his thesis work to the case of X-rays and discovered the concept of the 'reciprocal lattice'. While in Göttingen he met and married in 1913 Elisa (Ella) Philippson, a medical student. He returned to Munich where he did experimental and theoretical work on X-ray diffraction and early in 1914 he introduced the 'structure factor' concept.

Ewald had volunteered for the army before the the beginning of World War I but had been rejected as having a slight heart condition. However, having assisted at the university hospital in Munich using X-rays in medical diagnosis and enlisted in the army as an X-ray technician, he was sent to the Siemens factory in Berlin to learn more about X-ray tubes. He served on the Eastern war theatre, and in a quiet front near Dwinsk with little to do he conceived and developed the dynamic theory of X-ray diffraction, which when written up brought him the rank of 'Privatdozent'. When the war between Germany and Russia ended Ewald was sent to a laboratory in Kiel that tested torpedoes. In November 1918 he was able to join his wife and, by then, two children in Holzhausen, where his mother had built a house in 1911; Paul and Ella subsequently had two other children. In 1918 he became Privatdozent at the University of Munich.

In 1921 Ewald took up a professorial appointment at the Technical University of Stuttgart, and directed the theoretical physics department. He had a series of assistants for theory (including Fritz London and Hans Bethe) and also for experimental work. In 1932 Ewing was elected Rektor, equivalent to Vice-Chancellor in Britain, but in 1933, soon after the Nazis came to power he resigned this office because of his Jewish wife and his liberal opinions. Like many people in his position he hoped that the political situation could be improved and he stayed on in Germany. In 1936, when a young scientist, who was also a high-ranking Nazi, stated at a faculty meeting, 'Objectivity is no longer a valid or acceptable concept in science' Ewald walked out in protest, and soon after he was pensioned. Sir Lawrence Bragg obtained a research grant for him at Cambridge and Ewing left Germany in the autumn of 1937 followed later by his family. In August 1939 was appointed Lecturer at Queen's University, Belfast, where owing to war time circumstances the promise of promotion to a chair could not be implemented until 1945; in October of that year Ewald became a British subject. In 1949 he took up an appointment as Chairman of the Physics Department at Brooklyn Polytechnic Institute in the USA where a good research team was in place analysing the structures of complicated crystals, especially at low temperatures. He worked there until his retirement in 1959.

Ewald continued to work on the dynamical theory of diffraction all his life. He also made important contributions through his writings; in 1923 he published a book *Crystals and X-rays,* and in 1962 he edited the book *Fifty Years of X-ray Diffraction.* Much of his time was devoted to international cooperation among scientists and to providing research tools for X-ray crystallographers. Having been one of the editors of the *Zeitschrift für Crystallographie*, he was asked in 1946 to organise an international journal, which became *Acta Crystallographica*, and he served as Editor-in-Chief until 1960. Together with Carl Hermann as assistant he produced the first volume of the collection of crystal structure data, the *Strukturbericht*, covering the years 1913–28.

Following his retirement from Brooklyn in 1959 the Ewalds moved to Connecticut, and then in 1971 to Ithaca, New York. They travelled to many scientific meetings, and to Europe for holidays; his activities included hiking up steep mountain trails until his late eighties. In 1979 he gave the talk on von Laue at the conference in Berlin to commemorate the 100th anniversary of the birth of Einstein, von Laue, Hahn and Meitner. Ewing was the recipient of many academic honours.

BMFRS 1988, 34, 133–176. H.A. Bethe FMRS and G. Hildebrandt. DSB 17, 272–275 W. Kaiser.

SIR JAMES ALFRED EWING 1855–1935; FRS 1887

James Ewing was born in Dundee, Scotland, the youngest of three sons of James Ewing, a minister of the Free Church of Scotland, and Marjory née Ferguson. After schooling at West End academy and Dundee High School he entered Edinburgh University with a scholarship in engineering; during the summer vacations he participated in the work of three submarine cable-laying expeditions in Brazil and the River Plate. In 1878 he was appointed Professor of Mechanical Engineering and Physics at the Imperial University of Tokyo where he remained for five years. Here he carried out seismological work having established an observatory and also began his experimental research on magnetism.There followed in 1883 an appointment as Professor of Engineering in the new university college of Dundee, where he continued his work on magnetism. A move to Cambridge University occurred in 1890, as Professor of Mechanisms and Applied Mechanics; he became the founder of the Engineering School and the Mechanical Science tripos was established in 1892. He was elected to a Professorial Fellowship at King's College in 1898. Ewing became Director of Naval Education in 1903, developing a scheme of scientific and engineering training for naval officers. During the 1914–1918 war he was in charge of a group that intercepted and deciphered German messages. He became Principal and Vice-Chancellor of Edinburgh University in 1916, continuing in this post until his retirement in 1929.

In magnetism, his main field of research, Ewing made important discoveries, including the phenomenon of hysteresis. He also worked with Walter Rosenhain on plastic deformation, and invented a new type of extensometer for mechanical testing.

Ewing was twice married; first in 1879 to Annie Maria Thomasina Blackburn (died 1909), by whom he had a son and a daughter; in 1911 he married Ellen Lina Hopkinson and they had a son. His many honours included a Royal Medal (1895), CB (1907) and KCB (1911).

ONFRS 1932–1935, 1, 475–492. R.T. Glazebrook FRS. DSB 4, 500, 501. S. Dostrovsky. DNB 1931–1940, 264–266. E.I. Carlyle FRS.

SIR WILLIAM FAIRBAIRN 1789–1874; FRS 1850

William Fairbairn was born at Kelso, Roxburgshire one of five children of Andrew Fairbairn, a farm worker, who had married the daughter of a Jedburgh tradesman, named Henderson. William attended a parish school, and then a school at Mullochy; he also learnt some book-keeping. He started work at the age of 14, and in 1803 when the family moved to a farm near Newcastle upon Tyne, he was employed in a colliery; the next year he was apprenticed to a millwright, spending leisure time in studying mathematics and general literature. In 1811 he finished his apprenticeship, and after a period of work as a mill wright at Newcastle he went to London, where he was employed in the construction of a steam engine for digging. After various travels and further employment, including a short period in a foundry in Dublin, he went to work in Manchester in 1814. Later he set up in an engineering partnership and the business expanded. The partnership was disolved in 1832 and in 1835 he established a shipyard at Millwall. Other activities of Fairbairn included an interest in the Atlantic cable, membership of a committee on iron (1861–67) set up to advise on the armouring of ships, and extensive experimental investigations on the mechanical properties of iron and professional travels abroad. Fairbairn became a member of the Institute of Civil Engineers in 1830, and began to investigate the properties of iron boats, leading to activities in shipbuilding. In 1865 he published his *Treatise on Iron Shipbuilding.* A major part of his career was concerned with bridges and his engineering ability and inventiveness led to his earning a large sum by the time he was 40 years of age.

Fairbairn had become a friend of George and Robert Stephenson and when Robert designed the tubular Britannia and Conway bridges in North Wales he consulted Fairbairn and Eaton Hodgkinson. In addition to his role in the experiments at his large Millwall shipyard that led to the choice of the best form of girders, Fairbairn was associated with Stephenson in supervising the work for part of the time of the bridge building.

Fairbairn had seven sons and two daughters by his wife, Dorothy née Mar whom he married in 1916. Having declined a knighthood in 1861 he accepted a Baronetage in 1869. He received a Royal Medal in 1860, and in the same year gave the Bakerian lecture.
DNB. Gen. Ref. 20, v. IV. Ref. Ch. 9.14. A.I. Smith *The Engineer*, 1963, 216, 543–546.

MICHAEL FARADAY 1791–1867; FRS 1824
Michael Faraday who was born in Newington, Surrey (which became part of Southwark, London) was the third of four children. His father, James Faraday, was a blacksmith who had come to London from Yorkshire in 1791 to obtain work but with increasing poor health could only provide the family with the basic néeds; Michael's mother Margaret née Hastwell, played an important role in the family life. Both of the parents were members of the Sandemanian church, which stressed the love and sense of community of the early Christian church. Faraday continued in this faith throughout his life.

Having received only a rudimentary school education, Faraday, when about 13 years old, contributed to the family income by delivering newspapers for a Mr G. Ribeau, the owner of a shop selling and binding books. Beginning at the age of 14 he was apprenticed to the business until 1812, learning the art of bookbinding and living at Ribeau's; he developed the great manual dexterity which was evidenced in his later scientific experiments. He was also influenced by his wide reading, which included a book by Isaac Watts, a clergyman, on *The Improvement of the Mind;* also an article on electricity in the *Encylopaedia Britannica* aroused his scientific interest and led him to some experimental work on electrolytic decomposition, and to begin a serious pursuit of science. In 1810 he began to attend and to take notes at lectures at the City Philosophical Society, where he obtained a basic education in a range of subjects including electricity, optics, mechanics and chemistry. A book entitled *Conversations in Chemistry* was a further important influence. In 1812 a customer, impressed by Faraday, gave him tickets for four lectures by Humphry Davy at the Royal Institution; Faraday attended and sent the notes that he took to Davy. Contacts between Davy and Faraday led to Davy appointing him as an Assistant in the Laboratory of the Royal Institution in March 1813, with accommodation provided. His official post was 'Assistant in the Laboratory and Mineral Collection and Superintendent of the Apparatus'. There thus began a period in which Faraday's intellectual development was greatly influenced by Davy as he worked with him.

In October 1813 Davy and his wife set off to visit the Continent, Faraday having been asked to accompany them as secretary and scientific assistant. This provided Faraday with the opportunity to meet many leading scientists, including Volta, and to assist with experiments. Socially, while Davy treated Faraday with consideration, Lady Davy made the situation difficult treating Faraday as a menial. The itinerary included France and Italy and they returned in April 1815 to London where Faraday enthusiastically worked in chemistry. His first paper was published in 1816 and by 1820 his reputation as an analytical chemist was such that he acted as an expert witness in court cases. In 1821 Faraday married Sarah Barnard; the marriage although very happy was childless.

Election to the Royal Society came in 1824, although surprisingly not supported by Davy. Faraday's chemically-oriented work during the 1820s ranged widely. With James Stodart he carried out researches on alloy steels while his analytical chemistry work helped the finances of the Royal Institution. In organic chemistry his discoveries included benzene (1825) and vulcanisation of rubber. He was actively involved in the production of optical grade glass and also worked on the liquifaction of various gases, for example being the first to produce liquid chlorine. Also in this decade (1826) he instituted the Friday evening discourses at the Royal Institution, a series in which he was to give more than 100 up to his retirement in 1862. Faraday was an outstanding lecturer and in 1829 took an appointment as Professor of Chemistry at the Royal Military College, Woolwich, where over a period of some 20 years he gave 25 lectures a year. Also in 1829 he published a substantial and detailed book *Chemical Manipulation.* In 1833 Faraday became Fullerian Professor of Chemistry in the Royal Institution. Two years later he received a pension of £300 per year from the Government.

In physics Faraday discovered electromagnetic rotations in 1821 and early in the next decade (1831) there occurred his momentous discovery of electromagnetic induction. The decade of the 1830s saw the continuation of his outstanding scientific contributions in the overlapping fields of both chemistry and physics. In the period 1833–36 the outcome of his research included the laws of electrolysis, studies of fused salt electrolytes and of plasma chemistry. Faraday was an extremely hard worker, and often spent from 9 am to 11 pm in the laboratory. The strain resulting from his efforts caused a breakdown in 1840, and in 1841 he spent a period of eight months resting, involving a visit to Switzerland. When his health permitted his researches were resumed and during the period from 1845 there were further important discoveries, for example, in magneto-optics and magneto-chemistry, diamagnetism (1845) and colloidal metals (1857).

Faraday worked at the Royal Institution for somewhat more than 50 years and lived there most of that time. He worked in a basement laboratory, built his own equipment and often designed and constructed instruments. Of the approximately 160 papers that he wrote, he was sole author of nearly all of them. He became a leading figure in Victorian England and interacted with many eminent scientists. He declined the honour of a knighthood, but accepted a house near Hampton Court granted to him by Queen Victoria for his devotion to science. Recognition of his genius came to him by membership of many learned societies and other bodies. In the Royal Society he was Bakerian Lecturer on five occasions (1829, 1832, 1849, 1851 and 1857) and was awarded the Rumford Medal (1846) and on two occasions the Copley Medal (1832 and 1838) and the Royal Medal (1835 and 1846); however, he refused nomination to the office of President. His outstanding researches and discoveries surpassed those of any other physical scientist and also had a remarkable impact in industry and society. Helmholtz, in his Faraday lecture of 1881, said of him:

> It is in the highest degree astonishing to see what a large number of general theorems, the mathematical deduction of which requires the highest powers of mathematical analysis, he found by a kind of intuition, with the security of instinct, without the help of a single mathematical formula... With quite a wonderful sagacity and intellectual precision, Faraday performed in his brain the work of a great mathematician without using a single mathematical fomula...

Rutherford referred to him 'as one of the greatest scientific discoverers of all time'.
PRS 1869, 17, i–lxvii. HBJ. DSB 4, 527–540. L. Pearce Williams. DNB. Ref. Ch. 8.6, 9, 10, 21.

SIR WILLIAM SCOTT FARREN 1892–1970; FRS 1945

William Farren who was the Son of William Farren was born in Cambridge, where he became a pupil at the Perse School in 1901. He entered Trinity College, Cambridge in 1911 with an entrance scholarship. In 1912 he obtained a first class in Part I in the Mathematics Tripos, and obtained a senior scholarship at Trinity in 1913; in the following year he was awarded first class honours in the Mechanical Sciences Tripos. He then worked at British Thomson Houston at

Rugby until he moved in 1915 to the Royal Aircraft Factory at Farnborough, Hants, (which later became the Royal Aircraft Establishment), where he soon became Head of the Aeronautics Department. In 1916 he completed a flying course. In the following year he married Carol Erica Warrington and they had one daughter; after her death he married Mildred Alice Hooke. His work at Farnborough included the design of a seaplane. In 1920 he began to lecture in engineering and aeronautics at Cambridge, but was not appointed a university lecturer until 1926. He became a Fellow of Trinity College in 1933. He was joint author with G.I. Taylor of two papers through his role in designing apparatus for measuring strain and for measuring heat generation in research on plastic deformation. From 1937–1941 he held senior appointments at the Air Ministry, before being appointed Director of the Royal Aircraft Establishment in 1941. In 1947 he joined A.V. Roe; in 1951 he served as an assessor for the court of enquiry on the Comet disaster. He received the MBE in 1918, the CB in 1943 and was knighted in 1952.
BMFRS 1971, 17, 215–241. Sir George Thomson FRS and Sir Arnold Hall FRS. DNB 1961–1970, 347–349. Maria Morgan.

WILLIAM JAMES FEAST 1938–; FRS 1996
Univ. of Sheffield: BSc 1960. Univ. of Birmingham: PhD 1963. University of Durham: Lecturer 1965; Gillette International Research Fellow; Senior Lecturer 1976; Professor 1986; Courtaulds Professor of Polymer Chemistry 1989; Interdisciplinary Research Centre in Polymer Science and Technology, Director; 1994–.
Polymers.

GEOFFREY BERTRAM ROBERT FEILDEN 1917–; FRS 1959
Bedford School. Univ. of Cambridge: King's College, Scholar, MA. Lever Bros and Unilever Ltd, 1939–40. Power Jets Ltd, 1940–46. Ruston and Hornsby Ltd, 1946–59; Chief Engineer Turbine Department, 1949; Engineering Director, 1954. Hawker Siddeley Brush Turbines Ltd: Managing Director; Hawker Siddeley Industries Ltd; Director 1959–61. Davy Ashmore: Group Technical Director, 1961–68. British Standards Institution: Deputy Director General, 1968–70; Director, 1970–81. Feilden Associates: Principle Consultant, 1981–. Royal Society, Vice-President 1967–69. FREng 1976. CBE 1966.
Gas turbines; new materials including ceramics.

ENRICO FERMI 1901–1954; FMRS 1950
Enrico Fermi was born in Rome, the son of Alberto Fermi and Ida née de Gottis. He attended high school in Rome and entered the Scuola Normale Superiori at Pisa, where he obtained a Doctorate in 1922. He then studied at Göttingen and at Leiden before holding an appointment as a Lecturer in Mathematical Physics at Florence from 1924–26. In 1927 Fermi was appointed a Professor of Theoretical Physics in Rome. He married Laura Capon in 1928 and they had a son and a daughter. In the same year Fermi moved to the USA to take a professorial post at Columbia University. He studied chain reactions in uranium and a small atomic pile was built at Columbia in 1941. He then moved to Chicago University and started the construction of the nuclear reactor that produced the first controlled chain reaction in December 1942. In the following year Fermi moved to Los Alamos where he participated in the programme of development that led to the production of the atomic bomb. He returned to a chair at the University of Chicago in 1946. Fermi's earlier research included important contributions to the electronic theory of the solid state. In 1935 he was awarded the Nobel Prize for Physics.
BMFRS, 1955, 1, 69–76. E. Bretscher and J.D. Cockcroft. DSB, 4, 576–583. Emilio Segre.

SEBASTIAN ZIANI DE FERRANTI 1864–1930; FRS 1927
Sebastian de Ferranti was born in Liverpool, the second son of Cesar Ziani de Ferranti and Juliana née Scott. Having been educated in schools at Hampstead and at Ramsgate he obtained work in the Experimental Department in the works of Siemens Brothers at Charlton, near Woolwich. There he assisted William Siemens in experiments with an electric induction furnace for making steel and a took out a patent; he also, for a short time, attended evening classes at University College London. While at school he had made a dynamo, which he afterwards improved and patented in 1882 as the Ferranti alternator. In 1888 he married Gertrude Ruth Ince and they had three sons and four daughters. Ferranti had a distinguished career as an engineer and inventor; his inventions included an arc lamp and a meter for measuring electric current consumption; he took out a total of 176 patents between 1882 and 1927. In 1886 there began a period in which Ferranti played a major part in the development of power generation and distribution in London. From 1892 he devoted his attention to his private business as a manufacturing engineer. He established a factory in Lancashire in 1896 which became the headquarters of Ferranti Ltd. From 1895 he worked on the gas turbine and was involved in detailed invention and design. Ferranti, in his distinguished career as an engineer and inventor can be regarded as the father of the modern electrical engineering and supply industries.
PRS 1930 A, 127, xix–xxi. WHE. Notes and Records Roy. Soc. 1964, 19, 33–41. DNB 1922–1930, 301–302, E.I. Carlyle. Ref. Ch. 9.19.

JOHN EDWIN FIELD 1936–; FRS 1994
Univ. of London: BSc. Univ. of Cambridge: PhD, 1962; Magdalene College; Graduate Tutor, 1974–87; Assistant Lecturer, 1966–71; Lecturer 1971–90; Department of Physics; University Reader in Applied Physics, 1990– 94; Professor of Applied Physics, 1994–. Head of Physics and Chemistry of Solids Section, Cavendish Laboratory, OBE 1987.
Behaviour of materials under impact, fracture and erosion; moving cracks and explosive reactions.

LOUIS NAPOLEON GEORGE FILON 1875–1937; FRS 1910
Louis Filon was born at St Cloud, near Paris, the only son of Pierre Marie Augustin Filon and Marie Jeanne Madeleine née Poirel. When he was three years old his father, who had been tutor to the Prince Imperial, and had become blind, moved to England, settling in Margate. When he was about 12 years old, Louis attended Herne House School before entering University College London in 1894, where he graduated BA in 1896 with first class honours, and a gold medal in Greek. Having shown ability in mathematics he became a Demonstrator in Applied Mathematics with Karl Pearson, the Goldsmid Professor of Applied Mathematics and Mechanics. With an 1851 Exhibition he went to King's College, Cambridge in 1898. In 1904 he married Anne Godet, and they had one son and two daughters. Also in 1904 he moved back to University College London as a professor, and successor to Pearson. During the 1914–18 war he saw army service in France; later he became commandant to the 2nd Reserve Battalion of the London Regiment, and also held a technical post with the Admiralty Air Service. He served as Vice-Chancellor of London University from 1931 to 1935 and was awarded the CBE in 1935. As a mathematician his interests involved classical mechanics and the mechanics of continua. He collaborated with Ernest Coker, who also held a chair at University College, in the field of photoelasticity.
ONFRS 1936–38, 2, 501–509. G.B. Jeffery. DNB 1931–40, 270, 271. G.B. Jeffery.

GEORGE INGLE FINCH 1888–1970; FRS 1938

George Finch was born in Orange, New South Wales, the eldest son of C.E. Finch, Chairman of the Land Court of that State. Following his early education at Wolaroi College he was privately tutored in Europe. Having studied medicine for a short period at the École de Médecine in Paris he changed to physical science at the Eidgenössische Technische Hochschule, Zurich, studying from 1906 to 1911. He won the gold medal of the diploma course, and became a Research Assistant at the ETH and later at the University of Geneva. He spent a period as a manager in one of the subsidiaries of 'Badische Anilin and Sodafabrik'.

In 1912 Finch returned to England as a Research Chemist at the Royal Arsenal and in 1913 was appointed a Demonstrator in the newly formed Fuel Department at Imperial College, London; this later became the Department of Chemical Technology. In World War I he served with the Royal Field Artillery and the Royal Army Ordnance Division in France, Egypt and Macedonia with the rank of Captain; he was mentioned in despatches and was awarded the MBE. He returned as a Demonstrator to Imperial College in 1919 and was promoted to Lecturer in 1921. In the same year he married Agnes Isobel Johnson, a marriage from which there were three daughters. Promotion to Assistant Professor came in 1922, and then in 1936 he was appointed to a Chair of Applied Physical Chemistry, in which post he remained at Imperial College until his retirement in 1952. During that period he spent a year in Brussels (1937–38) as Franqui Professor. In the Second World War, Finch acted as scientific adviser to the Ministry of Home Security, being especially concerned with fire defence. He also led a group at Imperial College on the development of incendiary bombs. Following his retirement from Imperial College he became Director of the National Chemistry Laboratory in India where he served for five years.

Finch with his students carried out research on the effect of catalysts on the combustion of gases, epitaxial growth, wear and lubrication and electrodeposition; electron diffraction was used as an important technique. Finch's work brought him honours including the award of the Hughes Medal in 1944.

Finch became one of the most outstanding mountaineers of his generation. His enthusiasm for mountains dated back to his boyhood in Australia, and serious mountaineering began in 1917 in Zurich. With his critical scientific approach he brought safety, efficiency and meticulous planning and he had an excellent physique capable of strenuous activity. With an outstanding record as an Alpine climber he also made an outstanding contribution, with George Mallory, in the 1922 Mount Everest expedition, both as a climber and as a pionéer of the use of oxygen. He took an important part in the development of the type of apparatus which was used in the attempt on the summit, and was a co-leader of the second climb to 27,300 feet (8,300 m). His books on mountaineering included *The Making of a Mountaineer* (1924), in the writing of which he was assisted by his wife; this book helped to shape the attitude of a generation of Alpine climbers. In addition to mountaineering he enjoyed sailing; in 1932 he bought a yacht in which he spent active leisure.
BMFRS 1972, 18, 223–239. M. Blackman FRS. DNB 1961–1970, 358, 359. M. Blackman, FRS.

SIR HAROLD MONTAGUE FINNISTON 1912–1991; FRS 1969

Harold Montague Finniston was born into a family of Russian Jewish origin in Govanhill, Glasgow. The family home was a tenement in which each apartment had one room and a kitchen. Finniston, universally referred to as Monty, attended the local school, at Victoria, until he was 11. The family were able to move to a larger apartment in Crosshill. Monty won a scholarship for five years to Allan Glen's School and at the age of only 16 he obtained a university scholarship. He entered Glasgow University, Royal College of Science and Technology where he graduated in metallurgical chemistry. He then carried out research on the combustibility of coke obtaining a PhD and was a lecturer in metallurgy from 1933–35 and moved to Stewarts and Lloyds as a metallurgist, 1935–37. In 1936 he married Miriam Singer and they had a son and a daughter. He worked as chief research officer, Scottish coke committee, before joining the Royal Navy Scientific Service in 1940. Finniston worked at the naval ordnance inspection department in Sheffield and then at Chalk River, Canada, concerning the application of nuclear power to submarine propulsion.

After returning to Britain Finniston became Chief Metallurgist at the Atomic Energy Authority, Harwell, where he worked from 1948 to 1958, building up the activity aimed at developing civil nuclear power. In the Metallurgy Division, he organised, ran and inspired a highly successful team of very able, very active scientists dealing with new materials and new fabrication and behaviour problems. In addition to the responsibility for producing fuel elements important basic science was done. Monty knew everybody in the division by their first names; he made daily rounds, discussing and suggesting new ideas. In personal matters, he was thoughtful and generous. He worked extremely long hours, seven days a week.

In 1958 Finniston joined C.A. Parsons in Newcastle upon Tyne. There, he took charge of a large new laboratory to serve as the Nuclear Research Centre for Parsons and other member firms of the Nuclear Power Plant Company; he also became Parson's Technical Director. When investment in nuclear power in the UK slowed down he persuaded the Parson's Board in 1962 to convert it into a research company – International Research and Development Company (IRD) – to do contract research for government and industrial customers. Finniston was Managing Director of IRD and then Chairman, 1969–77. Examples of the work done at IRD include the construction of the world's first superconducting motor drive system and explosive joining techniques for underwater welding of petrochemical pipelines. Finniston's energy and enthusiasm were evident together with his belief in the potential contribution of advances in science and technology to the national interest.

In 1966 Finniston served on the government committee planning the future organization of the new re-nationalised steel corporation. Having joined the Corporation in 1967 as a Deputy Chairman he became Chief Executive in 1971 and Chairman in 1973. There were immense problems in achieving the necessary reorganisation and modernisation and although good progress was made there were international economic difficulties and political changes in England, and Finniston's contract was not renewed in 1976.

When he left the British Steel Corporation, Finniston continued to work as hard as ever. At various times he was chairman or director of some 14 companies and had many interests outside business, including serving on many government committees. Notably, he was Chairman of the Government Enquiry into the Engineering Profession, which produced the very valuable and wide-ranging document – *The Finniston Report*, 1980. It dealt with the critical issue of the quality of engineering in the recovery of Britain in world manufacturing and the need to improve the public perception of the status of engineers. However, its proposal to set up a statutory engineering authority concerned with the qualifications and registration of engineers was not implemented. Finniston gave much support to educational bodies; his many appointments in this sphere included university chancellorships and he strongly believed in the néed for more higher education and the importance of university research.

Finniston served as chairman of the body which became the Policy Studies Institute from 1974–84. He made constructive contributions to the work on employment policies, labour relations, comparisons with other countries, ethnic minorities and on the transport relationships of homes and jobs. Much time was devoted to charities, serving as chairman of a number; for example, he was first Chairman of the Prison Reform Trust. He was much in demand as a speaker and he became quite an established personality on TV and radio.

Finniston's many honours included election as FREng. He was Vice-President of the Royal Society from 1971–72 and was knighted in 1975. In metallurgy his remarkable achievements had spanned two major aspects of industry, nuclear power and iron and steel production, involving both process and physical metallurgy.

BMFRS 1992, 38, 133–144. N.J. Petch FRS.

MICHAEL ELLIS FISHER 1931–; FRS 1971

King's College, London: BSc 1951; PhD 1987. RAF: Flying Officer (educational), 1951–53. London Univ; Postgraduate Studentship, 1953–56; DSIR Senior Research Fellow, 1956–58. King's College, London: Lecturer in Theoretical Physics, 1958–62; Reader in Physics, 1962–64; Professor of Physics, 1965–66. Cornell Univ.: Professor of Chemistry and Mathematics, 1966–73; Horace White Professor of Chemistry, Physics and Mathematics, 1973–84; Chairman, Department of Chemistry, 1975–78. Univ. of Maryland: Distinguished University Professor and Regent's Professor. Institute for Physical Science and Technology, 1993–; Wilson H. Elkins Distinguished Professor, 1987–93. Royal Society: Member of Council and Vice-President, 1993–95. Bakerian Lecturer, 1979.

Theoretical aspects, interfaces and surfaces.

ALEXANDER FLECK (BARON FLECK OF SALTCOATS) 1889–1968; FRS 1955

Alexander Fleck was the only son of Robert Fleck, a coal merchant, and Agnes née Duncan. He was born in Glasgow and attended Saltwoods Public School and Hillhead High School. Because of family circumstances he left school at the age of 14 and became a laboratory assistant in the Chemistry Department at Glasgow University, where his keenness brought him to the attention of Frederick Soddy. Having studied at evening classes he became a full time student in the University and graduated in 1911 with an honours degree. He was an assistant to Soddy for two years, involved in research on radioactive elements. In 1913 he took a post with the Glasgow and West Scotland Radium Committee and was awarded a DSc in the same year.

In 1917 Fleck married Isabel Mitchell Kelly; there were no children. Also in 1917 he went to Wallsend as Chief Chemist to the Castner-Kellner Alkali Company, associated with Brunner Mond and Co. His work included improvements to the production of sodium and he was promoted to Works Manager in 1919. When Imperial Chemical Industries was formed by mergers in 1926 Fleck went to Billingham to be responsible for the planning and operation of a new works. In 1931 he became Managing Director of the General Chemical Division at Liverpool. He joined the ICI board in 1944, becoming Chairman two years later and remaining in this post until his retirement in 1960. During his career Fleck was involved in the production of a wide range of chemicals, including polymeric materials. Among his other activities he was involved in the nuclear industry, including service as Chairman of the Nuclear Safety Advisory Committee from 1960–65. In the Royal Society Fleck was Vice-President in 1960. He was made KBE in 1955 and Baron in 1961 for his services to the Ministry of Fuel and Power.

BMFRS 1971, 243–254. R. Holroyd. DNB 1961–1970, 363–365. C.M. Wright.

MARTIN FLEISCHMANN 1927–; FRS 1986

Worthing High School. Imperial College, London: ARCS 1947, BSc 1948, PhD 1951. Univ. of

Durham, Kings College: ICI Fellow, 1952–57. Univ. of Newcastle upon Tyne; Lecturer then Reader, 1957–67. Univ. of Southampton: Electricity Council Faraday Professor of Electrochemistry, 1967–77, Senior Fellow SERC 1977–82, Research Professor 1983–. Univ. of Utah; 1988–.
Electrochemistry.

SIR JOHN AMBROSE FLEMING 1849–1945; FRS 1892
John Fleming was born in Lancaster, the son of the Rev. George Fleming, a congregational minister, and Mary Ann Bazley née White. Having moved to London, he attended University College School and subsequently University College. He graduated in the first division in 1870 in spite of the fact that for two years he had been working as a clerk, studying only part time. During the period 1872–1874 he studied at South Kensington and he also taught as a science master. In 1877 he entered St John's College, Cambridge with an entrance exhibition, later holding a foundation scholarship. He worked under James Clerk Maxwell in the new Cavendish laboratory, and in 1880 was appointed a Demonstrator, having obtained the London DSc degree. In 1881 Fleming was appointed Professor of Physics and Mathematics at the newly constituted University College at Nottingham. After a year there he resigned to become a consultant to the Edison Electric Light Company in London. Then in 1885 he was appointed as Professor of Electrical Technology at University College, a post which he held for 41 years.

Fleming made important contributions in many fields including the design of transformers, the properties of materials at liquid air temperatures and electrical measurements. In his experiment with wireless telegraphy he collaborated with Guglielmo Marconi having a consultant post with the Marconi Wireless Telegraphy Company; he helped to design the transmitter used by Marconi in spanning the Atlantic in 1901. In 1904 he patented a 'thermionic valve', which used either platinum or aluminium as a cylinder surrounding the filament. In his family life he married Clara Ripley Pratt in 1887; she died in 1917 and he married Olive May Franks in 1933; there were no children. He was awarded the Hughes Medal in 1910 and was knighted in 1929.
ONFRS 1945–48, 5, 231–242 W.H. Eccles. DSB, 5, 32, 33, C. Susskind. DNB 1941–1950, 258–260. J.T. McGregor-Morris.

BRIAN HILTON FLOWERS (BARON FLOWERS OF QUEEN'S GATE) 1924–; FRS 1961
Bishop Gore Grammar School, Swansea. Univ. of Cambridge: Gonville and Caius College (Exhibitioner); MA. Anglo-Canadian Atomic Energy Project, 1944–46. Atomic Energy Research Establishment: research in nuclear physics and atomic energy, 1946–50. Univ. of Birmingham: Department of Mathematical Physics, 1950–52: DSc. AERE, Harwell: Head of Theorectical Physics Division, 1952–58. Univ. of Manchester: Professor of Theorectical Physics, 1958–61; Langworthy Professor of Physics, 1961–72 (on leave of absence as Chairman, Science Research Council, 1967–73). Imperial College of Science and Technology: Rector, 1973–85. Univ. of London: Vice-Chancellor, 1985–90. Chairman, Royal Commision on Environmental Pollution, 1973–76. Knighted 1969. Life Peer 1979.
Atomic physics.

MARTIN FOLKES 1690–1754; FRS 1714
Martin Folkes was born in Lincoln's Inn, London, the eldest son of of Martin Folkes, a bencher of Gray's Inn, and Dorothy née Hovell. Having attended the University of Saumur in France he entered Clare Hall, Cambridge in 1706, where he studied mathematics and matriculated in 1709;

he became a Cambridge MA in 1717. He married Lucretia Bradshaw in 1714 and they had one son and two daughters. Folkes had an interest in coins (including Roman) and antiquities and compiled a table of silver coins from the Conquest; his papers included the field of metrology. In the Royal Society he became Vice-President in 1722 and succeeded Newton as President in 1741 serving until 1752.
DSB 5, 53, 54. J.M. Rodney. DNB.

DAVID FORBES 1828–1876; FRS 1858
David Forbes was born at Douglas, Isle of Man, one of nine children of Edward Forbes and Jane née Teare. He showed an early interest in chemistry and after schooling in Brentwood, Essex he entered Edinburgh University. Having left university at the age of 19 Forbes spent some months in the metallurgical laboratory of John Percy in Birmingham. He then held the post of Superintendent of the mining and metallurgical works at Espedal in Norway for ten years. When a revolution threatened in 1848 he armed 400 of his men to assist the government, receiving the thanks of the king.

Entering into partnership with a firm of nickel smelters of Birmingham he went to South America in search of ores of nickel and cobalt; from 1857 to 1860 he travelled widely in Bolivia and Peru, making geological observations on the mineral and rock structures. After a return visit to England he travelled again in South America, adding to the large collection of minerals already made in Norway. Further travels followed in the South Sea Islands, and, in 1866, in Europe and Africa. Forbes's geological research included igneous and metamorphic phenomena and in Norway he experimented on the action of heat on minerals and rocks. He was among the first to apply microscopy to the study of rocks. Forbes settled in England and became Foreign Secretary of the Iron and Steel Institute, writing regular reports on progress in metal-working abroad.
DNB.

SIR HUGH FORD 1913–; FRS 1967
Northampton School. City and Guilds College, Imperial College, University of London: PhD, DSc(Eng). GWR locomotive works: practical training, 1931–36. Researches into heat transfer 1936–38. ICI Ltd. Northwich: Research and development engineering: 1939–42. BISRA: Chief Engineer Technical Department, 1945–47. Imperial College: Reader in Applied Mechanics, 1948–51, Professor, 1951–59; Head of Department of Mechanical Engineering, 1965–78; Professor of Mechanical Engineering, 1969–80; Pro-Rector, 1978–80. Sir Hugh Ford and Associates Ltd: Chairman, 1982–. FREng 1976. Knighted 1975.
Metal deformation in the elastic and plastic ranges and in experimental stress analysis; applied mechanics, rolling of metals.

JOHN SAMUEL FORREST 1907–1992; FRS 1966
James Forrest was the first of three children of Samuel Norris Forrest and Elizabeth Reed née Paterson. He was born in Hamilton, S.E. Glasgow where he studied at the Academy. In 1930 he graduated from Glasgow University with an ordinary BSc in pure science and honours in mathematics and natural philosophy. He was appointed as a Physicist at the Central Electricity Board in Glasgow in 1930 and a year later moved to London with the Board. His work was associated with the British Grid system. In 1940 Forrest married May Olding and in the same year he was involved in setting up a laboratory at Leatherhead. In 1947 the laboratory became part of the British Electrical Authority and by 1950 he had become Director. The activity later became the

Central Electricity Research Laboratory with Forrest as Director, serving until his retirement in 1973.

Forrest's work was particularly associated with electrical insulation and transmission lines at high voltages and included an investigation of the corrosion of the steel-cored aluminium conductors of the Grid. He was awarded the DSc degree of Glasgow Universty in 1950 and was elected FREng in 1976.

BMFRS 1994, 40, 107–126. T.E. Allibone FRS.

SIR CLEMENT LE NEVE FOSTER 1841–1904; FRS 1892

Clement Le Neve Foster was the second son of Peter Le Neve Foster, a barrister, and Georgiana Elizabeth née Chevallier. He was born in Camberwell, where he attended the Collegiate School. From the age of 12 to 16 he studied at the College Communal of Boulogne-sur-Mer, obtaining a BSc in the University of France in 1857. Entering the School of Mines in London in 1857, he obtained the Associateship in Mining, Metallurgy and Geology and was awarded the Edward-Forbes Medal and Prize. He spent a session at the Mining school of Freiberg, and then visited mining centres in Germany and Hungary.

From 1860 for five years Foster held an appointment with the Geological Survey of England and Wales. In 1865 he obtained the DSc degree of London University and became Lecturer to the Miners Association of Cornwall and Devon. In 1868 he carried out exploration in the Middle East and reported on the gold fields in Venezuela. In 1872 he married Sophia Chevallier, and they had one son and two daughters. From 1869–1872 he was an engineer to a gold mining company in Northern Italy. He became Inspector of Mines in Cornwall in 1872, then transferring to North Wales in 1880. Foster held the post of Professor of Mining at the Royal School of Mines, London, from 1890 until his death. He served on many important committees and was knighted in 1903.

ONFRS 1905, 75, 371–377. DNB 2nd Suppl., II, 42, 43. HTW.

GEORGE CAREY FOSTER 1835–1919; FRS 1869

George Foster, who was born at Sabden, Lincolnshire, obtained an honours degree in chemistry at University College London. He left England in 1858 to study at Ghent, Paris and Heidelberg. He became Professor of Natural Philosophy in Anderson's College, Glasgow in 1852. Three years later he was elected as the first Professor of Physics at University College London, a post which he held until 1898; from 1900 to 1904 he was Principal of the College. Foster's best known scientific work was connected with the use of the Wheatstone bridge for accurate measurements of electrical resistance. Also he served on the committee that was set up to test the resistance thermometer developed by William Siemens. Foster married Mary Anne Rosebank in 1868 and they had four sons and four daughters.

PRS A 1919–20, 96, xv–xviii. OJL. *Platinum Metals Review* 1980, 24, 3, 104–112. L.B. Hunt.

JEAN BAPTISTE JOSEPH FOURIER 1768–1830; FMRS 1823

Jean Fourier was born in Auxerre, the son of a tailor Joseph and his wife Edmée. Orphaned when he was young he was educated at a local military academy, where his interest in mathematics was aroused. Subsequently he attended a school in Benoît-sur-Loire, but when the French Revolution began he returned to Auxerre and taught at his former school. He was imprisoned in 1794 but was released and studied in Paris for a while at the École Normale. In 1795 he became an Assistant Lecturer at the École Polytechnique. He accompanied Napoleon on his Egyptian campaign in

1798 and conducted various diplomatic affairs. Back in France in 1801 he continued part time with mathematical studies, but also held administrative posts as a Prefect; in 1808 Napoleon conferred on him the title of Baron and subsequently of Count. Soon after he obtained an appointment at the Bureau of Statistics where he was able to focus his efforts on mathematics. As a mathematical physicist Fourier's important work included his theorem which could be used, for example, in the understanding of heat conduction.
DSB 5, 93–99. J.R. Ravetz and I. Grattan Guinness.

SIR RALPH HOWARD FOWLER 1889–1944; FRS 1925
Ralph Fowler was born at Fedsden, Royston, the eldest of two sons and a daughter of Howard and Ena Fowler. In 1902 he entered Winchester College as a scholar and six years later he entered Trinity College, Cambridge with a major scholarship, to take the Mathematical Tripos. He obtained first class honours in Part I (1909) and graduated BA as a wrangler in Part II, (1911) and MA (1915). After graduation Fowler commenced research in the field of pure mathematics and was elected to a fellowship of Trinity College in 1914. With the coming of war in 1914, Fowler was commissioned in the Royal Marine Artillery and served in Gallipoli, where he was severely wounded. Later in the war he took part in a range of applied projects, including gunnery, for the ministry of munition and was awarded the OBE in 1918. He returned to his fellowship at Trinity College in 1919 and became a Lecturer in Mathematics in 1924. In 1921 he married Eileen Rutherford and they had four children.

Fowler developed wide-ranging theoretical research, involving various collaborations, in areas of mathematical physics and physical chemistry. Themes included the high temperature dissociation of atoms at low pressures, statistical mechanics, intensities of spectral lines, statistical theory of strong electrolytes, thermionic and photoelectronic emission of electrons from metals, ferromagnetism and magnetostriction, and semiconductors. In 1932 he was elected to the Plummer Chair of Theoretical Physics. During the 1939–45 war he was involved with scientific liaison with Canada and the USA and he also did some work with the Admiralty. In the Royal Society Fowler gave the Bakerian Lecture in 1935 and was awarded a Royal Medal in the following year. He was knighted in 1942.
ONFRS 1945–48, 5, 67–78. E.A. Milne. DNB 1941–50, 268, 269. E.A. Milne.

ROBERT WERE FOX 1789–1877; FRS 1848
Robert Fox was born at Falmouth, Cornwall into a Quaker family; his father, a shipping agent, was also called Robert Were Fox, and his mother was Elizabeth née Tregelles. He was privately educated and showed a special interest in mathematics. In 1814 he married Maria Barclay. Fox began his scientific research in 1812 making experiments on high pressure steam, hoping to improve Watt's engines in use for pumping in Cornish mines. In 1815 he commenced investigations on the internal temperature of the earth aided by his access to the mines. He wrote around 50 scientific papers, the first of these (1819) being concerned with platinum alloys. The study of magnetic phenomena was one of his special interests, including the earth's magnetism and the electromagnetic properties of metalliferous veins in Cornish mines.
DNB.

SIR FREDERICK CHARLES FRANK 1911–1998; FRS 1954
Charles Frank was born in Durban, South Africa, the son of Frederick Frank and Medora Colin Emma née Read; his parents were English and returned to Suffolk when he was very young.

Following education at Thetford Grammar School and Ipswich School he entered Lincoln College, Oxford as a scholar where he obtained a BA and BSc in physics in 1933. There followed a period of research in the Dyson Perrins Laboratory and Engineering Laboratory, Oxford leading to a DPhil in 1937. From 1936–38 Frank was at the Kaiser Wilhelm Physics Research Institute in Berlin to work with Debye. From 1938–1940 he worked in the Colloid Science Laboratory at Cambridge. In 1939 he entered the Scientific Civil Service in which he served until 1946; for a period in 1940 he worked with the Chemical Defence Research Establishment, before moving to intelligence work in the Air Ministry where he made highly important contributions, for example, in the interpretation of aerial photographs. In 1940 he married Maida Asche; there were no children. In 1946 he went to the University of Bristol to the H.H. Wills Physics Laboratory, where he was appointed a Research Fellow in Theoretical Physics (1948) and a Reader in Physics (1954); from 1954–69 Frank was Professor of Physics, and from 1969–76 he held the appointments of Henry Overton Wills Professor of Physics and Director of the H.H. Wills Laboratory.

Frank's research ranged widely in the solid state physics of materials, including dislocations and plasticity in crystals, crystal morphology and growth, mechanical properties of polymers and liquid crystals; he also worked on cold fusion and the mechanics of the earth's crust. In the Royal Society he served as Vice-President from 1967–69, gave the Bakerian Lecture in 1973, and received a Royal Medal (1979) and the Copley Medal (1994). He was also awarded the OBE in 1946 and was knighted in 1977.

The Times, 27th April 1998. *Physics World*, 1998, **11**(6), 48. John Nye.

BENJAMIN FRANKLIN 1706–1790; FRS 1756

Benjamin Franklin was born in Boston, Massachusetts, one of 17 children of Josiah Franklin, who had come from Britain in 1683; his mother was Josiah's second wife Abiah ('Jane') née Folger. Benjamin had schooling up to the age of 10 and then assisted his father in his tallow and soap business. At the age of 12 he was indentured to his brother as a printer, but after a few years he moved to New York and then on to Philadelphia, from whence in 1724 he went to London where he became a skilled printer. Two years later he returned to Philadelphia and soon had his own business as a publisher and journalist. He became a major figure in the town and eventually in wider spheres, involved in civic affairs. He was Deputy Postmaster General for the British colonies in North America from 1753–1774; in 1757 he came to England as agent of the Pennsylvania Assembly to which he had been elected in 1751. He remained intermittently in England until 1775. Back in America he helped to draft the Declaration of Independence in 1776. Later he was involved in the peace negotiations between the American colonies and Britain and became President of Pennysylvania.

As a scientist he acquired a vey high reputation. From 1746 he spent seven years experimenting in electricity, making fundamental contributions. One of his earliest observations was the ability of a pointed metal object to discharge an electrified conductor and he invented the lightning conductor. He also developed the first plate condenser which consisted of glass sheets between lead plates. He was awarded the Copley Medal in 1753.

DSB 5, 129–141. R.V. Jenkins.

NORMAN LAURENCE FRANKLIN 1924–1986; FRS 1981

Norman (Ned) Franklin was born on the outskirts of Leeds, the only child of William Alexander and Beatrice Franklin. He entered Batley Grammar School in 1934, and then went to Leeds

University in 1942 where, in the Faculty of Chemical Engineering, he obtained first class honours in fuel technology in 1945. He was appointed a Technical Officer with the British Coke Research Association first in London, and then at Pontypridd, in South Wales where he was in charge of a research unit. In 1948 Franklin joined the lecturing staff at the University of Leeds in the Department of Chemical Engineering. In 1949 he married Bessie Copeland and they had two children.

Franklin joined the United Kingdom Atomic Energy Authority in the team at Risley in 1955 engaged on formulating the technical policy of the production group of the Authority. For a period during 1958 he became Special Assistant to the Managing Director. He also worked for a while as Technical Manager at the Chaplecross reactor site, and then returned to technical policy; he had special responsibility for fuel element design and evaluation and was rapidly promoted. When he was deputy manager of the production group he took a particular interest in the development of the ultra -centrifuge for uranium enrichment. Other appointments held by Franklin at various periods in his career were Managing Director and Chief Executive and Director of British Nuclear Fuels, member of the Board of the Atomic Energy Authority, Chairman and Managing Director of the National Nuclear Corporation and Chairman and Managing Director Nuclear Power Company. In 1984 when he left the nuclear industry he was appointed Professor of Nuclear Engineering at Imperial College. His honours included OBE (1963), CBE (1971) and FREng (1978).

BMFRS 1988, 34, 207–219. Sir John Hill FRS.

JACQUES FRIEDEL 1921–; FMRS 1988

École Polytechnic. École de Mines de Paris. Univ. of Bristol: Mining Engineer, École des Mines, Paris: 1948–56. Univ. de Paris: Maitre de Conferences, 1956– 59. Univ de Paris, (later Paris Sud): Professor of Solid State Physics, 1959– 89. Acad. of Sciences: Member 1977–; President 1991–94.

Dislocations. Solid state theory.

RICHARD HENRY FRIEND 1953–; FRS 1993

Rugby School. Univ. of Cambridge: Trinity College, MA, PhD; St John's College, Research Fellow; University Demonstrator in Physics, 1980–85; Lecturer, 1985–83; Reader in Experimental Physics, 1993–95: Cavendish Professor of Physics, 1995–.

Chemical physics; solid state physics; semi-conductors.

OTTO ROBERT FRISCH 1904–1979; FRS 1948

Otto Frisch was born in Vienna, the only son of Justinian Frisch and Auguste née Meitner (sister of Lise Meitner). He showed early talent for mathematics and also as a pianist. He was educated at the Piaristen Gymnasium and entered the University of Vienna in 1922 where he studied physics and was awarded the Dr.Phil in 1926. He spent a period in an industrial laboratory and then, in 1927 he took an appointment at the Physikalisch-Technische Reichsanstalt (the State Physical Laboratory) in Berlin. In 1930 he moved to Hamburg, where his worked included an investigation of the magnetic moment of the proton. When Hitler rose to power Frisch came to London in 1933 to Birkbeck College, but the following year he joined Bohr at Copenhagen and worked on nuclear physics. In 1938, when visiting his aunt Lise Meitner, he became aware of the letter she received from Otto Hahn concerning nuclear fission, and in 1939 he published a letter with Lisa

Meitner in *Nature* on the disintegration of uranium by neutrons. With the imminence of war Frisch came to England to the University of Birmingham where he and Rudolf Peierls wrote a memorandum concerning the possibility of the construction of a 'superbomb' based on a nuclear chain reaction in uranium. He moved to Liverpool in 1940 to James Chadwick's department to work on various aspects associated with chain reactions and uranium isotope separation. In 1943 became a naturalised British subject and went to Los Alamos to work on the atomic bomb project. Frisch returned to England in 1946 to the Atomic Energy Research Establishment at Harwell, where he was appointed as Head of the Nuclear Physics Division. From 1947–72 he was Jacksonian Professor of Natural Philosophy at the University of Cambridge; he was also a Fellow of Trinity College. In 1951 he married Ursula (Ulla) Blau and they had a daughter and a son. He was awarded the OBE in 1946.
BMFRS 1981, 27, 283–306. Sir Rudolf Peierls, FRS. DSB, 17, 320–322. J. Hendry. DNB 1971–80, 326, 327. Rudolf Peierls.

HERBERT FRÖHLICH 1905–1991; FRS 1951
Herbert Fröhlich was born at Rexington in the Black Forest; he was the second in a family of three children. The family moved to Munich and when he left school he apprenticed himself to a firm of textile manufacturers. In 1927 he entered Munich University and obtained a PhD in theoretical physics in 1930. He became a Privatdozent at Freiberg University in 1933. During 1934–35 he spent a period at Ioffe's Physico Technical Institute in Leningrad and later, in 1935, he was offered support by Tyndall to come to the Department of Physics at Bristol. During the war he carried out research at Bristol, and after the war was promoted to a Readership, and then was Professor of Physics at Liverpool from 1948–1973. After his retirement he was Professor of Solid State Physics at Salford University. He was twice married; his second wife was Fanchon Augst. Fröhlich's research included dielectrics, superconductors and biological systems. He published a book (in German) concerned with the application of quantum mechanics to electrons in metals (1936); the book included a section on semiconductors. In 1949 he published another book *Theory of dielectrics*.
BMFRS 1992, 38,147–162. Sir Nevill Mott FRS.

WILLIAM FROUDE 1810–1879; FRS 1870
William Froude, born at Darlington, Devonshire was the 14th son of the Archdeacon of Totnes, the Venerable Robert Hurrell Froude. He was educated at Westminster School and at Oriel College, Oxford where he obtained a first class honours BA degree in mathematics in 1832 and an MA in 1837. In his leisure he spent time studying chemistry and mechanics. In 1833 he was employed as a pupil of a distinguished civil engineer on some railway surveys. He joined the engineering staff of Isambard Kingdom Brunel in 1837, and was in charge of part of the construction of the Bristol and Exeter railway. More survey work was carried out in 1844, and soon after he went to live at Darlington with his father, until in 1859 following his father's death he moved to Torquay. His career involved important work as a naval architect, including employment by the Admiralty. As early as 1856, at Brunel's request he had investigated the laws of motion of ships; his work influenced the design of armour clad naval vessels. He received a Royal medal in 1876.
PRS 1879, 29, ii–vi. DNB.

JOHN FULLER ?–?; FRS 1704: JOHN FULLER ?–1755; FRS 1726: ROSE FULLER ?–1777; FRS 1732

The Fullers of Brightling Park, Sussex were iron masters. John, senior, married Elizabeth Rose and they had two sons. The elder son John served as Member of Parliament for Borough Bridge. The younger son Rose studied at Leyden (1729) and later was awarded the degree of MD of Cambridge; he served a Member of Parliament for New Romsey (1756–61), Maidstone (1761–68) and Rye (1768–77).
Bulloch's Roll.

DENNIS GABOR 1900–1979; FRS 1956

Dennis Gabor was born in Budapest, Hungary, the eldest of three sons; his father Berthalan Gabor was Director of Hungary's largest industrial firm, a coal mining company, and his mother was Adrienne née Kalman. Having attended school in Budapest he did military training near the end of World War I before commencing, in 1918, a course in mechanical engineering at the city's technical university. He left during the third year to go to Berlin where he completed his formal education at the Technische Hochschule, obtaining a diploma in electrical engineering in 1924 and a doctorate in 1927. Following industrial work in Germany and Hungary he came to England in 1933 and from 1934–1938 worked as a Research Engineer at the British Thomson-Houston Company in Rugby. In 1936 he married Marjorie Louise Butler; there were no children. From 1949 until his retirement in 1967 he was a member of the academic staff of Imperial College, London, culminating in his holding the Chair of Applied Electron Physics from 1958–1967. Gabor was renowned as an inventor. His invention of holography stemmed from his interest in electron microscopy and the challenge of developing an electron microscope that could 'see' atoms. He received the Nobel Prize for Physics in 1971.
BMFRS 1980, 26, 107–147. DSB Suppl., 17, 324–328. S.T. Keith.

SIR DOUGLAS STRUTT GALTON 1822–1899; FRS 1859

Douglas Galton who was born at Spring Hill, near Birmingham, was the second son of John Howard Galton and Isabelle née Strutt. His education included attendance at Rugby School and he then studied with distinction at the Royal Military Academy at Woolwich. In 1846 he was commissioned as a Second Lieutenant in the Royal Engineers, and was subsequently promoted to Lieutenant (1843), Second Captain (1851) and First Captain (1855). His army service included a period in the Mediterranean area, and appointment to the Ordnance Survey. In 1847 he became Secretary of the newly formed Railway Commission and was also Secretary to the Royal Commission on the Application of Iron to Railway Structures; he carried out experiments on the strength of iron. In 1854 he became Secretary to the Railway Department of the Board of Trade. Other activities and appointments in his career included service on several commissions, Assistant Permament Under-Secretary of State for nearly eight years beginning in 1862, Director of Public Works and Buildings in the Office of Works (commencing in 1869). In 1878 and 79 he presented papers on experiments on railway brakes. He married Marianne Nicholson in 1851 and they had two daughters. Galton was awarded the CB in 1865 and was knighted (KCB) in 1887.
DNB Suppl. up to 1901, 691–694. RHV.

WILLIAM ALEXANDER GAMBLING 1926–; FRS 1983

Univ. of Bristol: BSc, DSc. Univ. of Liverpool: PhD. Univ. of BC. National Research Council, 1955–57. Univ. of Southampton: Lecturer; Senior Lecturer; Reader, 1957–64; Dean of Engineer-

ing and Applied Science, 1972–75; Professor of Electronics, 1964–80; Head of Department, 1974–79; Professor of Optical Communications, 1980–95; Director Optoelectronics Research centre, 1989–95. City Univ. of Hong Kong: Royal Society Kan Tong Po Professor and Director Optoelectronic Research Centre, 1996–. FREng 1979.
Electronic and optical fibre communications; fibre fabrication.

JOSEPH-LOUIS GAY-LUSSAC 1778–1850; FMRS 1815
Joseph Gay-Lussac, born at St Leonard, France, was the eldest of five children of Antoine Gay, a lawyer, and Leonarde née Bourigner. In 1797 he entered the École Polytechnique, where he graduated in 1800; originally he studied engineering, but he became an assistant to Claude-Louis Berthollet, and took up chemistry. In 1808 he married Geneviève Marie Josephe Rojot, and they had five children. His appointments included Professor of Chemistry at the École Polytéchnique (1810), Professor of Physics at the Paris Faculty of Science (1808) and Professor of General Chemistry at the Musée National d'Histoire Naturelle. He carried out important research on gases, and also, in collaboration with Thénard he isolated and named boron. In politics Gay-Lussac served in the Chamber of Deputies.
PRS 1850, 76, 1013–1023. DSB 5, 317–327. M.P. Crosland.

GEOFFREY GEE 1910–1996; FRS 1951
Geoffrey Gee was the son of Thomas and Mary Ann Gee and was educated at New Mill Grammar School. He began to study chemistry at Manchester University in 1928, obtaining a BSc in 1931 and proceeded to research on colloid chemistry, being awarded an MSc in 1933. From 1933–38 he was associated with ICI Dyestuffs who financed him to work at Cambridge University, where he was awarded a PhD in 1936. In 1934 he married Marion Bowden and they had one son and two daughters. Gee became interested in polymers and joined the British Rubber Producers Association as a Research Chemist in 1938 on fundamental investigations on rubber, including elastic behaviour. In 1947 he was awarded a Cambridge ScD and became Director of BRPRA, a post which he held until he was appointed as Professor of Physical Chemistry at the University of Manchester in 1955; from 1955–77 he was Sir Samuel Hall Professor of Chemistry; he also served a Head of Department, Dean of the Faculty of Science(1963–65) and as Pro-Vice-Chancellor(1966–68) and (1972–77). At Manchester he built up an important research group in the physical chemistry of polymers. He was awarded a CBE in 1958.
The Independent, 18.1.97. C. Price and C. Booth.

PIERRE-GILLES DE GENNES 1932–; FMRS 1984
École Normale, Paris. Commissariat Énergie Atomique (Saclay): Research Engineer, 1955–69 Univ. Orsay: Professor, 1961–71. Coll. de France: Professor, 1971–. École de Physique et Chimie: Director, 1976–. Nobel Prize for Physics, 1991.
Theoretical physics: magnetism; superconductivity; liquid crystals; polymers.

CLAUDE-JOSEPH GEOFFROY 1685–1752; FRS 1715
Claude-Joseph Geoffroy, referred to as Geoffroy the Younger (le Cadet) was born in Paris, the second son of Matthieu Francois Geoffroy and Louise née de Vaux. Having qualified as an apothecary in 1703 he took over the family pharmacy in 1708 when his father died. Highly regarded in his profession he became Inspector of Pharmacy at the Paris Hospital, and in 1731 served as a Paris Alderman. His interests in chemistry developed and his research included work aiming to

find a chemical explanation for the colours of plants; his work on inorganic subjects was important in both the development of theory and practical significance. In the metallurgical sphere in 1725 he reported an investigation of the microscopically-observed fracture characteristics of a series of copper–zinc alloys, and related the behaviour to composition.
DSB 5, 351, 352. W.A.Smeaton. Bulloch's Roll. 'Observations sur un métal que résulte de l'alliage du cuivre & du zinc', *Histoire et Memoires de l'Académie des Sciences*, 1725, **39**, 57–66. Gen. Ref. 21.

JOSIAH WILLARD GIBBS 1839–1903; FMRS 1897
Josiah Willard Gibbs was born in New Haven, Connecticut where he lived throughout his life except for a period of study in Europe. He came from a distinguished and learned family. His father, also Josiah Willard Gibbs, was Professor of Sacred Literature in the Yale Divinity School, and his mother was Mary Anna Van Cleve. Gibbs had a distinguished career as a student at the University of Yale, obtaining prizes in mathematics and Latin and graduating in 1858 with high standing. He continued at Yale as a student of engineering in the new graduate school, and in 1863 he received one of the first PhD degrees awarded in America, for a thesis entitled 'On the form of the teeth of wheels in spur gearing'. He also did some other engineering work involving the design of an improved railway car brake (patented in 1866). Following his doctoral work he became a tutor at Yale, teaching Latin for two years and natural philosophy for another year. In 1866 he travelled to Europe where he studied in Paris (1866–67), in Berlin (1867–68), and in Heildeberg (1868–69), coming under the influence of some of the most distinguished mathematicians and physicists of the world. Having returned to New Haven his work included the design of a new form of governor for steam engines, and he turned his attention to the study of thermodynamics in the early 1870s. In 1871 he was elected Professor of Mathematical Physics at Yale, an appointment he held until his death. During the first nine years he was not paid a salary but lived on inherited income; only when he was invited to join the new John Hopkins University at Baltimore did Yale offer him a salary.

The contributions made by Gibbs through his theoretical work in thermodynamics became of outstanding importance in the development of physical chemistry and of metallurgy; in the latter field it was crucial in the work on phase diagrams that was a major and essential activity in providing an understanding of alloys. His work was presented in his memoir *On the equilibrium of heterogeneous substances,* the two parts of which were published in 1876 and 1878 respectively.

Gibbs never married, but lived a quiet life in the home of a colleague who had married one of his sisters, spending summers in the mountains, and attending church regularly. He gave generously time and thought to family and household problems. Although his work was carried out far from the European mainstream of science, it became widely recognised and he was invited to honorary membership of most of the leading learned societies and academies of both hemispheres in physics and mathematics. He received the Copley Medal in 1901. He was regarded by many as the greatest native-born American scientist.
ONFRS 1905, 75, 280–296 J.L. DSB, 5, 386–393 M.S. Klein. Chambers Biographical Encylopaedia of Scientists, Eds J. Daintith, S. Mitchell, E. Tostill, W. and R. Chambers Ltd, Edinburgh, 1981.

ALAN FRANK GIBSON 1923–1988; FRS 1978
Alan Frank Gibson was born in Calcutta, the only child of Heseltine Gibson, a metallurgist, and Ruby Margaret née Wilson. From 1931–37 he attended Stanley House School in Birmingham, for most of the time as a boarder; from 1937–42 he studied at Rydal School, Colwyn Bay. His parents were in India and he saw little of them until after the war. He entered Birmingham Uni-

versity with a State Bursary in Physics-with-Radio and after graduating in 1944 he joined the Telecommunications Research Establishment (TRE) at Malvern as a Junior Scientific Officer; he was assigned to a research project at Birmingham University and studied the luminescent properties of solids. In 1945 he married Judith Cresswell and they had one son and two daughters. From 1948–63 Gibson was at Malvern; in 1952 when a Transistor Physics Division was set up he was appointed as its Head. Collaborative research included the growth of single crystals of silicon and germanium. In 1953, when TRE became the Royal Radar Establishment, Gibson was promoted to Senior Principal Scientific Officer. He was appointed Superintendent of the Infra-red and Solid State Physics Group and in 1961 he was promoted to Deputy Chief Scientific Officer. In 1963 Gibson became the first Professor of Physics at the University of Essex. From 1977–83 he was Head of the Laser Division at the Rutherford Appleton Laboratory. His research was particularly concerned with the optical properties of semiconductors.
BMFRS 1991, 37, 221–244. R. Loudon FRS and E.G.S. Paige.

JOHN ANTHONY HARDINGE GIFFARD (HALSBURY, 3RD EARL) 1908–; FRS 1969
Eton. London Univ.: External BSc, 1st class honours, Chemistry and Mathematics, 1935. Lever Bros: 1935–42. Brown–Firth Research laboratories: 1942–47. Decca Record Co: Director of Research, 1947–49. National Research Development Corporation: Managing Director, 1949–59. Joseph Lucas Industries Ltd: Consultant and Director 1959–74. Head Wrightson Ltd, 1959–78. Brunel Univ.: Chancellor 1966–. FREng 1976.
Metallurgy and materials.

PERCY CARLYLE GILCHRIST 1851–1935; FRS 1891
Percy Gilchrist, the elder son of Alexander Gilchrist, a barrister, and his wife Anne née Burrows, was born at Lyme Regis; he had two sisters. He attended Felsted School and then from 1868–71 the Royal School of Mines where he trained as a metallurgist and analytical chemist and obtained the Murchison medal in 1870. He was an analytical chemist at an iron works at Cwm Avon in Glamorgan, and was appointed as Works Chemist at the Blaenavon Ironworks in Monmouthshire in 1876.

It was at Blaenavon that he collaborated with his cousin Sidney Gilchrist Thomas leading to a solution to the critical industrial problem of producing steel by the Bessemer process from pig irons high in phosphorus. This major advance reported in 1878 influenced steelmaking throughout the world. Although Gilchrist was overshadowed by his cousin his importance in the joint invention was clear; when Thomas received the Bessemer Medal of the Iron and Steel Institute in 1883 he acknowledged Gilchrist's 'unwearied exertions, conspicuous energy and ability' in their collaboration. Gilchrist married Nora Fitzmaurice in 1887 and they had a son and a daughter.
ONFRS 1936–38, 2, 19–27. H.C.H. Carpenter. DNB Missing Persons 251–252. W.H. Brock.

SIR WILLIAM HENRY GLANVILLE 1900–1976; FRS 1958
William Glanville was the only son and the second of three children of William Glanville and Amelia née Venning. He was educated at Kilburn High School, and after a brief period of service in the army at the end of the 1914–18 war he studied civil engineering at East London College (later Queen Mary College), London, graduating with first class honours in 1922. He then began a research career in the Scientific Civil Service by taking an appointment as an Engineering Assistant at the Building Research Station of the Department of Industrial and Scientific Research at Acton, where he began investigations of concrete. In 1925, he took charge of the first of a series of classical studies of reinforced concrete; important results were published in 1930

which were influential in relation to the establishment of codes of practice for the design of reinforced concrete structures. Glanville became Deputy Director of the DSIR Road Research Laboratory at Harmondsworth in 1936, where he maintained his interest in reinforced concrete, for example on the effect of the size of cracks on the corrosion of the reinforcing steel.

In 1939 Glanville succeeded Reginald Stradling as Director of the Road Research Laboratory. During the 1939–45 war the Laboratory carried out important research, which included rapid methods of constructing airfields, and the development of 'plastic protective plating' (a stone-filled bituminous material in a thin steel casing, which came into use to protect bridges and gun positions on allied merchant ships.) After the war Glanville's work on roads led his interest to extend into tar and bituminous materials, and problems of road safety such as vehicle skidding, including road surfaces and rubber tyres. Glanville was awarded the PhD and DSc degrees of London University in 1926 and 1930 respectively. He retired in 1965, was made a CBE in1944, CB in 1953, and was knighted in 1960. He married Millicent Patience Carr, and they had one son and one daughter.

BMFRS 1977, 23, 91–113, Lord Baker FRS. DNB 1971–80, 339–340, G. Grime.

SIR RICHARD TETLEY GLAZEBROOK 1854–1935; FRS 1882

Richard Glazebrook who was born in West Derby, Liverpool, was the eldest son of Nicholas Smith Glazebrook, a surgeon, and Sarah Ann née Tetley. He was educated at Dulwich College, Liverpool College and Trinity College, Cambridge (Scholar 1875, Fellow 1877). In the Mathematical Tripos he graduated Fifth Wrangler in 1876, and then began to study physics under James Clerk Maxwell, carrying out research in the Cavendish Laboratory. After becoming a demonstrator in 1880 he became a Lecturer and in 1891 was appointed Assistant Director of the Cavendish laboratory. He married Frances Gertrude Atkinson in 1883, and they had one son and three daughters. Also in 1883 he became Secretary of a British Association committee on Electrical Standards, and carried out accurate measurements. Having been appointed Senior Bursar of Trinity College in 1895 he demonstrated considerable business ability. In 1898 he became Principal of University College, Liverpool; a year later he was appointed as the first Director of the National Physical Laboratory at Teddington, then being formed under the management of the Royal Society. Following his retirement in 1919 he was Zaharoff Professor of Aviation and Director of the Aeronautics Department at Imperial College from 1923–26. In the Royal Society he was Foreign Secretary from 1926–29 and was the recipient of the Hughes Medal (1909) and of a Royal Medal (1931). He was honoured by the award of CB (1910), a knighthood (1917), KCB (1920) and KCVO (1934).

ONFRS 1936–38, 2, 29–56. Rayleigh. F.J. Selby. DNB 1931–40, 343, 344 W.C.D. Dampier.

EUGEN GLUECKAUF 1906–1981; FRS 1969

Eugen Glueckauf, the son of Bruno Glueckauf and Elas née Pretzfelder was born in Eisenhach, Thuringia. He entered the University of Berlin, but soon moved to the Technische Hochschule at Charlottenburg, where he was awarded the Dipl. Ing degree in 1930; he developed an interest in surface phenomena and continued at the Technische Hochschule, obtaining the Dr Ing degree in 1932; in 1933 he came to England and after a period of settling in he obtained an appointment as an Assistant at Imperial College, working on helium analysis in relation to meteorological investigations. In 1934 he married Irma Tepper, who had also come to England from Germany. Following a period of interment in 1940 he went to the Chemistry Department at the University of Durham with a two-year grant from DSIR on meteorological work, and in 1942 he was awarded a Research Studentship of the Royal Society. In 1945 he obtained an MSc. He moved to the Atomic Energy Research Establishement at Harwell in 1947 as a Principal Scientific Officer, and was promoted to Senior Principal Scientific Officer in 1949 and to Deputy Chief Scientist in 1952, having been awarded a London University DSc degree in 1951. He became Head of the Fission Product Technology Group and retired in 1971. Glueckhauf's contributions in the atomic energy field included projects on gaseous diffusion plant and the high temperature reactor programme.

BMFRS 1984, 30, 193–224, D.H. Everett FRS.

JONATHAN GODDARD c.1617–1675; FRS 1663

Jonathan Goddard was born in Greenwich, the son of Henry Goddard, a ship builder of Deptford. In 1632 he entered Magdalen Hall, Oxford, but left after 3 or 4 years without taking a degree. He studied medicine abroad, and in 1638 he graduated MB at Cambridge(Christ's College) and MD in 1643(Catherine Hall). He joined the College of Physicians from his practice in London and became a Fellow of the College in 1646. Goddard came to the attention of Cromwell and served in Ireland and Scotland as Physician-in-Chief of the Parliamentary Army. His subsequent career included the posts of Parliamentary Warden of Merton College, Oxford, Membership of Parliament and of the Council of State, Chancellor of the University of Oxford, and Professor of Physic at Gresham College (1655). He was a member of the first Council of the Royal Society, and in his laboratory made many experiments for the Society. One of his papers in *Philosophical Transactions* (1678) reported experiments on a method of refining of gold involving melting it with antimony.

The Record of the Royal Society. DNB.

VICTOR MORITZ GOLDSCHMIDT 1888–1947; FMRS 1943

Victor Goldschmidt was born in Zurich, Switzerland, the only child of Heinrich Jacob Goldschmidt, a distinguished physical chemist, and Amelie née Koehne. His earliest schooling was at Amsterdam and Heildeberg, and later at Oslo, Norway where his father was appointed Professor of Chemistry in the University. In 1905 he and his parents became Norwegian citizens. Goldschmidt entered the University in the following session where he studied mineralogy and geology, and inorganic and physical chemistry. He studied in Vienna in 1908–1909 and in Munich from 1911–1912. He gained his doctorate at Oslo in 1911 and was appointed in 1914 as Professor and Director of the Mineralogical Institute in the University. Apart from a period of six years at Göttingen (1929–1935) his work, involving students from many countries, was associated with his Oslo chair until 1942.

Goldschmidt's geological work included fundamental researches on rock metamorphism. His researches in geochemistry which began in the First World War continued over a long period and included the use of X-ray techniques for the analysis of rocks and minerals. He also carried out important work on metals establishing atomic radii.

After the German occupation of Norway during the 1939–45 war Goldschmidt was sent to a concentration camp in October 1942, and his property was confiscated; later that year there was a danger of his being deported to Poland, but he escaped to Sweden. In the spring of 1943 he came to Britain and worked under the auspices of the Agricultural Research Council. When he returned to Norway in 1946 he was very weak from prolonged illness.

Goldschmidt was a man of outstanding energy and fertility of ideas together with phenomenal knowledge of his fields. His many honours included honorary or foreign membership of most of

the leading scientific academies of Europe; he was made a Knight of the Royal Norwegian Order of St Olaf. In addition to the distinction he achieved in mineralogy, petrology and geochemistry his work in new areas of inorganic chemistry contributed to the fundamental understanding of metallic structures.
ONFRS 1948–49, 6, 51–66. C.E. Tilley. DSB 5, 456–458. E.D. Goldberg.

SIR CHARLES FREDERICK GOODEVE 1904–1980; FRS 1940

Charles Goodeve, one of a family of three girls and two boys, was born in the small town of Néepawa on the eastern edge of the Canadian prairies. His father was an Anglican clergyman and the family moved to Stonewall, north of Winnipeg when Charles was three, and then to Winnipeg when he was 10 years old; here he went to the Kelvin High School. In 1919 he entered Manitoba University as an arts student but transferred to science two years later, and passed the BSc examinations in chemistry and physics in 1925. To assist financially at university he worked part time and in the second year he was appointed as a Junior Laboratory Demonstrator. He held an Assistant Lectureship and worked on an electrolytic problem. In 1927 Goodeve obtained an MSc in electrochemistry and also was awarded 1851 Exhibition Scholarship to be held at University College, London. During his third year at university his sailing interests led him to enlist as a midshipman in the Royal Canadian Naval Volunteer Reserve.

In 1927 Goodeve came to London to the chemistry department at University College and began research including the subjects of unstable molecules and absorption spectra. This led him into photochemistry and the associated reaction kinetics and ultimately into his work on the physical chemistry of vision; during his time at University College up to 1939 he also worked on the oxides of chlorine and colloid chemistry. In 1932 Charles married Janet Wallace and they had two sons; she obtained a PhD degree in chemistry under Charles's supervision.

Goodeve became an Assistant Lecturer in 1928, Lecturer in Physical Chemistry in 1930 and Reader in 1937. He developed closer links with industry than was customary at that time, and obtained industrial funding. He maintained his naval interests through the Royal Naval Volunteer Reserve, went to sea in submarines and in minesweepers, and served in battleships and destroyers. In 1936 he was promoted to Lieutenant Commander and began to direct some of his research towards naval problems.

Goodeve's contributions in the 1939–1945 period were numerous and of great importance. At H.M.S.*Vernon* his work included developing methods for counteracting magnetic mines. He set up the Department of Miscellaneous Weapons Development and was awarded the OBE. In 1942 he was appointed Assistant (later Deputy) Controller Research and Development for the Admiralty and was created a Knight Bachelor in 1946.

When the war ended Goodeve became Director of the British Iron and Steel Research Association. He held this position until his retirement in 1969 and successfully built up a large research organisation complementary to the research and development groups in industry. Improved understanding and practice were achieved in areas such as blast furnaces, open hearth and arc furnaces and automatic gauge control in strip production. Goodeve also developed operational research.

After he retired Goodeve continued to work influentially in operational research, believing in the importance of the social dimensions and of human behaviour. During his career he was active in directorships and in other areas including government committee work. He was a Vice-President of the Royal Society from 1968–1970, and received many honours and distinctions.
BMFRS 1981, 27, 307–353. F.D. Richardson FRS. DNB 1971–1980, 349–350. E. Duckworth.

GEORGE GORE 1826–1908; FRS 1865

George Gore was born in Bristol where his father had a small business as a cooper. When he left school at about 13 years of age he began to work as an errand boy. From the age of 17 to 21 he was apprenticed to a cooper and worked at that trade also studying science and carrying out experiments in his leisure time. He was very interested in electrodeposition and in 1851 he moved to Birmingham, already the chief centre of electroplate manufacture, where he worked as a chemist for a local firm that manufactured phosphorus. He held classes on electroplating and on chemistry and from 1870 to 1880 taught physics and chemistry at King Edward's School.

From 1854 to 1963 Gore published a series of papers on the electrodeposition of metals. He acquired a reputation which gave him a leading position in the Birmingham as a consulting chemist for manufacturers. His discoveries were important in the early days of the art of electroplating and he wrote several textbooks on the subject. In other areas Gore carried out important work on anhydrous hydrofluoric acid. He studied the thermoelectric action of metals and liquids, and he also observed changes in properties associated with transformations during the cooling of iron. In about 1880 an ' Institute of Scientific Research' was established in Birmingham and he served as its Director until his death. He acted as an industrial consultant and continued research on electrolysis and voltaic cells. By his will his residuary estate was equally divided between the Royal Society and the Royal Institution to assist original scientific research.

PRS 1910–11 A, 84, xxi–xxii. JHP. DSB 5, 474. D.P. Jones.

HERBERT JOHN GOUGH 1890–1965; FRS 1933

Herbert Gough was born in Bermondsey, London, the second son of Henry James Gough, a civil servant, and Mary Anne née Gillis. Having attended primary school in Ealing he went to Regent Street Polytechnic Technical School from whence he won a studentship to University College School. Following a short period as a pupil teacher he served an apprenticeship at Messrs Vickers, Sons and Maxim from 1909–1913; he was then appointed a designer draughtsman with Vickers Ltd. During this period he obtained a BSc honours in engineering from London University and was later awarded a PhD and a DSc. From 1914–1938 Gough was on the staff of the engineering department at the National Physical Laboratory, becoming Superintendent in 1930. Gough's work at the NPL established him as a leading figure in research on metal fatigue, including aspects such as mechanisms and combined stress effects; he published a book *The Fatigue of Metals* in 1924. From 1914–19 Gough served in the Royal Engineers (Signals) becoming a Captain, receiving mention in dispatches twice and the MBE (Military) in 1919. In 1918 he married Sybil Holmes and they had one son and one daughter.

Gough became the first Director of Research in the War Office in 1938 ; four years later he was appointed Director-General of Scientific Research and Development at the Ministry of Supply. From 1945 until 1955 he was Engineer-in-Chief to Lever Brothers and Unilever Ltd. He was awarded the CB in 1942 and served on the Council of the Royal Society from 1939 to 40.

BMFRS 1966, 12, 181–194. S.F. Dorey. DNB 1961–1970, 445, 446. A. Kelly.

WILLIAM GOWLAND 1842–1922; FRS 1908

William Gowland, the son of George Thompson Gowland and his wife Catherine, was born in Sunderland. It was intended that he would enter the medical profession and he worked with a doctor at Sheffield for two or three years. However, attracted by scientific pursuits, he entered the Royal College of Chemistry in London in 1863. In the succeeding two years he obtained the Associateship in Mining and Metallurgy at the Royal School of Mines, with the award of the

Murchison Medal in Geology and the De la Beche Medal in Mining.

From 1870–1872 he was employed as Chemist and Metallurgist with the Broughton Copper Company in Manchester. Then he went to Japan in 1872 where he worked as Chemist and Metallurgist to the Imperial Mint at Osaka for several years. He became Assayer and Chief of the Foreign Staff and Adviser to the Imperial Arsenal in 1878; he made several expeditions into the mountains and also travelled in Korea, where he carried on work for the Japanese Government. In 1889 on his return to England he received the order of the Rising Sun. Gowland returned to the Broughton Copper Company as Chief Metallurgist in 1890 and in the same year he married Joanna Macaulay; following her death he married Maude Connacher.

In 1902 Gowland was appointed Professor of Metallurgy at the Royal School of Mines where he began to introduce microscopical methods of examining steel. Gowland's writings on metallurgical subjects were chiefly concerned with his observations in Japan and he was actively interested in the archaeology and applications of metals. He was the author of *Metallurgy of Nonferrous Metals* (1914). He served on the Council of the Royal Society from 1912–1914.
PRS 1922–23, A, 102 , xvi–xix. WAT.

GEORGE GRAHAM 1673–1751; FRS 1721

George Graham, who was born near Rigg, Cumberland, was brought up by a brother, William, following the death of his father. He apprenticed himself to a clockmaker in London in 1688, serving for seven years and obtaining the freedom of the Clockmakers' Company. Soon he obtained employment with Thomas Tompion the leading maker of clocks, watches and instruments at that time. In 1704 he married Elizabeth, Tompion's niece, and succeeded to the business in 1713. Among his inventions were the 'anchor escapement'. In addition to clocks and watches he made scientific instruments, including planetary models, barometers, quadrants, and micrometer screws for precise subdivisions. Graham was elected to the Council of the Royal Society in 1722. He was a member of the Society of Friends.
DSB 5, 490–492. E.A. Battison. DNB.

THOMAS GRAHAM 1805–1869; FRS 1836

Thomas Graham was born in Glasgow, the son of a merchant, and the eldest of a family of seven. After education at a preparatory school and the High School in Glasgow he entered the University in 1819, where he acquired scientific interests and decided to devote himself to science, instead of becoming a presbyterian minister as his father had wished. After receiving the MA degree in 1826 he spent nearly two years studying chemistry at the University of Edinburgh. Returning to Glasgow he taught in a private laboratory and then lectured on chemistry at the Mechanics' Institute in 1829; the next year he was appointed Professor of Chemistry at Anderson's College where he carried out research. There followed in 1837 his appointment to the Chair of Chemistry at University College London; in 1854 the Government appointed him as Master of the Mint. He had for many years acted as Non-Resident Assayer and he continued at the Royal Mint until his death. Graham remained unmarried.

As a chemist, Graham held ideas in advance of his contemporaries. His research was particularly concerned with the constitution of matter and the movement of atoms and molecules, involving important discoveries concerning diffusion of gases and the occlusion of gases in metals. Among his activities, he was appointed in 1847 by the Board of Ordnance to inquire into methods of casting guns. He was awarded a Royal Medal in 1838 and 1863 and the Copley Medal in 1850. He served for six years on the Council of the Society and was twice Vice-President. He

delivered the Bakerian Lectures in 1850 and 1854.
PRS 1869–70, 18, xvii–xxvi. RAS. DSB 5, 492–495. George B. Kauffman. DNB. Ref. Ch. 4.1.

GEORGE WILLIAM GRAY 1926–; FRS 1983
Univ. of Glasgow: BSc. Univ. of London: PhD. Univ. of Hull: Chemistry Dept Staff; 1946–; Senior Lecturer, 1960; Reader, 1964; Professor of Organic Chemistry, 1978; G.F. Grant Professor of Chemistry, 1984–90; Research consultant 1990–. Clifford Paterson Lecturer, 1985. Leverhulme Medal, 1987. CBE 1991.
Molecular structure and properties of liquid crystals; materials for liquid crystal electronic displays.

STEPHEN GRAY 1666–1736; FRS 1733
Stephen Gray was born in Canterbury, Kent. He followed his father into his trade as a dyer, but learned some latin and science, for the latter probably spending a period of study in London or Greenwich, perhaps with the Astronomer Royal. In 1707 and 1708 he spent a period in Cambridge, working on a project to set up a new observatory. In 1709 he became one of the Charterhouses's 'gentlemen pensioners'. His scientific work included important contributions on electricity. Gray was awarded the Copley Medal in 1731 and 1732.
DSB 5, 515–517. J.L. Heilbron. DNB.

GEOFFREY WILSON GREENWOOD 1929–; FRS 1992
Grange Grammar School, Bradford. Univ. of Sheffield: BSc, PhD, DMet. UKAEA, Harwell: Scientific and Senior Scientific Officer, 1953–60. Berkeley Nuclear Laboratories, Central Electricity Generating Board: Head of Fuel Materials Section, 1960–65. Scis. Div. Electricity Council Research Centre: Research Manager, 1965–66. Univ. of Sheffield: Professor of Metallurgy, 1966–94; Pro-Vice-Chancellor, 1979–83. FREng, 1990.
Metallurgy, materials science, physics, engineering; e.g materials for nuclear industry growth of fine precipitates in solids and of gas bubbles in irradiated metals; diffusional creep, grain shape anisotropy.

ALAN ARNOLD GRIFFITH 1893–1963; FRS 1941
Alan Griffith, a Londoner by birth, was the eldest of three children of George Griffith, explorer, journalist and author, and Elizabeth née Brierley. During his childhood the family moved to the Isle of Man where his father died. He received private tuition until 1906 when he went to the Douglas Secondary School. He was awarded a scholarship in 1910 and entered the School of Mechanical Engineering at the University of Liverpool. In 1914 he obtained the BEng degree with first class honours, the Rathbone Medal and the University Scholarship in Engineering tenable for a year of research. During that year he investigated the surface resistance to heat between metals and gases and obtained the MEng degree in 1917.

Griffith joined the Royal Aircraft Factory (subsequently known as the Royal Aircraft Establishment RAE) at Farnborough, Hampshire in 1915. Initially he received a general workshop training for several months and then held a series of posts during the next four years: Draughtsman, Technical Assistant and Senior Technical Assistant in the Physics and Instrument Department; he became a Senior Scientific Officer in 1920 and was awarded the D.Eng degree of Liverpool University in 1921. Griffith married Constance Vera Faulkner in 1925 and they had two daughters and a son.

In 1917 a joint paper by Griffith and G.I. Taylor on novel methods, using soap films, for the estimation of torsional stresses in sections of complicated shapes gained the Thomas Hawksley

Gold Medal of the Institution of Mechanical Engineers. Griffith continued with work using the soap film technique, and hence acquired the nickname of 'Bubble Griffith' at the RAE. In 1920 his paper 'Theory of rupture' made an important contribution to the science of fracture of materials .

Over a period of eight years as Senior Scientific Officer Griffith became involved in the study of the gas turbine, making great contributions to the science of aircraft propulsion. He became Principal Scientific Officer in charge of the Air Ministry Laboratory, South Kensington, London in 1928 and in the following year he evolved the contraflow principle for a gas turbine. In 1931 he moved back to the RAE to take charge of engine research. Griffith was associated with a wide variety of patents covering flame traps, ice indicators, de-icing, fuel vapourising and many piston engine features; his work greatly assisted the air effort during the second world war. Early in 1939 he was appointed as a Research Engineer with Rolls-Royce Limited where he designed a multi-axial compressor and turbine combined on the contra-flow principle. He also worked on the development of vertical take-off and landing, jet lift aircraft, including the first free flight of the 'Flying Bedstead' rig. Griffith received the CBE in 1948. After his retirement in 1960 he carried out some consulting for Rolls-Royce. He was one of the great figures in the science of aircraft propulsion and also contributed fundamentally to materials science.
BMFRS 1964, 10, 117–136. A.A. Rubbra, DNB 1961–70, 457–459. Kings Norton.

ERNEST HOWARD GRIFFITHS 1851–1932; FRS 1895
Ernest Griffiths was born at Brecknock, in Brecon, Wales, the son of Rev. Henry Griffiths. Following education at Owens College, Manchester he entered Sidney Sussex College, Cambridge in 1870 as a Whitworth Scholar, and took an ordinary degree in applied science in 1873. He then became a private tutor in Cambridge, but some time later decided to carry out scientific research and began in the field of heat. Griffiths was a colleague of Francis Neville at Sidney Sussex College, and in 1888 Heycock and Neville sought his advice in relation to the temperature measurements required in their alloy work. Griffiths married Elizabeth Clark in 1877 and his house became a meeting place for a circle of close friends, including Heycock and Neville.

Griffiths devoted spare time to research, initially in the Sidney Sussex Laboratory, but, after 1891, having found his measurements affected by traffic vibration, he built a laboratory in his house. He constructed a number of platinum resistance thermometers and carried out collaborative work with Heycock, Neville and Callendar.

In 1897 Griffiths was elected a Fellow of Sidney Sussex College. Then in 1901 he became Principal of the University College of South Wales and Monmouthshire at Cardiff, and also Professor of Experimental Philosophy. He served as Vice-Chancellor of the University of Wales, but also continued research activity in the field of exact physical measurement. In the Royal Society he served on the Council from 1909–1911 and in 1907 was awarded the Hughes Medal.
ONFRS 1932–35, 10, 15–18. WCDD. Ref. Ch. 4.1.

EZER GRIFFITHS 1888–1962; FRS 1926
Ezer Griffiths, who was born in Aberdare, Glamorgan, was the son of Abraham Lincoln Griffiths, a colliery mechanic, and Ann née Howells. There were nine children in the family and Ezer was the eldest of the six sons. Following his elementary education he attended the intermediate school at Aberdare. He then entered University College, Cardiff and obtained first class honours in physics. He was awarded a Research Studentship and later a Fellowship of the University of Wales. After research at Cardiff until 1915 he took an appointment at the National Physical

Laboratory,where he worked until his retirement in 1953. He received a DSc degree(Wales) and was awarded the OBE in 1950. Griffiths never married. In his research career Griffiths's main interests were in the field of heat in which he became a leading authority. Themes covered included heat insulation, heat transfer and evaporation and his research ranged from high temperatures to temperatures relevant to refrigeration. He investigated various thermal and other physical properties of a range of materials, including metallic and insulating types. His publications included a book on *Methods of Measuring Temperature* (1918) and one on *Pyrometers* (1925).
BMFRS 1962, 8, 41–48. Sir Charles Darwin. DNB 1951–1960, 459, 460. D.T. Lee.

SIR WILLIAM ROBERT GROVE 1811–1896; FRS 1840
William Grove was born at Swansea, the son of John Grove, a Magistrate and Deputy-Lieutentant of the County of Glamorgan, and Anne née Bevan. Following private education he entered Brasenose College, Oxford, where he graduated in 1832 and was awarded an MA in 1835. He was called to the Bar at Lincoln's Inn in 1835 but for several years ill health prevented him from actively practising his profession; during this period he carried out research on electrochemistry at home. In 1837 he married Emma Maria Powles and they had two sons and four daughters. To meet the financial néeds of his growing family he returned to the practice of law, but continued his scientific work.

In 1839 he devised a primary cell, which came to be called after him. From around 1840 he became Professor of Experimental Philosophy at the London Institution in Finsbury Circus and carried out research there for seven years. In a series of lectures, that he gave in 1842 he presented the concept of the conservation of energy (the first law of thermodynamics). Grove was also responsible for the origin of the fuel cell.

Grove was a Member of Council of the Royal Society in 1846 and 1847, and served as Secretary for the following two years; in 1847 he was awarded a Royal Medal. He played a leading part in the Society's reform movement. In later years Grove had a distinguished legal career, becoming a QC in 1853, a judge in 1871 and was knighted in 1872.
DSB 5, 559–561. E. L.Scott. DNB Suppl. II, 371–372 JNR. Ref. Ch. 4.1.

SIR JOSIAH JOHN GUEST 1785–1852; FRS 1830
Josiah Guest was the elder son of Thomas Guest, manager and part owner of Dowlais Ironworks, near Merthyr Tydfil, Wales, and Jemima née Phillips. Born at Dowlais he was educated at the Grammar Schools at Bridgnorth and Monmouth. He devoted himself to the Ironworks introducing various improvements by the application of chemical and engineering principles and the output greatly expanded. Having succeeded to sole management of the works in 1815 he was the sole proprietor from 1849 until his death.

Guest was twice married, first to Maria Elizabeth Ranken (died 1818) and then in 1833 to Charlotte Elizabeth Bertie, by whom he had 10 children. In politics he served in Parliament as the Member for Honiton from 1826–1831 while from 1832 until his death he represented Merthyr; he was created a Baronet in 1838. An enlightened employer, he was concerned with the interests of his employees and his charitable activities involved the founding of schools and places of worship.
DNB.

FREDERICK GUTHRIE 1833–1886; FRS 1871
Frederick Guthrie, a Scotsman by descent, was the son of Alexander Guthrie, a London trades-

man, and was born in Bayswater. After private tuition he attended University College School and then entered University College to study chemistry; one of his teachers was Thomas Graham. After graduating in arts at the age of 19 he went to Germany in 1854 where he received further chemical training under Robert Bunsen at the University of Heidelberg; then at Marburg he obtained a PhD degree. Having published a number of papers on organic chemistry he returned to England in 1855 and took his BA degree at London. The following year he was appointed Demonstrator in Chemistry at Owens College, Manchester. Later, in 1859, he was a Demonstrator under Lyon Playfair at Edinburgh. In 1861 he accepted an appointment as Professor of Physics in the Royal College in Mauritius where he continued scientific investigations.

He returned to England in 1868 as Lecturer in Physics at the Royal School of Mines in Jermyn Street; when the School was transferred to South Kensington he became Professor of Physics. Guthrie was an authority on science teaching. Among his scientific books was *Elements of Heat and Non-Metallic Chemistry.* He founded the Physical Society of London in 1873.

Of particular metallurgical interest is the important research that Guthrie carried out on the freezing points of solutions of salts and of metallic alloys, including the introduction of the term 'eutectic'.

DNB. Register of the RCS, RSM. T.G. Chambers, 1896. *Phil. Mag*., 1884 , **17**, 462–482.

LOUIS BERNARD GUYTON DE MORVEAU 1737–1816; FRS 1788

Louis Bernard Guyton, born in Dijon, Burgundy, was the son of a lawyer, Antoine Guyton, and Marguerite née Desaulle. Having been educated at the Jesuit College and the Faculty of Law in Dijon he practiced as an advocate from 1756–1762; he then served until 1782 in the Dijon Parliament (a royal court of law) as one of the public prosecutors, adding to his name 'de Morveau' from a family property. He was often referred to as Monsieur de Morveau until 1789; in the period of the French Revolution he became Guyton-Morveau, then Guyton, and then Guyton-Morveau again.

Guyton became interested in chemistry and this led to the installation of a laboratory in his house in 1769. He collaborated with various scientists including Lavoisier and was also involved in several industrial enterprises, e.g in glassmaking. His interests from the beginning of his scientific career included metallurgy and mineralogy. His research included an investigation of the use of coal in blast furnace practice, experiments on solidification and on etching, the fabrication and properties of platinum, pyrometry, and the tensile strength of metals.

In 1791 he went to Paris on his election to the National Assembly and was a Member of the National Convention which succeeded it in the following year, when France was declared a Rebublic. He became Secretary of the Committee of General Defence in 1793 and also served for a short period as President of the first Committee of Public Safety in 1793; after Robespierre's downfall the next year he again served on the latter Committee. Also in 1794 he was involved in the applications of science to the war. Having a decade earlier become associated with ballooning, testing various gases, he was one of the organisers of the military use of a captive balloon for observers to report on Austrian positions in a battle in 1794. He was active in the establishment of the École Polytechnique in 1794, where he taught and carried out research as a Professor of Chemistry from that year until 1811; he was also twice Director of the École. Guyton married Mme Picardet, a widow, in 1798; they had no children. In 1799 he was appointed by Napoleon Administrator of the Mints (nine altogether); when the Bourbon restoration occurred in 1814 he left this office, but resumed it until his retirement in 1815.

Guyton de Morveau's varied career included a turbulent period in European history, in which,

in addition to his political influence he made important contributions to a number of branches of chemistry, covering fundamental and industrial themes, including the metallurgy of various materials including iron and steel and platinum.
DSB 5, 600–604. W.A. Smeaton. Refs Ch. 2.8; 4.1.

SIR ROBERT ABBOTT HADFIELD 1858–1940; FRS 1909
Robert Hadfield was born at Attercliffe, then a village near Sheffield. He was the only son of Robert Hadfield and Marianne née Abbott. His father, who had gained experience in several branches of the steel industry, established a works in 1872 to produce steel castings, then a novelty in England. The firm also produced steel projectiles which until then had only been manufactured in France; Hadfield senior decided to develop their production independently rather than to buy foreign patents thus laying the foundation of what was to become one of the leading armament firms. Hadfield junior was educated at the Collegiate School in Sheffield, where he became interested in chemistry. He was allowed to use a basement in his father's house to carry out experiments with metals, and here he had a small melting furnace. At the age of between 16 and 17 years old he had the opportunity of studying at Oxford or Cambridge, but decided to enter the steel business; he spent a few months early in 1875 in the firm of Jonas and Colver, before entering the family works in Attercliffe, where he set up a laboratory and began to study alloys systematically. From the age of 24 as his father's health was failing Hadfield had to take over the administration of the firm and when his father died in 1888 a limited company was formed, of which he became Chairman and Managing Director.

Hadfield commenced a systematic study of alloys of iron with additions of manganese aiming to produce a hard steel with high wear resistance. This work led to the discovery of the remarkable properties of the alloy containing about 12–14 wt% manganese and 1 wt% carbon, which came to be known as Hadfield's manganese steel; this was patented in 1883 and became widely used e.g. in railway crossings and military helmets in the 1914–1918 war. Hadfield went on to make other investigations of alloy steels, including measurements of mechanical, electrical and magnetic properties; an alloy containing about 3 wt% silicon became important in electrical transformers. Collaboration by Hadfield with eminent physicists, including James Dewar, involved measurements of alloy properties at cryogenic temperatures. Among the activities of Hadfield's firm was the development of special steels for armour-piercing projectiles and for heat-resisting purposes. Hadfield was also involved in investigations on ingot solidification and on corrosion.

Keenly interested in the history of metallurgy Hadfield's studies included a collection of Sinhalese iron objects of the fifth century AD and the Delhi pillar. Of particular interest in the history of 19th century metallurgy was his detailed metallurgical examination of the alloys prepared by Michael Faraday and James Stodart.

Throughout his life Hadfield was one of the hardest and most well organised of workers. An enlightened employer, he was one of the first (1891) to introduce an eight-hour day into his works. He took a great interest in the national and international organisation of science and played a leading role in activities to form a central organisation of all engineering institutions. In 1925 he published *Metallurgy and its Influence on Modern Progress,* containing a review and an account of his own researches, with reflections on the needs of education and research.

Hadfield married Frances Belt Wickersham in 1894; they had no children. Soon after war broke out in 1914 the Hadfields founded a hospital at Wimereux. Many honours came to Hadfield from scientific and technical societies and from public bodies. He was knighted in 1908 and was

created a Baronet in 1917, in which year he also received the Freedom of the City of London. His reputation as a scientific investigator stems particularly from his discovery of manganese steel but he was also influential in the wide field of alloy steel development.
ONFRS 1939–41, 3, 647–664, C.H.Desch FRS. DSB, 6, 5. F. Greenaway. DNB 1931–1940 384–386, C.H. Desch.

JOHN HADLEY 1682–1744; FRS 1717
John Hadley was the son of George Hadley, Deputy Lieutenant and later High Sheriff of Hertfordshire, and Katherine Fitz née James. He became skilled in practical mechanics and by 1719 he had produced paraboloid mirrors of speculum metal superior to those produced by other London master opticians, achieving a mirror diameter of six inches (~15 cm). He became friends with the Scottish optician James Short. Hadley served on the Council of the Royal Society for many years and was Vice-President in 1728.
DSB 6, 5, 6. H.C. King. DNB.

OTTO HAHN 1879–1968; FMRS 1957
Otto Hahn was born in Frankfurt-on-the-Main, the youngest of three sons of Heinrich Hahn and, a widow, Charlotte Stutzmann. Otto was originally intended for a career in architecture and attended a technical school but he developed an interest in chemistry and studied the subject at the University of Marburg and obtained a doctorate in 1901. In the same year he joined an infantry regiment as a one-year volunteer and returned to Marburg in 1902. From 1904–05 he did chemical research in London under William Ramsay on radioactive elements. There followed a period when Hahn worked with Rutherford at McGill University, Montreal. In 1906 Hahn returned to Germany to work at the University of Berlin, in the Chemical Institute where he was joined by Lisa Meitner in 1907, beginning their collaborative work which was to extend over more than 30 years. Hahn became Head of the Radiochemistry Section of the new Kaiser Wilhelm Institute for Chemistry at Berlin-Dahlem, which opened in 1912. He married Edith Junghans in 1913 and they had one son. During the 1914–18 war Hahn served in the infantry on the Western, Eastern and Italian Fronts, including reluctant involvement with gas warfare. He was also involved in staff work and spent periods in Dahlem where he was able to continue some experimental research on radiochemistry with Meitner; they discovered the new radioelement protactinium in 1918. Hahn became Director of the Kaiser Wilhelm Institute in 1928.

During the 1930s Hahn's collaborative research on the neutron bombardment of uranium led to the proposal by Meitner and Frisch of 'nuclear fission'. During the 1939–45 war Hahn continued radiochemical work, but not within the programme of nuclear weapons research in Germany. In 1944 when the functions of the Kaiser Wilhelm Institute were taken over by the Max Planck Institute in Göttingen, Hahn became President. In the same year he was awarded the Nobel Prize for Chemistry, which he received in Stockholm in 1946. In 1945 Hahn was one of the group of German scientists brought to England for six months to a country house near Cambridge for questioning. Hahn became active in the cause for nuclear disarmament.
BMFRS 1970, 16, 279–313. R. Spence FRS. DSB 6, 14–17. Lawrence Badash.

EDMUND HALLEY 1656–1742; FRS 1678
Edmund Halley was born in Shoreditch, London, the son of Edmund Halley. He was educated at St Paul's School and at Queen's College, Oxford. Through Newton's influence he was appointed as Deputy Controller of the Mint at Chester in 1696, an office he held for two years. His subsequent career included the command of a Royal Navy warship, holding the Chair of Geometry at

Oxford, and appointment as Astronomer Royal in 1720.
DSB 6, 67–72. C.A. Ronan. DNB.

SIR WILLIAM BATE HARDY 1864–1934; FRS 1902

William Hardy was born at Erdington, Warwickshire and after attending Framlingham College he entered Gonville and Caius College, Cambridge. In 1888 he graduated first class in the Natural Science Tripos in Zoology. He was awarded a research scholarship and in 1892 became a College Fellow and later a Lecturer. In 1898 he married Alice Mary Finch and they had one son and two daughters. Hardy was appointed a University Demonstrator and then a Lecturer in Physiology in 1913. Around 1907 he became interested in molecular physics and set out to extend his mathematical knowledge. After the 1914–18 war his interests were extended to static friction and hence to lubrication. In the Royal Society Hardy was Biological Secretary (1915–25) , Vice-President (1914–15), received a Royal Medal (1926) and was Croonian Lecturer (1905) and Bakerian Lecturer (1925); his Bakerian Lecture was concerned with friction. He was knighted in 1925.
ONFRS 1932–35, 1, 327–333. FGH, FES. DNB 1931–40, 397, 398 A.V. Hill.

JOHN ALLEN HARKER 1870–1923; FRS 1910

John Harker was born at Alston, Cumberland, the son of the Rev. John Harker. He attended Stockport Grammar School, studied at Owens College, Manchester and was awarded a PhD at Tubingen. In his career he did important work on standards of temperature measurement using platinum resistance thermometers. He worked at Kew and at Sevres, and when the National Physical Laboratory was established in 1899 he became a member of staff; at the NPL his career included appointments as Head of the Heat Division and as Chief Assistant of the Laboratory. From 1916–21 he worked at the Ministry of Munitions, becoming Director of Research; he visited the USA on behalf of the Ministry and on the return voyage in 1918, the ship on which he was travelling was torpedoed. Harker was also associated with the firm J.T. Crowley, Consulting Engineers. He married Ada Richardson and they had two sons and three daughters. He was awarded the DSc degree and an OBE. Of notable metallurgical interest was his paper reported in 1905 on the construction of a furnace for achieving very high temperatures, involving the passage of electric current through the ceramic tube of the furnace.
PRS 1924 A, 105, xi–xiii. RTG.

JOHN EDWIN HARRIS 1932–; FRS 1988

Larkfield Grammar School, Chepstow. Univ. of Birmingham: BSc 1953, PhD 1956, DSc 1973. Associated Electrical Industries: 1956. CEGB 1959–80, seconded to Univ. of Sheffield, 1959–61. Berkeley Nuclear Laboratories: 1961–89; Section Head, 1966–69. Nuclear Electric plc: University Liaison Manager, 1988–90. Consultant and free lance Lecturer and writer, 1990–. Royal Society: Esso Gold Medal, 1979. FREng 1987. MBE, 1981.
Physical metallurgy of components of nuclear reactors under severe service conditions, deformation, creep and corrosion.

WILLIAM HARRISON 1728–1815; FRS 1765

William Harrison, like his father John, with whom he entered into partnership, was a watchmaker and clockmaker. In 1761 he sailed on a trial voyage to Jamaica for one of the marine chronometers, but although some successful measurements of longitude were made the rate of the watch

(loss or gain in timekeeping) was not known. There were disagreements between the Harrisons and the Board of Longitude, whose members were not satisfied with the outcome of the trial. In 1764 another voyage led to success. However, there were ongoing disputes before the full award was received by the Harrisons.
J. Betts. *John Harrison*, National Maritime Museum, Greenwich 1993.

MICHAEL HART 1938–; FRS 1982
Cotham Grammar School, Bristol. Univ. of Bristol: BSc, PhD, DSc. Cornell Univ.: Dept of Materials Science and Engineering; Research Associate, 1963–65. Univ. of Bristol: Lecturer in Physics; 1967–72; Reader in Physics, 1922–76. King's College, London: Wheatstone Professor of Physics and Head of Physics Dept, 1976–84; Daresbury Lab. SERC; Science Programme Coordinator (part time secondment) 1985–88; Univ. of Manchester: Professor of Physics, 1984–93. de Montford Univ.: Professor of Applied Physics, 1993–. CBE 1993. National Synchrotron Light Source, Brookhaven National Lab: Chairman 1995–.
X-ray optics, topography; defects in crystals; synchroton radiation.

SIR WALTER NOEL HARTLEY 1847–1913; FRS 1884
Walter Hartley was born in Lichfield, the son of Thomas Hartley and Caroline née Lockwood. He was educated privately prior to entering Edinburgh University in 1863, intending to study medicine. Having attended first year classes in chemistry he decided to pursue this subject and went to Germany to study at Marburg (1864-65). Following his return to England he worked at Manchester University and at the Royal Institution London; then in 1871 he became Senior Demonstrator at King's College, London. In 1879 he was appointed as Professor of Chemistry at the Royal College of Science in Dublin, and remained in this post until his retirement. He married Mary Lafley in 1882 and they had one son. Hartley's main field of research was in spectrography; his work included the interrelationships of the spectra of elements, including beryllium, on the basis of the Periodic classification. Of particular metallurgical interest was his investigation of the spectra of the flames seen during steelmaking by the Bessemer converter. Hartley was knighted in 1911.
PRS 1914 A, 90, vi–xiii. JYB.

CHARLES HATCHETT 1765–1847; FRS 1797
Charles Hatchett was the son of John Hatchett and his wife Elizabeth. John was a wealthy coachbuilder of Long Acre, London and gave his son a generous financial allowance. Charles married Elizabeth Colleck in 1786, and around 1800 he started a chemical manufacturing business near Chiswick, where soon after he took Thomas Brande as a young man into his laboratory. Hatchett carried out important work between 1796 and 1806, analysing a mineral, now known as columbite or niobite and reporting that it contained a previously unknown metal, which he named columbium. He acquired a reputation in Great Britain and on the Continent as a general analyst but also did important work on organic materials. Hatchett was awarded the Copley Medal in 1798.
DSB 6, 1972, 166, 167. E.L. Scott. DNB.

WILLIAM HERBERT HATFIELD 1882–1943; FRS 1935
William Hatfield, the son of Francis Hatfield and Martha née Sheppard, was a native of Sheffield. After leaving school he entered the laboratory of Messrs Henry Bessemer and Co., while attend-

ing evening classes in the University College, where John Arnold was Head of the Metallurgy Department. Hatfield was awarded the Mappin Medal in 1902 and later held a Carnegie Research Scholarship. In 1907 he married Edith Mariam Seagrave.

Hatfield worked with the ironfounding firm of John Crowley and Company, where he was largely concerned with malleable castings; in 1910 he became a Director and Joint Works Manager. A comprehensive study of the chemistry of cast iron was presented as a book, first published in 1912. In 1913 he was awarded the DMet degree of Sheffield University. In 1916 he became Director of the Brown–Firth Research Laboratories, a post in which he built up an active research centre covering a wide range of special or alloy steels, including armour plate, guns, marine forgings, high pressure boiler drums, turbine and aircraft materials, tool steels and various other products.

In alloy development Hatfield's name is particularly associated with the austenitic nickel–chromium (18:8) stainless steels, involving extensive investigations of properties. Creep behaviour was an area in which Hatfield did important work, including the introduction of the 'time-yield' test, a relatively short time test, which was used to provide an indication of longer term behaviour. He also did some work on the damping of engineering materials and light alloys for aircraft.

A prominent and successful role was played by Hatfield as chairman in committee work, stemming from the Advisory Council for Scientific and Industrial Research and the Iron and Steel Institute, and associated for example with important research programmes on the heterogeneity of steel ingots, alloy steels and corrosion respectively.

In the 1939–45 war Hatfield in his Company was involved in many problems relating to armaments and engineering. The work on research committees continued and he also served on various committees of government departments; he chaired the committee responsible for the examination of the construction and materials of enemy aircraft. During the whole period of the war he did not take a holiday; he experienced severe strain during a visit to the United States as one of a delegation of British metallurgists, and his death occurred soon after.

ONFRS 1942–45, 4, 617–625. C.H. Desch.

FRANCIS HAUKSBEE c.1666–1713; FRS 1705

Francis Hauksbee was one of the sons of Richard Hauksbee, a draper of Colchester. In 1678 he was apprenticed in the drapery trade to his elder brother. He finished his apprenticeship in 1687 and in the same year he married Mary, by whom he had several children. He became recognised as an instrument maker and skilled experimenter. From 1703–1713 Hauksbee reported his discoveries at meetings of the Royal Society, and his publications in *Philosophical Transactions* were brought together in *Physico-mechanical Experiments on Various Subjects* (1709). After 1704 he served as Curator of Experiments at the Royal Society. Of particular importance was his research on electricity, but he also worked on capillarity and measurements of specific gravity.

DSB 6, 169–175. H. Guerlac. DNB.

RENÉ-JUST HAÜY 1743–1822; FMRS 1818

René-Just Haüy was born in St-Just-en-Chaussée, France, the son of a poor weaver. Wth a scholarship to the Collège de Navaire in Paris, he obtained a classical and theological education and became a Regent there in 1764. He was ordained in 1770 and took a teaching post at the Collège Cardinal Lemoine. Haüy undertook studies in botany and then became interested in, and published papers on mineralogy. During the Revolution he experienced difficult times but became

Secretary of the Commission on Weights and Measures. In 1795 he started to teach physics and mineralogy courses at the École des Mines. His main work, *Traité de minéralogie,* published in 1801, presented his crystal theory and mineral classification. In 1802 he became Professor of Mineralogy at the Muséum d'Histoire Naturelle; appointment to the newly created Chair of Mineralogy at the Sorbonne took place in 1809.
DSB 6, 178–183 W. LeFanu.

SIR JOHN HAWKSHAW 1811–1891; FRS 1855
John Hawkshaw was born in Leeds, the son of Henry Hawkshaw and his wife née Carrington. He was educated at Leeds Grammar School and then became a pupil of an engineer with a large practice in Yorkshire. From 1832–34 he was in Venezuela, superintending some large copper mines. He married Ann Jackson in 1835. Hawkshaw worked initially in railway engineering and in 1845 became Engineer to the Manchester and Leeds Railway. In 1850 he came to London, where he set up as a consulting engineer. His numerous and wide ranging activities in civil engineering included the construction of the Charing Cross and Cannon Street railway bridges, consulting for the original Channel Tunnel Company, work on docks in London and Hull, foundations of forts at Spithead, and works abroad. Hawkshaw served as a Royal Commissioner on important enquiries, and also served on various government committees, including one relating to the use of steel in engineering structures. Hawkshaw was one of the foremost civil engineers of the 19th century. In the Royal Society he served on the Council in 1868–69, 1874–75 and 1881–82; he was knighted in 1873.
PRS 1891–1892, 50, i–iv. WP. DNB Suppl. v. II, 402–404. THB.

LIONEL HAWORTH 1912–; FRS 1971
Rondebosch Boys' High School. Univ. of Cape Town. Cape Town Corp's Gold Medal and Scholarship tenable abroad. BSc(Eng). Associated Equipment Co.: Graduate Apprentice, 1934. Rolls-Royce Ltd, Derby: Designer, 1936; Assistant Chief Designer, 1944; Deputy Chief Designer, 1951; Chief Designer (civil engines), 1954; Chief Engineer (prop. turbines), 1962. Bristol Siddeley Engines Ltd: Chief Design Consultant, 1963; Chief Designer, 1964; Director of Design; Aero Division, 1965. Rolls-Royce Ltd: Director of Design Aero Division, 1968–77. Lionel Haworth and Associates; Senior partner. FREng. 1976. OBE 1958.
Aero engines; engineering metallurgy.

SIR WALTER NORMAN HAWORTH 1883–1950; FRS 1928
Walter Haworth was born in Chorley, Lancashire, the second son and fourth child of Thomas and Hannah Haworth. He was educated at a local school and at the age of 14 began work at the linoleum factory which his father managed. He decided to become a chemist and, assisted by a private tutor, he passed the entrance examination to Victoria University, Manchester, where he obtained a first class honours degree in chemistry in 1906. He then worked in the field of organic chemistry, and spent a period at Göttingen with an 1851 Exhibition Scholarship; after a year back in Manchester he joined the chemistry staff of the Imperial College of Science and Technology in London in 1911. A year later he moved to St Andrews as a Lecturer. In 1920 he was appointed to the Chair of Organic Chemistry at Armstrong (later Kings) College at Newcastle upon Tyne. In 1922 he married Violet Chilton Dobbie and they had two sons. From 1925–1948 Haworth was Mason Professor of Chemistry at Birmingham University. His distinguished career as an organic chemist included research on sugars and carbohydrates. From 1939–45, as Chairman of the Brit-

ish Chemical Panel for Atomic Energy, he directed research on the production of uranium. In the Royal Society he was Vice-President in 1947; he was awarded the Davy Medal in 1934 and a Royal Medal in 1942. He also received the Nobel Prize for Chemistry in 1937 and was knighted in 1947.
ONFRS 1950–51, 7, 373–404. E.L. Hirst. DSB 6,184–186. Sheldon J. Kopperl. DNB 1941–50, 368, 369. L.L. Bircumshaw.

ALAN KENNETH HEAD 1925–; FRS 1988
Ballarat Grammar School. Scotch College. Univ. of Melbourne: BA, BSc, DSc. Univ. of Bristol: PhD. Commonwealth Scientific and Industrial Research Organization: Division of Aeronautics, Research Scientist, 1947–50. Aeronautical Research laboratories: 1953–57. CSIRO: Division of Tribophysics, 1957–81; of Chemical Physics, 1981–86; of Materials Science, 1987; Chief Research Scientist, 1969–90. Commonwealth Scientific Research Organisation, Australia: Hon. Research Fellow, 1990–.
Dislocations in crystals; surfaces and crystal anisotropy. Computer simulation of dislocation images in the electron microscope.

VOLKER HEINE 1930–; FRS 1974
Otago Univ: MSc, DipHons. Univ. of Cambridge: PhD; Demonstrator, 1958–63; Lecturer 1963–70; Reader in Theoretical Physics, 1970–76; Professor of Theoretical Physics 1976–1997. Royal Medal 1993.
Electron theory of solids; bonding and structure of metals and alloys; insulators and minerals.

WERNER KARL HEISENBERG 1901–1976; FMRS 1955
Werner Heisenberg was born in Wurzburg, Germany, the younger son of August and Anna Wecklein Heisenberg; his father taught in the university. Werner attended primary school in Wurzberg, and then in Munich when his family moved there in 1910. He attended the Maximilians-Gymnasium, where he showed outstanding ability in mathematics. In 1920 he entered the University of Munich and obtained a doctorate in theoretical physics under Arnold Sommerfeld in 1923. He then became assistant to Max Born at Göttingen and worked with Neils Bohr in Copenhagen from 1924–26. From 1927–41 he occupied the Chair of Theoretical Physics at Leipzig. In 1937 he married Elisabeth Schumacher and they had seven children. From 1942–45 Heisenberg was Director of the Max Planck Institute for Physics in Berlin. He was in charge of atomic research in Germany during the war and in 1946 became Director of the Max Planck Institute for Physics in Göttingen; the Institute was moved to Munich in 1958 and he was Director until 1970.

Heisenberg was the founder of quantum mechanics, and enunciated the uncertainty principle. His research included studies of ferromagnetism. He was the recipient of the Nobel Prize for Physics in 1932.
BMFRS 1977, 23, 213–251. Sir Nevill Mott FRS and Sir Rudolf Pieirls FRS. DSB Suppl., **17**, 394–403 D.C. Cassidy.

HENRY SELBY HELE-SHAW 1854–1941; FRS 1899
Henry Hele-Shaw was the eldest of 13 children of Henry Shaw, a solicitor, and Maria née Hele. He was born at Billericay, Essex and was privately educated. When he was 17 he was apprenticed at an engineering works at Bristol, where he worked long days and studied at evening classes. In 1876 he was awarded the first of a number of Whitworth Prizes and he became a student at University College, Bristol, where he graduated first on the list in 1880. He then became a Lec-

turer in Mathematics and Engineering at the College, and was appointed Professor in 1881. From 1885–1906 Hele-Shaw occupied the Chair of Engineering at the University College of Liverpool, and then spent two years in South Africa founding the Transvaal Technical Institute. He then worked as a consultant engineer. He married Ella Marion Rathbone in 1890 and they had one son and one daughter. Hele-Shaw's research included hydrodynamics, friction phenomena and automobile engineering; he was the inventor of a friction clutch.
ONFRS 1939–41, 3, 791– 811. H.L. Guy. DNB 1941–50, 373, 374. D.G. Christopherson.

THOMAS HETHERINGTON HENRY c.1816–1859; FRS 1846
Thomas Henry was the second son of Jacob Henry. For many years from 1847 as a chemist he was Superintendent at the brewery of Truman Hanbury and Buxton in Spitalfields, London. Later he set up as a consulting analytical chemist in Lincoln's Inn. His work included electroplating with platinum and palladium.
Bulloch's Roll. Ref. Ch. 4.1.

WILLIAM HENRY ?–1768; FRS 1755
William Henry probably came from Gloucestershire. He was educated at the University of Dublin, where he obtained an MA (1748) and subsequently (1750) the degrees of BD and DD. He was collated to the benefice of Killesher, County Fermanagh (1731); in 1734 he became Rector of Urney in County Tyrone, and in 1761, Dean of Kilaloe. He was a keen observer of natural history. Of metallurgical interest was his paper in *Philosophical Transactions* in 1751 concerning a copper mine in County Wicklow.
Bulloch's Roll.

SIR FREDERICK WILHELM (WILLIAM) HERSCHEL 1738–1822; FRS 1781
William Herschel was born in Hanover, Germany, the third of six surviving children in his family. His father Isaac Herschel, a military musician, arranged for William to join his regimental band as an oboist at the age of 15. In 1757 William came to England where he worked as a musician, first in Halifax and then in Bath, as an organist. His interests turned increasingly to astronomy, and for a decade from 1782, with his sister Caroline, he was active in this field, including the assembly of telescopes. He began grinding mirrors and used reflecting telescopes, and in order to increase the light-gathering power he set about making larger metal mirrors. In 1781 he was unsuccessful in making a mirror nearly 1 m in diameter, but some years later, with financial suppport from the King he produced one of nearly 1.2 m diameter. In 1782 he was appointed court astronomer. In 1788 he married Mary Pitt née Baldwin and in 1792 their only son John Frederick William Herschel was born. William's career included the discovery of the planet Uranus. He received the Copley Medal in 1781 and was knighted in 1816.
DSB 6, 328–336. M.A. Hoskin. DNB.

SIR JOHN FREDERICK WILLIAM HERSCHEL 1792–1871; FRS 1813
John Herschel was born at Slough, the only child of William and Mary Herschel. His education included a period at Eton and the study of mathematics at St John's College, Cambridge, where he entered as a scholar at the age of 17 and in 1813 was Senior Wrangler and first Smith's Prizeman. In the same year he was elected to a college fellowship and also as a Fellow of the Royal Society for his mathematical work. Herschel was also expert in chemistry, and in 1815 missed election to the Chair of Chemistry at Cambridge by just one vote. Unsure of his career choice he enrolled in the legal profession at Lincoln's Inn in 1814 where he developed his ac-

quaintance with William Wollaston. However, in 1815 he took a minor teaching post at St John's College, but left Cambridge the following year having taken his MA and after discussion with his father, who at 78 years of age wished to see his astronomical work continued. He married Margaret Brodie Stewart in 1829 and they had 12 children.

Herschel took over from his father various astronomical and instrumental technique and among his activities he made the 0.5 m mirror of speculum metal for his 7 m telescope. His research covered mathematics, and physical and geometrical optics, and birefringence of crystals. He travelled widely throughout the world, including journeys for meteorological and geological expeditions.

In the Royal Society Herschel was awarded the Coplcy Medal in 1821 and 1847, and a Royal Medal in 1833, 1836, and 1840; he served as Secretary from 1824 to 1827 and was elected President in 1845. His appointment as Master of the Mint took place in 1850 and lasted for five years. He was created a Baronet in 1838.

PRS 1871–1872, 20, xvii–xxiii. TRR. DSB 6, 323–328. D.S. Evans.

CHARLES THOMAS HEYCOCK 1858–1931; FRS 1895

Charles Heycock was born at Bourn, near Cambridge, the younger son of Frederick Heycock of Braunston, Oakham and received his early education at the Grammar Schools of Bedford and Oakham. After leaving school he spent a year at home, but managed to attend some lectures in chemistry at the University of Cambridge. Heycock spent the summer vacation of 1876 in the Cavendish Laboratory, where, with an undergraduate, he investigated the spectrum of indium, and was the joint author of a paper. Early in 1876 he was commissioned as a Lieutenant in the Third (Cambridge) Volunteer Battalion of the Suffolk Regiment. Having entered King's College, Cambridge as an Exhibitioner in 1877 he took the Natural Sciences Tripos in 1880. He then worked in association with Ernest Griffiths in a coaching partnership, and it was through Griffiths that he made contact with Francis Henry Neville, who became a lifelong friend and research colleague. For many years Heycock taught chemistry, physics and mineralogy for the Cambridge examinations, and in 1895 he was elected to a Fellowship at King's College. His research on alloys attracted the attention of the Goldsmiths' Company who endowed a Readership in Metallurgy at Cambridge; George Matthey played a leading role in this. Heycock was appointed in 1908 and held the post until his retirement in 1928. He was admitted to the Livery of the Goldsmiths' Company in 1900 and to the Court in 1913; he acted as Prime Warden during the year 1922–1923 and took a keen interest in the work of the Company's Assay Office. He married Caroline Sadler and they had one son and two daughters.

Heycock joined with Francis Neville in a comprehensive study of metals and their alloys; this partnership continued until the death of Neville in 1915. Their first joint paper in 1884 described a re-determination of the molecular weight of ozone. In the field of phase equilibria and microstructure in alloys, beginning with a publication in 1889, there ensued a remarkable series of papers in which novel directions of investigation were developed. Heycock was awarded the Davy Medal of the Royal Society in 1920. The major part of Heycock and Neville's experimental work was carried out in the small laboratory in Sidney Sussex College and as an addition to their many other duties. A large amount of valuable data stemmed from their high experimental skills and meticulous aim for accuracy.

Heycock was an excellent lecturer and had few equals as a teacher in the laboratory. He carried out much of the work of organising and planning the numerous extensions of the University Chemical Laboratories over some 25 years. He was active in the Volunteer Movement and during

the 1914–1918 war he was appointed Colonel of the Cambridge Regiment. He was appointed Deputy Lieutenant for Cambridge in 1921.
PRS 1932 A, 135, i–iii. W.J. Pope FRS. L.B. Hunt, *Heycock and Neville, Metallurgist and Materials Technologist*, July 1980, 392–394. Ref. Ch. 10.4.

JULIA STRETTON HIGGINS 1942–; FRS 1995
Univ. of Oxford: Somerville College, MA, DPhil, 1968. Mexborough Grammar School: Physics Teacher, 1966–68. Univ. of Manchester: Department of Chemistry, 1968–72. Centre de Recherche sur les Macromolecules, Strasbourg: Research Fellow, 1972–77. Inst. Laue–Langevin, Grenoble: Physicist, 1973–76. Imperial College, London: Department of Chemical Engineering, 1976–; Reader; Professor of Polymer Science, 1989–; City and Guilds College; Dean, 1993–96. Royal Society Council 1998–. CBE 1996. FREng 1999.
Polymers; neutron scattering.

SIR JOHN MCGREGOR HILL 1921–; FRS 1981
London Univ: King's College. Univ. of Cambridge: St John's College. Royal Air Force: Flight Lieutenant, 1941. Cavendish Laboratory, 1946. London University: Lecturer, 1948. United Kingdon Atomic Energy Authority: joined 1950; Member for Production, 1964–67; Chairman, 1967–81. British Nuclear Fuels plc: Chairman, 1971–83. FREng 1982. Knighted 1969.
Atomic energy.

RODNEY HILL 1921–; FRS 1961
Leeds Grammar School. Univ. of Cambridge: Pembroke College, MA, PhD, ScD. Armament Research Department, 1943–46. Cavendish Laboratory, 1946–48. BISRA, 1948–50. Bristol Univ.; Research Fellow, 1950–53; Reader, 1953. Univ.of Nottingham: Professor of Applied Mathematics, 1953–62; Professorial Research Fellow, 1962–63. Univ. of Cambridge: Gonville and Caius College, Berkeley bye-Fellow, 1963–69; Fellow 1972–; Reader in Mechanics,1969–72; Professor of Mechanics of Solids, 1972–79. Royal Medal 1993.
Theoretical mechanics of solids; plasticity; slip–line field theory-metal forming and machining; stress and strain in polycrystalline aggregates.

CYRIL HILSUM 1925–; FRS 1979
Raines School, London. Univ. of London: University College. Admiralty H.Q., 1945–47. Admiralty Research Laboratory, Teddington, 1947–50. Services Electronics Research Laboratory, Baldock: 1950–64. Royal Signals and Radar Establishment, Malvern: 1964–83. General Electric Co. (GEC): Research Laboratories: Chief Scientist, 1983–85. GEC Plc: Director of Research, 1985–92; Corporate Research Adviser, 1992–96. Royal Society: Clifford Paterson Lecturer, 1981; Philips Lecturer, 1989. FREng 1978. CBE 1990.
Semiconducting compounds; liquid crystals.

CHRISTOPHER HINTON (BARON HINTON OF BANKSIDE) 1901–1983; FRS 1954
Christopher Hinton was born in Tisbury, Wiltshire, the third of four children of Frederick Harvey Hinton, headmaster of the village school and Kate née Christopher. When the family moved to Chippenham he completed his schooling, leaving at the age of 16 to become an Engineering Apprentice, initially with a small firm and then with the Great Western Railway at Swindon. In addition to a long working week, at one stage 57 hours, he studied in the evening at Swindon Technical College. In 1923 he won a scholarship from the Institution of Mechanical Engineers

and entered Trinity College, Cambridge. There he won university and college awards obtaining a first class degree in mechanical science in two years, and spending his third year in research. Hinton then joined Brunner Mond, soon to become the Alkali Group of the new Imperial Chemical Industries, and at the age of 29 he became the Chief Engineer. In 1931 he married Lilian Boyer, and they had one daughter. He became Director of Ordnance Factory Construction at the Ministry of Supply in 1940, and other senior appointments in war production ensued.

After the war, Hinton embarked on a career in the Atomic Energy Programme set up by the Government; from his headquarters, which was in Risley, there developed the Industrial Group of the British Atomic Energy project of which he became Managing Director. In 1954 became Board Member for Engineering and Production of the Atomic Energy Authority. He served as Chairman of the Central Electrical Generating Board from 1957–64. He was Chancellor of the University of Bath from 1966–1980. Hinton's honours included the Rumford Medal of the Royal Society (1970), a knighthood in 1951, the award of the KBE in 1957, a life peerage on his retirement in 1964, and the Order of Merit in 1976.

BMFRS 1990, 36, 219–239. Margaret Gowing FRS. DNB 1981–85, 195–196. M. Gowing.

SIR PETER BERNHARD HIRSCH 1925–; FRS 1963

Sloane School, Chelsea. Univ. of Cambridge: St Catherine's College, BA, 1946; MA, 1950; PhD, 1951; Reader in Physics, 1964–66. Univ. of Oxford: Isaac Wolfson Professor of Metallurgy, 1966–92. Hughes Medal, 1973; UK AEA: 1982–89 (part time member 1982–84). Electricity Supply Council: Member 1969–82. Royal Medal, 1977. Knighted 1975.

Electron microscopy of thin crystals; imperfections in crystalline stuctures and their relationship to mechanical properties.

JAMES HOARE ?–1696; FRS 1664: JAMES HOARE ?–1679; FRS 1669

James Hoare (FRS 1664) was one of a group of distinguished goldsmiths; he established the sign of the Golden Bottle at Cheapside and was the founder of Hoare's bank, and banker to Oliver Cromwell. He became Surveyor, Warden and Controller at the Mint, while retaining his goldsmiths business at Cheapside. He had to appoint a Deputy at each of five temporary mints opened in county towns for the recoinage of clipped and obsolete silver, which started in 1696; at Chester he chose for Deputy Controller the astronomer, Edmund Halley. Hoare's son (also James) was a barrister and also became an FRS (1668) and held appointments at the Mint, including being joint Controller with his father.

Notes and Records of the Roy. Soc. 1964. Bulloch's Roll.

DOROTHY MARY CROWFOOT HODGKIN 1910–1994; FRS 1947

Dorothy Hodgkin was the daughter of John Winter Crowfoot and Grace Mary née Hood. She was born in Cairo where her father was working in the Egyptian education service. When the family returned to Britain she was educated at the Sir John Leman School at Beccles, Suffolk and went on to enter Somerville College, Oxford, where she studied chemistry and archeaology. Following her graduation she spent a period at Cambridge (1932–34) where she worked with Bernal on X-ray crystallography. She returned to Oxford as a Lecturer and Demonstrator in 1934. Three years later she married Thomas Hodgkin and they had two sons and one daughter. In 1956 Hodgkin was appointed Reader in X-ray Crystallography and from 1960–1977 she held the post of Wolfson Research Professor of the Royal Society. From 1970–88 she was Chancellor of Bristol University. Her distinguished research in X-ray crystallography was concerned with many complex biochemical molecules, and the structures that she elucidated included penicillin (1949), vitamin

B_{12} (1956) and insulin (1969). Hodgkin received a Royal Medal in 1956, and the Copley Medal in 1976; in 1964 she was awarded the Nobel Prize for Chemistry and in the following year was admitted to the Order of Merit.
Nature, 1984, 371, 20. Max Perutz. *The Daily Telegraph*, 1.8.94. *The Guardian*, 1.8.94. Guy Dodson.

EATON HODGKINSON 1789–1861; FRS 1841
Eaton Hodgkinson, the son of a farmer, was born at Anderton in the parish of Great Budworth, Cheshire. When Eaton was six years old his father died but his mother carried on the farm and he went to a school in Northwich, with a view to preparation for university and a career in the church; he disliked his classical studies and was moved to a private school in the same town where his natural ability in mathematics developed. He had to assist his widowed mother with the family farm and later in 1811, when they moved to Manchester, with the running of a pawnbroking business in which his mother invested her small amount of capital. In Manchester, where he extended his knowledge of mathematics and mechanics by private study he came to know John Dalton and other talented men then living there. Dalton guided and encouraged him and they studied together the work of Euler, Lagrange and Laplace. Hodgkinson was twice married, first to Catherine Johns and then to the daughter of Henry Holditch; there were no children.

Hodgkinson carried out important research on the mechanical properties of materials e.g. reported in his papers 'On the Transverse Strain and Strength of Materials' (1822) and 'On the Strength of Pillars of Cast Iron and other Materials' (1840); the latter led to the award of a Royal Medal in 1841. Together with William Fairbairn he assisted Robert Stephenson in experiments to determine the best form of girders for Menai Straits bridge. He edited the fourth edition of Tredgold's work on the strength of cast iron, and published a volume of his own in 1846 *Experimental Researches on the Strength and other Properties of Cast Iron.* From 1847 to 1849 he was one of the Royal Commissioners to inquire into the application of iron to railway structures. He also made the first sytematic investigation of metal fatigue. In 1847 he was appointed Professor of the Mechanical Principles of Engineering at University College, London. Through his research and publications on experimental and analytical aspects of the theory of elasticity and strength of materials Hodgkinson became the leading British authority in these fields.
PRS 1862–63, 12, xi–xiii. DSB 6, 452, H. Dorn. Gen. Ref. 21.

AUGUST WILHELM VON HOFMANN 1818–1892; FRS 1851
August von Hofmann was born in Giessen, Germany, the son of an architect, Johann Philipp Hofmann. Following his matriculation in 1836 he chose to study law and languages, but became attracted by Liebig's classes. He was awarded a doctorate in 1841 and later acted as personal assistant to Liebig prior to becoming Privadozent at the University of Bonn in 1845. Later that year he agreed to direct the Royal College of Chemistry in London, which was absorbed by the School of Mines in 1853; he also had an association with assaying activities at the Mint. Hofmann remained in London for 20 years until he took the Chair of Chemistry at Berlin. He was made a Baron on his seventieth birthday. He was married four times and had 11 children.

Hofmann was awarded the Copley Medal in 1875; he had a profound influence on chemistry in England and Germany, applying the method of science teaching by laboratory instruction established by Liebig, and also building up his own school of chemists, primarily in the experimental organic field and industrial applications of chemistry.
DSB, 6, 461–464 W.H. Brock.

ERNEST DEMETRIUS HONDROS 1930–; FRS 1984

Univ. of Melbourne: DSc, MSc. Univ. of Paris: Dr.d'Univ. CSIRO: Tribophysics Lab., Melbourne: Research Officer, 1955–59. Univ.of Paris: Lab de Chimie Minerale, Research Fellow, 1959–62. National Physical Laboratory Teddington: Senior Research Officer, Metallurgy Division, 1962–65; Principal Research Officer, 1965–68; Senior Principal Research Office (special merit), 1974; Superintendent Materials Applications Division. Petten Research Establishment Commission of European Communities, Joint Research Centre, Director,1985–1995. Institute of Advanced Materials, Petten (Netherlands) and ISPRA(Italy): 1988–95. CMG 1996.

Structure,chemical composition and energy of surfaces and interfaces of solids.

SIR ROBERT WILLIAM KERR HONEYCOMBE 1921–; FRS 1981

Geelong College, Australia. Univ. of Melbourne: Department of Metallurgy, Research Student, 1941–42. Council of Scientific and Industrial Research Organization, Australia: Research Officer, 1942–47. Univ. of Cambridge: Cavendish Laboratory: ICI Research Fellow, 1948. Royal Society Armourers' and Brasiers' Research Fellow 1949–51. Sheffield Univ.: Senior Lecturer in Physical Metallurgy, 1951–55; Professor, 1955–66. Univ. of Cambridge: Goldsmith's Professor of Metallurgy, 1966– 92; Clare Hall; President 1973–80. Royal Society: Vice-President, 1986–92; Treasurer, 1986–92. FREng 1980. Knighted 1990.

Plastic deformation; dislocation interactions in single crystals; micro-alloyed steels, microstructure and properties.

ROBERT HOOKE 1635–1703; FRS 1663

Robert Hooke was born at Freshwater on the Isle of Wight, the son of a clergyman, and intended for the ministry by his father; as a boy he was interested in 'mathematical magic', mechanical objects, sundials, clocks, water mills and model ships. In 1648 when his father died he inherited £100, and in view of his having shown some artistic talent his family sent him to London to serve an apprenticeship with Sir Peter Lely. Hooke however did not follow this interest and was befriended by the Master of Westminster School, who took him into his home. Here he learnt Latin, Greek and some Hebrew; he also read the first six books of Euclid in a week, and went on to apply geometry to mechanics; a further activity was to learn the organ.

He moved on to Oxford as a subsizar, entering Christ Church in 1653 as a chorister. Apparently he did not take a bachelor's degree but was nominated MA in 1663. He joined a group of brilliant men, including Robert Boyle, who met regularly to discuss scientific matters. He became assistant to Boyle. In 1658, at the time when he developed the air pump, he became interested in chronometers in connection with which he made important inventions.

When the Royal Society was founded in London he was appointed as the first Curator of Experiments in November 1662, with the task of providing each of the weekly meetings 'with three or four considerable experiments'. He was outstandingly inventive, generating a series of brilliant ideas, and providing experiments, demonstrations and discourses. In 1664 a lectureship in mechanics with an annual salary of £50 was founded for him and the Royal Society appointed him to the position of Curator for life at a salary of £30 and lodging at Gresham College, which later came to house the Royal Society; he remained in residence there until his death. For a period he was in charge of the Society's repository of rarities, and he also served as librarian. A further responsibility in 1665 was his appointment as Gresham Professor of Geometry, with a salary of £50. He was Secretary of the Royal Society from 1677–1682.

In 1666, after the great fire of London, Hooke's plan for a rectangular grid layout for the city was accepted although it was not implemented. Hooke was appointed as one of the three surveyors to reestablish property lines and to supervise the rebuilding. Hooke was able to use his artistic talents and also earned large sums of money. The period of 10 years following the fire was also a time of productive scientific work.

Hooke's special field of research field was mechanics, but during his life he worked in various other fields: the theory of light; combustion; respiration and the cellular structure of plants. A great microscopist, he constructed one of the most famous of early compound microscopes and produced in 1655 his most important book, *Micrographia,* a scientific milestone. His work on elasticity led to the law of proportionality of strain to stress that bears his name.

Hooke's activities contributed greatly to the success of the Royal Society. Unfortunately controversy with Newton was a feature of his career, initially stemming from work on the theory of light. He was a man of genius and has been described as the greatest experimental physicist before Faraday.

DSB 6, 1972, 481–488. R. S. Westfall. DNB. Gen. Refs 20 v. III, 25 II. Ref. Ch. 2.8.

SIR STANLEY GEORGE HOOKER 1907–1984; FRS 1962

Stanley Hooker was born on the Isle of Sheppey, the youngest of nine children of William Harry Hooker, inn keeper and later corn miller, and Ellen Mary née Russell. From Borden Grammar School Hooker went to Imperial College to study mathematics in 1926 with a Royal Scholarship. He was awarded an Armourers and Brasiers' Research Fellowship in 1930 and a DPhil at Brasenose College, Oxford in 1935. He then worked at the Admiralty Laboratories, Teddington, and at Woolwich Arsenal. In 1937 he married Margaret Bradbury, and they had one daughter; following the dissolution of the marriage he married Kate Maria Pope in 1950 and they also had a daughter.

Hooker joined Rolls-Royce at Derby in 1938. There he developed a supercharger for the Merlin 60 engine which enhanced the performance of the *Spitfire* fighter and played an important part in the winning of the air war. Later he became Chief Engineer of a team at Barnoldswick on the production of Whittle jet engines, and a squadron of RAF Meteors was delivered in July 1944. Hooker also worked on turbo-props. In 1948 he moved to Bristol Siddeley where his work included turbo-jets and he became Chief Engineer. After his retirement in 1970 he served for a period as Technical Director for Rolls-Royce. He was awarded the Royal Society Leverhulme Medal in 1981. He was appointed OBE in 1946, CBE in 1964 and was knighted in 1974.

BMFRS 1986, 32, 277–319. R.H.J. Young FRS, L. Haworth FRS, H. Pearson, G.L. Wilde and J.E. Ffowcs-Williams. DNB 1981–85 199, 200. L. Haworth.

BERTRAM HOPKINSON 1874–1918; FRS 1910

Bertram Hopkinson was born at Woodlea, Birmingham, the eldest son of John Hopkinson FRS and Evelyn née Oldenbourg. The family soon moved to London where his father became a consulting enginéeer and inventor. Bertram was educated at St Paul's School and then at the age of 17 he entered Trinity College, Cambridge with a major scholarship to study mathematics. Illness prevented him from taking the first part of the Tripos but in the final part he was placed in the first division of the first class. Having obtained his degree in 1895 he proceeded to read for the Bar to which he was called in 1897. However, when his father was killed in a climbing accident the following year he turned to engineering to take up as far as he could his father's unfinished work. Hopkinson acquired a considerable reputation as an engineer and was appointed to the Chair of Applied Mechanism and Applied Mechanics at Cambridge in 1903. In the same year he married

Mariana Siemens and they had seven daughters. Under his direction the School of Engineering at Cambridge expanded and its position was advanced.

Hopkinson's research activities included investigations on the fatigue of metals, the electrical and magnetic properties of iron and its alloys, and explosions in gases. In 1914 he became a Professorial Fellow of King's College and when the war began he was commissioned in the Royal Engineers, where he was involved in experimental and design work; for example, his 'pressure bar' for studying explosive detonations and impacts of bullets was adopted for testing at Woolwich. He served as secretary of a committee set up by the Royal Society to advise the Government on scientific matters relating to the war. He was appointed to the Department of Military Aeronautics, with experimental headquarters at Orfordness and later at Martlesham Heath, where he was responsible for the testing of aircraft; he rose to the rank of Colonel and was awarded the CMG. He learned to fly and in the course of his duties he often made solo flights on one of which an accident led to his death. At this time he was a Member of the Council of the Royal Society.

PRS 1918–1919, A, 95, xxvi–xxxvi. JAE. DNB 1912–1921, 268, 269. JAE.

JOHN HOPKINSON 1849–1898; FRS 1878

John Hopkinson was the eldest of 13 children of Alderman Hopkinson of Manchester and his wife who was a daughter of John Dewhurst. He entered Owens College, Manchester before he was 16 years of age. Having shown marked interest and ability for mathematics and physics, he went at the age of 18 with a scholarship to Trinity College, Cambridge, where he had a distinguished academic career. In 1871 he was Senior Wrangler and first Smith's Prizeman, having also obtained the London DSc degree, as well as a Whitworth Scholarship. He was elected to a Fellowship at Trinity, but left the University immediately after taking his degree to become an engineering pupil in the works of Messrs. Wren and Hopkinson, where his father was a partner. After a short time spent there he was appointed Engineering Manager in 1872 at Messrs Chance Brothers and Company, glassmakers of Birmingham. An important section of Chance's work related to lighthouse illumination, and in this field Hopkinson devised various improved and successful systems.

In 1878 Hopkinson moved from Birmingham to London to practise as a consulting engineer. In addition to his successful career as an electrical engineer, inventor and expert, he also made time for a remarkable amount of important purely scientific work, much of it relating to magnetism and the application of electricity and magnetism in engineering. He made important investigations of electrical and magnetic properties of various metals and alloys, including steels, for which he was awarded a Royal Medal in 1890. In the same year he was appointed as Professor of Engineering and head of the Siemens Laboratory at King's College, London. He served twice on the Council of the Royal Society.

In 1878 Hopkinson married Evelyn Oldenburg and they had six children. He was an accomplished climber and generally spent holidays in the Alps with his family; in 1898 he and three of his children were killed in a climbing accident at Evalona, Switzerland.

PRS 1898–99, 64, xvii–xxiv. JAE. DSB 6, 504. B. Dibner. DNB Suppl. v. II, 439–441. THB.

ARCHIBALD HOWIE 1934–; FRS 1978

Kirkaldy High School. Univ. of Edinburgh: BSc. California Institute of Technology, MS. Univ. of Cambridge: PhD; Trinity College, Research Scholar, 1957–60; Churchill College, Research Fellow, 1960–61; Cavendish Laboratory, ICI Research Fellow, 1960–61; Demonstrator in Phys-

ics, 1961–65; Lecturer, 1965–79; Reader 1979–86; Professor of Physics, 1986–; Head of Department, 1989–. Hughes Medal (jointly), 1988. CBE 1998.
Theory of electron diffraction and transmission electron microscopy applied to the study of lattice defects in crystals.

DEREK HULL 1931–; FRS 1989
Baines Grammar School, Poulton-le-Fylde. Univ. of Wales, Cardiff: PhD, DSc. Atomic Energy Research Establishment, Harwell and Clarendon Lab., Oxford: 1956–60. Univ. of Liverpool: Senior Lecturer, 1960–64; Henry Wortley Bell Professor of Materials Engineering, 1964–80; Dean of Engineering, 1971–74; Pro-Vice-Chancellor, 1983–84. Univ. of Cambridge: Goldsmiths Professor of Metallurgy, 1984–91. Senior Fellow 1991–. Univ. of Liverpool: Senior Fellow 1991. FREng 1986.
Dislocations, mechanical properties, composites.

WILLIAM HUME-ROTHERY 1899–1968; FRS 1937
William Hume-Rothery who was born at Worcester Park, Surrey, was the son of Joseph Hume Hume-Rothery and Ellen Maria née Carter; he had two sisters. Joseph had graduated with first class honours in physics in London University, had then been a scholar of Trinity College, Cambridge, and was 16th Wrangler, before becoming a barrister and patents lawyer. William's family moved to Cheltenham where he attended Suffolk Hall Grammar School from 1903 to 1912 then winning an entrance scholarship to Cheltenham College, where he studied until 1916. He became Head of his House and Cadet Lieutenant in the School Officers Training Corps. Having decided on a military career Hume-Rothery entered the Royal Military College, Woolwich in 1916 with a Prize Cadetship. However, in 1917 he contracted cerebrospinal meningitis which led to a long period in hospital and being left completely deaf. With an army career closed to him his parents thought of architecture as a career for him and he was sent for instruction in art. However, science attracted him and Magdalen College, Oxford accepted him in 1918. His parents moved to Headington Hill, Oxford, so that William, still in poor health, could live at home.

Hume-Rothery read chemistry, and in 1920, following the award of a Demyship by his College for an excellent examination performance, he proceeded to a first class performance in 1922. In his fourth year, a research year, he was introduced to optical microscopy. It seems that he had already a strong interest in intermetallic compounds, probably because they did not match the formal valency rules of inorganic chemistry, and that this was known to Frederick Soddy, then Dr Lee's Professor of Chemistry. On Soddy's advice, Hume-Rothery went to the Royal School of Mines, Imperial College, London where he worked under Harold Carpenter on the structure and properties of intermetallic compounds; this led to a London PhD in 1925 for a thesis which included the first mention of 'electron compounds'. Returning to Oxford in 1925 Hume-Rothery worked on the constitution of alloys, successively as Senior Demy of Magdalen, an Armourers and Brasiers' Company Research Fellow (1929–32) and a Warren Research Fellow of the Royal Society (1932–55). In 1931 he married Elizabeth Alice Fea and a daughter was born in 1934. Hume-Rothery was well known in Oxford, referred to as 'H.-R.', and he did not lack for research students. He remained in Oxford for the remainder of his life and was helped and sustained by the interest of Soddy and other distinguished chemists who succeeded Soddy.

Hume-Rothery was awarded a DSc (Oxford) in 1935 and became a Fellow of Magdalen College in 1938, a post which he held until 1943; also in 1938 he was appointed by the University as Lecturer in Metallurgical Chemistry. He continuously sought for recognition of his subject in

Oxford teaching. His early work had been done with very limited laboratory space and the establishment of a Metallurgy Department at Oxford was a long process involving assistance from industry and commerce. Developments included his election to a Readership in 1955, his election as the first Isaac Wolfson Professor of Metallurgy in 1958, (and holder of a professorial fellowship at St Edmund Hall) and the opening of a new building for an independent department in 1960.

Hume-Rothery published numerous original papers in scientific journals covering both experimental and theoretical topics with an emphasis on structure. The research showed the importance of relative atomic size and electronic factors and the 'Hume-Rothery' rules soon attracted much attention by metallurgists and solid state physicists. A wide range of alloys was investigated, including during the 1939–1945 war complex systems of direct industrial interest. X-ray diffraction was one of the important techniques and developments included experiments at high and low temperatures.

The success of Hume-Rothery in research was matched by his ability as an interpretative writer. He was the author, either alone or jointly, of seven books. His monograph *The Structure of Metals and Alloys* was published in 1936 and later editions were published with collaborating authors) gave the metallurgist the first time account of atomic structure, an outline of the theory of the metallic state, and an account of factors influencing alloy constitution. Atomic theory for students of metallurgy (1946) was presented as a series of articles in the form of a dialogue between an 'older metallurgist' and a 'young scientist'. Among the other books, *Metallurgical Equilibrium Diagrams,* published with J.W. Christian and W.B. Pearson presented the experience of his laboratory in accurate experimental methods.

Hume-Rothery overcame his deafness magnificently. In his family circle he could lipread well and he played his full part in committees with the aid of a 'scribe'. As a lecturer he became accomplished having learned to control his voice. He was a painter and a keen fisherman; he loved his home and garden, and he attended the church at Iffley regularly. 'H.-R.' made an outstanding contribution to the establishment of the scientific approach to the metallurgy of alloys which brought him international recognition and many honours, including the award of the OBE in 1951.

BMFRS 1969, 15, 109–139. G.V. Raynor FRS. DNB 1961–1970, 548, 549. G.V. Raynor FRS.

ROBERT HUNT 1807–1887; FRS 1854

Robert Hunt was born at Plymouth Dock (now Devonport), Devonshire; he was the posthumous son of a naval officer. He attended schools in Plymouth and Penzance and then worked with a surgeon in Paddington, London, where he learned some chemistry; there followed five years working with a physician and four years in charge of a medical dispensary. He then spent a period in Fowey, Cornwall where he established a Mechanics' Institute before returning to London to work for a chemical manufacturer. He carried out experiments on photography and also on electrical phenomena in mineral veins. Beginning in 1845 Hunt held for 36 years the government appointment of Keeper of Mining Records. He was appointed Lecturer in Mechanical Sciences at the School of Mines in London in 1851 and was also associated with work for the Great Exhibition. Two years later he began a period as Professor of Experimental Physics at the School of Mines. He was responsible for producing the *Mineral Statistics of the United Kingdom from 1855-1884*, and published other works on mining, including metalliferous mines; he was also editor of the *Dictionary of Arts, Manufactures and Minerals*.

PRS 1889–90, 47, i–ii. AG. DNB.

JAN INGENHOUSZ 1730–1799; FRS 1769

Jan Ingenhousz was born in Breda in the Netherlands, the second son of Arnoldus Ingenhousz, a leather merchant, and Maria née Beckers. After education at Breda Latin School he studied medicine taking his MD in Louvain in 1753. He then studied at Leyden prior to practising as a physician in his native town, where he began to experiment in physics and chemistry. In about 1765, having inherited from his father's estate, he came to Edinburgh and then to London, where he established a practice. Three years later, having played an important part in the early days of inoculation against smallpox, he was appointed Court Physician to the Empress Maria Theresa in Vienna. His salary enabled him to set up a laboratory and to experiment in physics and chemistry. Here he carried out experiments on platinum and its properties and through visits to Paris and London he encouraged other scientists to take an interest in this metal. He also attended meetings of the Lunar Society in Birmingham. Ingenhousz became famous for his pioneering work on plant physiology. He gave the Bakerian Lecture in 1778 and 1789.

DSB 7, 11–16. P.W. van der Pas. DNB.

SIR CHARLES EDWARD INGLIS 1875–1952; FRS 1930

Charles Inglis was the son of Dr Alexander Inglis and Florence née Feeney. The Inglis family of Auchindinny, about eight miles south of Edinburgh, had a history traced back to the 16th century, linked with an estate of about 730 acres bought in 1702. Charles's mother died soon after his birth and he was brought up by his father's unmarried sisters. He went to Cheltenham College at the age of 14 and in 1894 he entered King's College, Cambridge with a scholarship. In 1897 he was placed as 22nd Wrangler in the Mathematical Tripos, and in the following year having gained a first class in mechanical sciences he became a pupil of a consultant engineer in London.

After starting in the drawing office Inglis was soon assigned to work on the new extension to the Metropolitan Railway between Whitechapel and Bow. He was involved with the design and supervision of the nine bridges that had to be built. He also began a study on the subject of mechanical vibration. In 1901 he was made a Fellow of King's College, and returned to Cambridge as assistant to James Ewing. In the same year he married Eleanor Mary Moffatt and they had two daughters. Two years later, under Ewing's successor, he was appointed to a lectureship in engineering and continued his work on vibrations. In 1913 he published a paper on the stresses in a plate associated with cracks and sharp corners, which became important in relation to the theory of crack propagation.

At the outbreak of war in 1914 Inglis was commissioned in the Royal Engineers, and from 1916 to 1919 he was in charge of the department responsible for designing and supplying military bridges; he was made an OBE. Retiring with the rank of Major, Inglis returned to Cambridge in 1918, being elected in 1919 to the Chair of Mechanical Sciences, a post which he held until his retirement in 1943. His research contributions included important work on railway bridge stresses and other railway problems. Many honours came to him including the award of a knighthood in 1945. Inglis was a scholar and an engineer, appreciating the difficulties associated with engineering problems; he was also a great teacher of engineering.

ONFRS 1952–53, 8, 445–457. J.F. Baker.

GEORGE RANKIN IRWIN 1907–1998; FMRS 1987

George Irwin was born in El Paso, Texas, USA, the son of William Rankin Irwin and Mary Susan née Ross; he had an elder brother and a sister. His father died when he was very young and his mother took her children to her home town Rochester, Illinois. When Geroge was five years old the family moved to Springfield, Illinois, where he attended Springfield High School from 1921–25. The following year he began studies at Knox College in Galesburg and graduated with an AB degree in English in 1930; he had planned to become a journalist. However, in 1929, on a cycling tour in Europe, a visit to the Deutsches Teschnische Musem in Munich led to him deciding to study science and engineering. Accordingly he spent another year at Knox College studying physics, obtaining a physics degree in 1931.

Subsequently at the University of Illinois he was awarded an MA degree in physics in 1933 and a PhD in 1937. In 1933 he married Georgia Shearer and they had four children. From 1935–36 Irwin held the post of Associate Professor of Physics and then from 1936–37 he held an appointment in the Physics Department at the University of Illinois. There followed a long period of service from 1937–67 as a Physicist at the Naval Research Laboratory, Washington, DC; from 1937–53 he was Head of the Ballistics Branch of the Mechanics Division; in 1945 he served for several months in the Western Pacific in a technical capacity with the US Marine Corps. At the Naval Research Laboratory he was promoted to Associate Superintendent of the Division in 1948, followed in 1950 by appointment as Superintendent. His work in relation to World War II concerned laboratory investigations of penetration balllistics, combat damage to aircraft and the development of new armour materials. The group which he led were centrally involved in new non-metallic armours used extensively in the Korean and Vietman wars.

Following World War II Irwin's main personal research contributions were in the field of fracture behaviour, fracture testing and fracture control. He introduced fundamental concepts which led to the establishment of the field of 'fracture mechanics', a term that he introduced. Following his retirement in 1967 from the Naval Research Laboratory he was appointed as Boeing University Professor of Mechanics at Lehigh University and held this post until 1972; in this period he continued his work on fundamental and applied aspects of fracture mechanics relative to incorporation of the subject into the teaching of engineering science at universities. In 1972 he was appointed as Professor of Mechanical Engineering at the University of Maryland. where he continued his research, while maintaining links for a long period with Lehigh as adjunct research professor and consultant. Irwin's pioneering role in the field of fracture led to his being referred to as 'the father of fracture mechanics.'

H.P. Rossmanith ed., *Fracture Research in Retrospect*, An Anniversary Volume in Honour of George R. Irwin's 90th Birthday, A.A. Balkema/Rotterdam/Brookfield/1998. H.P. Rossmanith, *Technology, Law and Insurance*, 1997, 2, p 43–47. *Who's Who in America* v I, p. 209. *Marquis Who's Who*, Reed Reference Publishing Co, New Providence, NJ, 1995.

KENNETH HENDERSON JACK 1918–; FRS 1980

Tynemouth Municipal High School. Univ. of Durham: King's College, Newcastle upon Tyne, BSc, 1939, MSc 1944. Ministry of Supply: Experimental Officer, 1940–41. Univ. of Cambridge: FitzWilliam College, PhD, 1950; ScD 1978. Univ. of Durham: King's College; Lecturer in Chemistry, 1941–45, 1949–52; 1953–57. British Iron and Steel Research Association: Senior Scientific Officer, 1945–49. Univ. of Cambridge: Cavendish Laboratory, Research in Crystallography, 1947–49. Westinghouse Elec. Corp., Pittsburgh, Pa.: Research Engineer, 1952–53. Thermal Syndicate, Wallsend: Research Director, 1957–64. Univ. of Newcastle upon Tyne: Professor of Applied Crystal Chemistry, 1964– 84; Director of Wolfson Research Group for High Strength Materials, 1970–84; Royal Society, Armourers and Brasiers' company award, 1988. OBE 1997.

Interstitial structures important in metallurgy; nitrides and oxynitrides for high temperature applications.

SIR HERBERT JACKSON 1863–1936; FRS 1917

Herbert Jackson was born in Whitechapel the only surviving son of Samuel Jackson and Clementina Rebecca née Grant. Having attended King's College School he entered King's College London in 1879. He worked at the College successively as Demonstrator, Lecturer, Assistant Professor (1902), Professor of Organic Chemistry (1905) and Daniell Professor of Chemistry (1914). He married Amy Collister in 1900; there were no children from the marriage. Jackson's research ranged widely including phosphorescent phenomena and materials. He also worked on ceramics, including an interest in oriental ceramics and glasses. Microscopy was one of the techniques that he used. In the early part of the 1914–18 war he made important contributions to the industrial needs of the country by determining formulae for glasses for special purposes e.g. optical and heat resisting applications. In 1918 Jackson was appointed as the first Director of the British Scientific Instrument Research Association, a post which he held until his retirement in 1933. He was made KBE in 1917.

ONFRS 1936–38, 2, 307–314. H. Moore. DNB 1931–40, 469, 470. T. Martin.

NICOLAS JOSEPH DE JACQUIN 1727–1817; FRS 1788

Nicolas Jacquin was born in the Netherlands, the grandson of a French nobleman who had migrated there. His father, a distinguished cloth and velvet manufacturer, arranged for his son to have a thorough classical education at the Jesuit Gymnasium in Antwerp. Nicolas commenced theological studies and then changed to medicine in Paris. A friend of his father, who was director of the medical faculty at the University of Vienna, invited him there to continue studying with his support. He visited the West Indies and South America to enlarge the Imperial natural history collections (1754–1758). In 1763 he was appointed Professor of Practical Mining and Chemical Knowledge at the Mining School in Schemnitz, Hungary. Then 5 years later he obtained the Chair of Chemistry and Botany in the Medical Faculty at Vienna, a post which he held until 1796; in 1809 he became Rector of the University. His home, at which Mozart was a frequent visitor, played a role in Viennese social and scientific life. The honours that Jacquin received included being made a Baron (1806).

DSB 7, 57–59 Wilfried W. Oberhummer. Gen. Ref. 10.

SIR HENRY JAMES 1803–1877; FRS 1848

Henry James was the fifth son of John James and Jane née Hosken. He was born near St Agnes, Cornwall, received a Grammar School education in Exeter and studied at the Royal Military Academy, Woolwich. Embarking on a military career, he became a Probationer for the Corps of Royal Engineers in 1825 and a Second Lieutenant in 1826. The following year he was appointed to the Ordnance Survey and was promoted to Lieutenant in 1831 and Second Captain in 1842. There followed service with the Ordnance Survey in Ireland, and a period, commencing in 1846, with the Admiralty, superintending constructional work in Portsmouth Dockyard. In 1846, having been promoted to Captain, he was appointed a member of the Royal Commission enquiring into the application of iron in railway structures. His subsequent career included further service with the Ordnance Survey, in which he rose to become Director-General in 1857, a post which he held until 1875. Further military promotions occurred, culminating in his becoming a Lieutenant -General in 1874. James married Anne Watson, and they had two sons and one daughter. He was knighted in 1860.

DNB. Ref. Ch. 13.8.

REGINALD WILLIAM JAMES 1891–1964; FRS 1955

Reginald James was born in London, the elder son of William George Joseph James and Isabel Sarah née Ward. He attended school at Paddington and then, in 1903, studied at the Polytechnic Day School in Regent Street. He spent two post-matriculation years at the City of London school and won scholarships with which he entered St John's College, Cambridge in 1909. He obtained first class honours in the Natural Sciences Tripos in both Parts I and II, the latter in physics. In 1912 he commenced research in the Cavendish Laboratory under J.J. Thomson. He was invited in 1914 to join Shackleton's expedition to the Antarctic as a physicist. The enterprise turned out to involve conditions of severe hardship and danger from which he emerged with honour and respect. Having returned to England near the end of 1916 he was commissioned in the Royal Engineers in January 1917, and joined the experimental sound-ranging team which W.L. Bragg had established near Ypres; he played an important part in the work and became Head of the Sound-Ranging School with the rank of Captain.

James spent the period from 1919–1937 in the Physics Department at Manchester University under W.L. Bragg, initially as a Lecturer, then as a Senior Lecturer (1919) and then Reader in Experimental Physics (1934). He established an international reputation as an X-ray crystallographer. His collaborative research, included fundamental, quantitative investigations of X-ray reflection from various crystals, for example of aluminium; his work contributed importantly to investigations of complex crystal structures.

In 1936 James married Annie Watson and the following year they moved to South Africa where their two sons and one daughter were born and where he was Professor of Physics at Cape Town until his retirement in 1956.

BMFRS 1965, 11, 115–125. Sir Lawrence Bragg. DNB 1961–70, 576–578. W.H. Taylor.

CHARLES FREWEN JENKIN 1865–1940; FRS 1931

Charles Jenkin was born at Claygate, Surrey, the second son of Henry Charles Fleeming Jenkin and Anne née Austin. He was educated at Edinburgh Academy, Edinburgh University and Trinity College, Cambridge, where he was Second Optime in the Mathematical Tripos in 1886. Following graduation he entered the engineering workshops of Messrs Mather and Platt at Manchester. He then went to the London and North West Railway Company works at Crewe. Jenkin married Mary Oswald MacKenzie in 1889 and they had two sons and one daughter. In 1891 he became an Engineer at the Royal Gunpowder factory at Waltham Abbey. There followed practical works in other fields in Monmouthshire (1893–1898), Blackheath (1898–1903), Stafford (1903–1905) and London (1905–1908). He had an interest in research and was appointed to the new Chair of Engineering Science at Oxford University in 1908; he was a Fellow of New College and moved to Brasenose College in 1912. In 1914 he joined the Navy as a Lieutenant, RNVR and then worked for the Air Ministry in charge of aircraft materials investigations; in 1915 a Materials Testing Department was set up at Farnborough. He was promoted to the rank of Lieutenant-Colonel (RAF) and was awarded the CBE in 1919. Jenkin's interests included fatigue and corrosion fatigue. In 1929 he retired from his Oxford Chair and joined the Building Research Station at Garston, near Watford.

ONFRS 1939–41, 3, 575–585. R.V. Southwell. DNB 1931–40, 482, 483. R.V. Southwell.

HENRY CHARLES FLEEMING JENKIN 1833–1885; FRS 1865

Henry Jenkin was the son of Charles and Henrietta Camilla Jenkin. He was born near Dungeness and educated at Jedburgh and at Edinburgh Academy. In 1846 the family went to live abroad,

spending periods in Frankfurt-on-Main, Paris and Genoa. He studied physical sciences at the University of Genoa and graduated MA with first class honours. In 1851 he was apprenticed at Fairbairn's works at Manchester, involved in mechanical engineering and spent three years there. He did railway survey work in Switzerland in 1855 and worked as a draughtsman at Greenwich. In 1857 he entered the service of Messrs Newall in their submarine cable factory in Birkenhead, which was concerned with making the first Atlantic cable; he became Chief of Engineering and Electrical Staff. In 1859 he married Anne Austin and they had two sons. In the same year he came to know William Thomson and was associated with him on the conductivity and insulation of electric cables, making experimental measurements. From 1861 for a number of years he was in a partnership concerned with telegraphic and general engineering. Also he had continued association with William Thomson in various projects associated with cables. Jenkin was appointed Professor of Engineering at University College London in 1865 and three years later Professor of Engineering at Edinburgh University.
PRS 1885, 39, i–iii. WT. DNB.

STANLEY JEVONS 1835–1882; FRS 1872
Stanley Jevons was born in Liverpool, where his father was an iron merchant; at the age of 15 he was sent to London where he attended University College School, whence in 1851 he entered University College and commenced studies in arts and sciences. He won a gold medal in chemistry at the College and worked with a cousin Henry Roscoe. Towards the end of 1853 he received through Thomas Graham of the Mint the appointment of Assayer to the Australian Royal Mint in Sydney; he spent a period of training in assaying under Graham, and at the Paris Mint before he went to Australia in 1854. In 1859 he returned to England to resume his studies at University College. Having obtained the BA degree in 1860 he subsequently graduated as MA with the gold medal in logic, philosophy and political economy. His career interests turned mainly to the areas of political economy and logic. He held academic appointments in Manchester from 1863 to 1876, where he became a professor, and was married in 1867 to Harriet Ann Taylor. From 1876 to 1880 he was Professor of Political Economy at University College London.
PRS 1883, 35, i–xii, RH. DNB. Ref 10.9.

KENNETH LANGSTRETH JOHNSON 1925–; FRS 1982
Barrow Grammar School. Univ. of Manchester: MSc Tech, MA, PhD. Messrs Rotol Ltd, Gloucester: Engineer, 1944–49. College of Technology, Manchester: Assistant Lecturer, 1949–54. Univ. of Cambridge: Lecturer, then Reader in Engineering, 1954–77; Professor of Engineering, 1977–92. FREng 1987.
Contact mechanics;theoretical and experimental aspects.

PERCIVAL NORTON JOHNSON 1792–1866; FRS 1846
Percival Johnson was the son of John Johnson and Mary née Wight. At one time John was the only commercial assayer in London; in 1786 he had inherited the business, established in 1777 by his father, also named John Johnson. In 1807 Percival, who at that time had not yet begun to use the name of Norton, became an apprentice of the Worshipful Company of Goldsmiths, working with his father at his home in the City of London. The business consisted of the assaying of gold and silver, the analysis of ores and minerals and the buying and selling of materials containing the precious metals, including native platinum.

In 1817 Percival set up independently as 'Assayer and Practical Mineralogist' in the City and later in Hatton Garden, achieving an outstanding reputation as an assayer and metallurgist. He took up the refining business and a gold refinery was built in Hatton Garden where he successfully refined gold bars from certain Brazilian sources. He also commercially separated platinum group metals from the gold.

Following a visit to Germany Johnson set up production of German silver at Hatton Garden in 1829. In the mining field he was consulted and visited professionally nearly all the mines in England, Wales, Scotland, and Ireland, and many important ones elsewhere. He was involved with the then booming lead, copper, silver and tin mines of Devon and Cornwall, eventually becoming owner of several mines and he introduced improvements in mineral treatments. He aimed to improve social conditions of the miners and at his own expense he erected and became involved in supervising schools in the neighbourhoods of the mines. He introduced the use of the classification table and the ventilation of mines.

Johnson's greatest achievement however was in the platinum business, being the first to succeed in refining and manufacturing on a commercial scale, and playing a major role in extending the applications of platinum. His partnership with George Matthey as the firm of Johnson and Matthey began in 1851; thereafter Johnson, began to relinquish control of the business leaving Matthey largely in charge, until he finally retired in 1860 to live in his home at Stoke Fleming, Devonshire.

PRS 1867–68, 16, xxii–xxiv. Ref. Ch. 4.1. Donald McDonald, *Platinum Metals Review*, 1962, **6**(3), 112–114.

WILLIAM JOHNSON 1922–; FRS 1982

Central Grammar School, Nottingham. Manchester College of Science and Technology: BSc,Tech. BSc, Mathematics, London. Royal Electrical and Mechanical Engineers: war service, Lieutenant, UK and Italy, 1943–47. Home Civil Service: Assistant Principal Administrative Grade, 1948–50. Northampton Polytechnic, London: Lecturer, 1950–51. Univ. of Sheffield: Lecturer in Engineering, 1952–56. Manchester Univ.: Senior Lecturer in Engineering, 1956–60; Professor of Mechanical Engineering, 1960–75; Chairman, Department of Mechanical Engineering,1960–69. UMIST: Director of Medical Engineering, 1973–75. Univ. of Cambridge: Engineering Department, Professor of Mechanics, 1975–82. FREng 1983.

Mechanics of metal forming, novel manufacturing processes, impact engineering, medical and bioengineering, history of engineering mechanics.

JEAN FRÉDÉRIC JOLIOT 1900–1958; FMRS 1946

Frédéric Joliot was born in Paris, the youngest of the six children of Henri Joliot and Emilie née Roedener. In 1919 he entered the École Supérieure de Physique et Chimie Industrielle de la Ville de Paris. After he had graduated, there followed fifteen months of military service and a period of industrial work in a large steel mill in Luxembourg. He became an assistant to Marie Curie at l'Institut du Radium de Paris; in 1925 he married Marie Curie's elder daughter, Irene and they had a daughter and a son. In 1927 he published an account of electrochemical work on radioelements; polonium was deposited on a thin sheet of mica made conductive by gold. He went on to the study of some physical properties of thin metallic films. Joliot was awarded a PhD in 1930. Frédéric and his wife were awarded jointly the Nobel Prize for Chemistry in 1933 for their discovery of articially induced radioactivity. Frédéric was appointed Professor of Nuclear Physics at the Collège de France. In 1938 , following the work of Hahn and Strassman on uranium fission Joliot entered this field of research. In 1940 he obtained Norway's stock of heavy

water and this was smuggled to England; also seven tonnes of uranium oxide obtained from Belgium were hidden until after the war. During the occupation of France his work on chain reactions stopped but he was able to continue as Director of the Laboratory of Nuclear Chemistry, without doing military work; he experienced periods of arrest by the Gestapo. With the liberation of France in 1944 Joliot became Director of the Centre National de la Recherche Scientifique. Later he was Director of the Radium Institute and also for a period directed the French Atomic Energy Programme. He was awarded the Hughes Medal in 1947.
BMFRS 1960, 6, 87–105. P.M.S. Blackett. DSB 7, 151–157. Francis Perrin.

JOHN JOLY 1857–1933; FRS 1892
John Joly was born at Holywood in King's County, Ireland, the third son of the Rev. John Plunkett and Julia Anna Maria Georgina née Comtesse de Lusi. He was educated at Rathmines School, Dublin, and entered Trinity College Dublin in 1876, where his range of studies was wide. He obtained a first class in modern literature; then in 1883 having studied various science and engineering subjects he was awarded a first in engineering. An appointment as Assistant to the Professor of Engineering followed; in 1887 he became Professor of Geology in the University of Dublin, holding this appointment until his death. He remained unmarried.

Joly produced inventions and research results in a number of fields, including the development of a steam calorimeter with which he made the first determination of the specific heats of gases at constant volume. His geological work included an interest in applying radioactivity to the solution of problems of the earth's history. In mineralogy his work included accurate measurements of specific gravities and experiments on the fusion of minerals. He developed an electric furnace in which he reduced aluminium from topaz.In the Royal Society he was awarded a Royal Medal in 1910.
ONFRS 1932–35, 1, 259–286. DSB 7, 160, 161.V.A. Eyles. DNB 1931–40, 494–496.

HARRY JONES 1905–1986; FRS 1952
Harry Jones was born in Pudsey, Yorkshire, the seventh of eighth children of Ernest Jones and Hannah Elizabeth née Armitage. After attending a private school from 1910 to 1915 he attended Pudsey Grammar School. In 1923 he went to the University of Leeds and in 1926 obtained first class honours BSc in physics. Then he proceeded to do experimental research in Leeds on the loss of energy by electrons in gases, receiving a PhD degree in 1928. He obtained a senior award from the Department of Scientific and Industrial Research and turned to theoretical physics, going to Cambridge to work with R.H. Fowler.

In 1930 he moved to the University of Bristol as a Research Assistant to work on the behaviour of electrons in metals. He married Francis Molly O'Neill in 1931 and they had two girls and a boy. When Nevill Mott came to Bristol in 1933 to occupy the Chair of Theoretical Physics there followed an important period of collaboration up to 1937; the outcome included the book *The Theory of the Properties of Metals and Alloys*. Jones was awarded the PhD degree of Cambridge University in 1932.

In 1937 Jones became Assistant in Research at the Royal Society Mond Laboratory in Cambridge, and worked on the problems of liquid helium. In the following year he was appointed to a Readership at Imperial College, London, but from 1939–45 he held part time appointments with the Ministry of Home Security and later with the Ministry of Supply for research on explosives. Returning in 1946 to Imperial College he was appointed to a Chair as Head of the Mathematical Physics Group, and in 1955 to the Headship of the Department of Mathematics at a time

of university expansion; during the tenure of his Chair the Department doubled in size. For the last two years before his retirement (1970–1972) he was Pro-Rector and after retirement he remained a Research Fellow of the College until 1981.
BMFRS 1987, 33, 327–342. Sir Nevill Mott FRS.

BRIAN DAVID JOSEPHSON 1940–; FRS 1970
Cardiff High School. Univ. of Cambridge: BA, 1960, MA, PhD 1960. Univ. of Illinois: Research Assistant, Professor 1965–1966. Univ. of Cambridge: Assistant Director of Research in Physics, 1967–72; Reader in Physics, 1972–74; Professor of Physics 1974–. Nobel Prize for Physics, shared 1973.
Superconductivity.

JAMES PRESCOTT JOULE 1818–1889; FRS 1850
James Joule, whose ancestors were Derbyshire yeomen, was born at Salford near Manchester, the second of five children of Benjamin and Alice Prescott Joule. His first education was at home; later he and his elder brother received private teaching, including instruction from John Dalton in chemistry. He married Amelia Grimes in 1847, and they had two children before her death in 1854. Joule spent the rest of his life at various residences in the neighbourhood of Manchester. He carried out pioneering scientific experiments in laboratories installed at his own expense in his houses, or in a brewery. Later, having incurred financial losses he received some subsidies from scientific bodies, and was awarded a government pension in 1878.

In the period 1837–47 his creative work led to the general law of energy conservation and the establishing of the dynamical theory of heat. He achieved a position of great authority in the scientific community and continued to carry out a variety of skilful experimental investigations. Among aspects covered were the thermal effects of voltaic electricity(requiring highly accurate temperature measurements for which he constructed a special thermometer), the mechanical equivalent of heat, amalgams of various metals with mercury, and properties of materials. He became friends with William Thomson involving scientific collaboration, including an investigation on the thermal properties of gases, leading to the discovery of the Joule–Thomson effect, which became the basis of low temperature refrigeration. Also Joule, under Thomson's influence studied the thermal effects of strains on solids and liquids. He also carried out some welding experiments. Joule was awarded a Royal Medal in 1852 and the Copley Medal in 1870.
DSB 7, 180–182. L. Rosenfeld. Refs Ch. 8.6, 10, 22.

PIOTR LEONIDOVICH KAPITZA 1894–1984; FRS 1929
Piotr (Peter) Kapitza was born in Kronstadt, near St Petersburg. His father Leonid Petrovich Kapitza was a military engineer, and his mother Olga née Ieronimovna was an important figure in the literary world of St Petersburg. In 1912 he entered the Electrotechnical Faculty of the St Petersburg Polytechnical Institute. During the war for two years from 1914 he was an ambulance driver at the Russian front; then having graduated from the Institute in 1918 he was appointed to a teaching post there and began research which included magnetism. Soon after his demobilisation he had married Nadezhda Kyrillovna, but in 1919 in an influenza epidemic she and her two children died, as did Kapitza's father.

In 1921 Kapitza came to England as a member of a commission set up to renew scientific relations with other countries after the war. Rutherford accepted Kapitza to work in the Cavendish Laboratory at Cambridge, where he remained for 13 years. In his work on alpha particles he had the novel idea of producing large magnetic fields lasting only a very short time (impulsive magnetic fields). This work led him into pioneering research during the 1920s in solid state and low temperature physics, including the development of large machines to achieve high magnetic fields. He completed a PhD degree in 1923 and was admitted as a member of Trinity College. He was also awarded a Clerk Maxwell Scholarship.

Kapitza was skilled at constructing and handling delicate apparatus and with his new equipment he carried out important research on the magnetic behaviour of various materials, including metals. In 1925 he was appointed as Assistant Director of Magnetic Research, and in the same year was elected to a Research Fellowship at Trinity College. In 1927 he married Anna Alekseyevna Krylov, and they had two sons. In 1930 he was appointed to a Royal Society Messel professorship. The Royal Society Mond Laboratory was built to accommodate the high field magnetic equipment and to provide cryogenic facilities to extend his work to lower temperatures. After 1926 Kapitza visited the Soviet Union regularly, but in 1934 he was prevented from returning to England. He became the Director of a new Institute for Physical Problems in Moscow where with two interruptions he continued his scientific work until the end of his life. One of these interruptions, in the political climate of the post war period involved his temporary dismissal from his post as Director, and he demonstrated his personal courage in the difficulties of the situation. The main themes of the Institute were magnetism and low temperature physics, extended to include plasma physics, which became his chief personal interest. In 1965 Kapitza was allowed to travel again and was able to visit England. In his career he made outstanding contributions in science and technology and was awarded a Nobel Prize for Physics in 1978.

BMFRS 1985, 31, 327–374. D. Shoenberg FRS. Notes and Records of the Roy. Soc. 1988, 42, 205–228.

THEODORE VON KÁRMÁN 1881–1963; FMRS 1946

Theodore (Todor) von Kármán was born in Budapest, the third of five children of Maurice Kármán and Helene née Konn. He entered the Mintagimnazium, a school that his father had founded, at the age of nine. He proceeded to the Technical University of Budapest and graduated with highest honours as a mechanical engineer in 1902. He continued as an assistant (except for a period of military service) until 1906. Having obtained a two-year fellowship he studied with Ludwig Prandtl at the University of Göttingen. His doctoral dissertation in 1908 was concerned with buckling of columns. While at Göttingen he made a number of contributions in the field of mechanical behaviour and strength of materials. Von Kármán spent a semester at the Sorbonne where he became interested in aircraft. He became a Privatdozent at Göttingen and in 1911 he carried out important work on the theory of vortices, which became important in many engineering applications, including oscillations in bridges leading to collapse. In another field he was joint author of a paper in 1912 on the specific heat of solids. In the same year he returned to Hungary to the Chair of the Theory of Machines at the College of Mining Engineering in Selmecbanya.

In 1913 von Kármán moved to Aachen to become Professor of Aerodynamics and Mechanics and director of the Aeronautical Institute; he remained there until 1929 apart from the period 1914–1919 when he saw military service including work at an aircraft factory. After the war, von Kármán built up the Institute at Aachen to become an international centre and he consulted for leading aircraft companies. From 1930–49 he was Director of the Guggenheim Aeronautical Laboratory at the California Institute of Technology, which later became the NASA Jet Propulsion Laboratory; he was also and Director of Research of the Guggenheim Airship Institute at Akron, Ohio; he kept his position at Aachen until forced to resign in 1933 by the Nazis. He became a United States citizen in 1936. Von Kármán's research at Cal Tech included shapes of turbine blades in electricity generation and jet engines. Among other activities he became involved in rocket research in the 1930s and went on to play a major role in the American space programme in the 1950s and 60s.

Von Kármán never married, but for a long period lived with his mother and sister. He was the recipient of numerous honours from many countries in recognition of his wide ranging contributions to applied mechanics, aeronautics and astronautics and to wider spheres including education, and national and international affairs.

BMFRS 1966, 12, 335–365. S. Goldstein. Dictionary of American Biography Suppl., 7, 1961–65, 410–412. R.K. Gehrenbeck.

HENRY KATER 1777–1835; FRS 1814

Henry Kater was born in Bristol, where his father, of German descent, was associated with a firm of sugar boilers. Having spent two years in a lawyer's office, followed by study of mathematics, he became an ensign in the 12th Foot in 1799 serving in Madras. While in India he was promoted to Lieutenant, carried out survey work, and devised an improved form of pendulum. Following his return to England in 1808 he was promoted to command a company, and held various army posts. He was employed on pendulum experiments at the chief stations of the trigonometrical survey of Great Britain. In 1818 he produced a second pendulum of particularly high accuracy, involving the use of wootz steel. He lived mainly in London, where he gave a Bakerian Lecture in 1820 on 'the best kind of steel for compass needles'. Also in the Royal Society Kater received the Copley Medal in 1817 and served as a Member of Council and as Treasurer and Vice-President.

PRS 1834–35, 22, 350–354. HMC. DSB 7, 262, 263. H. Dorn. DNB.

CHRISTOPHER FRANK KEARTON
(BARON KEARTON OF WHITCHURCH) 1911–1992; FRS 1961

Christopher Kearton was born at Congleton, Cheshire, the only son of Christopher John Kearton and Lilian née Hancock. He attended Hanley High School and entered St John's College, Oxford, as an Open Exhibitioner in 1929. He graduated in natural sciences with first class honours in chemistry in 1933. He then went to ICI Billingham and became Shift Manager and Refinery Manager. In 1936 he married Agnes Kathleen Brander and they had two sons and two daughters. Kearton was chosen to join a team in the Tube Alloys Project concerning the atomic energy programme and in 1943 spent a period in the USA. From 1946–75 he was at Courtaulds. He headed a chemical engineering section (1946–52), was appointed to the Board (1952) and became Managing Director (1957). He became a director of many group companies and was involved in the Processing Division and the Synthetic Fibres Division. In 1975 he retired from the Board and his activities then involved an advisory role in the nuclear industry. Among his many responsibilities Kearton was Chancellor of the University of Bath from 1980–92. He received the OBE in 1945, was knighted in 1966 and became a Life Peer in 1970.

BMFRS 1995, 41, 221–241. Sir Norman Wooding.

JAMES KEIR 1735–1820; FRS 1785

James Keir was the youngest of 18 children of John Keir of Muiston Baxter and Queenshaugh, Stirlingshire and Margaret née Lind; both his parents were from prominent Scottish families. Following education at the Royal High School, Edinburgh, he attended medical school at Edinburgh University from 1754–55. With a wish to travel he purchased a commission in the Army, becoming an ensign in the 61st Regiment of Foot in 1757; he served during the Seven Years War, in the West Indies and in Ireland, and resigned with the rank of Captain in 1768.

Keir had retained an interest in chemistry from his time at Edinburgh and became a member of the Lunar Society of Birmingham. In 1770 he married Susanna Harvey and settled in West Bromwich, Staffordshire; they had one daughter. He began business as a glass manufacturer at Stourbridge in 1775 and a year later presented a paper to the Royal Society on observations of the crystallisation of glass during slow cooling. In 1778 he gave up his glass business to take charge of the Soho, Birmingham, engineering works of Boulton and Watt in Birmingham. Keir co-founded a chemical works in around 1780 in Tipton, where his activities included the development of the first commercially successful process for making synthetic alkalis. He also worked as an assistant to Priestley.

Keir's published work covered various fields, including military matters. In chemistry he translated, and augmented an important dictionary by Macquer, originally published in Paris in 1766. He wrote on experiments on the dissolution of metals in acids, including suggestions which contributed to the discovery of the electroplating process. As a pioneer industrial chemist Keir used his knowledge to improve the processes that he carried out, and did much to disseminate chemical knowledge.

DSB 7, 277, 278. E.L. Scott. DNB. Notes and Records of the Roy. Soc. 1967, 22, 144–154. B.M.D. Smith and J.L. Moillet.

ANDREW KELLER 1925–1999; FRS 1972

Andrew (born Andreas) Keller was born in Budapest, the only son of Imre Keller and Margit née Klein. During World War II he was required to work in a Jewish labour battalion in Ruthenia , but he escaped to the advancing Russian forces; he was subsequently in a displaced persons camp from which he escaped. At the end of the war he studied at Budapest University, and having graduated BSc, he came to England in 1948, just before the borders were closed by the communist authorities. After carrying out research at Bristol University, he joined ICI Dyestuffs in Manchester as a Technical Officer in 1948. In 1951 Keller married Eva Bulhack and they had one son and one daughter. Keller's interest in polymers developed and in 1955 he moved to the Physics Department at Bristol, with the encouragement of Charles Frank. From 1955–57 he was on a Ministry of Supply research appointment, and was then successively Research Assistant (1957–63), Lecturer (1963–65), Reader (1965–69), Research Professor of Polymer Physics (1969–91) and Emeritus Professor. Keller's important contributions to polymer science included the use of electron microscopy and electron diffraction to elucidate the structure of polymer crystals, and the production and study of ultra-stiff polymer fibres. He was awarded the Rumford Medal in 1994.

Physics World 1999, 12, 4, 44. A.H. Windle FRS. *Who's Who. The Guardian*, 13th May 1999. Peter Barham and Jeff Odell.

ANTHONY KELLY 1929–; FRS 1973

Presentation College, Reading. Univ. of Reading. Univ of Cambridge: Trinity College, BSc 1949. PhD, 1953. ScD, 1968. Univ. of Illinois: Research Associate 1953–55. Univ. of Birmingham: ICI Fellow, 1955. North Western Univ.: Assistant Associate Professor, 1950–59. Univ. of Cambridge: Lecturer, 1959–67. National Physical Laboratory: Division of Inorganic and Metallic Structure, Superintendent, 1967–69; Deputy Director, 1969–75 (seconded to ICI, 1974–75). Vice-Chancellor, Univ. of Surrey. FREng 1979. Bakerian Lecturer 1995. CBE 1988.

Structure of precipitation hardening alloys; fibre strengthening; composites; strong solids

MICHAEL JOSEPH KELLY 1949–; FRS 1993

Francis Douglas Memorial College, New Plymouth, NZ. Victoria University of Wellington. BSc, 1970, MSc, 1972. Univ.of Cambridge: Gonville and Caius College, PhD 1974; Trinity Hall, MA 1975, ScD 1990; Fellow 1974–81; 1989–92. Univ. of California, Berkeley: IBM Research Fellow, 1975–76. Cavendish Lab. SRC Research Fellow, 1977–81. GEC Hirst Research Centre: Member Research staff, 1981–92. Royal Society/SERC Research Fellow, 1989–91. Univ. of Surrey: Professor of Physics and Electronics, 1992–; Head of School of Electrical Engineering, Information Technology and Mathematics, 1997–. FREng 1998.

Low dimensional semiconductors: semiconductor physics for devices.

KEVIN KENDALL 1943–; FRS 1993

London Univ.: BSc Physics, external. Joseph Lucas: 1961–66. Univ. of Cambridge: Cavendish Laboratory, 1966–69, PhD. British Rail Research: 1969–71. Monash Univ.: 1972–74. ICI, Runcorn: 1974–93. Univ. of Keele: Professor of Materials Science, 1993–.

Adhesion, fracture, ceramics, material properties.

SIR ALEXANDER BLACKIE WILLIAM KENNEDY 1847–1928; FRS 1887

Alexander Kennedy who was born in Stepney was the eldest son of the Rev. John Kennedy and Helen Stodart née Blackie. He was educated at the City of London School and at the School of Mines (then at Jermyn Street). When he was 16 he began an apprenticeship which lasted for five and a half years with a firm in Millwall where he gained experience of marine engineering. In 1868 he was appointed leading draughtsman in a works at Jarrow-on-Tyne involved in marine engineering. Three years later he entered into partnership with a consultant marine engineer. In 1874 Kennedy married Elizabeth Verralls Smith and they had two sons and one daughter. Also in 1874 he was appointed Professor of Engineering at University College London. Between 1881 and 1892 his experimental work included the strength and elasticity of materials and the strength of riveted joints. He also designed steel and concrete structures. In 1889 he resigned his chair and went into partnership as a consulting engineer based in Westminster. He then turned his attention particularly to electrical engineering, including involvement in electric transport. During the 1914–18 war he was involved with technical military work. Kennedy was knighted in 1905.

ONFRS 1936–38, 2, 213–223. A. Gibb. DNB 1922–30, 464–466. E.I. Carlyle.

OLEG ALEXANDER KERENSKY 1905–1984; FRS 1970

Oleg Kerensky was born in St Petersburg, Russia, the elder of two sons of Alexander Federovitch, a barrister, and Olga Lvovna née Barenovsky. His father was a member of the Provisional Government after the February Revolution in 1917, but after the October Revolution he and his family went through difficult times. Olga and the boys came to England in 1920, where both Oleg and his brother studied engineering at the college which was then known unofficially as the Northampton Engineering College. Olag graduated in 1927 and soon afterwards joined Dorman Long and Company Ltd where he assisted in the project for designing and constructing the Sydney Harbour bridge. In 1928 he married Nathalie Bely and they had one son; after his wife's death he married Dorothy Harvey.

From 1937–46 Kerensky was associated with steel construction work with the firm of Holloway Brothers; during the 1939–45 war this work included the fabrication and assembly of pontoons for Bailey bridges and the Mulberry harbour. He was appointed Senior Designer for Holloway Brothers and became a British citizen in 1946. In the same year he joined Freeman Fox and

Partners as a Principal Bridge Designer. During his subsequent career he was associated with major civil engineering projects in England and abroad, including highways as well as bridges. In his work he used pioneering principles e.g. the use of high tensile steel grip bolts for site connections and high tensile rivets; also he introduced light weight concrete acting compositely with steel girders in a bridge deck. He was also involved in work on British Standards. He was awarded the CBE in 1964 and was elected FREng in 1976.
ONFRS 1986, 32, 321–353. M.R. Horne FRS. DNB 1981–85, 222–223. R. Horne FRS.

DAVID ANTHONY KING 1939–; FRS 1991
St John's College, Johannesburg. Univ. of Witwatersrand, Johannesburg: BSc, PhD (Rand). ScD (East Anglia). Imperial College, London: Shell Scholar, 1963–66. Univ. of East Anglia, Norwich: Lecturer in Chemical Physics, 1966–74. Univ. of Liverpool: Brunner Professor of Physical Chemistry, 1974–88. Univ. of Cambridge: Professor of Physical Chemistry, 1988–, and Head of Department of Chemistry, 1993–, Downing College: Master 1995–.
Interaction of gases with metal surfaces; spectrographic and other physical techniques.

GUSTAV ROBERT KIRCHHOFF 1824–1887; FMRS 1875
Gustav Kirchhoff, the son of a law councillor, was born in Germany at Konigsberg, where he studied at the University, graduating in 1847. In the same year he married Clara Richelot, and they had two daughters; following her death he married Luise Brommel. In 1848 in Berlin he obtained the qualification allowing him to lecture privately in a university, and two years later he was appointed Extraordinary Professor in Breslau. He became friends with Bunsen and in 1854 became a Professor of Physics at Heidelberg where he was associated with Bunsen in the work which gave rise to the discoveries of caesium and rubidium. He moved to Berlin as Professor of Mathematical Physics. His research work included spectroscopy, electricity, radiation and elasticity. He was awarded the Rumford Medal in 1862 and the Davy Medal in 1877.
PRS 1889, 46, vi–ix. DSB 7, 379–383. L. Rosenfeld.

MARTIN HEINRICH KLAPROTH 1743–1817; FRS 1795
Martin Klaproth was born in Wernigerode, Germany, the third son of Johann Julius Klaproth, a tailor. He was an apprentice in an apothecary shop, beginning in 1759, and moved as a journeyman to Hanover in 1766, where he became involved in chemistry and carried out some investigations. He eventually settled in Berlin in 1771, where he managed an apothecary shop until his marriage in 1780 brought him a dowry which enabled him to purchase his own shop. In 1782, having continued to pursue his interests in chemistry he was appointed to Prussia's highest medical board and obtained permission to lecture in chemistry. He advanced progressively through the medical bureaucracy to become High Councillor (1799–1817); also he held several teaching posts, including as Teacher of Chemistry at the Mining School (1784–1817) and as Professor of Chemistry in the University of Berlin.

Klaproth became the leading analytical chemist in Europe for a period from the late 1780s; this was the field in which his most important work was done. In 1789 he studied pitchblende from Joachimstal which was thought to be a mineral of zinc and iron; he concluded that it contained a new metal which he named uranium in honour of the newly-discovered planet Uranus. Working with minerals from many countries he was the discoverer or co-discovery of a number of other elements: zirconium, titanium, strontium, chromium and cerium in the period 1789–1803, and he confirmed the discoveries of tellurium and beryllium. In addition he contributed to

the theoretical development of chemistry. He was also consultant to the Berlin Porcelain Factory and in 1788 reported experiments on the use of platinum for decorating porcelain.
DSB 7, 394, 395. K. Hufbauer. Ref. Ch. 4.1.

JOHN FREDERICK KNOTT 1938–; FRS 1990
Queen Elizabeth's Hospital, Bristol. Sheffield Univ.: BMet 1959. Cambridge Univ. PhD, 1963, ScD, 1991. CEGB Laboratories, Leatherhead, Surrey: Research Officer 1962–67. Cambridge Univ.: Department of Metallurgy, Lecturer, 1967–81; Reader in Mechanical Metallurgy, 1981–90; Churchill College, Vice-Master 1988–90.Univ. of Birmingham: Feeney Professor of Metallurgy and Materials, 1990–, Head of the School of Metallurgy and Materials, 1990–96: Dean of Engineering 1995–. FREng 1988.
Physical metallurgy; fracture processes; metals and alloys; role of microstructure in crack initiation and propagation.

HANS KRONBERGER 1920–1970; FRS 1965
Hans Kronberger was born in Linz, Austria, the younger child and only son of Norbert and Olga Kronberger. He left Austria at the age of 18 when the German occupation occurred. He came to England and gained admission to King's College, Newcastle, to study mechanical engineering. When war came he was interned as a 'friendly enemy alien', first on the Isle of Man and then in Australia. In 1944 he returned to King's College and was awarded a BSc degree in physics in the same year. He then joined Francis Simon's team on the Tube Alloys project to work on isotope separation, based at Birmingham. In 1946 he joined the General Physics Division at the Atomic Energy Research Establishment at Harwell and was awarded a Birmingham PhD in 1948. In 1951 he went to the Capenhurst Laboratory of the Division of Atomic Energy (Production) and became its Head. Also in 1951 he married Joan Hanson, a widow, and they had two daughters. His career continued in atomic energy work; in 1958 he was appointed Director, Research and Development in the Industrial Group of the Atomic Energy Authority; subsequently he became Managing Director for Development and Scientist-in-Chief of the Reactor Group. In 1969 he was appointed a Member of the Atomic Energy Authority. Kronberger's work was particularly concerned with research and development of large scale separation of isotopes; his interests ranged from light elements such as hydrogen, helium and lithium to heavy elements such as uranium. He was also conerned with the development of reactor systems, including materials science aspects. He was awarded the Royal Society Leverhulme Tercentenary Medal (1964), an OBE (1957) and a CBE (1966).
BMFRS 1972, 18, 413–426. L. Rotherham FRS. DNB 1961–1970, 622, 623. Penney.

SIR HAROLD WALTER KROTO 1939–; FRS 1990
Bolton School. Univ. of Sheffield: BSc, PhD; Research Student 1961–64: NRCC, Postdoctoral Fellow. Bell Telephone Lab. , New Jersey: Research Scientist 1966–67. Univ. of Sussex: Research Fellow, 1967–68; Lecturer, 1968–77; Reader 1977–85; Professor of Chemistry, 1985–91; Royal Society Research Professor, 1991–. Nobel Prize for Chemistry, 1996. Knighted 1996.
Spectroscopic studies of unstable molecules; new carbon structures, fullerenes.

ADOLPH THEODOR VON KUPFFER 1799–1865; FMRS 1846
Adolph Kupffer, was born in Mitau in Courland, Russia, where his father was a merchant. When he was 16 he began to study medicine, first at the University of Dorpat, and then in 1816 at the

University of Berlin. However, he was not attracted to these studies and applied himself to mathematics, physical sciences and to mineralogy. In 1819 he went to the University of Göttingen, and the following year to Paris where he attended lectures in mineralogy. In St Petersburg in the winter of 1821–22 he lectured on crystallography, before obtaining an appointment in the spring of 1822 as Professor of Physics, Chemistry and Mineralogy at the University of Kasan. He took up his duties there in the following year, having visited Paris again. In 1828 having taken part in an geological expedition to the Urals he came to St Petersburg. His subsequent career included scientific exploration, and lecturing in various institutions. Then in 1843, a central observatory was established in St Petersburg, of which he was appointed Head.

Kupffer's research included a long and important series of experiments on the elasticity of metals and published over the period 1850–1861, concerning particularly vibrational constants and the effect of temperature. The Russian government had founded an Institute of Weights and Measures to which was also entrusted the task of investigating those properties of metals which can affect the standards of measurements, namely temperature and elasticity. Kupffer carried out experiments on the torsional vibrations of metals in the form of wire; iron, copper, platinum, silver and gold.

PRS 1866–67, 15, xlvi–xlvii. Ref. Ch. 13.8. VII pt I.

NICHOLAS KURTI 1908–1998; FRS 1956

Nicholas Kurti was born in Budapest, the son of Károly Kurti and Margit née Pintér. He was educated at the Minta-Gymnasium in the city, and obtained his first degree at the University of Budapest at the age of 18. Because of anti-Jewish laws he went abroad for further studies and graduated licencees sci. phys from the University of Paris two years later. He moved to Berlin in 1929 where his supervisor was Franz Simon; having been awarded a Dr.Phil degree Kurti moved with Simon to the Technische Hochschule in Breslau, where they worked together. With the rise to power of Hitler, they came in 1933 to the Clarendon Laboratory at the University of Oxford, through arrangements made by Frederick Lindemann; they continued to work together on low-temperature physics. Kurti developed new magnetic cooling techniques, achieving temperatures only a few hundredth of a degree above absolute zero; he identified various superconducting metals. Beginning in 1940, Simon and Kurti worked at Oxford on the British atomic bomb project, on the development of gaseous diffusion for the separation of U-235. Kurti worked on methods of producing diffusion membranes, and in 1941 he went with Simon and others to the USA to participate in the Manhattan project. In 1944 Kurti remained in the USA at Columbia University setting up diffusion membrane testing. In 1946 on his return to Oxford Kurti returned to Oxford and in the same year he married Giana Shipley; they had two daughters. Kurti continued to work in the field of cryophysics, developing techniques for cooling to increasingly low temperatures and investigating thermal and magnetic properties at low temperatures. In 1956, with his group, using nuclear techniques a temperature of 10^{-6} of a degree above absolute zero was achieved.

After his return from the USA Kurti held a series of appointments at Oxford: Demonstrator in Physics (1945–60); Reader (1960–67); Professor(1967–75); Director of the Mullard Cryogenic Laboratory in the Clarendon Laboratory; Professorial Fellow (1967–75) and then Emeritus Fellow. He was awarded an Oxford MA and was Senior Research Fellow at Brasenose College from 1947–67. In the Royal Society Kurti was a Member of the Council from 1964–67, and was Vice-President from 1965–67; he received the Hughes Medal in 1969. He was awarded the CBE in 1973.

The Times, 27 Nov. 1998. *The Independent*, 27 Nov. 1998. M.J.M. Leask. *The Guardian*, 28 Nov. 1988. Anthony Tucker.

JOSEPH LOUIS COMTE DE LAGRANGE 1736–1813; FRS 1791

Joseph Lagrange was born of a French father, Giuseppe Francesco Lodovico Lagrangia, and an Italian mother, Teresa Grosso in Turin. In 1755 he became Professor of Mathematics at the Royal Artillery School of Turin. He was appointed Director of the Berlin Academy of Sciences in 1766. In the following year he married his cousin Vittoria Conti; there were no children from the marriage. In 1787 at the invitation of Louis XVI he went to Paris as a Member of the French Royal Academy and became Professor of Mathematics at the École Polytechnique in 1797. A brilliant mathematician from his youth, having been largely self taught from his reading, his work included major contributions in the mechanics of solids and fluids.

DSB 7, 559–573. J. Itard.

SIR HORACE LAMB 1849–1934; FRS 1884

Horace Lamb was born at Stockport the son of John Lamb who was a foreman in a cotton mill, and Elizabeth née Rangeley. His father died when Horace was quite young; when his mother married again he was brought up by an aunt. He was educated at Stockport Grammar School, and at the age of 17 was awarded a classical scholarship to Queen's College, Cambridge. In view of his young age he was advised to spend a year at the Owens College in Manchester. In 1868 he obtained a minor scholarship at Trinity College, and in 1870 he became a scholar of that College. He was Second Wrangler in Mathematics in 1872 when he obtained the BA degree, and in the same year he won the second Smith's Prize and was elected a Fellow and Lecturer of his College. He was a very popular lecturer who taught various aspects of mathematics. In 1875 he married Elizabeth Foot and they had three sons and four daughters.

Lamb was appointed to a chair at Adelaide, Australia, commencing there in 1876. He returned in 1885 to Owens College as Professor of Pure Mathematics; later the title of the post was changed to Professor of Pure and Applied Mathematics. His research included the field of the theory of elasticity and the textbooks that he wrote included a massive treatise on hydrodynamics.

Having resigned his Manchester Chair in 1920 Lamb returned to Cambridge where he was made an Honorary Fellow of Trinity College, and an Honorary Lecturer in the University. He continued to work for 14 years, including in the wartime period on problems connected with the strength of aircraft, and from 1921–1927 he served on the Aeronautical Research Committee. Lamb served on the Council of the Royal Society and was Vice-President for two periods (1909–1910 and 1920–22); he received many distinctions, including a Royal Medal (1902), the Copley Medal (1923) and a knighthood in 1931.

ONFRS 1934–35, 1, 375–392. R.T. Glazebrook FRS. DSB, 7, 594, 595. K.E. Bullen, DNB 1931–40, 22, 523. S. Chapman.

FREDERICK WILLIAM LANCHESTER 1868–1946; FRS 1922

Frederick Lanchester was the fourth of eight children of Henry Lanchester and Octavia née Ward. He was born in Lewisham, near London, but when he was two years old the family moved to Hove, Sussex, where his father, an architect, set up business. In 1883, following education at a grammar school, he entered Hartley College, Southampton (which later became University College, Southampton); at Southampton his studies included metallurgy. He spent the period 1885–1888 at the Normal School of Science (later the Royal College of Science) at South Kensington, where his studies included mining. Subsequently at Finsbury Technical College he learnt workshop practice and he went on to become a highly skilled craftsman. Lanchester spent a short period as a Patent Office Draughtsman and then in 1889 he joined the Forward Gas Engine

Company at Saltley, Birmingham, manufacturers of gas-engines; initially he was Assistant Works Manager and Designer, and the next year he was appointed Works Manager. He set out to improve engine performance and to reduce costs also made important inventions. During the period 1889–90 he invented the pendulum governor and an engine starter.

In 1894 Lanchester decided to start out on his own account and formed a small private syndicate, setting up a workshop and aiming to develop a motor car incorporating revolutionary features such as a noiseless and vibrationless engine and transmission. Using only limited resources and facilities, a first, experimental, vehicle was produced in 1896, and a second model, two years later, won a gold medal. From 1899–1904 Lanchester produced cars for the market. For the Lanchester Engine Company, formed in 1899, he was for some time General Manager, Chief Designer and Works Manager, and in 1901 a vehicle with the innovative features was produced; however, there were financial difficulties. In 1905 the Lanchester Motor Company Ltd, was formed, but disagreements led to his resignation although he continued as Consultant Engineer and Designer until 1919. Around 1909 Lanchester became Consultant Engineer to the Daimler Motor Company and the Birmingham Small Arms Company.

From his earliest years Lanchester had a deep interest in the science of aeronautics. From 1892 onwards he experimented with model gliders to investigate the problems of lift and drag. He read an important paper in 1895 to the Birmingham Natural History and Philosophical Society on the theory of flight, but his theories were not published until 1915 when he presented them in his famous book *Aerial Flight*, in two volumes.

During World War I Lanchester's involvement in military work for the Government brought him to live in London. In 1919 Lanchester married Dorothea Cooper; they had no children. In 1924 they returned to Birmingham and a year later he founded laboratories intending to provide research and development services, but difficulties led to their closure in 1934. He became Consulting Engineer to Messr William Beardmore in 1928 for the construction of diesel engines, mainly for locomotive work. Lanchester as a pioneering inventor, combining abilities as a great scientist, engineer and designer, was a major fundamental contributor in the automotive industry and in aeronautics.

ONFRS 1945–48, 5, 757–766. H.R. Ricardo. DNB, 1941–50, 469–470. Henry R. Ricardo. Ref. Ch. 9.12 v.2. Hutchinsons Dictionary of Scientific Biography.

LEV DAVYDOVITCH LANDAU 1908–1968; FMRS 1960

Lev Landau was born in Baku where his father was an engineer and his mother a doctor; he had one older sister. He finished school at the age of 13 having shown exceptional talent, and being too young to go to university he went to a technical school where he studied economics. At the University of Baku which he entered in 1922 he studied concurrently in two departments, Physics and Chemistry. In 1924 Landau went to the Department of Physics at the University of Leningrad, where only two years later he published a thesis, and was appointed a Fellow. He spent one and a half years in Denmark with Bohr beginning in 1929 before moving to the Physical Chemistry Institute in Leningrad in 1931. The following year he went to Kharkov, in the Chair of Theoretical Physics in the Mechanical Machine Construction Institute, and in 1935 he occupied the Chair of General Physics at the University. In 1937 he moved to the Institute of Physical Problems in Moscow where Kapitza was working, and set up the Department of Theoretical Physics. In the same year he married K.T. Drobanzeva, and there was one son from the marriage. In 1938 Landau was joined by Eugenii Liftshitz, with whom there was scientific collaboration and close friendship. Landau's death occurred as a result of a car accident.

The distinguished and wide-ranging research of Landau in theoretical physics included magnetisation and ferromagnetic materials, superconductivity and second order phase transitions. He was awarded the Nobel Prize for Physics in 1962 for his work on the theoretical understanding and properties of liquid helium (superfluidity).
BMFRS 1969, 15, 141–158. P.L. Kapitza FRS and E.M. Lifshitz. DSB 7, 616–619. A.T. Grigorian.

ANDREW RICHARD LANG 1924–; FRS 1975
Univ. College of South West, Exeter. BSc, London, 1944, MSc, London, 1947. Univ of Cambridge: Cavendish Laboratory, Research Assistant, 1947–48; PhD, 1953. Lever Bros., Port Sunlight: Research Dept, 1945–47. North American Philips, Irvington-on-Hudson, NY: 1952–53. Harvard Univ: Instructor, 1953–54; Assistant Professor, 1954–59. Univ. of Bristol: Lecturer in Physics, 1960–66; Reader, 1966–79; Professor, 1979–87; Senior Research Fellow, 1995–.
X-ray diffraction and topography.

PAUL LANGEVIN 1872–1946; FMRS 1928
Paul Langevin, the second son of Victor Langevin, was born in Paris where he attended the École Lavoisier and the École de Physique et de Chimie Industrielles; at the latter Pierre Curie supervised his work. He entered the Sorbonne in 1891 where military service interrupted his studies for a year in 1893. The following year he entered the École Normal Supérieure. He won an academic competition which enabled him to go to the University of Cambridge in 1897 for a year where he studied under J.J. Thomson. In 1902 Langevin was awarded a PhD at the Sorbonne partly for his Cambridge work and partly for work under Curie. Langevin joined the Collège de France in 1902 and became Professor of Physics there two years later, a post which he held until 1909, when he moved to the Sorbonne to a similar position. During the 1914–1918 war he contributed to miltary research. During World War II in 1940 after the German occupation of France Langevin became Director of the École Municipale de Physique et de Chimie Industrielle, but he was arrested for his antifascist views and later placed under house arrest; he escaped to Switzerland in 1944 and returned to Paris later that year.

Langevin was the leading mathematical physicist of his time in France. His researches included investigations of the secondary emission of X-rays from metals exposed to radiation, and also magnetism. He received the Hughes Medal in 1915 and the Copley Medal in 1940.
ONFRS 1950–51, 7, 405–419. F. Joliot. DSB 8, 9–14. A.R. Weill-Brunschvicg.

IRVING LANGMUIR 1881–1957; FMRS 1935
Irving Langmuir's grandfather emigrated from Glasgow to Toronto, Canada, and his father Charles came to the United States where he married Sadie Comings. Irving was the third of four sons, and was born in Brooklyn, New York where he attended school. There was a period of three years during which his family moved to Paris where his father was in charge of the New York Life Insurance Agency. There, at a small boarding school, he was able to spend time in the school laboratory. He derived an interest in science from his brother Arthur who became an industrial chemist, and who encouraged him to have a workshop at the age of nine, and later a laboratory when he was only 12. After the years in Paris he went to schools in Philadelphia and in Brooklyn.

He studied at the School of Mines of Columbia University and graduated with the degree of metallurgical engineer in 1903. He then went to Germany for graduate studies in physical chemistry at the University of Göttingen where he carried out research under Walther Nernst on the dissocation of water vapour and carbon dioxide around glowing platinum wires; he obtained the

PhD degree in 1906. After Göttingen he became an Instructor in Chemistry at the Stevens Institute of Technology in Hoboken, New Jersey. In 1908 he analysed mathematically gas flow through heated vessels. He spent a period of summer employment at the General Electric Company, Schenectady, where the research laboratory was a leading centre of pure and of applied research, and began an association with W.R. Whitney, the Director, and with W.D. Coolidge and others. At the end of the period he accepted an invitation to stay at the laboratory.

Langmuir's extensive research included important investigations relevant to tungsten filaments in lamps, involving studies of gases at very low pressures and solid surfaces. Also fundamental studies were made of atomic structure, and of surfaces, including particularly in liquids, where his theory became a new basic approach to surface kinetics. He was also involved in the invention of the hydrogen welding torch.

During the 1914–1918 war he was concerned with the problems of submarine detection with a station on the New England coast. In World War II his studies of icing and his measurements of particle size led to his production of 'white smoke' as a smoke screen. Other work, which benefited from his hobby in meteorological topics, included the production of snow from supercooled clouds, the production of rain by a chain reaction in cumulus clouds at temperatures above freezing, and the control of precipitation from cumulus clouds by various seeding techniques. He continued his interest in meteorological research in the post-war period.

Three years after Langmuir went to Schenectady he married Marion Mersereau and they adopted a son and a daughter. Shared interests included mountain climbing in the Rockies and in Switzerland, and some small boat sailing. Langmuir loved to fly and was an excellent pilot.

World wide recognition came to Langmuir including the award of the Hughes Medal of the Royal Society in 1918 and of the Nobel Prize in 1932 for 'his discoveries and researches in the realm of surface chemistry'.

BMFRS 1958, 4, 167–184. Hugh Taylor. DSB 8, 22–25. C. Susskind. A.W. Hull, *Nature*, 1958, **181**, 148.

MAX THEODOR FELIX VON LAUE 1879–1960; FMRS 1949

Max Laue, who was born in Pfaffendorf near Koblenz, was the son of a civil official of the German military administration; when his father was raised to the hereditary nobility in 1913 Laue signed himself as M. v. Laue. During the winter of 1898–99 he began studying mathematics and physics in the University of Strasbourg while serving his compulsory year of military training. In the next year, studying in Göttingen, Laue turned to theoretical physics. He spent two years at Göttingen and a winter semester in Munich, and then moved to Berlin in 1902 where he worked with Max Planck on interference phenomena ; he was awarded the degree of DrPhil in 1903 '*magna cum laude*'. There followed a period of two years at Göttingen where he studied art and obtained the certificate required for high-school teaching.

In the autumn of 1905 Laue returned to Berlin as an assistant to Planck at the Institute for Theoretical Physics, where his work included the field of relativity; he became Planck's leading and favourite pupil and lecturer (Privatdozent) in 1906. In 1909, he became a Privatdozent at the Institute of Physics at the University of Munich with Arnold Sommerfeld. Initially Laue continued to work on relativity and his lectures were mainly on thermodynamics, optics, theory of radiation and relativity.

Laue began his work on the nature of X-rays and conceived the idea of X-ray diffraction in crystals in February 1912, linked with a discussion with Paul Ewald. There followed the experimental work carried out by Laue, Friedrich and Knipping which demonstrated the diffraction effect, and led to the opening up of the field of X-ray structural analysis of materials.

Laue accepted a Chair of Theoretical Physics at the University of Zurich in the summer of

1912. Two years later he accepted a chair at the University of Frankfurt, where he was during the 1914–18 war and for part of this period he worked with a group in the University of Wurzburg on the physics of electronic tubes. Laue was Professor of Theoretical Physics at the University in Berlin from 1919 to 1943. In addition he was Consultant at the Physikalisch Technische Reichsanstalt in Berlin-Charlottenburg, where Walther Meissner was working on superconductivity, and Laue also published in this field. Laue was also Deputy Director of the Kaiser Wilhelm Institut für Physik (KWI) in Berlin-Dahlem. Late in 1938 Otto Hahn and F. Strassmann working in the adjoining KWI for chemistry, discovered nuclear fission and the KWI for physics was taken over by the army. Laue refused to take responsibility for the nuclear work and only kept his own section. In 1943 he resigned from his professorship and the following year moved to a small town in Württemberg, Hechingen, where the Institute had been evacuated. He was taken prisoner by the English–American group carrying out the 'Mission Alsos'. and was one of 10 scientists taken to the mansion, Farmhall in Cambridgeshire, until in January 1946 he returned to Germany. He was at the forefront of the rebuilding of German science. For five years he lived in Göttingen, officially as acting Director of the Max Planck Institute for Physics. In 1951 Laue was appointed Director of the Fritz Haber Institute of Physical Chemistry in Berlin-Dahlem in which position he remained until his retirement in 1958. His death followed a motor car accident.

Laue's wide interests included sailing, mountaineering, skiing and motoring and many of the arts. He was a very influential figure in German physics, keenly interested in national and cultural politics. Laue showed his great political courage during the Nazi period when dissent was dangerous.

The scientific work of Laue spanned thermodynamics of radiation, optics, including X-ray optics, relativity, and superconductivity. The award of the Nobel Prize for Physics came in 1914 for his discovery of the diffraction of X–rays by crystals. He was one of the great physicists of the transition period from classical physics to quantum physics.

BMFRS 1960, 6, 135–156. P.P. Ewald. DSB 8, 1973, 50–53. A. Hermann.

ANTOINE-LAURENT LAVOISIER 1743–1794; FRS 1788

Antoine Lavoisier who was born in Paris was the son of Jean-Antoine Lavoisier, a lawyer, and Emilie née Punctis; he had a younger sister. Following his mother's death in 1748 he spent his childhood and the period up to his marriage with a maiden aunt in Paris. His formal education began at the Collège Mazarin in 1754; in 1761 he began to study in the faculty of law and graduated baccalaureate in 1763 and licentiate in 1764. His interest having turned to science he assisted in the preparation of a geological map of France. At the age of 23 he received an award for work on a new improved system of street lighting. From 1768 he carried out scientific work. Although Lavoisier came from a wealthy background he needed to work to support his research. Accordingly he took an administrative position in the Ferme Générale, which was a private agency that the government used to collect taxes (tax farming). In 1771 he married Marie Anne Pierette Paulze, who was 14 years of age and the daughter of a tax farmer; the marriage was childless. In 1775, he became inspector of gunpowder for the government. During the Revolution he, with others, came under suspicion, because of his membership of the Ferme Générale; having been arrested on wrongful charges he was executed in 1794.

Lavoisier's research which was based on careful measurements contributed greatly to fundamental knowledge, particularly in chemistry including the field of combustion. His fundamental discoveries made him one of the outstanding scientists of the 18th century. The publication of his book *Traite Elementaire de Chimie* in 1789 was a landmark in the history of chemistry.

DSB 7, 66–91. H. Guerlec. Ref. Ch. 4.1.

HENRI LOUIS LE CHATELIER 1850–1936; FMRS 1913

Henri Louis Le Chatelier was born in Paris, the eldest of six children. His father, Louis Le Chatelier, was closely associated as an engineer with the construction of the railways of France, Northern Spain, and Southern Austria; also he was France's Inspector-General of Mines and was associated with Saint Claire Deville in the establishment of an aluminium industry in France, and with W. Siemens in the construction of the first open-hearth steel furnace. Leading chemists in France visited his home. Henry Louis was thus influenced by the atmosphere of science and technology in which he grew up.

After attending a military academy for a short period Le Chatelier studied at the Collège Rollin, where he received a LittB in 1867 and a BS in 1868. He entered the École Polytechnique in 1869 and while still a student during the Franco-Prussian War he took part as a Sub-Lieutenant in the defence of Paris, before entering the École des Mines in 1871. He also followed courses at the Sorbonne and the Collège de France and was influenced by Deville, whose laboratory he had often visited with his father.

On graduating from the École des Mines in 1875 he went to Algeria as a member of an engineering mission, and after his return held a post as a Mining Engineer at Besancon. In 1877, however, he accepted the offer, which was based on his good record as a student, of the Chair of General Chemistry at the École des Mines. He remained at the École des Mines until his retirement in 1919. Le Chatelier received the degree of Doctor of Physical and Chemical Science in 1887, when his title was changed to Professor of Industrial Chemistry and Metallurgy. In the same year he accepted a chair at the Collège de France, and in 1907 he became Professor of General Chemistry at the Sorbonne in succession to Moissan and held this post until 1925.

When he took up his post at the École des Mines in 1877 he embarked on a long series of classic investigations on hydraulic cements, which were reported in his doctoral thesis in 1887. During his thesis work he developed the use of a platinum–platinum/rhodium thermocouple for high temperature measurements. Le Chatelier carried out important investigations on chemical reactions, including the enunciation of the principle (1884), which bears his name. In the metallurgical field pupils of Le Chatelier investigated the chemical reactions involved in the production of iron in the blast furnace, and reactions in steelmaking.

Although Le Chatelier had declined to include a metallurgy course in his lectures at the École des Mines, he became interested in the nature of metallic alloys as a subject requiring study. His investigations of the physical properties of series of alloys and their relation to the composition, also using microscopical methods, gave greater precision to the ideas of Matthiesen and others, which were little known in France. His contact with Floris Osmond, who had adopted Sorby's methods of microscopical examination, led him into the field of transformations in solid alloys during heating and cooling. He demonstrated the existence of intermetallic compounds and contributed to metallography by his improvements in apparatus and in experimental techniques.

Le Chatelier was greatly interested in the organisation of science, especially with reference to its industrial applications. In the training of men for industry, his policy was to confine academic courses to fundamental instruction, leaving practical procedures to be learned in the laboratory and in the factory. In 1904 he founded the *Revue de Metallurgie* to which he devoted much time and energy. He held important positions on many commissions and boards, for example advising the French government during World War I. Many honours and official positions came to Le Chatelier, among them the Davy Medal in 1916.

Le Chatelier married Genevieve Nicolas in 1876 and their descendants comprised seven children, thirty-four grandchildren, and six great-grandchildren. At the commemoration held at the

Sorbonne, presided over by the President of the Republic (one of his former students) delegates attended from many countries. He exerted both directly and through his numerous pupils an outstanding influence in France.
BMFRS 1936–38, 2, 251–259. C.H. Desch. FRS. DSB 8, 116–120. Henry M. Leicester.

PETER GEORGE LeCOMBER 1941–1992; FRS 1992
Peter LeComber who was born at Ilford, Essex was the son of George Henry LeComber and Florence née Peck. He attended Becontree Heath Primary School and, at the age of 11, went to the South-East Essex Technical College, with a scholarship. In 1959 he entered the University of Leicester where he graduated in 1962 with an upper second degree in physics. He joined the research group in solid state physics, where in the 1950s research was carried out on the transport and optical properties of low mobility solids, both crystalline and non-crystalline. LeComber worked on cadmium sulphide and zinc sulphide on a topic involving piezoelectric phenomena. Soon after commencing his PhD research he married Joy Smith and they had a son and daughter. In 1965 LeComber took an appoinment as a Research Associate at Purdue University in Indiana working on photoconductivity. In 1967 Lecomber returned to Leicester with an SRC Research Fellowship. He moved to Dundee as Lecturer in Physics and carried out research on disordered semiconductors, particularly silicon in amorphous form. During the 1970s important work on doping was carried out and then in the 1980s LeComber's research moved towards the applications of amorphous materials in devices and he was closely associated with important developments. He was appointed to a Personal Chair in 1986 and a few years later became Harris Professor of Physics. From 1987–1990 he served as the first Head of the merged Physics and Engineering Departments; subsequently he led the amorphous semiconductor group.
BMFRS 1994, 39, 215–225. W.E. Spear FRS.

ANTHONY LEDWITH 1933–; FRS 1995
Univ. of London: BSc (external) 1954. Univ. of Liverpool: PhD, 1957; DSc, 1970, Chemistry Department: 1956–80; Lecturer, Senior Lecturer, Reader and Professor; Campbell Brown Professor of Industrial Chemistry and Head of the Department of Inorganic, Physical and Industrial Chemistry, 1980; Dean Faculty of Science, 1980–84. Pilkington plc,: Deputy Director Group R. and D. 1984–88; Head of Group Research. 1988–1996. Univ. of Sheffield: Professor and Head of Department of Chemistry, 1996–99. EPSRC: Chief Executive, 1999–. CBE 1995
Photochemical initiation processes for curing; design of poymeric materials for xerography; photo-processes in polymer systems: semiconductors.

CHARLES HERBERT LEES 1864–1952; FRS 1906
Charles Lees was born in Oldham, Lancashire, the second of three sons of John Lees and his wife Jane née Ogden. In 1870 he entered Highfield Academy in Oldham; then in 1879 he entered the office of the Gas Superintendent of the town and attended evening classes in the Oldham School of Science and Art. He obtained an exhibition in 1882 to enter Owens College in Manchester and had a very successful career as a student; for example, in 1886–87 he studied higher pure mathematics with Horace Lamb. In 1887 he went to Strasbourg to study at the University. In the following year he was elected to a fellowship at Manchester where he lectured on thermodynamics. He spent a period at the City and Guilds College in London in 1891; in the same year he became Senior Assistant Lecturer and Demonstrator in Physics at Owens College and was awarded a DSc in 1895. In 1900 he was appointed Lecturer in Physics and Assistant Director of the Physi-

cal Laboratories at the University of Manchester. He married Evelyn May Savidge in 1902 and they had three sons and two daughters. Lees was appointed Professor of Physics at East London College (which became Queen Mary College), Professor of Physics in the University of London in 1912, and Vice-Principal of his College in 1917, serving in this position until 1930.

Lees possessed a combination of great practical and mathematical ability, and in his research was best known for his careful and valuable measurements of thermal conductivities, which he carried out over a long period beginning at Strassburg. He gave the Bakerian Lecture in 1908 presenting work on thermal and electrical conductivities of metals and alloys, including low temperature measurements.

ONFRS 1952–53, 8, 523–528. Wm Wilson. *Who was Who* 1951–60.

HENRY BEAUMONT LEESON 1800–1872; FRS 1849

Henry Leeson was educated at Caius College, Cambridge and obtained an MA. Subsequently he studied medicine at St Thomas's Hospital, London and obtained the qualifications MD (Oxon 1840), MRCP (1840) and FRCP (1846). He became Physician and Lecturer in Forensic Medicine and Chemistry at St Thomas's Hospital. His research included electrodeposition in which area he filed a patent in 1846 covering a range of metals.

Bulloch's Roll. Ref. Ch. 4.1.

ANTONI VAN LEEUWENHOEK 1632–1723; FRS 1680

Antoni van Leeuwenhoek was born in Delft, Netherlands. His father Philips Thoniszoon, a basket maker, who was married to Margriet Jacobsdochter van den Berch, died in 1638, and Antoni was sent to the Grammar School of Warmond, near Leiden. He moved to Amsterdam in 1648 where he was apprenticed to a cloth merchant. Subsequently, he returned to Delft and became a shopkeeper, draper and haberdasher. He married, in 1654, Barbara de May, daughter of a serge merchant from Norwich, England and there were five children from the marriage. His wife died in 1666 and five years later he married Cornelia Swalmius; the only child of this marriage died in infancy.

Leeuwenhoek began a career as a civil servant in the municipality of Delft in 1660. Later he was appointed as Chief Warden of Delft (1677) , Inspector of Weights and Measures for the City (1679) and Surveyor to the Court of Holland (1669); in 1666 he visited England. His appointments, and the pension that he received in his old age for his scientific achievements made him financially secure.

In his research Leeuwenhoek showed a keen intellect and great manual dexterity. His chief area of work, developing from the microscopes that he began to produce in about 1671 was the microscopical investigation of organic and inorganic structures, in particular in the field of biology, in which he made important discoveries. This work included aspects of metallurgical interest.

The Royal Society, with which Leeuwenhoek exchanged regular correspondence, encouraged him to investigate new fields and prepared English translations or summaries of his letters for the *Philosophical Transactions.* His achievements, over half a century, through his observations of fundamental importance led to his becoming famous in his own country and abroad; his many notable visitors included kings and princes, among them James II.

DSB 8, 126–130. J. Heniger. Roy. Soc. Notes and Records. 9, 36. Ref. Ch. 2.8.

SIR JOHN EDWARD LENNARD-JONES 1894–1954; FRS 1933

John Jones was born in Leigh, Lancashire, the eldest son of Hugh Jones and Mary Ellen née Rigby. At Leigh Grammar School he specialised in classics, then at Manchester University he studied mathematics, obtaining first class honours in 1915. In the 1914–1918 war he gained his wings in the Royal Flying Corps, saw service in France and also did some work on aerodynamics at the National Physical Laboratory. In 1919 he returned to Manchester, before moving to Cambridge with an 1851 Senior Exhibition at Trinity College and working on the forces between atoms and molecules. Moving to Bristol he became Reader in 1925 and Professor of Theoretical Physics in 1927. In 1925 he married Kathleen Mary Lennard and they took the name of Lennard-Jones; they had one son and one daughter. He studied quantum mechanics at Göttingen in 1929, and was mainly responsible for introducing the new theory to the group of physicists at Bristol that R.M. Tyndall was gathering; here he began his work on molecular orbitals. He was appointed to the Plummer Chair of Theoretical Chemistry at Cambridge in 1932. From 1942 he served as Chief Superintendent of Armament Research, then becoming Director-General, Scientific Research Defence in 1945. He returned to academic life in 1946 and was a Professorial Fellow at Corpus Christie College at Cambridge; he was appointed KBE in the same year. In 1953 he became Principal of the University College of North Staffordshire at Keele. In the Royal Society he received the Davy Medal in 1953.

BMFRS 1955, 1, 175–184. N.F. Mott. DSB 8, 185–187. G. Brush. DNB 1951–60, 621–622, N.F. Mott.

FRANK MATTHEWS LESLIE 1935–; FRS 1995

Queen's College, Dundee, Univ. of St. Andrews: BSc, DSc, 1995. Univ. of Manchester, PhD, 1996; Assistant Lecturer in Mathematics, 1959–61. MIT: Research Associate in Mathematics, 1961–62. Newcastle Univ: Lecturer in Mathematics, 1962–68. Strathclyde Univ: Senior Lecturer, 1968–71; Reader, 1971–79; Personal Professor, 1979–82; Professor of Mathematics, 1982–. Justice of the Peace.

Theory of liquid crystals.

GILBERT NEWTON LEWIS 1875–1946; FMRS 1940

Gilbert Lewis was born in Weymouth, Massaschusetts, the son of Frank Wesley Lewis, a lawyer, and Mary Burr Lewis. He was educated at the Universities of Nebraska and Harvard, graduating from the latter with a bachelor's degree in 1896; he spent a year teaching in Andover and then returned to Harvard where he obtained an MA. In 1899 he was awarded a PhD for a thesis on electrochemical and thermochemical relations of zinc and cadmium amalgams. His career then included a period of three years involving an appointment as an instructor at Harvard, study in Europe and a post in the Bureau of Science in Manila. In 1905 he took a research appointment at MIT. In 1912 he married Mary Hinckley Sheldon and there were two sons and a daughter from the marriage. In the same year Lewis became Chairman of the Chemistry Department at the University of California, Berkeley. In 1917 he was commissioned as a Major in the Gas Service and went to France; he was awarded the Distinguished Service Medal and was promoted to Lieutenant-Colonel. He returned to Berkeley and remained there until his death. His research included the experimental determination of free energies and in 1923 he was co-author with M. Randall of *Thermodynamics and the Free Energy of Chemical Substances.* He also introduced the concept of activity. Another important field of his work was the electronic theory of valency. He was awarded the Davy Medal in 1929.

ONFRS 1945–48, 5, 491–506. J.L. Hildebrand. DSB 8, 289–294. R.E. Kohler.

WILLIAM LEWIS 1708–1781; FRS 1745

William Lewis, son of John Lewis, was a London physician and lecturer, who had studied medicine first at Christ Church, Oxford and then at Emmanuel College, Cambridge (BA 1734, MA 1737, MB 1741, MD 1745). He then settled in London and became established as a public lecturer on chemistry, and on the improvement of pharmacy and the manufacturing arts. By 1745 he was well enough known as a physician, practical chemist and lecturer to be elected as a Fellow of the Royal Society. Two years later he moved to Kingston where he rented a spacious house, and equipped a large laboratory. Lewis's interests ranged over many aspects of engineering, metallurgy and chemistry. In the metals field he was mainly interested in gold, platina and iron. On the basis of his acquaintance with Brownrigg he was allowed to select a small sample of platina from the samples that had been presented to the Royal Society but in 1754 he obtained from Spain a large enough supply for full chemical and physical examination. Lewis was awarded the Copley Medal in 1754.

DSB 8, 297–299. J. Ekland. DNB. *Platinum Metals Review* 1963, **7**(2), 66–69. Ref. Ch. 4.1.

JUSTUS VON LIEBIG 1803–1873; FMRS 1840

Justus Liebeg was born in Darmstadt, the second of nine children of Johann Georg and Maria Karoline Moserin Liebig and went to school at the Gymnasium of the town. He acquired an interest in chemistry through the work of his father whos business involved a small laboratory supplying colours and drugs; he studied chemistry texts in the local library and carried out chemical experiments. He decided to devote himself to chemistry and after a period working with an apothecary he went, in 1820, to university, first at Bonn, and then at Erlangen, where he received a doctorate and published his first chemical paper in 1822. He obtained a travelling stipend from the Grand Duke of Darmstadt to study in Paris. There he entered the laboratory of Louis Thénard; later through a recommendation from Baron Alexander von Humbolt he went to work with Joseph Gay-Lussac.This led to Gay-Lussac recommending him to the Grand Duke Ludwig who appointed him Professor Extraordinary at the University of Giessen in 1824. Two years later Liebig was promoted to be Professor of Chemistry and remained at Giessen until in 1851 he moved to the University of Munich. In 1826 he married Henriette Moldenhauer and they had five children.

Liebig was awarded the Copley Medal in 1840. During his time at Giessen he was made a Baron (1845); his reputation rapidly increased and when young chemists from all parts of Europe came to work for him, the State eventually decided to build him a large chemical laboratory outside Giessen. Many eminent investigators, teachers, and practical men for various industries came from his school.

PRS 1875–76, 24, xxvii–xxxvii. DSB 3, 329–350. F.C. Holmes.

EVGENII MIKHAILOVICH LIFSHITZ 1915–1985; FMRS 1983

Evgenii Lifshitz was born in Kharkov, the elder of two sons of Mikhail Ilyich Lifshitz and Berta née Evzorovna; his father was a doctor and Professor of Medicine at the University. When he left secondary school in 1929 he studied for two years at the Chemical College. He proceeded to enter the Physics and Mechanics Faculty of the Kharkov Mechanics and Machine Building Institute, where he graduated in 1933. He then became a Graduate Student under Landau at the Ukrainian Physicotechnical Institute where he obtained a PhD the following year. He continued to work at the Institute until 1938 as a Senior Research Scientist and was awarded the DSc of Leningrad State University in 1939. From 1939 Lifshitz worked at the Academy of Sciences Institute of Physical Problems in Moscow; he also taught at a number of other institutions. He was married and had a son.

Lifshitz carried out important work in theoretical physics, some involving close collaboration with Landau. His investigations included the fields of ferromagnetism and second order phase transitions.

BMFRS 1990, 36, 337–357. Ya. B. Zel'dovich, FMRS and M.I. Kaganov.

FREDERICK ALEXANDER LINDEMANN (VISCOUNT CHERWELL) 1886–1957; FRS 1920

Frederick Lindemann was born in Baden-Baden, Germany. His father, A.F. Lindemann, was an engineer who emigrated to Britain fom Alsace, and his mother, Olga Noble, was American. He went to school in Scotland and Germany and obtained a PhD at the University of Berlin in 1910. Lindemann was a tennis player of international standard and won the European championship in Germany in 1914. In the same year he came to Britain in 1914 and worked on aircraft stability at the Royal Aircraft Establishment, Farnborough, where he acted as a test pilot for his theory of how to escape from a spin. Lindemann became Professor of Experimental Philosophy at Oxford in 1919, where he built up the research activity in the Clarendon Laboratory with a team that included notable Jewish émigrés from Germany during the 1930s. Appointed as Scientific Advisor to Winston Churchill in 1940 and serving as Paymaster-General from 1942–1945 and from 1951–1953, he played an important role concerning scientific research during the war time period and atomic energy research leading to the creation of the Atomic Energy Research Authority after the war. Lindemann was at Oxford from 1945–1951 and 1953–1956. In his early research with Nernst he made an important contribution to quantum theory and made low temperature measurements of specific heats. He was awarded the Hughes Medal in 1956 and was appointed PC in 1943, became Baron Cherwell in 1941, CH in 1953 and Viscount Cherwell in 1956.

BMFRS 1958, 4, 45–71. G.P. Thomson. DSB 8, 368, 369. R.V. Jones.

HENRY SOLOMON LIPSON 1910–1991; FRS 1957

Henry Lipson's grandparents were Polish Jewish émigrés who had settled in Liverpool. He was the youngest child, and only son, of the three children of Isaac and Sarah Lipson. From a background of shopkeeping Henry's father later became a steelworker at Shotton, near Liverpool. Henry was educated at Hawarden County School and at the age of 17, with a scholarship and a Flintshire County Exhibition, he entered Liverpool University to study physics. In 1927, having obtained a first class honours BSc, he was made a Demonstrator in the Physics Department.

With another graduate student, Arnold Beevers, Lipson worked on the determination of crystal structures by X-ray diffraction; with some initial guidance from W.L. Bragg's group at Manchester they went on to introduce a procedure (the Lipson–Beever strips) for aiding the interpretation of complex structures. Lipson was awarded an MSc in 1931 (the PhD degree was not favoured in the Department); he continued his research holding the post of Demonstrator until in 1933 he was awarded the Oliver Lodge Fellowship which he held until 1936.

In 1936 Bragg obtained a grant from the Department of Industrial and Scientific Research for Lipson to work with Albert Bradley on the structure of metals and he became unofficial supervisor of research students in Bradley's group. Both Bradley and Lispon moved with Bragg to the National Physical Laboratory in 1937, and then to Cambridge the following year when Bragg became Cavendish Professor. Lipson was awarded a Liverpool DSc in 1939 and a Cambridge MA in 1942. As a Research Assistant at Cambridge Lipson's work included collaboration with A.J.C. Wilson. In 1937 Lipson married Jane Rosenthal and they had three children.

At the end of the war Lipson was appointed Head of the Physics Department in the Manchester College of Technology, which later became the University of Manchester Institute of Science and Technology. The department was largely concerned with service teaching, and Lipson immediately began to build up the existing crystallographic research to make the department an important centre. Working with a young assistant lecturer, Charles Taylor, they undertook research exploiting the analogy between optical and X-ray diffraction to assist in the solution of crystal structures. Lipson was appointed Professor in 1954 and retired when he was 67 but continued his research in the Electrical Engineering Department. He was awarded the CBE in 1976. Henry Lipson as a scientist made important contributions to the field of structure of alloys and he was an enthusiastic and effective teacher.
BMFRS 1994, 39, 229–244. M.M. Woolfson FRS.

GEORGE DOWNING LIVEING 1827–1924; FRS 1879
Born in Nayland, Suffolk George Liveing was the eldest son of Edward Liveing, a surgeon, and Catherine née Downing. He entered St John's College, Cambridge in 1847 and graduated as 11th Wrangler in the Mathematical Tripos in 1850. He then read for the newly established Natural Sciences Tripos, and obtained distinctions in chemistry and mineralogy in 1851. There followed a period of study in Berlin before returning to Cambridge, where in 1852 he started a course of practical chemistry for medical students. Downing was elected as a Fellow of St John's College in 1853, and the College also founded a lectureship for him and built him a laboratory. In 1860 he married Catherine Ingram; they had no children. In the same year he was appointed Professor of Chemistry at the Staff College at the Military College at Sandhurst, while continuing to teach at Cambridge. He was appointed as Professor of Chemistry at Cambridge in 1861; two years later the University started building laboratories, initiating a great development in experimental science, and during his subsequent career he played an important role in the development of chemistry in the University.

When James Dewar was appointed Jacksonian Professor of experimental philosophy in 1875, there began a long collaboration with Liveing; between 1878 and 1900 they published nearly 80 joint papers based on their spectroscopic investigations, which included metallic vapours. In addtion to spectroscopy Liveing's other interests included thermodynamics. In 1889 he was appointed a Professorial Fellow at St Johns. In 1908 he retired from the Chair of Chemistry, at the age of 81, but continued to maintain contact with his research. Liveing became President of St John's College in 1911. For some years he was engaged in research in the Department of Metallurgy. In the Royal Society he served on the Council from 1891–92 and 1903–14, and was awarded the Davy Medal in 1901.
PRS 1925, A, 109. xxviii–xxix. CTH. DNB 1922–30, 510–512, W.C.D. Dampier.

JOSEPH LOCKE 1805–1860; FRS 1838
Joseph Locke, who was born at Attercliffe, near Sheffield, was the fourth and youngest son of William Locke, a colliery manager. He was educated at Barnsley Grammar School, and in 1823 was articled to George Stephenson; when his articles were completed he continued to work with Stephenson on the construction of the Manchester–Liverpool railway. In his highly influential career as a civil engineer he was extensively associated with railway construction in England and on the Continent. He set a new standard of accuracy in estimating the costs of projects by supplying to contractors detailed specifications and by closely controlling the progress of the work. Together with Robert Stephenson and Isambard Kingdom Brunel he formed the triumvirate of

the railway engineering world. In 1834 he married Phoebe McCreery. In 1847 he moved to Honiton in Devonshire and served, until his death, as the Member of Parliament for the Borough of Honiton.
DNB.

SIR BEN LOCKSPEISER 1891–1990; FRS 1949

Ben Lockspeiser was born in London to Jewish parents Leon and Rose Lockspeiser, who had recently come from Eastern Europe. He attended the Grocer's School at Hackney, and went on to Sidney Sussex College, Cambridge, with an open scholarship. He obtained a first in the Natural Science Tripos, Part I and proceeded to the Mechanical Science Tripos in 1913. There followed a short period of study at the Royal School of Mines, London before he enlisted in the RAMC in 1914 and was sent to Gallipoli; dysentery led to him being invalidated out. In 1921 Lockspeiser joined the Royal Aircraft Establishment at Farnborough, and continued to work in the Scientific Civil Service until his retirement. He married Elsie Shuttleworth, and they had a son and two daughters.

At Farnborough Lockspeiser was in a section headed by A.A. Griffith, with whom he worked on stress concentrations in plastic crystals and on metal fatigue phenomena. Among other research projects two new types of manometers for the measurement of the vapour pressure of solids were described in 1930. In 1939 he became Assistant Director of Scientific Research at the Air Ministry; in 1941 he was appointed Deputy Director of Research at the Ministry of Aircraft Production where in 1943 he became Chief Scientist and in 1945 Director General. He was Chief Scientist at the Ministry of Supply from 1946–1949 and then became Secretary of DSIR. After retirement he joined the board of many companies. Lockspeiser was elected FREng in 1976 and was created a knight bachelor in 1946 and KCB in 1950.
BMFRS 1994, 39, 247–261. A.P.J. Edwards.

HEINZ LONDON 1907–1970; FRS 1961

Heinz London was born in Bonn, the son of Franz London, Professor of Mathematics at the University, and his wife Luise. Heinz's father died when he was nine, and his education was guided by his elder brother, Fritz, who became a leading figure in quantum chemistry. As was the custom in Germany, Heinz attended several universities: Bonn (1926–27); then, after a period with the chemical firm W.C. Heraeus, Berlin Technische Hochschule (1927); Munich (1929–31); at Munich he attended lectures by Planck, von Laue and Sommerfeld. He then carried out research in low temperature physics in the University of Breslau under Franz Simon and obtained a PhD in 1933 for a thesis which contained important new ideas on superconductivity. In 1933, with the advent of Hitler to power, he left Germany and went to the Clarendon Laboratory at Oxford where Simon, with others from Germany, including Fritz London had already come. Heinz continued to develop his work on superconductivity, before moving to Bristol in 1936, where his research included work on the superconductivity of thin metal films. In 1939 he married Gertrude Rosenthal, but following a divorce he married Lucie Meissner in 1945, by whom he had two sons and two daughters.

Early in the war years London became involved in work on isotope separation relating to the atomic bomb project, and worked at several centres including Bristol, Birmingham and Imperial College. He became a naturalised British citizen in 1942. After the war London moved to the Atomic Energy Research Establishment at Harwell, initially as a Principal Scientific Officer then as Senior Principal Scientific Officer (1950) and Deputy Chief Scientist (1958). Ideas and inventions were a notable feature of his career, including, for example, ideas for the technical exploi-

tation of superconductivity.
BMFRS 1971, 17, 441–461. D. Shoenberg FRS. DSB 8, 479–483. C.W. Everitt and W.M. Fairbank. DNB 1961–70, 672, 673. D. Shoenberg.

DAME KATHLEEN LONSDALE (NÉE YARDLEY) 1903–1971; FRS 1945
Kathleen Yardley was born in Newbridge, Ireland, the youngest of 10 children of the local post-master, Harry Frederick Yardley and Jesse née Cameron.In 1908 the family moved to England to Seven Kings, Essex, where Kathleen won a scholarship from the elementary school to the County School for Girls in Ilford. She went to Bedford College for Women in London at the age of 16 and having studied physics for a year changed to mathematics. In 1922 she graduated as the top student in the University and joined W.H. Bragg's research group at University College London, and later at the Royal Institution. She married Thomas Lonsdale in 1927 and moved to Leeds where they had three children and she worked in the Physics Department of the University. Returning to London in 1931 Lonsdale carried out research at the Royal Institution for 15 years. In 1945 she was one of first two women to be elected Fellows of the Royal Society. The following year she became Professor of Chemistry and Head of the Department of Crystallography at University College London. Lonsdale became DBE in 1956. Her research in X-ray crystallography, paticularly relating to organic molecules, made her one of the leading scientists in the field.
ONFRS 1975, 21, 447–484. Dorothy M.C. Hodgkin. OM. FRS. DSB 8, 484–486 J.M. Robertson.

GILBERT GEORGE LONZARICH 1945–; FRS 1989
University of Cambridge: Department of Physics, Cavendish Laboratory; Professor in Condensed Matter Physics; Fellow of Trinity College.
Metallic magnetism.

HENDRIK ANTOON LORENTZ 1853–1928; FMRS 1905
Hendrik Lorentz was born in Arnhem, the Netherlands, the son of Gerrit Frederik Lorentz and Geertruide née van Ginkel. He received his primary and scondary education in Arnhem, before entering the University of Leiden at the age of 17. Two year later he returned to Arnhem as a teacher, and studied alone for his doctorate which he received in 1875. In 1877 he was appointed to the new Chair of Theoretical Physics at Leiden. He married Aletta Catharina Kaiser in 1881 and they had two sons and one daughter. In a series of publications beginning in 1892 he made a major contribution in the field of electron theory. Lorentz moved to Haarlem in 1912 as Director of the Teyler Institute. In 1925 he gave an address to the Institute of Metals on 'The Motion of Electricity in Metallic Bodies'. Among honours recognising his achievements were the Nobel Prize for Physics (shared) in 1902, the Rumford Medal (1908) and the Copley Medal (1918).
PRS 1928, A, 121, xx–xxviii. OWR. DSB, 8, 487–500 R. McCormack.

AUGUSTUS EDWARD HOUGH LOVE 1863–1940; FRS 1894
Augustus Love was born at Weston-Super-Mare, the second of three sons of John Henry Love, a surgeon. He also had a sister, who later in his life kept house for him as he remained unmarried. The family moved to Wolverhampton, and Love entered the Grammar School in 1874. He entered St John's College, Cambridge, with a sizarship in 1882. After uncertainty as to whether to read classics or mathematics, he gradually chose the latter. He was elected Scholar of the College in 1884 and was Second Wrangler in the Mathematical Tripos of 1885. He was elected into a Fellowship at St John's in 1886 which he retained until 1899; for most of this period he was

College Lecturer in Mathematics. In 1899 he went to Oxford to the Sedleian Chair of Natural Philosophy, a position which he held until his death. In 1927 he was elected a Fellow of Queen's College, Oxford.

Love's mathematical research field, in which he made fundamental discoveries, was the theory of elasticity of solids and its application to problems of the earth's crust; his work also involved problems of hydrodynamics and electromagnetism. He was also a great expositor, in which capacity his work included *A Treatise on the Mathematical Theory of Elasticity* (first published in two volumes in 1892 and 1893). Among his honours were a Royal Medal in 1909 and the Sylvester Medal of the Royal Society in 1937.

ONFRS 1939–41, 3, 467–482, E.A. Milne. DSB 8, 516, 517. K.E. Bullen. DNB 1931–1940 545–546; E.A. Milne.

ANDREW McCANCE 1889–1983; FRS 1943

Andrew McCance, who was the third of four children of John McCance and Janet Ferguson née McGaw, was born at Cadder, near Glasgow. His father was then a cloth buyer but later worked for a shipping company. Andrew was educated at Morrison's Academy in Crieff, and then at Allan Glen's School in Glasgow, where science was encouraged. He studied at the Royal School of Mines in London where he graduated in metallurgy in 1910.

McCance obtained unpaid experience in steelmaking at Beardmores in Glasgow, where, six months later, he was given an appointment at 30 shillings a week. He became Assistant Armour Manager and worked for the company until 1919. In 1913 he was seriously injured in an accident involving a crane and a period of five months was spent in recovery. During his time at Beardmores McCance held considerable managerial responsibilities but also sought proper scientific understanding of the technical problems. Experimental work was carried out at the Royal Technical College and at the University of Glasgow to supplement his investigations in the works. McCance's early papers were concerned with some classical metallurgical problems, including the allotropy of iron and the microstructures of heat-treated steels. He made an important contribution also in the field of non-metallic inclusions in steel. In 1916 he was awarded the London DSc degree.

In 1919 McCance started the Clyde Alloy Steel Company, Motherwell. During the 1920s he worked long and hard, involving management and technical expertise. However, it was also in the early 1920s that McCance made his most important scientific contribution which concerned the physical chemistry of steelmaking, a field in which he had begun work at Beardmores.

In 1930 McCance became general manager of Colvilles at the invitation of John Craig and the partnership between Craig and McCance shaped the development of the steel industry in Scotland. McCance played a major role during the war in maintaining Colvilles as one of the largest producers of armour plate. By the middle of 1943 when attention was turned to post-war planning, the Colville policy became increasingly determined by McCance. He became Vice-Chairman and joint Managing Director of the Company in 1944 and was knighted in 1947. In 1956 Sir Andrew succeeded Craig as Chairman, and during his tenure of this post the average annual trading profit of Colvilles was £8 million. He retired in 1965, two years before nationalisation to which he was strongly opposed.

During the 1930s and 1940s although McCance had little opportunity for personal research, he read widely and thought. His work sought a quantitative approach and he produced a number of review lectures. In the strong research department at Colvilles in which he provided a general influence on the work done and investigations were made particularly on brittle fracture, creep and weldability problems.

In 1936 Sir Andrew married Joya Harriet Gladys Burford and they had two daughters. McCance was active on a number of scientific bodies, and served in various positions such as Deputy Lieutenant for Lanarkshire and Chairman of the Royal Technical College, Glasgow. He possessed a strong personality, with a management style involving consulting and listening, but then making up his own mind. His interests included music and the history of art. McCance played a dominant role in the development of the steel industry during the period 1930–65, combining commercial and technical leadership.

BMFRS 1984, 30, 389–405. N.J. Petch FRS and L. Barnard. DNB 1981–1985, 253–254. Monty Finniston.

JAMES DESMOND CALDWELL McCONNELL 1930–; FRS 1987

Queens' Univ. of Belfast: BSc, MSc. Univ. of Cambridge: MA, 1955, PhD 1956; Demonstrator, 1955; Lecturer, 1960; Reader, 1972–82. Schlumberger Cambridge Research: Head of Department of Rock Physics, 1983–88. Univ. of Oxford: Department of Earth Sciences; Professor of Physical Chemistry of Minerals, 1986–95; Head of Department, 1991–95.

Structural transformations in minerals.

ALAN LINDSAY MACKAY 1926–; FRS 1988

Wolverhampton Grammar School. Oundle School. Univ. of Cambridge: Trinity College, BA, MA. Univ. of London: BSc, PhD, DSc. Birkbeck College, Department of Crystallography, 1957–91; Lecturer, Reader, Professor 1986–91.

Crystallography: geometry and symmetry of crystalline and quasi-crystalline materials: five-fold symmetry.

ALLAN ROY MACKINTOSH 1936–1995; FRS 1991

Allan Mackintosh was born in Nottingham, the younger of two sons of Malcolm Roy Mackintosh and Alice née Williams. He was educated at Haydn Road Primary School and then at Nottingham High School, beginning in 1947. In 1954 he entered Peterhouse College, Cambridge as a scholar to read physics. After graduating BA he carried out doctoral research in the Cavendish laboratory, supervised by Sir Brian Pippard, investigating the Fermi surface of lead and tin using ultrasonic attenuation. He married Jette Stannow in 1958 and they had two daughters and one son. In 1960 he was appointed Associate Professor of Physics at Iowa State University where he became a leading expert in the new field of the physical properties, particularly magnetic behaviour, of rare-earth metals. Having spent a sabbatical in 1963/64 at the Riso National Laboratory in Denmark, he moved to Denmark in 1966 to the Technical University at Lyngby, and then in 1970 to the University of Copenhagen where he became Professor of Physics. From 1971–76 Mackintosh was Director of the Danish National Laboratory at Riso, and then returned to his Copenhagen chair until his death. Also from 1986–1989 he was Director of the Nordic Institute for Theoretical Physics in Copenhagen. In 1975 he was made a Knight of the Dannebrog Order.

BMFRS, 1997, 43, 321–331. Brebis Bleaney CBE, FRS. *Physics World* 1996, March, 67, 68. K. McEwan.

JOHN CUNNINGHAM McLENNAN 1867–1935; FRS 1915

John McLennan was born in Ingersoll, Ontario, Canada, the son of David McLennan, who had come from Scotland in 1865, and Barbara née Cunningham; he had one sister. He was educated at the Collegiate Institutions at Clinton and Stratford, and taught in various country schools to obtain money for his university education. He entered the University of Toronto in 1888 and graduated first in the class , with first class honours in physics in 1892. He then spent six years as a Demonstrator at Toronto, and was at the University of Cambridge from 1898–99 working with

J.J. Thomson. There followed eight years at Toronto as Demonstrator, Associate Professor (1902), Director of the Physics Laboratory (1904) and Professor of Physics (1907). He married Elsie Monro Ramsay in 1910; there were no children from the marriage. In 1917 he came to London to spend a period working with Rutherford on anti-submarine work and received the OBE. He was appointed Dean of the Graduate School at Toronto in 1930. McLennan's research included the fields of radioactivity, spectroscopy and ionisation potentials of various metals, and low temperature physics, for example, measurements of superconductivity and magnetism for certain metals and alloys. In the Royal Society he was awarded a Royal Medal in 1927, gave the Bakerian Lecture in 1928, and served as Vice-President from 1931–34. He was knighted (KBE) in 1935.
ONFRS, 1932–35, 4, 577–583. A.S. Eve. DSB, 18, 586, 587. L. Pyenson. DNB, 1931–40, 584, 585. J.A. Stevenson.

HEINRICH GUSTAV MAGNUS 1802–1870; FMRS 1863

Heinrich Magnus, who was born in Berlin, was the son of Johann Matthias Magnus, a prosperous founder of a London trading firm. Following a private education in mathematics and natural sciences he entered the University of Berlin in 1822. His first publication stemmed from an investigation of pyrophoric iron, cobalt and nickel. Following his doctorate dissertation (1827) on tellurium, he went in the following year to study with Berzelius. On returning to Berlin his habilitations-schrift on mineral analysis (1831) enabled him to begin lecturing on technology at the university; he was appointed Extraordinary Professor in 1834 and rose to be Professor of Technology and Physics(1845) and Rector (1861, 1862). His travels included industrial visits in England in 1835. Magnus married Bertha Humblot in 1840. His wide research interests, in addition to mineral analysis, included inorganic chemistry and organic agricultural chemistry. His main achievements were in physico-chemical and physical investigations, for example, in heat, electrolysis, thermoelectric currents, magnetism and mechanics.
PRS 1871–72, 20, xxvii–xxix. DSB 9, 18, 19. G.B. Kauffman. Ref. Ch. 4.1.

ROBERT MALLET 1810–1881; FRS 1854

Robert Mallet was born in Dublin, Ireland, the son of John Mallet of Devonshire, who settled in Dublin as an iron, brass and copper founder. He attended Bective House School until he was nearly 16; then following travel on the Continent he entered Trinity College, Dublin in 1826, where his education included special training in mathematics. After leaving College on graduation as BA in 1830 he spent much time in his father's works, in addition to visiting engineering establishments in England and taking instruction in surveying and levelling. In 1831 he made a tour of the Continent which led to a paper on glacier mechanisms. In the same year he married Cordelia Watson, by whom he had three sons and three daughters; his wife died in 1854, and in 1861 he married May Daniel.

In 1832 he became a full partner with his father in the firm J. and R. Mallet, and was in charge of the Victoria Foundry, with the fitting and machining shops.The early 1830s saw the dawning of the railway era, with an associated increase in engineering activity in Ireland. Under Mallet the foundry expanded into a major concern with large contracts for railway plant, permanent way materials, cast iron bridges, cranes and other ironwork. Mallet's career involved many aspects of engineering construction, including swivel bridges (1836) and railway stations and engine sheds (1848–1849). By the 1860s engineering work in Ireland having become scarce, the foundry was closed and Mallet moved to London in 1861 where he established himself as a consulting engineer. He became Honorary Master in Engineering of Trinity College in 1862.

In addition to his civil engineering activities Mallet carried out extensive scientific research; much of this was geological, e.g. concerning glaciers and the columnar structure of basalts, but also including a considerable metallurgical activity. From an early age he had developed an interest in chemistry, having a small room in his home as a laboratory. He was an early worker in the field of corrosion.

As early as 1832, on a visit to the Continent he had become interested in heavy ordnance, and over the years he carried out important work on this topic reported in his book *Physical Conditions Involved in the Construction of Artillery* (1856).

Mallet's metallurgical work ranged quite widely; in casting technology his observations and conclusions, including the recognition of the importance of planes of weakness associated with the meeting of different crystal growth fronts, illustrate his considerable ability as an experimental scientist, contributing to the development of metallurgical knowledge at a formative time in the 19th century.
PRS 1881–82, 33, xix–xx. DSB 9, 60, 61. W. Fischer. DNB. Refs Ch. 9.9, 13.

SIR ERNEST MARSDEN 1889–1970; FRS 1946

Ernest Marsden was born at Rushton, near Blackburn, the second son in a family of four sons and one daughter of Thomas Marsden and Phoebe née Holden. He was educated at Queen Elizabeth's Grammar School, Blackburn and at the University of Manchester, where he graduated with an honours degree in physics in 1906. Working with Rutherford on radioactivity he investigated the scattering of alpha particles from metal surfaces and found that the effect was dependent on the atomic weight of the metal. For a short period he taught at the East London College, before returning to Manchester as John Harling Fellow and working with Hans Geiger. He became a Lecturer and Research Assistant in 1912 and then went to Victoria University College, Wellington, New Zealand. In 1913 he married Margaret Sutcliffe and they had one daughter and one son; following his wife's death he married Joyce Winifred Chote in 1958. Marsden was commissioned in the New Zealand Territorial Force and volunteered for service overseas in the 1914–18 war. He was seconded to the Royal Engineers Sound Ranging Section in France and was mentioned in dispatches and awarded the MC (1919); he resumed Territorial Service in New Zealand, and retired with the rank of Major. His career included senior administrative posts in New Zealand, including serving as Permanent Secretary of the Department of Scientific and Industrial Research. During World War II he was scientific adviser to the New Zealand fighting services. His honours included CBE(1935) and CMG(1946); he was made a knight bachelor in 1958.
BMFRS, 1971, 17, 463–496. C.A. Fleming FRS. DSB, 18, 595–597 C.A. Fleming. DNB 1961–70, 731–733. C.A. Fleming.

WALTER CHARLES MARSHALL
(BARON MARSHALL OF GORING) 1932–1996; FRS 1971

Walter Marshall was born in Rumney, Cardiff, the youngest of the three children of Frank Marshall and Amy née Pearson. His grammar school education was at St Illtyd's College in Cardiff. In 1949 he entered Birmingham University with a Major County Scholarship; having been awarded a first class honours degree in mathematical physics, he proceeded to obtain a PhD degree at the age of 22 for a thesis on 'Antiferromagnetism and neutron scattering from ferromagnets'; his supervisor was Rudolf Peierls. Marshall then joined the Theoretical Physics Division at the Atomic Energy Research Establishment at Harwell in 1954 as a Scientific Officer. Following some work on plasma physics he returned to condensed matter physics and spent two periods of study in the

USA during 1957 and 1958, with visits to Berkeley, Livermore and Harvard. He married Ann Vivienne Sheppard in 1955, and they had a son and a daughter.

In 1959 Marshall became the Leader of the Solid State Theory Group at Harwell; promotion occurred to Head of the Theoretical Physics Division in 1960, a post which he held until 1966. His personal contributions in solid state theory concerned magnetic properties, including transition metals. he interacted strongly with experimentalists and directed programmes concerned with obtaineing improved understanding of real materials.

Marshall was promoted to the post of Deputy Director of Harwell in 1966, and two years later he became Director, serving until 1975. He set about developing Harwell to become a laboratory for improving industrial performance generally and for meeting specified government requirements. Marshall was increasingly concerned with energy policy during the 1970s, and held a number of senior appointments including Headship of the Research Group of the UKAEA from 1969–1975, with responsibility for the Culham and Harwell Laboratories. In 1972 he was appointed to the Board of the UKAEA , serving until 1982 and becoming Deputy Chairman in 1975 and holding the post of Chairman from 1981–1982. From 1974–1977 he was Chief Scientist to the Department of Energy on a part-time basis; in 1977 he was dismissed from this post by the Secretary for Energy and returned full time to the UKAEA. Marshall was a strong advocate of nuclear power and played a major role in the debate on the choice of reactor for the next series of civil reactors; he favoured the pressurised water type (PWR), but there were concerns as to the integrity of the pressure vessels. Marshall set up a study group which made a thorough analysis and reported in 1976; there ensued further work, with a group including more academic members, and this led to a second report in 1980 and the adoption of the PWR.

In 1983 Marshall was appointed as the Chairman of the Central Electricity Generating Board in which capacity he became the target of groups opposed to nuclear energy. During the miners strike in 1985–86 he was responsible for and successful in maintaining the electricity supply to the United Kingdom. In 1989 in the light of political decisions concerning the future of CEGB Marshall resigned as Chairman. He was appointed as Executive-Chairman of a new international organisation called the World Association of Nuclear Operators (WANO); this was set up in response to the Chernobyl explosion. Marshall headed the organisation until 1993. In 1995 Marshall became Chairman of an insurance syndicate specialising in nuclear risks.

Marshall was an outstanding theoretical physicist; his scientific publications included two technical books *Thermal neutron scattering* (1984) and *Nuclear power technology* (1984). He was also a distinguished public servant, and a gifted public speaker. His achievements were recognised by the award of many honours, including a CBE (1973), a knighthood (1982) and a life peerage (1985).

BMFRS 1998, 44, 299–312. D. Fishlock OBE and L.E.J. Roberts, CBE, FRS; *The Times* 23.2.96. *The Guardian* 23.2.96. *The Independent* 26.2.96.

THOMAS NELSON MARSHAM 1923–1989; FRS 1986

Thomas Marsham was the son of Captain Thomas Brabban Marsham, of the Merchant Navy, and his wife née Nelson. He was educated at Crosby Grammar School and at Merchant Taylor's School, Crosby, Lancashire. In 1941 he left school and became a sea-going radio officer, seeing service in the Far East. When he returned in 1946 he became a student in the City of Liverpool Technical College and then in the University; he studied physics, chemistry, mathematics and oceanography. He was awarded a first class degree in general science in 1949, followed by a 2.1 honours degree in physics the following year. With the award of an Oliver Lodge Research Fellowship he carried out research in nuclear physics, obtaining a PhD in 1953. Marsham then entered the atomic energy industry as a physicist working at the Technical Policy Branch at Risley on uranium enrichment and diffusion plant problems. In 1955 he was appointed Reactor Manager of Calder Hall Nuclear Power Station, where his work included the commissioning and operation of two MAGNOX reactors. He returned to Risley as Deputy Chief Physicist Reactors in 1957, and then in the following year he returned to Windscale as Deputy Works General Manager, Windscale and Calder Works. In 1958 he married Sheila Margaret Griffin and they had two sons.

Marsham's responsibilities at Windscale included the commissioning of the prototype advanced gas-cooled reactor. He was appointed as Director of Technical Operations at the headquarters of the Reactor Group of AEA in 1965, and, in 1969, he became Deputy Managing Director of the Group while remaining Director of Technical Resources. His contributions included work on the fast-neutron fission reactor. In 1977 he was appointed Managing Director of the Northern Division of the Atomic Energy Authority, and two years later he became a full-time member of the Board for Reactor Development. From 1975–1980 he was also a Non-Executive Director of the Nuclear Power Company, while in 1979 he became a Non-Executive Director of Britich Nuclear Fuels plc. Marsham was awarded the OBE (1964) and CBE (1976). He was the recipient of the Royal Society Esso Energy Medal in 1978 and was elected FREng in 1986.
BMFRS 1992, 38, 231–246. L.E.J. Roberts, FRS.

RAGHUNATH ANANT MASHELKAR 1943–; FRS 1998
Council of Scientific and Industrial Research, New Delhi, India: Director General. National Chemical Laboratory, India: Director.
Polymer engineering;polymerisation reactors; diffusion.

SIR RONALD MASON 1930–; FRS 1975
Universities of Wales and London. British Empire Cancer Campaign: Research Associate 1953–61. Imperial College, London: Lecturer, 1961–63. Sheffield Univ.: Professor of Inorganic Chemistry, 1963–71. Ministry of Defence: Chief Scientific Adviser, 1977–83. Sussex Univ.: Professor of Chemistry, 1971–88; Pro-Vice-Chancellor, 1977. British Ceramic Research Ltd: Chairman, 1990–. University College Hospitals NHS Trust: Chairman, 1993–. KCB 1980.
Surface science: transition metal surfaces-diffraction and spectroscopic studies; chemisorption; catalysis. Ceramics.

SIR HARRIE STEWART WILSON MASSEY 1908–1983; FRS 1940
Harrie Massey was born in St. Kilda, a suburb of Melbourne, Australia. He was the only child of Harrie Stewart Massie and Eleanor née Wilson. After attending the local school at Hoddles Creek he obtained a scholarship to the University High School in Melbourne. At Melbourne University, which he entered in 1925 with a scholarship, he took the honours courses in both chemistry and natural philosophy in three years, obtaining a first class honours BSc degree in 1927. He continued his studies and obtained a first class BA degree in mathematics in 1929; also at the beginning of 1928 he began work leading to an MSc degree in the Department of Natural Philosophy; the experimental aspect of the project was concerned with the reflection of soft X-rays from metal surfaces. In 1925 he married Jessica Elizabeth Barton-Bruce and they had one daughter. With an Aitchison Scholarship Massey went to the University of Cambridge, Trinity College in 1929 and worked in the Cavendish Laboratory; he was awarded an 1851 Senior Research Studentship in

1931 and a PhD in 1932. His research was on atomic collisions, a subject on which he published a book in 1933, and on which he continued to work throughout his career. He held an appointment as a Lecturer in Mathematical Physics at Queen's University, Belfast from 1933–38. Massey then came to University College London as Goldschmid Professor of Mathematics.

At the end of 1939 Massey began work with the Admiralty on counter-measures to the German magnetic mines; he led a team of young physicists and in 1941 he was appointed Deputy Chief Scientist concerned with mine design, and in 1943 he became a Chief Scientist. In the same year, he was appointed to lead a group at Berkeley, California concerned with the large-scale separation of U-235. He returned to University College London in 1945 and in 1950 was appointed as Quain Professor of Physics, a post which he held until 1972; he served as Vice-Provost from 1969–73. His activities included space research and he became President of the Council of the European Space Research Organization. In the Royal Society Massey served as a Council Member (1949–51, 1959–60), as Physical Secretary (1969–78) and as Vice-President (1969–78); he was awarded the Hughes Medal (1955) and a Royal Medal (1958). He was knighted in 1960.
BMFRS 1984, 30, 445–511. Sir David Bates FRS, Sir Robert Boyd FRS and D.G. Davis. DNB 1981–85, 271, 272. Nevill Mott.

GEORGE MATTHEY 1825–1913; FRS 1879

George Matthey was one of the sons of John Matthey, a wealthy stock broker and foreign exchange dealer, whose father had come from Switzerland in 1790 to settle in London. George was educated at Arragon House, Twickenham. John Matthey was a great friend of Percival Johnson, and in 1838 they made an agreement that he would provide capital to expand Johnson's business, and that his sons, George and Edward Matthey, would be apprenticed to the business. George, apprenticed under the Goldsmith's Company, at once began his work in the assay laboratory. At this time the firm was known as Johnson and Cock and besides gaining the excellent training from Percival Johnson, George had the benefit of William James Cock (son of Thomas Cock) as a mentor. When William Cock retired in 1838, George Matthey was put in charge of the small scale platinum refining these operations. In 1851 Johnson took Matthey into partnership the firm becoming Johnson and Matthey, and with Matthey largely in charge. Matthey's activities included an important role in providing high purity metals for work on international standards. By 1880 the platinum business was a prosperous activity.

In 1853 Matthey married Charlotte Ann Davies and they had 11 children. Matthey was a leading figure in the Worshipful Company of Goldsmiths, and was Prime Warden in 1872 and in 1894. Also he took an active interest in technical education in London. In 1891 Johnson Matthey was incorporated as a limited company and George Matthey became Chairman, but retained supervision of the platinum refining. He finally retired in 1909, following a career that had spanned just over seventy years; the development of the platinum industry was largely due to his efforts.
Platinum Metals Review 1979, **23**(2), 68–77. Ref. Ch. 4.1.

AUGUSTUS MATTHIESEN 1831–1870; FRS 1861

Augustus Matthiesen was a Londoner by birth, the son of a merchant who died while the boy was quite young. A paralytic attack during infancy resulted in a disability in his right hand, and although he showed an interest in chemistry as a boy his guardians sent him to learn farming as an occupation suited to his situation. He then became interested in agricultural chemistry a subject that was attracting wide attention. When he came of age he went to Giessen to continue his education in chemistry and was awarded a PhD in 1853. Following this he spent nearly four years at Heildeberg with Robert Bunsen under whose direction he prepared significant quantities of lithium, strontium, magnesium and calcium using Bunsen's electrolytic method. Then in the laboratory of Gustav Kirchhoff he studied the electrical conductivity of these and other metals. On returning to London in 1857 he studied organic chemistry for a short period with Augustus Hofmann at the Royal College of Chemistry.

Matthiesen was a skilled experimentalist. Working in his laboratory in his home at Torrington Place in London, he carried out a series of investigations on the physical properties of pure metals and alloys which became classical. From 1862 to 1865 he was a Lecturer in Chemistry at St Mary's Hospital and in 1868 he took up a similar appointment at St Bartholomew's Hospital. He carried out work for the British Association Committee on Electrical Standards, and prepared a number of standards from various metals and alloys. He experienced severe nervous strain and his death by suicide occurred in 1870, only a year after he was awarded a Royal Medal.
DSB 9, 178. D.P. Jones. DNB. Gen. Ref. 21.

JAMES CLERK MAXWELL 1831–1879; FRS 1861

James Maxwell was the son of John Clerk Maxwell and Francis née Cay; he was born in Edinburgh, where he was educated at the Royal Academy from 1841 to 1847 and then at the University. In 1850 he went to Cambridge to Peterhouse and then Trinity College, where he graduated in 1854 as Second Wrangler and Smith's Prizeman; he became a Fellow of Trinity the following year. Two years later he became Professor of Natural Philosophy at Marischal College, Aberdeen. Maxwell married Katherine Mary Dewar in 1858; there were no children from the marriage. He took the post of Professor of Natural Philosophy and Astronomy at King's College, London in 1860. Five years later he returned to his family home in Scotland where he carried out research. In 1871 he became the first Professor of Experimental Physics at Cambridge and planned and developed the Cavendish Laboratory which opened in 1874; he remained at Cambridge until 1879.

Maxwell was awarded the Rumford Medal in 1860 and was the Bakerian Lecturer in 1866. His research included major fundamental contributions in the fields of electromagnetism and the kinetic theory of gases; he is widely regarded as the greatest theoretical physicist of the 19th century.
PRS 1881–82, 33, i–xvi. DSB 9, 198–230. C.W.F. Everitt.

JOHN MAYOW 1640–1679; FRS 1678

John Mayow was born at Bray, near Looe, the second son of Phillip Mayowe. He matriculated at Wadham College, Oxford in 1658, and was received as a commoner in 1659 and admitted as a scholar in 1660; he was elected to a Fellowship at All Souls in 1660. He graduated bachelor of common law in 1670. He also studied medicine and when he left Oxford in 1670 he practised in Bath. During the 1670s he spent time in London and became known for his investigations of atmospheric composition, respiration and combustion.
DSB 9, 242–247. Theodore M. Brown.

LISA MEITNER 1878–1968; FMRS 1955

Lisa Meitner was born in Vienna, the third of eight children of the Jewish family of Dr Philipp Meitner, a lawyer. Before beginning her studies in physics at the University of Vienna in 1901 she qualified as a teacher of French. She obtained a PhD in 1905 for research on thermal conduction and two years later went to Berlin to the University to attend Planck's lectures; from 1912–

15 she was assistant to Planck at the Institute of Theoretical Physics. Also in Berlin she carried out experimental work with Otto Hahn on radioactivity in the Chemical Institute of the University, beginning in 1907. In 1912 when the new Kaiser Wilhelm Institute for Chemistry opened with Hahn as Head of the Radiochemical Section she went to work with him as a member of the Institute; she became joint Director with Hahn and Head of the Physics Department in 1917. During the 1914–18 war Meitner served as an X-ray nurse in Austrian Army Field Hospitals and Hahn was in the army, but there were opportunities for some research in Berlin on radioactive elements and their properties, during times of leave from their military duties. They discovered the new radioelement protactinium in 1918. Meitner was appointed Docent at the University of Berlin in 1922 and Extraordinary Professor of Physics in 1926.

During the 1930s Meitner worked with Hahn on uranium bombarded with neutrons. In 1938 Austria was annexed by Germany and, since her Austrian citizenship no longer protected her from danger, she escaped to Holland, and then to Denmark, before taking a post at the Nobel Institute in Stockholm, where a cyclotron was being built. Having received from Hahn the results he had obtained on the neutron bombardment of uranium she discussed them with her nephew Otto Frisch and they wrote a paper on 'nuclear fission'. In 1947 a laboratory was established for her at the Royal Institute of Technology, Stockholm by the Swedish Atomic Energy Commission and she became a Swedish citizen two year later. When she retired in 1960 she came to England and settled in Cambridge.

BMFRS 1970, 16, 405–420. O.R. Frisch FRS.

JOSEPH WILLIAM MELLOR 1869–1938; FRS 1927

Joseph Mellor was born in Huddersfield, but at the age of 10 years his parents moved to New Zealand. There as a young man he worked in a boot factory and took evening classes at the Dunedin Technical School. Having matriculated at the age of 23 he entered the University of Otago a year later. Following his graduation he held a post as a science lecturer at an agricultural college. The award of an 1851 Exhibition enabled him to return to England and enter the University of Manchester in 1899 where he stayed until 1904. There, at Owens College, he worked on problems in both inorganic and organic chemistry; his main research project concerned the combination of hydrogen and chlorine. On leaving Manchester he moved to North Staffordshire, where for a short period he was a science master at Newcastle (Staffs) High School, before being appointed as a Lecturer in Pottery Manufacture in Stoke-on-Trent. His research in ceramics ranged widely, including refractory materials and glazes. Mellor achieved a considerable reputation as a teacher of ceramics, and when the North Staffordshire Technical College was built in 1914 he served as Principal of the Pottery Department until he retired in 1934. He was actively involved in the work of the Ceramic Society, from its early days, holding the office of Honorary Secretary from 1905 until his death.

Mellor recognised the importance of the durability of refractory materials in industries concerned with high temperature manufacturing processes. He was influential in the setting up the Refractories Committee of the Institution of Gas Engineers. When the British Refractories Association was formed in 1919, under the aegis of the Department of Scientific and Industrial Research, Mellor was appointed as the first Director and held the post until 1937, rendering notable service to the industry. Mellor had great literary ability and was the author of several books, including the monumental work (16 volumes) *A Comprehensive Treatise on Theoretical and Inorganic Chemistry* (1937). Mellor's achievements were widely recognised and he received the CBE in 1937.

ONFRS 1936–38, 2, 573–575. A.T. Green.

SIR HARRY WORK MELVILLE 1908–; FRS 1941

George Heriot's School, Edinburgh. Edinburgh Univ.: Carnegie Research Scholar, PhD, DSc. Univ. of Cambridge: Trinity College, 1851 Exhibitioner; Fellow, 1933–44; DSc; Colloid Laboratory, Assistant Director, 1938–40. Univ. of Aberdeen: Professor of Chemistry, 1940–48. Ministry of Supply: Scientific Adviser to Chief Superintendent, Chemical Defence, 1940–43. Radar Research Station: Superintendent, 1943–45. Univ. of Birmingham: Mason Professor of Chemistry, 1948–56. Department of Scientific and Industrial Research: Secretary, 1967–76. Queen Mary College, University of London: Principal, 1967–76. Davy Medal 1955. Bakerian Lecturer, 1956. KCB 1958.

Chain reactions involving free radicals. Kinetics of polymerisation.

DMITRI IVANOVICH MENDELEEV 1834–1907; FMRS 1892

Dmitri Ivanovich Mendeleev was the youngest of 14 children of Ivan Pavlovitch Mendeleev, Director of the Gymnasium at Tobolsk, Siberia. His mother was Marie Dimitrievna Kornilova, from an old Russian family, manufacturers of paper and glass. When he was 15 he went to Moscow to continue his education, and a year later to St Petersburg, where he mainly studied physical sciences. He then took teaching appointments at Simferopol in the Crimea and at Odessa. In 1856 he returned to the University at St Petersburg as Privatdozent. From 1859–1861 he worked in a private laboratory in Heidelberg studying physical constants of chemical compounds. Back in St Petersburg he obtained his doctorate, and soon afterwards became Professor of Chemistry in the Technological Institute. In 1866 he became Professor of General Chemistry in the University, while also holding the Chair of Organic Chemistry. He was often employed by the government in relation to important technical matters, and in the department of weights and measures; the latter work brought him to England several times.

Mendeleev was awarded the Davy Medal in 1882 and the Copley Medal in 1905. The work with which his name is primarily connected, i.e. the development of the Periodic Law, was communicated to the Russian Chemical Society in March 1869 as a paper on 'The relations of the properties to the atomic weights of the elements'. This law was one of the most fertile concepts in the development of modern chemistry.

PRS 1910–11, A, 84, xvii–xx. WAT. DSB 9, 286–295. B.M. Kedrov.

KURT ALFRED GEORG MENDELSSOHN 1906–1980; FRS 1951

Kurt Mendelssohn was the only child of Ernst Moritz Mendelssohn and Elisabeth née Ruprecht. He was born in Berlin-Schoenberg and from the age of 6 to 19 he was educated at the Goethe Schule in Berlin. In 1925 he entered the University of Berlin where he studied physics, chemistry, mathematics and psychology. He commenced research in 1927 at the Physikalisch Chemische Institut, his supervisor being his cousin Franz Simon, who had recently achieved the liquifaction of helium by a new technique. In 1930 when Simon moved to the Chair of Chemical Physics at the Technische Hochschule, Berlin, Mendelssohn went with him. In 1933 Mendelssohn married Jutta Zarniko, one of Simon's students, and they had five children.

In 1933 he came to England to escape from Germany (followed later by his wife), and obtained an appointment with Lindemann at Oxford, supported by Imperial Chemical Industries. Mendelssohn was appointed an ICI Fellow at Oxford in 1945; two years later he became a Demonstrator and then from 1955 until his retirement in 1973 he was a Reader in Physics. In 1965 he was elected an Ordinary Fellow of the newly founded Iffley College (which was renamed Wolfson College in 1966); he later became a Professorial Fellow. His research at Oxford in the field of low temperature physics included

the superconductivity of alloys (and also their thermal conductivity) and the superfluidity of liquid helium. He travelled widely and had many research students and visiting scientists from abroad in his group. Mendelssohn was awarded the Hughes Medal of the Royal Society in 1967.
BMFRS 1983, 29, 361–398. D. Shoenberg.

SIR JAMES WOODHAM MENTER 1921–; FRS 1966
Dover Grammar School. Univ. of Cambridge: Peterhouse College, PhD, 1949, ScD, 1960. Admiralty: Experimental Officer, 1942–45. Cambridge Univ.: Research 1956–68; ICI Fellow 1951–54. Tube Investments Research laboratories, Hinxton Hall: 1954. TI Ltd: Director Research and Development, 1965–76. TI Labs: Director, 1961–68. Royal Society: Vice-President, 1971–76; Treasurer, 1972–76. Knighted 1973.
Electron microscopical investigations of crystal lattices.

FRANK BRIAN MERCER 1927–1998; FRS 1984
Brian Mercer was a member of a family which, for generations, had been involved in the textile manufacturing industry. It was intended that he should take a City and Guilds course to become a Chartered Textile Technologist before beginning university studies. However, with the coming of war in 1939 he enlisted in the Royal Horseguards, but as a result of an accident he was invalided out, and worked in the family business. When Mercer was 23 he was appointed as Managing Director of the Mercer Group. In 1954 he founded British Tufting Machinery Ltd based in Blackburn. Later he sold this company and founded Netlon Ltd in Blackburn in 1959. This followed his novel and highly significant invention of the Netlon process, which enabled plastics to be extruded in a one-stage process to form a net-like product. He was Chairman until 1975 and then President. Netlon products were manufactured, and patents were taken out representing improvements; licensing to many international companies and in many countries took place. In the late 1970s and early 1980s Mercer developed another mesh-like plastic product, Tensor; this was designed to be used in the construction industry, for stabilisation and reinforcement of soil structures.

In the Royal Society Mercer was awarded the Mullard Medal in 1974 and gave the Philips Lecture in 1986. He was elected FREng in 1986 and was awarded the OBE in 1981.
Materials World, 1999, February, 98.

CHRISTOPHER MERRET 1614–1695; FRS 1663
Christopher Merret (or Merrett) was the son of Christopher Merret, and was born at Winchomb, Gloucestershire. In 1631 he became a student at Gloucester Hall, Oxford, then moved to Oriel College two years later, graduating BA in 1635. Returning to Gloucester Hall he studied medicine and graduated MB in 1636; he was awarded an MD in 1643. By the time he had begun practice in London around 1640 he had married Ann Jenour. He became a Fellow of the College of Physicians in 1651. He was appointed the first Keeper of the Library and Museum of the College. Following the Great Fire of London in 1666, in which the College Museum and Library were destroyed, the College decided that it no longer needed his services. The subsequent dispute led to his losing his Fellowship of the College in 1681. Merret was a significant writer, whose work included natural history. He was also the translator of the book by Antonio Neri on glassmaking (*L'arte vetraria*, 1662); he added his own extensive observations on glassmaking to this book, and also reported some metallurgical information on English practice in making brasses.
DSB 9, 312, 313. Leonard M. Payne. DNB. Gen. Ref. 19. v III.

HENRY ALEXANDER MIERS 1858–1942; FRS 1896

Henry Miers was born in Rio de Janeiro, Brazil, the third son of Francis Charles Miers, a civil engineer, and Susan Fry Miers. He won scholarships to Eton and to Oxford, where he graduated in mathematics in 1881. Through working visits at Oxford and Strasbourg in the area of crystallography he prepared for and obtained a post as assistant, working on crystal forms, at the British Museum. In 1886 he began to give a course in crystallography at the recently opened City and Guilds of London Institute, where his most famous pupil was William Jackson Pope. In 1895 he was appointed Waynefleet Professor of Mineralogy at Oxford. After holding the post of Principal of the University of London commencing in 1908, there followed a series of administrative posts, including the Vice-Chancellorship of Manchester University, beginning in 1915. He was knighted in 1912.

ONFRS 1942–44, 4, 369–380. Sir Thomas Holland. DSB 9, 379–38. W.T. Holser.

DAVID ANDREW BARCLAY MILLER 1954–; FRS 1995

St Andrews Univ.: BSc 1976. Heriot-Watt Univ.: PhD physics, 1979; Research Associate, 1979; Lecturer, 1980–81. AT and T Bell labs, Holmdel, NJ: Technical Staff, 1981–87; Head Advanced Photonic Research Dept (formerly AT and T Bell) Labs: 1987–96. Stanford Univ.: W.M. Keck Foundation Professor of Electrical Engineering, 1997–; Professor of Electrical Engineering, 1996–; Director E.L. Ginzton Laboratories and Solid State Photonics Laboratory, 1997–.

Optical properties of semiconductors.

STEWART CHRICHTON MILLER 1934–1999; FRS 1996

Stewart Miller was the son of William Young Chrichton Miller and Grace Margaret née Finley. He was educated at Kirkaldy High School and at Edinburgh University, where he graduated with a BSc degree. He joined Rolls Royce in 1954, and after a period of training he held a series of technological appointments from 1956–76. Subsequently, Miller was Chief Engineer for the RB 211–535 Engine Work (1977–84), Director of Engineering (1985–90), Managing Director Aerospace Group (1991–92), and Director of Engineering and Technology (1993–96). His activities including directing work on the hollow titanium fan blade for aeroengine compressors. In 1960 he married Catherine Proudfoot McCourtie, and they had two sons and two daughters. He was elected FREng in 1987 and was awarded the CBE in 1990.

Who's Who.

WILLIAM ALLEN MILLER 1817–1870; FRS 1845

William Miller was born at Ipswich and at the age of 15 he was apprenticed to his uncle, a Birmingham surgeon. Five years later he entered the medical department of King's College, London, where he obtained the degrees of MB and MD in the University of London in 1841–42. During the period of his medical studies he was awarded a prize in theology, and also spent a few months in 1840 working in the laboratory of Justus Liebig at Giessen; on his return he was appointed as a Demonstrator in Chemistry at King's College, London. Having become Assistant Lecturer to John Daniell in 1841 he co-operated with him in his research. He married Eliza Forrest in 1842 and there were three children from the marriage. On Daniell's death in 1845, Allen succeeded him to the Chair of Chemistry at King's College. His research included spectrum analysis, astronomical physics and electrochemistry. Among his other activities he aided in the chemical testing of the stone used in building the Houses of Parliament, and was Assayer to the Mint and to the Bank of England. In the Royal Society he served on the Council; from 1861

until his death he was Treasurer and from 1855–57 and 1861–70 he was Vice-President.
PRS 1870–71, 19, xix–xxvi. CT. DNB.

WILLIAM HALLOWES MILLER 1801–1880; FRS 1838

William Miller came from a family with military associations and was born at Velindre, near Llandovery. After education at private schools he entered St John's College, Cambridge and graduated as Fifth Wrangler in the Mathematical Tripos in 1826. He was elected to a college fellowship in 1829 and to the Chair of Mineralogy in 1832. In accordance with the statutes he proceeded in 1841 to the degree of MD in order to retain his Fellowship. However, he vacated this when he married Harriet Minty in 1844; there were two sons and four daughters from the marriage.

Miller had an exceptionally wide knowledge of natural philosophy. It was in crystallography that he acquired his great reputation, with the introduction of a new system incorporating the indices which carried his name. Miller had a share in the reconstruction of the standards of weight and length which had been destroyed in 1834 when the Houses of Parliament were burnt. He was Secretary of the Royal Society from 1856–1873 and was awarded a Royal Medal in 1870.
PRS 1880–81, 31, ii–vii.

SIR EDGAR WILLIAM JOHN MITCHELL 1925–; FRS 1986

Kingsbridge Grammar School. Univ. of Sheffield: BSc, MSc. Univ. of Bristol : PhD. Metropolitan Vickers Research Department: Research Physicist. Univ. of Bristol: 1948–1950. Univ. Reading: 1957–78 : Professor of Physics, 1961–78; Dean, Faculty of Science, 1966– 69; Dept Vice Chancellor, 1976–78. Univ. of Oxford: Dr Lees Professor of Experimental Philosophy, 1978–88; Professor of Physics 1988–89. Science Research Council: Chairman, 1967–70. CBE 1976. Knighted 1990.
Solid state physics.

JOHN WESLEY MITCHELL 1913–; FRS 1956

Canterbury Univ. College, Christchurch, N.Z.: BSc, 1934; MSc 1935. Univ of Oxford: PhD 1938, DSc, 1960. Univ. of Bristol: 1945–59; Reader in Experimental Physics. Univ. of Virginia: Professor of Physics, 1959–63. National Chemical Lab.: Director, 1963–64. Univ. of Virginia: William Barton Rogers Professor of Physics, 1968–79; Senior Research Fellow 1979–.
Plastic deformation of crystals, photographic sensitivity.

HENRY KEITH MOFFATT 1935–; FRS 1986

George Watson's College, Edinburgh. Edinburgh Univ.: BSc. Univ. of Cambridge: BA, PhD, ScD; Lecturer in Mathematics and Director of Studies in Mathematics at Trinity College; 1961–76. Univ. of Bristol: Professor of Applied Mathematics; 1977–80. Univ. of Cambridge: Professor of Mathematical Physics; 1980–; Head of Department and Applied Mathematics and Theoretical Phsics; 1983–91; Isaac Newton Institute for Mathematical Science: Director 1996–.
Dynamics of fluids and plasmas: metallurgical magnetohydrodynamics.

(FERDINAND-FREDERIC-HENRI) HENRY MOISSAN 1852–1907; FMRS 1905

Henry Moissan was born in Paris to parents of modest means; in 1864 his parents moved to Meaux where he attended the Municipal College, but did not complete his studies. Returning to Paris he worked for two years as a pharmacy apprentice. Then in 1872 he went to work at the

Muséum d'Histoire Naturelle where he carried out research on plant physiology. To obtain more formal academic education he studied in Paris, being awarded the preliminary degree of bachelier in 1874, licencié in 1877, pharmacien de premiere classe in 1879 and docteur ès sciences physiques in 1880. He married Leonie Lugan of Meaux in 1882 and received financial support from her father which helped him to pursue his scientific work. In 1886 he was appointed to a Chair in Toxicology at the École Supériere de Pharmacie and three years later he became Professor of Inorganic Chemistry there; in 1900 he held the Chair of Inorganic Chemistry at the Faculty of Sciences.

Moissan decided on inorganic chemistry as his main interest. His first research, on the oxides of the iron group of metals, particularly with compounds of chromium, attracted the attention of Deville and Debray. In 1884 he commenced his remarkable programme of work on compounds of fluorine, which culminated in the isolation of this element. The preparation of two gaseous fluorides of carbon led Moissan to attempt to remove the fluorine, in the hope that the carbon would be liberated in the form of diamond; this work led to his investigation of the artificial production of diamond.

For this investigation Moissan set about dissolving carbon in molten iron followed by quenching from the liquid state aiming to achieve high internal pressures. To obtain the necessary high temperatures he invented an electric arc furnace. The liquid iron–carbon alloys were quenched into cold water, and small particles, having the characteristics of diamond were chemically extracted from the metal. Work carried out some decades later indicated that these particles consisted of another type of inorganic compound. However, using the furnace, Moissan prepared a large number of high melting point metals (for example, tungsten and titanium) and compounds (for example SiC, and other carbides and borides). The furnace provided substantial impetus to the development of high temperature chemistry and found important metallurgical applications in industry.

Moissan was the recipient of numerous honours, including honorary membership of academies in many countries and the Davy Medal (1896). He had a remarkable influence on inorganic chemistry and in 1906 he was awarded the Nobel Prize for Chemistry.
PRS 1907–1908A, 80, xxx–xxxvii. W.R. DSB 9, 450–452. A. Berman.

ALFRED MORITZ MOND (FIRST BARON MELCHETT) 1868–1930; FRS 1928

Alfred Mond, the younger son of Ludwig Mond and Frida née Loewenthal, was born at Farnworth, Lancashire. Following education at Cheltenham College he entered St John's College, Cambridge, where he failed in the Natural Science Tripos. He then attended Edinburgh University, and was called to the Bar in 1894, subsequently practising for a while. In 1894 he married Violet Florence Mabel Goetze, and they had one son and three daughters. In 1895 he became a director in his father's business (Brunner, Mond and Company) and then Managing Director. He succeeded his father as Chairman of the Mond Nickel Company. He achieved prominence through the many important enterprises with which he was associated, and the amalgamations which in 1926 were consolidated as Imperial Chemical Industries Ltd. Alfred Mond had a successful political career, beginning in 1906 as Member of Parliament for Chester; from 1910 to 1923 he represented Swansea, and from 1924 to 1928, Carmarthen. He also held ministerial posts, and was created a Baronet in 1910 and a Privy Councillor in 1913, becoming Baron Melchett in 1928.
PRS 1931 A, 131, ii–v. EFA. DNB 1922–1930, 603–605. H. Withers.

LUDWIG MOND 1839–1909; FRS 1891

Ludwig Mond was born in Cassel in Germany where his father Moritz Mond was a well-to-do merchant; his mother was Henrietta née Levinsohn. Following work at Marburg in 1855, he began the next year to study with Robert Bunsen at the University of Heidelberg. In his industrial career which began in 1859 he worked in various chemical plants in Germany.

In 1862 he came to England where he took out an English patent for the recovery of sulphur from alkali waste, and in 1863 he went to a works at Widnes to perfect the process. In 1864 he moved to Utrecht to construct and manage a soda works, and returned to England in 1874, later (1880) becoming a British citizen. Mond married his cousin Frida Loewenthal in 1866, and they had two sons, Robert Ludwig and Alfred Moritz. Mond continued in industrial work eventually entering into partnership with John Tomlinson Brunner, leading to the setting up of a works in Winnington to operate the Solvay soda-ammonia process. The firm became Brunner, Mond and Company, a large alkali manufacturers; they were among the first to adopt an eight hour day and to provide model housing and playing fields for their employees.

Mond moved London in 1884 where he set up a laboratory for research in the stables of his home. It was here in 1889, with his assistant Karl Langer, that the discovery was made of the reaction of finely divided nickel with carbon monoxide to form a volatile compound, nickel carbonyl. This discovery led to Mond's development of the important industrial process for the production of nickel by the decomposition of the carbonyl and to the formation of the Mond Nickel Company.

Ludwig Mond was a man of great scientific achievements and energy, with a genius for perceiving industrial possibilities of scientific discoveries. His interest beyond his science included music and art, and he acquired a remarkable collection of pictures. His benefactions included the sum of £100,000 given in 1896 to found and equip the Davy–Faraday Laboratory in a house next to the Royal Institution.

DSB 9, 466, 467. W.A. Campbell. DNB 2nd Suppl., II, 1912, 631–634. PJH. Ref. Ch. 6.3.

SIR ROBERT LUDWIG MOND 1867–1938; FRS 1938

Robert Ludwig Mond was the elder son of Ludwig Mond and Frida née Lowenthal, and was born at Farnworth, Lancashire. After attending Cheltenham College, he was educated at Cambridge University (Peterhouse College), Zurich Polytechnium, Edinburgh University and at Glasgow University, where he was an assistant to Lord Kelvin. He worked in his father's factory at Winnington, Cheshire and carried out research on metal carbonyls, including collaboration with his father on investigations on nickel carbonyl. His work was of great importance in the commercial development of the carbonyl process and he also played a role in investigations leading to improvements in zinc production by electrolysis of zinc chloride. After his work at Winnington he joined the board of Brunner, Mond and Company, and later that of the Mond Nickel Company, of which he eventually became Chairman. In 1898 he married Helen Edith Levis who died in 1905 and by whom he had two daughters; in 1922 he married Marie Louise Le Manach, the widow of Simon Guggenheim. Robert Mond had a close association with the Davy–Faraday Research Laboratory at the Royal Institution. He was a generous benefactor in scientific and charitable activities and in Egyptology, in which subject he had a keen interest. He was knighted in 1932.

ONFRS 1936–38, 2, 627–632. J.F. Thorpe. DNB 1931–1946, 622, 623. R. Robertson.

MICHAEL ARTHUR MOORE 1943–; FRS 1989

Huddersfield New College. Univ. of Oxford, Oriel College: BA 1964, DPhil 1967: Magdalen College; Prize Fellow, 1967–71. Univ. of Illinois: Research Associate, 1967–69. Univ. of Sussex: Lecturer in Physics, 1971–76. Univ. of Manchester: Professor of Theoretical Physics, 1976–.
Statistical mechanics, including polymer physics and spin glasses.

HENTON MORROGH 1917–; FRS 1964

George Dixon Grammar School, Birmingham. British Cast Iron Research Association: Senior Research Officer, 1942; Research Manager, 1945; Deputy Director, 1957; Director, 1959. Univ. of Technology, Loughborough: Department of Industrial Engineering and Management, Visiting Professor, 1967–72. FREng 1979. CBE 1969.
Microstructure and solidification of cast irons; development of ductile cast iron.

HENRY MOSELEY 1801–1872; FRS 1839

Henry Moseley, the son of Dr William Willis Moseley and Margaret née Jackson, was born in Newcastle-under-Lyme where his father had a private school. Henry attended the Grammar School in the town before going to a school at Abbeville, followed by a short period at a naval school in Portsmouth. He studied mathematics at St John's College, Cambridge, graduating as Seventh Wrangler in 1826. In 1827 he was ordained Deacon, and in the following year priest, in the Anglican Church. At West Moulton, Taunton, in addition to his responsibilities as a curate, he pursued his interests in mathematics. King's College, London appointed him as Professor of Natural and Experimental Philosophy and Astronomy in 1831. He married Harriet Nottidge in 1835 and they had one son. In 1853 Moseley became Canon Residentiary at Bristol Cathedral, and in the next year vicar at Olveston, Gloucestershire; an appointment as Chaplain in Ordinary to Queen Victoria followed in 1855. Moseley's writings ensuing from his deep interest in applied mathematics were influential in engineering, for example in the development of wrought iron bridge construction using box girders.
DNB. Notes and Records of the Roy. Soc. 1976, 30, 169–179. T.M. Charlton.

SIR NEVILL FRANCIS MOTT 1905–1996; FRS 1936

Nevill Mott was born in Leeds, the son of Charles Francis Mott, a former director of education for the city and Lilian Mary née Reynolds. Both his mother and father had been research students in the Cavendish Laboratory under J.J. Thomson. Mott attended Baswick House Preparatory School and Clifton College and then went to St John's College, Cambridge with an open scholarship. He was classed as Wrangler with distinction in the Mathematical Tripos in 1926, and proceeded to three years research in applied mathematics. In autum 1928 he spent a period with Bohr in Copenhagen. He became a Lecturer at Manchester University in 1929 and came back to Gonville and Caius College, Cambridge in the following year as a Fellow and Lecturer. He married Ruth Horder in 1930 and they had two daughters. In 1933 he became Melville Wills Professor in Theoretical Physics at Bristol University, where in 1948 he was appointed as Henry Overton Wills Professor of Physics and Director of the Henry Herbert Wills Physical Laboratory. During the 1939–1945 war his work included a period as Superintendent of Theoretical Research in Armaments at Fort Halstead, Kent, begining in 1943. From 1954 to 1971 he was Cavendish Professor of Physics at Cambridge University; in 1959 he was elected additionally as Master of Caius College, serving until 1966. From 1971–1973 he was a Senior Research Fellow at Imperial College, London.

Mott was a leading authority in the field of solid state physics. When he went to Bristol he worked on metals and alloys and later research themes included semiconductors and insulators. He possessed the ability to explain mathematics in physical terms and he was the author or co-author of important books including the book with H. Jones entitled *The Theory of the Properties of Metals and Alloys* (1936). His honours included the award of Royal Society Medals: Hughes (1941), Royal (1953) and Copley (1972); Bakerian Lecturer (1953); election as FREng (1987); a knighthood (1972): a Nobel Prize (joint) for Physics (1977) and appointment as CH (1995).
BMFRS 1998, 44, 315–328. Sir Brian Pippard FRS. *The Times*, 12 August 1966. *Who's Who* 1996.

JOSEPH MOXON 1627–1691; FRS 1678
Joseph Moxon was born in Wakefield, Yorkshire, the son of James Moxon, a printer who worked in Holland from 1637–1643. In 1643 he began a three year partnership with his father in a printing business in London, near Bishopsgate. Joseph married Susan Marson in 1648, and they had a daughter; his wife died about 9 years later and he married Hannah Cooke, by whom he had a son. From 1649 Moxon spent several years learning to make globes, maps, and mathematical instruments, which led to his main business from 1753 onwards dealing in these items, and in mathematical books; he had a shop initially at Cornhill but subsequently at other locations in London. Soon after 1660 he was appointed hygrographer i.e.map and chart printer and seller, to Charles II. Moxon was closely associated with Fellows of the Royal Society from the time of its foundation, and was probably the first tradesman to be elected as a Fellow. His various technological activities included a substantial interest in metallurgical matters.

Moxon's book, begun in 1677, *Mechanick Exercises or the Doctrine of Handy-works applied to the art of smithing, joining, carpentry, turning, bricklaying etc* included practical metallurgical information. Another important book dealt with printing, *Mechanick Exercises on the whole art of Printing* (1683) and Moxon had a special interest in type casting and the alloys used for this.

Moxon also published in other fields: astronomy, geography, hydrography and architecture. For example, he gave an account of the construction of timber floors, showing that it was recognised that the strength of a beam is more seriously weakened by cutting a mortice across the wood fibres than it is by cutting even a fairly long slot with the fibres. He thus made important contributions in various practical aspects of metallurgy and strength of materials.
DNB. Notes and Records Roy. Soc. 1995, 49, 2, 193–208. Refs Ch. 2.1, 5.

ALEXANDER MUIRHEAD 1848–1920; FRS 1904
Alexander Muirhead was born at Salthoun, East Lothian, the second son of John Muirhead, a farmer, and Margaret née Lauder. His father took up a post as superintendent with the Electric Telegraph Co. in London, when Alexander was three years old, and moved to London. Having been educated by a private tutor up to the age of 15, he attended University College School. He was an outstanding pupil, although he suffered from partial deafness following an accident. In 1869 he graduated with honours in chemistry at University College and then studied natural sciences at St Bartholomew's Hospital, obtaining the DSc degree in 1872. He became scientific adviser to the firm of telegraph engineers founded by his father in partnership with J.L. Clark in around 1869, working on the development of techniques for duplexing signals in submarine cables and making precise measurements of resistance and capacity. Muirhead set up a private research laboratory in London and he established the manufacturing business of Muirhead and Co. in 1885. In 1893 he married Mary Elizabeth Blomfield; there were no children.
PRS 1921–1922A, 100, viii–ix. DJL. DNB Missing Persons 484, 485. R.M. Birse.

PETRUS VAN MUSSCHENBROEK 1692–1761; FRS 1734

Petrus van Musschenbroek came from a well known family of brass founders and instrument makers who settled in Leyden, the Netherlands in the latter part of the 16th century. His father, Johan, married Maria van der Straeten and they had two sons, Jan and Petrus. When Johan died, Petrus was not yet 15 years old and he owed his further education to his brother. Studies at the University of Leiden led to his receiving a doctorate in medicine, which profession he practised in Leiden, after a visit to London, during which he is reported to have studied with Isaac Newton. In 1719, having received a doctoratc in philosophy, Musschenbroek was appointed as a Professor in Mathematics and Philosophy at Duisburg where in 1721 he also became Extraordinary Professor of Medicine. From 1723 to 1740, he held the Chair of Natural Philosophy and Mathematics at Utrecht, together with, from 1732, the Chair of Astronomy. His final academic appointment was at Leiden where he taught from 1740 until his death. Marriage to Adriana van de Water took place in 1724, and following her death in 1732, he married Helen Helena Alstorphius in 1738.

Musschenbroek wrote books describing his varied experimental work, including the topics of mechanics of rigid bodies, heat, cohesion, capillarity, magnetism and electricity; among his discoveries he is generally credited with originating the Leyden jar. Of metallurgical interest is his pyrometer (1731) based on the expansion of a metal bar. Concerning mechanical properties he made an extensive series of experiments to determine the behaviour of building materials.
DSB 9, 594–597. D.J. Struik. Bulloch's Roll. Gen. Ref. 20, v. III.

APOLLOS MUSSIN-PUSHKIN 1760–1805; FRS 1799

Apollos Mussin-Pushkin was a Russian Count, a member of the Imperial Court of Catherine the Great in St Petersburg, in which city he founded the Mining College. He developed a method of compressing platinum sponge with mercury, followed by heating and forging. He was the founder of the St Petersburg Mining Institute. He travelled widely in Europe on behalf of the mining authorities, meeting many other scientists. In 1802, after the annexation of Georgia by Russia, he was sent there to study the mineral resources, and he continued his work on platinum in a laboratory in Tiflis.
Ref. Ch. 4.1.

FRANK REGINALD NUNES NABARRO 1916–; FRS 1971

Nottingham High School. Oxford Univ.: New College, MA, BSc. DSc, Birmingham. Ministry of Supply: Senior Experimental Officer, 1941–45. Bristol Univ.: Royal Society Warren Research Fellow, 1945–49. Birmingham Univ.: Lecturer in Metallurgy, 1949–53. Univ. of Witwatersrand: Professor and Head of Department of Physics, 1953–77; City of Johannesburg Professor of Physics, 1970–77; Dean Faculty of Science, 1968–70; Deputy Vice-Chancellor, 1978–80. South African Council for Scientific and Industrial Research: Professional Research Fellow, 1994–. MBE 1946.
Solid state physics, physics of creep, crystal dislocations.

EDWARD NAIRNE 1726–1806; FRS 1776

Edward Nairne was probably a member of the Nairne family of Sandwich, Kent, and at an early age he became interested in science. After an apprenticeship beginning in 1741 he established a shop in London at 20 Cornhill as an 'optical, mathematical and philosophical instrument maker' in around 1748, and he obtained royal patronage. From 1771 he began to contribute papers to *Philosophical Transactions*; over a period he reported on experiments in optics, pneumatics, and, in particular, electricity. In 1772 he invented an improved form of electrostatic machine. His shop

provided a range of items: microscopes, telescopes, equipment for navigation, surveying and measuring, electrical machinery and vacuum pumps. In 1801 he gave up his business and moved to Chelsea.
DSB 9, 607–608. Roderick S. Webster. DNB.

THOMAS NEALE 1641–1699; FRS 1664
Thomas Neale was appointed Master and Worker of the Mint by Charles II and held office also under James II and William III until 1699. He put forward a proposal for amending the silver coins of England. He was associated with the organisation of various large scale lotteries.
Notes and Rec R. S. DNB. 1964, 19, 156–157 Sir John Craig.

LOUIS EUGÈNE FÉLIX NÉEL 1904–; FMRS 1966
Lycée du Parc, Lyon. Lycée Saint-Louis Paris and École Normale Superieure, Paris. Professor: Universities of Strasbourg, 1937–45: Grenoble, 1945–76. Scientific Adviser to the Navy, 1952–82. President Inst. Nat. Polytechnique, Grenoble, 1971–76. Nobel Prize for Physics, 1970.
Magnetism.

JAMES BEAUMONT NEILSON 1792–1865; FRS 1846
James Neilson who was born in Shettleson, near Glasgow was the son of a millwright. Having left school before he was 14 he worked on a winding engine at the Govan Colliery where he showed an aptitude for mechanics. He was then apprenticed to his elder brother, John, an engineman, and spent leisure time in improving his education. When his apprenticeship was completed Neilson worked for a time as a journeyman to his brother, and then at the age of 22 he was appointed engine-wright of a colliery at Irvine. A year later he married Barbara Montgomerie; her dowry enabled them to manage during a period of unemployment, and they migrated to Glasgow. Here at the age of 25 he obtained a post as Foreman of the Gasworks, the first of its kind to be established in the city. After five years he became Manager and Engineer of the Works, where he introduced several important improvements into both the manufacture and the utilisation of the gas. In these early successes he had the benefit of the knowledge of physical and chemical science which he learned as a student at the Anderson's College, Glasgow. With the aid of the directors of the gas company he established a workman's institution, with a library, lecture-room, laboratory and workshop.

Around 1835 he commenced the work which led to his important discovery of the advantages to be achieved by using hot blast in the ironmaking process, which eventually resulted in successful industrial application. Neilson retired from the managership of the gasworks in 1847 and went to live on the Isle of Bute. In 1851 he moved to an estate in the Stewarty of Kircudbright, where he promoted local improvements, and founded an institution similar to that which he had established for workmen in Glasgow.
DNB.

HERMANN WALTHER NERNST 1864–1941; FMRS 1932
Walther Nernst, who was born at Briesen, East Prussia, was the third child of Gustav Nernst, a judge, and Ottilie née Nerger. He graduated from the Gymnasium in Grandenz and studied at the universities of Zurich, Berlin, Wurzburg, and Graz. In 1891 he became an Associate Professor of Physics at Göttingen, and in the following year he married Emma Lohmeyer; there were two sons and three daughters from the marriage. Nernst was appointed Professor of Physical Chem-

istry, at Göttingen in 1894 and then in 1905 Professor of Physical Chemistry in Berlin. From 1922–1924 he was President of the Physikalisch-Technisches Reichsanstalt. He then returned to the University as Professor of Physics and Director of the Physical Laboratory. In 1933 he retired, out of favour with the Nazis, to a life of farming and agriculture on his estate.

Nernst carried out important work in electrochemistry and also in thermodynamics (he formulated the third law) and was awarded the Nobel Prize for Chemistry in 1920. By the turn of the century he had made a large fortune by selling a type of electric light, which though superior to the Edison carbon filament became obsolete the with advent of the tungsten filament lamp. He was awarded the Copley Medal in 1886.

ONFRS 1942–44, 4, 101–112. Lord Cherwell and F. Simon. DSB 15, 432–453. E.N. Hiebert.

FRANZ ERNST NEUMANN 1798–1895; FMRS 1862

Franz Neumann was born in Joachmimsthal, Germany and attended the Berlin Gymnasium. During service as a volunteer in the Prussian army he was wounded in 1815 at Ligny. Having left the army a year later he studied theology at the University of Berlin (1817) but went to Jena in 1818 to start studies in science. Back in Berlin in 1819 he studied mineralogy and crystallography and obtained a doctorate in 1825. Following appointment as Privatdozent at the University of Konigsberg, he became a Lecturer and then Professor of Mineralogy and Physics in 1829. His interests included the theory of elasticity of crystals and his work on crystallography included a method of spherical projection. He received the Copley Medal in 1886.

PRS 1896–97, 60, viii–xi. AS. DSB 10, 26–29. J.G. Burke.

FRANCIS HENRY NEVILLE 1847–1915; FRS 1897

Born at Exeter, Devonshire, where he attended school, Francis Neville entered Sidney Sussex College, Cambridge in 1867, graduating BA in 1871 as 15th Wrangler in the Mathematical Tripos. He became a Fellow of the College in the same year, and MA in 1874; he was Senior Proctor (1878–88) and College Lecturer in Chemistry and Physics. His interests were in experimental science rather than mathematics and early in 1880 he took over management of the college chemical laboratory which consisted of a series of sheds. This responsibility he maintained until he retired in 1908, when he moved to Letchworth where he set up a laboratory in his own home and continued collaborative work with Charles Heycock. The college laboratory was left empty and two years later was burnt as a result of a celebratory bonfire lit by the college rowing eight; although some apparatus was saved the research records were destroyed.

Neville, in his important research partnership with Heycock brought experimental skill and mathematical precision; he was also gifted in other ways – a man of wide reading and interests, an excellent scholar of French, German and Italian, and an authority on Italian history. He was active in international correspondence with leading scientists, including Le Chatelier on the subject of intermetallic compounds and Roozeboom on the Phase Rule in relation to the results on the copper–tin system that he and Heycock obtained.

L.B. Hunt: 'Heycock and Neville', *The Metallurgist and Materials Technologist* 1980, 392–394. *Nature* 1915, 432. Ref. Ch. 10.4.

ROBERT STIRLING NEWALL 1812–1889; FRS 1875

Robert Newall was born in Dundee, where his father placed him in a mercantile office. However, he soon went to London, where he carried out experiments with an engineering firm and spent two years promoting the interests of the firm in America. In 1840 he took out a patent for the

invention of wire rope. With partners he established works at Gateshead-on-Tyne for the manufacture of his product, which quickly became used world-wide. He continued to improve the manufacturing technique and produced a new machine in 1885. Newall played an important role in the production of submarine cables by surrounding the gutta percha insulation with strong wires. He produced the first successful cable from Dover to Calais in 1851, and in 1853 he invented the 'brake drum' for laying cable in deep seas. He was directly involved in many cable-laying projects in the seas around England and elsewhere; half of the first Atlantic cable was made at his Works. His interests extended to scientific pursuits including astronomy. He married Mary Pattinson and they had four sons and one daughter.
PRS 1889, 46, xxxiii–xxxv. JNL. DNB.

RONALD CHARLES NEWMAN 1931–; FRS 1998
Tottenham Grammar School. Imperial College, London: BSc, Physics, PhD. AEI Central Research Lab., Aldermaston Court, Berks; Semiconductor Section: 1955–63. AEI Research Lab, Rugby: Senior Research Scientist: 1964. Univ. of Reading: Lecturer, J.J.Thomson Physical Lab, 1964–69; Reader, 1969–75; Professor, 1975–88. Imperial College: Professor and Associate Director IRC for Semiconductor Materials; 1989–.
Epitaxial growth. Semiconductor science.

SIR ISAAC NEWTON 1642–1727; FRS 1671
Isaac Newton came from a family belonging to the yeoman class of manorial England and was born at Woolsthorpe in Lincolnshire. His father, also Isaac, whom Newton never knew died in 1642. When the boy was 3 years old his mother Hannah remarried the Rev. Barnabas Smith (Rector of nearby North Witham). Smith owned a small library containing theological books and these passed to his stepson. In his early life Newton showed practical talents in mechanics, and skill with his hands; in later life these talents were demonstrated in experimentation in practical optics and chemistry, including the ability to turn wood on a lathe, to melt metals in a furnace, to cut screw threads and to do simple blacksmith's work. Newton began to attend Grantham Grammar School in 1654 and in 1661 he entered Trinity College, Cambridge as a Subsizar. He was elected a Scholar in 1664 and obtained a BA degree in January 1665.

Newton became interested in mathematics and obtained from his reading and note-taking a good grounding in areas of algebra and geometry. He then moved on to become an independent investigator, working intensively, making outstanding contributions in his mathematical work and in his work on astronomy and optics, including the development of the reflecting telescope. He was formally admited to the Fellowship of the Royal Society when he visited London in 1666, having been elected three years earlier.

He also carried out extensive work in chemistry in which his activities of metallurgical interest included investigations on the production of speculum metal mirrors for telescopes, low melting point alloys, thermometry and pyrometallurgy.

Newton served the State in various capacities for 30 years, including being for two periods a University Member of Parliament. A particularly notable sphere of work was the Mastership of the Royal Mint, which commenced in 1699, following his previous appointment as Warden in 1696.

Newton became President of the Royal Society in 1703, and was annually re-elected to this position for 25 years. In 1705 Queen Anne, on a visit to Cambridge, knighted Newton at Trinity College, the first mathematician and philosopher to be honoured in this way. His later years

included a period of controversy over claims of priority with Leibnitz regarding the invention of the calculus. He turned again to theological interest that had intensely concerned him in his younger years. In the five years prior to his death his health began to deteriorate, and in his last months he gave up attendance at the Mint to his deputy, although remaining as Master. Newton presided for the last time over the Council of the Royal Society early in March 1727; he died before the end of the month and was buried in Westminster Abbey. Newton, with his outstanding discoveries as a mathematician and natural philosopher has been described as 'the greatest mind in British history'.
DSB 10, 43–101. I. B. Cohen. DNB. Refs Ch. 2.3, 4, 12.

FRANK NICHOLLS 1699–1778; FRS 1728
Frank Nicholls, who was born in London, was the second son of John Nicholls, a barrister. After attending Westminster School he entered Exeter College, Oxford in 1714, where he studied classics and physics. Having obtained the degrees of BA (1718) and MA (1721) he proceeded to graduate in medicine, MB (1724) and MD (1729); before he completed his medical studies he lectured as a Reader in Anatomy at Oxford. After practising medicine for a short time in Cornwall, he settled in London. He married Elizabeth Mead and they had five children. He gave the Croonian Lecture in 1739. In addition to his medical practice and research, Nicholls had an interest in, and published papers on, minerals.
DNB.

SIR ROBIN BUCHANAN NICHOLSON 1934–; FRS 1978
Oundle school. Univ. of Cambridge: St Catherine's College; BA, 1956; PhD, 1959, MA, 1960; University Demonstrator in Metallurgy, 1960; Lecturer in Metallurgy, 1964. Manchester Univ.: Professor of Metallurgy, 1966. INCO Europe Ltd, Director of Research Laboratory, 1972; Director, 1975; Managing Director, 1976–81. Chief Scientific Adviser to Cabinet Office: 1983–85 (Central policy review staff, 1981–83). Pilkington plc (formerly Pilkington Bros plc): Director, 1986–1996. Pilkington Optronics Ltd: Chairman 1991–98. Royal Society: Philips Lecturer, 1968. FREng 1980.Knighted 1985.
Electron microscopy; strengthening of crystals, precipitation hardening; structure/property relationships.

JOHN FREDERICK NYE 1923–; FRS 1976
Stowe School. Univ. of Cambridge: Kings College, Major Scholarship, MA, PhD, 1948; Cavendish Laboratory, 1944–49: University Demonstrator in Mineralogy and Petrology, 1949–51. Bell Telephone Laboratories, New Jersey, USA: 1952–53. Bristol Univ.: Lecturer, 1953; Reader, 1965; Professor of Physics, 1969–88; Melville Wills Professor of Physics, 1985–88.
Physical properties of crystals; glaciology; optics; catastrophe theory.

SIR CHARLES WILLIAM OATLEY 1904–1996; FRS 1969
Charles Oatley was born in Frome, Somerset, the son of William Oatley and Ada Mary née Dorrington. After attending a local council school he went on to Bedford Modern School as a boarder. He won an Exhibition to St John's College, Cambridge which he entered in 1922 and took the Natural Sciences Tripos, with physics in Part II. He developed an interest in electronics and obtained a post with Radio Accessories in Willesden in 1925. Two years later he was appointed as a Demonstrator at King's College, London, subsequently becoming a Lecturer. In 1930 he married Enid West and they had two sons. With the outbreak of war he joined a team at

the Air Defence Experimental Establishment at Christchurch, Dorset to work on radar. The Establishment moved to Malvern and became the Royal Radar Research Establishment and in 1944 he was appointed Acting Superintendent. After the was Oatley returned to Cambridge with a Fellowship at Trinity College and a Lectureship in Electrical Engineering and he began to build up a research group. In 1954 he was appointed Reader and from 1960–71 he was Professor of Electrical Engineering; under his leadership the teaching of electronics was modernised.

Oatley became interested in scanning electron microscopy, which had been the subject of inconclusive work in Germany and the USA in the 1930s. He began work with a research student, Dennis McMullan, in 1948 leading to the production of a working instrument in 1951. Further development work led to improved instruments and, in 1965, to the Steroscan instrument manufactured by the Cambridge Instrument Company. Oatley's awards included a Royal Medal (1969), the Mullard Medal (1973), FREng (1976), OBE (1956), and a knighthood (1974).

BMFRS 1998, 44, 331–347. K.C.A. Smith. *The Times*, 1.4.96. *The Independent*, 21.3.96, A.N. Broers and K.C.A. Smith. *Physics World*, July, 1996, H. Ahmed. *Nature*, 25.4.96 D. McMullan.

WILLIAM ODLING 1829–1921; FRS 1859

William Odling was born in Southwark, London, the only son of George Odling, a London doctor. Born and educated in London, he entered Guy's Hospital as a student at the age of 16 and graduated MD in 1851. He attended Hofmann's course at the Royal College of Chemistry for a semester. In his career he had several teaching posts at Guy's Hospital, was Medical Officer of Health for Lambeth (1856–1872) and taught chemistry at St Barthomolew's Hospital (1863–1870). He was appointed Fullerian Professor of Chemistry at the Royal Institution in 1867; then from 1872 to 1912 he was Waynefleet Professor of Chemistry at Oxford University. He married Elizabeth Mary Smee in 1872 and they had three sons. In the Royal Society he was Vice-President in 1869 and President from 1873–75. In his research Odling had a lifelong interest in classifying chemical compounds and between 1857 and 1865 he studied relationships between the chemical elements, producing a scheme which was one of the forerunners of Mendeleev's Periodic Table.

PRS 1921–22, A, 100, i–vii. HBD. DSB 10, 177–179. W.H. Brock.

HANS CHRISTIAN OERSTED 1777–1851; FMRS 1821

Hans Oersted was born at Rudkobing, Langeland, Denmark, the elder son of Soren Christian Oersted and Karen née Hermansen. He entered the University of Copenhagen in 1794 where he studied astronomy, physics, mathematics, chemistry, and pharmacy; he graduated in pharmacy in 1797 and obtained a doctorate two years later. He worked as a pharmacist and also travelled in Europe from 1801–03 for further studies in science. In 1806 Oersted became Professor of Physics at Copenhagen; then in 1829 he was appointed Director of the Polytechnic Institute in Copenhagen where he remained until his death.

In 1820 Oersted made his important discovery in electromagnetism of the deflection of the needle of a magnetic compass resulting from the passage of an electric current through a wire above the compass. He also investigated the compressibility of gases and liquids and did work on thermoelectricity. Of metallurgical interest was his report in 1825 of his method for producing aluminium by the reduction of the chloride. He was awarded the Copley Medal in 1820.

DSB 14, 474–478. R. Keen.

SIR MARK (MARCUS LAURENCE ELWIN) OLIPHANT 1901–; FRS 1937

Unley and Adelaide High Schools. Univ. of Adelaide, S. Australia. Univ. of Cambridge: Trinity College; 1851 Exhibition, Overseas, 1927, Senior 1929; PhD 1929. Messel Research Fellowship of the Royal Society, 1931; St John's College, Fellow and Lecturer 1934; Cavendish Laboratory, Assistant Director of Research, 1935; Univ. of Birmingham: Poyting Professor of Physics, 1937–50. ANU, Canberra: Institute of Advanced Studies; Professor of Physics and Ionized Gases, 1964–67. Governor of S. Australia 1971–76. KBE 1959.

Nuclear physics; surface properties.

HEIKE KAMERLINGH ONNES 1853–1926; FMRS 1916

Heike Kamerlingh Onnes, born in Groningen, the Netherlands, was the son of a well-known manufacturer. In 1870 he entered the University of Groningen to study physics and mathematics. The following year he spent some time at Heidelberg with Bunsen and Kirchhoff. Returning to Groningen in 1873 he took his doctorate in 1876. Two years later he was appointed Assistant at the Polytechnic School at Delft (later the Technical University of Delft) and in 1882 he became Professor of Experimental Physics at Leiden, a post which he held until 1923; in 1894 he founded the cryogenic laboratory which was to become world famous.

Onnes was a highly skilled experimentalist; he had a particular interest in the properties of matter at low temperatures and in 1908 he liquified helium. He discovered the phenomenon of superconductivity in 1911, received the Rumford Medal in 1912 and in the following year was awarded the Nobel Prize for Physics.

PRS 1927, A, 113, i–vi. FGD. DSB 7, 220–222. J. van den Handel.

SIR HENRY JOHN ORAM 1858–1939; FRS 1912

Henry Oram was born at Plymouth, Devonshire, the eldest son of John Joseph Oram and Jane née Hall. He was educated at private schools before entering the Royal Naval Engineering College at Keyham, Devonport in 1873. Entry to the Royal Navy as an Assistant Engineer occurred in 1879. He then spent three years at the Royal Naval Engineering College, Greenwich. He married Emily Kate Bardens in 1881 and they had two sons and two daughters.

In 1882 Oram became a Junior Engineer on troopships; he returned to London in 1884 as Assistant Engineer in the Department of the Engineer-in-Chief at the Admiralty where he spent more than 33 years. He experienced rapid promotion: Chief Engineer (1889), Staff Engineer (1893), Fleet Engineer and also Inspector of Machines (1897), Chief Inspector of Machines (1901, in 1903 the title was changed to Engineering Rear-Admiral), Deputy Engineer-in-Chief (1902), Engineer-in-Chief of the Fleet with the rank of Engineering Vice-Admiral (1907); he retired in 1917. The period during which he was Engineer-in-Chief was one of active change for the Navy with the introduction of the water-tube boiler, oil fuel and steam turbines and he played an important role; he served on a Royal Commission on Oil Fuel and Engines, and as President of the Institute of Metals in 1914. Oram received a CB in 1906 and was knighted (KCB) in 1910.

ONFRS 1939–41, 3, 155–158. E.H. Tennyson d'Eyncourt. DNB 1931–40, 660. W.H. Whayman and H.H. Brown.

EGON OROWAN 1902–1989; FRS 1947

Egon Orowan was born in Obuda, Budapest. His father Berthold Orowan was a mechanical engineer, and his mother was Josze Spitzer Sagvari. Orowan studied in the Staatsober Gymnasium and also in the University of Vienna, from 1920–1922, where his subjects were physics, chemistry, mathematics and astronomy. In 1928 he went to the Technical University of Berlin, where, in the physics department he began a project involving the preparation of single crystals of zinc, tin etc., and investigating their plasticity at liquid air temperatures. His observations of jerky flow of a zinc crystal led him to the concept of dislocations.

Orowan began a doctoral thesis on the cleavage of mica in1931 which led to the award of the PhD the following year. In 1936 there began a period when Orowan could not find employment, during which he spent time re-thinking the results of his experiments of the previous three years. Beginning in 1936 he worked with the Tungsram Research Laboratory, where he played an important role in the large scale separation of krypton from air by fractional distillation. Then in 1937 he moved to the Physics Department of the University of Birmingham, where his main interest was the theory of fatigue. From 1939 until 1950 he worked in the Cavendish Laboratory at Cambridge. He married Jolan Schonfel in 1941 and they had a daughter.

During Orowan's period in Cambridge he led a group whose research included X-ray studies of deformed metals, the rolling of metals, notch brittleness, yield point phenomena and dislocation effects in precipitation hardening. He also contributed significantly in the field of glaciology, and introduced models, based on the physical flow properties of ice. This and his interest in rock mechanics led on to his studies in earthquake mechanisms and flow in the earth's mantle.

In 1950 Orowan became George Washington Professor in the Mechanical Engineering Department at the Massachusetts Institute of Technology. Subsequently he became Head of the Materials Division. He continued to work on problems such as mechanisms of crystal plasticity, brittle and ductile fracture, fatigue and the application of these to geology. Orowan as a teacher had particular ability in clarifying complex concepts. Following his retirement in 1958 he became involved in other fields such as the stability of Western industrial economies and ageing of societies. From the mid 50s until 1980 he had many links with industry and was a very effective industrial consultant. Orowan was internationally recognised by many honours and awards for his important contributions to the deformation and fracture behaviour of materials.

BMFRS 1995, 41, 315–340. F. R. N. Nabarro FRS and A.S. Argon. Gen. Ref. 22.

SIR FREDERICK WILLIAM PAGE 1917–; FRS 1978

Rutlish School, Merton. Univ. of Cambridge: St Catherine's College, MA. Hawker Aircraft Co.: 1938. English Electric: 1945; Chief Engineer, 1950. English Electric Aviation: Director and Chief Executive (aircraft), 1950. BAC: Managing Director Aircraft Division, 1965–72: Chairman, 1967; Managing Director Aircraft and Chairman Commercial Aircraft Division, 1972. SEPECAT (Anglo-French Jaguar Programme); Joint Chairman, 1966–73. British Aerospace plc; Aircraft Group: Chairman and Chief executive, 1977–82. FREng 1977. CBE 1961. Knighted 1979.

Aircraft materials.

EDWARD GEORGE SYDNEY PAIGE 1930–; FRS 1983

Univ. of Reading:BSc, PhD. Royal Radar Establishment: Junior Research Fellow to DCSO, 1955–77. Univ. of Oxford: Professor of Electrical Engineering and Fellow St John's College, 1977–97.

Electron transport; semiconductors.

DENIS PAPIN 1647–c.1712; FRS 1682

Denis Papin was born in Blois, France, the son of Denis Papin and Magdaleine née Pineau. He studied medicine and qualified MD in 1669. Intent on a scientific career he came to Paris as assistant to Christiaan Huygens. In 1675 he moved to London and assisted Robert Boyle in his

air pump experiments. He spent some time in Vienna and returned to London in 1684, when he demonstrated to the Royal Society a device for casting medallions. In 1687 Papin became Professor of Mathematics at Marburg; some 10 years later he went to Cassel and returned to London in 1707. Papin attempted to use steam as a source of power.
DSB 10, 292, 293. P.P. MacLachlan. DNB.

SIR CHARLES ALGERNON PARSONS 1854–1931; FRS 1898
Charles Parsons was the youngest of six sons of William Parsons, third Earl of Rosse, and Mary née Field. Born in London, Charles was brought up at the family seat, Birr Castle, Parsonstown, Ireland. After a private education he spent two years studying mathematics at Trinity College, Dublin, before entering St John's College, Cambridge. There he attended lectures on mechanism and applied mechanics and studied mathematics, graduating BA as 11th Wrangler in 1877. He then commenced a four year apprenticeship at the Elswick Works of Sir William Armstrong and Company. In 1883 he married Katherine Bethell and they had a daughter and a son. From 1881–1883 Parsons worked in an engineering firm in Leeds prior to becoming a junior partner in 1884 in Clarke, Chapman & Co. at Gateshead. There he worked on the development of a steam turbine to drive dynamos. The first Parsons turbo-dynamo, built in 1884, was successful, developing an output of 7.5 kW. Five years later he founded C.A. Parsons and Co. at Heaton, near Newcastle upon Tyne. Later he formed the Parsons Marine Turbine Company, achieving great success with the development of turbine-propelled vessels. His activities also involved optics, including the production of searchlight reflectors and glass for optical purposes. Like Moissan, Parsons also had an interest in attempting to produce diamonds under conditions of high temperature and pressure.

Parsons was the recipient of numerous honours. In the Royal Society he was Vice-President (1908), Bakerian lecturer (1918), and recipient of the Rumford Medal (1902) and Copley Medal (1928). He was appointed CB in 1904, KCB in 1911 and admitted to the Order of Merit in 1927. Parsons took out over 300 patents, and was considered to be the most original British engineer since John Watt.
PRS 1931, A, cxxxi, v– xxv. JAE. DNB 1931–1940, 672–678. C.D. Gibb. Ref. Ch. 9.19.

WILLIAM PARSONS (THIRD EARL OF ROSSE) 1800–1867; FRS 1831
William Parsons, who was born in York, was the eldest son of Lawrence Parsons, the second Earl of Rosse. Following private education at Birr Castle, Ireland, the family seat, he spent a year at Trinity College, Dublin before going to Magdalen College, Oxford, where he graduated with first class honours in mathematics in 1822. Prior to his father's death he held the title Lord Oxmanton. While he was an undergraduate he became Member of Parliament for King's County in 1821, serving until 1834. In 1831 he became Lord Lieutenant of County Offaly, in which Birr was situated, and in 1841 he was elected an Irish representative peer in the House of Lords. In addition to his achievements as an administrator and public servant he made important contributions in astronomy. He used reflecting telescopes and produced large bronze mirrors in a foundry which he set up in Birr Castle. A three foot (0.9 m) diameter speculum was produced in around 1839, followed a few years later by six foot (1.8 m) diameter products using novel casting techniques. He had an interest also in building iron armoured ships. Parsons married Mary Field in 1836 and they had four sons. Following the potato famine in Ireland in 1856 Parsons devoted most of the rent from his Irish properties to alleviating local poverty. In the Royal Society he was President from 1848–1854. He was awarded a Royal Medal in 1851.
PRS 1867–68, 16, xxvi–xlii. DSB 10, 328, 329. J.D. North and C.A. Ronan. *Phil Trans*, 1840, 130, 503–527.

DONALD WILLIAM PASHLEY 1927–; FRS 1968

Henry Thornton School, London. Imperial College, London: BSc 1947, PhD 1950: Research Fellow, 1950–55. TI Research Laboratories, Hinxton Hall: Research Scientist, 1956–61; Group Leader and Division Head, 1962–67; Assistant Director, 1967–68; Director, 1968–79 (also Director of Research, TI Ltd, 1976–79. Imperial College: Professor of Materials, 1979–92; Head of Department of Materials, 1979–90; Dean Royal School of Mines, 1986–89; Senior Research Fellow, 1992–.

Electron microscopy of thin crystals, electron microscopy and diffraction, thin films, epitaxy.

SIR CHARLES WILLIAM PASLEY 1780–1861; FRS 1816

Charles Pasley was born in Eskdalamuir, Dumfriesshire. He was educated at Langholm and then entered a school in Selkirk in 1794 and proceeded to the Royal Military Academy at Woolwich in 1796. He was commissioned Second Lieutenant in the Royal Artillery in 1797 and in the following year transferred to the Royal Engineers. He received a series of promotions: First Lieutenant (1799), Second Captain (1805); First Captain (1807); Brevet Major (1812); Brevet Lieutenant Colonel (1813); Regimental Lieutenant Colonel (1814); Brevet Colonel (1830); Regimental Colonel (1831); Major General (1841); Lieutenant General (1851); Colonel Commandant of the Royal Engineers (1853). During his career he saw extensive active service, for example in Spain and at Walcheren, and he was wounded. In 1811 he commanded the Plymouth Company of Royal Military Artificiers, and his work and the system he introduced to improve the practice of military engineering led to his appointment in 1812 as Director of the Chatham Establishment of the Royal Engineers for Field Instruction; he served in this appointment for nearly 30 years. Beginning in 1836 he work on materials such as limes, mortars, cements and concretes, and his experiments led to the large scale production of certain types of cements. In 1841 Pasley was appointed Inspector General of Railways. Pasley married Harriet Spencer Cooper in 1814; following her death he married Martha Matilda Roberts and they had six children. He received the CB in 1831 and was knighted in 1846 (KCB).

DNB.

HUGH LEE PATTINSON 1796–1858; FRS 1852

Born at Alston, Cumberland, Hugh Lee Pattinson was the son of Thomas Pattinson, a retail trader and his wife Margaret née Lee; both his parents were members of the Society of Friends. He received his early education at school in his native town, and in his youth, having attended a lecture on chemistry, he became committed to science, and studied electricity and chemistry with the help of books and rudimentary apparatus. In 1815 he married Phoebe Walton and took the additional name of Lee. In about 1821 he became clerk and assistant to a soap-boiler in Newcastle upon Tyne, an appointment which gave him facilities for pursuing his studies.

Four years later he obtained the post of Assay Master to the Lords of the Manor at Alston; this involved inspecting ores levied as royalties from extensive lead mines in his native district. In 1829 Pattinson discovered an easy way of separating the silver from a lead ore and subsequently, as Manager of a lead smelting and refining works, he implemented the process. Pattinson also made other industrial developments. Contemporaneously with William Armstrong, Pattinson discovered the electrical effect associated with effluent steam. In 1858 he retired from business, and in order to practise astronomy he studied mathematics and physics and built a telescope at his residence near Gateshead.

PRS 1859, 9, 534, 535. DNB.

WOLFGANG ERNST PAULI 1900–1958; FMRS 1953

Wolfgang Pauli was born in Vienna, the son of Joseph Pauli and Bertha née Schutz; he had a younger sister. In 1918 he entered the University of Munich, and worked under Arnold Sommerfeld, obtaining a PhD in 1922. Sommerfeld also gave him the task, at the age of 19, of preparing a comprehensive review of work on relativity, and this proved to be an important monograph. He then spent a period at Gottingen before moving to Copenhagen to work with Bohr; in 1923 he became a Privatdozent at Hamburg, and in 1928 he was appointed as Professor of Experimental Physics at the Eidgenössiche Technical University at Zurich; he held this post until the end of his life, although during the 1939–45 war he was at the Institute for Advanced Studies at Princeton in the USA. In 1934 he married Francisca Bertram. In his research career his proposal of the principle in quantum mechanics known as the Pauli principle, was of outstanding importance and he was awarded the Nobel Prize for Physics in 1945.

BMFRS 1959, 5, 175–192. R.E. Peierls. DSB 10, 422–425. M.Fierz.

LINUS CARL PAULING 1901–1994; FMRS 1948

Linus Pauling was the eldest of three children of a pharmacist Herman Henry William Pauling and Lucy Isabelle née Darling. He was born in Portland, Oregon, and following his high school education entered Oregon Agricultural College (later Oregon State University) in Corvallis in 1917. He majored in chemical engineering, chemistry and physics; he was appointed assistant and in his senior year this work involved chemistry, mechanics and materials. He graduated in chemical engineering in 1922 and in the same year married Ava Helen Miller: there were three sons and a daughter from the marriage. Pauling went to the California Institute of Technology and obtained a PhD (*summa cum laude*) in physical chemistry for research involving the X-ray determination of crystal structures; in 1925 he was also a teaching fellow. There followed two years of research at Cal Tech. During 1926 and 1927 Pauling held a Guggenheim Fellowship, and visited Europe to work with Arnold Sommerfeld, Niels Bohr, Erwin Schrodinger and W.H. Bragg. In 1927 he was appointed Assistant Professor of Chemistry at Cal Tech, becoming Associate Professor in 1929 and Professor in 1931. He became Chairman of the Department of Chemistry and Chemical Engineering in 1936. During World War II he did defence research.

Pauling's distinguished research covered a variety of topics in chemistry and biology, including chemical bonding and molecular structure. He applied X-ray diffraction, electron diffraction and magnetic effects to study the structure of molecules, and made significant contributions to applying quantum mechanics to bonding phenomena in compounds. He received the Davy Medal in 1947. In 1954 he was awarded the Nobel Prize for Chemistry, while in 1962 he was the recipient of the Nobel Peace Prize.

BMFRS 1996, 42, 315–338. Jack D. Dinitz FRS. *Nobel Prize Winners*. An A.H. Wilson Biographical Dictionary. H.W. Wilson Co. NY 1987.

DAVID NEIL PAYNE 1944–; FRS 1992

Univ. of Southampton: BSc, PhD. English Elec. Co: Commissioning Engineer, 1962. Univ. of Southampton: Research Assistant,1969; Junior Research Fellow, 1971; Pirelli Research Fellow: 1972; Senior Research Fellow, 1978; Principal Research Fellow, 1981; Pirelli Reader, 1984; Optoelectronics Research Centre; Deputy Director, 1989–95; Director, 1995–.

Optoelectronics.

GEORGE PEARSON 1751–1828; FRS 1791

Born in Rotherham, George Pearson was the son of John Pearson, an apothecary. He studied at Edinburgh University, where he obtained the MD degree in 1773, having also studied chemistry under Joseph Black. After a brief period as St Thomas's Hospital, London, he spent two years in Europe before establishing a medical practice in Doncaster, where he remained for six years. Eventually he moved to London, where he became Chief Physician of St George's Hospital. His research covered medicine and chemistry, and included work on the nature of wootz steel in association with Stodart. He served on the Council of the Royal Society for many years and gave a Bakerian Lecture in 1827.

DSB 10, 445–447. E.L. Scott. DNB. Ref. Ch. 8.11.

KARL PEARSON 1857–1936; FRS 1896

Karl Pearson was the younger son and second of three children of William Pearson KC and Fanny née Smith. He was born in London, where at the age of nine he entered University College School. In 1873 he went to King's College, Cambridge with a scholarship and graduated BA as Third Wrangler in the Mathematical Tripos in 1879. He became a Fellow of the College in 1880. A period in Germany was spent at Heidelberg and Berlin on a wide range of studies including physics, engineering, law and history. When Pearson returned to Cambridge he worked for a while in the engineering workshops. He then went to London where he studied law at Lincoln's Inn Fields and was called to the Bar in 1881, also obtaining the LLB and MA degrees of Cambridge University in the following year. He did not practice law but in 1884 was appointed as the Goldsmid Professor of Applied Mathematics and Mechanics at University College London; he was also Lecturer in Geometry at Gresham College, London from 1891–94. In 1911 he gave up the Goldsmid Chair and was elected as the Galton Professor of Eugenics, a post which he held until 1933. Pearson married Maria Sharpe in 1890 and they had one son and two daughters; he married Margaret Child in 1929. Pearson's research included elasticity and he was the editor of Todhunter's book on the history of elasticity. He also made important contributions to statistics. He was awarded the Darwin Medal of the Royal Society in 1898.

ONFRS 1936–38, 2, 73–110. L.N.G. Filon. DSB, 10, 447–473. C. Eisenhart. DNB 1931–40, 681–684. M. Greenwood.

RENDEL SEBASTIAN PEASE 1922–; FRS 1977

Bedales School. Univ. of Cambridge: Trinity College, MA, ScD. Ministry of Aircraft Production at ORSU, Unit HQ, Bomber Command: Scientific Officer, 1942–46. AERE, Harwell: Research 1947–61. UKAEA, Culham Labs for Plasma Physics and Nuclear Fusion: Divisional Head, 1961–67. UKAEA Research Group; Assistant Director, 1967; Culham Labs, Director, 1968–81. UKAEA: Programme Director for Fusion; 1981–87. Royal Society: Council, 1985–87; Vice-President 1986–87.

Plasma physics; fusion power.

SIR RUDOLF ERNST PEIERLS 1907–1995; FRS 1945

Rudolf Peierls was born in Oberschöneweide near Berlin, the youngest of the three children of Heinrich Peierls, managing director of a cable company, and Elisabeth née Weigert. He attended the local Gymnasium and then studied physics in Berlin (1925–26) under Max Planck and Walther Nernst, wave mechanics with Sommerfeld at Munich (1926–28) and electron theory of metal structure with Heisenberg at Leipzig (1928–29, PhD 1929). He was an assistant to Wolfgang

Pauli at the Federal Institute of Technology at Zurich in 1929–1932, where he applied quantum electron dynamics to investigate electron behaviour in metals. In 1931 he married Genia Nikolae Una Kannegiser whom he met in Odessa in 1930, when he was in Russia as a Visiting Lecturer at Leningrad; they had one son and three daughters. He was Rockefeller Fellow in Rome and Cambridge between 1932 and 33, and Research Fellow under W.L. Bragg from 1933–35 at Manchester University where he began to study nuclear physics. From 1935–37 he was Assistant in Research in the Royal Mond Laboratory in Cambridge in magnetism and low temperature physics. In 1937 he became Professor of Applied Mathematics and subsequently Professor of Mathematical Physics at Birmingham.

Peierls became a naturalised British subject in 1940 and was involved in the atomic energy project, first in Birmingham from 1940–43 and then from 1943–45 in the USA, leading a group at Los Alamos. He was appointed as Wykham Professor of Physics and Oxford, and a Fellow of New College (1963–74). From 1975–77 he was Professor of Physics at the University of Washington, Seattle. His honours included a Royal Medal (1959), the Copley Medal (1986), CBE (1946) and a knighthood (1968). He maintained a close involvement with the science and politics of nuclear weapons control and was one of the leaders of the Pugwash Movement. His research included fundamental contributions to solid state theory, including the subject of dislocation movement in crystals.

Physics World 1995, Nov., 63, 64. J. Paton. Ref. Ch. 18.18.

JOHN BRIAN PENDRY 1943–; FRS 1984

Univ. of Cambridge: Downing College, BA 1965, MA, PhD 1969; Research Fellow, 1969–72. Bell Laboratories, USA: Member of Technical Staff, 1972–73. Univ. of Cambridge: Cavendish Laboratory, Senior Assistant in Research, 1973–75. Daresbury Laboratory: Senior Principal Scientific Officer, Head of Theory Group, 1975–81. Imperial College, London: Professor of Theoretical Solid State Physics, 1981–; Dean Royal College of Science, 1993–96; EPSRC Senior Research Fellow, 1997–; Head Department of Physics, 1998–.

Surface science; photonic materials.

SIR WILLIAM GEORGE PENNEY
(BARON PENNEY OF EAST HENDRED) 1909–1991; FRS 1946

William Penney was the son of William A. Penney and Blanche Evelyn née Johnson; he was born in Gibraltar where his father was stationed with the Army. In 1914, while his father was in the war, the family lived near Sheerness where he attended primary school. He then attended school in Colchester before spending two years (1924–26) at Sheerness Junior Technical College. When he left school he took a job as a laboratory assistant, but with scholarship support, he entered the Royal College of Science, Imperial College, London in 1927 to study mathematics, having been granted exemption from the first year of the course. Penney's brilliance led to a first class honours and the award of the Governor's Prize in 1929. He continued with graduate work and obtained a PhD in 1931. Two years were then spent at the University of Wisconsin with a Commonwealth Fund Fellowship; he worked with Van Vleck and obtained an MA degree. He then entered Trinity College, Cambridge, with an 1851 Exhibition Senior Studentship and obtained a PhD degree. His main research in the 1930s was concerned with the applications of quantum mechanics to fundamental problems in the theory of solids and the theory of the chemical bond. He was awarded a London University DSc in 1935 and in 1936 he spent a short time as Stokes Student at Pembroke College, Cambridge. He then returned to Imperial College as Assistant Professor and

Reader in Mathematics. In 1935 he married Adele Minnie Elms and they had two sons. Following the death of his wife in 1945 he married Eleanor Joan Quennell.

With the coming of war in 1939 Penney continued to teach at Imperial College until in 1940 he was called up by the Ministry of Home Security, where he worked on the blast effects of bombing; transfer to the Admiralty occurred where his work included Mulberry Harbour design. In 1944 he was appointed to the Los Alamos laboratory to work on the atomic bomb project. He took part on a flight to witness the dropping of the second bomb and then visited Japan to investigate the blast effects of the bombs.

At the end of the war, although he wished to return to academic life, he was persuaded to remain in government service. In 1946 he accepted apppointment as Chief Superintendent of Armaments Research under the Ministry of Supply, based at Fort Halstead. and then from 1951 directed the work at the Atomic Weapons Research Establishment at Aldermaston leading to the construction and testing of atomic weapons. In 1954 when the UK Atomic Energy Authority was formed Penney became Member for Weapons Development, and in 1959 he became Member for Research; in 1961 he was apppointed Deputy Chairman, and three years later Chairman. He returned to Imperial College as Rector in 1967, serving in this post until 1973. In the Royal Society Penney was Treasurer from 1956–60 and Vice-President from 1958–59. His honours included the award of an OBE (1946), KBE (1952), a life peerage (1967) and the OM (1969).
BMFRS 1994, 39, 283–302. Lord Sherfield FRS.

MICHAEL PEPPER 1942–; FRS 1983

St Marylebone Grammar School. Univ. of Reading: BSc, Physics, 1963; PhD Physics, 1967. MA, 1987, ScD 1989 Cantab. Mullard Ltd: Research in Solid State Physics, 1967–69. The Plessey Co, Allen Clark Research Centre: 1969–73. Univ. of Cambridge: Cavendish Lab. (in association with the Plessey Co. 1973–82); Warren Research Fellow of the Royal Society, 1978–86. GEC plc Hirst Research Centre: Principal Research Fellow, 1982–87. Univ. of Cambridge: Professor of Physics; 1987–. Hughes Medal 1987.
Solid state chemistry; semiconductors and solid state physics.

WILLIAM HASLEDINE PEPYS 1775–1856; FRS 1808

Born in London William Pepys was the son of a cutler and maker of surgical instruments. He was a Quaker, and appears to have succeeded to his father's business. He had a remarkable skill and ingenuity, his interest being more in the invention than the use of apparatus. His most important researches were carried out with William Allen, and he took considerable interest in platinum and in attempts to melt it. He showed that soft iron when heated in contact with diamond produced steel. Another activity was the supervision of the construction of a battery of 2,000 plates for the Royal Institution. He participated in the founding of the Askesian and other societies, and was actively involved with the Royal Institution of which he was President in 1816.
DNB. (Also DNB George Children). L.B. Hunt. *Platinum Metals Review* 1987, **31**(1), 26–41.

JOHN PERCY 1817–1889; FRS 1847

John Percy was the third son of a solicitor, Henry Percy, and was born in Nottingham. After private schooling he attended chemical lectures at a local school of medicine.He wished to be a chemist but, accepting his father's intention that he should graduate in medicine he was taken by his brother Edmund to Paris in 1834 at the age of 17 to begin medical studies. He attended lectures given by Gay-Lussac and Thénard on chemistry, and also lectures on botany. During travels in Switzerland and the south of France in 1836 he made a large collection of mineralogical and botanical specimens. In the same year he entered the Medical School at the University of Edinburgh where he graduated MD at the age of 21, receiving a gold medal for a thesis on the presence of alcohol in the brain.

There followed a period of his life based in Birmingham. In 1839 he married a wealthy cousin, Grace Piercy of Warley Hall, Birmingham. He was appointed as a Physician at the Queen's Hospital, although he did not establish a practice. For a period he also held an appointment as Professor of Organic Chemistry at Queen's College, Birmingham. The metallurgical works in the neighbourhood stimulated his interest in metallurgy, and in addition to publishing medical papers he also published on industrial subjects, including work on crystallised slags with David Forbes and William Hallowes Miller. His medical research soon gave place to a systematic study of metallurgy. In 1848, the year after he was elected to the Royal Society on the basis of his medical research, he presented a paper on a method of extracting silver from its ores.

In 1851 he was appointed Lecturer in Metallurgy at the newly founded Institute which later became the Royal School of Mines; the post later became a Professorship. In this period of his career he had a profound influence on the progress of British metallurgy. Percy was an excellent lecturer and teacher, innovative and methodical and most English metallurgists of his time were his pupils.

He contributed to knowledge of metallic alloys, and he discovered the alloy of copper and aluminium(aluminium bronze.) His important work on iron and steel included the supervision of the extensive series of British ores and the question of the need to remove phosphorus in the Bessemer steel making process. Percy also made a very large collection of metallurgical specimens to illustrate points relating to the manufacture and use of metals. Percy's role as an author was particularly significant through his great treatise on *Metallurgy,* which consisted of a series of volumes, the first of which was published in 1861. In 1879 the Government decided to complete the removal of the Royal School of Mines from Jermyn Street to South Kensington. Objecting strongly to this, Percy offered to rebuild the metallurgical laboratory in Jermyn Street, but his offer was refused and he resigned his chair in 1879.

In addition to his position at the Royal School of Mines Percy was appointed Lecturer on Metallurgy to the artillery officers at Woolwich around 1864, and retained this post until his death. Another appointment was as Superintendent of Ventilation, Warming and Lighting of the Houses of Parliament. Also he served the Government in various ways, including membership of the Secretary for War's commission on the application of iron for defensive purposes (1861), and on Royal Commissions (in 1871 and 1875) on matters relating to coal. In the Royal Society he was a Council Member from 1857–1859. He took an interest in social and political questions, on which he wrote many letters to *The Times* under the signature 'Y'.

Percy saw clearly the importance of metallurgy in the national scene. and also through his career he played an important role in transforming metallurgy from a collection of practices into a scientific discipline.

PRS 1889, 46, xxxv–xl. WCR-A. DSB, 10, 511–512. F. Greenaway. DNB. S. Cackett *JHMS*, 1989, 23, 2, 92–98.

SIR JOSEPH ERNST PETAVEL 1873–1936; FRS 1907

Joseph Petavel was born in London, the younger son of Emmanuel Petavel and Susanna née Olliff. The family moved to Geneva in 1876 and then to Lausanne, where Joseph studied engineering at the University. In 1893 he entered University College London to study science and engi-

neering. Then with an 1851 Exhibition Scholarship he spent three years at the Royal Institution with Dewar and Fleming, where he made accurate measurents of physical properties of materials at low temperatures. From 1901–1903 he held a John Harling Research Fellowship at Manchester University. He designed a gauge for measuring pressures in exploding gas mixtures and used it in constructing electric furnaces for studying chemical reactions at high pressures and temperatures. He spent a period at the St Louis International Exhibition erecting and demonstrating a reproduction of the Royal Institution plant for gas liquifaction. Petavel returned to Manchester in 1905, where in 1908 he became Professor of Engineering. During the 1914–18 war the work of his laboratory included testing of materials. In 1919 he was appointed Director of the National Physical Laboratory and was knighted (KBE) in 1920. He remained unmarried.
ONFRS 1936–38, 2, 183–203. Robert Robertson. DNB 1931–40, 69, 690. R. Robertson.

NORMAN JAMES PETCH 1917–1992; FRS 1974
Norman Petch was born in Glasgow, the third child in the family of three boys and one girl of George and Jane Petch. He had an English father who was a commercial traveller and a Scottish mother. The family moved to England, where most of his youth was spent in the West Riding of Yorkshire. Having attended the village school at Aberford he studied at the Grammar School in Tadcaster, and then at the West Leeds High School, when the family moved to Leeds. Petch in 1935 won a scholarship to Queen Mary College, London, and obtained a BSc in chemistry two years later. Further studies followed at the University of Sheffield where he was awarded a BMet at the age of 21.

In 1939 when Queen Mary College was evacuated to Cambridge Petch entered the crystallography group at the Cavendish Laboratory. His first research, under the supervision of Harold Lipson led to the elucidation of the complex crystal structure of cementite and later he deduced the positions of the interstitial carbon atoms in austenite and in martensite. In 1942 he joined the Metallurgy Division of the Royal Aircraft Establishment at Farnborough where his responsibilities, as head of a newly created Crystallography Section included radiography and electron microscopy, and mainly the investigation of failures of aircraft components and related equipment.

In 1946 Petch returned to the Cavendish Laboratory as a member of the Metal Physics Research Group headed by Egon Orowan. He carried out work supported by the British Iron and Steel Research Association (BISRA) concerned with the yield and fracture stresses of steels at low temperatures. This research, which proceeded in parallel with that of Eric Hall led to the relationship between lower yield stress and grain size which came to be known as the Hall–Petch equation. From 1948–49 Petch was BISRA Research Associate at Cambridge, and BISRA planned to appoint him as Head of Physical Metallurgy at a new laboratory to be built at Sheffield. However, before he could take up this position he was appointed Reader in Metallurgy at Leeds University in 1949; he was promoted to be the first Professor of Metallurgy in Leeds in 1956, where he established a full department in the subject. At Leeds much of Petch's research activity, with various co-workers, concentrated on aspects of the Hall–Petch equation, and on the ductile-brittle transition.

Appointment as Cochrane Professor of Metallurgy at the University of Durham, King's College, at Newcastle upon Tyne occurred in 1959. A major project on the brittle fracture of steels was carried out and also work on the brittle fracture of alumina. From 1968–1971 Petch was a Pro-Vice-Chancellor in the University at Newcastle.

Petch was appointed to the Chair of Metallurgy at Strathclyde University in 1972. During his

10 years tenure of this post his research interest was mainly in the field of deformation and fracture of steels. Following his retirement in 1965 he spent most of his time on the Black Isle, while keeping contact with former colleagues.

Following the end of his marriage to Marion Blight, Petch married Eileen Allen in 1949, and in the early 1950s two daughters were born. From 1955, when his wife contracted poliomyelitis which tragically confined her to a wheel chair until her death in 1975, Petch gave devoted care. In 1976 he married Marjorie Jackson.

Petch's career included service on important committees and the honours he received included election as FREng in 1979; he served on the Council of the Royal Society from 1979–81. His pionéering work on mechanical properties led to substantial beneficial consequences for industrial practice, particularly in ferrous metallurgy.

BMFRS 1995, 41, 341–357. Sir Robert Honeycombe FRS.

JOHN BERNARD PETHICA; FRS 1999

Univ. of Oxford: Professor of Materials Science.

Atomic-scale mechanics. Nanoindentation. Scanning tunnelling microscopy.

DAVID GODFREY PETTIFOR 1945–; FRS 1994

Univ. of Witswatersrand: BSc, 1967. PhD, Cantab, 1970. Univ. of Dar-as-Salaam: Department of Physics, Lecturer, 1971. Univ. of Cambridge: Research Assistant, 1974. Bell Laboratories, USA: visiting Research Scientist, 1978. Imperial College, London: Lecturer and Reader in Mathematics, 1978–88; Professor of Theoretical Solid State Physics, 1988–92. Univ. of Oxford: Isaac Wolfson Professor of Metallurgy, 1992–; Head of Department of Materials, 1994–95.

Predictive models of cohesion and structure in alloys based on theories of atomic structure. Magnetism of iron under pressure and temperature.

SIR WILLIAM PETTY 1623–1687; FRS 1663

William Petty was the eldest surviving child of Anthony Petty, clothworker and tailor, and his wife Francesca Denby Petty. He was born in Romsey, Hampshire, and at the age of 13 he joined a merchant ship as a cabin boy; later he served in the Navy. In 1643 he studied medicine in the Netherlands and in 1645 spent a period in Paris. Having returned to England he resumed his anatomical studies and by 1649 he was at Oxford University. In 1650 he received the doctorate of physic and became Professor of Anatomy and Vice-Principal of Brasenose College. He also became Professor of Music at Gresham College, London. He was appointed Physician General to Cromwell's army in Ireland in 1652 and carried out land survey and mapping. Petty became wealthy and was knighted by Charles II in 1661. He married Elizabeth Fenton née Waller and they had two sons and one daughter. His scientific work included an interest in elasticity.

DSB 10, 564–567. Frank N. Egerton III. DNB.

LEONARD BESSEMER PFEIL 1898–1969; FRS 1951

Leonard Bessemer Pfeil was born in London, the third of four children; his father was an accountant, descended from a Frankfurt banker, and his mother came from a family with industrial connections. The name Bessemer derived from his birth occurring two days before the death of Sir Henry Bessemer. Pfeil was educated at St Dunstan's College, Catford, London where he showed an interest in chemistry. When he left school during the 1914–1918 war, he worked as a metallurgical chemist in South London, when he acquired the manipulative skill that later marked his experimental researches, and he also attended evening classes in London. In 1918 he was accepted as an Airforce Cadet for training, but was demobilised a month after the Armistice, and went immediately to the Royal School of Mines to study metallurgy under Professor Harold Carpenter. A brilliant student, he graduated BSc in 1921 with first class honours, and won the Bessemer Medal and the Murchison Medal. He married Brenda Beatrice Butler in 1924 and they had two sons.

After graduation, Pfeil was appointed to be a Junior Lecturer in Metallurgy under Professor Charles Edwards at Swansea, where Pfeil began by working on metallographic aspects of steel relating to technological problems of the local industry. He also carried out fundamental work involving the growth of single crystals of iron by the strain – anneal method, and the study of their mechanical properties; this led to an important investigation on the mechanism of oxide scale formation on iron during heat treatment. Pfeil was a very hard worker, and an excellent lecturer. His research led to the award of the London DSc in 1927.

In 1930 Pfeil was appointed Assistant Manager in charge of the Research and Development Department of the Mond Nickel Company in Birmingham, which was then located in a few rooms in the works of Henry Wiggin and Company Ltd. A range of industrial problems was investigated, but the main objective was to build up a staff capable of developing all the applications of nickel. Extensive programmes were carried out on the functions of nickel in steels and cast irons, and on the corrosion resistance of nickel alloys generally. A new laboratory in Birmingham, built in 1934–1936, soon became one of the principal metallurgical research laboratories in the country. Projects included the cooling transformations of alloy steels and the welding of the stronger, alloyed plate steels. Work also concerned special materials problems of the telecommunications industries and of the aircraft engine, in which Pfeil himself showed a special interest. The major wartime contribution of the laboratory was to develop the highly successful nickel-chromium based 'Nimonic' series of heat resisting alloys, with the result that the British aircraft industry had a valuable lead in the post-war period. Pfeil's role was crucial role in the development of high temperature alloys that enabled the Whittle jet engine to be successfully produced.

Pfeil also became increasingly involved in matters external to the Birmingham Laboratory: work was done on fibrous alloy structures in platinum tungsten alloy wires used in sparking plug electrodes for aircraft engines, and on the production by powder metallurgy of diffusion membranes for separating isotopes of uranium.

At the end of the war Pfeil became Manager of the Research and Development Department of the Mond Nickel Company Ltd, and moved to London. His work came to concentrate increasingly on the technical problems of production policy. He became a Director of Henry Wiggin and Company Ltd in 1946 and of The Mond Nickel Company Ltd in 1951. In this period he served on many official and institutional advisory committees. In 1955 he relinquished his control of development and research in his organisation to give more attention to wider problems. From 1960 until his retirement in 1963 he was Vice-Chairman of International Nickel Ltd. Pfeil was awarded the OBE in 1947. In the course of his influential career he made unusually varied and significant contributions to physical metallurgy.

BMFRS 1972, 18, 477–487. N.P. Allen FRS.

SIR LIONEL ALEXANDER (ALISTAIR) BETHUNE PILKINGTON 1920–1995; FRS 1969

Lionel Pilkington was the second son of Lionel Pilkington, an engineer and his wife Evelyn. (The Christian name Alistair was one that he adopted.) He was born in Calcutta where his father worked for a few years. From the mid 1920s he was in England, where he attended Sherborne

School. At Cambridge, which he entered in 1938 he studied mechanical sciences and was a reserve Artillery Officer. He saw war service from 1939–1945 in Egypt, Greece and also Crete, where he became a prisoner of war. He resumed his studies at Cambridge in October 1945. His father's interest in the family tree led to a contact with the St Helen's glass-making family of Pilkington. There was no direct family link but the contact led to his taking an appointment with the Pilkington Company in 1947. He held a series of appointments involving rapid promotion, being appointed a Board Member in 1953, Director 1955–58, Deputy Chairman, 1971–73, Chairman 1973–80, Non-Executive Director 1980–85, and President in 1985. Alistair Pilkington was responsible for the invention and development of the float glass process, which came to ultimate success in the early 1960s. He was awarded the Mullard Medal in 1968 and gave the Philips Lecture in 1990.

From his marriage in 1945 to Patricia Elliott there were a son and a daughter; following the death of his wife he married Kathleen Eldridge Haynes in 1978. He was active in philanthropic work, and also in higher education, serving as Pro-Chancellor of Lancaster University (1980–90) and Chancellor of Liverpool University, from 1994.
The Times, May 6th, 1995. *Who's Who* 1995.

COLIN TREVOR PILLINGER 1943–; FRS 1993
Kingswood Grammar School. Univ. College, Swansea: Univ of Wales: BSc Hons Chemistry, 1965, PhD 1968. Univ. of Bristol: Department of Chemistry; Postdoctoral Fellow, 1968–72; BSC Fellow: 1972–74; Research Associate, 1970–76 Univ. of Cambridge: Department of Earth Sciences: Associate and Senior Research Associate, 1976–84. Open Univ.: Department of Earth Sciences: Senior Research Fellow, 1984–90; Professor of Planetary Science, 1990–.
Mineral physics of diamonds. Mass spectrometry.

SIR ALFRED BRIAN PIPPARD 1920–; FRS 1956
Clifton College. Univ. of Cambridge: Clare College, BA 1941, MA 1945, PhD 1949, ScD 1966. Radar Research and Development Establishment, Great Malvern: 1941–45. Univ. of Cambridge: Pembroke College, Stokes Student, 1945–46; Demonstrator in Physics, 1946; Lecturer, 1950; Reader 1959; John Humphrey Plummer Professor of Physics, 1961; Cavendish Professor of Physics, 1971–82; President Clare Hall, 1966–73. Hughes Medal, 1959. Knighted 1975.
Superconductors: conduction in surface layers; magnetoresistance.

MAX KARL ERNST LUDWIG PLANCK 1858–1947; FMRS 1926
Max Planck was born at Kiel, the fourth child of Johann Julius Wilhelm von Planck, Professor of Civil Law, and Emma née Patzig. The family moved to Munich in 1867 where he attended the Maximilian Gymnasium before entering the University of Munich in 1874. He studied mathematics and physics and turned his attention particularly to the latter, and spent a period in Berlin in 1877 and 1878; he obtained a PhD at Munich in 1879 for work on the second law of thermodynamics and became a Lecturer in the following year. In 1885 he became Extraordinary Professor of Physics at Kiel and in 1888 later he moved to Berlin to become Assistant Professor of Physics and Director of the Institute for Theoretical Physics. From 1892 until 1926 Planck held the post of Professor of Physics at Berlin. In 1930 he became President of the Kaiser Wilhelm Institute, but resigned in 1937 in the situation of the Nazi's treatment of Jewish scientists. The Institute moved to Göttingen in 1945 and was renamed the Max Planck Institute; he was appointed its President, a position which he held until his death.

In his family life Planck experienced great tragedy, through the death of his first wife, Marie Merck, and also of the two sons and two daughters of his second wife, Marga von Hoesslin; a further loss occurred when most of his books and manuscripts were destroyed in an air raid in the Second World War.

Planck's research included the fields of thermodynamics and radiation. In 1900 he proposed the revolutionary idea that radiation consists of indivisible particles, an idea that developed into quantum mechanics. He was awarded the Nobel Prize for Physics in 1918.

ONFRS 1948–49, 6, 161–188. M. Born. DSB 11, 7–17. H. Kangro.

JOSEPH ANTOINE FERDINAND PLATEAU 1801–1883; FMRS 1870

Joseph Plateau, born in Brussels was the son of an artist, who died when Joseph was 14. Having received his early education in Brussels he entered the University of Liege to study law in 1822; however, he developed an interest in science and after he had received his law diploma, he became a candidate for an advanced degree in physical sciences and mathematics. In 1829 he became Docteur es Sciences, and returned to Brussels the following year as Professor of Physics at the Institut Gaggia, a leading teaching institution. He became Professor of Experimental Physics at the State University of Ghent in 1835, and then from 1844 until his retirement in 1872 Professeur Ordinaire. In 1843 he completely lost his sight, having some years earlier stared into the sun in an experiment. However, Plateau continued his productive career. His earlier work was concerned with physiological optics and visual perception. Other important work was on thin films formed within wire contours dipped into a mixture of soapy water and glycerine. His theoretical work, which led him to conclude that the shapes always had minimum surfaces, was confirmed experimentally and was of fundamental interest in relation to surface and interfacial tension effects in metallurgical systems. Plateau had an exceptional gift for visualising physical results and physically interpreting geometrical results.

DSB 11, 20–22. E. Koppelman.

SIR LYON PLAYFAIR (BARON PLAYFAIR OF ST ANDREWS) 1818–1898; FRS 1848

Lyon Playfair was born in Chunar, India, the son of George Playfair, a Medical Officer and Janet née Ross. He was brought up by relatives in Scotland and received his early education at the parish school in St Andrews. In 1832 he entered the University of St Andrews and in 1835 he began to study for the medical profession, entering Thomas Graham's chemistry classes at the Andersonian Institution in Glasgow. He went to Edinburgh University in 1837 to complete his medical studies, but had to discontinue for health reasons. For a short period he went to Calcutta to enter a business house, but then returned to England to study chemistry and became Graham's assistant in London. From 1839–1840 he worked in Liebeg's laboratory at Giessen and obtained a PhD. Then, after work in an industrial post in Lancashire, he was appointed Professor of Chemistry at the Royal Institution, Manchester, where he spent three years, including collaboration with Joule, and also, on blast furnace gases, with Bunsen. In 1845 he became Chemist to the Geological Survey and subsequently Professor of Chemistry in the newly founded School of Mines in London. Playfair married Margaret Eliza Oakes in 1846 and they had one son; following his wife's death Playfair married Jean Ann Milligan in 1857, and then, after her death, he married Edith Russell in 1878.

Playfair was the Chief Adviser to Albert, Prince Consort, for the Great Exhibition of 1851 and was influential in advancing science and technology. He was appointed to a Chair of Chemistry at Edinburgh in 1858. He was active in political life, serving an a Member of Parliament for the

universities of St Andrews and Edinburgh from 1868–1885, and for South Leeds from 1885–1892; he was appointed as Postmaster-General in 1873 and was Deputy Speaker of the House of Commons from 1880–1883. He was knighted (KCB) on retirement from the latter office and was raised to the peerage as Baron Playfair of St Andrews in 1892.
PRS 1898–99, 64, ix–xi. ACB. DSB 11, 36, 37. W.V. Farrar. DNB Suppl. III, 270–272. AH–n.

SIMÉON DENIS POISSON 1781–1840; FMRS 1818
Siméon Poisson was born at Pithiviers, Loiret, France. His health was weak and his mother arranged for the care of a nurse. His father, formerly a soldier, after retiring from the army obtained a low-ranking administrative post, but with the coming of the Revolution he became President of the District. Poisson, at the age of 14 had been placed in the care of his uncle, a surgeon at Fontainebleau, but he had little interest in entering the medical profession. In 1796 he became a pupil of the École Centrale of Fontainebleau, where he showed great ability and progressed rapidly in mathematics. Two years later he obtained admission by competitive examination to the École Polytechnique, Paris.

Poisson's progress was such that in 1800 he presented a memoir which cleared up a very difficult and obscure point of analysis. He continued his mathematical work, and was promoted to full Professor at the Polytechnique in 1806, also later holding other important appointments. In 1817 Poisson married Nancy de Bardi, an orphan, born in England to émigré parents. He became a Baron in 1837. He was awarded the Copley Medal in 1832.

Poisson's life style was simple and he spent long hours working. In his career he made important contributions to the applications of mathematics to physical science, including electricity, magnetism , astronomy and mechanics. His research impacted on metallurgy through his contributions to elasticity theory remembered especially through 'Poisson's ratio'. He was a leading mathematical physicist and influential in the scientific life of France.
PRS 1840, 4, 269–272. DSB 15, Suppl. I, 480–490. P. Costabol.

MICHAEL POLANYI 1891–1976; FRS 1944
Michael Polanyi was born in Budapest, Hungary one of a family of five children of Mihaly Polyanyi and Cecilia née Wohl. His father was a civil engineer and entrepreneur involved in the planning and development of railways in Europe, but lost all his fortune when Michael was eight years old. In 1905 when his father died Michael assisted the family finances by tutoring other high school students.

When he graduated from high school in 1909 Polanyi entered the University of Budapest to study medicine. He became increasingly interested in research and wrote a paper of medical interest the next year, followed by two other papers. His interest in other fields had been developing and soon after qualifying in medicine in 1913 he entered the Technische Hochschule in Karlsruhe, Germany to study chemistry. On the outbreak of war in 1914 Polanyi had to return to Austria–Hungary. He served in the army as a medical officer, but his interest in physical chemistry continued and he wrote several papers. With the break up of Austria–Hungary at the end of the war Polanyi remained in Budapest to complete his doctoral degree in chemistry. However, in view of the the political situation he returned to Karlsruhe in 1919, where he became interested in the rate of chemical reactions. Here also he met his future wife, Magda Kemeny, whom he married in 1921 in Berlin; they had two sons.

Late in 1920 Polanyi moved to Berlin, to the Kaiser Wilhelm Institut für Faserstoffchemie (fibre chemistry) where he worked on the X-ray analysis of fibrous structures, in particular of cellulose, but also of metals, and on new methods of X-ray analysis in general. This then led him to study the structure and properties of crystals principally those of metals. Polanyi and his collaborators grew crystals, mainly metallic, and investigated their properties under stress. He proposed the theory that the lower strengths of real material as compared with theoretical strength result from lattice imperfections, thus making him one of the originators of the concept of dislocations.

In 1923 Polanyi was appointed as one of the department heads at the Institut für Physikalische Chemie und Electrochemie. He maintained for several years his interests in crystallography and crystal structure but turned his attention increasingly to the theme of the rate of chemical reactions. With the influence of the Nazi party, even before Hitler became Chancellor in 1933, there occurred the dismissal of several Jewish scientists. Polanyi was greatly concerned by this and soon gave up his position at the Institute and accepted the Chair of Physical Chemistry at the University of Manchester, starting work there in the autumn of 1933.

Polanyi's work turned to the field of economics in 1932 and he led a small team which produced a pioneer educational film on economics and unemployment. During the 1940s he decided to concentrate on philosophy and in 1948 he received a Personal Chair in Social Studies. He moved to Merton College, Oxford in 1958 as a Senior Research Fellow, and during the ensuing 15 years he travelled extensively and published papers on a wide range of subjects: scientific, political and aesthetic.

BMFRS 1977, 23, 413–448. E.P. Wigner FMRS and R.A. Hodgkin. DSB 18, 718–719 F. Szabadváary. DNB 1971–80 677–688 R.A. Hodgkin.

WILLIAM POLE 1814–1900; FRS 1861

William Pole was born in Birmingham. His career included engineering practice in England, for example involvement in the Britannia bridge project. His appointments included Lecturer at the Royal Engineer Establishment at Chatham, Professor of Civil Engineering at Elphinstone College, Bombay, and Professor of Civil Engineering at University College London (1855–1896). He served the government in providing technical advice on various matters e.g on heavy artillery, and was a member of several Royal Commissions. In the Royal Society he was a Council Member. In 1846 he married Matilda Gauntlett. Pole was a distinguished musician, receiving the DMus degree from Oxford University in 1867.

PRS 1905, 75, 117–120. JWS. Notes and Records of the Roy. Soc. 1976, 30, 169–179. T.M. Charlton. *Who was Who* 1897–1915.

JEAN VICTOR PONCELET 1788–1867; FMRS 1842

Jean Poncelet was born in Metz, France, the illegitimate, but later recognised son of Claude Poncelet, a rich landowner. After his studies, which included a period at the École Polytechnique in Paris he entered the École d'Application of Metz as a Sub-Lieutenant of Engineers. In 1812 he assisted in the construction of defensive works on Walcheren Island. In the same year he saw action in the campaign against Russia as a Military Engineer and having been taken prisoner, experienced great hardship. He returned to France in 1814 and became a Captain of Engineers at Metz. His work included the theory of machines and industrial mechanics; he was the author of the book *Introduction a la Mécanique Industrielle Physique* (1829). His appointments included Chairs in Mechanics and from 1848–50 he was Commandant of the École Polytechnique, with the rank of Brigadier-General. In 1842 he married Louise Palmyre Gaudin.

PRS 1869–70, 18, i–ii. DSB 11, 76–82. R. Taton.

SIR WILLIAM JACKSON POPE 1870–1939; FRS 1902

William Pope was born in London, one of eight children of William Pope and Alice née Hall. In 1885 he won an entrance scholarship to the Finsbury Technical College and the City and Guilds of London Institute at South Kensington. In 1897 he became Head of the Chemical Department at the Institute of the Goldsmiths at New Cross, London. Four years later he was appointed Head of the Chemical Department at the Municipal School of Technology and Professor of Chemistry at Manchester. From 1908 until his death he was Professor of Chemistry at Cambridge, working in the organic field. Arising from his association with W.H. Barlow he suggested to W.L. Bragg a possible crystal structure for sodium chloride which Bragg confirmed in 1913 by X-ray diffraction. Pope received the Davy Medal in 1914, the CB in 1918 and was knighted (KBE) in 1919.
BMFRS 1939–41, 3, 291–324. Charles S. Gibson. DSB 1, 84–92. F.G. Mann.

JOHN ANTHONY POPLE 1925–; FRS 1961

Bristol Grammar School. Univ. of Cambridge: Trinity College, MA, PhD; Fellow, 1951–58; Lecturer in Mathematics, 1954–58. National Physical Laboratory: Superintendent of Basic Physics, 1958–64. Carnegie-Mellon Univ.: Professor of Chemical Physics, subsequently John Chalmers' Warner University Professor of Natural Science, 1964–93. North Western Univ. adj Professor of Chemistry, 1986–93. Trustees Professor of Chemistry, 1993–. Davy Medal, 1988.
Molecular physics and theoretical chemistry.

ROBERT PORRETT 1783–1868; FRS 1848

Robert Porrett was born in London where his father was Ordnance Storekeeper in the Tower of London and resided there. Robert was employed as assistant to his father, and later succeeded him as head of the department from 1795 to 1850. In his leisure he pursued scientific interests especially in chemical research. He also had interests in antiquarian matters, particularly in arms and armour. He was a bachelor.
PRS 1869–70, 18, iv, v. DNB. Bulloch's Roll. J. Chemical Soc. of London 1869, VII, vii.

ALBERT MARCEL GERMAIN RENÉ PORTEVIN 1880–1962; FMRS 1952

Born in Paris, Albert Portevin was brought up by his mother after his father's early death. In 1889 he entered the École Centrale des Arts et Manufactures, one of the Grandes Écoles d'Ingenieurs Francaises, where he graduated in 1902 after a brilliant career as a student. After a period with Credit Lyonnais and later with the Societe Métallurgique de la Bonneville he joined the De Dion-Bouton factories as Chief Chemist. For more than 40 years he was connected with the teaching staff of the École Centrale, which he joined in 1910 to create the laboratory of microscopical metallography and where he became professor in 1925. He also taught at other science schools in Paris – the École Supérieure de Fonderie and at the École Supérieure de Soudure. The application of science to industry was an important part of his work and he contributed his deep knowledge to many organisations, concerned for example, with electrochemistry, electrometallurgy and ferrous metallurgy.

Portevin's research activities ranged very widely, including heat treatment of steels, casting, welding, corrosion, trace elements and inclusions. His research output was represented by about 800 papers in association with many collaborators. Portevin became Editor in Chief of the *Revue de Métallurgie* in 1907. Among his other appointments he was a consultant to the Science Committee of the UNESCO national commission in France. A well known lecturer, he was in demand all over the world. His work was characterised by his great clarity in the establishment of programmes of research and interpreting results, and in his great ability to communicate results. In education he was influential through the many engineers that he taught. He was also a man of exceptionally wide culture and was interested in obtaining deep knowledge of classical authors as well as the natural sciences.

Portevin's contributions to science brought numerous honours from learned societies in France and other countries, and honorary doctorates from universities. Other awards of national recognition came from many countries, including that of CBE in 1957. He was a respected teacher to his pupils and one of the most prestigious leaders of French metallography.

BMFRS 1963, 9, 223–235. P. Bastien. JISI 1962, 200, 558, 1070. 1071. P. Bastien. DSB 11, 100, 101. G. Chaudron.

ALAN RICHARD POWELL 1894–1975; FRS 1953

Alan Powell was educated at the City of London School which he left at the age of 19 to join the firm of G.T. Holloway of Poplar, consulting chemists and metallurgists. This firm, in which by 1916 Powell had become Chief Chemist, had an established reputation in the assay of minerals and in the chemistry of the rare earth metals, thorium and other, then rare, metals. Through a professional friendship with a mining engineer he was recommended to Johnson, Matthey and Co. Ltd when the company conceived the idea of setting up a research department. Powell's appointment began in 1918.

Powell's first laboratory, where for three years he had one young laboratory assistant, was in a back room in Hatton Garden. During this period he was a co-author with Dr W.R. Schoeller of a textbook on the *Analysis of Minerals and Ores of the Rarer Elements.* His work included the study of the chemistry and metallurgy of niobium and tantalum. In 1920 the Department was equipped with a chemical laboratory, a balance room, fume-cupboards and a furnace room; two scientific colleagues were appointed and Powell became formally Head of Department. Over a period of 14 years Powell was associated with Schoeller in research on tantaloniobic minerals, mostly carried out at the Sir John Cass Technical Institute.

Varied work was undertaken over the years, to which Powell brought a comprehensive knowledge and technical skill. This included investigations to overcome the problem of the instability of platinum–rhodium thermocouples. The laboratory became responsible for precious metal analysis, which involved further increases in staff, and which laid the foundations of the analytical laboratory. As the work of the Department grew, sectionalisation and further increases in staff were required; however, the whole staff could be mobilised to deal with an important and urgent problem, as in the work begun in 1926 which overcame the difficulties in extracting precious metals from certain South African ores. A new laboratory was built in Wembley in 1938. Work included problems arising in the field of metallic electrical contacts, the development of special solders and fluxes, the design and development of high temperature electric furnaces, and the development of new alloys of the precious metals for specialized uses. Powell gave lectures at the Royal School of Mines during the 1940s. In 1954 he retired from the Managership of the Research Department, but until 1960 he was a Research Consultant, and long after he had finally retired he continued to pursue his interest of personal experimentation at the laboratory bench which was reserved for him.

Away from chemical and metallurgical work Alan Powell was a keen and successful gardener. He was married in 1914 to Marguerite Tremmel and their family consisted of one son and three daughters; Marguerite died in 1956, and he married Mildred Mary Coleman in 1960.

Alan Powell did not publish much of his research results. He was primarily an experimentalist,

who could work equally readily on the scale of a few milligrams of metal or with tonnes of ore. Possessing an unrivalled knowledge of inorganic chemistry, he conveyed his imaginative thinking to his colleagues in discussion. He was a research leader of great distinction.
BMFRS 1976, 22, 307–318. G.V. Raynor FRS.

HENRY POWER 1623–1668; FRS 1663

Henry Power matriculated at Cambridge, as a Pensioner of Christ's College in 1641 and graduated BA in 1644, MA in 1648 and MD in 1655. It seems that he practised medicine at Halifax for some time, but then moved to New Hall near Elland. In 1664 he published *Experimental Philosophy,* in three books, on microscopy, atmospheric pressure, and magnetism.
DSB 11, 121, 122. G.W. O'Brien. DNB. Gen. Ref. 20, v III.

LUDWIG PRANDTL 1875–1953; FMRS 1928

Ludwig Prandtl was born in Freising, Germany, the only child of Alexander Prandtl, an Engineering Professor, and Magdalene née Ostermann. He entered the Technische Hochschule at Munich in 1894 where he specialised in engineering. He graduated in 1898 and two years later completed a doctoral thesis 1900 on the lateral instability of beams in bending. He then worked in Maschinenfabrik Augsburg-Nurnberg where he developed an interest in fluid flow. Prandtl was appointed to a Chair at the Technische Hochschule in Hanover in 1901; he moved to Göttingen in 1904 as Professor of Applied Mechanics and built up an important centre for aerodynamics. He married Gertrude Foppl and they had two daughters. Prandtl's work included supersonic flow, boundary layer theory and turbulent flow; in the early 1920s he carried out research on plastic deformation.
BMFRS 1959, 5, 193–205. A. Busemann. DSB 11, 123–125. J.H. Lienhard.

SIR WILLIAM HENRY PREECE 1834–1913; FRS 1881

William Preece was born at Bryn Helen, Caernarvon, the eldest son of Richard Matthias Preece, stockbroker and Jane née Hughes. After attending King's College School and King's College, London he went to the Royal Institution where he worked with Faraday and became interested in the problems to be solved in applied electricity and telegraphy. After entering the office of a civil engineer in 1852 he was appointed the following year to the Electric and International Telegraph Company, becoming Superintendent of its Southern District in 1852. Preece was Engineer to the Channel Islands Telegraph Company from 1858 to 1862, and then in 1870 began a long association with the Post Office. He rose to be Chief Engineer in 1892, and following his retirement in 1899 he continued consulting. He was responsible for many improvements and inventions in radio-telegraphic communications. He introduced into England the first telephone receivers as patented by Alexander Graham Bell. In 1864 he married Anne Agnes Pocock and they had four sons and three daughters. He was made CB in 1894 and knighted (KCB) in 1899.
DNB 1912–1921, 442, 443. APMF.

JOSEPH PRIESTLEY 1733–1804; FRS 1766

Joseph Priestley was born at Fieldhead, Birstal, near Leeds the son of Jonas Priestley and his wife Mary née Smith, who died when Joseph was seven. His father was a weaver and cloth dresser and a congregationalist. Joseph was educated at a non-conformist seminary at Daventry from 1752–1755 and then became a Presbyterian Minister at Needham Market. In 1758 he moved to Nantwich and began to teach languages at Warrington Academy. He married Mary Wilkinson from an ironmaster's family in 1762. On a visit to London he met Benjamin Franklin, which led him to take up the study of electricity, and eventually to write a history of the existing knowledge of the subject. This, together with several new electrical experiments, established his reputation and led to his election as an FRS.

In 1767 Priestley became a minister of a chapel at Leeds and began chemical experiments and among his discoveries was that of oxygen, announced in 1775, and of carbon monoxide. His most productive work was done between 1773 and 1780, when he was librarian and literary companion to Lord Shelbourne, whose estate was at Calne, Wiltshire. He was a minister in Birmingham and in 1780 he was a member of the Lunar Society. There were serious riots in Birmingham in 1791, and, as a dissenter, his home was broken into by a mob destroying his books and papers. In 1794 he went to the USA to join his sons. His research included combustion and respiration, gaseous diffusion, conduction of heat, the effect of electrical discharge on gases at low pressures. He received the Copley Medal in 1772. In addition to his scientific work he was a theologian and philosopher.

DSB 11, 139–147. R.E. Schofield.

JAMES PRINSEP 1799–1840; FRS 1828

James Prinsep, the seventh son of John Prinsep, commenced architectural studies at the age of 15, but because of an eyesight problem relating to drawing work, he trained to be an assayer. At the age of 20 he became Assistant Assay Master at the Mint in Calcutta, subsequently being appointed Assay Master in Benares, and then Deputy Assay Master at Calcutta (1830); from 1832 until 1838, when he returned to England for health reasons, he was Assay Master and Secretary to the Mint Committee at Calcutta. He was the author of a reform of weights and measures of India. His metallurgical activities included pyrometry, and the preservation of iron from rusting. In India he also carried out important building work, for example canals and bridges. He became well-known for his literary work, and studied antiquities in India. He married Harriet Aubert and they had one daughter.

PRS 1837–43, 4, 259, 260. DNB.

MAURICE HENRY LECORNEY PRYCE 1913–; FRS 1951

Royal Grammar School; Guildford. Univ. of Cambridge: Trinity College; Faculty Asst. Lecturer, 1937–39. Univ. of Liverpool: Reader in Theoretical Physics, 1939–45: Admiralty Signal Establishment: Radar Research, 1941–44; National Research Council of Canada, Montreal: Atomic Energy Research, 1944–45. Univ. of Cambridge: University Lecturer in Mathematics and Fellow Trinity College, 1945–46. Univ. of Oxford: Wykeham Professor of Physics, 1946–54. Univ. of Bristol: Henry Overton Wills Professor of Physics: 1954–64. Univ. of Southern California: Professor of Physics, 1964–68. Univ. of British Columbia: Professor of Physics, 1968–78.

Theoretical physics.

SIR ALFRED GRENVILE PUGSLEY 1903–1998; FRS 1952

Alfred Pugsley was the son of H.W. Pugsley, a civil servant in the Admiralty. He was educated at Rutlish School and then from 1920–1923 at Battersea Polytechnic, where he graduated in engineering with a London BSc degree. He was a civil engineering apprentice at the Royal Arsenal, Woolwich from 1923–26 and from 1926–1931 he worked as a Technical Officer at the Royal Airship Works at Cardington, Bedford in a team working on the design of the R101 airship. In 1928 he married Kathleen M. Warner; there were no children from the marriage. When the

Cardington team was disbanded in 1930 Pugsley moved to the Royal Aircraft Establishment at Farnborough, Hants where he worked on aeroplane flutter. He remained at the RAE throughout the war, becoming Head of the Structural and Mechanical Engineering Department in 1941 and receiving the OBE in 1944. During the later part of his period at the RAE he became interested in problems of repeated loading, leading to fatigue damage of aircraft. Pugsley was appointed to the Chair of Civil Engineering at Bristol University in 1945, an appointment which he held until his retirement in 1968. He served as Pro-Vice-Chancellor from 1961–64. Pugsley's interests ranged widely over engineering science and he extended his work into the field of safety of engineering structures, in which he contributed to understanding of key design factors. His activities included serving as Chairman of the Aeronautical Research Council from 1952–57. He was knighted in 1956.
The Times, March 9th, 1998.

PETER NICHOLAS PUSEY 1942–; FRS 1996
St Edwards School, Oxford. Univ. of Cambridge: Clare College, MA. Univ. of Pittsburgh: PhD 1964. IBM Yorktown Heights, New York: Postdoctoral Fellow, 1969–72. Royal Signals and Radar Establishment: SPSO, later Grade 6, 1972–91. Univ. of Edinburgh: Professor of Physics 1991–; Head of Department of Physics and Astronomy.
Colloidal and polymeric systems.

SIR WILLIAM RAMSAY 1852–1916; FRS 1888
William Ramsay was born in Glasgow, the only child of William Ramsay, a civil engineer and businessman, and Catherine née Robertson. Having received his secondary education at Glasgow Academy he matriculated at the University in 1816 and studied classics. His interests in science developed and he attended lectures in chemistry and physics (the latter, in 1870, given by William Thomson). In 1869 he commenced a period of 18 months as apprentice to a local analyst. Then, in 1871, he went to the Organic Chemistry Laboratory at the University of Tübingen where he obtained a PhD in 1872. Having returned to Glasgow he became an Assistant at Anderson's College, prior to becoming an Assistant at the University in 1874. In 1881 he became Professor of Chemistry at University College, Bristol also holding the post of Principal. He married Margaret Buchanan in 1881 and they had two children. In 1887 he was appointed to the Chair of Chemistry at University College London, a post which he held until his retirement in 1912. Ramsay's research included the notable discovery and isolation of the inert gases, and investigations on radioactivity. His earlier interest in liquids and vapours included work on mercury; later he worked on the extraction of radium from pitchblende. He was made KCB in 1902 and received the Davy Medal in 1895 and the Nobel Prize for Chemistry in 1904.
PRS 1916–17 , 93, xlii–liv, JNC. DSB 11, 277–284. T.J. Trenn. DNB 1912–1921, 444–446. FGD. Morris S. Travers, *William Ramsay*, Edward Arnold Ltd, The Royal Society, London, 1956.

JESSE RAMSDEN 1735–1800; FRS 1786
Jesse Ramsden was born in a suburb of Halifax, the son of an innkeeper. He was apprenticed at the age of 16 to a cloth worker and subsequently became a clerk in a cloth warehouse. However, in 1758 he apprenticed himself to a mathematical instrument maker, and opened his own shop in London four years later. He became the most skilful and capable instrument maker of the 18th century, with interests in mathematics, optics and mechanics. During his career he made many instruments including theodolites, micrometers, barometers, balances and screw cutting lathes;

also he devised a method of measuring the expansion of metal bars against a standard. He received the Copley Medal in 1795.
DSB 11, 284–285. R.S. Webster. DNB. Gen. Ref. 26.

SIR JOHN TURTON RANDALL 1905–1984; FRS 1946

John Randall was the only son and the eldest of three children of Sidney Randall and Hannah née Turton. He was born at Newton-le-Willows near Manchester, and attended the Grammar School at Ashton-in-Makerfield. He entered Manchester University where he obtained a first class honours degree in physics, with a graduate prize in 1925, and an MSc in 1926. He then joined the scientific staff of the General Electric Company Ltd at the laboratories in Wembley. Two years later he married Doris Duckworth and they had one son. Randall's collaborative work at Wembley included X-ray powder examination of deposits on the inner glass surfaces of lamps and valves at various stages of manufacture. He also made fundamental investigations of the structure of glasses and of liquids, including some metals.He also played a leading role in developing luminescent powders for discharge lamps.

In 1937, recognised as a leader in the field of luminescence, Randall moved to the Physics Department at Birmingham as a Warren Research Fellow. He was awarded the DSc degree by Manchester University in the following year. With the coming of war in 1939 he joined the group in the Department working on radar and became the joint inventor of the vital device – the cavity magnetron. He moved to St Andrews University in 1944 to the Chair of Natural Philosophy and entered the field of biophysics. Two years later Randall moved to King's College London as Wheatstone Professor of Physics; the Medical Research Council set up the Biophysics Research Unit, which he directed; research themes included investigations of muscle, DNA and collagen. When he retired in 1970 he moved to Edinburgh University where he continued research. The honours received by Randall included a knighthood (1962) and the Hughes Medal (1946).
BMFRS 1987, 33, 493–535. M.H.F. Wilkins FRS. DNB 1981–1985, 330, 331. M.H.F. Wilkins.

WILLIAM JOHN MACQUORN RANKINE 1820–1872; FRS 1853

Born in Edinburgh, William Rankine was the son of David Rankine, an Army Engineer Officer, and Barbara née Graham. He was educated at Ayr Academy and Edinburgh High School prior to studying studied natural philosophy from 1836–38 at Edinburgh University. He assisted his father who was Superintendent of the Glasgow and Dalkeith Railway. Also he was involved as a civil engineer in various engineering schemes, including a period in Ireland apprenticed to a civil engineer on railway projects. He made an investigation in the early 1840s of the fracture of railway axles. In 1855 he was appointed to the Chair of Civil Engineering and Mechanics at Glasgow University, in which post he remained until his death. In 1859 he was commissioned in the University Rifle Volunteers. His researches included the theory of heat engines and thermodynamics. He remained unmarried.
PRS 1872–73, 21, i–iv. DSB 11, 291–295. E.M. Parkinson. DNB.

CHINTAMANI NAGESA RAMACHANDRA RAO 1934–; FRS 1982

Univ. of Mysore. Univ. of Purdue, USA: PhD. Univ. of California, Berkeley: Research Chemist, 1958–59. Indian Institute of Science: Lecturer, 1959–63. Indian Institute of Science and Technology, Kanpur: Professor, 1963–76; Head of Chemistry Department, 1964–68, Dean of Research, 1967–72; Jawaharlal Nehru Fellow, 1973–75. Indian Institute of Science, Bangalore: Solid State and Structural Chemistry Unit and Materials Research Laboratory, Chairman 1970–

84; Director, 1984–94. Jawaharlal Nehru Centre for Advanced Scientific Research: President, 1994–.
Solid state chemistry of ionic substances. Chemical spectroscopy. Oxides. Superconductors.

RUDOLPH ERIC RASPE 1737–1794; FRS 1769
Rudolph Raspe, who was born in Hanover, Germany was the son of an accountant in the Department of Mines and Forests; he developed an early interest in geology through contact in the Harz with some of the mining communities. In 1756 he began to study law at the University of Göttingen; a year later the went to Leizig to study where his interest ranged from science to literature, and where he obtained a master's degree in 1760. He obtained a post as a Clerk in the University Library at Hannover, where he made translations of various works and published works of his own. In 1767 Raspe became Professor of the Collegium Carolinum in Cassel and Curator of the collection of gems and medals belonging to the Landgrave of Hesse; soon afterwards he became Librarian of Cassel. He began writing on natural science and his election to the Royal Society came from a paper in 1769 to *Philosophical Transactions* proposing the previous existence of elephants, or mammoths, in the northern regions of the world. In 1771 he was married and there were two children from the marriage.

Having stolen valuable coins from the collection at Cassel he fled to England in 1775. There the Royal Society removed his name from the list of Fellows. He eventually turned to mining and prospecting as a source of finance. Matthew Boulton offered him a post as Assay Master for some mines at Dolcouth in North Cornwall and he worked there from 1783–1786. It was during this period that he wrote *Baron Munchausen's Narrative of His Marvellous Travels and Campaigns in Russia.* In 1787 he moved to Edinburgh to work in a post cataloguing a collection of gems. During the period 1789–1792 he carried out mineralogical surveys of the Highlands. In 1791 Raspe published a translation of Baron Inigo Born's treatise on his new process of of amalgamation of gold and silver ores. Following Raspe's involvement in a financial irregularity associated with a venture for mineral exploitation in Scotland, he went to County Donegal in Ireland in 1793 to work in a copper mine.
DSB 11, 302 –305. A.V. Carrozzi. DNB.

JOHN URPETH RASTRICK 1780–1856; FRS 1837
John Rastrick, who was the eldest son of John Rastrick, was born at Morpeth, Northumberland. When he was about 15 years old he was articled to his father who was an engineer and machinist. In around 1801 he entered the Ketley Ironworks in Shropshire to obtain experience in the use of cast iron for machinery. Soon after he entered into a partnership in Bridgnorth, in which he was associated with an iron foundry. He practised independently as a civil engineer, including railway engineering. In 1815–16 he built a cast iron bridge over the River Wye at Chepstow and in 1817 became the managing partner in a firm of iron founders and manufacturers of machinery at Stourbridge, Worcestershire, playing a major role in the design and construction of rolling mills, steam engines etc. Rastrick also designed ironworks at Chillington, near Wolverhampton and at Strutt End, near Stourbridge. In his later career he became prominent in various aspects of railway engineering; he retired from active work in 1847.
DNB.

EDWARD PETER RAYNES 1945–; FRS 1987
St Peters School, Yorks. Univ. of Cambridge: Gonville and Caius College, MA, PhD. Royal

Signals and Radar Establishment: 1971–92; SPSO 1981, DCSO 1998–92. Sharp Labs of Europe Ltd Oxford. Chief Scientist, 1992–95; Director of Research, 1995–98. Univ. of Oxford: Professor of Optoelectronic Engineering, 1998–.
Liquid crystals; their physics, Chemistry and applications.

GEOFFREY VINCENT RAYNOR 1913–1983; FRS 1959
Geoffrey Raynor was born in Nottingham, the youngest of three sons of Alfred Ernest, a lace-dressers' manager, and Florence Lottie née Campion. From the local infant and primary school at West Bridgford he won a scholarship to Nottingham High School in 1925. He was also good at sport, with a particular interest in rowing. When his father died in 1927, a time of industrial depression, his mother thought that he should take a job in a bank. The school, however, persuaded her to let him enter the sixth form and in 1931 he won a Nottingham County Major Scholarship with which he entered Keble College, Oxford in 1932. In the final honour School of Natural Science he took Part I in 1935; for his Part II research he worked in the inorganic field in the group led by William Hume-Rothery and was awarded first class honours in 1936. During his undergraduate career he achieved considerable success in university rowing.

Raynor joined Hume-Rothery's research group and was awarded the DPhil in 1939, as well as the MA; he was financially supported by various research assistantships and also became a University Demonstrator in Inorganic Chemistry. He made a wide-ranging investigation of Brillouin zone effects in alloys, showing great interpretative skills in dealing with masses of complex experimental facts; he had a great capacity for work. During the war Raynor tried to join the Army, but was refused release from the scientific work for the war effort to which the Hume-Rothery and Raynor group had turned, including studies of industrially important alloys. In 1943 he married Jean Brockless, who had read physical chemistry and carried out research for the DPhil; the first of their three sons was born in 1944.

In 1945 Raynor accepted an ICI Research Fellowship in the Department of Metallurgy at Birmingham University. He became Reader in Theoretical Metallurgy in 1947 (and was awarded an Oxford DSc in 1948), Professor of Metal Physics (1949–54), Professor of Physical Metallurgy (1954–55), and Feeney Professor of Physical Metallurgy (1955–69) and Head of the Department in 1955. He built up and maintained for many years a large and brilliant research group on alloy constitution. His lectures to undergraduates were notable for their excellence and inspiration.

From 1966 to 1969 Raynor was Dean of the the Faculty of Science and Engineering at Birmingham at a time of expansion and constitutional change. He was much involved in 1969 with the formation of the School of Metallurgy, which incorporated the two Departments of Physical Metallurgy and of Industrial Metallurgy. From 1969 until 1973 he was Deputy Principal of the University. Having retained a Personal Chair in the Metallurgy Department he continued research throughout his life.

Raynor's prolific writings included a number of books; two of these: *The Structure of Metals and Alloys* (with Hume-Rothery) first published in 1944, and *An Introduction to the Electron Theory of Metals and Alloys*, first published in 1947, became classical metallurgical texts.

The critical evaluation of published alloy phase diagrams was an ongoing interest for Raynor and beginning in 1943 he had published a series of annotated diagrams of several of the major binary alloy systems. In 1976, when the Metals Society formed an Alloy Phase Diagram Data Committee, he became its Chairman, serving in this capacity for six years; he also contributed substantially in the critical assessment of important ternary systems. Raynor was the world leader

and most senior figure in the science of alloy phases and was the recipient of many awards.
BMFRS 1984, 30, 547–563. Sir Alan Cottrell. FRS. DNB 31, 332 R.E. Smallman FRS.

RICHARD EDMUND REASON 1903–1987; FRS 1971

Richard Reason was the son of a professional picture engraver, whose practical skills he showed in his own career; he acquired his interest in music from his mother, née Wolton. He was educated at Tonbridge School and entered the Royal College of Science in 1922 where he studied physics and developed a particular interest in optical instrumentation, a theme which was to be his career work. In 1925 he joined the firm of Taylor, Taylor and Hobson (Kapella) at Leicester and in the course of time became Research Manager. This firm had been founded in 1886 by William Taylor (1865–1937; FRS 1934, a distinguished designer of scientific instruments) and his brother Thomas; (the firm eventually became Rank Taylor Hobson Ltd). The qualities that Reason brought to his work included his use and understanding of materials, optics and kinematics; also his love of music brought him into contact with electronics, through his interest in high fidelity reproduction. His background in these areas was utilised in his first instrumental development, namely the electrolimit gauge, in which he designed the mechanism which translated the movement from the probe to the transducer, and also the electrical circuits.

Reason went on to invent and develop many important items of optical equipment. He became interested in metrology, particularly surface measurements, and this led to a series of devices, beginning with the 'Talysurf', ('taly' from the Greek – to measure); this instrument, using a stylus and giving a graphical representation of a surface became available to industry in the early 1940s. Reason and his co-workers went on to develop the 'Talyrond' for roundness measurements. Another highly important development was the 'Talystep' instrument, conceived for the measurement of thin films and representing a breakthrough in sensitivy; originally envisaged for quantifying the deposition of thin film of glass it became highly important in the semiconductor industry e.g. for metallised films. Up to the age of 78, after his formal retirement, he continued his association with Rank Taylor Hobson Ltd. He married Jane Eve and they had two daughters and one son.

Reason took out more than 90 patents, many of them in his sole name. He is regarded as the father of surface metrology in the United Kingdom and to a large extent in the world.
BMFRS 1990, 36, 437–462. David J. Whitehouse.

RENÉ-ANTOINE FERCHAULT DE RÉAUMUR 1683–1757; FRS 1738

René-Antoine de Réaumur came from the Ferchault family, who were prosperous in trade, and bought the ancient Réaumur estate early in the 17th century. He was born at La Rochelle, France, where his father Rene Ferchault was a judge of appeal; his mother Genevieve Bouchel was the daughter of a magistrate. Before he reached the age of two his father died. His early education was probably with the Jesuits, and he subsequently went to Bourges for three years where he studied law. In 1703 he went to live in Paris, where he met a cousin who was a 'student geometer' of the Academy of Sciences and from whom he took lessons. Pierre Varignon a great mathematician, became his friend and teacher, and nominated him to be his 'student geometer' at the Academy. Réaumur communicated work to the Academy on geometrical subjects in 1708 and 1709, but late in 1709 he changed the direction of his scientific career by reading a paper on the growth of animal shells. From then on he carried out exceptionally varied research in natural history (in which he published extensively and became one of the greatest naturalists of his age), technology, biology, physics and metallurgy.

The Academy of Sciences, to which Réaumur was elected Pensionnaire Mecanicien in 1711, was given the task by Louis XIV's Finance Minister, of collecting a description of all the arts, industries and professions, to form a kind of industrial encylopaedia to facilitate the examination and improvement of industrial technology. Réaumur began his task in 1713 with a paper on the art of drawing gold into thread and wire. Two years later he published his investigations of the arts concerned with precious stones. His findings were presented to the Academy in a series of Memoirs in 1720, 1721 and 1722. These were collected and published under the title '*L'Art de convertir le fer forgé en acier, et l'art d'adoucir le fer fondu, ou de faire desouvrgaes de fer fondu aussi finis que de fer forgé*'. The French government took considerable interest in Réaumur's work, and he was granted a pension which subsidised his researches.

In ferrous metallurgy Réaumur's research included investigations of fractures, the cementation process, the heat treatment of steel and the malleabilising of white cast iron. Hardness testing and mechanical property assessment of metals were other themes. He also made an extensive investigation of the porcelain industry and was very well known for the thermometer scale that he invented, and which bears his name.

Réaumur never married, but devoted his time to his scientific and academic career. He was perhaps the most prestigious member of the Academy of Sciences during the first half of the 18th century, and held the office of Director twelve times. Two years before his death he inherited the castle and lordship of La Bermondiere in Maine, and it was there that he died in a riding accident. He attained enormous authority and reputation in the European scientific community, recognised by his election to the Academies of Science of several countries and to the Royal Society, and he was one of the founders of modern scientific ferrous metallurgy.

Bulloch's Roll. DSB 11, 1975, 327–335. J.B. Gough. Gen. Refs 17, 21, 22.

SIR EDWARD JAMES REED 1830–1906; FRS 1876

Edward Reed was born in Sheerness, the son of John Reed. Having served apprenticeships with shipwrights in Sheerness he entered the School of Mathematics and Naval Construction established at Portsmouth. He married Rosetta Barnaby in 1851. During 1852 he served as a draughtsman in Sheerness and then turned to technical journalism. In 1854 he submitted to the Admiralty a design for a fast armour clad frigate, but this was rejected. In 1860 he was appointed Secretary to the newly formed Institution of Naval Architects. Two years later Reed proposed designs for the conversion of wooden men-of-war into armour clads, and in 1863 he was appointed Chief Constructor of the Navy, and introduced a new system of framing. On resigning from the Admiralty in 1870, he joined Sir Joseph Whitworth at his Ordnance Works in Manchester. The following year he began practise as a naval architect in London and designed ships for several foreign navies. Reed entered Parliament in 1874 as Member for Pembroke Boroughs, subsequently representing Cardiff from 1880–95 and 1900–1905. He was awarded the CB in 1868 and was knighted in 1880.

DNB 2nd Suppl. III, 171–173. LGCL.

HENRI VICTOR RÉGNAULT 1810–1878; FMRS 1852

Henri Régnault was born in Aix-la-Chapelle; his father, who was an officer in Napoleon's army died in the Russian campaign in 1812, and his mother also died soon afterwards. His education was supervised by a friend of his father, who obtained employment for him in a draper's shop in Paris. Despite financial hardships he took lessons for the entrance examination of the École Polytechnique, where he was admitted in 1830 and graduated with distinction two years later.

There followed two years of study at the École des Mines in Paris, before extensive travel in Europe to study mining and metallurgical processes. After short periods of research at Giessen and at Lyons he returned to the École Polytechnique as assistant to Gay-Lussac in 1836, and then, in 1840, became his successor to the chair of chemistry. His interests turned to physics, of which subject he became Professor at the Collège de France in 1841; there, for the next 13 years he carried out his most important experimental work. From 1854 he was Director of the famous porcelain works at Sèvres. In 1870 all his instruments and books were destroyed by Prussian soldiers; this loss and the death of his son clouded his last years .

In chemistry his main reputation was in the organic field. In physics, encouraged by Dumas, his activity included a systematic experimental study of the specific heats of a wide range of solids and liquids. He received the Rumford Medal (1848) and the Copley Medal (1869).
DSB 11, 352–354. Robert Fox.

GEORGE RENNIE 1791–1866; FRS 1822

George Rennie was the elder son of John Rennie, and was born in Blackfriars, London. Having attended St Paul's School and the University of Edinburgh, he entered his father's engineering office in 1811. In 1818 he reported data on the crushing strength of metal blocks, including cast iron. Also in 1818 he was appointed, on the recomendation of James Watt and Sir Joseph Banks, as Inspector of Machinery and Clerk of the Irons (dies) at the Royal Mint, a post which he held for nearly 8 years. He was also involved in the manufacture of a great variety of machinery; in conjunction with Messr Boulton and Watt he supplied machinery for a number of foreign mints. When his father died he entered into partnership with his younger brother, John, and was involved in many major engineering projects in England and abroad. In 1828 he married Margaret Anne Jackson, and they had two sons and two daughters. In the Royal Society he contributed papers, including work on the friction of metals and other materials. He was Treasurer (1845–1850) and Vice-President (1830–32; 1842–43; 1845–51).
PRS 1867–68, 16, xxxiii–xxxv. Notes and Records of the Roy. Soc. 1964. DNB.

JOHN RENNIE 1761–1821; FRS 1798

John Rennie was born in Phantassie, Huntingdonshire, the youngest son of James Rennie, a farmer. He attended school in Dunbar and studied at the University of Edinburgh from 1780–1783. He became a millwright who introduced cast iron pinions instead of wooden trundles in the machinery. Early in life he married Martha Mackintosh, and they had two children, George and John (later Sir John). In 1784 he visited James Watt at his Soho works in Birmingham, and worked there for a while before moving to London to work at a flour mill for which Boulton and Watt were building a steam engine. In 1791 he set himself up as an engineer in London, and subsequently was involved in many major construction projects, including canals, docks, harbours and bridges. He was called in to make an inspection of the Mint, and was consequently involved in the building of a new mint.
Notes and Records of the Roy. Soc. 1964. DNB.

SIR JOHN RENNIE 1794–1874; FRS 1823

John Rennie, the second son of John Rennie, was born in Blackfriars, London. He entered his father's business, and assisted him in his engineering work. When his father died in 1821 he entered into partnership with his brother George, and caried out many civil engineering projects; he erected the machinery for several foreign mints. His work included railway engineering, and

he was Engineer to the Admiralty, associated with harbours in England and Ireland. He was knighted in 1831.
PRS 1874–75, 23, x–xi. DNB.

OSBORNE REYNOLDS 1842–1912; FRS 1877
Osborne Reynolds was born in Belfast, the son of the Rev. Osborne Reynolds and Jane née Hickman. In 1861 he worked in the workshops of a mechanical engineering firm at Stony Stratford. He then entered Queen's College, Cambridge,where he graduated in mathematics as Seventh Wrangler in 1867. He was awarded a Fellowship and also worked for a short period with a civil engineer in London. In 1868 he was appointed to the newly instituted Professorship of Engineering at Owens College, Manchester a post which he held for 37 years. In the same year he married Charlotte Chadwick; following her death he married Anne Wilkinson in 1881 and they had three sons and one daughter. His research in the field of fluid mechanics included investigations on the condensation of steam on metal surfaces, of heat transfer between metal surfaces and liquids, of turbulent and laminar flow, defining in 1883 the dimensionless number (Reynolds number) determining the type of flow, and the development of pumps and turbines. He received a Royal Medal in 1888 and was Bakerian Lecturer in 1887.
PRS 1913 A, 88, xv–xxi. HL. DSB 11, 392–394. R.H. Kargon. DNB 1912–1921, 456–457. HC.

THEODORE WILLIAM RICHARDS 1868–1928; FMRS 1919
Theodore Williams was born in Germantown, Pennsylvania; the son of William Trost Richards and Anna Matlack Richards. Following education at home he entered Haverford College in 1882 and graduated three years later. He then proceeded to Harvard University and obtained a chemistry degree, *summa cum laude*, in 1886; he carried out research and obtained a Harvard PhD in 1888. With a travelling fellowship Richardson then visited several European countries, before returning to Harvard, where he was appointed an assistant in 1889. There followed a series of posts at Harvard: Instructor (1891), Assistant Professor (1894), Full Professor (1901), Chairman of the Chemistry Department (1903–1911), Erving Professor of Chemistry and Director of the Wolcott Gibbs Memorial Laboratory (1912–1928). There followed visits in Europe again in 1895, before appointment as Professor of Chemistry at Harvard, a post which he held for the rest of his career. Richards married Miriam Stuart Thayer in 1896 and they had one daughter and two sons. Richards's research was particularly concerned with precise measurements of atomic masses of many elements, and he also investigated physical properties such as atomic volumes. He was awarded the Davy Medal in 1910 and the Nobel Prize for Chemistry in 1914.
PRS 1928, A21, xxix–xxxiv. HBD. DSB 11,416–418. Sheldon J. Koppel.

FREDERICK DENYS RICHARDSON 1913–1983; FRS 1968
Denys Richardson, as he was generally called by his friends, was born in Streatham, London, the third in a family of three sons and one daughter of Charles Willerton Richardson and Kate Harriet née Bunker. His father died when Denys was three, and when his mother died in 1919, his aunt became the family guardian and and he went to a local kindergarten; at the age of eight he went as a boarder to University School.

Richardson entered University College London to take the honours chemistry course; the lectures given by Charles Goodeve showed him value of applying classical thermodynamics to the study of chemical reactions and stimulated his interest in the study of the kinetics and equilibrium conditions of reacting systems. He became much involved in the Artillery Unit of the Officers' Training Corps.

After graduation Richardson carried out research with Goodeve for a PhD degree on a project concerned with the preparation of Cl_2O_6 from ClO_2 and O_3. Working with these unstable and explosive compounds required apparatus to be built entirely of glass, and he became adept in glass blowing. He obtained his PhD in 1936, and after this worked with Goodeve on a project supported by the Admiralty. He then applied for one of the Commonwealth Fund Fellowships (now Harkness Fellowships). To combine his interests in chemistry and natural history he proposed a project in the field of agricultural chemistry, and was awarded a fellowship to spend two years at Princeton University.

Richardson returned from America in 1939 and early in 1940 he joined Goodeve, as a Sub-Lieutenant in the Royal Naval Volunteer Reserve. They worked at the Admiralty Department, H.M.S. *Vernon*, at Portsmouth in the Department of Miscellaneous Weapons Development; Richardson remained in this Department until the end of the war becoming Deputy Director (1943–1946), having received the rank of Commander in 1942. The work covered a wide range of projects anti-aircraft weapons, rockets, and anti-submarine techniques, and his invention of the star shell. In 1942 Richardson married Irene Austin, a graduate in engineering and town planning, who joined the Admiralty Scientific Research Department; they had two sons.

At the end of the war, at the invitation of Goodeve, Richardson joined the British Iron and Steel Research Association (BISRA) to build up the chemistry department and laboratories, established at Battersea. In collaborative work he made an important fundamental contribution to chemical metallurgy through the preparation of diagrams showing free energy changes for the formation of various compounds – oxides, sulphides etc. – as a function of temperature.

In 1950 Richardson moved to the Metallurgy Department at Imperial College with a Nuffield Foundation Fellowship, and built up what became the Nuffield Research Group in Extraction Metallurgy. In 1955 he was awarded the DSc degree of London University. His work at Imperial College brought him many academic honours, including election as FREng in 1976. The research, involving many students and other research colleagues, was wide ranging, covering both ferrous and non-ferrous themes: for example, the vacuum degassing of steel; thermodynamics and structures of slag solutions and kinetics of reactions involving levitation melted drops. He recognised the importance of, and obtained support for, engineering aspects of metal processing. He had a major influence on the development of a new curriculum for the undergraduate degree course in metallurgy. As an author, his important book *The Physical Chemistry of Melts in Metallurgy*, was published in 1974. Following his retirement in 1976 he became a Senior Research Fellow.

Denys Richardson had a wide range of interests beyond his scientific work. He was an accomplished painter in water colours, a connoiseur of fine art and a gifted conversationalist. His care for the welfare of those who worked with him was unending, and through a world-wide network of ex-students and colleagues his influence as a leader in chemical metallurgy remains.

BMFRS 1985, 31, 495–521. J.H.E. Jeffes. DNB 1981–1985, 339, 340. J.H.J. Jeffes.

SIR OWEN WILLANS RICHARDSON 1879–1959; FRS 1913

Owen Richardson was the eldest of three children of Josiah Henry Richardson and Charlotte Maria née Willans. He was born in Dewsbury, Yorkshire where he attended school before entering Batley Grammar School. At Cambridge, where he entered Trinity College with a scholarship, he took the Natural Sciences Tripos and obtained a BA with first class honours in 1900, having studied chemistry and physics in Part II. He became a fellow of Trinity College in 1902 and Clerk Maxwell Scholar in 1904, also obtaining a DSc of London University in 1904. In the Cavendish Laboratory he carried out important fundamental work under J.J. Thomson including the subject of current from hot platinum filaments in vacuum. Richardson married Lilian Maude Wilson in 1906 and they had two sons and one daughter; following his wife's death he married Henrietta Maria Rupp in 1948.

In 1906 Richardson was appointed to the Chair of Physics at Princeton University, where he carried out research on various phenomena related to thermionic emission. He returned to England in 1914 to occupy the Wheatstone Chair of Physics at King's College, London; in the same year he published the book *The electron theory of matter*. His research also included spectroscopy and metallic conduction. In 1924 he was appointed Yarrow Research Professor and retired in 1944.

Richardson's research made important contributions to fields of physics applied to developments in radio, telephony, television and X-ray technology. In the Royal Society he served as a Foreign Secretary and was awarded the Hughes Medal (1920) and a Royal Medal (1930). He was awarded the Nobel Prize for Physics in 1928 for his work on thermionic phenomena and was knighted in 1939.

BMFRS 1959, 5, 207–215. Wm Wilson. DSB 11, 419–423. Loyd S. Swenson Jr. DNB 1951–60, 839–871. E.W. Foster.

SIR ERIC KEIGHTLEY RIDEAL 1890–1974; FRS 1930

Eric Rideal was born at Sydenham, Kent, the eldest of four children of Samuel Rideal, a leading analyst and consulting chemist, and Elizabeth née Keightley. When the family moved to Elstead, Surrey, Eric attended Farnham Grammar School, and later went to Oundle School. In 1907 he entered Trinity Hall, Cambridge with an Open Scholarship in Natural Sciences, graduating BA with first class honours in 1910. He carried out research at Aachen and then at Bonn, where he was awarded a PhD in 1912, for a thesis on electrochemical aspects of uranium. He then returned to England where he worked with Ulick Evans for two years on electrochemical consulting work in his father's rooms in Westminster. After a visit to Ecuador on a consultancy project he came back to England and joined the Artists' Rifles; later he was made a Captain in the Royal Engineers and went to the Somme on technical duties. Having been invalided out of the army in 1916 he worked for the Munitions Invention Department in a team at University College, London.

From 1919–20 Rideal was a visiting professor at Urbana, Illinois, and was then appointed to a Lectureship and a Fellowship at Trinity Hall, Cambridge. He married Margaret ('Peggy') Jackson and they had one daughter. Rideal remained at Cambridge from 1921–44, being appointed to the Chair of Colloidal Physics in 1930, transferring in the following year to the new Chair of Colloid Science. In 1946 Rideal as appointed Fullerian Professor of Chemistry and Director of the Davy-Faraday laboratory at the Royal Institution. From 1950–55 he occupied the Chair of Chemistry at King's College, London.

Rideal's research career as a physical chemist involved the fields of electrochemistry, colloid and surface chemistry, chemisorption and analysis, and spectroscopy. In 1951 he received the Davy Medal and was Bakerian Lecturer; he was also knighted in the same year.

BMFRS 1976, 22, 381–413. D.D. Eley FRS. DSB 18, 739–743. W.H. Brock. DNB 1971–80, 723. D.D. Eley.

BRIAN KIDD RIDLEY 1931–; FRS 1994

Univ. of Durham: BSc. PhD. Mullard Research Lab., Redhill: Research Physicist, 1956–64. Univ. of Essex: Lecturer in Physics, 1954–67; Senior Lecturer, 1967–71; Reader, 1971–84; Professor, 1984–91; Research Professor of Physics, 1991–.

Quantum processes in semiconductors. Electrons and phonons in semiconductor multi layers.

ALFRED EDWARD RINGWOOD 1930–1993; FRS 1977

Alfred ('Ted') Ringwood was born in Kew, a suburb of Melbourne; his father, Alfred Edward Ringwood, suffered from his army experiences in France in the 1914–18 war; consequently the family was in straitened financial circumstances during the Depression and Ted's mother, Wilhelmena Grace, née Robertson,worked to provide financial support. Ted, an only child, attended Hawthorn West State School and then in 1943 obtained a scholarship to Geelong Grammar School as a boarder. He was awarded scholarships, with which he entered Melbourne University to study geology, and while at university his interests broadened to include metallurgy and materials science. Having graduated in 1950 with first class honours in geology he was awarded a scholarship to study for an MSc degree, for which his field project work involved field-mapping and petrology; this brought the opportunity to explore an exhausted silver mine, from which he arranged for discarded galena-rich tailings to be used for feedstock for the Melbourne shot tower. Following the award of an MSc, with honours, in 1953 , Ringwood began research for a PhD. Initially he made experimental studies on the origin of metalliferous ore deposits, but his interests moved to the application of geochemistry to obtaining understanding of the structure of the Earth.

By the time he was awarded his PhD he had already published important papers and had pioneered methods for studying the interior of the Earth, utilising crystal chemistry concepts and experimental investigations. In 1957 and 1958 for 15 months he worked at Harvard University as a Melbourne Research Fellow obtaining experience of high-pressure experimental techniques Ringwood moved to the new Department of Geophysics in the Research School of Physical Sciences at the Australian National University, Canberra. He was appointed Senior Research Fellow in 1959, Senior Fellow in 1960 and Personal Professor in 1963. His work included investigations of meteorites and the construction of a Bridgman-anvil high-pressure apparatus. The meteorite work involved visits to Sweden where he met and married, in 1960, Gun Iver Carlson; they had a son and a daughter.

At Canberra, Ringwood's research included important investigations of polymorphic phase transitions of various minerals under high pressures; later, in the 1960s, collaborative work was done using apparatus which reproduced pressure-temperature conditions to depths of around 150 km within the Earth. In 1977, in work on the Earth's outer core of molten iron he carried out investigations to demonstrate that high pressures and temperatures increase the solubility of oxygen in molten iron. Nuclear waste disposal was another area in which he worked. In the 1980s, in association with a major mining company the availability of small industrial-grade diamonds led him to invent and patent a new diamond-based cutting tool material; this was suitable for drilling hard rock and for machining ultra-hard ceramics. From 1978–83 he served as Director of the Research School of Earth Sciences, which had been formed in 1972. A leading figure in Australian science, Ringwood received many awards in recognition of his work, including being the Bakerian Lecturer in 1983.

BMFRS 1998, 44, 351–362. D.H. Green FRS.

SIR DEREK HARRY ROBERTS 1932–; FRS 1980

Manchester Central High School. Manchester Univ.: BSc. Plessey Company's Coswell Research Laboratory: 1953, Plessey Semiconductors: General Manager, 1967; Allen Clark Research Centre, 1969. Plessey Microelectronics Division: Managing Director, 1973; Technical Director, 1983–85. GEC: joint Deputy Managing Director (Technical), 1985–88. University College London: Provost, 1989–99. Clifford Paterson Lecturer, 1980. FREng 1980. CBE 1983. Knighted 1995.

Advanced electronic components particularly silicon integrated circuits, opto-electronic devices and memory systems.

SIR GARETH GWYNN ROBERTS 1940–; FRS 1984

University College of North Wales, BSc, PhD, DSc, MA(Oxon) 1987. Univ. of Wales: Lecturer in Physics, 1963–66. Xerox Corp. USA. Research Physicist, 1966–68. New Univ. of Ulster: Senior Lecturer, Reader, Professor of Physics, 1968–76. Univ. of Durham: Professor of Applied Physics and Electronics and Head of Optical Physics Division, 1976–85. Thorn EMI plc: Chief Scientist, 1985; Director of Research, 1986–90. Univ. of Sheffield: Vice-Chancellor, 1991. Clifford Paterson Lecturer, 1987. Knighted 1997.

Physics of semiconductors; insulating films on semiconductors.

SIR GILBERT ROBERTS 1899–1978; FRS 1965

Gilbert Roberts was born in Hampstead, the son of William Roberts; he had an elder sister and a half brother. He attended school in Bromley and studied engineering at Gresham College. In the 1914–18 war he joined the Royal Flying Corps, under age, and was commissioned as a Second Lieutenant. He served in France and was wounded in 1918. With an army scholarship he entered the City and Guilds College of the Imperial College of Science and Technology in 1919 and obtained a BSc first class degree in 1922. Roberts then joined Sir Douglas Fox and Partners (later named Freeman, Fox and Partners) to work on the design of the Sydney Bridge. In 1925 he married Elizabeth Nada Hora; they had two daughters. He became Assistant Engineer on a railway project in Lancashire in 1925. Then he joined the Bridge Department of Dorman Long and Company in charge of various projects. In 1936 he was appointed Engineer-in-Charge of Development and Construction with Sir William Arroll and Co. in Glasgow. In 1943 he became a Director and Chief Engineer of the Company. He became a partner in Freeman Fox in 1948. In his distinguished career as a civil engineer Roberts was associated with many major bridge projects, including the Severn Bridge and the Forth Bridge; during the 1939–45 war his work included the construction of welded landing craft. He was associated with the design of the Dome of Discovery for the Festival of Britain (1951). Also he was involved in the design and fabrication of heavy industrial plant and structures. Roberts was a pioneer in the use of high tensile steel and of welded constructions. He was elected FREng in 1976 and received a Royal Medal in 1968. His was knighted in 1965.

BMFRS 1979, 25, 477–503. O.A. Kerensky FRS, FREng.

LEWIS EDWARD JOHN ROBERTS 1922–; FRS 1982

Swansea Grammar School. Univ. of Oxford: Jesus College; MA, DPhil; Clarendon Laboratory, 1944. Chalk River Research Establishment, Ontario, Canada: Scientific Officer; 1946–47. AERE, Harwell: 1947 Principal Scientific Officer, 1952. Univ. of California, Berkeley: Commonwealth Research Fellow, 1954–55. AERE, Harwell: Deputy Head Chemistry Division, 1966; Assistant Director 1967; Director 1975–86. UKAEA: Member, 1979–86. Univ. of East Anglia: Wolfson Professor of Environmental Risk Assessment, 1986–90. CBE 1978.

Nuclear energy.

SIR WILLIAM CHANDLER ROBERTS-AUSTEN 1843–1902; FRS 1875

William Chandler Roberts was born in 1843, his father, George Roberts being of Welsh descent, while his mother, Maria Louisa Chandler, belonged to the Kentish families of Austen and Chandler. In 1885 he became the heir of his uncle Nathaniel Austen of Haffenden and Camborne in

Kent, and obtained royal licence to add the name of Austen. At the age of eighteen, after private education, he entered the Royal School of Mines, South Kensington, London, with the intention of becoming a mining engineer; however, when he obtained the Associateship of the School in 1865 he was employed by Thomas Graham, then Master of the Mint, as a private assistant.

Following Graham's death in 1869 and reorganisation at the Mint Roberts-Austen was appointed in 1870 to a new post, that of 'Chemist of the Mint'. His tour of 13 European mints in the same year was the beginning of his international influence and interests. In 1882 he became 'Chemist and Assayer'; he also held the post of acting Deputy-Master for several months preceding his death. In 1883 he visited the principal mints and assay offices in the USA. During his career at the Mint Roberts-Austen was responsible for the composition of coinage to the value of £150 million of gold and £31 million of silver.

Roberts-Austen always valued the recollection of his early association with the Royal School of Mines, and in 1880 he was appointed to the Chair of Metallurgy in succession to John Percy, while still retaining his appointment at the Mint, where most of his research was done. He was an excellent teacher and enlarged the scope of research at the School. During the 22 years that he held his chair he trained a succession of men who later held important positions in Britain and in many parts of the world.

Roberts-Austen's scientific papers practically all relate to metallurgical problems, or were connected with his work at the Mint. They dealt with topics such as the spectroscopic characteristics of alloys, the physical and chemical nature of alloys, the structure of metals and the connection between the properties of metals and the periodic law, and the nature of hydrogen occluded by palladium.

An important part of his research began in 1890, at the request of the Institution of Mechanical Engineers; this was a fundamental investigation of the effects of small additions of certain elements on the mechanical and physical properties of the common metals and their alloys. He was innovative in pyrometry and developed the use of photomicrography. The five reports he submitted to the Institution, which included studies of the iron-carbon phase diagram, represented a major contribution to physical metallurgy.

Much of Roberts-Austen's work was influenced by his strong artistic sense and by his high regard for beauty of form or colour; this was evidenced in his work at the Mint, especially in medal striking and he lectured widely on the colours of metals for art work. He had an interest in oriental metallurgy and historical metallurgy, investigating with Arthur Gilbert, the sculptor, the revival of 15th century methods of casting.

On scientific and technical operations on coinage Roberts-Austen became the leading world authority. An original, skilful and resourceful experimenter, he excelled in the then new field of physical metallurgy, where his wide-ranging research, much of it concerned with alloys, contributed greatly to the scientific development of the subject. His teaching was characterised by its thoroughness and in the enthusiasm to illuminate the scientific problems of physical metallurgy. His very influential book *Introduction to the Study of Metallurgy* (1891), which ran to six editions, aimed to treat the subject as a whole. The book proved to be a masterly guide to a knowledge of the principles of metallurgy.

Roberts-Austen had a high reputation as a lecturer and demonstrator, and was socially popular. He married Florence Maud Allridge in 1876; they had no children. At Chilworth, near Guildford, Surrey, where he had a residence he was a good neighbour, and erected a chapel for the benefit of the district. In the Royal Society he served on the council from 1890–1892 and was Bakerian Lecturer for 1896. He was called on to serve on many committees, councils and commissions,

including the Treasury Committee in 1897 which led to the establishment of the National Physical Laboratory at Teddington, Middlesex. He was the recipient of a number of awards, including honorary degrees. In 1888 he was made a CB and he received a knighthood, KCB, in 1899.
PRS 1905, 75, 192–198. DSB 11, 485, 486. F. Greenaway. DNB 1901–1911 212–3. TEJ. Refs Ch. 10.5, 9.

ANDREW ROBERTSON 1883–1977; FRS 1940

Andrew Robertson was born in Fleetwood, Lancashire, the youngest of six children of James and Elizabeth Robertson. On leaving school he was apprenticed at the age of fourteen to his father's marine engineering business. Then with a grant he entered the Engineering Department of Owens College, Manchester in 1902 and graduated with first class honours in 1905. He became a Demonstrator and did research in the Department, obtaining an MSc in 1909. Four years later he was appointed Vulcan Research Fellow. While he was at Manchester he married Mabel Bailey; they had no children. With the outbreak of war in 1914 Robertson went via the Royal Naval Volunteer Reserve to the Royal Aircraft Factory (later the Royal Aircraft Establishment) at Farnborough. where he became Chief Assistant in the Materials Testing Department; by the end of the war he had been promoted to the rank of Major. Robertson had been awarded a Manchester DSc in 1915 and was appointed as Professor of Mechanical and Mining Engineering at Bristol in 1919. Robertson's research included important investigations of the strength of thick cylinders under internal pressure, and of the transition from the elastic to the plastic state in mild steel, including the phenomenon of the yield drop. He carried out work on struts in relation to major bridge projects. Robertson was awarded the DSc of Manchester University in 1915.
BMFRS 1978, 24, 515–528. Sir Alfred Pugsley FRS. DNB 1971–1980, 727, 728. Baker.

BENJAMIN ROBINS 1707–1751; FRS 1727

Benjamin Robins was the only son of John Robins, a Quaker, and was born in Bath. He showed ability in mathematics and when he left school he came to London where he studied languages and higher mathematics in order to prepare for teaching. He taught particularly pure and applied mathematics and physical science, but gradually gave this up and became an engineer. Robins was involved in the construction of works such as mills, bridges and harbours. He studied the principles of gunnery and of fortifications, and in 1742 he published a book on gunnery. He invented the ballistic pendulum for measuring the velocity of projectiles. He became involved in projects overseas, for example, he went to India as Engineer General to repair the forts of the East India Company. He was unmarried.
DSB 11, 493–494. J. Morton Briggs Jr. DNB.

PETER NEVILLE ROBSON 1930–; FRS 1987

Univ. of Cambridge: BA. Univ. of Sheffield: PhD. Metropolitan Vickers Electrical Co., Manchester: Research Engineer, 1954–57. Univ. of Sheffield: Lecturer, 1957–63; Senior Lecturer, 1963–66. Stanford Univ.: Research Fellow, 1966–67. University College London: Reader 1967–68. Univ. of Sheffield: Professor of Electronic and Electrical Engineering, 1968–96. FREng 1983. OBE 1983.
Semiconductor devices and magnetic theory.

JOHN ROEBUCK 1718–1794; FRS 1764

John Roebuck, born in Sheffield, was the third son of John Roebuck and Sarah née Roe. Following his education at Sheffield Grammar School and at a school in Northampton, he did not enter his father's prosperous business as a cutler, manufacturer and merchant. Instead he studied medicine at Edinburgh, and then went to Leiden to complete his medical studies, obtaining his MD in 1742, and presenting a medical dissertation the following year. Roebuck set up as a physician in Birmingham, but also developed his interest in chemistry and its industrial applications. He established a precious metal refinery, acted as a consultant for local industry and also introduced the use of lead vessels instead of glass vessels in thc manufacture of sulphuric acid. Roebuck also became involved in the production of ceramics. Around 1746 he married Ann Ward, and there were children from the marriage.

With partners he founded the Carron Ironworks near Falkirk, Stirlingshire, in 1759. Roebuck's industrial activities were extended by leasing coal mines and salt works in West Lothian, where he brought in John Watt to improve the pumping engine facilities. Unfortunately, he eventually encountered financial difficulties with his various business activities, but his creditors retained him in some management responsibilities, and made him an annual allowance. To his various occupations he added farming on a substantial scale and he was a successful agriculturalist.

Roebuck's honours included receiving the Freedom of the City of Edinburgh. In his varied career as an industrialist his establishment of the Carron works and the improvements he introduced into iron manufacture gave an impetus to the then much néeded Scottish industrial enterprise.

DSB 11, 499, 500. A. Clow. DNB.

SIR HENRY ENFIELD ROSCOE 1833–1915; FRS 1863

Henry Roscoe was born in London where his father, Henry Roscoe was a barrister, who became a Judge in Liverpool; his mother, Maria née Fletcher was the daughter of a Liverpool merchant. His father died when he was four years old, and he was educated at a preparatory school and at the High School of the Liverpool Institute. He entered University College London in 1848, where one of his teachers was Thomas Graham. Having obtained a BA degree with honours in chemistry in 1852 he carried out research with Bunsen in Heidelberg obtaining a PhD in 1854; he then continued with Bunsen, working on the chemical action of light.

In 1856 he returned to London, where he set up a private laboratory, lectured at an army school in Eltham and did some analytical work. In the following year he was elected as Professor of Chemistry at the recently founded Owens College in Manchester. He married Lucy Potter in 1863 and they had one son and two daughters. Roscoe was Vice-Chancellor of London University from 1896–1902. His most important contribution to chemistry was the preparation of pure vanadium and his study of its oxides and chlorides. Roscoe was knighted in 1884, was elected MP for South Manchester in the following year, serving until 1895, and was made a Privy Councillor in 1909. In the Royal Society he was a Council Member and Vice-President, and was awarded a Royal Medal in 1873.

PRS 1916–17 A, 93, i–xxi. TET. DSB 11, 536–539. R.H. Kargon. DNB 1912–1921, 478, 479. HBD.

WALTER ROSENHAIN 1875–1934; FRS 1913

Walter Rosenhain was born in Berlin, the only son and youngest child of Moritz and Friedericke Rosenhain. The family emigrated to Australia when he was five and lived in Geelong. His early education was at Wesley College in Melbourne, from whence he entered Queen's College in the University of Melbourne in 1891, graduating in physics and engineering in 1897. He then came to England as the holder of an 1851 Exhibition Scholarship, and entered St John's College, Cambridge, with the intention of studying civil engineering.

The direction of his life's work was determined when he consulted Professor (later Sir Alfred)

Ewing as to a suitable subject for research and proceeded to work with him from 1898–1899; their classic investigation established the process of slip as fundamental in the mechanism of plastic deformation.

On leaving Cambridge Rosenhain entered the glass works of Messrs Chance Bros of Smethwick, near Birmingham in 1900. The following year he married Louise Monash, the sister of Sir John Monash and they had two daughters. He published papers on optical glass and a textbook on glass manufacture. During the 1914–18 war he was called on to resume work on optical glass and this led him to the study of refractory materials of high purity. While at Smethwick he continued his work on the structure of metals in a small private laboratory. He trained his wife in metallurgical specimen preparation and she shared his work. He published work on the deformation of iron and steel and on the mechanism of fracture. In 1908 he was awarded the DSc degree of Manchester University.

Rosenhain established a reputation as a brilliant investigator in metallurgy, and in 1906, he succeeded Harold Carpenter as Superintendent of the Department of Metallurgy and Metallurgical Chemistry at the National Physical Laboratory. He held this post for 25 years during which period a stream of important researches issued from it under his direction. The staff, four in number initially increased to about 70, and the space and facilities were also expanded greatly. Rosenhain aimed from the beginning to undertake fundamental research, with the highest available degree of accuracy, and at the same time to serve the growing needs of industry.

The work of the Alloys Research Committee of the Institution of Mechanical Engineers continued at the NPL during Rosenhain's period of office; important alloy research was done during the 1914–18 war, including aluminium-base systems. Beginning in 1921 investigations of the constitution of the binary alloys of iron were made under the auspices of the Alloys of Iron Research Committee. Rosenhain continued his activity on fundamental aspects of physical metallurgy; for example, the modes of deformation of metals as a function of temperature, and the mechanism of solid solution hardening. Rosenhain also contributed greatly to metallurgy by his improvements in metallographic techniques; for example as a skilled microscopist he was very interested in microscope design. His text book *Physical Metallurgy*, in its three editions, became the standard work in the field.

As a leader of a team of investigators, Rosenhain was able to retain their confidence and their affection. His power of exposition was great, and he was a formidable controversialist. He played a leading part in the organisation of metallurgists, his fluency in French and German making him a link with foreign metallurgical bodies. In 1931 he retired from the National Physical Laboratory, and until shortly before his death practised as a consulting metallurgist (e.g. for Messrs J. Stone) with an office in London. His research contributions during his career, covering both ferrous and non-ferrous metallurgy, established him with a world-wide reputation.

ONFRS 1932–35, 1, 353–359. C.H. Desch FRS. DSB 11, 548–550. R.W. Cahn. DNB 1931–40, 749, 750 C.H. Desch. Ref. Ch. 10.6.

SIR JOSEPH ROTBLAT 1908–; FRS 1995

Univ. of Warsaw, Poland. MA, DSc, Warsaw. Radiological Laboratory of Scientific Society of Warsaw: Research Fellow; 1933–39. Atomic Physics Institute of Free Univ. of Poland: 1937–39. Univ. of Liverpool: Oliver Lodge Fellow, 1939–40; PhD; Lecturer, Senior Lecturer, Department of Physics, 1940–49; Director of Research in Nuclear Physics, 1945–49; work on atomic energy at Liverpool Univ. and at Los Alamos. St Bartholomew's Medical College, London Univ: Professor of Physics; 1950–76. DSc. Nobel Peace Prize 1995. CBE 1965. Knighted 1998, (KCMG).
Nuclear physics and radiation biology.

LEONARD ROTHERHAM 1913–; FRS 1963

Strutt School, Belper. University College London. Firth Brown Research Laboratories: Physicist, 1935–46. Royal Aircraft Establishment, Farnborough, Hants: Head of Materials Department, 1946–50. UKAEA Industrial Group, Risley: Director, Research and Development, 1950–58. CEGB: Member for Research, 1958–69. Electricity Suppply Industry and Electricity Council: Head of Research, 1965–69. Bath Univ.: Vice-Chancellor, 1969–76. DSc. FREng, 1976. CBE 1970.

Creep of metals. Manufacture of new metals. Development of nuclear power.

JOHN MARTIN ROWELL 1935–; FRS 1989

Univ. of Oxford: Wadham College, BSc, MA; DPhil 1961. Bell Tel. Labs: 1962–84. Bell Communications Research: 1984–89. Conductors Inc, 1989–95. North Western Univ.: Materials Research Institute; Professor, 1997–.

Low temperature physics; electron transport in metals; superconductivity.

PHILIP CHARLES RUFFLES 1939–; FRS 1998

Sevenoaks School, Kent. Univ. of Bristol: BSc. Rolls Royce: Graduate Apprentice, 1961–63; Engineer Prelininary Design, 1963–68; Project Development Engineer RB211, 1968–74; Project Engineer, 1974–75; Manager JJ IoD team, E. Halifax, 1976: Chief Engineer RB211, 1977–80; Head of Engineering Helicopters, 1981–83; Director of Technology, 1984–97; Director of Design Engineering, 1987–89; Technical Director, 1989–90; Director of Engineering, 1991–96; Director of Engineering and Technology Aerospace Group, Derby, 1997–. FREng 1988.

Gas turbine engineering: applications of fluid dynamics and materials science.

PRINCE RUPERT OF THE RHINE 1619–1682; FRS 1665

Prince Rupert was the grandson of James I; his mother the Winter Queen of Bohemia, was Charles I's sister, Elizabeth the wife of King Frederick of Bohemia. At the age of 22 Prince Rupert was given command of the Royalist Cavalry in the English Civil War. He acquired a major reputation in the miltary field. In his early years he had an artistic interest, which later extended into science. He did not marry, but left two natural children.

After nearly 40 years involved in warfare he spent his retirement in scientific work, which was assisted by his appointment in 1668 as Constable of Windsor Castle, where he set up a laboratory, workshop, forge and studio. His facility at drawing enabled him to appreciate the art of design, and this led him into the development of mechanical inventions. He also made a close study of weapons and gunnery, including ballistics. Although he was apparently not a regular attender at Royal Society meetings he continued to send in a great variety of inventions, many related to military matters. Among these were a tiny automatic pistol, a more powerful type of gunpowder, a method of making iron ductile and a method for producing lead shot. He also introduced the art of mezzotinting into England.

DNB. P. Mirrah, *Prince Rupert of the Rhine*, Constable, London 1976. Ballistics in the 17th c A.R. Hall p.11. Camb. Univ. Press, 1952. K. Dewhurst, 'Prince Rupert as a scientist', *British J. Hist. Soc.*, 1962–63, **1**, 365–373. Gen. Ref. 20. v. III.

JOHN SCOTT RUSSELL 1808–1882; FRS 1849

John Russell, the eldest son of David Russell a Scottish clergyman and Agnes née Clark, was born at Parkhead near Glasgow. Originally intended for the church , instead he entered an engineering workshop. He began studies at St Andrews when he was 13 and in the following year he

matriculated to Glasgow University where he graduated MA in 1825. He then worked as a mathematics teacher in Edinburgh and also gave lectures to medical students. In 1832 he was elected temporarily to the Chair of Natural Philosophy at Glasgow, where he carried out research on the nature of waves. In 1836 he married Harriette Osborne and they had three daughters and one son. For a period he was manager of a large ship building yard at Greenock. Having moved to London in 1844 he was for many years a shipbuilder on the Thames. He constructed the iron vessel the *Great Eastern,* a project in which Brunel was involved. He was a strong advocate of iron clad warships and was joint designer of the *Warrior,* the first sea-going armoured frigate.
PRS 1882–83, 34, xv–xvii. PS. WHB. DNB.

SIR ERNEST RUTHERFORD
(BARON RUTHERFORD OF NELSON) 1871–1937; FRS 1903

Ernest Rutherford was born near Nelson, New Zealand, the fourth child in the family of twelve children of James Rutherford, wheelwright and farmer, and his wife Martha née Thompson. He entered Nelson College in 1887 and then went to Canterbury College two years later, where he obtained a BA in 1804. He began a study of mathematics and physics obtaining an MA in 1893 and a BSc in 1894. In 1895 he entered the Cavendish Laboratory in Cambridge with a scholarship as J.J. Thomson's first research student and began work in the field of atomic physics. He was appointed Professor of Physics at McGill University, Montreal in 1898 and in 1907 moved to Manchester as Professor of Physics. During World War I he carried out work for the Admiralty on locating submarines. In 1919 he became Professor of Physics and Director of the Cavendish Laboratory. In addition he was Professor of Natural Philosophy at The Royal Institution from 1921. His research was outstanding in contributing to the founding of nuclear physics. Among his achievements were the first explanation that radioactivity is produced by the disintegration of atoms, the discovery that alpha particles consist of helium nuclei, and showing that atoms consist of central nuclei surrunded by electrons. Rutherford was the recipient of the Nobel Prize for Chemistry in 1908 and of the Copley Medal in 1922; he was Bakerian Lecturer in 1904 and 1920. He was President of the Royal Society from 1925–30, was knighted in 1914, raised to the peerage as Baron Rutherford of Nelson in 1921, and received the Order of Merit in 1925.
ONFRS 1936–38, 1, 395–423. A.S. Eve and J. Chadwick. DSB 12, 25–36. L. Badash. DNB 1931–40, 765–774. H.T. Tizard.

JOHANNES ROBERT RYDBERG 1854–1919; FMRS 1919

Johannes Rydberg was born in Halmstad, Sweden, the son of Sven and Maria Andersons Rydberg. Following education at the Halmstad Gymnasium he entered the University of Lund in 1873, where he graduated in mathematics in 1875, and obtained a doctorate in 1879. His career continued at the University where he was successively Lecturer in Mathematics (1889), Lecturer in Physics (1880) and Professor of Physics (1897). In 1886 he married Lydia E.M. Carlsson and they had two daughters and one son. In his research his interest in the relation of the properties of the elements with their position in the Periodic Table led him into the field of spectroscopy and the formula that bears his name (1890).
DSB 12, 42–45 C.L. Maler.

JOHN WALTER RYDE 1898–1961; FRS 1948

John Ryde was born in Brighton, the son of Walter William Ryde and Hannah née Buckland. He was educated at St Paul's School, London and at the City and Guilds Technical College at Finsbury.

In 1916, having enlisted in the Royal Engineers he served in France and was commissioned in 1918. He then joined the scientific staff of the General Electric Company Ltd Laboratories at Wembley where he spent his scientific career, becoming Chief Scientist in 1953. He married Dorothy Ritchie in 1930 and they had one one son. Ryde's research included electrical discharge in gases, thermionic emission and spectrographic analysis; he and his team invented the high pressure mercury vapour lamp in the early 1930s.
BMFRS 1962, 8, 105–117. R. Whiddington. DNB 1961–70, 910, 911. B.S. Cooper.

SIR BERNHARD SAMUELSON 1820–1905; FRS 1881
Bernhard Samuelson was the eldest of six sons of Samuel Henry Samuelson, a merchant and Sarah née Hertz; he was born in Hamburg where his mother was on a visit. His father settled in Hull and Bernhard attended a private school at Skirlaugh, Yorkshire. He entered his father's office at the age of 14, but was soon apprenticed to Rudolph Zwilchenhart and Company, a Swiss firm of merchants at Liverpool where he spent six years. In 1837 he was sent by the firm to Warrington to purchase locomotive engines for export and this led him to seek knowledge of engineering. He became manager of Messrs Sharp, Stewart and Co., engineers of Manchester in 1842. He married Caroline Blundell in 1844 and they had four sons and four daughters; following his wife's death he married Leila Mathilda Serena in 1889.

Samuelson bought a small factory of agricultural implements at Banbury in 1848 and greatly expanded the works. In 1853 he became aware of the future of the Cleveland iron trade, following the development of the seam of Cleveland ironstone in 1851. He erected blast furnaces at South Bank, Middlesbrough, near the works of Bolckow and Vaughan. He operated the plant until 1863 when he sold it and built more extensive works near Newport. His interest in the practical applications of science grew and he studied the construction of large blast furnaces. By 1870 there were eight furnaces and by 1872 2,500–3,000 tonnes of pig iron a week were being produced. In 1887 the firm of Sir B. Samuelson and Co. was formed and he served as Chairman until 1895. Also in 1870 he built the Britannia Ironworks at Middlesbrough. He experimented unsuccessfully with the Siemens–Martin process aiming to make steel starting from the Cleveland ore.

Samuelson's home was in Banbury and he served as a Member of Parliament for Banbury and subsequently for North Oxfordshire for many years (although not continuously) being first elected in 1859 and retiring in 1895, when he was made a Privy Councillor. In the House of Commons he gave advice on industrial matters; he was also a great advocate of technical education, and served on a royal commission on Scientific Instruction in 1870, and in 1881 was Chairman of a Royal Commission on Technical Instruction. Samuelson became a Baronet in 1884.
DNB 2nd Supplement, v. III, 258–260. WFS.

FÉLIX SAVART 1791–1841; FMRS 1839
Félix Savart was born at Mezieres, France, the son of Gerard Savart, an engineer at the military school of Metz. He studied medicine, first at the military hospital at Metz and then at the University of Strasbourg, where he received a medical degree in 1816. Following service as a surgeon in the French army, and a period of medical practice in Strasburg he then moved to Paris. Savart had an interest in the physics of the violin, and met Jean-Baptiste Biot, who helped him find a teaching post in physics. He obtained a post in the Collège de France, and taught acoustics, a subject in which he investigated many phenomena. His research activities also covered magnetism, the physics of the violin and aspects of voice and hearing. Savart was highly regarded as an experi-

menter. He used vibrations to obtain information about the structure of metals from the dependence of elasticity, covering a range of metals and alloy. His work covered features such as preferred orientation, static and dynamic modulii, and aspects of internal friction, that became important themes in modern physical metallurgy.
PRS 1841, 50, 352–354. DSB 12, 129–130. S. Dostrovsky. Bulloch's Roll. Gen. Ref. 21.

THOMAS SAVERY c.1650–1715; FRS 1706
Thomas Savery was born at Shilstone, near Modbury, Devonshire, the son of Richard Savery. He became a military engineer, and occupied his spare time in mechanical experiments. In 1698 he patented and constructed what was known as the 'fire engine', thus being the first to utilise fuel as a practical means of performing mechanical work. In 1702 he became a Captain in the Engineers and subsequently held other posts. Of metallurgical interest he invented a hand bellows which produced temperatures in a wood or coal fire sufficient to melt metals.
DNB.

ERWIN SCHRÖDINGER 1887–1961; FMRS 1949
Erwin Schrödinger was born in Vienna, the only child of Rudolf Schrödinger and his wife, who was a daughter of Alexander Bauer. His early education was mainly from a private tutor. At the age of 11 he went to the Akademische Gymnasium in Vienna before entering the University in 1906 to study physics. In 1910 he was awarded a doctorate and then joined the staff. During the 1914–18 war Dirac served as an officer in the fortress artillery and then returned to Vienna. In 1920 he married a girl from Salzburg. He also went to Germany where he spent short periods at Jena, Stuttgart and Breslau. He was then appointed Professor of Physics at Zurich. There followed a move to Berlin in 1927 as Professor of Theoretical Physics. In 1933 with the rise of the Nazis Dirac came to Oxford as a Fellow of Magdalen College. However, he moved to Graz in 1936 until the German take-over of Austria; he then went to the Institute for Advanced Studies in Dublin. In 1956 he returned to Austria to a Chair at the University of Vienna. Dirac's early research covered both experimental and theoretical work; beginning in 1910 his publications included dielectrics, magnetism and X-ray diffraction in lattices. He was the founder of wave mechanics and shared, with Dirac, the Nobel Prize for Physics in 1933.
BMFRS 1961, 7, 221–228. W. Heitler. DSB 12, 217–223. Armin Hermann.

GLENN THEODORE SEABORG 1912–1999; FMRS 1985
Glenn Seaborg was born in Ishpeming, Michigan into an immigrant Swedish family; his parents were H. Theodore and Selma Seaborg. When Glenn was 10 years old the family moved to Los Angeles, where he graduated from High School in 1929. He entered the University of California, where he was awarded a science degree in in 1934. In 1937 he obtained a PhD degree in chemistry, having worked with Gilbert Lewis, with whom he continued for two years as a Research Associate. From 1939–41 Seaborg held the post of Instructor; he was appointed Associate Professor in 1941. He married Helen L. Griggs in 1942 and they had four sons and two daughters. From 1942–46 he was on leave of absence from the University as head of the plutonium work for the Manhattan Project at the University of Chicago Metallurgical Laboratory. In 1945 he was appointed Professor of Chemistry in the University of California, a post which he held until 1971. He served as Associate Director of the Lawrence Radiation Laboratory at Berkeley from 1954–61 and from 1971–75. He was Chancellor of the University of California from 1958–61 and then held the appointment, on leave of absence, as Chairman of the US Atomic Energy

Commission from 1961–1971. Beginning in 1975 he was again Associate Director of the Lawrence Radiation Laboratory.

Seaborg was renowned for his research on the synthetic transition elements, the actinides, from elements 93 onwards in the Periodic Table; this work included the discovery of plutonium in 1940. Together with Edwin McMillan he was awarded the Nobel Prize for Chemistry in 1951. He was a foreign member of 10 academies of science and received 50 honorary degrees.

R. Porter, *The Hutchinson Dictionary of Scientific Biography*, Helicon, 1994. David Millar, Ian Miller, John Millar and Margaret Millar, *The Cambridge Dictionary of Scientists*, Cambridge University Press, 1996. *The International Who's Who*.

JOHN SENEX ?–1740; FRS 1728

John Senex became known as a cartographer, engraver and globemaker. By 1719 he had established a bookselling establishment in Salisbury Court, Fleet Street, London.

DNB.

IAN ALEXANDER SHANKS 1948–; FRS 1984

Dumbarton Academy. Univ. of Glasgow: BSc. Glasgow College of Technology: PhD. Scottish Colorfoto Labs: Projects Manager, 1970–72. Portsmouth Polytechnic: Research Student, 1972–73 (liquid crystal displays). RSRE, Malvern: Displays and L-B films) 1973–82. Unilever Research, 1982; Principal Scientist, 1984–86 (electronic biosensors). Thorn EMI plc: Chief Scientist, 1986–94. Unilever Research: Divisional Science Adviser, Research and Engineering Division, 1994–. FREng 1992.

Liquid crystal displays; optical and optoelectronic devices.

DAVID SHOENBERG 1911–; FRS 1953

Latymer Upper School, London. Cambridge Univ.: Trinity College, Scholar, Research in Low Temperature Physics, PhD 1935; 1851 Exhibition Senior Student, 1936–39; in charge of Royal Society Mond Laboratory, 1947–73; Univ. Lecturer in Physics, 1944–52; Reader 1952–73; Professor of Physics and Head of Low Temperature Physics Group, Cavendish Laboratory, 1973–78. Hughes Medal 1995. MBE 1944.

Low temperature physics and magnetism.

JAMES SHORT 1710–1768; FRS 1737

James Short was the son of William Short, a joiner in Edinburgh, where he was born. After education at Edinburgh High School he entered the University in 1726; he also obtained an MA at St Andrews. He became a protégé of the professor of mathematics, and also started to make mirrors for reflecting telescopes, first using glass and then speculum metal. The perfecting of metal mirrors became his life's work, and he referred to himself as an 'Optician, solely for reflecting telescopes'. His skill led to immediate recognition and he went on to achieve international reputation and a fortune. His success was the outcome of his control of alloy composition and casting, and also on the techniques of polishing; he was the first to give the specula a true parabolic form. He settled permanently in London in 1738 with a workshop in Surrey Street, off the Strand. In the Royal Society he served as a Member of Council.

DSB 12, 413, 414. G.L'E. Turner. DNB. Notes and Records Roy. Soc. 1969, 24, 109–112.

KARLE MANNE GEORG SIEGBAHN 1886–1978; FMRS 1954

Karle Siegbahn, who was born in Orebo, Sweden, was the son of Nils Reinhold George Siegbahn, a station master and Emma née Zetterberg. The family moved to Lund, where in 1906 he entered the University of Stockholm, where he obtained an MA and a DSc. In 1910 he married Karin Evelina Hogbom. He worked at Lund until 1923, apart from periods of study at several universities in Europe; he became Professor of Physics in 1920. From 1924–1937 he was Professor of Physics at the University of Uppsala. In 1937 he took up an appointment as the first Director of the Nobel Institute of Experimental Physics, a post which he held until his retirement in 1964. He was the recipient of the Nobel Prize for Physics in 1924 for his research on X-ray spectroscopy. BMFRS 1991, 37, 429–444. H. Atterling.

SIR CHARLES WILLIAM SIEMENS 1823–1883; FRS 1862

Born at Lenthe, Hannover, in Germany, Charles Siemens, was one of the eight sons of Christian Ferdinand Siemens and Eleonore née Deichmann; he was baptised Carl Wilhelm, but since one of his brothers was named Carl he was always known as William. His father, who was a farmer of government lands, died in 1839. The eldest son Werner, then 23 years old and an officer in the Prussian artillery, took the place of a father to the younger brothers, superintending their education and helping their start in life. Of the eight sons, four, – Werner, William, Frederick and Carl – were closely involved in practical applications of science or in managing great industrial enterprises which stemmed from their scientific inventions.

William was educated at the Gymnasium at Lubeck, the Polytechnic School of Magneburg and then at the University of Göttingen, where he studied natural sciences under Friedrich Wohler. On leaving university in 1842 he spent a year as a pupil in the engine works of Count Stolberg, adopting the profession of engineer. He considered that an engineer should be involved in the practical development of the results of science, in order to serve mankind and should be able to adapt the results of science to the néeds of life. In his own career he exemplified this and also made important contributions in pure science.

In 1843 William Siemens came to England for the first time with a joint invention that he and his brother Werner had made in Germany in electrogilding; persevering through the difficulties, with little knowledge of the English language, he proved the usefulness of the invention and brought it into practical effect. William gave up his position in the Stolberg factory in 1844, and came to London with two other inventions, a chronometric governor for steam engines (jointly with Werner) and the process of anastatic printing developed by the brothers from a previously available invention. During the 10 years from his arrival in England till about 1856 he was employed on various engineering works and in railway work, and in 1859 he became naturalised in England. In the same year he married Anne Gordon; they had no children.

Following earlier work on the regeneration principle in the economic use of heat, William Siemens worked with his brother Frederick to apply this principle in metallurgy. This led to the regenerative gas furnace, which was applied in steelmaking. In the steel works which he founded in Birmingham in 1865 he achieved practical success in the open-hearth steelmaking process, which came to bear his name. Siemens also made pioneering contributions and built up another reputation in the field of the electric telegraph. Werner, William and Carl established the works of Siemens Brothers. Cable laying and repairing activities of the firm were world wide. William played a major role in relation to machines and processes. In the field of transport William was one of the first to apply electric power to railway locomotion. In addition to his inventiveness he was a shrewd and successful man of business. During the last 15 years of his life he gave active support to the development of the engineering profession and its societies. He stimulated public interest in fuel conservation, reduction of air pollution, and the importance of electric power in engineering applications.

Among the many honours conferred on William Siemens for his contributions to pure and applied science were a knighthood in 1883 and recognition from the Emperor of Brazil, the Shah of Persia, and from France. In the Royal Society he twice served on the Council. An article in *The Times* said of him 'Looking back along the line of England's scientific worthies, there are few who have served the people better than their adopted son – few, if any, whose life's record will show so long a list of useful labours'.

PRS 1884, 37, i–x. DSB, 12, 424 R.A. Chapman. DNB. Ref. Ch. 7.6.

SIR FRANCIS (FRANZ) EUGEN SIMON 1893–1956; FRS 1941

Franz Simon was born in Berlin, the only son and second of three children of Ernst Simon, a wealthy man and Anna Philibert née Mendelssohn. Following education at the Kaiser Friedrich Reform Gymnasium, which he entered in 1903, he studied physics, chemistry and mathematics at the University of Munich for a year, beginning in 1912. He then spent a period at Göttingen before commencing military service in 1913. During the 1914–18 war, he served as a Field Artillery Officer with the rank of Lieutenant; he was twice wounded, became a poison gas casualty, and was awarded the Iron Cross, First Class. In 1919 he went to the University of Berlin to study physics where he was influenced by Planck, von Laue and by Nernst, under whom he obtained his PhD in 1921. In the following year he married Charlotte Munchhausen and they had two daughters. He also became assistant to Nernst at the Physical Chemical Institute of the University of Berlin; he became a Privatdozen in 1924 and Associate Professor in 1927. Simon was appointed to the Chair of Physical Chemistry at the Technical University in Breslau in 1931. However, in 1933 with the rise to power of Hitler, he came to work at the Clarendon Laboratory, Oxford, at the invitation of Frederick Lindemann. He became a Reader in Thermodynamics in 1935. During the 1939–45 war he worked on the separation of uranium isotopes by gaseous diffusion for the atomic bomb project. In 1945 Simon was appointed Professor of Thermodynamics at Oxford. He was appointed Dr Lee's Professor of Experimental Philosophy in 1956, but died within a few weeks.

Simon's achievements included the development of methods of achieving extremely low temperatures and the building up of the Clarendon Laboratory as a major school of low temperature physics. The research led to the discovery of new superconductors and new magnetic phenomena. He received the Rumford Medal in 1948, the CBE in 1948 and was knighted in 1955.

BMFRS 1958, 4, 225, 256. N. Kurti. DSB 12, 437 439. K. Mendelssohn. DNB 1951–60, 890–892. N. Kurti.

HERBERT WAKEFIELD BANKS SKINNER 1900–1960; FRS 1942

Herbert Skinner was born in Ealing, near London, the son of George Herbert and Mabel Elizabeth Skinner. Following education at Durston House School in Ealing, he entered Rugby School in 1914? Beginning in 1919 he studied at Trinity College, Cambridge, where he obtained a first class in Part I mathematics, followed in 1922 by a first class in natural sciences. From 1922 to 1927 he carried out research in physics in the Cavendish Laboratory; he held an 1851 Exhibition Senior Studentship, and for a period worked with Peter Kapitza. In 1927, with a Henry Herbert Wills Fellowship, Skinner moved to the Physics Department at Bristol University, where Tyndall was the Head of Department. He carried out research there from 1927–39, including work on the excitation potential of light metals; during the time of his appointment at Bristol he was Rockefeller Fellow at the Massachusetts Institute of Technology from 1932–33. In 1931 he married Erna

Wurmbrand and they had one daughter.

In the war-time period Skinner did radar work at the Telecommunications Research Establishment from 1940–44. He then went to work from 1944–45 on atomic energy as part of a group of British physicists at Berkeley, California; he was involved in the electromagnetic separation of uranium isotopes. In 1946, he joined the British atomic energy project at Harwell, and established and led the General Physics Division, whose activities included X-ray and neutron diffraction worked. From 1946–50 he was Head of General Physics at the Atomic Energy Research Establishment, Harwell. In 1949 he was appointed to the Lyon Jones Chair of Physics at Liverpool University. Skinner carried out important work in soft X-ray spectroscopy, obtaining direct experimental confirmation of fundamental aspects of electron theory in metals.

BMFRS 1960, 6, 259–268. H. Jones. *Who was Who* 1951–60.

HENRY SLINGSBY c.1621–c.1688; FRS 1663

Henry Slingsby was the son of Sir Henry Slingsby, a Royalist. He became Master of the Mint in 1663 and also served in Parliament.

Bulloch's Roll.

RAYMOND EDWARD SMALLMAN 1929–; FRS 1986

Rugeley Grammar School. Univ. of Birmingham: BSc, PhD, DSc. AERE, Harwell: 1953–58. Univ. of Birmingham: Lecturer, Department of Physical Metallurgy, 1958; Senior Lecturer, 1963; Professor of Physical Metallurgy, 1964; Head of Department of Physical Metallurgy and Science of Materials, 1969–81, of Metallurgy and Materials, 1981–88; Dean Faculty of Science and Engineering, 1984–85, of Faculty of Engineering, 1985–87; Pro-Vice-Chancellor and Vice-Principal, 1987–92; Feeney Professor of Metallurgy and Materials Science, 1969–. FREng 1991. CBE 1992.

Defect analysis in electron microscopy; relationship of microstructure to properties.

JOHN SMEATON 1724–1792; FRS 1753

John Smeaton was born at Austhorpe, near Leeds, the eldest son of William Smeaton, an attorney, and Mary née Stones. He showed considerable mechanical ability as a boy and made a small lathe and other tools, doing his own casting and forging. Following his education at Leeds Grammar School he entered his father's office. In 1742 he came to London to continue his legal studies at Westminster Hall, but he abandoned the law and, after entering the employment of an instrument manufacturer he opened a shop of his own in 1750. His father provided him with a generous allowance. In 1756 he married Anne and they had two daughters. He regularly attended the meetings of the Royal Society and contributed various papers. From about 1752 he was mainly occupied by engineering problems, and he became one of the leading British engineers and a man of science. He travelled, in 1754 through the Low Countries to study canal and harbour systems, and later constructed a lighthouse for erection on the Eddystone Reef off Plymouth; he also carried out bridge-building and canal work. In the metallurgical field he was consulting engineer for the Carron Ironworks founded by James Roebuck and his partners. In the Royal Society he was the recipient of the Copley Medal in 1759.

DSB 12, 461–463. H. Dorn. DNB.

ALFRED SMEE 1818–1877; FRS 1841

Alfred Smee, who was born in Camberwell, was the second son of William Smee, accountant-

general to the Bank of England. He entered St Paul's School in 1829 and in 1834 he became a medical student at King's College, London, where he was awarded the silver medal and prize for chemistry in 1836 and the silver medals for anatomy and physiology in 1837. He then moved to St Bartholomew's Hospital and obtained a prize in surgery. For most of his student life he lived in his father's official residence within the Bank of England, and here he carried out his research on chemistry and electrometallurgy which afterwards made him famous. He began to practise as a consultant surgeon in Finsbury Circus in 1840, particularly concerned with diseases of the eye; however, he also spent much time with chemical problems and in the study of electrical science. 'Smee's battery' was an outcome of this work. In 1841 Smee was appointed Surgeon to the Bank of England, a post especially created for him, in which he invented a durable writing-ink and was involved in perfecting a system of printing bank notes and cheques.
DNB.

EDWIN SMITH 1931–; FRS 1996

Chesterfield Grammar School. Univ. of Nottingham: BSc. Univ. of Sheffield PhD. Univ. of Manchester: MSc. AEI Research Laboratory, Aldermaston: 1955–61. CEGB Research Laboratory, Leatherhead: 1961– 68. Univ. of Manchester: Professor of Metallurgy, UMIST Materials Science Centre, 1968– 88; Dean, Faculty of Science 1983–85; Pro-Vice-Chancellor 1984– 88; Consultant and Honorary Fellow, 1988–.
Physical metallurgy, fracture.

SIR FRANK EWART SMITH 1897–1995; FRS 1957

Frank Smith was born in Loughton, Essex, where his father Richard Sidney Smith, was a pharmaceutical chemist and optician; his mother was Kathleen Winifred née Dawes. The family moved to Hastings soon after his birth? and following private education he entered Uckfield Grammar School in 1906. At the age of 12 he obtained a scholarship to Christ's Hospital and in 1915 he was awarded a scholarship in mathematics, physics and chemistry to Sidney Sussex College, Cambridge. However, he left school in 1916 to join the Royal Horse Artillery as an Officer Cadet. Having received his commission he transferred to a heavy battery of the Royal Garrison Artillery; he saw active service in France and was mentioned in despatches. He was appointed Assistant Adjutant to the 48th Brigade of the RGA and returned to Cambridge in 1919 to read for the Mechanical Engineering Tripos. Following the award of a first class honours in 1921, after 7 terms, he was awarded a scholarship for research. Stemming from his interest in metallurgy he investigated the expansion of iron, including the highest purity grade available, and steel, through the allotropic transformations up to 1000 °C; he used vacuum conditions for this work and was awarded a PhD.

In 1923 Smith joined the staff of Fertilizers and Synthetic Products Ltd (later to beome a division of ICI) at Billingham. In the following year he married Kathleen Winifred Davies and they had a son and a daughter. At ICI his initial work included the design of a pressure seal for a plant producing ammonia at high pressures. He also worked for a period in the Research Department and in 1928 was appointed Deputy Chief Engineer. From 1923–30 his work covered research, design, construction and production tasks and in 1931 he became Chief Engineer at the Billingham site. In the 1939–45 wartime period Smith's work included appointment in 1942 as Chief Engineer and Superintendent Armaments Design in the Ministry of Supply; the work, which included a strong metallurgical activity, involved responsibility for the design and development of armaments for all three Services. Among his particular interests was the assessment of

German secret weapons.

Smith returned to ICI in 1945 and was appointed Technical Director with responsibility for all engineering and related work; his policy involved applying all available technology to the development of new equipment and techniques for the control of operating plant. He retired from ICI in 1959. In the Royal Society he served on the Council from 1963–64, and in 1976 he was elected FREng as a founder member.

BMFRS 1996, 42, 421–431. G.F.Whitby FREng. *Who's Who.*

GEORGE DAVID WILLIAM SMITH 1943–; FRS 1996

St Benedicts School, Aldershot. Salesian College, Farnborough. Univ. of Oxford: Corpus Christi College, Scholar, BA Hons Metallurgy 1965, MA, DPhil 1968; SRC Research Fellow, 1968–70; Junior Research Fellow 1968–72; Research Fellow, Wolfson College, 1972–77; University Research Fellow, Department of Metallurgy, 1970–75; Senior Research Fellow, 1975–77; Lecturer in Metallurgy 1977–92; George Kelley Reader in Metallurgy, 1992–96; Professor of Materials Science, 1996–.

Field ion/atom probe microscopy; phase transformations.

ROBERT ALLAN SMITH 1909–1980; FRS 1962

Robert (Robin) Smith was the elder of two sons of George J.T. Smith and Elisabeth née Allan. He was born at Kelso, Roxburgshire, where he attended the High School. In 1926 he entered Edinburgh University and studied mathematics and natural philosophy, obtaining an MA, first class, in 1930. He went to Emmanuel College, Cambridge with a Maclaren Scholarship and took the Mathematical Tripos, Part II, obtaining an MA in 1932, becoming a Wrangler. He carried out research on atomic collisions at the Cavendish Laboratory obtaining a PhD in 1935. In 1934 he married Doris Marguerite Lousie Ward; they had two daughters and one son. Smith held a Carnegie Teaching Fellowship at St Andrews University until 1938, and then became a Lecturer in Mathematics at the University of Reading. In 1939 his wartime science service began at the Bawdsey Research Station; from 1942 he was at the Telecommunications Research Establishment at Malvern. He continued to work at Malvern until 1962 including the period when the title was changed to Royal Radar Establishment. From 1961–62 Smith was Professor of Physics at Sheffield University. During the period 1962–67 he was Director of the Materials Science Centre at MIT, where his work included investigations of impurity bands in semiconductors. From 1968–74 he was Principal and Vice-Chancellor of Heriot-Watt University. His publications included the book *Wave mechanics of crystalline solids* (1961).

BMFRS 1982, 28, 479–504. S.D. Smith FRS.

GEORGE JAMES SNELUS 1837–1906; FRS 1887

George Snelus was born in Camden Town, North London, the son of James and Susannah Snelus; his father, a master builder, died when George was seven. He was trained as a schoolteacher at St John's College, Battersea, but subsequently, when he was teaching at a school in Macclesfield he attended science lectures at Owens College, where he was influenced by Henry Roscoe. In 1864, having won a Royal Albert Scholarship, he entered a three year course at the Royal School of Mines, London, where he obtained the Associateship in Metallurgy and Mining, with the De la Beche Medal for Mining. In 1867 he married Lavinia Whitfield, and they had three sons and three daughters. He was appointed Chemist to the Dowlais Ironworks, a post which he held for four years.

In 1871 he was commissioned by the Iron and Steel Institute to go to the USA to investigate the chemistry of the Dank's rotary puddling process. It was during the course of this investigation that he thought of the possibility of eliminating phosphorus from molten phosphoric pig iron by oxidation in a basic lined vessel. He carried out experimental trials showing that phosphorus could be almost entirely removed but there remained some practical difficulties, and it was not until 1879 that Thomas and Gilchrist developed a successful industrial process. For his important contribution to the crucial technological achievement, he was awarded a gold medal at the Paris exhibition in 1878, and the Bessemer Gold Medal with Thomas; also he received a share of profits from the British and American rights from the invention. Another important contribution made by Snelus to metallurgical chemistry was in the analytical determination of phosphorus in steel. In 1872 Snelus became Works Manager and, later, General Manager of the West Cumberland Iron and Steel Company at Workington, where he remained until 1900; he was also director of several mining concerns in Cumberland. Snelus was an enthusiastic member of the Volunteer Force from 1859–1891, retiring with the rank of Honorary Major; he was one of the best rifle shots in the country.
PRS 1906–07, 78 A, lx–lxi, JES. DNB, 2nd Suppl. III, 353, 354. WAB.

FREDERICK SODDY 1877–1956; FRS 1910
Frederick Soddy was born at Eastbourne, Surrey, the youngest of seven sons of Benjamin Soddy, a London corn merchant and Hannah née Green. Following his education at Eastbourne College and a period at University College, Aberystwyth, he studied at Merton College, Oxford, where he obtained first class honours in chemistry in 1898. He was appointed as a Demonstrator at McGill College, Montreal, where he collaborated with Ernest Rutherford on radioactive investigations of thorium salts. Returning to London in 1903 he worked with Sir William Ramsay. Then from 1904–05 he was a Lecturer in Physical Chemistry and Radioactivity at Glasgow University. In 1908 he married Winifred Moller Beilby; there were no children. He was appointed to the Professorship of Chemistry at Aberdeen in 1914; then from 1919 to his retirement in 1936 he was Dr Lee's Professor of Chemistry at Oxford. In 1921 he received the Nobel Prize for Chemistry.
BMFRS 1957, 3, 203–216. A. Fleck. DSB, 12, 504–509. T.J. Trenn. DNB 1951–1960, 904, 905. Fleck.

LASZLO SOLYMAR 1930–; FRS 1995
Technical Univ. Budapest: Hungarian equivalents of BSc and PhD in Engineering; Lecturer, 1952–53. Research Inst. for Telecommunications, Budapest: Research Engineer, 1953–56. Standard Telecom Labs, Harlow: 1956–65. Univ. of Oxford: Brasenose College; Fellow in Engineering, 1966–86; Lecturer, 1971–86; Donald Pollock Reader in Engineering Science, 1986–92; Professor of Applied Electromagnetics, 1992–97; Hartford College, Professorial Fellow.
Electrical properties of materials; superconducting tunnelling; photorefractive materials.

ARNOLD JOHANNES WILHELM SOMMERFELD 1868–1951; FMRS 1926
Arnold Sommerfeld was born at Konigsberg, East Prussia, the son of Franz Sommerfeld, a physician and Cäcile née Matthias. He was educated at the Alsträdtisches Gymnasium from 1875–1876 and at the University of Konigsberg where he obtained a PhD. He then spent a year preparing for the examinations to qualify as a Gymnasium teacher in mathematics and physics. There followed a year of compulsory military service before he became an assistant at the Mineralogical Institute in Göttingen in 1893; then in 1895 he became a Privatdozen in Mathematics at the University. An appointment as Professor of Mathematics at the Mining Academy in Clausthal

began in 1897. In 1900 he became Professor at the Technische Hochschule in Aachen. A move to Munich as Director of the Institute of Physics occurred in 1906. Sommerfeld retired in 1940, but returned to his post after World War II. His research included (in 1916) a proposed important modification to Bohr's model of the atom. Sommerfeld married Johanna Höpfner.
ONFRS 1952–53, 8, 275–296. Max Born. DSB 12, 525–532. P. Forman. A. Hermann.

THOMAS SOPWITH 1803–1879; FRS 1845
Thomas Sopwith was born in Newcastle upon Tyne, the son of Jacob Sopwith, a builder and cabinet maker, and Isabella née Lowes. He took up surveying and engineering, and became involved in mining and surveying work, and also in railway engineering. In 1845 he was appointed Chief Agent for Lead Mines in Northumberland and Durham and worked as a mining engineer. He lived at Allenheads, and was also involved with building houses, schools and libraries. Sopwith married three times; his first two wives died, and from his third marriage to Anne Potter in 1858 he had one daughter.
Roger Burt. *The British Lead Mining Industry*. Dyllanson Truran, Cornwall, 1984. DNB.

HENRY CLIFTON SORBY 1826–1908; FRS 1857
Henry Sorby was the son of Henry Sorby and Amelia née Lambert. His father was a partner in the firm of John and Henry Sorby, edge tool manufacturers. The family had been connected with the main industry of Sheffield since the 16th century. One of Henry's ancestors was the first Master Cutler of the Cutlers' company; his grandfather also became Master Cutler. Henry was born at Woodbourne, near Sheffield, an estate which belonged to his father, and which he inherited. He received his early education at the Sheffield Collegiate School where he obtained as a mathematics prize a book entitled *Readings in Science* which interested him in research and experiment. During the next four years he had a full-time private tutor, who, having had a medical training introduced Sorby to anatomy and chemistry, besides supervising his mathematical studies. Sorby's father died in 1847 and left him a modest fortune, which enabled him to to devote himself entirely to science, continuing to live in Sheffield for the rest of his life. He continued to study mathematics, optics, chemistry and anatomy. Also he found time for water colour drawing – 'I worked' he said 'not to pass an examination, but to qualify myself for a career of original investigation'.

Sorby decided to remain at Sheffield, where there were opportunities for conducting experimental research. He never married but lived with his mother until her death in 1874. Subsequently he bought a yacht and for many years spent the summer months dredging and making biological and physical observations in the estuaries and inland waters of the east of England. The winters were spent in Sheffield, carrying on his experiments, much of the work being done in his own house.

His earliest publication appears to have been in the field of agricultural chemistry, probably stemming from some of the chemical studies with his tutor. Sorby's most influential scientific work was done between 1849 and 1864 on the application of the microscope to geology (he came to be known as the father of microscopical petrography) and to metallurgy; he also carried out important work in biological microscopy, microspectrographic analysis, archaeology of masonry, flow and sedimentation in river and ancient natural history.

In 1863 Sorby turned his attention to metals, pioneering the technique of polishing and etching metallic samples and then examining them by microscopy using reflected light. In a paper published in 1864 he referred to the constituents present in steel, and in this and later publications made fundamental contributions to the understanding of the microstructure of steels, including the constituent which became known as pearlite.

Much of Sorby's time from 1880 was involved as President of Firth College, Sheffield, supporting its incorporation as part of the University of Sheffield. The last four years of Sorby's life involved physical disability, and for long periods confinement to bed, but he spent time reviewing his earlier geological work, and submitted a paper for publication a few days before he died. Sorby was awarded a Royal Medal in 1874. He was gifted with great originality, perseverance and mechanical ingenuity. His career as a full-time independent research scientist of extraordinary versatility led to his publishing more than 150 papers; his achievements left their mark in many fields.

PRS 1908, B, v.80, lvi–lxvi. AG. DSB, 12, 542–546. C.S. Smith. DNB 2nd Suppl. 355, 356 TGB. Gen. Refs 21, 22.

SIR RICHARD VYNNE SOUTHWELL 1888–1970; FRS 1925

Richard Southwell was born in Norwich, the son of Edwin Baterbee Southwell and Annie née Vynne. In 1898 he entered King Edward VI School in Norwich where at the age of 16 he changed from the study of classics to mathematics. He won an Exhibition at Trinity Hall, Cambridge, but entered Trinity College the following year (1907) as a Commoner. At Cambridge he won scholarships and prizes and obtained first class honours in both parts of the Mechanical Sciences Tripos (in 1909 and 1910). In 1912 he was awarded a Fellowship at Trinity College and in 1914 the College offered him a Lectureship. However, he volunteered for the army at the outbreak of war and was commisioned in the Army Service Corps going to France before the end of the year. In 1915 he was called to serve as a Lieutenant in the part of the Royal Naval Air Service responsible for the development of non-rigid airships. Southwell was promoted to Lieutenant Commander in 1917 and transferred to the Royal Air Force as a Major in 1918. He was on the staff of the Royal Aircraft Establishment at Farnborough in charge of the aerodynamics and structural departments until his demobilisation in 1919. In the previous year he had married Isabella Wilhemina Warburton Wingate and they had four daughters.

Southwell returned to Trinity College for a period before moving to the National Physical Laboratory in 1920 as Superintendent of the Aeronautics Department. He returned to Trinity College in 1925 as a Fellow and Faculty Lecturer in Mathematics. In 1939 he was appointed as Professor of Engineering Science at Oxford University. From 1942 to his retirement in 1948 he was Rector of Imperial College, London.

Southwell's research in applied mechanics and aeronautics included elasticity, the theory of frameworks and the impact testing of materials. In the Royal Society Southwell served on the Council from 1930–31. He was knighted in 1948.

BMFRS 1972, 18, 549–565. D.G. Christopherson. DNB 1961–70, 967–969. A.G. Pugsley.

WALTER ERIC SPEAR 1921–; FRS 1980

Musterchule, Frankfurt /Main. Univ. of London: BSc 1947, PhD 1950, DSc 1967, Univ. of Leicester: Lecturer, 1953, Reader 1967 in Physics. University of Dundee: Harris Professor of Physics, 1968–90. Bakerian Lecturer, 1988. Rumford Medal, 1990.

Electronic and transport properties in crystalline solids, liquids and amorphous semiconductors.

ROBERT SPENCE 1905–1976; FRS 1959

Robert Spence, the only child of Robert and Rebecca Spence, was born at South Shields, County

Durham. He was educated at the Stanhope Road Elementary School from 1911–1917 and then at Westhoe Grammar School. At the age of 17 he entered Armstrong College (now King's) in Newcastle, then part of the University of Durham. He was awarded a first class BSc degree in chemistry (1926), an MSc and a PhD(1930); later he obtained a DSc degree. In 1928 Spence was awarded a Commonwealth Fund Fellowship at Princeton University, USA, where he worked in the Department of Physical Chemistry. He returned to England in 1931 to a lectureship at Leeds University. In 1936 he married Kate Lockwood and they had three children. Spence was commissioned in the Territorial Army in 1936 to command the University Officers Training Corps; he was promoted to Captain in 1939, and then to Major and Lt-Col. in 1941. Spence sought transfer to active service with a tank regiment, but the army seconded him to the Ministry of Aircraft Production and sent him in 1942 to the Middle East as Chemical Warfare Advisor, HQ RAF, with the rank of Hon. Wing Commander. Subsequently he served in Italy, before returning to Britain near the end of 1944.

Spence then went to Chalk River, Montreal, Canada in June 1945 to join the atomic energy team headed by Cockcroft, where he worked to develop separation processes for plutonium and fission products and to produce plutonium. Spence and his team, using only the milligram quantity of plutonium available, carried out work which led them on to evolve the industrial plant constructed in Britain for the separation process, which came into operation in February 1952. In 1950 Spence was appointed Chief Chemist at the Atomic Energy Research Establishment at Harwell and he was also Head of the Chemistry Division. In 1960 he became Deputy Director of Harwell, and from 1964–68 he served as Director. Spence was a key figure in the successes achieved in the British atomic energy programme His research in the field of atomic energy ranged widely, including the production of uranium from sea water, the basic chemistry of polonium and the isolation and purification of protactinium. From 1968–73 Spence was Professor of Chemistry and Master of Keynes College at the University of Kent at Canterbury. He was awarded the CB in 1953.

BMFRS 1977, 23, 501–528. E. Glueckauf FRS.

ANTHONY JAMES MERRILL SPENCER 1929–; FRS 1987

Queen Mary's Grammar School, Walsall. Univ. of Cambridge: Queens College, MA, PhD, ScD. Brown Univ., USA: Research Associate, 1955–57. UKAEA: Senior Scientific Officer, 1957–60. Univ. of Nottingham: Lecturer; Reader; Professor, 1960–94; Professor of Theoretical Mechanics, 1965–94.

Deformation of fibre reinforced materials; continuum theory of mechanics of fibre reinforced composites.

SIR JOHN STANLEY 1663–1744; FRS 1698

John Stanley was the son of John Stanley of Grange Gorman Co., Dublin. He went to Trinity College, Dublin in 1676 and obtained an MA in 1684. He was Secretary to several Lord Chamberlains of the Household and also was appointed Warden of the Mint in 1699. He was created Baronet in 1698.

Bulloch's Roll.

THOMAS EDWARD STANTON 1865–1931; FRS 1914

Thomas Stanton, the son of Thomas Stanton, was born at Atherstone, Warwickshire, where he attended the Grammar School. At the age of nineteen he was apprenticed to the firm of Gimson and Co., Leicester, general engineers and millwrights. In 1888 he entered Owens College, Manchester, and obtained a first class honours degree in engineering from the Victoria University of Manchester in 1891. He then became a Demonstrator in the Whitworth Laboratory at Owens College until 1896. There followed an appointment as Senior (Assistant) Lecturer in Engineering at University College, Liverpool until in 1899 he became Professor of Civil and Mechanical Engineering at University College, Bristol. In 1901 Stanton joined the National Physical Laboratory as Superintendent of the Engineering Department, a post which he held until 1930 and where his research interests included the fatigue of metals. He married Martha Grace Child in 1912 and they had one son and one daughter. He was awarded the CBE in 1920 and was knighted in 1928.

PRS 1932 A, 135, ix–xv. JEP. FJS.

JOHN EDWARD STEAD 1851–1923; FRS 1903

John Stead was born at Howden-on-Tyne, Northumberland. He was educated privately by his father, the Reverend W. Stead, because of weakness of his spine, and at the age of sixteen he became apprenticed to John Pattinson, the public analysts of Newcastle upon Tyne. After serving three years he became analyst to the Tharsis Sulphur and Copper Company at Hebburn-on-Tyne. From there he went as chemist to Messrs. Bolckow, Vaughan and Company, a large firm of iron manufacturers, who had purchased the works of a Lancashire steel company at Garston, and subsequently he transferred to the Middlesborough Works of the Company. While in Manchester he attended evening classes at Owens College. At the age of 25 he became a partner of John Pattinson, and remained associated with the firm for the remainder of his life, becoming known as one of the best analysts in England. In 1887 Stead married L.M. Livens and they had one son.

Stead's research arose out of the metallurgical problems he encountered in his commercial practice. Over a period of of 46 years he covered a very wide range of subjects in the metallurgy of iron and steel. The influence of phosphorus on iron was a subject in which he had a particular interest. Specific topics on which he did pionéering research were brittleness in steel, overheated steel and burnt steel, the nature of blast furnace 'bears', and 'ghosts' in steel forgings. Other studies included the influence of arsenic and sulphur in steel, the welding up of blowholes and cavities in steel ingots, phosphide-containing eutectics, and the crystallisation of electrodeposited iron. Stead also played a very important part in the early days of the basic Bessemer process being the first to propose the correct explanation that phosphorus was removed during the 'afterblow'.

Stead was a born investigator. Although the equipment of his laboratory was simple, he had access to resources which were not open to any other investigator since all the works in the Middlesbrough district were available to him for experiments and he could thus study phenomena on a large scale. He generously gave young metallurgists the benefit of his great knowledge and experience, and was highly regarded by the industry with which he was associated.

PRS 1924 A, 106, i–v. H.C.H. Carpenter FRS. DSB 1976, 13, 7, 8. F. Greenaway.

JOHN WICKHAM STEEDS 1940–; FRS 1988

University College, London: BSc, 1961. Univ. of Cambridge: PhD, 1965; Selwyn College, Research Fellow, 1964–67; Fellow, 1967; IBM Research Fellow, 1966–67. University of Bristol: Lecturer in Physics, 1966–77; Reader 1977–85; Research Professor in Physics and Head of Microstructural Group, 1985–; Director of Interface Analysis Centre, 1990–.

Electron diffraction and electron microscopy of materials: convergent beam analysis: dislocation arrangements: precipitates.

JOHN STENHOUSE 1809–1880; FRS 1848

John Stenhouse was born in Glasgow the son of William Stenhouse, a calico printer and his Elizabeth née Currie. After a grammar school education he studied at the University. His early interests were more literary than scientific but because of poor eyesight he had to give up literature as a pursuit. He took up the study of chemistry, including a period with Thomas Graham at Andersonian Institution. Subsequently he carried out chemical research with Liebig at Giessen for about two years. Following his return to Scotland in 1839 the failure of the Glasgow Commercial exchange led to the loss of the fortune left to him by his father. In 1850 he received the degree of LLD from Aberdeen University. Stenhouse moved to London in 1851 and was appointed Lecturer in Chemistry at St Bartholomew's Hospital, but resigned in 1857 as a result of an attack of paralysis. For several years he lived in Nice and then returned to England, set up a laboratory and resumed his scientific research, despite his physical handicap.

In 1865 he succeeded August Hofmann as one of the Non-Resident Assayers to the Royal Mint, continuing until this post was abolished in 1870. Much of his research was concerned with organic chemistry and the lichens; he was the holder of many patents in areas including charcoal airfilters and respirators, and the preparation of materials for treating textile fabrics. He was awarded a Royal Medal in 1871.

PRS 1880–81, 31, xix–xxi. DNB.

ROBERT STEPHENSON 1803–1859; FRS 1849

Robert Stephenson was the only son of George Stephenson (1781–1848) one of the leading pioneers of the railway engineering era, who did not accept a knighthood or election as a Fellow of the Royal Society. He was born at Willington Quay, near Newcastle, and when the family moved to Killingworth in 1804, where his mother died two years later, Robert began his elementary education in the village school at Long Benton. In 1814 his father sent him to Bruce's Academy in Newcastle. When he left school in 1819 he became an apprentice at a colliery and in 1821 assisted his father in railway survey work. In 1822 he studied natural philosophy, chemistry and natural history for six months at Edinburgh University, and then managed the locomotive factory his father established in Edinburgh in 1823. After a period of ill health he spent three years in Colombia, South America, where he supervised the working of some gold and silver mines. From 1827–33 he assisted his father in railway engineering, and became an engineer from 1838 in this field. His achievements included the famous *Rocket*, engine built under his direction. His large scale construction work included the Conway and Menai bridges in North Wales and a bridge over the St Lawrence river in Canada. Stephenson married Francis Sanderson; there were no children. From 1847 until his death he served as the MP for Whitby, Yorkshire, and was the recipient of many distinctions. However, like his father, he declined the offer of a knighthood. His engineering work brought him substantial wealth.

DNB. Ref. Ch. 9, 17.

JAMES STODART 1760–1823; FRS 1821

James Stodart was a maker of surgical instruments and a cutler, with his business at 401 Strand, London. His trade card, dated 1800, referred to fine razors, varieties of knives, scissors, etc, tempered by the thermometer. A card from about 1820 referred to razors and other cutlery, made from wootz, a steel from India, preferred by him to the best steel in Europe after years of comparative trial.

Stodart had acquired a reputation as a man of science, and had published metallurgical papers in the period 1802–1805: namely 'Account of an experiment to imitate the Damascus sword blades', 'Methods of gilding upon steel by immersion in a liquid' and 'Precipitation of Platina as a covering or defence to polished steel, and also to brass'. He also worked on the tempering of steel, and deduced that temper colours resulted from oxidation. At the request of Sir Joseph Banks, Stodart participated in the investigation of the properties of wootz. He had an important association with Faraday, leading to joint publications in work on the alloying of iron carried out at the Royal Institution.

English Cyclopaedia, 733–735, Charles Knight. Bradbury and Evans, London 1858.

SIR GEORGE GABRIEL STOKES 1819–1903; FRS 1851

George Stokes was born at Skreen, County Sligo, Ireland, the youngest of six children of Gabriel Stokes, rector of the parish and Elizabeth née Haughton. Having received his early education from his father and the parish clerk, he attended schools in Dublin and Bristol. A student of Pembroke College, Cambridge from 1837, he graduated in 1841 as Senior Wrangler and Smith's Prizeman, and obtained a College Fellowship. He was Lucasian Professor at Cambridge from 1849 until his death; since the Chair was not well endowed he also taught at the Government School of Mines in London in the 1850s to augment his income. In 1857 he married Mary Robinson, and they had two sons and one daughter.

Stokes's research, in which he showed both mathematical ability and experimental skill, included the fields of elasticity and hydrodynamics; he derived the relationship (Stokes's law) giving the force resisting the motion of a spherical body through a fluid. He served as a Member of Parliament for Cambridge University from 1887–1891, and was knighted in 1889. In 1902 he became Master of Pembroke College. In the Royal Society he received the Rumford Medal (1852) and the Copley Medal (1893); he served as Secretary for more than 30 years , and was President from 1885–1890.

PRS 1905, 75, 195–216. DSB 13, 74–79. E.M. Parkinson. DNB 2nd Suppl., 3, 421–424. JL.

ARTHUR MARSHALL STONEHAM 1940–; FRS 1989

Univ. of Bristol: BSc, 1961, PhD, 1964. UKAEA, Harwell: 1964–; Group Leader, 1974; individual merit promotions to band level, 1974 and Senior level, 1979; Head of Materials Physics and Metallurgy Division, 1989–90. UKAEA, Harwell: Head of Technical Area, Core and Fuel Studies, 1988–90; Chief Scientist, 1993–96. AEA Industrial Technology: Director of Research, 1990–. University College, London. Massey Professor of Physics and Director, Centre for Materials Research, 1995–.

Theory of defects in insulators and semiconductors, particularly those which control optical, electrical and chemical properties.

EDMUND CLIFTON STONER 1899–1968; FRS 1937

Edmund Stoner was born at East Molesey, Esher in Surrey, the son of Arthur Hallett Stoner, a cricketer, and Mary Ann née Fleet. He was brought up by foster parents and was educated in South Shields and then at Bolton, where he obtained a scholarship to the Grammar School. For health reasons he was not called up for military service and in 1918 he obtained scholarships to Cambridge University at Emmanuel College. In the Natural Sciences Tripos Part I (botany, chemistry and physics) he obtained a first class, and a year later also a first class in Part II (physics). From 1921 to 1924 he held a DSIR maintenance grant and carried out research in the Cavendish Laboratory. With Rutherford's agreement he worked on the absorption of X-rays by a selected set of elements, but the progress of the work was delayed by illness.

Stoner obtained a Lectureship in Physics at Leeds University in 1924; he was promoted to a Readership in 1927 and to a Chair in Theoretical Physics in 1939, having been awarded a Cambridge DSc in 1938. In 1951 he married (Jean) Heather Crawford; there were no children from the marriage. Stoner continued his career at Leeds, becoming Head of Department in 1951 and retiring in 1963. Stoner's research included extensive theoretical work in the field of magnetism; his first book, published in 1926 was *Magnetism and Atomic Structure.*

BMFRS 1969, 15, 201–237. L.F. Bates. DSB 18, 876, 877. G. Cantor. DNB 1961–70 984, 985. E.P. Wohlfarth.

MICHAEL JAMES STOWELL 1935–; FRS 1984

St Julian's High School, Newport, Wales. Bristol Univ.: BSc, 1957, PhD, 1961. Ohio State Univ.: Research Fellow, 1962–63. TI Research Laboratories: Research Scientist and Group Leader, 1960–78. TI Research: Materials Department, Research Manager, 1978–88. Univ. of Minnesota: Research Fellow, 1970. Alcan International Ltd: Principal Consulting Scientist, 1989–90; Research Director, 1990–94.

Physical metallurgy, electron microscopy, epitaxy, nucleation theory, superplasticity; ferrous and non-ferrous alloys.

SIR REGINALD EDWARD STRADLING 1891–1952; FRS 1943

Reginald Stradling was born in Bristol, the second of three children of Edward John Stradling and Sarah Mary née Bennet. He was educated at Bristol Grammar School from whence he entered Bristol University in 1909, with a scholarship. In 1912 he graduated with a BSc in civil engineering. There followed practical training with a consulting engineer and work with firms in Bolton and Birmingham. He volunteered in 1914 for service with the Royal Engineers, was commissioned and went to France in 1915. He reached the rank of Captain and Adjutant, Divisional Engineers, and served continuously until the end of 1917, when he was invalided out of the Army having been awarded the MC and twice mentioned in dispatches. In 1918 he married Inda Pippard and they had one daughter and one son. Stradling became a Lecturer in Civil Engineering at Birmingham University; he did research on the properties of cement and concrete, and was awarded a PhD in 1922. In the same year he was appointed as Head of the Civil Engineering and Building Department at Bradford Technical College.

In 1924 Stradling became Director of Building Research in the Department of Scientific Research at the Building Research Station, located initially at Acton and later at Garston, near Watford. He was instrumental in 1925 in setting up the Steel Structures Committee, which over a period of five and a half years carried out important work relating to the design of steel framed buildings; Stradling served as the Executive Officer for the Committee. He was awarded the DSc of Bristol University in 1935. With the coming of war in 1939 Stradling was appointed Chief Adviser to the Ministry of Home Security. His subsequent appointments included that of Chief Scientific Adviser to the Ministry of Works (1944–49). He was appointed CB in 1934 and knighted in 1945.

ONFRS 1952–53, 8, 197–307, A.J.S. Pippard. DNB 1951–60, 933–934. A.J.S. Pippard.

JOHN WILLIAM STRUTT (THIRD BARON RAYLEIGH) 1842–1919; FRS 1873

John Strutt was born at Langford Grove, Maldon, Essex the eldest son of John James Strutt (second Baron) and Clara Elizabeth La Touche née Vicars. His education included short periods at Eton and Harrow, but he suffered from ill-health. He entered Trinity College, Cambridge in 1861 and graduated senior wrangler and Smith's prizeman in 1865. After graduation, having visited the USA, he purchased equipment and set up a laboratory to carry out experiments at the family seat at Terling Place. In 1871 he married Evelyn Georgina Mary Balfour, the sister of Arthur James Balfour, the statesman; there were three sons from the marriage. Having succeeded to the title in 1873 he took up serious work in his laboratory. He was Cavendish Professor of Experimental Physics at Cambridge from 1879–1884. He also held a professorship at the Royal Institution from 1877–1905, although this involved spending only relatively short periods in London. He was Chancellor of the University of Cambridge from 1908–1919.

Strutt's scientific research ranged widely, including optics, acoustics, elasticity and capillarity. With Ramsay he discovered and isolated argon and was awarded the Nobel Prize for Physics in 1904. In the Royal Society he served as Secretary (1885–1896) and President (1905–1908) and was the recipient of three medals: Royal (1882); Copley (1899) and Rumford (1914). He was awarded the Order of Merit in 1902.

PRS 1920–21 A, 98, iv–l Arthur Schuster. DSB 13, 100–107. R.B. Lindsay. DNB 1912–1921, 514–517. JHJ.

WILLIAM STRUTT 1756–1830; FRS 1817

William Strutt was the eldest son of Jedediah Strutt, a cotton spinner, from whom he inherited his mechanical skills. He became a prominent mill owner in Derby and Belper. His work included devising a system of ventilation and heating of large buildings, and improving methods of constructing stoves. He was also concerned with the construction of 'fire-proof' buildings. He married Barbara Evans and they had one son and three daughters.

Bulloch's Roll. DNB (*see* Jedediah Strutt).

WILLIE SUCKSMITH 1896–1981; FRS 1940

Willie Sucksmith was born at Low Moor, a suburb of Bradford, Yorkshire; his father was then secretary of a small firm of chemical manufacturers nearby. His early upbringing, following a change in his father's work was mainly in the village of Hipperholme, near Halifax, where he attended the Grammar School from the ages of 10 to 18. Entering the University of Leeds in 1914 he studied mathematics, physics and chemistry, together with education. In 1915 he volunteered for the army, joining the West Yorkshire Regiment; service in France began in 1916 where he was transferred to the divisional signals unit, remaining there until demobilisation in 1919.

Study at the University of Leeds was resumed, leading to first class honours in physics in 1921, followed by an appointment as an Assistant Lecturer in the Bristol University in the Physics Department under A.M. Tyndall. In 1924 he married Minnie Sykes Thornton and they had one daughter. In the University, following his early collaboration with A.P. Chattock and Leslie Bates he continued to work in the field of magnetism including the application of measurements of magnetic properties to the study of alloy constitution. During his time at Bristol he spent a period in 1933 at the Technische Hochschule in Zurich with a Rockefeller Foundation Fellowship. He was promoted from Lecturer to Reader in 1940 and in the same year he was appointed Professor of Physics at Sheffield University.

War-time conditions led to an emphasis on applied research when new magnetic materials were needed e.g. for radar. Much of the industrial work required was carried out in the Sheffield neighbourhood; Sucksmith formed an important link between the armed forces and the manufacturers and had contacts with the Permanent Magnet Association. In the period after the war

Sucksmith built up a strong research school in experimental magnetism.
BMFRS 1982, 28, 575–587. J. Crangle and H.S. Lipson FRS.

SIR JOSEPH WILSON SWAN 1828–1914; FRS 1894

Joseph Swan, the second son of John Swan and Isabel née Cameron, was born near Sunderland. He received his education locally, and left school at the age of 12. From his school books he obtained knowledge of Dalton's atomic theory and of laboratory apparatus. In the autumn of 1842 he was apprenticed to a firm of druggists, where he obtained considerable experience in chemistry. Before the end of his apprenticeship he joined the business of a Newcastle chemist, his future brother-in-law, where he continued his experimental researches and later became a partner.

Swan was interested in photography and made important advances in processing. However, his name became more widely known in connection with the development of incandescent electric lighting, after his successful production of fine filaments from carbon. His first lamp was shown in 1879, and in that year his own house and that of Sir William Crookes were lit by those lamps. Swan also invented the cellular lead plate for secondary batteries.

Swan was twice married, first, in 1862 to Francis White, who died in 1878, and then in 1881 to Hannah, sister of his first wife; he had four sons and three daughters. He received many distinctions, including the Hughes Medal of the Royal Society in 1904, and a knighthood in the same year.

PRS 1919–20, A, 96, ix–xiv. W.G. DNB 1912–1921, 518–519. WBF. Ref. Ch. 9.19.

SIR JAMES SWINBURNE 1858–1958; FRS 1906

James Swinburne was the third of six sons of Thomas Anthony Swinburne, a naval officer, and Anne née Fraser. In 1934 he succeeded to a baronetcy. He was born in Inverness, Scotland and spent much of his childhood on the small island of Eilean Shona in Loch Moidart. In 1870 he went to Clifton College, Bristol where his interests in engineering and science developed. Soon after leaving school he was apprenticed to a locomotive works in Manchester where he developed his inherent and remarkable manual skill. Subsequently he moved to an engineering works in Tyneside and became interested in the rising electrical industry. In 1881 he was engaged by Joseph Swan to established a lamp factory in Paris and in the following year he went to Boston, Massachussetts on a similar assignment. In about 1885 he joined Colonel R.E.B. Crompton in his dynamo works as Technical Assistant, and later became Manager. Swinburne married Ellen Wilson in 1886 and they had three sons; following her death he married Lilian Gilchrist Carey in 1898; they had two daughters. In 1894 he moved to Victoria Street, London as a consultant with a well equipped workshop and chemical and physical laboratory.

Swinburne made many important contributions in electrical engineering e.g in the theory of dynamo design. He also had a great interest in applying physics and chemistry to the problems of industry and in the development of new materials. His numerous patents included subjects relevant to metallurgy, for example the extraction of metals, batteries, and electrolysis of fused salts. Among his interests was the metallurgy of safety razor blades in which subject he was an authority.He was highly skilled as an expert witness in legal cases. In the field of plastics, stemming from his work on developing a commercial product from the reaction of pheonl and formaldehyde. His work was anticipated by Baekeland's patent (Bakelite) in 1907. However, Swinburne did much to pioneer the development of the modern plastics industry.

BMFRS 1959, 5, 253–268. F.A. Freeth. DNB 1951–60, 945, 946. F.A. Freeth.

SIR CHARLES SYKES 1905–1982; FRS 1943

Charles Sykes, the only son of Sam and Louisa Sykes, was born in the village of Clowne in Derbyshire. He went to the Clowne Council School at the age of five. His father, who had been the manager of the Cooperative Society at Bakewell before going into a green grocery business on his own account in the village, often kept him at home on one day a week to help in the business; this practice continued throughout his time at Grammar School. At the age of 11 he went to the Staveley Netherthorpe Grammar School with a county scholarship. He won a county major scholarship and scholarships to Leeds, Liverpool and Sheffield Universities in 1922.

Having chosen Sheffield, Sykes continued to live at home. He gained first class honours in physics in the final BSc, and also was good at sports. He registered for the PhD in physics and was awarded a grant from the Department of Scientific and Industrial Research to study 'The excitation of band-spectra by low-voltage electrons'. However, after one year the Metropolitan Vickers Electrical Company of Trafford Park, Manchester offered him a two-year research studentship to work on the alloys of zirconium. This work had been begun by T.E. Allibone, who moved to the Cavendish Laboratory, Cambridge in 1926 and recommended that Sykes should be appointed to succeed him, working under Desch for a PhD in metallurgy. Having made the transfer, Sykes continued Alibone's work, involving vacuum melting on some alloys of zirconium. Then he went on to make other alloys studying their electrical and magnetic properties. Sykes was awarded the PhD degree in metallurgy in 1928 and was appointed to the Physics Section of the Metropolitan-Vickers Co., where he ultimately became Section Leader and stayed till 1940. In 1930 he married Norah Staton, and they had a son and a daughter.

Sykes continued his metallurgical work at Manchester and collaborated in the design and construction of larger induction melting furnaces. Among the steels and other alloys produced, were some from the iron-aluminium system; Sykes, with colleagues, investigated the properties of alloys of various aluminium contents, and demonstrated the occurrence of lattice ordering. Other work during the 1930s included the production of heating elements, and the powder metallurgy production of cutting materials,

In 1940 Sykes moved to the National Physical Laboratory, Teddington as Superintendent of the Metallurgy Division. He supervised work on armour-piercing shot and was given the additional responsibility of Superintendent of the Technical Ballistics Department at the Armaments Research Establishment at Fort Halstead, Kent. He was appointed Director of the Brown-Firth Research Laboratory, Sheffield, in 1944, where his work included contributions to the special creep resistant steels being developed for high temperature gas turbines. Other projects concerned the properties of low alloy steels, 'hair line' cracks and the role of hydrogen in the failure of steel castings and forgings. In 1951 he became Managing Director of Thomas Firth and John Brown Ltd, in 1962, Deputy Chairman, and in 1964, Chairman; following his retirement in 1967 he remained on the board for another six years.

Sykes played an active role in various spheres of the community, including service as a magistrate and as Pro-Chancellor of Sheffield University. His awards included the CBE in 1956 and a knighthood in 1964. After retirement he devoted himself to the art of gardening. He was an outstanding example of a graduate in pure science who moved into the metallurgical profession and reached its heights.

BMFRS 1983, 29, 553–583. T.E. Allibone, FRS. DNB 1981–93, 388, 389. T.E. Allibone.

MICHAEL SZWARC 1909–; FRS 1966

Warsaw Inst. of Tech; Chemical Engineering, 1933. Hebrew Univ. Jerusalem: PhD 1942. Univ.

of Manchester: Lecturer, 1945–52; PhD, Physical Chemistry, 1947, DSc, 1949. State Univ. College of Environmental Sciences at Syracuse, NY, 1952–82; State Univ. of NY, Distinguished Professor, 1960–79.
Polymers.

DAVID TABOR 1913–; FRS 1963
Regent Street Polytechnic. Universities of London and Cambridge. London, BSc, 1934. Cambridge, PhD, 1939, ScD, 1956. Univ. of Cambridge: Reader in Physics, 1964–73; Professor of Physics, 1973–81; Head of Physics and Chemistry of Solids, Cavendish Laboratory, 1969–81. Royal Medal 1992.
Friction and wear between solids; meechanical and physical aspects of moving parts in contact; adhesion; lubrication. Hardness.

SIR GEOFFREY INGRAM TAYLOR 1886–1975; FRS 1919
Geoffrey Taylor was born at St John's Wood, London; he was the elder of two sons of Edward Ingram Taylor, an artist and Margaret née Boole. He attended a preparatory school in Hampstead, where he was attracted to science. With his brother he attended Christmas lectures at the Royal Institution in 1897–98; this led him, with friends to construct a Wimshurst machine to which he connected a small X-ray bulb to generate X-rays of low intensity and to make X-ray photographs. From1899 to 1905 he was at University College School, where he was particularly interested in mathematics; also he developed his practical skills and designed and built a sailing boat. He was awarded a scholarship by Trinity College, Cambridge and began to read mathematics in 1905. He was placed 22nd in Part I of the Mathematical Tripos in 1907, and then took Part II of the Natural Sciences Tripos in physics in 1908, obtaining first-class honours, and receiving a major scholarship at Trinity.

From the suggestions for research projects made by J.J. Thomson, Taylor selected an investigation of interference fringes formed by light waves of very small intensity. Also he made a theoretical study of the structure of shock waves, for which he was awarded a Smith's Prize at Cambridge. In 1910 he was elected to a Prize Fellowship at Trinity College, which enabled him to spend all his time on research for six years. His first major investigation was carried out, following his appointment in 1911 to the Schuster Readership in Dynamical Meteorology at Cambridge. He made important contributions in turbulent flow, and in 1912 he served as the meteorologist on an expedition to observe the path of icebergs in the North Atlantic subsequent to the sinking of thc Titanic.

Early in the 1914–1918 war Taylor was recruited by the Royal Aircraft Factory at Farnborough; he joined a group of civilian scientists working for the Royal Flying Corps helping to put the design and operation of aeroplanes on a scientific basis. Since the early work was concerned with instruments or devices for use in aircraft, Taylor concluded that his contribution would be more effective if he could obtain direct experience of flying. He was allowed to join the Royal Flying Corps as an ordinary pupil, and when he had obtained pilot's certificate in 1915 he was transferred back to Farnborough. In the same year he carried out an important investigation with A.A. Griffith of the stress distribution in cylindrical shafts under torsion relating to the weakening effect of keyways; they used an analogue technique which depended on the displacement of soap films.

In October 1919 Taylor returned to Cambridge as a Lecturer and Fellow in Mathematics at Trinity College. In the same year Rutherford became Cavendish Professor of Physics, and it appears that he and Taylor soon became friends. Rutherford provided Taylor with a room where, during the next 35 years, he did his experimental work. In 1923 Taylor was appointed to the Royal Society Yarrow Research Professorship, which he held at Cambridge until his retirement in 1952. Walter Thompson, a technician, worked with him for about 40 years.

Taylor, in 1925 married Stephanie Ravenhill; they had no children. Sailing was a common interest and they made several notable voyages including one to the Lofoten Islands; Taylor designed an original design of anchor, which became popular with small boat owners.

In the inter-war years Taylor was an adviser on government committees, and had a close connection with aeronautical research. He made outstanding contributions in two major research areas, namely the turbulent motion of fluids and the deformation of crystalline materials. In the latter area the work included investigations of plastic flow of single crystals and polycrystalline materials, and he was a pioneer in introducing the concept of dislocations.

During the 1939–45 war, Taylor's advice and help were much sought after by military and civil authorities and he was an outstandingly effective consultant on many problems. His work included the detonation of high explosives and propagation of blast waves, and in 1944–5 he visited Los Alamos, New Mexico in connection with the atomic bomb project.

In 1952 he reached retirement age but continued his work in the Cavendish Laboratory. For around 20 years he worked on fluid mechanics exploring a remarkable range of problems, he writing nearly 50 papers and travelling abroad to many conferences and meetings. A severe stroke in 1972 left him seriously physically handicapped.

Geoffrey Taylor's numerous honours included the Copley Medal (1944), a knighthood in 1944, and the Order of Merit in 1969. In a career of more than 60 years, in various fields, including metal physics, he made highly original and important contributions; he was one of the outstanding scientists of the 20th century.

BMFRS 1976, 22, 565–633. G.K. Batchelor FRS. DSB, 18, 896–898. G. Battimelli. DNB 1971–1980, 22, 833–834. G.K. Batchelor FRS.

THOMAS TELFORD 1757–1834; FRS 1827

Thomas Telford was born in Westerkirk, Eskdale, Eastern Dumfriesshire, Scotland. His father, James Telford, a shepherd, died when he was in infancy. His mother, Janet née Jackson earned a little by farming, and when Thomas was old enough he herded cattle. Having attended the parish school at intervals he was apprenticed, at the age of 15, to a stone mason at Langholm. His intelligence, hard work and love of reading attracted the attention of a Langdale lady who gave him access to her library. When he had finished his apprenticeship he worked as a journeyman mason. In 1780 Telford moved to Edinburgh, where he worked as a mason. After two years he moved to London, where he worked squaring and shaping blocks of Portland stone at Somerset House. From 1784 he spent two years in Portsmouth, where he was commissioned to oversee building work. There ensued a move to Shropshire, where he became Surveyor of Public Works; then in 1793 he was appointed sole Agent, Surveyor and Architect for the Ellesmere canal. He also carried out work overseas. Telford remained unmarried. His bridge building activities involved the use of cast iron, in which he was the first and greatest master, and also wrought iron. One of the world's greatest bridge builders, his work included the design and construction, from 1819–1825, of a suspension bridge over the Menai Straits in North Wales. He declined the offer of a knighthood.

DNB. Ref. Ch. 9.18.

SMITHSON TENNANT 1761–1815; FRS 1785

Smithson Tennant was born at Selby, Yorkshire, the son of Calvert Tennant, vicar of Selby, and Mary Daunt Tennant. Following education in the Grammar Schools at Tadcaster and Beverley he studied medicine at Edinburgh, where he attended the lectures of the chemist, Joseph Black. In 1782 he became Pensioner and then Fellow Commoner at Christ's College, Cambridge, where he studied chemistry and botany . He travelled in Denmark and Sweden in 1784, and the following year he was elected to the Royal Society while he was still an undergraduate. He moved to Emmanuel College in 1785, and in the following year graduated MB. During the following years he travelled in Europe, and on his return took up residence in London and in 1796 obtained the MD degree at Cambridge. Tennant became interested in agriculture and farmed land in Somerset, although he lived for most of the year in London. He occupied himself with literary and scientific studies, including his work with Wollaston involving a business partnership. The award of the Copley Medal came in 1804, when he published his discovery of iridium and osmium, and in 1813 he was appointed Professor of Chemistry at Cambridge. His death resulted from an accident in France.

DSB 13, 280, 281. D.C. Goodman. DNB. Ref. Ch. 4.1.

LOUIS JACQUES THÉNARD 1777–1857; FMRS 1824

Born in the village of La Louptiere, Aube, France, Louis Thénard was one of seven children of Etienne Amable Thenard and Cecile née Savourat, peasant farmers. Having received an elementary education from a priest, he went the collège at Sens when he was 11 years old. With the ambition to become a pharmacist he went to Paris at the age of 17 during the Reign of Terror. He attended lectures by Nicholas Vauquelin and was taken into the Vauquelin household as a servant. Later, Thénard was allowed to deputise for Vauquelin as a lecturer, and, in 1798, he was appointed as a Demonstrator at the École Polytechnique. When Vauquelin retired from his chair at the Collège de France in 1804, Thénard succeeded him. In 1814 he married the daughter of Arnould Humblot-Conté. His career involved a number of appointments, including Professorships of Chemistry, Dean of the Paris Faculty of Sciences (1822) and Chancellor of the University of France (1845–1852).

Thénard's wide-ranging research included themes of metallurgical interest. In collaboration with Guy Lussac he prepared potassium by fusing potash with iron filings; also he isolated boron. He discovered hydrogen peroxide (reported in 1818) and investigated the effects of finely divided metals, including the platinum group metals and gold and silver, on this important chemical. With Pierre Dulong he carried out important research on the catalytic effects of platinum and other platinum group metals. Other topics included oxides and the oxidation of metals, and chemical analysis. Thénard was the author of an important and large chemistry textbook *Traite de chimie elementaire.*

Thénard became involved in political life, being elected to the Chamber of Deputies in 1827 and 1830; in 1832 he was nominated as a peer and served in the upper chamber, contributing to debates on technical and scientific matters. Among his many honours, he was given the title of Baron in 1825.

DSB 13, 309–314, M.P. Crosland. Ref. Ch. 4.1.

SIR JOHN MEURIG THOMAS 1932–; FRS 1977

Gwendraith Grammar School. Queen Mary College, London. UKAEA: Scientific Officer, 1957–58. University College of North Wales, Bangor: Assistant Lecturer, 1958–59; Lecturer, 1959–65;

Reader, 1965–69 in Chemistry. University College of Wales, Aberystwyth, 1967–78: Professor and Head of Department of Chemistry. Univ. of Cambridge: King's College, Fellow, 1978–86. Royal Institution of Great Britain and Faraday Research Laboratories, London: Resident Professor of Chemistry, 1986–88; Fullerian Professor of Chemistry, 1988–94. University of Wales: Deputy Pro-chancellor, 1991–94. Univ. of Cambridge: Peterhouse College, Master, 1993–. Knighted 1991.
Solid state and surface Chemistry;catalysis and influence of crystalline imperfections.

ROBERT KEMEYS THOMAS 1941–; FRS 1998
Radley College, Abingdon. Univ. of Oxford: St Johns College, MA, DPhil 1968; Royal Society Pickering Fellow, 1970–75; Merton College Fellow, 1975–78; University College Fellow, 1978–. Lecturer in Physical Chemical Laboratory, 1978–.
Wet interfaces; X-ray and neutron scattering.

SIR BENJAMIN THOMPSON (COUNT RUMFORD) 1753–1814; FRS 1779
Benjamin Thomson was born in Woburn, Massachussets, the only son of Benjamin Thompson and Ruth née Simonds. His education was largely by study with friends and clergymen. He moved to Concord in 1772 where he married a wealthy widow, Sarah Walker Rolfe, and they had one child. He was commissioned as a Major by the Royal Governor of New Hampshire; after the fall of Boston during the American Revolution he fled to England where he advanced in government circles to become Under-Secretary of State for the Colonies. He had a brief military career and retired from the British army as a Colonel at the age of 31. He was knighted by George III, and joined the court of the Elector of Bavaria, where he rose to be head of the army. In 1793 he was made a Count of the Holy Roman Empire, taking the name Count Rumford. He returned to London briefly around 1800 but settled later in Paris where he married the widow of Lavoisier; this marriage, as did his first, ended in separation. During his career he carried out some work in science and technology, for example, he measured the heat generated in boring cannon. Also of great importance for the world of science he founded the Royal Institution in London. Rumford received the Copley Medal in 1792 and also was the first recipient of the Rumford Medal when it was instituted in 1800.
DSB 13, 350–352. S.C. Brown. DNB.

HANS PETER JÖRGEN JULIUS THOMSEN 1826–1909; FMRS 1902
Julius Thomsen was born in Copenhagen, the son of Thomas Thomsen, a bank auditor and Jensine Friederike née Lund. He worked in the chemistry laboratory of the university of Copenhagen and studied successfully to enter the Polytekniske Laereanstalt, where he obtained an MSc in 1843 in applied natural sciences. Between 1847 and 1856 he held appointments in the Polytekniske and did a study tour in France and Germany. Other appointments later in his career included Director of Weights and Measures at Copenhagen (1856–59) and teaching physics at the Danish Military College (1859–66). Between 1864 and 1901 he held academic posts at the University (including Professor of Chemistry), while also serving as Professor of Chemistry at the Polytekniske. He also held the offices of Rector of the University, Principal of the Polytekniske, and President of the Royal Danish Academy of Sciences and Letters (1888–1909). In his technical work he developed a method for preparing soda from cryolite but his main activities were in pure science. Thomsen conducted over a period of 30 years a programme of thermochemical studies, during which he personally carried out more than 3,500 calorimetric measurements; the data were later

assembled in a four-volume work. His work also included electrochemical research. He was awarded the Davy Medal in 1883.
PRS 1910–11, A 84, xxiii–xxix. TET. WCU. DSB 13, 358, 359. S. Veibel.

SIR GEORGE PAGET THOMSON 1892–1975; FRS 1930
George Paget Thomson was born in Cambridge, the son of Joseph John Thomson Rose Elizabeth née Paget. Following education at King's College Choir School and the Perse School, Cambridge, he entered Trinity College where he obtained first class honours in mathematics and physics and was Smith's prizeman in 1916. In 1913 he became a Fellow and Lecturer in Mathematics at Corpus Christi College. He saw war service from 1914–1919, first with the Queen's Regiment in France, and then on secondment to the Royal Flying Corps, based at Farnborough, Hampshire, working on aeronautical problems. Having returned to Cambridge in 1919 he was appointed, in 1922, to the Chair of Natural Philosophy at Aberdeen. In 1924 he married Kathleen Buchanan Smith and they had two sons and two daughters.

In 1925 and 1927 Thomson demonstrated that a beam of electrons passing through a thin gold foil was diffracted. He became Professor of Physics at Imperial College, London in 1930, and in 1937 shared the Nobel Prize for Physics with Clinton J. Davisson for their work on electron diffraction. From around 1932 Thomson had become interested in nuclear physics; in 1939 he alerted the Air Ministry to the feasibility of a uranium bomb and he was involved in the wartime atomic energy project in Britain. He was knighted in 1943. He received the Hughes Medal in 1939 and a Royal Medal in 1949 .
BMFRS 1977, 23, 528–556, P.B. Moon FRS. DSB Suppl. 18, 908–912. J. Hendry. DNB 1971–1980, 842. P.B. Moon FRS.

JAMES THOMSON 1822–1892; FRS 1877
James Thomson was born in Belfast, the eldest son of James Thomson and Margaret née Gardiner; James's elder brother was William (later Lord Kelvin). His early education was supervised by his father, and he attended school for only a short period. At the age of 10 he began to attend Glasgow University where he matriculated in 1834 and graduated MA in 1839, with honours in mathematics and natural philosophy. In 1841 he spent six months in the engineering department of a spinning mill in Glasgow, and subsequently became a pupil at the Horsley Ironworks at Tipton, Staffordshire and in the works of Fairbairn and Co at Millwall. He settled as a civil engineer in Belfast in 1851, and two years later became resident engineer to the water commissioners. In 1853 he married Elizabeth Hancock and they had one son and two daughters. Thomson was appointed Professor of Civil Engineering at Queen's College, Belfast in 1857; then in 1873 he moved to a similar post at Glasgow University. His research included fluids, pumps and turbines. Also he worked on the strength of materials, including the elasticity and strength of spiral springs and bars subjected to torsion. He also worked on the elasticity of ice and crystallisation phenomena. He was interested in safety aspects of engineering structures e.g. boilers, and the principles on which their strength could be estimated and proved.
PRS 1893, 53, i–x, JTB. DNB.

SIR JOSEPH JOHN THOMSON 1856–1940; FRS 1884
Joseph Thomson was born at Cheetham Hill, near Manchester, the elder son of a bookseller, Joseph James Thomson and Emma née Swindells. At first he was intended for an engineering career and his father sent him to Owens College when he was 14. Following his father's death

two years later, the family could not afford the premium for training as an engineer, but he continued at the college studying science. He won a scholarship to Trinity College, Cambridge, where he commenced his studies in the Mathematical Tripos in 1876. He graduated as Second Wrangler in 1880, was awarded a fellowship by Trinity College in 1881, and remained a member of the college for the rest of his life culminating in being appointed Master in 1918. He carried out work in the Cavendish Laboratory and his appointment to the Cavendish Chair of Physics took place in 1884. In 1890 he married Rose Elizabeth Paget and they had a son, George Paget Thomson, and a daughter.

Of outstanding importance, in the period following the discovery of X-rays, was Thomson's discovery of the electron in 1897. Great interest had been stimulated in the cathode rays that produce X-rays; Thomson demonstrated the deflection of these waves by magnetic and electric fields, determined the ratio of electric charge to mass of the rays, and established that the rays were negatively charged particles, much smaller than atoms. Thomson's awards included medals of the Royal Society: Copley (1914), Royal (1894); Hughes (1902), and he gave the Bakerian Lectures in 1887 and 1913. He received the Nobel Prize for Physics in 1906. He was knighted in 1908, and received the Order of Merit in 1912.

ONFRS 1939–41 3, 587–609. Rayleigh. DSB 13, 362–372. J.L. Heilbron. DNB 1931–1940, 857–863. O.W. Richardson. *Physics World* 1997, **10**(4), 33–36. G. Squires.

THOMAS THOMSON 1773–1852; FRS 1811

Thomas Thomson was born in Crieff, Scotland, the son of John Thomson and Elizabeth née Ewan. He studied classics, mathematics and natural philosophy at the University of St Andrews. In 1791, considering medicine as a career, he attended various lecture courses at Edinburgh University, and graduated MD in 1799. However, in 1795–96 he decided to take up chemistry. From 1796–1800 he wrote articles for the *Encylopaedia Brittanica* on chemistry and mineralogy. Subsequently his occupations included lecturing, text book writing and research. He was the author of a history of the Royal Society (1812), which included aspects of mining and metallurgical interest. In 1816 he married Agnes Colquhoun and they had a son and a daughter. He was appointed Regius Professor of Chemistry at the University of Glasgow in 1818. His research included co-operation with Humphry Davy on temper colours.

DSB 13, 372–374. J.B. Morrell. DNB. Ref. Ch. 8.5.

SIR WILLIAM THOMSON (BARON KELVIN OF LARGS) 1824–1907; FRS 1851

William Thomson was born in Belfast, the second son and fourth child of James Thomson and Margaret née Gardiner. His father was Professor of Mathematics at Belfast and became Professor of Mathematics at Glasgow in 1832. William matriculated at the University of Glasgow at the age of 10, and made his mark in physics and mathematics in his studies at the university. He did not take a degree at Glasgow but entered Peterhouse College, Cambridge in 1841. He graduated second wrangler and first Smith's prizeman in 1845, and then spent a few months in Paris working in Regnault's laboratory. At the age of 21 he had published a dozen original papers, and in 1845 he was made a Fellow of Peterhouse College. He was appointed as Professor of Natural Philosophy at Glasgow in 1846, a post which he held for 53 years. He was Chancellor of Glasgow University from 1904 to 1907. In 1852 he married a second cousin, Margaret Crum; following her death in 1870 he married Frances Anna Blandy in 1874; there were no children.

Thomson's research included the field of thermodynamics, the strength of materials including elasticity, electricity and magnetism. The utilisation of science for practical objectives was his

ambition. His reputation and interests brought him to the attention of a consortium of British industrialists who in the mid 1850s proposed the laying of a submarine telegraph cable between Ireland and Newfoundland; his involvement in the project led to his being knighted in 1866. His main researches were concerned with electromagnetism and thermodynamics. In the field of metals, he carried out research on thermoelectric, thermoelastic and thermomagnetic behaviour, which formed the subject of his Bakerian Lecture in 1856. His work on the homogeneous partitioning of space later became of interest in relation to grain structures of metallic crystalline aggregates. He was the author of over 700 publications. In the Royal Society he received a Royal Medal (1856) and the Copley Medal (1883), and was President from 1890–94. He was made a peer in 1892, received thc Order of Merit in 1902 and became a Privy Councillor.
PRS 1908, A, 81, iii–lxxvi. JL. DSB 13, 374–388. J.Z. Buchwald. DNB 2nd Suppl. III, 508–517. SPT.

SIR THOMAS EDWARD THORPE 1845–1925; FRS 1876
Thomas Thorpe was born at Harphuray, near Manchester; he was the son of George Thorpe, a cotton merchant, and Mary née Wilde. He attended Hulme Grammar School and entered Owens College, Manchester in 1863 under the guidance of Henry Roscoe. In 1870 he married Caroline Emma Watts; they had no children. In the same year he went to Heidelberg to study with Bunsen; he obtained a PhD and then spent a period in Bonn. Thorpe held a series of professorships of chemistry at the Andersonian College, Glasgow (1870–74), the Yorkshire College of Science, Leeds (1874–85), and the Royal College of Science, London (1885–94 and 1909–1912). From 1894–1909 he held an appointment as the first Government Chemist. In his work on inorganic chemistry he shared in Roscoe's work on vanadium; also, with students he made accurate determinations of atomic weights of several metals. Thorpe received the CB in 1900 and was knighted in 1909. In the Royal Society he received a Royal Medal in 1889, was joint Bakerian Lecturer in 1889, and was Foreign Secretary from 1899–1903.
PRS 1925, A, 109, xviii–xxiv. AEHT. DSB 13, 389, 390. S.J. Kopperl. DNB 1922–1930, 842, 843. J.C. Philip.

SIR WILLIAM AUGUSTUS TILDEN 1842–1926; FRS 1880
William Tilden, who was born in London, was the eldest son of Augustus Tilden. He began an apprenticeship to a pharmacist in 1857, and served the last year of this at the Royal College of Science. Beginnning in 1863 he was a Demonstrator at the Pharmaceutical Society for nine years, while obtaining a BSc of London University (1868); in 1871 he was awarded a DSc. In 1869 he married Charlotte Pither Bush; following her death he married Julia Mary Ramie in 1907. Tilden took a post as a science teacher at Clifton College, Bristol, and continued his research. In 1880 he became the first Professor of Chemistry at Mason College, Birmingham, and in 1894 moved to the Royal College of Science, London as Professor of Chemistry, where he remained until his retirement in 1909. His main research was in organic chemistry, but later in his career he worked on the relationship between specific heat and atomic weight. He was Bakerian Lecturer in 1900 and was awarded the Davy Medal in 1905.
PRS 1928, A, 117, i–v. JCP. DSB 13, 410, 411. G.R. van Hecke.

STEPHEN PROKOFIEVITCH TIMOSHENKO 1878–1972; FMRS 1944
Stephen Timoshenko was born in the village of Shpotovka in Ukrania. He graduated from the real-Gymnasium in the town of Romni in 1896 and entered the Institute of Ways of Communication to study engineering; following graduation in 1901 he did a year of compulsory military

training. In 1902 he married Alexandra Archangelskaya, a medical school student. His first job was as an assistant at the mechanics laboratory of the Institute, where he made experiments on the strength of rails, structural steel and cement. From 1903–1906 he was an instructor at the St Petersburg Polytechnical Institute, where he began creative scientific work utilising mathematics to solve engineering problems. He spent the summers in Germany, including work with Ludwig Prandtl on beams; he also applied energy methods to stability problems. In 1906 he was selected in open competition for a professorship in strength of materials at Kiev Polytechnic, and three years later was elected Dean of the Engineering School. The political situation in the country led to difficulties resulting in Timoshenko and the other Deans being dismissed. After two years as a part-time teaching assistant he obtained a post as consultant to shipyards relating to structural strength. He was able to travel as a result of a prize that he obtained, and this involved a visit to the mathematical congress in Cambridge. In 1913 he was reinstated as a professor at the Institute of Ways of Communication, where he continued his scientific work, solving new problems in the theory of elasticity. After 1914 conditions worsened for him in Russia, and he left for Yugoslavia to take the Chair of Strength of Materials in Zagreb.

In 1992 Timoshenko accepted the offer of a post with the Vibration Specialty Company in Philadelphia in the USA. A year later he joined the Westinghouse Company in Pittsburgh in their research group in mechanics. Among his projects he developed photoelastic equipment for stuying stress distributions, and he also worked with a new pendulum hardness tester. An important project in the Company was a theoretical and experimental study of rails, and he led a group of engineers on this work; he also began an activity teaching elasticity theory to young company engineers. In 1928 Timoshenko went to the University of Michigan where he proved a very popular teacher, and his scientific work there was largely confined to the writing of engineering textbooks. e.g *Strength of Materials and Theory of Elasticity*. A move occurred in 1936 to Stanford University where he continued similar activity to that at Michigan, including publishing further books. Although formal retirement occurred in 1944 Timoshenko continued some teaching until 1955. In 1965 he moved to Germany to live with his daughter.

Timoshenko was an outstanding scientist, distinguished engineer and inspiring teacher, whose work advanced mechanics as a science and promoted its application to practical engineering problems.

BMFRS 1973, 19, 679–694. E.H. Mansfield FRS and D.H. Young.

ISAAC TODHUNTER 1820–1884; FRS 1862

Isaac Todhunter was the second son of George Todhunter and Mary née Hume. He was born at Rye, Sussex, and having attended school in Hastings he became an assistant master at a school in Peckham and attended evening classes at University College, London. He obtained a London BA degree in 1842, and proceeded to obtain an MA, with a gold medal award. Concurrently he taught at a school in Wimbledon. In 1844 he entered St John's College, Cambridge, where, in 1848 he obtained the senior wranglership and the Smith's prize. Having become a Fellow of the College in 1849 he then lectured and tutored, and compiled an important mathematical treatise. In 1864, he resigned his fellowship, when he married Louisa Anna Maria Davies. He served on the Council of the Royal Society from 1871–1873, and received the Adams Prize in 1871. His publications, included the field of mechanics, and of particular interest in the metallurgical field was the very detailed and substantial work *A History of Elasticity and of the Strength of Materials*, published posthumously in 1886. This work includes much information on the research relevant to the development of understanding of the mechanical properties of metals, and of practi-

cal engineering applications.
PRS 1884, 37, xvii–xxxii. EJR. DSB 13, 426–428. M.E. Baron. DNB.

SAMUEL TOLANSKY 1907–1973; FRS 1952

Samuel Tolansky was born in Newcastle upon Tyne, the second child in a family of two boys and two girls. His father Barnet Turlausky (the surname was changed) a tailor, married to Moise Chaiet, of Lithuanian Jewish origin had migrated from Odessa at about the turn of the century. Having attended Snow Street Primary School his secondary education was at Rutherford College from 1919–25; scholarships enabled him to enter Armstrong College, then part of Durham University. He obtained first class honours and then spent a year in the education department. From 1929–31 Tolansky studied spectroscopy at Berlin with a travelling scholarship; also two years were spent with an 1851 exhibitition at Imperial College. He married Ottilie Pinkasovich in 1935, and there was a son and daughter from the marriage. In 1934 he had become an Assistant Lecturer under W.L. Bragg at Manchester University where there followed a series of promotions: Lecturer (1937); Senior Lecturer (1945); Reader (1946). During the war he worked at Manchester on an aspect of the Tube Alloys project (atomic energy). In 1947 Tolansky became Professor of Physics at Royal Holloway College, London University. His research included the use of multibeam interference of light to investigate surface microtopography. He examined a range of materials, metals and diamonds, revealing important surface features at high resolution, and became a leader in this field..
BMFRS 1974, 20, 429–455. R.W.D. Ditchburn FRS and G.D. Rochester FRS. DNB 1901–80 851, 852. R.W.D. Ditchburn.

SIR JOHN SEALY TOWNSEND 1868–1957; FRS 1903

John Townsend was born in Galway, Ireland, the second son of Edward and Judith Townsend. He was educated at Corrig School, and at Trinity College, Dublin, which he entered in 1885 and where he studied mathematics, mathematical physics and experimental science. In 1888 he won a scholarship in mathematics and in 1890 he graduated BA.For the next four years he was fellowship prizeman, and in 1895 he entered Trinity College, Cambridge as an advanced student. He did research with J.J. Thomson and in 1898 he was elected Clerk Maxwell scholar. In 1899 he was elected a fellow of Trinity College, and was appointed as an Asistant Demonstrator. Townsend became the first holder of the Wykeham Chair of Experimental Physics at Oxford in 1900 and held a fellowship at New College. In 1911 he married Mary Georgiana Lambert and they had two sons. During the 1914–18 war Townsend carried out wireless research in the Royal Naval Air Service, with the rank of Major. Townsend's research at Cambridge included investigations of the electron properties of gases and secondary X-rays. At Oxford he introduced the concept of an 'electron gas'.
BMFRS 1957, 3, 257–272. A. von Engel. DSB 13, 445–447. Thaddeus J. Trenn. DNB 1951–60, 983–985. A. von Engel.

MORRIS WILLIAM TRAVERS 1872–1961; FRS 1904

Morris Travers was born in Kensington, London, the second son of William Travers, a doctor surgeon, and Anne née Pocock. Having attended schools at Woking and Ramsgate he entered Blundells School at Tiverton at the age of 12. In 1889 he entered University College London and obtained an honours BSc degree in chemistry two years later. He initially specialised in organic chemistry and spent two years at Nancy. He returned to University College to William Ramsay's

laboratory as a demonstrator. In 1895 he worked with Ramsay on the discovery of helium and subsequently collaborated with him in the discovery of krypton, neon and xenon. Travers became Assistant Professor of Chemistry at University College in 1898. He moved to University College, Bristol in 1904 as Professor of Chemistry. In 1906 he went to India as Director of the proposed Institute of Science at Bangalore, which came into being in 1911. He had married Dorothy Gray in 1909 and they had a son and daughter. Returning to England in 1914, he joined Baird and Tatlock at the beginning of the war and made important contributions in the building of glass furnaces and in the production of scientific glass ware. He collaborated in the founding of Travers and Clark in 1920, a firm specialising in the construction of glass furnaces and plant for coal gasification. Travers became an Honorary Professor and Research Fellow at Bristol and Reader in Applied Chemistry in 1927. In 1937 he retired from his university post, but continued research. During World War II he worked as a consultant on explosives for the Armament Research Establishment of the Ministry of Supply. Travers took up a post with the National Smelting Company, Ltd at Avonmouth in 1945.
BMFRS 1963, 9, 301–313. C.E.H. Bawn. DNB 1961–70, 1014, 1015. C.E.H. Bawn.

MARTIN TRIEWALD 1691–1747; FRS 1731
Martin Triewald was born in Sweden. Intended for a career in the law, he came to England for this purpose. However, he was employed by the proprietor of some coal mines near Newcastle to superintend his machinery and works. In 1726 he returned to Sweden where he constructed a steam engine and lectured on philosophy. Triewald was made Director of Machinery by the king of Sweden; also he became Captain of Engineers and Inspector of Fortifications, in which connection he invented various machines.
Bulloch's Roll. Gen. Ref. 20. v. IV.

EDWARD TROUGHTON 1753–1835; FRS 1810
Edward Troughton was born in Corney, Northumberland, the third son of Francis Troughton, a 'husbandman'. In 1770 he was apprenticed in London to his eldest brother, John, who had set up as a 'mechanician'. When his brother died he carried on the business. Troughton remained unmarried. He became the finest scientific instrument maker of his day in England, including astronomical instruments, and made a wide range of inventions. He was the recipient of the Copley Medal in 1809.
PRS 1834–35, 3, 355. DSB 13, 470, 471. R.S. Webster. DNB.

WILLIAM ERNEST STEPHEN TURNER 1881–1963; FRS 1938
William Turner was born in Wednesbury, Staffordshire into a working class family; he was the eldest son and the second of seven children of William George Turner and Emma Blanche née Gardner. Having attended a school at Smethwick he obtained a scholarship to King Edward VI Grammar School at Birmingham. With a further scholarship he studied at Mason University College, and was awarded a first class honours BSc (external) by the University of London in 1902. He was then a research scholar at Birmingham, obtaining an MSc in 1904. In 1904 Turner became a Junior Demonstrator and Lecturer at University College, Sheffield. His interest in the application of science to industrial problems was shown by the course in physical chemistry that he gave to metallurgy students. In 1908 an honours course in physical chemistry was set up and began to supervise a group of research students. Also in 1908 Turner married Mary Isobel Marshall in 1908 and they had two sons and two daughters; following his wife's death in 1939 he married

Annie Helen Nairn Monro in 1943. In 1911 he was awarded the DSc degree of London University.

Realising that the local industries might welcome help from university scientists Turner proposed the formation of a technical advisory committee; this came about in 1914, with him as secretary. Initially advice was given on problems in the field of metallurgy, but glass plants sought advice, and he drew up a report on the glass industry of Yorkshire. A department with the title of glass manufacture was set up in the University of Sheffield in 1915 but soon Turner introduced the term 'glass technology'. In 1916 he founded the Society of Glass Technology. Turner became a professor and carried out extensive research covering fields such as the resistance of chemical glass to reagents, and furnace design. Requirements of other industries led to topics such as the design of glass-metal seals for the electrical industry. Following his retirement in 1945, Turner set up a consulting firm with a friend. Although physically handicapped by the effects of poliomyelitis in childhood he was a vigorous walker, even in the Alps. Among his honours was the award of the OBE in 1918.

BMFRS 1964, 10, 325–355. R.W. Douglas. DSB 13, 504–05. F. Greenaway. DNB 1961–70, 1023–24. R.W. Douglas.

JOHN TYNDALL 1820–1893; FRS 1852

John Tyndall was born at Leighton Bridge, County Carlow, Ireland. Having worked on the ordnance survey for Ireland and England, he spent three years as a railway engineer. He studied chemistry with Bunsen at Marburg in 1848, obtaining a PhD in 1850; in the following year he studied in Magnus's laboratory in Berlin, before returning to England. Tyndall was appointed Professor of Natural Philosophy at the Royal Institution, London in 1867, where his research included heat, the behaviour of crystalline bodies in a magnetic field, and the scattering of light by small particles suspended in air. In 1876 he married Louisa Hamilton. In the Royal Society he delivered the Bakerian Lecture in 1855 and received the Rumford Medal in 1864.

PRS 1894, 55, xviii–xxxiv. EF. DNB.

ALFRED RENÉ JEAN PAUL UBBELOHDE 1907–1988; FRS 1951

Paul Ubbelohde was born in Antwerp; he was the third son in the family of four sons and one daughter of Francis Christian Ubbelohde and Angele née Verspreeuwen. His father was a merchant who, with his family, moved to London in 1914. Paul contracted polymyelitis as a child and as a result had a paralysed right arm. He was educated at Richmond County School and Colet Court and then for six years at St Paul's School. He was a scholar at Christ Church, Oxford from 1926–1930, graduating with first class honours in chemistry. He became a naturalised British citizen in his mid-twenties. Following his graduation he began research in the Department of Thermodynamics at Oxford. He also took the BSc special chemistry degree of London University as an external student in 1928, obtaining first class honours; he spent the year 1931–32 at Göttingen, and from 1933–35 he held a senior research award from the Department of Scientific Research. From 1936–1940 Ubbelohde held the Dewar Fellowship at the Royal Institution, London, where he used X-ray diffraction to study phase transitions. Ubbelohde was Principal Experimental Officer at the Ministry of Supply establishment in Swansea carrying out work on detonations and explosives from 1940–45; he was awarded a DSc, Oxford, in 1941. He was appointed Professor of Chemistry and Head of the Department of Chemistry at Queen's University, Belfast. In 1954 he became as Professor of Thermodynamics to the Department of Chemical Engineering and Chemical Technology at Imperial College, London; from 1961 until his retire-

ment in 1975 he was Head of the Department. Following his retirement he became a Senior Research Fellow. His research ranged widely, including investigations of carbon, graphite and its compounds, phase transitions and ionic melts. He remained unmarried. Ubbelohde was elected FREng as a founder member; he was awarded the CBE in 1963.
BMFRS 1990, 35, 381–402. F.J. Weinberg FRS.

ANTONIA DE ULLOA DE LA TORRE GIRAL 1716–1795; FRS 1746
Antonia de Ulloa, born in Seville, Spain, became a mariner, astronomer and mathematician. He was commissioned to accompany an expedition (1736–1745) sent by the Paris Academy of Sciences to America to measure a meridian of latitude. The ship was captured by the English en route back to Spain, and during his subsequent stay in England he was befriended by Martin Folkes then President of the Royal Society, and was elected as a Fellow. After his return to Spain he published an account of the expedition which included a description of platina found in Colombia in 1736. Ulloa was sent by Ferdinand VI on a mission throught Europe to learn about recent scientific discoveries. His subsequent career included an appointment in 1758 as General Manager of the Mines of Huancavelica, Peru, and later high positions in the Navy; he was also the first Spanish Governor of Louisiana (1766–1768).
DSB 13, 530, 531. J. Vernet.

WILLIAM CAWTHORNE UNWIN 1838–1933; FRS 1886
William Unwin was the eldest child of the Rev. William Jordan Unwin, a congregational minister and Eliza née Davey. He was born at Woodbridge, Suffolk, and educated at the City of London School (1848–1854) and then at New College, St John's Wood, where he spent a year. He came to know the engineer, William Fairbairn leading to a friendship that continued until Fairbairn's death. Fairbairn, then 67 years of age, took Unwin, at the age of 18, into his employment in Manchester. Some of Unwin's earliest experience in practical engineering were concerned with the testing of materials. He helped Fairbairn in his work on the effects of impact, vibrating action and long continued cyclic change of load on the strength of wrought iron girders. Unwin obtained the BSc degree of London University in 1861 by study in the evenings.

There followed industrial appointments as Works Manager in 1862 of Messr Williamson Bros. of Kendal manufacturing water tubing and millwork, and as Manager of the Engine Department at Fairbairn Engineering in 1866. Unwin sought an academic career and gave lectures to Royal Engineer Officers at Chatham. He was appointed Instructor at the School of Naval Archictecture and Marine Engineering in South Kensington in 1861. Other academic appointments ensued including his appointment as Professor of Civil and Mechanical Engineering at the Institute of City and Guilds of London, which was incorporated into Imperial College in 1900. Unwin's research in his engineering laboratories at the College continued to involve mechanical testing and properties of materials; it included the publication of a treatise entitled *The Testing of Materials for Construction* (1883), and investigations on steels and copper-based alloys. From 1896 to 1900 he served on a Board of Trade committee enquiring into the loss of strength of steel rails caused by prolonged use.

Unwin remained unmarried throughout his life; for fifty years he lived in the same London flat, much of this period with his only sister. His hobbies were photography, fishing and climbing. During his long retirement from professional duties he took a lively interest in the work of his old students.
ONFRS 1934–35, 1, 167–178. J.S. Wilson. DNB 1931–1940, 878, 879. E.G. Walker.

HAROLD CLAYTON UREY 1893–1981; FMRS 1947

Harold Urey was born in Walkerton, Indiana. After graduating from high school in 1911 he taught for three years in country schools. In 1914 he entered the University of Montana, Missoula, obtaining a BS degree, majoring in biology with a minor in chemistry, in 1917. He then went as Research Chemist to the Barrett Chemical Company in Philadelphia. From 1919–21 he was an Instructor in Chemistry at Montana State University before entering the University of California at Berkeley, where he worked on heat capacities and entropies of gases, obtaining a PhD in 1923. He went to the University of Copenhagen for a year as an American-Scandinavian Research Fellow and studied atomic physics with Bohr. Then for five years he was an Associate in Chemistry at John Hopkins University in Baltimore, followed by appointments at Columbia University, New York City as Associate Professor of Chemistry (1929–33) and Ernest Kempton Fellow (1933–36); he was appointed Professor of Chemistry in 1934. In 1931 Urey with colleagues, discovered deuterium (heavy hydrogen). Other research concerned liquid-solid ion exchange systems for the separation of isotopes,such as those of carbon, nitrogen oxygen and sulphur; his group provided the basic information for the separation of the fissionable uranium-235 isotope from uranium-238.

During World War II Urey was Director of Research of the Substitute Alloys Materials Laboratories at Columbia, which became part of the Manhattan project to develop the atomic bomb. In 1940 Urey as the recognised world leader in isotope separation began investigations of various methods for the separation of the uranium isotopes: thermal diffusion, chemical separation and centrifugal fractionation column methods. Urey had broad responsibilities in the whole research programme on uranium. In 1945 he was appointed Professor of Chemistry at the Institute of Nuclear Studies at the University of Chicago. He became Professor-at-large of Chemistry at the University of California at la Jolla in 1958. Urey's research included the subject of the evolution of the earth; he proposed a theory that the earth had formed from small, planetar particles, mainly metallic. Urey received the Nobel Prize for Chemistry in 1934. He married Frieda Daum and they had three daughters and one son.

BMFRS 1983, 29, 623–659. K.P. Cohen, S.K. Runcorn, H.E. Suess and H.G. Thode FRS. DSB, 18, 943–948. Joseph N. Tutarewicz.

JOHN HASBROUCK VAN VLECK 1899–1980; FMRS 1967

John Van Vleck was a tenth generation American, originating from the Van Vleck family in the Netherlands. He was the only child of Edward Burr Van Vleck and Hester Laurence née Raymond. His father was a faculty member at Wesleyan University at the time when John, was born at Middletown. In 1906 his father moved to Madison as Professor of Mathematics at the University. John's education was mainly in Madison, where he also attended the University. He obtained an AB in physics in 1920 and then an AM at Harvard in 1921. His doctoral thesis at Harvard submitted in 1922 concerned quantum theory. In 1923 he became Assistant Professor at the University of Minnesota, being promoted to Associate Professor in 1926 and Full Professor in 1927. He married Abigail Pearson in 1927; there were no children from the marriage. They moved to Madison in 1928, and he became a Professor at the University of Wisconsin. He moved to Harvard in 1934, where after a year as Associate Professor, he became a Full Professor; from 1951–1969 he held the Hollis Chair of Mathematics and Natural Philosophy. During World War II his work included a period at the Radiation Laboratory at MIT, where he was concerned with radio countermeasures.

During his time at Minnesota Van Vleck was the first to apply quantum mechanics to the

theory of electric and magnetic susceptibilities Then from 1928–1941 at Wisconsin and Harvard he established the detailed theory of paramagnetism, including studies of rare earths, on a firm quantum mechanical basis. In this work he developed all the fundamental formulae for the interactions between magnetic electrons and the surrounding ions in the crystal attice. In 1932 he published *The theory of electric and magnetic susceptibilities*. At Harvard his work spanned widely the field of magnetism, including ferromagnetism and he is considered to be the father of modern magnetism. He was awarded the Nobel Prize for Physics in 1977, jointly with P. Anderson and M.F. Mott.
BMFRS 1982, 28, 627–665. B. Bleaney FRS.

JACOBUS HENRICK VAN'T HOFF 1852–1911; FMRS 1897

Jacobus van't Hoff was born in Rotterdam, the third of seven children of Jacobus Henricus van't Hoff, a physician, and Alida Jacoba née Kolff. Having studied technology at Delft from 1869–1871, he entered the University of Leiden, where he studied mainly mathematics. His subsequent career included study in Paris and a PhD degree in the Netherlands (1874). He held lecturing posts at the state veterinerary school in Utrecht, and at the University of Amsterdam, the latter in theoretical and physical chemistry. In the period beginning 1878 at Amsterdam he was Professor of Chemistry, Mineralogy and Geology, and Head of the Department of Chemistry. He married Johanna Francisca Mees in 1878 and they had two daughters and two sons. In 1896 he became a Professor at the Prussian Academy of Sciences in Berlin.

Van't Hoff's research covered chemical thermodynamics and affinities. He showed that the laws of thermodynamics were valid not only for gases but for dilute solutions, and proposed a theory of reaction rates. The honours that he received included the Davy Medal (1893) and the Nobel Prize for Chemistry (1901) for his work on osmotic pressure and the laws of chemical thermodynamics.
PRS 1911–12, A, 86, xxxix–xliii. FGD. DSB 13, 575–581. H.A.M. Snelders.

CROMWELL FLEETWOOD VARLEY 1828–1883; FRS 1871

Cromwell Varley, the only son of Cornelius Varley was named after two of his ancestors, Oliver Cromwell and General Fleetwood. Born in Kentish Town, London, he was educated at St Saviour's School, Southwark. After leaving school he studied telegraphy and, through the influence of W.F. Cooke he was employed in 1846 by the Electric and International Telegraphy Company, where he remained until 1868. He then retired into private life but continued to produce inventions. His contributions included a method of localising faults in submarine cables and an important role in connection with the Atlantic cable. He was twice married and had two sons and two daughters by this first wife.
DNB.

NICOLAS LOUIS VAUQUELIN 1763–1829; FMRS 1823

Nicolas Louis Vauquelin, the son of Nicolas Vauquelin, an estate manager, and Catherine le Chartier, was born in St Andre d'Hebertot, Normandy. He became an assistant to a pharmacist in Rouen when about 14, and then moved to Paris to work for a pharmacist. In around 1784 he became assistant to the chemist Fourcroy, and later his collaborator. For a period, after leaving Fourcroy's laboratory, he practiced pharmacy; then in 1793 he was sent by the government to organise the production of saltpetre which was urgently required for gunpowder. On returning to Paris the following year, he worked again with Fourcroy, as Assistant Professor of Chemistry at the institute which later became the École Polytechnique. During his career he held various academic appointments, including Professor of Assaying at the École des Mines, Professor of Chemistry at the Collége de France (1801), Director of the École de Pharmacie (1803) and Professor of Applied Chemistry at the Muséum d'Histoire Naturelle (1804) where in 1811 he succeeded Fourcroy as Professor of Chemistry. Other appointments included Official Assayer of Precious Metals for Paris, and Inspector of Mines. In 1828 he was elected to the Parliament. He remained unmarried.

Vauquelin's research included numerous analyses of minerals and the discovery of chromium. He also carried out work on platinum and other platinum group metals, including commercial enterprises in refining and fabrication of platinum.
DSB 13, 596–598. W.A. Smeaton.

CHARLES BLACKER VIGNOLES 1793–1875; FRS 1855
Charles Vignolles was born in Woodbrook, County Wexford, Ireland. His father, Charles Henry Vignoles, a descendant of a Huguenot family, was an army officer; his father, and also his mother, Camilla née Hutton, died soon after Charles's birth, and he was placed in the care of his grandfather in England. After commencing training for the law, he went to Sandhurst in 1810, which led to his embarking on a military career in various regiments and countries; he maintained a connection with the army until 1833, while also being involved in civil engineering. He married Mary Griffiths in 1817 and, after her death he remarried; there were four sons. In his engineering career Vignoles worked in Great Britain and abroad, including railway construction projects; in 1837 he introduced the flat-bottomed rail. His work involved association with John Rennie and his brother George, and also with George Stephenson. In 1841 Vignoles became the first Professor of Civil Engineering at University College London.
DNB.

JOHN HENRY VIVIAN 1785–1855; FRS 1823
John Vivian was born in Truro, Cornwall, the second son of John Vivian and Betsy née Cranch. His father was a Cornish mine owner who became involved in copper smelting in the county. In 1810 John and his elder brother brother Richard partnered their father in establishing the Hafod copper smelting works at Swansea, South Wales and in 1823 he published a paper dealing with the extraction and refining techniques used. He was Member of Parliament for Swansea from 1832–1855, and was also a Justice of the Peace and a Major in the Royal Stannary Artillery. In 1816 he married Sarah Jones and they had a son Henry Hussey (later Sir Henry) who managed the Hafod works and played a major role in extraction metallurgy in Swansea, concerned with copper, zinc and other non-ferrous metals.
Bulloch's Roll. Gen. Ref. 23.

SIR RICHARD HUSSEY VIVIAN (FIRST BARON VIVIAN) 1775–1842; FRS 1841
Richard Vivian was the eldest son of John Vivian and Betsy née Cranch. He was born in Truro where he was educated at the Grammar School, before entering Harrow. He spent two terms at Exeter College, Oxford, and then went to France in 1791 to study the language. Having returned to England he became an Ensign in the Infantry and embarked on a distinguished military career with a series of promotions reaching the rank of Lieutenant-General in 1830; he saw active service in Flanders, in Spain and at Waterloo and was severely wounded. He was a partner with his father in the Hafod copper works. For various periods he served as Member of Parliament for

Truro, Windsor and East Cornwall. He also served as Major-General of the Ordnance, in which post he recognised the importance of science for the national welfare. He was appointed KCB (1815), Bt (1828), Baron (1841) and Privy Councillor (1835). Richard Vivian married Eliza Champion de Crespigny in 1804 and they had daughters and two sons; in 1833, following Eliza's death he married Letitia Webster and they had one daughter. Of interest in the metallurgical field he was a partner with his father and with his brother in the Hafod copper smelting works at Swansea.
PRS 1837–43–43, 4, 415. DNB.

WOLDEMAR VOIGT 1850–1919; FMRS 1913
Woldemar Voigt was born in Leipzig, Germany, where he graduated from the Nickolauschule in 1868. Having entered the University his studies were interrupted from 1870–71 by service in the Franco–Prussian war. He then went to Konigsberg to continue his studies. With a great interest in music he considered it as a careeer, but decided to be a physicist. At Konigsberg he was influenced by Franz Neumann and in 1874 he completed a dissertation on the elastic properties of rock salt. In 1875 he became an Assistant Professor at Göttingen and was appointed to a Chair of Theoretical Physics in 1883, a post which he held until 1919. Elasticity was a main field of his research, including measurements of various crystals, including rock salt, and studies of anisotropy; he also worked on other physical properties of crystals. Voigt made frequent visits to England.
PRS 1921, A, 99, xxix–xxx. H.L. DSB 14, 61–63. S. Goldberg.

ALLESANDRO GIUSEPPE ANTONIA ANASTASIO VOLTA 1745–1827; FRS 1791
Alessandro Volta was born in Como in Italy from a Lombard family enobled by the municipality of Como. His father Filippo, with a Jesuit background, married Maddelena de' conti Inzaghi in 1773; Alessandro was the youngest of seven children; he was only seven years of age when his father died and an uncle took charge of his education, part of which was at a Jesuit college. The mid 1770s saw the beginning of his career. He became Professor of Experimental Physics at the state Gymnasium in Como, and travelled to centres of learning, becoming Professor of Experimental Physics at Pavia university in Austria. In 1794 he married Teresa Peregrini and they had three sons. His career was influenced by the complex historical events of the time, and Napoleon became his patron, making him a count. His scientific work included the field of galvanism and his most famous invention was that of the voltaic pile. His honours included the award of the Copley medal in 1794.
DSB 14, 69–82 FRS 1791. J.L. Heilbron.

SIR BARNES NEVILLE WALLIS 1887–1979; FRS 1945
Barnes Wallis was born at Ripley in Derbyshire, the second son of Charles, a general practitioner, and Edith née Eyre. His father moved to a South-East London practice in 1891, but illness left him disabled and family finances were difficult. Barnes obtained a competitive place at Christ's Hospital school, which he entered at the age of 12. He left school just before he was 17 and took an engineering apprenticeship at the Thames Engineering Works in Blackheath, but having decided to become a Marine Engineer he transferred to John Samuel White's shipyard at Cowes, Isle of Wight. He attended evening classes and passed the London matriculation examination. He joined the Vickers Company in 1911 and in 1913 he entered the design drawing office in London to work on the design of a new airship, the R 9. He served for a period in the Artists Rifles and was then moved into the Royal Naval Air serevice and allocated to Vickers. Wallis became Chief

Designer in the airship department of Vickers at Barrow-in-Furness. In 1921 his department at Vickers was closed and he found himself unemployed; however, Vickers gave him a retainer, which enabled him to study and to obtain, in 1922, an external degree in engineering at London University. In 1923 he returned to Vickers, and over a period he held various appointments: Chief Designer for a subsidiary of Vickers concerned with airships (1916–22); Chief Designer of Structures at Vickers Weybridge works (1923–1945); Chief of Aeronautical Research and Development at the British Aircraft corporation division at Weybridge (1945–1971) . His contributions to airship design and construction included the R100 (beginning in 1924). In 1925 he married a distant cousin, 'Molly' Frances née Bloxam and they had two sons and two daughters.

Wallis was an expert in the use of Duralumin alloys in aircraft construction; he evolved the geodesic system for fixed wing aircraft and was involved in the Wellington bomber airframe. His inventions in ballistics included the 'bouncing bomb' used for attacking the Ruhr dams. He received the CBE in 1943 and was knighted in 1968. In the Royal Society he was awarded a Royal Medal in 1975.

BNFRS 1981, 27, 603–627. Sir Alfred Pugsley FRS and N.E. Rowe. DNB 1971–80, 899–901. A.G. Pugsley.

SIR ALAN WALSH 1916–; FRS 1969

Darwen Grammar School, Australia. Manchester Univ.: BSc 1938, MSc Tech 1946, DSc 1960. British Nonferrous Metals Research Association: 1939–42; 42–46. Ministry of Aircraft Production: 1943. CSIRO, Melbourne, Division of Physics: 1946–77; Assistant Chief of Division, 1961–77; Consultant Spectroscopist–; Royal Medal, 1976. Knighted 1977.

Spectroscopy.

KENNETH WALTERS 1934–; FRS 1991

University College of Swansea: BSc 1956, MSc 1957, PhD 1959, DSc 1984. University College of Wales, Aberystwyth: Department of Mathematics; Lecturer, 1960–65; Senior Lecturer, 1965–70; Reader, 1970–73; Professor of Applied Mathematics, 1973–.

Theoretical and and experimental rheology, measurement and analysis of properties of non-Newtonian fluids such as elastic liquids; application in manufacturing industry.

HENRY WARBURTON 1784?–1858; FRS 1809

Henry Warburton was the son of John Warburton, of Eltham, Kent, a timber merchant. He was educated at Eton and at Trinity College, Cambridge, which he entered in 1802. He graduated BA as 12th wrangler in 1806, and MA in 1812. After working for some years in the timber trade, he pursued his interest in science and in politics and abandoned commercial life. He was the closest friend of William Wollaston, who, during his final illness, dictated to him the details of the process for producing malleable platinum that Wollaston had developed; this information was presented by Warburton as a Bakerian Lecture in 1828 on behalf of Wollaston. Warburton was a Member of Parliament for many years, being first elected for Bridport in 1826 and later represented Kendal.

PRS 1857–59, 9, 555, 556. DNB. Ref. Ch. 4.1.

IAN MACMILLAN WARD 1928–; FRS 1983

Royal Grammar School, Newcastle upon Tyne. Univ. of Oxford: Magdalen College, MA, DPhil. ICI Fibres: Technical Officer, 1954–61; seconded to Division of Applied Mathematics, Brown University, USA, 1961–62. ICI Fibres: Head of Basic Physics Section, 1962–65. ICI Research

Associate 1964. Univ. of Bristol: Senior Lecturer in Physics, 1965–69. Univ. of Leeds: Chairman Department of Physics, 1975–78; Cavendish Professor 1987–94; Research Professor and Technical Director Centre for Industrial Polymers; 1994–. IRC in Polymer Science and Technology–Universities of Leeds, Bradford and Durham: Director, 1989–94.
Physical properties of the solid state of polymerized materials, invention of the technology for the production of very strong versions of these materials.

RICHARD WATSON 1737–1816; FRS 1769
Richard Watson, who was born in Heversham, Westmoreland, was the son of Thomas Watson, a clergyman. He was educated at school in Heversham and entered Trinity College, Cambridge as a Sizar, provided for by a legacy from his father. He obtained a scholarship in 1757 and distinguished himself in mathematics, graduating BA as second wrangler in 1759; in the following year he became a Fellow of his college, and obtained an MA in 1762. Watson was elected Professor of Chemistry in 1764, a subject which he had not previously studied; he studied intensively and after 14 months began a series of lectures which became popular. In 1771 he obtained the king's mandate for a DD degree and was elected as Professor of Divinity. In 1773 he married Dorothy Wilson and they had six children. His church appointments included a period as Archdeacon of Ely beginning in 1774, prior to becoming Bishop of Llandaff in 1782. He published, beginning in 1781 *Chemical Essays,* as a series of volumes, which included metallurgiclal topics, such as the smelting of lead ore.
DSB 14, 191, 192. E.L. Scott. DNB.

SIR WILLIAM WATSON 1715–1787; FRS 1741
William Watson, the son of a tradesman, was born in the Smithfield part of London and in 1726 he entered the Merchant Taylors' School. In 1730 he was apprenticed to an apothecary and had a strong interest in botany. In 1738 he married and had a son and daughter. He set up in business for himself and became distinguished for his scientific knowledge; he contributed to the *Philosopical Transactions* many papers, concerning natural history, electricity and medicine. William Brownrigg passed to Watson results of experiments that he had made on platinum, which Watson communicated to the Royal Society in 1750. He was awarded the Copley Medal in 1745 for his electrical research. His work included experiments on electrical discharge in vacuo. His experiments became famous outside scientific circles, and George III (then Prince of Wales) and others fashionable people went to see them at his house in Aldersgate Street. He obtained the Licentiateship of the Royal College of Physicians in 1759 and was elected Fellow in 1784, having practiced as a physician. He was knighted by George III in 1786, and from 1774–76 and 1782–87 was Vice-President of the Royal Society.
DSB 14, 193–196. J.L. Heilbron. DNB. Ref. Ch. 4.1.

JAMES WATT 1736–1819; FRS 1785
James Watt was born at Crawfordsdyke near Greenock, the son of James Watt and Agnes née Muirhead; his father was a man of varied occupations including that of a builder and a merchant. He received schooling in Greenock and worked as a boy in his father's shop making models, demonstrating manual dexterity. At the age of about 17 he spent a period in Glasgow before going to London in 1755 to become apprenticed as a mathematical instrument maker. After a year there he returned to Greenock, and became mathematical instrument maker at Glasgow University, where he became interested in and invented a steam engine representing an improve-

ment on Thomas Newcomen's engine. His marriage to a cousin Margaret Miller took place in 1763 and they had two sons and two daughters; following the death of his wife in 1773 he married Ann MacGregor, by whom he had a son and daughter.

Best known for his role in the invention and development of the steam engine, Watt in 1765 made a separate condenser for the Newcomen engine and patented it in 1769; he developed it commercially, initially with John Roebuck, and subsequently with Matthew Boulton. Roebuck had founded the Carron works in Scotland and it appears that Watt succeeded Smeaton as engineer advisor to the Carron foundry. In 1774 he moved to Birmingham, and the following year entered into partnership with Matthew Boulton, who had founded the Soho Engineering Works there. Their partnership transformed the prospects for the steam engine; two full scale engines were built at the Soho works with cast iron cylinders. Watt retired from the firm of Boulton and Watt in 1800. Recognition for his achievements extended to countries abroad; for example in 1781 he received an invitation, which he declined, to go to Russia to be the Director of Mines, Metallurgy and Ordnance Castings. Watt corresponded with scientists in Europe, visited Paris with Boulton in 1786, and knew Lavoisier and Berthollet well. In his interest in chemistry he had close contacts with such men as Priestley, Keir and Wedgwood, members of the Lunar Society.

DSB 14, 196–199. H. Dorn. DNB. Ref. Ch. 2.7.

WILLIAM WATT 1912–1985; FRS 1976

William Watt was the only child of Patrick Watt and Flora née Corsar. He was born in Edinburgh where he was educated at North Merchiston Primary School and at George Heriot's School. In 1928 he became a Laboratory Assistant with the (then) Edinburgh and East Scotland College of Agriculture. Also he studied for four years at Heriot-Watt College in Edinburgh and obtained the Diploma and Associateship of the College. Concurrently he obtained a first class honours BSc in chemistry in the University of London external examinations. In 1936 he joined British Celanese as a works chemist, but after six months he became an Assistant in the Chemistry Department of the Royal Aircraft Establishment at Farnborough. Hants, where he worked on various aircraft materials. From 1941–48 he was a member of a team concerned with fire and explosion hazards and was promoted to Technical Officer. Watt was transferred to the Metallurgy Department in 1944 to a team carrying out exploratory research on high temperature materials, including titanium alloys; Watt's work was in the non-metallic field, involving sintered oxides, such as alumina, and carbides, such as zirconium carbide. He became a Senior Scientific Officer in 1946 and a Principal Scientific Officer in the following year. In 1946 he married Irene Isabel Corps and they had two daughters.

Beginning in the early 1950s Watt worked on pyrolitic graphite and on impermeable graphite fuel cans for the nuclear industry. In the field of fibres in collaborative work he produced a polycrystalline carbon fibre of high strength and high modulus; his approach was based on a polyacrilonitrile (PAN) textile fibre and a process was patented in 1962. In 1969 Watt received the OBE. He retired in 1975 from the RAE, and was appointed Senior Research Fellow in the (then) Department of Metallurgy and Materials Technology at the University of Surrey at Guildford.

BMFRS 1987, 33, 643–667. W.M. Nair and E.H.

DENIS LAWRENCE WEAIRE 1942–; FRS 1999

Belfast Royal Academy. Univ. of Cambridge, Harkness Fellowship: 1967–69; Clare College, Fellow 1967–69. Yale Univ.: Instructor; Associate Professor, 1970–74. Heriot-Watt Univ.: Senior

Lecturer, Professor, 1974–79. Univ. College, Dublin: Chair of Experimental Physics, 1980–84. Trinity College, Dublin: Erasmus Smith's Professor of Natural and Experimental Philosophy, 1984–; Dean of Science, 1989.
Solid state theory; materials science.

WILHELM EDUARD WEBER 1804–1891; FMRS 1850
Wilhelm Weber was born in Wittenberg, Germany, one of 12 children of Michael Weber, Professor of Theology at Wittenberg University. The family moved to Halle, where Weber entered the University in 1822; he obtained a Doctorate in 1822, became a Lecturer and then, in 1828, an Assistant Professorship. In 1831 Wilhelm became Professor of Physics at Göttingen University; in 1843 he was appointed Professor of Physics at Leipzig, returning to his former post at Göttingen in 1849. Weber's research included electrodynamics and the electrical structure of matter. During the period 1836–41 he investigated the dependence of magnetism on temperature. He constructed accurate measureing instruments and did work on electrical and thermal conductivity and elasticity. He was the recipient of the Copley Medal in 1859.
DSB 14, 203–209. A.E. Woodruff.

RICHARD WECK 1913–1986; FRS 1975
Richard Weck was born in Franzensbad in Bohemia, then part of Austria. His father managed a small restaurant and his mother died young, as did his only sister. With the responsibility for looking after his younger brother he supplemented the finances by giving private lessons in mathematics and physics and by earning money in the hotel business.

He entered the Technical University of Prague in 1931 to study civil and structural engineering; during the vacations he spent time working on building sites and graduated as Ingenieur in 1936. He joined a civil engineering construction company in Prague for a year, working as a designer and clerk of the works. In 1937 he began research on the plastic theories of structural analysis and design at the Technical University. When Czechoslovakia was relinquished to Germany in 1938, Weck who was involved in the anti-fascist movement had to leave Prague and to come to Britain. His wife Katie née Bartl, whom he had married in 1933 joined him at the beginning of 1939 and they settled in Byfleet and Weybridge; they became British citizens in 1946.

Weck was employed in the design of electric arc furnaces and the layout of ancillary equipment for steel foundries working for the Electric Furnace Company, and Campbell and Gifford, Consulting Engineers, equipping foundries for war production. In 1943 he joined a small team working for the Welding Research Council of the Institute of Welding, which was the forerunner of the British Welding Research Association (BWRA). In 1943 Weck was commissioned to edit a handbook for welded steel structural steelwork. When J.F. Baker (later Lord Baker of Windrush, FRS) was appointed to the Chair of Engineering at Cambridge in 1943 he obtained the secondment of Weck to initiate research on the critical problem of brittle fracture in ship steels; Weck made important contributions, including in the area of weld fracture.

In 1946 Weck continued his association with Professor Baker, who became one of the founding members of the BWRA. Metallurgical and welding process work was conducted at premises in London, but the initial engineering researches were in Cambridge University engineering laboratory. Weck was awarded a PhD in 1948. He was Principal Scientific Officer at BWRA from 1947–1951 and head of its new fatigue laboratory.

As the work of BWRA expanded Weck set out to find a suitable new facility, and Abington Hall, near Cambridge was obtained as a site. Starting with an area of neglected park land and some wooden hutments, experiencing spartan conditions in the first winter (1946–47) Weck con-

tributed greatly to the creation of the organisation, for example in the construction of the fatigue testing laboratory. Leaving BWRA in 1952 he returned to Cambridge University until 1957 to organise and run the postgraduate course on structures which had a new emphasis on the application of welding. He maintained his close interest in the work of BWRA and in 1957 he became its Director of Research, serving until his retirement in 1977. When BWRA merged with the Institute of Welding he became the first Director-General of The Welding Institute; he played a major role, including through his travel to establish links. In the development and expansion of the Institute to become an organisation of high international status. From 1967 for six years Weck was a visiting Industrial Professor at Imperial College, London, in the Departments of Mechanical Engineering and of Mctallurgy; for nine years, beginning in 1975 he held a complementary appointment in the Department of Civil Engineering. Among the honours that Weck received was the CBE in 1969 and he was elected FREng in 1976.
BMFRS 1986, 32, 631–647. A.A. Wells FRS and E.G. West. DNB 1986–1990, 473, 474. A.A. Wells.

SIR JOSIAH WEDGWOOD 1730–1795; FRS 1783
Josiah Wedgwood was born in Burslem, Staffordshire, the 13th and youngest son of Thomas Wedgwood and Mary née Stringer. Josiah's father, as well as several uncles and cousins, were potters. When his father died he began work in the pottery with his eldest brother at the age of nine; later he served an apprenticeship with his brother, and obtained an insight into many aspects of the potter's art, also developing an interest in original experimenting. There followed experience in other potteries, until he set up business on his own account in Burslem as a master potter in 1758. His business expanded and he improved the works practice, introducing better organisation and cleanliness, aiming at greater efficiency and precision and improved decorative aspects. Wedgwood adopted or improved many of the processes that had been developed in the ceramics industry during the period 1720–60, adding novel features from his own experimental work. Thus he evolved an improved stoneware and termed it Queen's ware, having obtained royal patronage; this ware was soon being made by many other potteries, and in time became one of the major products of the Potteries. Wedgwood married Sarah Wedgwood, a distant cousin in 1764 and they had three sons and four daughters. In 1766 he acquired a site for a new factory and built a village there, which he called Etruria, for his workmen.

Wedgwood, was influenetial in pottery manufacture. His worl included investigations of ceramic chemistry, looking for new clays, glazes and enamels; he carried out analyses of clays and examined metallic ores. Healso carried out other work relevant to metallurgy. He was probably the first to develop a practical technique for measuring high temperatures. Another significant contribution made was to supply equipment (pyrometers, retorts, crucibles, tubing) to various scientists, and he also engaged in scientific correspondence. Through his membership of the Lunar Society of Birmingham he was associated with the foremost British scientists of the day and he was one of the most progressive industrialists of the 18th century.
DSB 14, 213–214. H. Dorn. DNB.

ALAN ARTHUR WELLS 1924–; FRS 1977
City of London School. Univ. of Nottingham. Univ. of Cambridge: Clare College, PhD. British Welding Research Association: 1956; Deputy Director (Scientific), 1963. Queen's University of Belfast: Professor of Structural Science, 1964; Head of Civil Engineering Department, 1970–77; Dean Faculty of Applied Science and Technology, 1973–76; Professorial Fellow, 1990. Welding Institute: Director General, 1977–88. FREng, 1978. OBE 1982.
Welding Technology; fracture mechanics.

SIR CHARLES WHEATSTONE 1802–1875; FRS 1836

Charles Wheatstone was born into a family of musical instrument makers and dealers; he was born in Gloucester and was privately educated. In 1816 he was apprenticed to his uncle in the music business in London. When he was twenty-one he set up in business in London as a musical instrument maker; he also carried out experiments on sound and acoustics and on light. He was appointed Professor of Experimental Physics at King's College, London, where he held this position until the end of his life. He carried out experiments on the rate of transmission of electricity along wires and his inventions included instruments for electrical measurements. Wheatstone entered into partnership with W.F. Cooke which led to important inventions in the commercial development of telegraphy. Wheatstone had an extraordinary ability to decipher hieroglyphics and cipher despatches. In 1847 he married Emma West and they had five children. He was knighted in 1868. In the Royal Society he received a Royal Medal in 1840 and 1843, the Copley Medal in 1868 and gave the Bakerian Lecture in 1843 and 1852.

PRS 1875–76, 24, xvi–xxvii. DSB 14, 289–291. S. Dostrovsky. DNB.

MICHAEL JOHN WHELAN 1931–; FRS 1976

Farnborough Grammar School. Univ. of Cambridge: Gonville and Caius College; Fellow 1958–66; Demonstrator in Physics, 1961–65; Assistant Director of Research, 1965–66. Univ. of Oxford: Reader, Department of Materials, 1966–72; Professor of Microscopy of Materials,1992–. Hughes Medal (jointly)1988.

Electron microscopy and diffraction.

WILLIAM WHEWELL 1794–1866; FRS 1820

William Whewell was born in Lancaster, the eldest of seven children of John Whewell, a master carpenter, and Elizabeth née Bennison. He was educated in Lancaster and at Heversham, prior to entering Trinity College, Cambridge in 1812 with an exhibition. He was second wrangler and second Smith's prizeman in 1816. He obtained a college fellowship in 1817 and became assistant tutor a year later. He was awarded an MA in 1819 and a DD in 1844. He was mathematical tutor at Trinity College and was also ordained in the Church of England (deacon in 1825, priest in 1826). Whewell published papers on crystallography and was elected to the Chair of Crystallography at Cambridge in 1828; he resigned from this post in 1832. In 1837 he received a Royal Medal for his work on tides. He became Master of Trinity College in 1841. In the same year he married Cordelia Marshall , and following her death he married Lady Evering Affleck in 1858. Whewell was Vice-Chancellor of Cambridge University in 1842 and 1845.

Of metallurgical interest was his correspondence with Faraday concerning Faraday's discoveries in electricity which led to the adoption of terms which became established in electrochemistry.

PRS 1867–68, 16, l–lxi. JFWH. DSB 14, 292–295. R.E. Butts. DNB.

JOHN WILLIAM WHITE 1937–; FRS 1993

Newcastle High School, Australia. Sydney Univ.: MSc. Oxford Univ.: Lincoln College; 1851 Scholar, 1959, MA, DPhil; ICI Research Fellow, 1962; University Lecturer, 1963–85; Australian National Univ., Canberra: Professor of Physical and Theoretical Chemistry, 1985–; Dean, Research School of Chemistry 1995–; Pro-Vice-Chancellor and Chairman, Board of Institute of Advanced Studies, 1992–94. CMG 1981.

Neutron scattering methods for structure studies; molecular crystals.

SIR WILLIAM HENRY WHITE 1845–1913; FRS 1888

William White, who became a leading naval architect, was born at Devonport, the youngest child of Richard White, currier, and Jane née Matthews. Following an apprenticeship in the naval dockyard at Devonport, he, and other apprentices, were appointed by the Admiralty in 1864 for training at the then newly founded School of Naval Architecture and Marine Engineering at South Kensington, London. In 1867 he obtained highest honours at the school and was appointed to the Admiralty staff. It was a time when many warships with iron hulls were being constructed in place of vessels with wooden hulls. White collaborated with the Chief Constructor of the Navy in some publications, including a book on '*Ship building in iron and steel*' (1869) and became influential in naval ship construction, including battleships and cruisers, for example, the cruiser, *Iris,* laid down in 1875, which was the first steel vessel built for the Navy.

Between 1883 and 1885 White was Director and Manager to Armstrong and Company at Elswick, involved in designing and building ships. He then became Director of Naval Construction, in which post he made improvements incorporating advances including in the quality of materials. He also served as Assistant Controller to the Navy. From 1870–1873 he also lectured on naval design in South Kensington, and he continued (until 1881) to lecture when the school was transferred to the Royal Naval College, Greenwich. In 1875 he married Alice Martin; following her death in 1886 he married Annie Marshall in 1890; there were three sons and one daughter. White was knighted (KCB) in 1895.

PRS 1913–14 A, 89, xiv–xix, PW. DNB 1912–1921, 574–576. PW.

JOHN WHITEHURST 1713–1788; FRS 1779

John Whitehurst was born in Congleton, Cheshire, the son of John Whitehouse, a clock and watch maker, who taught his son the trade. In 1736 he entered business for himself at Derby, including making instruments such as thermometers and barometers. He was widely consulted for his skill in mechanics, pneumatics and hydraulics. In 1775, when an act was passed for the better regulation of gold coinage, he was appointed stamper of the money weights. He moved later to London, where his house became a place visited by many men of science. He married Elizabeth Gretten in 1783, but there were no surviving children. Celebrated as a clockmaker, Whitehurst was also a pioneer in geology.

Notes and Records Roy. Soc. 1954, 19, 156, 157. DSB 14, 311, 312. J. Challinor. DNB.

SIR FRANK WHITTLE 1907–1996; FRS 1947

Frank Whittle was born in Coventry, the child of Moses and Sara Alice Whittle. His father was a designer and craftsman who ran his own workshop. From an early age Whittle showed a practical aptitude and he became interested in aircraft. When he was 10 the family moved to Leamington Spa, where he won a scholarship from Milverton Council School to Leamington College. Whittle was determined to enter the Royal Air Force, but the financial situation did not make it possible for him to go to the RAF College at Cranwell. However, he was accepted for an RAF apprenticeship in 1923. His ability having been recognised by the RAF he was given a cadetship to Cranwell. For the thesis that formed part of the course he wrote on 'Future developments in aircraft design'; he concluded that to attain very high speeds with long range very high altitude flight would be needed with low air density. A naturally gifted pilot he passed out second at Cranwell in 1928 with the award of a prize and was posted as a Pilot Officer to a fighter squadron. In the following year, while taking a flying instructor's course, he had the idea of using a gas turbine for the jet propulsion of aircraft and a patent was taken out in 1930. However, no official interest was shown in the invention and the patent was not protected by a security classification, so that in 1932 it was published. Whittle married Dorothy May Lee in 1930 and they had two sons. He served as a flying instructor and then was a test pilot for seaplanes at Felixstowe. In 1932 he was sent to the RAF School of Aeronautical Engineering at Henlow; in 1934 the RAF arranged for him to go to Peterhouse College, Cambridge where two years later he graduated with first class honours in the Mechanical Sciences Tripos; he then continued on to a postgraduate course for a year, during which time he was able to work on his turbojet project.

In 1935, with some support and finance, Whittle founded Power Jets as a company to develop aircraft gas turbines for jet propulsion. Work was done at the British Thomson Houston company at Rugby and Whittle was transferred in 1936 to the special duty list of the RAF for secondment to this work. In 1937 the first run of a gas turbine jet engine was achieved, but it was not until 1939 that the Air Ministry decided to test fly one of the engines. The Gloster aircraft company was commissioned to build an experimental plane and in May 1941 the Gloster Whittle E 28/39 first flew at Cranwell. However, it was three years before the Gloster Meteor, the production model, was coming into service. During 1943 Whittle was on a course at the RAF staff college.

The way in which the government handled the Power Jets company led to Whittle resigning early in 1946; he went to the Ministry of Supply as a technical adviser on engine design and production. In 1948 he resigned from the RAF on grounds of ill-health, having been appointed CBE in 1944 and CB in 1947 and reaching the rank of Air Commodore. Shortly after he had left the RAF he was knighted KBE in 1948. He also received a financial award (£100,000) for his invention. Between 1961 and 1970 Whittle worked as a consultant to Bristol Siddeley Engines and to Rolls-Royce on the development of a turbo drill for deep sea exploration.

Whittle's first marriage was dissolved in 1976, but in the same year he married Hazel Hall, an American, and moved to the USA where he became a Research Professor at the US Naval Academy in Annapolis. He was the recipient of numerous honours including the Rumford Medal in 1950. The award of the Order of Merit in 1986 marked the achievements resulting from his inventive skills and determination that contributed outstandingly to the revolution in aircraft propulsion that has occurred during the latter part of the 20th century.

BMFRS 1998, 44, 435–452. G.B.R. Feilden CBE, FREng, FRS and Sir William Hawthorne CBE, FREng, FRS. *The Times*, August 10th 1996.

SIR JOSEPH WHITWORTH 1803–1887; FRS 1857

Joseph Whitworth was born at Stockport, Lancashire, the son of Charles Whitworth, a schoolmaster and Sarah née Hulse. He was educated at his father's school and at William Vint's Academy at Idle, near Leeds; having left school at the age of 14 he entered his uncle's cotton mill in Derbyshire to learn the business. However, showing more interest in machinery than in the commercial aspects, he mastered the construction of every machine. In 1821 he went to Manchester to work as a mechanic in a firm of machinists. Then, soon after his marriage to Fanny Ankers in 1825, he moved to London and entered the firm of Maudslay and Co., in the Westminster Bridge Road, where his exceptional talent was soon recognised; here he developed a method of preparing truly plane surfaces. On leaving Maudsley's, his experience in London included a period in the workshop of John Clement, where the calculating machine invented by Charles Babbage was currently being constructed.

The year 1833 saw Whitworth return to Manchester, where he rented a room with steam power and set up as a tool maker, displaying a sign 'Joseph Whitworth, Tool Maker, from London'. His business and reputation rapidly increased. The next 20 years were occupied mainly with the improvement of machine tools, which were displayed at the Great Exhibition of 1851, where he was the outstanding machine-tool maker; he maintained this leading position also at the International Exhibition of 1862. Whitworth developed a machine capable of very precise measurement, and a system of standard measures and gauges, notably his uniform system of screw threads. His work was characterised by the highest standards of workmanship and materials.

In the field of armaments he made important contributions, including the invention of a rifled gun. Over the years Whitworth had built up an extensive works at Openshaw, near Manchester, and, in 1874, with his employees as shareholders he converted it into a limited liability company, which in 1897 was united with Armstrongs of Elswick.

Whitworth was created Baronet in 1869, and following the death of his first wife in 1870, he married Mary Louisa a year later; there were no children. Around that time he purchased a seat and an estate near Matlock. In 1868 he founded 30 scholarships for the theory and practice of engineering, and through his will these became permament; also over a period of years a total of nearly £600,000 was distributed from his estate to various educational and charitable causes.
PRS 1887, 42, ix–xi. JH. DNB.

GUSTAV HEINRICH WIEDEMANN 1826–1899; FMRS 1884
Gustav Wiedemann was born in Berlin, into a merchant family. His father died when he was two, and when his mother died when he was 15 he was brought up by grandparents. Having studied at the Köllnische RealGymnasium in Berlin from 1838–1844 he studied natural sciences at the University of Berlin for three years. In 1847 he obtained a Doctorate and became a Lecturer in Theoretical Physics. He married Clara Mitscherlich in 1851, and they had one daughter and two sons. Wiedemann held a number of academic appointments: a Professorship of Physics at the University of Basel (1854) and posts at Polytechnische Schules in Brunswick (1863) and Karlsruhe (1866). In 1871 he was appointed to Germany's first Chair of Physical Chemistry at Leipzig, followed in 1887 by appointment as Professor of Physics. His research included investigations of the electrical and thermal conductivities of metals and of the relationship between magnetization and mechanical phenomena.
PRS 1905, 75, 41–43. GCF. DSB, 14, 329–341. H.-G. Korber. Ref. Ch. 13.8.

EUGENE PAUL WIGNER 1902–1995; FMRS 1970
Eugene Wigner was the son of Antal Wigner, a businessman, and Elisabeth née Einhorn. He was born in Budapest where he was educated at the Lutheran Gymnasium. He entered the Technische Hochschule in Berlin where he studied chemical engineering and obtained a PhD in 1925; he then held an appointment as Assistant Lecturer. After a period at the University of Göttingen (1927–28) he returned to Berlin until he moved to the USA in 1930 and initially took a part-time post as Professor of Mathematical Physics at Princeton. In 1936 he moved to the University of Wisconsin where he was Professor of Physics from 1936–38. He became a naturalised American citizen in 1937. Wigner returned as Professor of Mathematical Physics to Princeton; he remained there until his retirement in 1971 apart from a period on leave of absence from 1942–45 at the Metallurgical Laboratory at Chicago, working with Fermi on nuclear chain reactions, and the construction of the first pile, and a further period from 1946–47 as Director of the Clinton Laboratories at Oak Ridge, Tennessee. In 1938 Wigner married Amelia Frank; following her death he married Mary Wheeler in1941, and they had one son and one daughter; Mary died in 1977 and Wigner married Eileen Hamilton in 1979.

In his research career, following some early work on chemical reactions and the spectroscopy of compounds Wigner made fundamental contributions in quantum and nuclear physics; in reactor technology he anticipated the heating of the graphite moderator resulting from neutron bombardment – an important phenomenon referred to as the 'Wigner effect'. In 1963 he shared the Nobel Prize for Physics.
The Independent, 13.1.95. V. Telegidi.

WILLLIAM LIONEL WILKINSON 1931–; FRS 1990
Univ. of Cambridge: Christ's College, MA, PhD, ScD.; Salters Research Scholar, 1953–56. Univ. of Wales: University College, Swansea: Lecturer in Chemical Engineering, 1956–59. UKAEA Production Group: 1959–67. Univ. of Bradford: Professor of Chemical Engineering, 1967–79. British Nuclear Fuels Ltd: Deputy Director, 1979–84; Technical Director, 1984–86; Deputy Chief Executive, 1986–92; Non-Executive Director, 1992–. Allied Colloids plc: Director, 1990–, Deputy Chairman, 1992–. FREng 1980. CBE 1987.
Fluid mechanics; heat transfer; polymer processing; process design.

EVAN JAMES WILLIAMS 1903–1945; FRS 1939
Evan James was born in Cwmsychpant near Llanbyther in rural Wales. He was the youngest of three sons of James Williams and Elizabeth née Lloyd. He attended Llanwenog National School and then Llandyssul County School. In 1919 he won a scholarship to Swansea Technical College, which a year later became a constituent college of the University of Wales. In 1923 he obtained first class honours in physics and the following year an MSc. He was awarded a Fellowship of the University of Wales and went to Manchester to work with W.L. Bragg, obtaining a PhD in 1926. With the award of an 1851 Exhibition Scholarship Williams he went to Cambridge in 1927 to work with Sir Ernest Rutherford in the Cavendish Laboratory. In 1929 he obtained a PhD at Cambridge, and the following year a DSc (Wales). From 1929–1934 and 1934–36 he carried out teaching and research at Manchester; he spent a year (1933–34) working under Niels Bohr in Copenhagen. In 1937 he was appointed Senior Lecturer at Liverpool, but in the following year he moved to the University College of Wales at Aberystwyth to occupy the Chair of Physics.

Williams was a theoretical physicist and also a skilled experimentalist, his main research being in the field of atomic and nuclear physics. His work also included subjects of metallurgical interest: at Swansea he investigated the electrical conductivity of mercury with various added elements, while at Cambridge in 1934 and 1935, he collaborated with Bragg on a theory of the effect of thermal agitation on the atomic arrangement in alloys, including ordering phenomena.

From 1939–41 Williams was a Principal Scientific Officer at the Royal Aircraft Establishment Farnborough. where he was initially concerned with developing apparatus for detecting submarines from the air. From 1941–42 he was Director of Operational Research at RAF Coastal Command , and then from 1943–45 he was Scientific Adviser on anti-U Boat warfare at the Admiralty, and served as Assistant Director of Naval Operations from 1944–45.
ONFRS 1945–48, 5, 387–406. P.M.S. Blackett.

JAMES GORDON WILLIAMS 1938–; FRS 1994
Imperial College, London: BSc (Eng), PhD, DSc. Royal Aircraft Establishment, Farnborough: 1956–61. Imperial College: Assistant Lecturer, 1962–64; Lecturer, 1964–79; Reader, 1970–75; Professor of Polymer Engineering, 1975–90; Professor and Head of Department of Mechanical

Engineering, 1990–. FREng 1982.
Polymeric materials: deformation, fatigue, impact behaviour, cracks and fracture; applications in machinery design.

SIR PETER MICHAEL WILLIAMS 1945–; FRS 1999
Hymers College, Hull. Univ. of Cambridge: Trinity College, MA, PhD; Selwyn College, Mullard Research Fellow, 1969–70. Imperial College, London: Department of Chemical Technology; Lecturer, 1970–75. VG Instrument Group, 1975–82, (Deputy Managing Director, 1979–82). Oxford Instruments Group plc., 1982–; Managing Director, 1983–85; Chief Executive, 1985–98; Chairman 1991. Particle Physics and Astronomy Research Council: Chairman 1994–. FREng 1996. CBE 1992. Knighted 1998.
Solid state physics. Instrument design.

ROBERT HUGHES (ROBIN) WILLIAMS 1941–; FRS 1990
Bala Boys Grammar School. Univ. College of North Wales, Bangor: BSc, PhD, DSc. Univ. of Wales Research Fellow, 1966–68. New Univ. of Ulster: Lecturer then Reader and Professor, 1968–83. Univ. of Wales, College of Cardiff: Professor and Head of Department of Physics, later Physics and Astronomy, 1984–94; Deputy Principal, 1993–94. Univ. of Wales, Swansea: Vice-Chancellor and Principal 1994–.
Solid state physics and semiconductor devices; metal/semiconductor contacts.

JOHN RAYMOND WILLIS 1940–; FRS 1992
Imperial College, London: BSc, PhD; Assistant Lecturer, 1962–64. New York Univ.: Research Associate, 1964–65. Univ of Cambridge: Senior Assistant in Research, 1965–67; MA(Cantab), 1966; Assistant Director of Research, 1968–72. Univ. of Bath: Professor of Applied Mathematics, 1972–94. Univ. of Cambridge: Professor of Theoretical Solid Mechanics, 1994–.
Mechanics and physics of solids. Composite materials.

ROBERT WILLIS 1800–1875; FRS 1830
Robert Willis, the son of Robert Darling Willis, was born in London. He received part of his education at King's Lynn, before entering Gonville and Caius College, Cambridge in 1822 as a Pensioner. Having graduated BA in 1826 as ninth wrangler, he became Frankland fellow of his college, in the same year. He was ordained Deacon and Priest in 1827, and became a Foundation Fellow of his college in 1829. In 1832 he married Mary Anne Humfrey. He focussed his work on the study of mechanisms (e.g. gear teeth) and from 1837 until his death he held the appointment of Jacksonian Professor of Natural and Experimental Philosophy at Cambridge University. His activities included membership of the commission concerned with iron in railway structures. In 1853, soon after the Government School of Mines had been established in London, he lectured there on applied mechanics. Willis also had a major interest in the field of architecture and archaeology.
DSB 14, 403, 404. B. Roth. DNB.

SIR ALAN HARRIES WILSON 1906–1995; FRS 1942
Alan Wilson was the son of H. and A. Wilson. Following his education at Wallasey Grammar School he studied at Emmanuel College, Cambridge. He was awarded the Smith's Prize (1928) and the Adams Prize (1931). From 1929–1933 he was a Fellow of Emmanuel College, 1929–33,

and then fellow of Trinity College and University Lecturer in Mathematics from 1933–45. In 1934 he married Margaret Constance Monks and they had two sons. His chief interest was in atomic physics, in which field he worked on the elctronic theory of metals, insulators and semi-conductors. Wilson made important advances in the theory of the optical properties of metals, of diamagnetism, and of thermal conductivity and thermoelectric effects at low temperatures. His publications included *The theory of metals* (1936) and *Semiconductors and metals* (1939). Following his achievements in the academic sphere he joined the textile company Courtaulds in 1945, where he rose to become Managing Director in 1945 and Deputy Chairman (1957–62). He joined the board of Glaxo Group Ltd and became Chairman in the same year, serving for 10 years. He served on a wide range of public and other bodies and was knighted in 1961.
The Times, Oct 5, 1995 p 23a. *Who was Who*, 1995.

ARTHUR JAMES COCHRAN WILSON 1914–1995; FRS 1963
Arthur Wilson was born in Springhill, Nova Scotia, Canada, the only son of Arthur A. Wilson and Hildegarde Gretchen née Geldert. His father was a doctor who while serving in the Canadian army during the 1914–18 war was reported missing. Arthur and his sister were brought up by their mother. Following attendance from 1922–28 at King's Collegiate School, Windsor, Nova Scotia he entered Dalhousie University, Halifax in 1930 where he obtained the B.Sc degree in physics with high honours, and a university medal in 1934. He was awarded an MSc from Dalhousie University, Halifax in 1936 for work on the temperature dependence of the heat capacity of a number of metals. Two years he was awarded a PhD at the Massachusetts Institute of Technology. From 1937–1940 he was an 1851 Exhibition Scholar, and then a Research Assistant at the Cavendish Laboratory, Cambridge from 1940–1945; in 1942 he obtained the PhD degree from the University of Cambridge for research involving accurate measurements of the thermal expansion of aluminium and lead, using X-ray diffraction for the precision measurement of lattice parameters. In 1946 he married Harriett Charlotte Friedeberg and they had a son and a daughter.

Wilson was appointed a Lecturer in the Physics Department at University College, Cardiff, Wales, in 1945; he became a Senior Lecturer 1946, and was Professor from 1954–1965. From 1965–1982 he was Professor of Crystallography in the Department of Physics at Birmingham University. He held visiting professorships at Georgia Institute of Technology in 1965, 1968 and 1971; Tokyo University, 1972 and was Emeritus Professor at the University of Birmingham and emeritus fellow of the crystallographic data centre at Cambridge University.

X-ray investigations of crystal imperfections and X-ray statistics were Wilson's main areas of research, and his work included important contributions in X-ray powder diffractometry. His publications included books: *X-ray Optics*, 1949; *Mathematical Theory of X-ray Powder Diffractometry*, 1963; *Elements of X-ray Crystallography*; he was editor of *Acta Crystallographica* from 1960–1977 and of the *International Tables* for X-ray Crystallography.
BMFRS 1997, 43, 521–535. M.M. Woolfson FRS. *Materials World*, 1995, **3**(10), 451.

ALAN HARDWICK WINDLE 1942–; FRS 1997
Whitgift School, Croydon. Imperial College, London: 1960, BSc (Eng) 1963, ARSM 1963, Univ. of Cambridge: Trinity College, PhD 1966. Imperial College: ICI Research Fellow 1966–67; Lecturer in Metallurgy, 1967–75. Univ. of Cambridge: Lecturer in Metallurgy and Materials Science, 1975–92; Professor of Materials Science, 1992–; Head of Department 1996–.
Polymers; polymer glasses, diffusion, liquid crystalline polymers, polymer modelling.

COLIN GEORGE WINDSOR 1938–; FRS 1995

Beckenham Grammar School. Oxford Univ.: Magdalen College; BA, Physics, DPhil 1963; Clarendon Lab.; Magnetic Resonance Research, 1963. Yale Univ.: Research Fellow, 1964. Harwell: Neutron Scattering Research, 1964–96. AEA Technology; National Non-Destructive Testing Centre: Senior Scientist, 1988–96. UKAEA Fusion: Programme Area Manager, 1996–.

Applications of neutron beams to the study of materials.

FRIEDRICH WÖHLER 1800–1882; FMRS 1854

Friedrich Wöhler was born in a village, Eschersheim near Frankfort-on-the-Main, Germany. His father, Anton August Wohler became a leading citizen in the city; his mother was Anna Katharina née Schroder. He showed an early interest in experimenting and in collecting minerals. In 1814 he went to the Frankfort Gymnasium; then at the age of 20 he commenced studies in medicine, initially at Marburg and then, a year later, at the University of Heidelberg, where he spent as much time as his medical studies allowed working with Leopold Gmelin on chemistry. His main aim was to qualify in medicine and he obtained the degree of doctor in medicine, surgery and midwifery. Subsequently, however, Gmelin advised him to give up medicine and to concentrate on chemistry. Accordingly he went to work with Berzelius as his student in Stockholm; there he assisted in the discovery of silicon, boron and zirconium. He became a life long friend of Berzelius and translated his works into German.

After leaving Sweden he held teaching appointments in Berlin (1825–1831) and Cassel (1831–1836), carrying out important research in both inorganic and organic chemistry. This included a pioneer role in the discovery of aluminium and beryllium, and work on the platinum group metals. He also published many papers on minerals, meteorites and their analysis. In organic chemistry, in addition to work carried out in collaboration with Liebig, he made an important investigation in which he produced an organic compound, urea, from inorganic sources.

His later career included appointments as Professor of Chemistry to the Medical Faculty in the University of Göttingen (1836–1882) and Inspector-General of Pharmacies in the kingdom of Hanover. Among his many honours and distinctions, he receive the Copley Medal in 1872. Wohler was twice married. His first wife, Franziska, whom he married in 1830 died in 1832, leaving him a son and a daughter; in 1834 he re-married Julie Pfeiffer, by whom he had four daughters.

The school of chemistry grew steadily during the 46 years Wöhler spent at Göttingen. Students came from many countries, including America, to study under him, and his lectures were heard by around 8,000 students.

PRS 1883, 35, xii–xx. DSB 14, 474–479. R. Keen. *Platinum Metals Review*, 1985, **29**(2), 81–85.

WILLIAM HYDE WOLLASTON 1766–1828; FRS 1793

William Wollaston was born at East Dereham, Norfolk, the third son of Francis Wollaston, a vicar and Fellow of the Royal Society, and Althea née Hyde. His family was well known through their interests in science and theology. Following education at Charterhouse School from 1774–1778, he entered Caius College, Cambridge as a medical student in 1782, first as a pensioner and then a scholar; his studies included botany and astronomy, and he acquired an interest in chemistry; the latter was further stimulated by Smithson Tennant who was also studying medicine. Wollaston graduated MB in 1788 and MD in 1793, having been appointed to a Senior Fellowship in 1787, which he retained till his death. He also held the offices of Greek and Hebrew Lecturer. On leaving Cambridge he went as a physician to Huntingdon, and thence to Bury St Edmunds. In 1795 he was elected a Fellow of the Royal College of Physicians.

Wollaston's committment to various branches of natural science had been increasing, and in 1800 he decided to retire from medical practice. He pursued a very wide range of interests including physiology, chemistry, optics, mineralogy, electricity, mechanics and botany. He took a house in Fitzroy Square, London, where he set up a laboratory and carried out chemical research. Within five years, during which time he also worked in association with Smithson Tennant, he developed a process for making platinum malleable, which brought him a substantial fortune. Wollaston's work also led to the discovery of two other new elements palladium and rhodium. He also did research on optics and his reputation as an inventor of optical apparatus was high; an invention through which he did important work in crystallography was the reflecting goniometer, which first made possible the exact measurement of crystals and determination of minerals.

In 1802 he was awarded the Copley Medal, and in 1828 a Royal Medal; he was Secretary of the Royal Society from 1804–1816 and served also as Vice-President. He gave the Croonian Lecture in 1809, and was the Bakerian Lecturer in 1802, 1805, 1812 and 1828. In 1820 he was President of the Society for an interim period before the election to that post of Humphry Davy. From 1800 he began to suffer in health, and in 1828, when a serious medical condition had developed, he continued to direct experiments in a room next to his sick-room up to near the time of his death.

Wollaston was amongst the foremost men of science in Europe; his talents included extraordinary experimental dexterity, with the ability to work with unusually small quantities, and he was regarded as the most skilful chemist and mineralogist of his day.
DSB 14, 484–494. D.C. Goodman. DNB.

GRAHAM CHARLES WOOD 1934–; FRS 1997

Bromley Grammar School. Univ. of Cambridge: Christ's College, 1953, BA 1953, PhD 1959, MA 1960, ScD, 1972; Postdoctoral Fellow, 1959– 61. Univ. of Manchester Institute of Science and Technology (UMIST): 1961–72, successively Lecturer, Senior Lecturer and Reader in Corrosion Science; Professor of Corrosion Science and Engineering, 1972–97; Head of Corrosion and Protection Centre, 1972–82; Deputy Principal, 1983; Vice-Principal, 1992–94; Pro-Vice-Chancellor, 1994–. FREng 1990.
Corrosion and oxidation.

SIR MARTIN FRANCIS WOOD 1927–; FRS 1987

Gresham's. Cambridge Univ.: Trinity College MA. Imperial College; Royal School of Mines: BSc. Oxford Univ.; Christ Church: MA. National Service; Bevin Boy, South Wales and Derbyshire Coalfields, 1945–48. NCB: 1953–55. Oxford Univ., Clarendon Lab: Senior Research Officer, 1959–69. Chairman, National Committee for Superconductivity, SERC DTI: 1987–92. Oxford Instruments plc: Founder, 1959, Deputy Chairman; Chairman until 1983. Mullard Medal, 1982. Clifford Paterson Lecturer, 1996. Knighted 1986.
Technology of high magnetic fields; development of cryomagnetic equipment.

STEPHEN ESSLEMONT WOODS 1912–1994; FRS 1974

Stephen Woods was born in Cardiff, the fifth of eight children of William Woods and Annie née Whyte, and his father was a shipmaster. From 1923–1929 Stephen received his secondary education at Canton High School, Cardiff. He was awarded a Welsh foundation scholarship in 1929 and entered Jesus College, Oxford to study chemistry in the honour school of natural science; in 1931 he was awarded a Goldsmiths' Company Exhibition. Two years later he obtained a first class honours BSc, having carried out research on the decomposition of nitrogen trichloride in the Part II course. Following graduation he was a Demonstrator in the Jesus College Laboratory, and also received a grant from the Department of Scientific and Industrial Research for the doctorate research that he carried out from 1933–1935 on the same theme as his undergraduate project. In 1937 he received the MA and DPhil degrees.

Woods joined the Research Department of the then National Smelting Company at Avonmouth, Bristol, as a Chemist in 1935. During his career he held a series of appointments: Assistant Research Manager (1948); Deputy Research Manager (1953); Development Manager (1954); Process Development Manager (1963); Head of Research and Process Development (1963); Head of Research (1964); Technical Advisor to the Managing Director, Imperial Smelting Corporation (1966); Senior Research Associate (1970) and Research Secretariat – Chief Scientist (1971). In addition he was a Director of the Imperial Smelting Processes Ltd from 1964 and of the National Smelting Corporation from 1965–1970.

At Avonmouth, Woods played an important role in the team that successfully developed the zinc blast furnace process that achieved continuous production of zinc on a large scale and with improved fuel economy. Stephen Woods was widely known as 'Doc'; his advice was sought by people throughout the Rio Tinto Zinc organisation, and he applied himself to technical problems of RTZ world wide. He had the rare ability to have productive scientific discussions with a wide range of his colleagues, both theoretical specialists and experimentalists. Encouragement and guidance were given from his analytical approach and wisdom, and he showed remarkable scientific insight. His hobbies were reading, book collecting and walking. He married Ivy Oliver in 1939 and they had two daughters and a son.

Although Woods did not publish widely in the literature, he received in conjunction with S.W.K. Morgan, L.J. Derham and B.G. Perry, the Mullard Award of the Royal Society in 1970. As a scientist he will be remembered for his work in metals production, particularly his decisive contribution to the successful development of the zinc blast furnace process.

BMFRS 1988, 44, 455–469. A.K. Barbour. S.W.K. Morgan, 'Development of the zinc-lead blast-furnace as a research project', *Trans Inst. Min. Metall.*, (Section C) 1983, **92**, C 192–200.

CHARLES ROMLEY ALDER WRIGHT 1844–1894; FRS 1881

Charles Alder Wright was born at Southend, Essex, the son of Romley Wright, a civil engineer. During his boyhood, his parents moved to Manchester, where he attended Owens College, with a view to entering his father's profession. However, he developed a liking for experimental chemistry influenced by Henry Roscoe. He took the London matriculation examination, gaining the top place in 1863, followed the next year by taking the intermediate examinations and obtaining an exhibition in chemistry and natural philosophy.

His first experimental investigation under Roscoe was published in 1866. Soon after, in a post as a Chemist in the Runcorn Soap and Alkali Company he carried out investigations on the manufacture of sulphuric acid and alkalis. Moving to London, he became an Assistant in a laboratory attached to St Thomas's Hospital, and afterwards joined Augustus Matthiessen in organic chemistry investigations. From 1871 until his death he was a Lecturer at St Mary's Hospital. Wright also carried out metallurgical work (1869) in association with Lowthian Bell on the theory of iron smelting, the results being published in Bell's book *On the Principles of Manufacture of Iron and Steel.*

In physical metallurgy Wright collaborated with William Roberts-Austen on measurements of specific heat. Also he conducted extensive and important collaborative experiments which made

a pioneering contribution to the study of binary and higher order phase diagrams.
PRS 1894–95, 57, v–vi. TET. PRS 1890–91, 49, 156–173. A. Prince, *Alloy Phase Diagrams,* Elsevier, Amsterdam 1966.

SIR JOHN WROTTESLEY 1798–1867; FRS 1841

John Wrottesley was the eldest son of Sir John Wrottesley, Wrottesley Hall, Staffordshire, and Caroline née Bennet. He was educated at Westminster School (1810–1814)and at Christ Church, Oxford, where he graduated BA in 1819. In 1821 he married Sophia Eizabeth Gifford and they had five sons and two daughters. Beginning at Lincoln's Inn, London in the same year he was called to the bar in 1823. At Blackheath, from 1829 he constructed an astronomical observatory, which in 1841 he transferred to Wrottesley Hall. He served on several Royal Commissions of a scientific nature, including that dealing with iron in the railway industry. In the Royal Society he was Vice-President from 1844–46; 1852–53 and 1855–60 and President from 1854–57.
PRS 1867–68, 16, lxiii–lxiv. DNB.

SIR ALFRED FERNANDEZ YARROW 1842–1932; FRS 1922

Born in London, Alfred Yarrow was the elder son of Edgar William Yarrow and Esther née Lindo. He was educated at University College School, which he left at the age of 15 to serve a five year apprenticeship to Messrs Ranhill Salkeld and Co., a firm of marine engine builders. With a friend he attended lectures by Faraday at the Royal Institution and he and his friend carried out experiments in engineering and took out some patents. When his apprenticeship was completed he was engaged as their London representative by Messrs Coleman and Sons of Chelmsford. In 1866 Yarrow entered into a partnership to establish a small works on the Isle of Dogs, London. There in 1868 he set about improving the design and performance of steam launches; between 1869–75, 350 small vessels were made, including some constructed by pre-fabrication to allow transport of parts for later assembly. Subsequently, he concentrated on smaller war vessels, and in 1876 produced his first torpedo boat for the Argentine fleet. In the following year the firm of Yarrow was constructing these vessels for the British, French, Greek and Russian Navies; he continued to develop the design of torpedo boats and in 1894 the Russian Navy was supplied with a this type of vessel with a speed of 30 knots.

In 1917 after 40 years at his works on the Thames he moved to the River Clyde in Scotland to a new shipyard at Scotstoun. The following year his first destroyer was launched. Having retired to Hampshire in 1913, he offered his services to Lord Fisher with the coming of war in 1914. They planned the rapid construction of additional destroyers, and he was responsible for the construction of shallow draft gunboats for service in Mesopotamia. During the four years of war Yarrow's shipyard constructed 29 destroyers.

Yarrow married Minnie Florence Franklin in 1875 and they had three sons and two daughters; following her death he married Eleanor Cecilia Barnes. He was created a Baronet in 1916 in recognition of his war services. Among his generous benefactions was the sum of £100,000 to the Royal Society for the provision of research professorships. His achievements as a shipbuilder and marine engineer led to the creation of one of the most famous engineering firms in the world.
ONFRS 1932–35, 1, 7–11 E.H.T.-D'E. DNB 1931–40, 926–928. E.H. Tennyson-D'Eyncourt.

WILLIAM YOLLAND 1810–1885; FRS 1859

William Yolland was born at Plympton St Mary, Devonshire, the son of John and Priscilla Yolland. He attended Trueman's mathematical school, Exeter, and the Royal Military Academy at Woolwich.

In 1828 he was commissioned as a Second Lieutenant in the Royal Engineers; he received a series of promotions, culminating in the rank of Brevet Colonel in 1858. His career after initial commissioning included a period in Canada (1831–35) and appointments in ordnance and in surveying. In 1843 he married Ellen Catherine Rainier and they had five daughters and one son. Yolland was appointed an Inspector of Railways under the Board of Trade in 1854; he retired from military service in 1863 but continued his work with the Board of Trade. From 1877 until his death he served as Chief Inspector of Railways. In 1880 he was a member of the commission enquiring into the Tay bridge disaster. He was awarded the CB in 1881.
DNB.

THOMAS YOUNG 1773–1829; FRS 1794

Thomas Young was born in Milverton, Somerset, the eldest son of Thomas Young, a mercer and banker and Sara née Davis and was raised as a member of the Society of Friends. He showed extraordinary ability at an early age, having learnt to read fluently by the age of two and beginning the study of Latin by the time he was six. Although largely self educated in languages and natural philosophy, he attended two boarding schools from 1780 to 1786, where he learned elementary mathematics, acquired a reading knowledge of Latin, Greek, French and Italian, and also skill in drawing and turning. His exceptional linguistic ability included a particular interest in ancient languages of the Middle East, and he began to study independently eight of these, including Hebrew, Arabic, Persian, and Turkish. By 1792 he had become proficient in Greek and Latin, as well as having advanced his mathematical knowledge and having read extensively, including Newton's *Principia* and *Opticks*, and Lavoisier's *Elements of Chemistry*.

In 1792, guided by a great-uncle, a distinguished London physician, he began to study medicine at St Bartholomew's Hospital, London; he continued these studies in Edinburgh and in Göttingen, where he graduated MD in 1796. During the period in Edinburgh he began to enjoy music, dancing and the theatre, and to move away from the practices of the Society of Friends; later in his life he formally became a member of the Church of England. From 1797–1803 he was a Fellow-Commoner student at Emmanuel College, Cambridge, partly to fulfil the requirements of the Royal College of Physicians, and partly to comply with the wish of his uncle, who sponsored both his medical education and his election to the Royal Society, at the age of 21. While at Cambridge he was known as 'Phenomenon Young', and began original scientific studies; he was awarded an MB in 1803 and an MD five years later. His uncle also introduced him to influential people such as William Herschel. When his uncle died in 1797, Young was the beneficiary of his London house, his library and paintings, and the sum of £10,000.

Having moved from Cambridge to London in 1800 Young attempted to establish a medical practice, but this venture was not very successful. Consequently he had time to attend Royal Society meetings regularly and became known to the President, Sir Joseph Banks and to Benjamin Thompson, Count Rumford, founder of the Royal Institution. This led to Young's appointment as 'Professor of Natural Philosophy, editor of the journals and Superintendent of the House', at the Royal Institution at an annual salary of £300. The lectures that he was required to give on natural philosophy and the mechanical arts did not prove to be very successful because they were not sufficiently 'popular' in their approach. Young, having resigned his post in 1803, resumed medical practice in London, and also, in the summers, at Worthing, a fashionable sea side resort. In 1804 he married Eliza Maxwell, who was related to the Scottish aristocracy. His medical career continued, including election to the Royal College of Physicians in 1809, lecturing at the Middlesex Hospital, and, in 1811 election to a permanent position as physician to St George's Hospital.

Young's early research was concerned with the physiology of vision, and with optics. In the latter he discovered the principle of interference, and obtained experimental proof, illustrating the wave nature of light. In mechanics he established important concepts in a course of lectures published in 1807; a notable contribution was made in his definition of a 'modulus of elasticity'. Another significant contribution, published in 1805, was a theory of capillary action.

Resulting from his uncle's influence and his marriage Young occupied a high social position. With his ability and influential situation he followed careers as a physician, writer, natural philosopher, administrator of science and Egyptologist. In the Royal Society he held the office of Foreign Secretary from 1804 to his death, and was a member of its Council from 1806. He gave the Bakerian Lecture in 1800, 1801 and 1803, and the Croonian Lecture in 1808. He was a consultant to the Admiralty, Secretary of the Royal Commission on Weights and Measures, and Secretary of the Board of Longitude. His wide ranging writings included coverage of optics and mechanics. From 1815 onwards he published papers in Egyptology, mainly concerned with the reading of tablets involving hieroglyphics; he was one of the first to interpret the writings on the Rosetta Stone. Possessing extraordinary talents and insight, Young initiated important investigations that others were able to pursue and complete.

Bulloch's Roll. Roy. Soc. Journal Book XLV, 530–536. DSB 14, 562–572. E.W. Moore. DNB.

JOHN MICHAEL ZIMAN 1925–; FRS 1967

Hamilton High School, NZ. Victoria Univ. Coll., Wellington, NZ. Oxford Univ.: Balliol College: Junior Lecturer in Mathematics; 1951–53; Pressed Steel Co Ltd., Research Fellow, 1953–54. Cambridge Univ.: Lecturer in Physics, 1954–64. Bristol Univ.: Professor of Theoretical Physics, 1964–69; Melville Wills Professor of Physics, 1969–76; Director H.H. Wills Physical Lab, 1971–81. Henry Wills Overton Professor of Physics, 1971–82; Emeritus Professor 1988–.

Solid state physics: transport phenomena; electronic properties of liquid metals.

APPENDIX 1

Some Aspects of the History of Metallurgical Education of University Rank in Great Britain, Relating to Fellows of the Royal Society (Up to mid-20th century)

Fellows of the Royal Society have exerted a major influence on higher education in metallurgy in Great Britain in the past century and a half, through holding chairs and headships of University departments, designing undergraduate courses, building up research schools, and writing important books both at undergraduate and higher levels. The thematic chapters of the book, and the biographies (Part II) contain information on the roles played, and a brief summary of part of the historical background is given here.

In 1924, as part of the Proceedings of the Empire Mining and Metallurgical Congress held in London, Professor Henry Cort Harold Carpenter, (later Sir Harold, 1875–1940; FRS 1918) presented a detailed review of metallurgical education of university rank in Great Britain (A1.i). Some of the courses discussed were founded at Colleges which subsequently became university institutions. His review, supplemented by other sources, (A1.ii, iii) presents an interesting insight into historical and academic aspects.

Advanced teaching in metallurgy in Great Britain was an offspring of the Great Exhibition held in 1851 which created a demand for scientific technical education. A technical school for mining and other branches of industry, especially in the chemical field, was formed in 1851 in connection with the Museum of Practical Geology in Jermyn Street, London, and known as The Government School of Mines and of Science Applied to the Arts. In succeeding years changes occurred which led to the Institution concentrating on mining,metallurgy and geology, and in 1863 the Institution became The Royal School of Mines. John Percy (1817–1889; FRS 1847) was the first holder of the chair in Metallurgy, and his successor in 1880 was William Chandler Roberts-Austen (later Sir William, 1843–1902; FRS 1875). For a period of about 30 years the Royal School of Mines was the only institution in Great Britain providing courses leading to qualifications equivalent to a degree in metallurgy. However, in 1892 three other centres were listed as offering 'advanced instruction' in metallurgy: Camborne School of Mines, Cornwall; Sheffield Technical College; and Durham College of Science, Newcastle upon Tyne. Regular instruction in metallurgy was also offered at King's College, London.

During the early years of his professorship Roberts-Austen was actively involved with the Science and Art Department in organising metallurgy teaching throughout the country; as the examiner in metallurgy under the Department he was influential in guiding the teaching. Moreover, his text-book for students of metallurgy entitled *An Introduction to the Study of Metallurgy*, first published in 1891 quickly acquired a world-wide reputation, and was followed by later editions, the fifth just before his death. The use of the word *study* in the title is significant. The preface to the first edition commented '*The literature of metallurgy is rich, but those beginning to study it need guidance to a knowledge of the principles on which the art is rightly practiced.*'... A reviewer wrote: '*There has been a distinct want of a systematic exposition of the general principles of metallurgy and of clear statements as to the physical characteristics of metal and alloys.... The evident purpose of the volume is to meet this want, the author having deliberately subordinated details of smelting operations in order that he might deal at length with the physical properties of metals and the constitution and characteristics of alloys, ... The book will hardly be popular with the class of students who merely attempt to 'cram'* '.. Roberts-Austen's fundamental approach is illustrated by his far-sighted advice to students that success would '*depend on the facility with which you are able to use the tools of thought furnished by chemistry, physics and mechanics.*'

Between 1884 and 1894 five important metallurgical centres of higher education were established: at Sheffield (1884), Birmingham (1886), Glasgow (1886), Newcastle upon Tyne(1891) and Cardiff (1894). Their establishment and development were linked with important metallurgical activities in the respective localities. After an interval of twelve years Departments of Metallurgy were established at Manchester (1906), Leeds (1907) and Cambridge (1908), for reasons specific to the local situations. At Manchester, a major centre of engineering activity, the title of the chair, Professor of Metallurgy and Metallography reflected the rapid development of metallography (essentially physical metallurgy) and its relevance to engineering and its problems. Leeds also was a centre of engineering rather than metallurgy, and at the University there was a close relation between fuel and metallurgy,expressed in the title, Department of Coal Gas and Fuel Industries with Metallurgy. At Cambridge the establishment of instruction was associated with the foundation of a Readership by the Worshipful Company of Goldsmiths, the first incumbent being Charles Thomas Heycock (1858 –1931; FRS 1895) special attention was paid to the study of precious metals and assaying, and the main emphasis was postgraduate. After a further period of twelve years, professorships of metallurgy were established in 1920 at the University of Liverpool, and at the University College of Swansea. In Liverpool, where there was a strong engineering faculty, the emphasis was on engineering aspects. Swansea, historically a famous metallurgical centre, already had a technical college giving metallurgical instruction.

Carpenter (A1.i) referred to the transformation that had gradually occurred of metallurgy from an art into a science rendering it a subject suitable for university instruction. He referred to the eleven departments in existence in 1924 as having their own special character,with every branch of the subject represented. Also he expressed the view 'Since metallurgical departments are expensive to found and maintain, any further resources which may become available for education and research should be devoted, not to establishing new departments, but to strengthening and extending existing ones'.

However, the subsequent history of higher education shows that his view was not adopted; beginning some three decades later following World War II there was a substantial increase in advanced metallurgical education and research both in terms of numbers of departments and of undergraduate and postgraduate students, particularly during the period of general expansion of university education in Britain. The role of Fellows of the Royal Society has continued strongly throughout this later period, during which period the transition from metallurgy to materials science and engineering has been a key feature.

REFERENCES

A1.i. H.C.H. Carpenter, *Metallurgical Education of University Rank in Great Britain*, Proc. Empire Mining and Metallurgical Congress 1924 Part V, Edited by the General Secretaries and G. Shaw Scott, Published by the Offices of the Congress, London, 1925, 43–79, Discussion 80–90.

A1.ii. J.H. Merivale, 'The Education of Mining Engineers', *Trans. Inst. Min. Eng.*, 1892–93, v5, 626–630.

A1.iii. H. O'Neill, 'University metallurgical training in Britain', *Bull. Instn. Metallurgists*, 1948, **6**, 22–29.

Note: Interesting information concerning University Metallurgy Departments and Fellows of the Royal Society referred to in the thematic chapters is contained in the book:

Michael Sanderson, *The Universities and British Industry 1850–1970*, Routledge and Kegan Paul, London 1972.

APPENDIX 2

Royal Society Awards

ROYAL SOCIETY AWARDS

Fellows and Foreign Members with associations with metallurgy and materials, and referred to in the book.

(Note: in most cases the awards were not made for metallurgical research.)

Several medallists referred to in the thematic chapters, but not elected FRS or FMRS, are included in the following lists.

The citations for medallists since 1940 are quoted in abridged or complete form from the *Royal Society Record*, and the *Supplement 1940–89*.

THE COPLEY MEDAL (1731)

This medal is the premier award of the Royal Society, given annually for outstanding achievements in research in any branch of science.

1731 Stephen Gray
1732 Stephen Gray
1734 John Theophilus Desaguliers
1736 John Theophilus Desaguliers
1741 John Theophilus Desaguliers
1744 Henry Baker
1749 John Harrison (not elected FRS)
1753 Benjamin Franklin
1754 William Lewis
1758 John Dollond
1759 John Smeaton
1766 William Brownrigg
Hon. Henry Cavendish
1768 Peter Woulfe
1772 Joseph Priestley
1781 William Herschel
1792 Benjamin Count Rumford
1794 Alessandro Volta
1795 Jesse Ramsden
1798 Charles Hatchett
1802 William Hyde Wollaston
1803 Richard Chenevix
1804 Smithson Tennant
1805 Sir Humphry Davy
1809 Edward Troughton
1813 William Thomas Brande

1817 Captain Henry Kater
1820 Hans Christian Oersted
1821 John Frederick William Herschel
1825 Dominique (Francois) Jean Arago
Peter Barlow
1832 Michael Faraday
Baron Siméon Denis Poisson
1836 Jöns Jacob Berzelius
1837 Antoine-César Becquerel
1837 John Frederic Daniell
1838 Michael Faraday
1840 Justus von Liebig
1843 Jean Baptiste André Dumas
1847 Sir John Frederick William Herschel
1859 Wilhelm Eduard Weber
1860 Robert Wilhelm Bunsen
1862 Thomas Graham
1868 Sir Charles Wheatstone
1869 Henri Victor Régnault
1869 James Prescott Joule
1872 Friedrich Wöhler
1875 August Wilhelm von Hofmann
1883 Sir William Thomson
1886 Franz Ernst Neumann
1893 Sir George Gabriel Stokes
1899 Lord Rayleigh
1900 Pierre Eugène Marcellin Berthelot
1901 Josiah Willard Gibbs
1904 Sir William Crookes
1905 Dmitri Ivanovich Mendeleev
1914 Sir Joseph John Thomson
1916 Sir James Dewar
1918 Hendrik Antoon Lorentz
1922 Sir Ernest Rutherford
1923 Horace Lamb
1925 Albert Einstein
1928 Sir Charles Parsons
1930 Sir William Henry Bragg
1935 Charles Galton Darwin
1938 Niels Henrik David Bohr
1940 Paul Langevin: for his pioneer work on the electron theory of magnetism, his fundamental contributions to discharge of electricity in gases, and his important work in many branches of theoretical physics .
1944 Sir Geoffrey Ingram Taylor: for his many contributions to aerodynamics, hydrodynamics and the structure of metals, which have had a profound influence on the advance of physical science and its applications.

1950 James Chadwick: for his outstanding work in nuclear physics and the development of atomic energy, especially for his discovery of the neutron.
1952 Paul Adrien Maurice Dirac: in recognition of his remarkable contributions to relativistic dynamics of a particle in quantum mechanics.
1966 Sir William Lawrence Bragg: in recognition of his distinguished contributions to the development of methods of structural determination by X-ray diffraction.
1972 Sir Nevill Francis Mott: in recognition of his original contributions over a long period to atomic and solid state physics.
1976 Dorothy Mary Crowfoot Hodgkin: in recognition of her outstanding work on the structure of complex molecules , particularly penicillin, vitamin B_{12} and insulin.
1986 Sir Rudolf Ernst Peierls: in recognition of his fundamental contributions to a wide range of theoretical physics , and signal advances in proposing the probable existence of nuclear chain reactions in fissile molecules.
1994 Sir Frederick Charles Frank: in recognition of his fundamental contributions to the theory of crystal morphology , in particular to the source of dislocations and their consequences in interfaces and crystal growth ; to fundamental understanding of liquid crystals and the concept of disclination, and to the extension of crystallinity concepts to aperiodic crystals. He has also contributed through a variety of remarkable insights into a great number of physical problems.
1995 Sir Alan Howard Cottrell: in recognition of his contributions to the understanding of mechanical properties of metals and related topics through his pioneering studies on crystal plasticity, dislocation impurity interactions, fracture and irradiation effects.

THE RUMFORD MEDAL (1800)

The medal is awarded biennially in recognition of an outstandingly important discovery in the field of thermal or optical properties of matter made by a scientist working in Europe, noting that Rumford was concerned to see recognized discoveries that tended to promote the good of mankind.

1800 Sir Benjamin Thompson (Count Rumford)
1816 Sir Humphry Davy
1832 John Frederic Daniell
1840 Jean Baptiste Biot
1846 Michael Faraday
1848 Henri Victor Régnault
1850 Dominique Francois Jean Arago
1852 George Gabriel Stokes
1860 James Clerk Maxwell
1862 Gustav Robert Kirchhoff
1864 John Tyndall
1872 Anders Jöns Ångström
1894 James Dewar
1900 Antoine-Henri Becquerel
1902 Charles Algernon Parsons
1904 Ernest Rutherford
1906 Hugh Longbourne Callendar

1908 Hendrik Antoon Lorentz
1912 Heike Kamerlingh Onnes
1914 Lord Rayleigh
1916 William Henry Bragg
1930 Peter Joseph Wilhelm Debye
1935 Charles Vernon Boys
1936 Ernest George Coker
1940 Karl Manne Georg Siegbahn: for his pioneer work in high precision X-ray spectroscopy and its applications.
1948 Franz Eugen Simon: for his outstanding contributions to the attainment of low temperatures and to the study of the properties of substances at temperatures near the absolute zero.
1950 Frank Whittle: for his pioneering contributions to the jet propulsion of aircraft.
1954 Cecil Reginal Burch: for his distinguished contributions to the techniquefor the production of high vacua and for the development of the reflecting microscope.
1956 Francis William Bowden: in recognition of his distinguished work on the nature of friction.
1968 Dennis Gabor: in recognition of his distinguished contributions to optics, especially by establishing the principles of holography.
1970 Christopher Hinton: in recognition of his outstanding contributions to engineering and of his leadership of engineering design teams in the chemical and atomic energy industries and in electricity generation.
1974 Alan Howard Cottrell: in recognition of his contributions to physical metallurgy and particularly in extending knowledge of the role of dislocations in the fracture of metals.
1990 Walter Eric Spear: for discovering and applying techniques for depositing and characterising thin films of high quality amorphous silicon and for demonstrating that these can be doped to give useful electronic devices, such as cost-effective solar cells and large arrays of thin film transistors, now used in commercial, flat-panel, LED colour TV systems.
1994 Andrew Keller: in recognition of his contributions to polymer science, in particular his elucidiation of the basis of polymeric crystallisation, to methods of making strong fibres and to the understanding of polymer solutions which underlie this technology.
1998 Richard Henry Friend: in recognition of his leading research in the development of polymer-based electronics and opoelectronics.

THE ROYAL MEDALS (1825 and 1965)

Three Royal Medals are awarded annually by the Sovereign upon the recommendation of the Council, two for the most important contributions to the advancement of Natural Knowledge (one to each of the two great divisions) and the other for distinguished contributions in the applied sciences.

1826 John Dalton
1827 Sir Humphry Davy
1828 William Hyde Wollaston
1833 Sir John Frederick William Herschel
1838 Thomas Graham
1835 Michael Faraday

1836 Sir John Frederick William Herschel
1837 Rev William Whewell
1838 Thomas Graham
1840 Charles Wheatstone
Sir John Frederick William Herschel
1841 Eaton Hodgkinson
1842 John Frederic Daniell
1843 Charles Wheatstone
1846 Michael Faraday
1847 William Robert Grove
1850 Thomas Graham
1851 William Parsons (third Earl of Rosse)
1852 James Prescott Joule
1854 August Wilhelm von Hofmann
1856 William Thomson
1860 William Fairbairn
1869 Augustus Matthiesen
1870 William Hallowes Miller
1871 John Stenhouse
1873 Henry Enfield Roscoe
1874 Henry Clifton Sorby
1875 William Crookes
1876 William Froude
1877 Frederick Augustus Abel
1882 Lord Rayleigh
1888 Osborne Reynolds
1890 John Hopkinson
1894 Joseph John Thomson
1895 James Alfred Ewing
1896 Charles Vernon Boys
1902 Horace Lamb
1909 Augustus Edward Hough Love
1910 John Joly
1926 Sir William Bate Hardy
1930 Owen Willans Richardson
1927 John Cunningham McLennan
1931 Sir Richard Tetley Glazebrook
1933 Geoffrey Ingram Taylor
1936 Ralph Howard Fowler
1942 Walter Norman Haworth: for his fundamental contributions to organic chemistry, particularly to the constitution of the sugars and the structures of complex polysaccharides.
1945 John Desmond Bernal: for his work on the structure of proteins and other substances by X-ray methods and for the solution of many problems requiring a physical approach.
1946 William Lawrence Bragg: for his distinguished researches on the sciences of X-ray structure analysis and X-ray spectroscopy.
1949 George Paget Thomson: for his distinguished contributions to many branches of atomic physics, and especially for his work in establishing the wave properties of the electron.

1953 Nevill Francis Mott: in recognition of his eminent work in the field of quantum theory and particularly in the theory of metals.

1954 Sir John Douglas Cockcroft: in recognition of his distinguished work on nuclear and atomic physics.

1956 Dorothy Mary Crowfoot Hodgkin: in recognition of her distinguished work in the elucidation of the structures of penicillin, vitamin B_{12} and other important compounds by the methods of X-ray crystallography.

1958 Harrie Stewart Wilson Massey: in recognition of his distinguished contributions to physics and particularly for his experimental and theoretical studies of collision phenomena in gases.

1959 Rudolf Ernst Peierls: in recognition of his distinguished work on the theoretical foundations of high energy and nuclear physics.

1968 Gilbert Roberts: in recognition of his distinguished contributions to civil engineering, and in particular to the design and construction of long-span suspension bridges.

1969 Charles William Oatley: in recognition of his distinguished work in the development of radar and latterly for the design and development of a highly successful scanning electron microscope.

1970 John Fleetwood Baker: in recognition of his fundamental and applied work on the plastic behaviour and design of framed structures which is now being used throughout the world.

1975 Barnes Wallis: in recognition of the originality of his ideas and the determination with which he pursued them.

1976 Sir Alan Walsh: in recognition of his distinguished contributions to emission and infra-red spectroscopy and his origination of the atomic method of quantitative analysis

1977 Sir Peter Bernhard Hirsch: in recognition of his distinguished studies of defects in crystals and especially of his elucidation of the process of work hardening.

1979 Frederick Charles Frank: in recognition of his outstanding contributions to the theory of crystal growth, dislocations, phase transformations and polymers with wide application in physics, chemistry and geology.

1979 Vernon Ellis Coslett: in recognitition of his outstanding contributions to the design and development of the X-ray microscope, the scanning electron microprobe analyser, the high voltage and ultra-high resolution (2.5 nm) electron microscopes and their applications in many disciplines.

1986 Eric Albert Ash: in recognition of his outstanding researches on acoustic microscopy, leading to wholly new techniques and substantial improvements in resolution of acoustic microscopes

1992 David Tabor: distinguished for his seminal contributions to the basic study of friction and wear between solids, of considerable relevance to the design of machinery.

1993 Rodney Hill: for his outstanding contributions to the theoretical mechanics of solids and especially the plasticity of solids. He has made many diverse contributions to our understanding of the complex properties and behaviour of solid materials

1993 Volker Heine: in recognition of his contributions to solid state theory, in particular the bonding and structure of solids. He has been one of the most influential figures in the recent history of solid state theory .

1994 Sivaramakrishna Chandrasekhar: for his many new discoveries on the understanding of liquid crystals , for a synthesis of the subject in his seminal book *The invention of discotic liquid crystals and for elucidating their remarkable properties* .

1998 Donald Charlton Bradley: in recognition of his pioneering work on the molecular chemistry of metal alkoxides and metal amides, their synthesis, structure and bonding, and for his studies of their conversions to metal oxides and metal nitrides, processes which now find commonplace applications in materials science, especially in the fields of microelectronics and chemical vapour deposition.

1999 Archibald Howie: in recognition of his oustanding contributions to the development and application of electron microscopy of materials, and to the underlying theories of electron scattering, in particular his extensive contributions to inelastic scattering theory, his systematic high resolution microscope studies of amorphous materials, his introduction of the concept of coherence volume for hollow cone dark field imaging and his pioneering use of a high angle annular dark field detector to image small catalyst particles.

THE DAVY MEDAL (1877)

The award is made annually for an outstandingly important recent discovery in any branch of chemistry made in Europe or North America.

1877 Robert Wilhelm Eberhard Bunsen
Gustav Robert Kirchhoff
1882 Dmitri Ivanovich Mendeleev
Lothar Meyer (not elected FRS)
1883 Marcelin Berthelot
1883 Julius Thomsen
1887 John Alexander Reina Newlands (not elected FRS)
1888 William Crookes
1892 François Marie Raoult
1893 Jacobus Henricus van't Hoff
1895 William Ramsay
1896 Ferdinand Frederic-Henri (Henry) Moissan
1901 George Downing Liveing
1902 Svante August Arrhenius
1903 Pierre Curie and Madame Curie (not elected FRS)
1909 Sir James Dewar
1910 Theodore William Richards
1914 William Jackson Pope
1916 Henri Louis Le Chatelier
1920 Charles Thomas Heycock
1929 Gilbert Newton Lewis
1940 Harold Clayton Urey: for his isolation of deuterium, the heavy hydrogen isotope, and for his work on this and other isotopes in following the detailed course of chemical reactions.
1947 Linus Carl Pauling: in recognition of his distinguished contributions to the theory of valency and their application to systems of biological importance.
1951 Eric Keightley Rideal: for his distinguished contributions to the subject of surface chemistry.
1955 Harry Work Melville: in recognition of his distinguished work in physical chemistry and in polymeric actions.
1957 Kathleen Lonsdale: for her distinguished studies in the structure and growth of crystals.

1970 Charles Alfred Coulson: in recognition of his distinguished work in theoretical chemistry.
1973 John Stuart Anderson: in recognition of his distinguished contributions to chemistry, especially on the structural investigations of imperfect surfaces and non-stoichiometric materials.
1984 Samuel Frederick Edwards: in recognition of his distinguished contributions to the theoretical basis of thermodynamic aspects of polymer chemistry.
1988 John Anthony Pople: in recognition of his wide-ranging contributions to theoretical chemistry, especially his development and application of techniques for the computation of molecular wave-functions and properties.
1994 Sir John Meurig Thomas: for his pioneering studies of solid state chemistry and for the major advances he has made in the design of new materials for heterogeneous catalysis.

THE HUGHES MEDAL (1902)
The award is made annually in recognition of an original discovery in the physical sciences, particularly electricity and magnetism or their applications.

1902 Joseph John Thomson
1904 Sir Joseph Wilson Swan
1907 Ernest Howard Griffiths
1909 William Tetley Glazebrook
1910 John Ambrose Fleming
1915 Paul Langevin
1918 Irving Langmuir
1920 Owen Willans Richardson
1921 Niels Henrik David Bohr
1922 Francis William Aston
1931 William Lawrence Bragg
1932 James Chadwick
1934 Karl Manne Georg Siegbahn
1935 Clinton Joseph Davisson (not elected FMRS)
1938 John Douglas Cockcroft (joint)
1939 George Paget Thompson
1941 Nevill Francis Mott: for his fertile application of the principles of quantum theory to many branches of physics, especially in the field of nuclear and collision theory, in the theory of metals and in the theory of photographic emulsions.
1942 Enrico Fermi: in recognition of his outstanding contributions to the knowledge of the electrical structure of matter, his work on quantum theory and his experimental studies of the neutron.
1944 George Ingle Finch: in recognition of his fundamental contributions to the study of the structure and properties of surfaces and for his important work on the electrical ignition of gases.
1946 John Turton Randall: for his distinguished research into fluorescent materials and into the production of high frequency electromagnetic radiation.
1947 Jean-Frederic Joliot: in recognition of his distinguished contributions to nuclear physics, particularly the discovery of artificial radioactivity and of neutron emission in the fusion process.

1950 Max Born: for his contributions to theoretical physics in general and to the development of quantum mechanics.
1955 Harrie Stewart Wilson Massey: in recognition of his distinguished contributions to atomic and molecular physics, particularly in regard to collisions involving the production and recombination of ions.
1956 Frederick Alexander Lindemann (Lord Cherwell): in recognition of his distinguished work in many fields: the melting point formula and theory of specific heats: ionisation of stars; meteors and temperature inversion in the stratosphere.
1958 Edward Neville da Costa Andrade: in recognition of his distinguished contributions to many branches of classical physics.
1959 Alfred Brian Pippard: in recognition of his distinguished contributions in the field of low temperature physics.
1961 Alan Howard Cottrell: in recognition of his distinguished work on the physical properties of metals, particularly in relation to mechanical deformation and to the effects of irradiation.
1962 Brebis Bleaney: in recognition of his distinguished studies of electrical and magnetic phenomena and their connection with atomic and molecular properties .
1967 Kurt Alfred Georg Mendelssohn: in recognition of his distinguished contributions to cryophysics , expecially his discoveries in superconductivity and superfluidity.
1969 Nicholas Kurti: in recognition of his distinguished work in low temperature physics, and in thermodynamics.
1972 Brian David Josephson: in recognition particularly of his discovery of the remarkable properties of junctions between superconducting materials.
1973 Peter Bernhard Hirsch: in recognition of his distinguished contributions to the development of the electron microscopy of thin film techniques for the study of crystal defects and its application to a very wide range of problems in materials science and metallurgy.
1978 William Cochran: in recognition of his pioneering contributions to the science of X-ray crystallography.... for his original contributions to lattice dynamics and its relation to phase transitions.
1987 Michael Pepper: in recognition of his many important experimental investigations into the fundamanetal properties of semiconductors, especially low dimensional systems, where he has elucidated some of their unusual properties like electron localization and the quantum Hall effect.
1988 Archibald Howie and Michael John Whelan: in recognition of their contributions to the theory of electron diffraction and microscopy and its application to the study of lattice defects in crystals.
1995 David Shoenberg: in recognition of his work on the electronic structure of solids, in particular by exploiting low temperature techniques, particularly the De Haas van Alphen effect defining the Fermi surface of many metals .
1997 Andrew Richard Lang: in recognition of his fundamental work on X-ray diffraction physics and for his development of the techniques of X-ray topography, in particular in studying defects in crystal structures.

THE LEVERHULME MEDAL

The Leverhulme Trust Fund marked the occasion of the Tercentenary of the Royal Society by the award of a gold medal; this award is made triennually for an outstandingly significant contribution in the field of pure or applied chemistry or engineering.

1969 Hans Kronberger: in recognition of his many distinguished contributions to nuclear reactor research and development and for outstanding contributions in all branches of this field.
1981 Stanley George Hooker: in recognition of his outstanding work on superchargers of the *Merlin* engine, the development of the first Rolls-Royce jet engines, then Bristol engines including that for the jump jet and, later, the final development of the Rolls-Royce RB211.
1987 George William Gray: for his many contributions to the technologically important field of liquid crystals.

THE MULLARD MEDAL (1967)
The award was provided by a gift by Mullard Ltd to recognise an outstanding contribution or contributions to the advancement of science or engineering or technology leading directly to national prosperity in the United Kingdom.

1968 Lionel Alexander Bethune Pilkington: for his outstanding advances in the technology of glass manufacture and, in particular, for his invention and development of the float glass process.
1970 Stephen Esslemont Woods (shared): in recognition of outstanding contributions to the concept and development of the Imperial Smelting zinc blast furnace process.
1973 Charles William Oatley: in recognition of his outstanding contribution over an extended period to the design and development of the scanning electron microscope in which he played a significant and continuous part.
1974 Frank Brian Mercer: in recognition of his invention of the Netlon net process – an extrusion process for the manufacture of integral or knotless plastic net – which was of great ingenuity and simplicity with an extremely wide range of applications.
1982 Sir Martin Francis Wood (shared): in recognition of the development, manufacture and marketing of advanced superconducting magnet systems as a result of which Oxford Instruments Ltd was established as the leading supplier of these systems throughout the world.

THE ESSO ENERGY MEDAL (1974)
The award was instituted following an agreement between the Royal Society and Esso Petroleum Company (now Esso UK plc) for outstanding contributions to the advancement of science or engineering or technology leading to the more efficient mobilisation, use or conservation of energy resources.

1978 Thomas Nelson Marsham (shared): for their joint contributions, variously through the United Kingdom Atomic Energy Authority and the Nuclear Power Company, to the development of the successful Prototype Fast Reactor at Dounreay which was commissioned in 1976.
1979 John Edwin Harris (shared): for outstanding contributions to major improvements in nuclear fuel utilisation in the MAGNOX reactors of the Central Electicity Generating Board.
1994 Alan Arthur Wells (shared): for work on the design and development of shoreline wave power stations.

THE ARMOURERS AND BRASIERS' MEDAL (1984)

The award is for excellence in materials science and technology.

1985 Michael Farries Ashby
1988 Kenneth Henderson Jack (shared)
1993 James Derek Birchall (shared)
1997 Harshad Kumar Dharamshi Hansraj Bhadeshia
1999 David Godfrey Pettifor

ARMOURERS AND BRASIERS' RESEARCH FELLOWS

1924–29 Constance Flig Elam (not elected FRS)
1929–31 William Hume-Rothery
1932–36 Ulick Richardson Evans
1949–51 Robert William Kerr Honeycombe

WARREN RESEARCH FELLOWS

1932–38 Albert James Bradley
1932–42 William Hume-Rothery
1937–44 John Turton Randall
1943–55 William Hume-Rothery
1945–49 Frank Reginald Nunes Nabarro
1948–66 Cecil Reginald Burch
1977–80 Brebis Bleaney
1978–86 Michael Pepper

APPENDIX 3

Bakerian Lectures

The Bakerian Lecture is the premier lecture of the Royal Society in physical sciences. It originated in 1775 through a bequest by Henry Baker FRS for an oration or discourse which was to be spoken or read yearly by one of the Fellows.

The lectures listed here were those which relate to metallurgical, or related, themes covered in the book and presented by Fellows listed in the book.

METALLURGY AND MATERIALS

1775 Peter Woulfe, 'Experiments made in order to ascertain the nature of some Mineral Substances, and in particular how far the Acids of Sea-Salt, and of Vitreol contribute to Mineralise Metallic and other Substances, Part I', *Journal-book*, vol. xxix, p. 135.

1788 Tiberius Cavallo, 'On an Improvement in the Blow Pipe', *Journal-book*, vol. xxxiii, p 257.*

1790 Tiberius Cavallo, 'A description of a new pyrometer', *Journal-book*, vol. xxxiv, p. 208.

1807 Sir Humphry Davy, 'On some new Phenomena of Chemical Changes produced by Electricity, particularly the Decomposition of the fixed Alkalis, and the Exhibition of the new Substances which constitute their Bases', *Phil. Trans.*, 1808.

1809 Sir Humphry Davy, 'On some new Electro-Chemical Researches, on various objects, particularly the Metallic Bodies from the Alkalis and Earths; and on some combinations of Hydrogen', *Phil. Trans.*, 1810.

1812 William Hyde Wollaston, 'On the Elementary Particles of certain Crystals', *Phil.Trans.*, 1813.

1813 William Thomas Brande, 'On some new Electro-Chemical Phenomena', *Phil. Trans.*, 1814.

1819 Robert John Strutt, 'A structure of the line spectrum of sodium as excited by fluorescence', *Proc. A*, vol 96.

1820 Captain Henry Kater, 'On the best kind of Steel, and form, for a Compass Needle', *Phil. Trans.*, 1821.

1828 William Hyde Wollaston, 'On a method of rendering platina malleable', *Phil. Trans.*, 1829.

1829 Michael Faraday, 'On the Manufacture of Glass for Optical Purposes', *Phil. Trans.*, 1830.

1833 Samuel Hunter Christie, 'Experimental Determination of the Laws of Magneto - Electric Induction in different masses of the same metal , and of its intensity in different metals', *Abstracts of Papers*, vol. iii, 177.

1850 Thomas Graham, 'On the Diffusion of Liquids', *Phil. Trans.*, 1850.

*Tiberius Cavallo was Bakerian lecture in thirteen successive years (from 1780–1792), covering a range of subjects, including thermometry, electricity and magnetism. Only two of these lectures are listed here. Sir Humphry Davy gave seven Bakerian Lectures in the period 1805–1826, of which two are listed here.

1855 John Tyndall, 'On the Nature of the Force by which Bodies are repelled from the Poles of a Magnet; to which is added a prefixed account of some experiments on Molecular Influence', *Phil. Trans.*, 1855.

1856 William Thomson, 'On the Electrodynamic Qualities of Metals', *Phil. Trans.*, 1856.

1857 Michael Faraday, Experimental Relations of Gold (and other metals) to Light', *Phil. Trans.*, 1857.

1868 Henry Enfield Roscoe, 'Researches on Vanadium', *Phil. Trans.*, 1868.

1871 Charles William Siemens, 'On the Increase of Electrical Resistanc in Conductors with Rise of Temperature, and its Application to the Measure of Ordinary and Furnace Temperatures ; also on a simple Method of measuring Electrical Resistances', *Proc. vol. 18*

1883 William Crookes, 'On Radiant Matter Spectroscopy : the Detection and wide Distribution of Yttrium', *Phil. Trans.*, 1883.

1896 William Chandler Roberts-Austen, 'On the Diffusion of Metals', *Phil. Trans.*, 1896.

1897 Osborne Reynolds and W.H. Moseley, 'On the mechanical equivalent of heat', *Phil. Trans. A*, vol .190.

1899 James Alfred Ewing and Walter Rosenhain, 'The Crystalline Structure of Metals', *Phil. Trans. A*, vol. 193.

1900 William Augustus Tilden, 'On the Specific Heat of Metals and the Relation of Specific Heat to Atomic Weight', *Phil.Trans. A*, vol. 194.

1903 Charles Thomas Heycock and Francis Henry Neville, 'On the Constitution of the Copper-tin Series of Alloys', *Phil. Trans. A*, vol. 202.

1904 Ernest Rutherford, 'The Succession of Changes in Radioactive Bodies', *Phil. Trans. A*, vol. 204.

1907 Thomas Edward Thorpe, 'The Atomic Weight of Radium', *Proc. A*, vol. 80.

1908 Charles Herbert Lees, 'The Effect of Temperature and Pressure on the Thermal Conductivities of Solids', *Phil. Trans. A*, vol. 208.

1915 William Henry Bragg, 'X-rays and Crystals', *Phil. Trans. A*, vol. 215.

1918 Sir Charles Parsons, 'Experiments on the Artifical Production of Diamond', *Phil. Trans. A*, vol. 220.

1919 Robert John Strutt, 'A study of the line spectrum of sodium as excited by fluorescence', *Proc. A*, vol. 96.

1920 Sir Ernest Rutherford, 'Nuclear constitution of atoms', *Proc. A*, vol. 87.

1923 Geoffrey Ingram Taylor and Constance F. Elam, 'The Distortion of an Aluminium Crystals during a Tensile Test', *Proc. A*, vol. 102.

1925 Sir William Bate Hardy and Ida Bircumshaw, 'Boundary Lubrication – plane surfaces and the limitations of Amonton's law', *Proc. A*, vol. 108.

1927 Francis William Aston, 'A New Mass-Spectrograph and the Whole Number Rule', *Proc. A*, vol. 115.

1933 James Chadwick, 'The Neutron', *Proc. A*, vol. 142.

1934 William Lawrence Bragg, 'The Structure of Alloys', *Proc. A*, vol. 145.

1935 Ralph Howard Fowler, 'The anomalous specific heats of crystals with special reference to the contribution of molecular rotations', *Proc. A*, vol. 151.

1950 Percy Williams Bridgman, 'Physics above 20,000 kg/cm^2', *Proc. A*, vol. 203.

1951 Eric Keightley Rideal, 'Reactions in monolayers', *Proc. A*, vol. 209.

1953 Nevill Francis Mott, 'Dislocations, plastic flow and creep in metals', *Proc. A*, vol. 222.

1956 Harry Work Melville, 'Addition polymerization', *Proc. A*, vol. 237, 149.

1962 John Desmond Bernal, 'Structure of liquids', *Proc. A*, vol. 280, 147.
1963 Alan Howard Cottrell, 'Fracture', *Proc. A*, vol. 276, 1.
1973 Frederick Charles Frank, 'Crystals imperfect ....'
1974 John Meurig Thomas, 'New microcrystalline catalysts', *Phil. Trans. A*, vol. 333, 173.
1979 Michael Ellis Fisher, 'Multicritical points in magnets and fluids a review of some novel states of matter'.
1983 Alfred Edward Ringwood, 'The earth's core: its composition, formation and bearing upon the origin of the earth', *Proc. A*, vol. 394, 1.
1995 Anthony Kelly, 'Composites: towards intelligent materials design', *Phil. Trans. A*, 1996, vol. 354, 1841.
1988 Walter Eric Spear, 'Amorphous semi-conductors: a new generation of electronic materials'.
1999 Peter Day, 'The molecular chemistry of magnets and superconductors'.

APPENDIX 4

Nobel Prize Winners (Fellows and Foreign Members referred to in the Biographical Section)

P – Physics   C – Chemistry

1901 Jacobus Henricus van't Hoff C
1902 Hendrik Antoon Lorentz P
1903 Antoine-Henri Becquerel P
1903 Svante Arrhenius C
1904 John William Strutt, third Baron Rayleigh P
1904 William Ramsay C
1906 Joseph John Thomson P
1906 Henri Moissan C
1908 Ernest Rutherford C
1913 Heike Kammerlingh Onnes P
1914 Max von Laue P
1914 Theodore William Richards C
1915 William Henry Bragg P
1915 William Lawrence Bragg P
1918 Max Karl Ernst Planck P
1920 Walther Hermann Nernst C
1921 Albert Einstein P
1921 Frederick Soddy C
1922 Niels Henrik David Bohr P
1922 Francis William Aston C
1924 Karl Manne Georg Siegbahn P
1928 Owen Willans Richardson P
1929 Louis Victor de Broglie P
1932 Walter Karl Heisenberg P
1932 Irving Langmuir C
1933 Paul Adrien Maurice Dirac P
1933 Erwin Schrödinger P
1934 Harold Clayton Urey C
1935 James Chadwick P
1935 Jean-Frederic Joliot C

1936 Peter Joseph Wilhelm Debye C
1937 George Paget Thomson P
1937 Walter Norman Haworth C
1938 Enrico Fermi P
1944 Otto Hahn C
1945 Wolfgang Pauli P
1946 Percy Williams Bridgman P
1951 Sir John Douglas Cockcroft P
1951 Glenn Theodore Seaborg C
1954 Max Born P
1954 Linus Carl Pauling C
1956 John Bardeen P
1962 Lev Davidovich Landau P
1963 Eugene Paul Wigner P
1964 Dorothy Mary Crowfoot Hodgkin C
1967 Hans Albrecht Bethe P
1970 Louis Eugène Felix Néel P
1971 Dennis Gabor P
1972 John Bardeen P
1973 Brian David Josephson P
1977 Philip Warren Anderson P
1977 Sir Nevill Francis Mott P
1977 John Hasbrouk van Vleck P
1978 Piotr Leonidovich Kapitza P
1991 Pierre-Gilles de Gennes P

APPENDIX 5

Fellows of the Royal Society – Presidents of British Institutes of Metallurgy and Materials

Note in some cases election as Fellows occurred subsequent to the date of service as Presidents.

THE IRON AND STEEL INSTITUTE 1869–1973

1869–71 William Cavendish (Seventh Duke of Devonshire)
1871–73 Sir Henry Bessemer
1873–75 Sir Lowthian Bell
1877–79 Sir William Siemens
1883–85 Sir Bernard Samuelson
1885–87 John Percy
1891–93 Sir Frederick Augustus Abel
1899–1901 Sir William Chandler Roberts-Austen
1905–07 Sir Robert Hadfield
1920–22 John Edward Stead
1935–37 Sir Harold Carpenter
1946–48 Cecil Henry Desch
1948–50 Sir Andrew McCance
1961–62 Sir Charles Goodeve
1968–69 John Hugh Chesters

THE INSTITUTE OF METALS 1908–1973 and 1985–1992

1908–10 Sir William Henry White
1912–13 William Gowland
1914–16 Sir Henry John Oram
1916–18 Sir George Thomas Beilby
1920–22 Sir Henry Cort Harold Carpenter
1928–30 Walter Rosenhain
1938–40 Cecil Henry Desch
1954–55 Stanley Fabe Dorey
1957–58 Leonard Bessemer Pfeil
1963–64 Sir Hugh Ford
1965–66 Leonard Rotherham
1967–68 Sir Monty Finniston
1985–87 Sir Hugh Ford

THE METALS SOCIETY 1974–1984

1974–75 Sir Harold Montague Finniston
1976–77 Sir James Menter
1980–81 Sir Robert William Kerr Honeycombe

THE INSTITUTE OF MATERIALS 1993–

1995–96 Sir Ronald Mason
1996–97 Anthony Kelly
1997–98 Sir Robin Buchanan Nicholson

THE INSTITUTION OF METALLURGISTS 1945–1984

1953–54 Leonard Bessemer Pfeil
1961–62 Norman Percy Allen
1964–65 Leonard Rotherham
1967–68 Henton Morrogh
1975–76 Sir Harold Montague Finniston
1977–78 Sir Robert William Kerr Honeycombe

THE INSTITUTION OF MINING AND METALLURGY 1892–

1934–35 Sir Henry Cort Harold Carpenter
1975–76 Frederick Denys Richardson

THE PLASTICS INSTITUTE 1931–1975

1937–38 Sir James Swinburne, Bart
1970–74 Sir Harry Melville

THE PLASTICS AND RUBBER INSTITUTE 1975–1992

1975 Sir Harry Melville
1990–92 Sir Geoffrey Allen

THE BRITISH CERAMICS SOCIETY 1939–1986

1951–52 John Hugh Chesters

THE INSTITUTE OF CERAMICS 1954–1993

1961–63 John Hugh Chesters

APPENDIX 6

Fellows and Foreign Members of the Royal Society – Recipients of Premier Awards of British Institutes of Metallurgy and Materials

All of the awards listed, with the exception of the Gold Medal of the Institution of Mining and Metallurgy are now associated with the Institute of Materials, which was formed in 1993 by the merger of the Institute of Metals, the Plastics and Rubber Institute, the Institute of Ceramics and the Composites Society. (see the historical summary below)

The medals/awards referred to here are predominantly for metallurgical work, representing mainly 'premier' awards of the Institutes originally founded in the field of metallurgy. The Griffiths award (initiated in 1965) covers all types of materials and the list includes some who received the award for work in the non-metallics field. (In the earlier years up to around 1985 the award was associated with the Materials Science Club). The premier awards on polymers and rubbers, originally established by the Institutes specifically concerned with these materials, are also listed together with information on the Mellor Memorial Lectureship, which is the premier award associated with the former Institute of Ceramics.

Note: in some cases election to the Royal Society occurred subsequent to the date of the Awards.

BESSEMER GOLD MEDAL (FERROUS METALLURGY)

1874 Sir Isaac Lowthian Bell
1875 Sir Charles William Siemens
1878 John Percy
1889 Sir Joseph Whitworth
1883 George James Snelus
1891 William George Armstrong (Baron Armstrong)
1897 Sir Frederick Augustus Abel
1901 John Edward Stead
1904 Sir Robert Abbott Hadfield
1905 John Oliver Arnold
1911 Henry Louis Le Chatelier
1929 Sir Charles Algernon Parsons
1930 Walter Rosenhain
1931 Sir Henry Cort Harold Carpenter

1933 William Herbert Hatfield
1935 Albert Marcel Germain René Portevin
1938 Cecil Henry Desch
1940 Sir Andrew McCance
1956 Charles Sykes
1962 Sir Charles Frederick Goodeve
1965 Norman Percy Allen
1966 John Hugh Chesters
1968 Frederick Denys Richardson
1973 Sir James Woodham Menter
1974 Sir Harold Montague Finniston
1977 Henton Morrogh

PLATINUM MEDAL (NON-FERROUS METALLURGY)
1938 Sir William Lawrence Bragg
1939 Sir Henry Cort Harold Carpenter
1941 Cecil Henry Desch
1949 William Hume- Rothery
1950 Albert Marcel Germain René Portevin
1959 Leonard Bessemer Pfeil
1964 Sir Geoffrey Ingram Taylor
1965 Sir Alan Howard Cottrell
1967 Norman Percy Allen
1976 Sir Peter Bernhard Hirsch
1980 Ulick Richardson Evans
1982 Sir Robin Buchanan Nicholson
1984 John Wyrill Christian
1986 Alan Arthur Wells
1989 Raymond Edward Smallman
1992 Anthony Kelly
1997 Frank Reginald Nunes Nabarro
1998 Michael Farries Ashby

A.A. GRIFFITH MEDAL
(MATERIALS SCIENCE – ALL TYPES OF MATERIAL – FIRST AWARDED 1965)
1965 Sir Alan Howard Cottrell
1967 Sir Frederick Charles Frank
1968 David Tabor
1969 Sir Geoffrey Ingram Taylor
1970 Sir Hugh Ford
1973 Sir Nevill Francis Mott
1974 Anthony Kelly
1975 Sir Harold Montague Finniston
1976 John Hugh Chesters
1978 Sir Lionel Alexander (Alistair) Bethune Pilkington
1979 Sir Peter Bernhard Hirsch

1981 Michael Farries Ashby
1982 Ian McMillan Ward
1983 Robert Wolfgang Cahn
1985 Derek Hull
1987 Ernest Demetrius Hondros
1989 Kenneth Anderson Jack
1995 Geoffrey Wilson Greenwood
1999 John Frederick Knott

COLWYN MEDAL (RUBBER – FIRST AWARDED 1928)
1952 Geoffrey Gee
1957 Harry Work Melville
1962 Leslie Clifford Bateman
1973 Cecil Edwin Henry Bawn

SWINBURNE AWARD (PLASTICS – FIRST AWARDED 1960)
1960 Geoffrey Gee
1966 Cecil Edward Henry Bawn
1974 Andrew Keller
1980 Sir Geoffrey Allen
1986 James Gordon Williams
1988 Ian MacMillan Ward
1992 Alan Hardwick Windle
1994 William James Feast

PRINCE PHILIP AWARD
(POLYMERS IN THE SERVICE OF MANKIND – FIRST AWARDED 1977)
1977 John Charnley
1987 Frank Brian Mercer

MELLOR MEMORIAL LECTURERS (CERAMICS)
1957 William Ernest Stephen Turner
1960 John Hugh Chesters
1973 Kenneth Henderson Jack
1984 James Derek Birchall
1991 Ernest Demetrius Hondros

GOLD MEDAL (INSTITUTION OF MINING AND METALLURGY 1892–)
1909 William Gowland
1927 Sir Alfred Mond
1931 Sir Henry Harold Cort Carpenter
1972 Frederick Denys Richardson

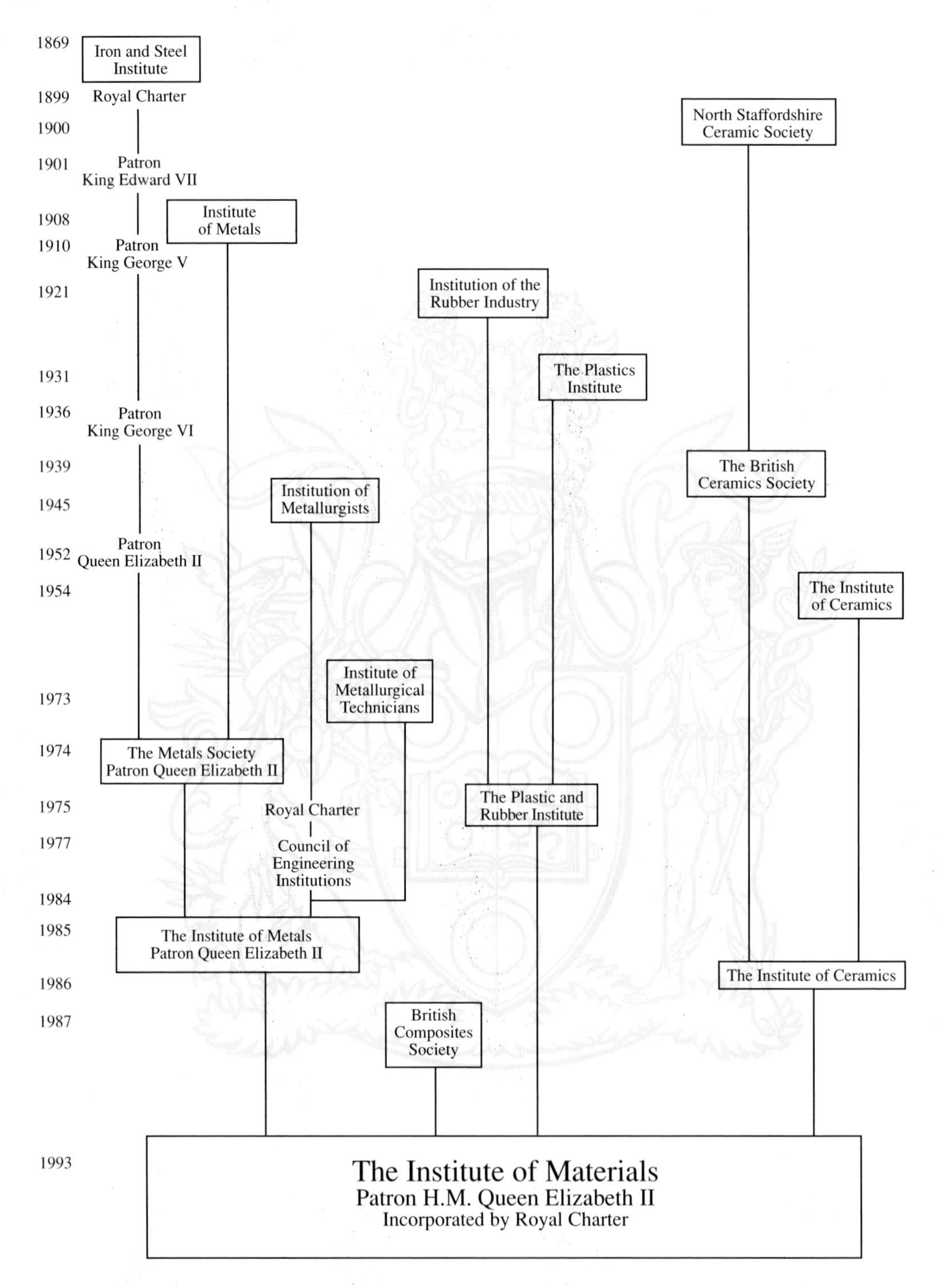

Genealogy of The Institute of Materials

APPENDIX 7

Fellows and Foreign Members of the Royal Society (referred to in the Biographical Section) Chronological List of Election Dates*

1663 Ashley; Boyle; Croone; Goddard; Hooke; Merret; Petty; Power; Slingsby
1664 Hoare, J.; Neale;
1665 Prince Rupert
**** ****
1667 Brown
1668 Hoare, J. (Senior)
**** ****
1672 Newton
**** ****
1678 Halley; Mayow; Moxon
**** ****
1680 Leeuwenhoeck
**** ****
1682 Papin
**** ****
1698 Stanley
**** ****
1704 Fuller, John (senior)
1705 Hauksbee; Savery
**** ****
1714 Desaguliers; Folkes
1715 Geoffroy
**** ****
1717 Hadley
1718 Conduitt
**** ****
1720 Graham, G.
**** ****
1726 Belidor

*All dates of election are quotedin the 'NewStyle' in relation to the Calendar (New Style) Act of 1751.

1727 Fuller, John; Robins
1728 Nicholls; Senex
**** ****
1731 Triewald
1732 Fuller (Rose)
1733 Gray
**** ****
1734 Musschenbroek
**** ****
1737 Short
**** ****
1738 Ellicott; Réaumur
**** ****
1740 Buffon
1741 Baker, H.; Watson, W.
1742 Brownrigg
**** ****
1745 Lewis, W.
1746 Ulloa
1747 Euler
**** ****
1750 Bernouilli, D.
**** ****
1753 Smeaton
**** ****
1755 Henry, W.; Ascanius
1756 Franklin, B.
**** ****
1759 Wright, E
1760 Cavendish, H.
1761 Dollond, J.; Boscovich
**** ****
1764 Berthoud; Pallas; Roebuck
1765 Bergman; Harrison, W.
1766 Banks; Priestley
1767 Woulfe
**** ****
1769 Ingenhousz; Raspe; Watson, R.
**** ****
1776 Nairne
**** ****
1779 Cavallo; Thompson (Rumford)
**** ****
1781 Herschel, W.
**** ****
1783 Wedgwood, J

**** ****
1785 Boulton; Keir; Tennant; Watt
1786 Ramsden
**** ****
1788 Crell; Guyton de Morveau; Jacquin; Lavoisier
1789 Berthollet
**** ****
1791 Lagrange; Pearson, G.; Volta
**** ****
1793 Wollaston
1794 Young
1795 Klaproth
**** ****
1797 Hatchett
1798 Rennie, John
1799 Mussin-Pushkin
**** ****
1801 Chenevix
**** ****
1803 Davy, H
1804 Cockshutt; Dillwyn
**** ****
1807 Allen,W.; Children, J.
1808 Pepys, W.H.
1809 Bingley; Brande; Warburton
1810 Troughton
1811 Thomson, T.
**** ****
1813 Berzelius (FMRS); Herschel, J.F.W.
1814 Brunel (M.I.); Daniell; Kater
1815 Biot (FMRS); Gay-Lussac (FMRS)
1816 Babbage; Pasley
1817 Strutt, W.
1818 Arago (FMRS); Haüy (FMRS); Poisson (FMRS)
1819 Dollond,G.
1820 Whewell
1821 Oersted (FMRS); Stodart
1822 Dalton; Rennie, G.
1823 Barlow, P.; Fourier (FMRS); Rennie, Sir John; Vauquelin (FMRS); Vivian, J.H.
1824 Faraday; Thénard (FMRS)
**** ****
1826 Christie; Davy, E; Dulong (FMRS)
1827 Ampère (FMRS); Telford
1828 Prinsep
1829 Cavendish, W. (Duke of Devonshire)
1830 Brunel, I.K.; Cubitt; Guest; Willis

1831 Parsons, W. (Earl of Rosse)
1832 Cauchy (FMRS)
**** ****
1836 Graham; Wheatstone
1837 Becquerel, A.-C. (FMRS); Clark, W.T.; Rastrick
1838 Donkin; Miller, W.H.; Locke
1839 Moseley, H.; Savart (FMRS)
1840 Dumas (FMRS); Grove; Liebig (FMRS)
1841 Hodgkinson; Smee; Vivian, R.H.; Wrottesley
1842 Poncelet (FMRS)
**** ****
1845 Miller, W.A.; Sopwith
1846 Armstrong; Bird; de la Rive (FMRS); Henry; Johnson, P.N.; Kuppfer (FMRS); Neilson
1847 Percy
1848 Fox; James, H.; Playfair; Porrett; Stenhouse
1849 Leeson; Russell; Stephenson, R.
1850 Barlow, W.H.; Fairbairn; Joule; Weber (FMRS)
1851 Hofmann; Stokes; Thomson (Kelvin)
1852 Pattinson; Tyndall; Regnault (FMRS)
1853 Rankine
1854 Hunt; Mallet; Wöhler (FMRS)
1855 Hawkshaw; Vignoles
**** ****
1857 Sorby; Whitworth
1858 Bunsen (FMRS); Forbes
1859 Galton, D.; Odling; Yolland
1860 Abel
1861 Matthiesen; Maxwell; Pole
1862 Neumann (FMRS); Siemens, C.W.; Todhunter
1863 Crookes; Magnus (FMRS); Roscoe
**** ****
1865 Gore; Jenkin, H.C.F.
**** ****
1867 Parsons, W.
**** ****
1869 Foster, G.C.
1870 Angström (FMRS); Froude; Plateau (FMRS)
1871 Guthrie; Varley
1872 Jevons
1873 Bramwell; Strutt, J.W. (3rd Baron Rayleigh)
1874 Bell
1875 Newall; Roberts-Austen
1876 Reed; Thorpe
1877 Berthelot (FMRS); Dewar; Reynolds; Thomson, James
1878 Hopkinson, J.
1879 Bessemer; Liveing; Matthey

1880 Tilden
1881 Preece; Samuelson; Wright
1882 Chamberlain, J.; Glazebrook
**** ****
1884 Hartley; Lamb; Thomson, J.J.; Wiedemann (FMRS)
**** ****
1886 Unwin
1887 Ewing; Kennedy; Kirchhoff (FMRS); Snelus
1888 Andrews; Becquerel, A.E. (FMRS); Bottomley; Boys; Ramsay; White, W.H.
1889 Clark
1890 Baker, B.
1891 Anderson, W.; Gilchrist; Mond, L.
1892 Fleming; Foster, C. le N.; Joly; Mendeleev (FMRS)
**** ****
1894 Calendar; Love; Swan
1895 Griffiths, E.H.; Heycock
1896 Elgar; Miers; Pearson, K;
1897 Amagat (FMRS); Gibbs (FMRS); Neville; van't Hoff (FMRS)
1898 Parsons, C.A.
1899 Barrett; Hele-Shaw
**** ****
1902 Hardy; Pope; Thomsen (FMRS)
1903 Rutherford; Stead; Townsend
1904 Muirhead; Travers
1905 Lorentz (FMRS); Moissan (FMRS)
1906 Beilby; Lees; Swinburne
1907 Bragg,W.L.; Petavel
1908 Barlow,W; Becquerel, A.-H. (FMRS); Gowland
1909 Hadfield
1910 Arrhenius (FMRS); Filon; Harker; Hopkinson, B.; Soddy
**** ****
1912 Arnold; Oram
1913 Dalby; Le Chatelier (FMRS); Richardson, O.W.; Rosenhain; Voigt (FMRS)
1914 Stanton
1915 MacLennan
1916 Coker; Onnes (FMRS)
1917 Bairstow; Jackson
1918 Carpenter
1919 Richards (FMRS); Rydberg (FMRS); Taylor
1920 Chattock; Lindemann
1921 Aston; Bragg, W.L.; Einstein (FMRS)
1922 Darwin, C.G.; Lanchester; Yarrow
1923 Desch
**** ****
1925 Fowler R.H.; Southwell
1926 Bohr (FMRS); Griffiths, Ezer; Planck (FMRS); Sommerfeld (FMRS)

1927 Baldwin; Chadwick; Ferranti; Mellor
1928 Haworth; Langevin (FMRS); Prandtl (FMRS); Mond, A.M. (Lord Melchett)
1929 Kapitza
1930 Dirac; Edwards, C.A.; Inglis; Rideal; Thomson, G.P.
1931 Jenkin, C.F.
1932 Nernst (FMRS)
1933 Crompton; Debye (FMRS); Gough; Lennard-Jones
**** ****
1935 Andrade; Hatfield; Langmuir (FMRS)
1936 Cockcroft; Mott
1937 Bernal; Hume-Rothery; Oliphant; Stoner
1938 Bengough; Chamberlain, A.N.; Finch; Mond, R.L.; Turner
1939 Born; Bradley, A.J.; Williams
1940 Cook; Goodeve; Lewis, G.N. (FMRS); Massey; Robertson; Sucksmith
1941 Grifitth, A.A.; Melville; Simon
1942 Skinner; Wilson, A.H.
1943 Goldschmidt (FMRS); McCance; Sykes; Stradling
1944 Burch; Polanyi; Timoshenko (FMRS)
1945 Anderson J. (Waverley); Farren; Lonsdale; Peierls
1946 Joliot (FMRS); Kármàn (FMRS); Marsden; Penney; Randall
1947 Hodgkin; Orowan; Urey (FMRS); Whittle
1948 Allibone; Bowden; Constant; Dorey; Frisch; Pauling (FMRS); Ryde
1949 Bailey; Bridgman (FMRS); Evans, U.R.; Laue (FMRS); Lockspeiser; Schrödinger (FMRS)
1950 Bates; Bleaney; Coulson; Fermi (FMRS)
1951 Frohlich; Mendelssohn; Pfeil; Pryce; Ubbelohde
1952 Akers; Bawn; Jones, H.; Portevin (FMRS); Pugsley; Tolansky
1953 Anderson J.S.; de Broglie (FMRS); Pauli (FMRS); Powell; Shoenberg
1954 Frank; Hinton; Siegbahn (FMRS)
1955 Cottrell; Fleck; Heisenberg (FMRS); James, RW; Meitner (FMRS)
1956 Allen, N.P.; Baker, J.F.; Barrer; Gabor; Kurti; Mitchell, J.W.; Pippard
1957 Bethe (FMRS); Hahn (FMRS); Lipson
1958 Ewald
1959 Feilden; Raynor; Spence
1960 Landau (FMRS)
1961 Hill,R.; Flowers; Kearton; London, H.; Pople
1962 Blackman; Cochran; Hooker; Smith, R.A.
1963 Hirsch; Rotherham; Tabor; Wilson, A.J.C.
1964 Bamford; Morrogh
1965 Kronberger; Roberts
1966 Edwards, S.; Forrest; Menter; Néel (FMRS); Szwarc
1967 de Bruyne; Bunn; Ford; van Vleck (FMRS); Ziman
1968 Bateman; Dyson; Pashley; Richardson
1969 Chesters; Finniston; Gifford, (Lord Hailsbury); Glueckhauf; Oatley; Pilkington; Walsh
1970 Addison; Kerensky; Josephson; Marshall; Reason; Wigner (FMRS)
1971 Fisher; Haworth; Nabarro
1972 Coslett; Keller

1973 Bacon; Bardeen (FMRS); Kelly, A.
1974 Eshelby; Heine; Petch; Woods
1975 Charnley; Christian; Lang; Mason; Weck
1976 Allen,G; Elliott; Nye; Watt; Whelan
1977 Ash; Bilby; Duncumb; Pease; Ringwood; Wells
1978 Gibson; Howie; Nicholson; Page
1979 Ashby; Crossland; Hilsum
1980 Anderson, P.W. (FMRS); Basinski; Bradley, D.C.; Jack; Roberts, D.H.; Spear
1981 Franklin,N.; Hill, J.; Honeycombe
1982 Birchall; Brown; Hart; Johnson, K.L.; Johnson, W.; Lifshitz (FMRS); Rao; Roberts, E.J.
1983 Chandrasekhar; Gambling; Gray, G.W.; Paige; Pepper; Ward
1984 De Gennes (FMRS); Desty; Hondros; Mercer; Pendry; Roberts, G.G.; Shanks; Stowell
1985 Bullough; Seaborg (FMRS)
1986 Broers; Day; Fleischmann; Marsham; Mitchell, E.W.J.; Moffatt; Smallman
1987 Dowson; Irwin (FMRS); McConnell; Raynes; Robson; Spencer; Wood, M.F.
1988 Cowley; Evans, T.; Head; Friedel (FMRS); Harris; Mackay; Steeds
1989 Ball; Hull; Lonzarich; Moore; Rowell; Stoneham
1990 Collier; Knott; Kroto; Williams, R.H.W.; Wilkinson, W.
1991 Cahn; Coles; King; Mackintosh; Walters
1992 Greenwood; LeComber; Payne; Willis, J.R.
1993 Burdekin; Dearnaley; Friend; Kelly, M.J.; Kendall; Pillinger; White, J.
1994 Cheetham; Field; Pettifor; Ridley; Williams J.G.
1995 Higgins; Ledwith; Leslie; Miller, D.; Rotblat; Solymar; Windsor
1996 Adams; Dowling; Feast; Miller, S.C.; Pusey; Smith, E.; Smith, G.
1997 Eaves; Windle; Wood
1998 Atkinson; Bhadeshia; Bowen; Mashelkar; Newman; Ruffles; Thomas
1999 Cockayne; Donald; Pethica; Weaire; Williams, P.

APPENDIX 8

Bibliography

The planning for the compilation of biographies aimed initially to rely dominantly on information from Royal Society sources. From 1955 to the present day *Biographical Memoirs of Fellows of the Royal Society* (BMFRS) were published, while from 1932–1954 information was presented as Obituary Notices (ONFRS); these Memoirs and Notices, usually written by other Fellows, are often very extensive and comprehensive concerning family background, career, achievements, scientific research (typically with a list of publications) and other aspects such as character and personality. For an earlier period of about 100 years (1830–1932) obituary notices of many, but not all, deceased Fellows and Foreign Members were published in the Proceedings of the Royal Society. Initially these formed an integral part, albeit sometimes brief, of the President's Anniversary Address, but from 1860 the information appeared as a supplement at the end of the Proceedings Volumes. In 1905, volume 5, the last issued before the A/B division of the Proceedings, consisted entirely of biographical information, chiefly covering the period 1898–1904. Photographic portraits began to appear in 1895, 1896.

For the early part of the 18th century back to the founding of the Society, there are other sources accessible in the Royal Society Library, including *Bulloch's Roll* and the *Journal Book*. These provide information on Fellows such as dates of birth, death and election, offices held in the Society and awards of Medals; also other biographical sources are listed, and sometimes, brief details of careers. The Record of the Royal Society (in two parts: Sir Henry Lyons ed., *The Record of the Royal Society*, 4th issue, 1940; and John Rowlinson FRS and Norman Robinson eds, *Record of the Royal Society – Supplement 1993*) includes much essential information, including full names and dates of elections, lists of recipients of medals, and historical detail of the Society. Also the *Notes and Records of the Royal Society* include significant information, as do the annual issues of the Year Book.

In compiling the biographical information required for this book, particularly for the earlier periods, it became necessary to extend the sources used beyond those of the Royal Society publications; also it was decided to refer to other sources for Fellows in the more recent periods to supplement the Royal Society information. In this connection the *Dictionary of Scientific Biography* (DSB) and the *Dictionary of National Biography* (DNB) with their various Supplements have proved invaluable. The former incorporates much information about Foreign Members as well as Fellows of the Royal Society. Also the DSB accounts often contain extensive bibliographies. Other biographical compilations have been used in some cases.

In the Biographical Section, details of the DSB sources are quoted; this is the case also for DNB sources for the 20th century period.

Further important historical sources have been extensively consulted. For example, works by Cyril Stanley Smith (1903–1991) and Robert Franklin Mehl (1898–1976), who themselves were leading figures in the development of metallurgical science in the 20th century, have provided detailed information on the contributions of individual scientists, and/or historical context. Other important sources are the publications of Robert Wolfgang Cahn FRS on historical aspects of metallurgy and materials science. The volumes of *A History of Technology*, Editors Charles Singer, E.J. Holmyard, A.R. Hall and Trevor I. Williams are also essential sources. Another major source used in the present work is *A History of Platinum and its Allied Metals* by Donald McDonald and Leslie B. Hunt. The subject of platinum metallurgy is one in which activities of Fellows can be traced over more than two centuries. The books of Margaret Gowing FRS are invaluable sources for the history of atomic energy in Britain. For a relatively small number of Fellows, for example, Newton and Faraday, detailed biographies have been consulted.

The bibliography presented below is a wide-ranging, but nevertheless selective list, from the very extensive literature available relevant to the subject of metallurgical history. The list designated 'General References' contains sources which are relevant to various thematic chapters and also to the Biography Section. For each of the thematic Chapters some of the particularly relevant sources are listed here. In the Biographical Section, abbreviations are used for specific references e.g. Gen. Ref. 1 and Ref. Ch. 1.1.

GENERAL REFERENCES

1. J. Bernal, *Science in History, v.2. The Scientific and Industrial Revolutions*, Pelican, 1969.
2. William H. Brock, *The Fontana History of Chemistry*, Fontana Press, London, 1992.
3. R.W.Cahn 'The Historical Development of Physical Metallurgy' c.1., *Physical Metallurgy*, North Holland Publ.Co., Amsterdam, 1965, 1–31.
4. Donald Cardwell, *The Fontana History of Technology*, Fontana Press, London, 1994.
5. J.A. Charles and C.H. Long, *History of Metallurgy*, Brittanica, Macropoaedia, v 21, 15th ed., 1993, 417–422.
6. Ronald W. Clark, *Works of Man*, Century Publishing Ltd, London, 1985.
7. Joan Day and R.F. Tylecote, eds, *The Industrial Revolution in Metals*, The Institute of Metals, 1991.
8. T.K. Derry and Trevor I. Williams, *A Short History of Technology from the earliest times to AD 1900*, Clarendon Press, Oxford, 1960.
9. A. Feldman and P. Ford, *Scientists and Inventors*, Bloomsbury Books, London, 1989.
10. D. Goodman and C.A. Russell, eds, *The Rise of Scientific Europe 1500–1800*, Hodder and Stoughton, The Open University, 1991.
11. Sir Robert Hadfield, *Metallurgy and its Influence on Modern Progress*, Chapman and Hall Ltd, London, 1925.
12. A. Rupert Hall, *The Revolution in Science 1500–1750*, Longman, London, 1983.
13. Alexander Hellemans and Bryan Bunch, *The Timetables of Science*, Sidgwick and Jackson, London 1989.
14. John Hudson, *The History of Chemistry*, MacMillan Press Ltd, Basingstoke, 1992.
15. Bernard Jaffe, *Crucibles: The Story of Chemistry*, 4th Edn, Dover Publications Inc., New York, 1976.
16. Ian McNeil, ed., *An Encylopaedia of the History of Technology. Part I Materials*; 1. S. Darling, *Non-Ferrous Metals*. 2. W.K.V. Gale, *Ferrous Materials*, Routledge, London and New York, 1990.
17. R.F. Mehl, *A Brief History of the Science of Metals*, Amer. Inst. Min. and Met. Eng., N.Y., 1948.
18. C. More, *The Industrial Age*, Longman, London, 1989, 111–128.
19. J.R. Partington, *A History of Chemistry*, v. II. 1961, v. III. 1962, v. IV. 1964. MacMillan and Co. Ltd, London.

20. Charles Singer, E.J. Holmyard, A.R. Hall and Trevor I. Williams, eds, *A History of Technology*, v. III, *From the Renaissance to the Industrial Revolution c 1650–c 1750*; 1957: v. IV *The Industrial Revolution c 1750 – c 1850*; 1958: v. V *The Late Nineteenth Century c. 1850 – c. 1950*, 1958. Clarendon Press, Oxford.
21. C.S. Smith, *A History of Metallography*, University of Chicago Press, Chicago, 1960.
22. C.S. Smith, ed., *The Sorby Centennial Symposium on the History of Metallography*, Amer. Inst. Min. Met. and Petr. Eng., Gordon and Breach Sc. Publ., N.Y., 1965.
23. R.F. Tylecote, *A History of Metallurgy*, Metals Society, London, 1976.
24. Trevor I. Williams, ed., *A History of Technology*, v. VI *The Twentieth Century c. 1900–c. 1950*, Part I; v. VII, Part II, Clarendon Press, Oxford, 1978.
25. A. Wolf, *A History of Science, Technology and Philosophy in the 16th and 17th Centuries v. I and II*, George Allen and Unwin Ltd, London, 1968.
26. A. Wolf, *A History of Science, Technology and Philosophy in the 18th Century, v. I and II*, George Allen and Unwin Ltd, London, 1968.

CHAPTER 1

1. L. Aitchison, *A History of Metals*, New York, Interscience, 1960.
2. G. Agricola, *De Re Metallica*, 1556, with Introduction and Epilogue by Herbert C. Hoover and Lou Henry Hoover, New York, Dover, 1950.
3. R. Ashton, *The English Civil War: Conservation and Revolution 1603–1649*, Weidenfield and Nicolson, London, 1978.
4. J.D. Bernal, *Science in History, Vols 1–4*, London, Pelican, London, 1969.
5. M. Boas, *Nature and Nature's Laws*, Macmillan, London, 1970.
6. R. Boyle, *The Sceptical Chymist*, 1661, Dent, London, 1911.
7. G. Clark, *Science and Social Welfare in the Age of Newton*, Oxford Univ. Press, London, 1937.
8. R. Cotterill, *The Cambridge Guide to the Material World*, Cambridge University Press, Cambridge, 1985.
9. A.C. Crombie, *Augustine to Galileo: Vol 2 Science in the Later Middle Ages and Early Modern Times, l3th–l7th Centuries*, Mercury, London, 1961.
10. J.C. Crowther, *Francis Bacon: The First Statesman of Science*, The Cresset Press, 1960.
11. Lewis Feuer, *The Scientific Revolution*, Basic Books Inc., New York, 1963.
12. W Gilbert, *De Magnete*, 1600, translated by P.F. Mottelay, Dover, New York, 1958.
13. D Goodman and C A Russell, eds, *The Rise of Scientific Europe, 1500–1800*, Hodder and Stoughton, The Open University, 1991.
14. A. Rupert Hall, *The Revolution in Science 1500–1750*, Longman, London, 1983.
15. A. Rupert Hall, *From Galileo to Newton 1630–1720*, Fontana, London, 1963.
16. M. Boas Hall, *All Scientists Now: The Royal Society in the Nineteenth Century*, Cambridge University Press, Cambridge, 1984.
17. W. Haller, *The Rise of Puritanism*, University of Pennsylvania Press, 2nd Edn, 1984, Philadelphia, 1984.
18. R. D. Haynes, *From Faust to Strangelove: Representation of the Scientist in Western Literature*, John Hopkins University Press, Baltimore, Press, 1994.
19. C. Hill, *Puritanism and Revolution*, Mercury Books, London,1958.
20. C. Hill, *The Century of Revolution, 1603–1714*, Reinhold (UK), London, 1961.
21. R. Hooke, *Micrographia*, 1665, Dover Publications Inc., New York, 1961.

22. R.S. Kirby, S. Withington, A.B. Darling and F.G. Kilgour, *Engineering in History*, McGraw-Hill, New York, 1956.
23. R. F. Mehl, *A Brief History of the Science of Metals*, AIME, New York, 1948.
24. Robert K. Merton, *Science, Technology and Society in the Seventeenth Century*, Osiris, 1938, reprinted Harper, New York, 1970.
25. W.R. Owens and A. Hughes, eds, *Seventeenth Century England – A Changing Culture, Vols I & II*, Ward Lock/Open University, London, 1980.
26. W. Pagel, *Paracelsus: An Introduction to Philosophical Medicine in the Era of the Renaissance*, Karger, Basel and New York, 1958.
27. T.A. Ricard, *Man and Metals*, McGraw-Hill, New York, 1932.
28. C. Russell, ed., *The Origins of the English Civil War*, Macmillan, London, 1973.
29. C. Russell, ed., *Science in Europe 1500–1800: A Primary Source Anthology*, The Open University, Milton Keynes, 1991.
30. C.J. Schneer, *The Evolution of Physical Science*, Grove Press, New York, 1960.
31. C.S. Smith, *A History of Metallography*, University of Chicago Press, Chicago, 1960.
32. C.S. Smith, *A Search for Structure: Selected Essays on Science, Art and History*, MIT Press, Cambridge, Mass., 1982.
33. T. Sprat, *History of the Royal Society*, 1667, Washington University Press, 1959.
34. R.F. Tylecote, *A History of Metallurgy*, The Institute of Metals, London, 1992.
35. Charles Webster, *The Great Instauration: Science, Medicine and Reform 1626–1660*, Gerald Duckworth and Co. Ltd, 1975.

CHAPTER 2

1. H. Davis and H. Carter, *Moxon's Mechanick Exercises on Printing*, Oxford Univ. Press, 1962.
2. *Engineering Heritage* v. 1 The Institution of Mechanical Engineers, 1963. v. 2. E.G. Semler, ed., Heinemann Education Books Ltd, London on behalf of the Institution of Mechanical Engineers, 1966.
3. A. Rupert Hall, *Isaac Newton, Adventurer in Thought*, Blackwell, Oxford, 1992.
4. F. Manuel, *A Portrait of Newton*, Harvard University Press, 1968.
5. J. Moxon, *Mechanical Exercises or the Doctrine of Handy-works applied to the art of smithing, joining, carpentry, turning, bricklaying etc.* Fac. reprint, Praeger Publishers, London, 1970.
6. *Réamur's Memoirs on steel and iron*, a translation from the original, printed in 1722, by Anneliese Grunhaldt Sisco, with an Introduction and Notes by Cyril Stanley Smith, Chicago: University of Chicago Press, London, Cambridge University Press? 1956.
7. Robert E. Schofield, *The Lunar Society of Birmingham*, Oxford, Clarendon Press, 1963.
8. C.S. Smith, *A History of Metallography*, University of Chicago Press, Chicago, 1960.
9. T. Thomson, *History of the Royal Society from its institution to the end of the eighteenth century*, Robert Baldwin, London, 1812.
10. S.P. Timoshenko, *History of Strength of Materials*, McGraw-Hill Publishing Co., London, 1953, 3.
11. R.F. Tylecote, *A History of Metallurgy*, The Institute of Metals, London, 1992.
12. Richard S. Westfall, *The Life of Isaac Newton*, Cambridge University Press, London, 1993.

CHAPTER 3

1. Donald Cardwell, *The Fontana History of Technology*, Fontana Press, London, 1994.
2. Sir Robert Hadfield, *Faraday and His Metallurgical Researches*, Chapman and Hall Ltd, London, 1931.
3. John Hudson, *The History of Chemistry*, MacMillan Press Ltd, Basingstoke, 1992.
4. Bernard Jaffe, *Crucibles: The Story of Chemistry*, 4th Edn Dover Publications Inc., New York, 1976.
5. Frank A.I. James, 'Michael's Faraday's Work on Optical Glass', *Phys Educ.*, 1991, 296–300.

CHAPTER 4

1. D. McDonald and L.B. Hunt, *A History of Platinum and its Allied Metals*, Johnson Matthey, London, 1982.

CHAPTER 5

1. Sir John Craig, 'The Royal Society and the Royal Mint', *Notes and Records of the Royal Society*, 1964, **19**, 156–167.
2. John Craig, *The History of the Mint*, Cambridge University Press, 1953.

CHAPTER 6

1. William Crookes and Ernst Rohrig, *A Treatise on Metallurgy*, (1868) based on the German edition of Kerl's *Metallurgy*.
2. Joan Day and R.F. Tylecote, eds, *The Industrial Revolution in Metals*, The Institute of Metals, 1991.
3. F.B. Howard-White, *Nickel: An Historical Review*, Methuen and Co. Ltd, London, 1963.
4. John Percy, *Metallurgy, Fuels, Refractories etc.*, 1861 (revised edition 1875); *The Metallurgy of Lead*, 1870; *Silver and Gold, Part 1*, 1880; John Murray, London.

CHAPTER 7

1. *Sir Henry Bessemer FRS An Autobiography*, Office of Engineering, London, 1905, Strand. Published in 1989 by The Institute of Metals, London.
2. C. Bodsworth, ed., *Sir Henry Bessemer: Father of the Steel Industry*, IoM Communications, London, 1998.
3. Joan Day and R.F. Tylecote, eds, *The Industrial Revolution in Metals*, The Institute of Metals, 1991.
4. J.S. Jeans, *Steel its History, Manufacture and Use*, Earl F.N. Spons, London, NY, 1880.
5. John Percy, *Metallurgy, Fuels, Refractories etc.*, 1861 (revised edition 1875); *Iron and Steel*, 1864; John Murray, London.
6. W. Pole, *The Life of Sir William Siemens*, John Murray, London, 1888.

CHAPTER 8

1. B. Bowers, *A History of Electric Light and Power*, Peter Peregrinus, Stevenage, 1982.
2. B. Bowers, *Michael Faraday and Electricity*, Priory Press, 1974.
3. B. Bowers, *Sir Charles Wheatstone*, HMSO, 1975.
4. D. Burns, *The Political Economy of Nuclear Energy*, Inst. of Economic Affairs, London, 1967.

5. I.C. Byatt, *The British Electricity Industry 1875–1914*, Oxford University Press, 1979.
6. Geoffrey Cantor, *Michael Faraday, Sandemanian and Scientist: A Study of Science and Religion in the Nineteenth Century*, Macmillan Press, 1991.
7. Gwendy Caroe, *The Royal Institution – An Informal History*, John Murray, Albermale Street, London, 1985.
8. R.W. Clark, *Edison: The Man who made the Future*, Macdonald and Janes, 1977.
9. J.C. Crowther, *British Scientists of the Nineteenth Century, Volume I*, Penguin Books, Allen Lane, Pelican Books, 1940.
10. Donald S.Cardwell, *James Joule: An Autobiography*, Manchester University Press, Manchester and New York, 1989.
11. Sir Robert Hadfield, *Metallurgy and its Influence on Modern Progress*, Chapman and Hall Ltd, London, 1925.
12. L. Hannah, *Electricity Before Nationalization: A Study of the Electricity Supply Industry in Britain to 1948*, MacMillan, 1979.
13. L. Hannah, *Engineers Managers and Politicians: The First Fifteen Years of Nationalised Electricity Supply in Britain*, MacMillan, 1982.
14. F. Ledger and H. Sallis, *Crisis Management in the Power Industry: An Inside Story*, Routledge, 1995.
15. N.C. Parsons and W.G. Scaife, *Materials and the Development of the Turbine*, 2nd Parsons International Conference, Institute of Metals, London, 1989, 1–4.
16. W. Pole, *The Life of Sir William Siemens*, John Murray, London, 1888.
17. J.D. Scott, *Siemens Brothers 1858–1958 An Essay in the History of Industry*, Weidenfeld and Nicolson, London, 1957.
18. W.G.F. Scaife, *Charles A. Parsons 1884 patents and his prototype turbo-generator*, 1st Parsons Int. Conference, Institute of Metals, London, 5–12.
19. C.S. Smith, *A History of Metallography*, Univ of Chicago Press, Chicago, 1960.
20. J Surrey, *The British Electricity Experiment Privatization: The Record, The Issues, The Lessons*, Earthscan, 1996.
21. John Meurig Thomas, *Michael Faraday and the Royal Institution*, Adam Hilger, Bristol, Institute of Physics Pub., 1991.
22. *The Scientific Papers of James Prescott Joule*, Physical Society, London, 2 parts 1884 and 1887.
23. G.F. Williams, *Birmingham and the beginnings of industrial electrometallurgy*, Special Publication No. 80, Royal Society of Chemistry, 1989.
24. L.R. Williams, *Michael Faraday*, Cambridge University Press, 1965.
25. L.P. Williams (ed.), *The Selected Correspondence of Michael Faraday: Vol 1 1812–48, Vol 2 1849–66*, Cambridge University Press, 1971.

CHAPTER 9

1. Peter Barlow FRS, *A Treatise on the Strength of Materials*, A New Edition revised by his sons P.W. Barlow FRS and W.H. Barlow FRS, to which is added a summary of experiments by Eaton Hodgkinson FRS, William Fairbairn FRS and David Kirkaldy; Lockwood and Co., London, Stationers' Hall Court, 1867.
2. W.H. Barlow, F.J. Bramwell, J. Thomson, D. Galton, B. Baker, W.C. Unwin, A.B.W. Kennedy, C. Barlow, H.S. Hele-Shaw, W.C. Roberts-Austen, A.T. Atchison (Secretary), *Report on the endurance of metals under repeated and varying stresses and the proper working stresses*

on railway bridges and other structures subjected to varying loads, Report British Association, 1887, 424–438. John Murray, London, 1888.
3. Derrick Beckett, *Stephenson's Britain*, David and Charles (Publishers) Ltd, Newton Abbot, 1984.
4. Asa Briggs, *From Iron Bridge to Crystal Palace: Impact and Images of the Industrial Revolution*, Thomas Hudson Ltd, 1979.
5. Asa Briggs, *Victorian Things*, Penguin Books, London, 1988.
6. Isambard Brunel, *The Life of Isambard Kingdom Brunel, Civil Engineer*, (with an Introductory Chapter by L.T.C. Rolt) David and Charles Reprints, 1971. (Originally published by Longmans, Green and Company, 1870).
7. R.A. Buchanan, *The Engineers: a history of the engineering profession in Britain, 1750–1914*, Kingsley, London, 1989.
8. T.M. Charlton, 'Contributions to the Science of Bridge-Building in the Nineteenth Century by Henry Moseley Hon LLD, FRS and William Pole DMus, FRS', *Notes and Records of the Royal Society*, 1976, **30**, 169–179.
9. R.Cox, *Robert Mallet FRS, Centenary Seminar Papers*, The Institution of Engineers of Ireland and the Royal Irish Academy, 1982.
10. Chester H. Gibbons, 'History of Testing Machines for Materials', *Trans. Newcomen Soc.*, 1934–35, **15**, 169–184.
11. Eaton Hodgkinson, 'Experimental researches on the Strength of Pillars of Cast Iron, and other Materials', *Phil. Trans.*, 1840, **130**, 385–456.E
12. *Engineering Heritage* v. 1. The Institution of Mechanical Engineers, 1963. v.2. E.G. Semler, ed., Heinemann Education Books Ltd, London on behalf of the Institution of Mechanical Engineers, 1966.
13. Robert Mallet, *On the Construction of Artillery (On the physical conditions involved in the construction of artillery with an investigation of the relative and absolute values of the materials principally employed and of some hitherto unexplained causes of the destruction of cannon in service)*, Longman, Brown, Green, Longman and Roberts, 1856.
14. W. Pole, *The Life of Sir William Fairbairn*, Longman, Green and Co. Ltd, 1877.
15. *Report of the Commission appointed to inquire into the application of iron to railway structures*, HMSO, London, 1849.
16. L.T.C. Rolt, *Isambard Kingdom Brunel*, Penguin Books, Pelican Books, Harmondsworth, Middlesex, 1970.
17. L.T.C. Rolt, *George and Robert Stephenson – The Railway Revolution*, Penguin Books, Pelican Books, Harmondsworth, Middlesex, 1978.
18. L.T.C. Rolt, *Thomas Telford*, Penguin Books, Pelican Books, Harmondsworth, Middlesex, 1979.
19. L.T.C. Rolt, *Victorian Engineering*, Penguin Books, The Penguin Press, 1970.
20. A.I. Smith, 'William Fairbairn and Mechanical Properties of Materials', *The Engineer*, 1963, **216**, 543–545.

CHAPTER 10

1. Robert W. Cahn, 'Physics of Materials', *Twentieth Century Physics v. III*, Laurie M. Brown, Abraham Pais, Sir Brian Pippard, eds, Institute of Physics Publishing and American Inst. of Physics Press, 1995, Ch. 19, 1505–1564.
2. Sir Harold Carpenter and J.M. Robertson, *Metals, vols I and II*, Oxford University Press,

London, 1939.
3. C.H. Desch, *Metallography*, Longmans, Green and Co, London, 6th Edn, 1944.
4. A.L. Greer, *Sidney Sussex Quartercentenary Essays*, Cambridge, D.E.Beales and A. Nisbet, eds, Boydell and Brewer, 1996.
5. F.X. Kayser and J.W. Paterson, 'Sir William Chandler Roberts-Austen – His role in the development of binary phase diagrams and modern physical metallurgy', *J. Phase Equilibria*, 1998, **19**(1), 11–18.
6. A. Kelly FRS, 'Walter Rosenhain and Materials Research at Teddington', *Rosenhain Centenary Conference*, Royal Society, London, 1976, 5–36.
7. W.C. Roberts-Austen, *An Introduction to the Study of Metallurgy*, Charles Griffin and Co. Ltd, London, 1898.
8. W. Rosenhain, *Introduction to Physical Metallurgy*, Constable and Co. Ltd, London, 1914.
9. S.W. Smith, *Roberts-Austen, A Record of His Work*, Charles Griffin and Co. Ltd, 1914.

CHAPTER 11

1. Robert W. Cahn, 'Physics of Materials', *Twentieth Century Physics v. III*, Laurie M. Brown, Abraham Pais, Sir Brian Pippard, eds, Institute of Physics Publishing and American Inst. of Physics Press, 1995, Ch. 19, 1505–1564.
2. Sir Harold Carpenter and J.M.Robertson, *Metals, vols I and II*, Oxford University Press, London, 1939.
3. J.G. Crowther, *The Cavendish Laboratory*, MacMillan, London, 1970.
4. C.H. Desch, *Metallography*, Longmans, Green and Co, London, 6th Edn.
5. Sir Robert Hadfield, *Metallurgy and its Influence on Modern Progress*, Chapman and Hall Ltd, 1925.
6. Sir Robert Hadfield, *Faraday and His Metallurgical Researches*, Chapmand and Hall, Ltd, London, 1931.
7. William C. Roberts-Austen, *Reports of the Alloy Steels Committee*, First report, 1891; Second report, 1893; Third report, 1895; Fourth report, 1897; Fifth report, 1899; Sixth report (completed by William Gowland) 1904, Institution of Mechanical Engineers, London.
8. W.C. Roberts-Austen, *An Introduction to the Study of Metallurgy*, Charles Griffin and Co. Ltd, London, 1898.
9. W. Rosenhain, *Introduction to Physical Metallurgy*, Constable and Co. Ltd, London, 1914.
10. S.W. Smith, *Roberts-Austen, A Record of His Work*, Charles Griffin and Co. Ltd, 1914.

CHAPTER 12

1. Robert W. Cahn, 'Physics of Materials', *Twentieth Century Physics v. III*, Laurie M. Brown, Abraham Pais, Sir Brian Pippard, eds, Institute of Physics Publishing and American Inst. of Physics Press, 1995, Ch. 19, 1505–1564.
2. Gwendy Caroe, *The Royal Institution - An Informal History*, John Murray, Albermale Street, London, 1985.
3. J.G. Crowther, *The Cavendish Laboratory*, MacMillan, London, 1970.

CHAPTER 13

1. Robert W. Cahn, 'Physics of Materials', *Twentieth Century Physics v. III*, Laurie M. Brown, Abraham Pais, Sir Brian Pippard, eds, Institute of Physics Publishing and American Inst. of Physics Press, 1995, Ch. 19, 1505–1564.

2. Robert W. Cahn, 'Historical perspective on the development of aluminides.' *Nickel and Iron Aluminides Processing, Properties and Applications*, Cincinnati, Oh, USA, 1996.
3. Sir Harold Carpenter and J.M.Robertson, *Metals,* vols I and II, Oxford University Press, London, 1939.
4. C.H. Desch, *Metallography*, Longmans, Green and Co, London, 6th Edn, 1944.
5. A. Kelly FRS, 'Walter Rosenhain and Materials Research at Teddington', *Rosenhain Centenary Conference*, Royal Society, London, 1976, 5–36.
6. W. Rosenhain, *Introduction to Physical Metallurgy*, Constable and Co. Ltd, London, 1914.
7. S.P. Timoshenko, *History of Strength of Materials*, McGraw-Hill Publishing Co., London, 1953.
8. I. Todhunter, ed., and completed by K.Pearson, *A History of the Theory of Elasticity and of the Strength of Metals, vol 1, 1886; vol 2, Parts I and II*, 1893, Cambridge University Press.

CHAPTER 14

1. W.H. Barlow, F.J. Bramwell, J. Thomson, D. Galton, B. Baker, W.C. Unwin, A.B.W. Kennedy, C. Barlow, H.S. Hele-Shaw, W.C. Roberts-Austen, A.T. Atchison (Secretary), *Report on the endurance of metals under repeated and varying stresses and the proper working stresses on railway bridges and other structures subjected to varying loads*, Report British Association, 1887, 424–438. John Murray, London,1888.
2. Sir Harold Carpenter and J.M.Robertson, *Metals, vols I and II*, Oxford University Press, London 1939.
3. R. Cazaud, *Fatigue of Metals*, Chapman and Hall Ltd, London, 1953.
4. C.H. Desch, *Metallography*, Longmans, Green and Co., London, 6th Edn,
5. N.E. Frost, K.J. Marsh, and L.P. Pook, *Metal Fatigue*, Clarendon Press, Oxford, 1974.
6. H.J. Gough, *The Fatigue of Metals*, Scott, Greenwood and Son, London, 1924.
7. M.S. Loveday, M.F. Day, B.F. Dyson, *Measurement of High Temperature Mechanical Properties of Materials*, HMSO, London, 1982.
8. J.Y. Mann, *Bibliography on the Fatigue of Materials, Components and Structures, 1838–1950*, Pergamon Press, 1970.
9. W. Pole, *The Life of Sir William Fairbairn*, Longman, Green and Co., Ltd, 1877.
10. P. Rossmanith, ed., *Fracture Research in Retrospect*, A.A. Balkema/Rotterdam/Brookfield, 1997.
11. A.I.Smith, 'William Fairbairn and Mechanical Properties of Materials', *The Engineer*, 1963, **216**, 543–545.
12. S.P. Timoshenko, *History of Strength of Materials*, McGraw-Hill Publishing Co., London, 1953.

CHAPTER 15

1. *Sir Henry Bessemer, FRS An Autobiography*, Office of Engineering, London, 1905, Strand. Published in 1989 by The Institute of Metals, London.
2. R.Cox, *Robert Mallet FRS, Centenary Seminar Papers*, The Institution of Engineers of Ireland and the Royal Irish Academy, 1982.
3. *Engineering Heritage v. 1.* The Institution of Mechanical Engineers, 1963. *v. 2.* E.G. Semler, ed., Heinemann Education Books Ltd, London on behalf of the Institution of Mechanical Engineers, 1966.
4. Robert Mallet, *On the Construction of Artillery (On the physical conditions involved in the*

construction of artillery with an investigation of the relative and absolute values of the materials principally employed and of some hitherto unexplained causes of the destruction of cannon in service), Longman, Brown ,Green, Longman and Roberts, 1856.

CHAPTER 16

1. Robert W. Cahn, 'Physics of Materials', *Twentieth Century Physics v. III*, Laurie M. Brown, Abraham Pais, Sir Brian Pippard, eds, Institute of Physics Publishing and American Inst. of Physics Press, 1995, Ch. 19, 1505–1564.
2. A.L.Greer, *Sidney Sussex Quatercentenary Essays*, Cambridge, D.E. Beales and A. Nisbet, Boydell and Brewer, 1996.
3. F.J. Humphreys and M. Hatherly, *Recrystallisation and Related Annealing Phenomena*, Pergamon, 1995.
4. F.X. Kayser and J.W. Paterson, 'Sir William Chandler Roberts-Austen – His role in the development of binary phase diagrams and modern physical metallurgy', *J. Phase Equilibria*, 1998, **19**(1), 11–18.
5. A. Kelly FRS, 'Walter Rosenhain and Materials Research at Teddington', *Rosenhain Centenary Conference*, Royal Society, London, 1976, 5–36.
6. A. Kelly and R.B. Nicholson, *Progress in Materials Science*, 1963, **10**, 151.
7. D. McLean, *Grain Boundaries in Metals*, Clarendon Press, Oxford, 1957.
8. W.C. Roberts-Austen, *An Introduction to the Study of Metallurgy*, Charles Griffin and Co. Ltd, London, 1898.
9. S.W. Smith, *Roberts-Austen, A Record of His Work*, Charles Griffin and Co. Ltd, 1914.

CHAPTER 17

1. Robert W. Cahn, 'Physics of Materials', *Twentieth Century Physics v. III*, Laurie M. Brown, Abraham Pais, Sir Brian Pippard, eds, Institute of Physics Publishing and American Inst. of Physics Press, 1995, Ch. 19, 1505–1564.

CHAPTER 18

1. H. Bolter, *Inside Sellafield: Taking the Lid off the World's Nuclear Dustbin*, Quarter Books, 1996.
2. A. Brown, *The Neutron and the Bomb: A Biography of Sir James Chadwick*, Oxford University Press, 1997.
3. D. Burns, Nuclear Power and the Energy Crisis: Politics and the Atomic Industry, MacMillan, 1978.
4. R.W. Clark, *The Greatest Power on Earth: The Story of Nuclear Fission*, Sidgwick and Jackson, 1980.
5. R.W. Clark, *The Birth of the Bomb*, Scientific Book Club, 1961.
6. B.L. Cohen, *The Nuclear Energy Option: An Alternative for the 90s*, Plenum, 1990.
7. H.A. Cole, *Understanding Nuclear Power: A Technical Guide to the Industry and its Processes*, Gower, 1988.
8. J.H. Fremlin, *Power Production: What are the Risks?*, Hilger, 1989.
9. O. Frisch, *What Little I Remember*, Cambridge University Press, 1979.
10. M. Gowing, *Britain and Atomic Energy 1939–1945*, MacMillan, 1964.
11. M Gowing, *Independence and Deterrence: Britain and Atomic Energy 1945–1952, Vol 1 Policy Making*, and *Vol 2 Policy Execution*, MacMillan, 1975.

12. J. Harris and E. Sykes, eds, *Physical Metallurgy of Reactor Fuel Elements*, Metals Society, 1973
13. Bernard Jaffe, *Crucibles: The Story of Chemistry*, 4th Edn, Dover Publications Inc., New York, 1976.
14. R. Jungk, *Brighter than 1000 Suns*, Penguin 1960.
15. D.J. Littler,ed., *Properties of Reactor Materials and the Effects of Radiation Damage*, Butterworths, 1961.
16. A. McKay, *The Making of the Atomic Age*, Oxford University Press, 1984.
17. W.C. Patterson, *Nuclear Power*, Penguin, 1976.
18. R. Peierls, *Bird of Passage*, Princeton University Press, 1985.
19. M. Perutz, *Is Science Necessary?*, Oxford University Press, 1991.
20. T. Price, *Political Electricity: What Future for Nuclear Energy?*, Oxford University Press, 1990.
21. S.F. Pugh, *The Harwell Metallurgy Division, 1946–54*, UKAEA, 1986. H86/1009. AERE 12109.
22. R. Rhodes, *The Making of the Atomic Bomb*, Penguin, 1986.
23. J. Rigden, *Rabi, Scientist and Citizen*, Basic Books, 1987.
24. S. Ripon, *Nuclear Energy*, Heinemann, 1984.
25. S. Zuckerman, *Nuclear Illusion and Reality*, Collins, 1982.
26. S. Zuckerman, *Scientists and War*, Hamish Hamilton, 1966.

CHAPTER 19

1. C. Barnett, *The Audit of War: The Illusion and Reality of Britain as a Great Nation*, Papermac, 1987.
2. C. Barnett, *The Lost Victory: British Dreams, British Realities 1945–1950*, MacMillan, 1995.
3. H. Bessemer, *An Autobiography*, 1905 and reprinted by Institute of Metals, 1989.
4. J.D. Bernal, *The Social Function of Science*, Routledge, London, 1939.
5. J.D. Bernal, *Science in History: Volume 2 The Scientific and Industrial Revolutions*, Penguin, 1954.
6. E.G. Bowen, *Radar Days*, Hilger, 1987.
7. D.E.S. Cardwell, *The Organisation of Science in England*, Heinemann, 1957.
8. R.W. Clark, *The Rise of the Boffins*, Phoenix, 1962.
9. F. Habashi, *A History of Metallurgy*, Oxford University Press, 1994.
10. M. Boas Hall, *All Scientists Now: The Royal Society in the Nineteenth Century*, Cambridge University Press, 1984.
11. W Hutton, *The State We're In*, Cape, 1995.
12. R.V. Jones, *Most Secret War: British Scientific Intelligence 1939–1945*, Hamish Hamilton, 1978.
13. A. Koestler, ed., *Suicide of a Nation: An Enquiry int the State of Britain Today*, Hutchinson 1963.
14. B. Lovell, *Echoes of War: The Story of H2S Radar*, Hilger, 1991.
15. M. Moseley, *Irascible Genius: A Life of Charles Babbage*, Inventor, Hutchinson, 1964.
16. M. Nicholson, *The System: The Misgovernment of Modern Britain*, Hodder and Stoughton, 1967.
17. H. Rose and S. Rose, *Science and Society*, Penguin, 1969.
18. A. Rowe, *One Story of Radar*, Cambridge University Press, 1948.

19. A. Sampson, Anatomy of Britain Today, Hodder and Stoughton, 1965.
20. Michael Sanderson, *The Universities and British Industry 1850–1970*, Routledge and Kegan Paul, London, 1972.
21. Thomas Sprat, *History of the Royal Society*, 1667.
22. R.F. Tylecote, *A History of Metallurgy*, The Metals Society, 1976.
23. J.M. Ziman, *Reliable Knowledge: An Exploration of the Grounds for Belief in Science*, Cambridge University Press, Cambridge, 1978.

DICTIONARIES, ENCYLOPAEDIAS AND OTHER BIOGRAPHICAL SOURCES

1. *Dictionary of National Biography and Supplements*, Smith, Elder and Co. Ltd, London. 1875–1912: Oxford University Press, 1912–.
2. Charles Coulston Gillespie, Editor-in-Chief, *Dictionary of Scientific Biography, volumes 1 –16* ; Fredericci L. Holmes, Editor-in-Chief. Supplement, Biographies, and Topical Essay. Index v, 17, 18, Suppl. II 1990 Charles Scribners' Sons, New York.
3. *Dictionary of American Biography and Supplements*, Charles Scribner and Sons, New York.
4. F.W. Boase, *Modern English Biography*, Franck Cass & Co. Ltd, London, 1945.
5. John Daintith, Sarah Mitchell, Elizabeth Tootill, Derek Gjertsen, *Biographical Encylopedia of Scientists*, 2nd Edition, vol I and II, Institute of Physics Publishers, Bristol and Philadelphia, 1994.
6. David Millar, Ian Millar, John Millar and Margaret Millar, *The Cambridge Dictionary of Scientists*, Cambridge University Press, Cambridge, 1996.
7. R.Porter, Consultant Editor, *The Hutchinson Dictionary of Scientific Biography*, Helicon Publishing Ltd, Oxford, 1994.
8. Trevor Williams, ed., *Collins Biographical Dictionary of Scientists*, Harper Collins Publishing, Glasgow, 1994.
9. *Who's Who*, Adam and Charles Black (Publishers) Ltd, London.
10. *The International Who's Who*, Europa Publications Ltd, London.

Name Index

Bold type corresponds to Fellows and Foreign members for whom biographical information is contained in the Biographies Section.

Academic and Industrial Index

Subject Index